T0190280

OCEAN ENGINEERING MECHANICS

Ocean Engineering Mechanics provides an introduction to water waves and wave-structure interactions for fixed and floating bodies. The author provides a foundation in wave mechanics, including a thorough discussion of linear and nonlinear regular waves, and he presents methods for determining the averaged properties of random waves. He then explains applications to engineering situations in coastal zones. This introduction to the coastal engineering aspects of wave mechanics includes an introduction to shore protection. The book also covers the basics of wave-structure interactions for situations involving rigid structures, compliant structures, and floating bodies in regular and random seas. The final chapters deal with the various analytical methods available for the engineering analyses of wave-induced forces and motions of floating and compliant structures in regular and random seas. An introduction to soil-structure interactions is also included. This book can be used for both introductory and advanced courses in ocean engineering mechanics.

Michael E. McCormick is currently the Corbin A. McNeill Professor in the Department of Naval Architecture and Ocean Engineering at the U.S. Naval Academy. He is a Fellow of the Marine Technology Society, the American Society of Mechanical Engineers, and the American Society of Civil Engineers. In 1976, he became the co-editor of the journal *Ocean Engineering* and remained so for thirty years. Prior to that, he was the editor of the *Marine Technology Society Journal.* Professor McCormick was the first recipient of the U.S. Naval Academy Alumni Award for Research Excellence, and he was also awarded the U.S. Navy Meritorious Civilian Service Award and the U.S. Navy Superior Civilian Service Award. He is the author of *Ocean Engineering Wave Mechanics* and *Ocean Wave Energy Conversion.* Professor McCormick received a Ph.D. in mechanical engineering from the Catholic University of America and a Ph.D. in civil engineering and a Sc.D. in engineering science from Trinity College, Dublin.

Ocean Engineering Mechanics

WITH APPLICATIONS

Michael E. McCormick
United States Naval Academy

CAMBRIDGE
UNIVERSITY PRESS

32 Avenue of the Americas, New York NY 10013-2473, USA

Cambridge University Press is part of the University of Cambridge.

It furthers the University's mission by disseminating knowledge in the pursuit of education, learning and research at the highest international levels of excellence.

www.cambridge.org
Information on this title: www.cambridge.org/9781107427556

First published 2010
First paperback edition 2014

A catalogue record for this publication is available from the British Library

Library of Congress Cataloguing in Publication data

McCormick, Michael E., 1936–
Ocean engineering mechanics : with applications / Michael E. McCormick.
p. cm.
Includes bibliographical references and index.
ISBN 978-0-521-85952-3 (hardback)
1. Ocean engineering. 2. Hydraulic structures. 3. Hydrodynamics. I. Title.
TC1645.M317 2009
620'.4162–dc22 2009014172

ISBN 978-0-521-85952-3 Hardback
ISBN 978-1-107-42755-6 Paperback

I dedicate this book to my dear wife, Mary Ann, and to my family for their love and support, and to my dear friend and colleague, Professor Rameswar Bhattacharyya, for his never-ending support and encouragement, and to the late Professor Manley St. Denis for all he taught me in the early days of my career.

Contents

Preface

It has been more than three decades since my first book on ocean engineering, *Ocean Engineering Wave Mechanics*. My purpose in writing that book was to give ocean engineering students and ocean technologists an introduction to the mechanics of water waves, and to present and demonstrate the analytical techniques used in wave-structure interaction problems. Since the 1973 publication of that book, ocean technology has been one of the most rapidly advancing engineering fields. The purpose of this book is to present both fundamental and advanced techniques in the analyses of both water waves and wave-structure interactions. The classical analytical works in the areas of wave mechanics are discussed in detail so that the reader can follow the lines of thought of the masters who produced these classic analyses.

Most of the material presented herein is for readers with a basic education in applied mechanics, including fluid mechanics or hydraulics and applied mathematics. The material is presented so that the reader can immediately apply the various analytical techniques to problems of interest. To this end, examples are presented in each section. Certain topics, such as the cnoidal theory, are of an advanced analytical nature and, as such, are more appropriate for postgraduate education. Following these topics are examples designed to demonstrate the application of these advanced analytical methods.

I would like to express my thanks to Dr. David R. B. Kraemer of the University of Wisconsin, Platteville, for his help and advice during the preparation of much of this book. Dr. Kraemer's expertise in computational techniques in fluid and applied mechanics and his willingness to share that expertise were of great value. In addition, my sincere appreciation goes to Mr. Jeffrey Cerquetti of Johnson, Mirmiran & Thompson, Inc. and Dr. Patrick J. Hudson of the Applied Physics Laboratory of Johns Hopkins University for their advice and expertise in numerous areas of hydrodynamics.

Two of my friends and former students at the U.S. Naval Academy have made my situation conducive to book writing. Those are Mr. Robert Murtha and Mr. Bernard Bailey. Each of these fine gentlemen knows how they have contributed.

A special thanks is given to Professor Jacek Mostwin of Johns Hopkins Medical Institutions. Because of his consideration and skills, I was able to complete this book.

My sincere appreciation is given to my long-time friends and colleagues, Dr. Ronald Gularte and Mrs. Alice Gularte, for proofreading the manuscript. They provided guidance and editorial comments that were invaluable.

Finally, I would like to express my appreciation to my dear friend and colleague, Professor Rameswar Bhattacharyya of the U.S. Naval Academy, for his suggestions and encouragement. It has been my good fortune to be able to work closely with Professor Bhattacharyya for more than thirty-eight years, and I have profited greatly from the experience.

Michael E. McCormick
Annapolis, Maryland
December 2008

Notation

General

a_w	added mass (kg)
a	cylindrical radius (m)
A_w	added-mass moment of inertia (N-m-s^2/rad)
A	area (m^2)
$\mathrm{b}_{\mathrm{r,P,v}}$	linear radiation, power take-off, and viscous damping coefficients (N-s/m)
b	half-breadth and crest width (m)
b_{v}	nonlinear viscous damping coefficient (N-s^2/m^2)
B	breadth of a structure into the page, or beam of a floating body (m)
$\mathbf{c}$	celerity vector (m/s)
c_g	group velocity (m/s)
C_d	drag coefficient
D	diameter (m)
d	draft of a fixed or floating structure (m)
e	2.7182818...
f	frequency (Hz)
F	force (N)
g	gravitational constant ($\simeq$9.81 m/s^2)
h	water depth (m)
H	traveling wave height (m)
$H_n^{(1,2)}(\,)$	Hankel function of the first and second kinds
H	standing wave height (m)
i	$(-1)^{1/2}$
i,j,k	x,y,z-unit vectors
I_{e}	second moment of area with respect to the e-axis (m^4)
I_{e}	body mass-moment of inertia with respect to the e-axis (N-m-s^2/rad)
I_n	modified Bessel function of the first kind
$J_n(\,)$	Bessel function of the first kind
KC	Keulegan-Carpenter number
$K_n(\,)$	modified Bessel function of the second kind
K_r	refraction coefficient in eq. 6.85
K_R	reflection coefficient in eq. 6.23

K_S	shoaling coefficient in eq. 3.78
k	wave number, $2\pi/\lambda$ (1/m)
L	body length (m)
m	body mass (kg)
M	moment (N-m)
n	order of Bessel function and index number
$\boldsymbol{n}$	normal unit vector
p	pressure (N/m^2)
P	energy flux (N-m/s)
Q	volume flow rate (m^3/s)
r	radial coordinate (m)
$\boldsymbol{r}$	position vector (m)
R	radius (m)
$R_{e\ell}$	Reynolds number based on length ℓ
s	local coordinate (m)
SPM	*Shore Protection Manual* (U.S. Army, 1984)
SWL	still-water level
t	time (s)
T	line tension (N)
T	wave period (s)
u,v,w	x,y,z-velocity components (m/s)
U	nominal speed (m/s)
V	velocity (m/s)
W	body weight (N)
x,y	inertial horizontal coordinates (m)
X,Y	local horizontal coordinates (m)
$Y_n(\,)$	Bessel function of the second kind
z	inertial vertical coordinate (m)
Z	local vertical coordinate (m)
$\Re$	real part of a quantity
$\Im$	imaginary part of a quantity
η	free-surface displacement (m)
θ	angular coordinate (radians, degrees)
λ	wavelength (m)
μ	dynamic viscosity (N-s/m^2)
ν	kinematic viscosity (m^2/s)
ρ	mass density (kg/m^3)
φ	velocity potential (m^2/s)
ψ	two-dimensional stream function (m^2/s)
ω	circular wave frequency (rad/s)

Subscripts

avg	average value
o	amplitude
max	maximum value
0	deep-water wave properties

Chapter 1

$E(T_o)$ wave-energy spectral density (m^2/s)
F fetch (m, km)
F_{min} minimum fetch (m, km)
T_o modal period (s)
U wind speed (m/s, km/hr)
X_D developing length of a wind-generated sea (m, km)

Chapter 2

C_p pressure coefficient in Figure 2.11
$f(t)$ see eq. 2.70
f_V vortex-shedding frequency (Hz)
F_B buoyant force (N)
F_r Froude number
j index number
$L_{m,p}$ model and prototype lengths (m)
m,n indices
n_ς ς-scale factor ($\varsigma = F, L, p, P, t, V$)
$M_{+,-}$ three-dimensional source and sink strengths (m^3/s)
$\mathrm{M}_{+,-}$ line source and sink strengths (m^2/s)
N maximum index number
$R_{o,i}$ outer and inner diameters (m)
R,β,Θ spherical coordinates
$S_{t\ell}$ Strouhal number based on length ℓ
$\hat{\mathbf{S}}$ safety factor
V_0 upstream velocity (m/s)
$\vee$ volume (m^3)
γ weight density (N/m^3)
Γ circulation (m^2/s)
$\sigma_{1,2,u}$ axial, hoop, and ultimate stresses (N/m^2)
Φ three-dimensional velocity potential (m^2/s)
Ψ three-dimensional stream function (m^3/s)

Chapter 3

α,β arbitrary phase angles in eqs. 3.12 and 3.13
C arbitrary constant
E total energy (N-m)
E_p potential energy (N-m)
E_k kinetic energy (N-m)
$F_{A,B}(\lambda)$ wavelength functions in eq. 3.34
K_S shoaling coefficient
$\boldsymbol{N}$ sea-bed normal unit vector
P energy-flux vector (N-m/s)
$T(t)$ time function in eq. 3.9
U,W standing-wave horizontal and vertical particle velocity components (m/s)
$X(x)$ spatial function in eq. 3.9

$Z(z)$ spatial function in eq. 3.9
Δ_F see eq. 3.35
ε power-conversion efficiency
$\xi(x,t)$ translating horizontal displacement (m)
$\zeta(x,t)$ translating vertical displacement (m)
Φ standing-wave velocity potential (m^2/s)
Ψ standing-wave stream function (m^3/s)

Subscripts

wc water column

Chapter 4

A integration constant in eq. 4.93
B perturbation constant in eq. 4.94
C free-surface constant value (m)
E' energy per crest width (N-m/m)
$E(\,)$ complete elliptic integral of the second kind
f_j, F_j see eqs. 4.77 and 4.78
$K(\,)$ complete elliptic integral of the first kind
K total energy per unit volume (N-m/m^3)
ℓ height of the trough above the sea (m)
S bottom coordinate ($= z + h$)
U_R Ursell parameter; see Figure 4.1
U,W Stokian horizontal and vertical velocity components (m/s)
U_{con} convective velocity (m/s)
α crest angle from the vertical (radians, degrees)
Γ free-surface function in eq. 4.88
ε perturbation constant in eqs. 4.8 and 4.9
η free-surface displacement from the wave trough (m)
M parameter
ξ horizontal particle convection length (m)

Subscripts

b breaking condition
c at a wave crest
α angular component
R radial component
0 properties at an origin of a coordinate system

Chapter 5

a wave amplitude (m)
A Weibull parameter in eq. 5.27
A generic spectral parameter in eq. 5.42
A_o coefficient in eq. 5.89
B Weibull parameter in eq. 5.27

B	generic spectral parameter in eq. 5.42
E	energy per free-surface area (N-m/m^2)
$\underline{E}$	energy intensity (N-m/m)
F	fetch (m, km)
$G(\,)$	spreading function in eq. 5.92
–	wave period index bed (m)
I	maximum wave period index
j	wave height index
J	maximum wave height index
M	dimensionless wave height ratio
m	Weibull parameter in eq. 5.27
m_H	shape factor in eq. 5.102
m_T	shape factor in eq. 5.103
m	generic spectral parameter in eq. 5.42
n	generic spectral parameter in eq. 5.42
$n_{i,j}$	number of observed waves corresponding to the index j
N	expected number of observed waves
$p(\,)$	probability density function
P( )	cumulative frequency of occurrence
$P(\,)$	cumulative probability of occurrence
R	radial coordinate from a wave crest (m)
s	spreading parameter
$S(T)$	wave spectral density (m^2/s)
t_D	duration (hours)
$U_{10,19.5}$	wind speed at heights of 10 m and 19.5 m above the still-water level (km/hr)
Z	arbitrary variable
α	equivalent Mach angle in Figure 5.17 (radians, degrees)
β	wind angle from onshore direction (radians, degrees)
Γ_2	the gamma function evaluated at (m+2)/m
δ	boundary layer thickness in Figure 5.16 (m)
θ	angle from wind direction in the horizontal plane (radians, degrees)
Θ	angle from true north (radians, degrees)

Subscripts

avg	averaged
B	Bretschneider spectral density
DF	critical duration
fds	fully developed sea
h	at a finite water depth
$H_{j,J}$	property of the j or J wave height
I	wave component index
J	generic wave spectral density
J	direction index
JON	JONSWAP spectral density
LT	long-term
rms	root-mean-square
s	significant wave

z	zero up-crossing period
$+-$	maximum and negative wave amplitudes

Chapter 6

$\mathrm{A}_{I,R}$	incident and reflected amplitude coefficients
A,B	coefficients in eq. 6.132
b	wave crest width (m)
$\mathrm{B_B}$	boundary value of amplitude function
$\mathrm{B}_n(\,)$	generic Bessel function in eq. 6.99d
B	breadth of structure (m)
B_F	complex coefficient in eq. 6.111
B_G	complex coefficient in eq. 6.113
B_{FG}	$B_F B_G$
C	line-integration path
C,D	coefficients in eq. 6.133
$C(u_L)$	Fresnel integral
e_n	energy per unit crest length over the *n*th step of the shoal in eq. 6.78
$\mathrm{E}(\,)$	amplitude function in eq. 6.37
$E(\,)$	see eq. 6.163
$F(\,)$	arbitrary spatial function in eq. 6.101
$G(\,)$	arbitrary spatial function in eq. 6.103
f_μ	linearized friction factor in eq. 6.31
H'	pure shoaling wave height (m)
K	frequency parameter in eq. 6.50
$\mathrm{K}^{(A,B)}$	see eq. 6.171
K_A	absorption coefficient in eq. 6.24
K_D	diffraction coefficient in eq. 6.23
ℓ	length of Region B in Figure 6.7 (m)
m	slope of structural face
N	wall porosity
N	number of *quasi*-steps on the shoal in Figure 6.13
$\boldsymbol{N}$	normal unit vector on the sea bed
$\mathrm{P}(r)$	defined in eq. 6.99
$P_{0,S}$	points in Figure 6.21
r_S	relative position vector in Figure 6.21
$\mathrm{R_u}$	runup (m)
s	arbitrary dependent variable in eq. 6.51
s,S	coordinates in eqs. 6.141 and 6.142 (m)
$\boldsymbol{s}$	displacement vector in the direction of wave travel (m)
Q_{mn}	see eq. 6.169
$Q(x)$	separation of variables function in eq. 6.192
$S(u_L)$	Fresnel integral
S_{mn}	see eq. 6.165
$\mathrm{T}(t)$	time function in separation-of-variables solution in eq. 6.46
$\mathrm{X}(x)$	spatial function in separation-of-variables solution in eq. 6.46
$Y(y)$	separation-of-variables function in eq. 6.192
Y_0	alongshore distance over the deep-water contour in Figure 6.17 (m)
$Z(x,z)$	see eq. 6.181

Δ	differential operator, defined in eq. 6.185
ε	angle between wave direction and wall (radians, degrees)
ε	phase angles defined in eq. 6.140 (radians, degrees)
θ	angle measured from the leeward side of the seawall (radians, degrees)
Θ	angle measured from the normal on the leeward side of the seawall (radians, degrees), as in Figure 6.21
Λ	alongshore component of the wavelength in eq. 6.20 (m)
μ	see eq. 6.160
ν	see eq. 6.161
σ	phase angle in eq. 6.37 (radians, degrees)
Q_{sed}	volume-rate of sediment transport
Σ_N	defined in eq. 6.80
φ	spatially dependent velocity potential (m^2/s)
Φ	spatially and temporally dependent velocity potential (m^2/s)

Subscripts

a	properties upwave of the shoal
b	properties downwave of the shoal (m^3/s)
A	property in Region A
A	absorbed
B	property in Region B
C	property in Region C
D	diffraction properties
I	incident properties
m	summation index
R	reflected properties
T	transmitted properties

Chapter 7

a	slope-dependent coefficient in eq. 7.6
A	parameter in eq. 7.2
b	slope-dependent coefficient in eq. 7.7
B	parameter in eq. 7.3
$C_{y,\varepsilon}$	constants associated with stresses
C_p	porosity factor in eq. 7.15, equal to $1\text{-}\forall_{void}/\forall_{total}$
D	local rate of energy dissipation ($N\text{-}m^{-1}\text{-}s^{-1}$)
D_{50}	mean sediment diameter (m, cm, mm)
$E_{1,2}$	constants in eqs. 7.71 and 7.72
f_μ	friction factor in eq. 7.58
H'	pure shoaling wave height (m)
K	proportionality constant in eq. 7.41
m	slope of the sea bed
P_x	energy flux line intensity in eq. 7.51 (N/s)
R_u	runup in Figure 7.1 (m)
$[s]$	equivalent radiation stress matrix (N/m)
$s_{XX,YY}$	equivalent components of radiation stress in eq. 7.29 (N/m)
S_{XX}	principal component of radiation stress in eq. 7.20 (N/m)

S_{YY} transverse component of radiation stress in eq. 7.23 (N/m)
[$\boldsymbol{S}$] radiation stress matrix (N/m)
T_ε function of the eddy viscosity in eq. 7.54 (N/m)
U,V,W particle velocity components with respect to the wave direction in Figure 7.7 (m/s)
V velocity vector in Figure 7.7
V_l alongshore (or longshore) velocity in Figure 7.6 (m/s)
$\underline{V}_s$ V_l/V_b
x_S maximum set-up in eq. 7.48 (m)
$\underline{x}$ seaward coordinate from swash line in Figure 7.13 (m)
X,Y,Z wave coordinate system in Figure 7.7 (m)
γ experimental proportionality constant in eq. 7.44
κ parametric constant in eq. 7.49 (1/m)
μ_ε eddy viscosity in eq. 7.59 (N-s/m^2)
ξ surf similarity parameter in eq. 7.9
τ_y time-averaged bed shear stress (N/m^2)
τ_{xys} radiation stress (N/m^2)
τ_ε effective eddy shear stress in eq. 7.61 (N/m^2)
χ $\underline{x}/x_b$
$\mathscr{F}$ momentum flux in eq. 7.35 (N/m)

Subscripts

b breaking condition
ℓs longshore property
s surf-zone conditions
S maximum set-up
sed sediment property
ε eddy viscosity property

Chapter 8

B_T cap width of the trunk (m)
f expected number of failures
F_r Froude number
H' pure shoaling wave height (m)
h_T height of the breakwater (m)
k_Δ layer coefficient in eq. 8.5
$K_{D\,T}$ stability coefficient in eq. 8.1
$L_{0,1..}$ alongshore separation distance between groins 0 and 1, 1 and 2,… (m)
L_g groin length in Figure 8.2 (m)
m slope of the sea bed
m shape parameter in eq. 8.14
n number of primary stone layers
n_L length scale factor in eq. 8.11
n_t time scale factor in eq. 8.13
n_V velocity scale factor in eq. 8.13
N number of cap stones of the breakwater trunk
N number of armor stones of a breakwater

N_{100} expected number of observed waves over 100 years
P probability of failure (= 1 – **R**), as in Example 8.4
Q_{sed} volume rate of sediment transport (m^3/s)
r_T total thickness of the primary armor stone layer (m)
R reliability, as defined in Example 8.4
V_ℓ alongshore (or longshore) velocity in Figure 8.2 (m/s)
W_{1T} armor stone weight (N)
W_{2T} shield stone weight (N)
W_{3T} foundation stone weight (N)
W_{4T} toe stone weight (N)
ρ_{stone} mass density of the stone material (kg/m^3)
ε angle of the breakwater weather face with respect to the horizontal (radians, degrees)

Subscripts

D design condition
avg average value
max maximum value
ref reference value
rms root-mean-square value
T breakwater trunk properties

Chapter 9

a semi-length of a Lewis form (m)
a radius of a circle or circular cylinder (m)
a_e equivalent radius in eq. 9.81 (m)
A semi-length of a rectangular caisson (m)
A_d projected area (m^2)
A_m see eq. 9.141
A_1 Lewis transformation constant (m^2)
A_3 Lewis transformation constants (m^4)
b semi-width of a Lewis form (m)
B semi-width of a rectangular caisson (m)
B_{mn} see eq. 9.141
B_1 $2Y_{max}|_{a=1}$
C contour enclosing an area S (m)
C_i inertial coefficient in eq. 9.26
C_d drag coefficient
C_M mass coefficient defined in Figure 9.17 and eq. 9.79 for a circular caisson
C_{M} mass coefficient defined in eq. 9.80 for a rectangular caisson
$f(\ )$ arbitrary function
E_m constant associated with the index m in eq. 9.63
F_d drag force (N)
F_w wave-induced pressure force on the wall (N)
F_W wave-induced pressure force on the wall, excluding higher-order terms in η_w in eq. 9.5 (N)
$\digamma$ non-dimensional force defined in eq. 9.151

$\mathcal{F}_{1mn}$	Fourier series in eq. 9.111
$G_\gamma(z)$	see eq. 9.138
$\boldsymbol{i}_r$	unit vector in the radial direction
K_d	drag parameter in eq. 9.173
K_i	inertial parameter in eq. 9.175
K_n	depth-draft parameter in eq. 9.108
KC	Keulegan-Carpenter number in eq. 9.45
ℓ	length of a cross-brace (m)
N_M	expected number of waves over M years
$p(\,)$	wave-height probability density function in eq. 9.180 (1/m)
p_w	wave-induced wall pressure (N/m^2)
P	point in Figure 9.27
P_{1m}	see eqs. 9.109 and 9.110
$P(\,)$	wave-height probability in eq. 9.179
$q(z)$	see eq. 9.94
$Q(z)$	empirical function in eq. 9.94
$Q(\beta)$	separation of variables function in Section 9.2H(3)
$R(r)$	separation of variables function in Section 9.2H(3)
S	area of a fluid enclosed by a contour C (m^2)
$S(f)$	wave spectral density (m^2-s)
$T(f)$	transfer function
u_{max}	horizontal particle motion at a wave crest (m/s)
U_R	Ursell parameter or number in eq. 9.54
V_0	body velocity (m/s)
V_∞	free-stream velocity at $x = \pm\infty$ (m/s)
$\vee_{disp}$	displaced volume (m^3)
$w_{\zeta,z}$	complex potentials in ζ- and z-planes (m^2/s)
X	x at a
X,Y,Z	inertial coordinates on the sea bed (m)
Y	y at a (m)
z	$x + iy$ (m)
$Z(z)$	separation of variables function in Section 9.2H(3)
Z_w	depth of the center of pressure (m)
α	eigenvalue in eq. 9.113
β	angle from the direction of motion in the x-y plane (radians, degrees)
γ	angle measured positively from the ξ-axis in the ζ-plane, and angle measured from the wave direction to a ray in Figure 9.28 (radians, degrees)
$\Gamma_{m\alpha}$	see eq. 9.141
$\delta_{\alpha\gamma}$	Kronecker delta in eq. 9.140
ε_m	Neumann's symbol
ζ	$\xi + i\varepsilon$ (m)
κ	dispersion parameter in eq. 9.117
$\Lambda(ka)$	MacCamy-Fuchs amplitude function in eq. 9.72
ν	kinematic viscosity (m^2/s)
Φ	displacement potential in eq. 9.99
$\sigma(ka)$	MacCamy-Fuchs phase angle in eq. 9.73

Subscripts

B	Bretschneider wave spectral density
$brace$	property on a cross-brace
cp	property at the center of pressure
CL	property on the centerline of the cylinder
G	Garrett force and moment
I	incident wave property
j,k,l	caisson indices
m,n	summation indices
MF	MacCamy-Fuchs property
PM	Pierson-Moskowitz wave spectral density
S	scattered wave property
s	significant wave property
w	property at the wall
X,Y	properties in the x- and y-directions
□	properties associated with the rectangular caisson

Chapter 10

a_P	damping plate radius in Figure 10.10 (m)
ALP	articulated-leg platform
A_d	projected area for drag (m^2)
A_{wp}	waterplane area for drag (m^2)
b_{cz}	heaving critical damping coefficient (N-s/m)
b_p	power take-off damping coefficient (N-s/m)
b_r	radiation damping coefficient (N-s/m)
b_v	equivalent linear viscous damping coefficient (N-s/m)
b_z	combine damping coefficient (N-s/m) in eq. 10.25
b_v	nonlinear viscous damping coefficient (N-s^2/m^2)
C	constant in eq. 10.55
C_d	drag coefficient
$\mathcal{H}(\omega)$	amplitude response function in eq. 10.51
$\mathcal{H}(\omega)*$	complex conjugate of (ω)
k_S	spring constant (N/m)
l_S	relaxed mooring line length (m)
N	power of the velocity in eq. 10.1
N	number of mooring lines
N_M	number of observations over M years
$p(Z)$	probability density (1/m)
$P(\ )$	probability of an event
P_z	power absorbed (N-m/s)
$S_J(T)$	wave spectral density in eq. 10.55 (m^2-s)
$S_z(T)$	response spectral density in eq. 10.55 (m^2-s)
T	time interval (s)
T_S	mooring line tension (N)
T_{nz}	natural heaving period (s)
V_z	heaving velocity vector of a body (m/s)
w_{cw}	capture width in eq. 10.46 (m)

$Z(\omega)$ frequency-dependent heaving amplitude (m)
α_z phase angle between the incident wave and the wave-induced force (radians, degrees)
γ 0.5772157 (Euler's constant)
δd change in draft (m)
Δ_z damping ratio in eq. 10.21
ε_z phase angle between the heaving response and the wave-induced force (radians, degrees)
ω_{nz} natural heaving frequency (rad/s)
$\omega_{1,2}$ bounds of the half-power frequency bandwidth (rad/s)

Subscripts

ABS absolute value
avg averaged property
b damping property
B Bretschneider spectral density
dyn dynamic pressure
j amplitude number
J generic spectral density
n natural frequency property
N wave index in eq. 10.60
rms root-mean-square value
S property of mooring line spring effect
x,*y*,*z* motion directions

Chapter 11

a radius of a circle (m)
a_w total added mass of a floating body (kg)
A area in eq. 11.185 (m^2)
A_w added-mass moment of inertia (N-m-s^2/rad)
A_0 Lewis parameter (m^0)
A_1 Lewis parameter (m^2)
A_2 Lewis parameter (m^3)
A_3 Lewis parameter (m^4)
b_Z linear damping coefficient for heaving motions (N-s/m)
b_{wZ} *quasi*-linear damping term in eqs. 11.82 and 11.84 (N-s/m)
b_ξ waterline semi-breadth of a fixed or floating body at a distance ξ from the center of gravity (m)
B center of buoyancy
B′ displaced center of buoyancy
B_θ linear damping coefficient for pitching motions (N-m-s/rad)
$B_\xi(\xi)$ breadth of a body at ξ (m)
c linear restoring coefficient (N/m)
ci( ) cosine integral
C angular restoring coefficient (N-m/rad)
C_{area} sectional area coefficient in eq. 11.43
C_{max} (length) maxima coefficient in eq. 11.42

C_{SF} scale factor in eq. 11.40 when $a = 1$
$C_{SF\text{-}a}$ scale factor in eq. 11.40 when $a \neq 1$
C_{smith} Smith correction factor in eq. 11.63b
d,e,f angular coupling terms in eq. 11.15 (see Table 11.1)
d_ξ draft at ξ (m)
D_ξ diameter of a semicircular section at ξ (m)
D,E,F linear coupling terms in eq. 11.16 (see Table 11.1)
f( ) see eqs. 11.96 through 11.98
F(k) see eq. 11.159
F_a inertial reaction force in eq. 11.29 (N)
F_B buoyant force (N)
F_r radiation damping force in eq. 11.31 (N)
F_W total wave force in eq. 11.25 (N)
g( ) see eqs. 11.96 through 11.98
G center of gravity
GM metacentric height (m)
k_2 shape parameter in eq. 11.144
k_4 frequency coefficient in Table 11.3
K keel
$I_{x,y}$ second moment of area with respect to the x- or y-axis (m^4)
$L_{i,j}$ operators defined in eqs. 11.113 through 11.116, where i = 1,2 and j = 1,2
L waterplane ship length (m)
ℓ_{aft} distance from stern to G in the waterplane (m)
ℓ_{fwd} distance from the bow to G in the waterplane (m) pitching motions (N-m-s/rad)
ℓ freeboard of a floating body (m)
M metacenter
M_o two-dimensional source strength (m^2/s)
M_a inertial reaction moment in eq. 11.30 (N-m)
M_r radiation damping moment in eq. 11.31 (N-m)
M_W total wave moment in eq. 11.26 (N-m)
O origin of the ship coordinate system, X, Y, Z
P point on the strip in Figure 11.5
P point on the strip in Figure 11.24a
r radius of a circle (m)
r,β polar coordinates in the Y-Z plane, as in Figures 11.7b, 11.7c, and 11.12
R_Z amplitude ratio in eqs. 11.88 and 11.89
R_ξ radius of a semicircular section at ξ (m)
si( ) sine integral
s curvilinear coordinate in eq. 11.185 (m)
s_o amplitude in eq. 11.178 (m) S(Y,Z,t) strip envelope geometry in Figure 11.22 (m)
S_ξ strip area at ξ (m^2)
S_{body} spatial portion of strip envelope geometry in eq. 11.178 (m)
T_e period of encounter (s)
U ship's forward speed (m/s)
$V_w(t)$ vertical speed of the free surface (m/s)
V_z heaving body speed (m/s)
V_{wz} vertical water particle velocity (m/s)

$V_{w\eta}$	vertical water particle velocity at $z = \eta$ (m/s)
w_Z	complex velocity potential in eq. 11.51
x,y,z	coordinate system attached to the calm-water free-surface (m)
X,Y,Z	coordinate system at the center of gravity of a body (m)
X_B	see Figure 11.1
Y_{max}	maximum half-breadth of a Lewis form (m)
Y_ξ	half-breadth at ξ (m)
W	body weight (N)
z	$x + iy$
$\underline{\mathrm{z}}$	iz in eq. 11.49 and Figure 11.10b
Z_Z	heaving magnification factor in eq. 11.140
Z_θ	pitching magnification factor in eq. 11.141
Z_o	amplitude of vertical body motion (m)
Z_{max}	maximum half-height of a Lewis form (m)
Z_o	source location on the vertical axis in Figure 11.23 (m)
Z_{ref}	reference draft in eq. 11.89 (m)
Z_{stat}	static displacement in eq. 11.131
Z_ξ	half-height at ξ (m)
α	angular coordinate measured from the negative Z (or z) direction (radians, degrees)
γ	angular coordinate in Figure 11.7 (radians, degrees)
Δ_Z	damping ratio for heaving motions
Δ_θ	damping ratio for pitching motions
$\varepsilon, \in$	complex variables Figure 11.7a
ε_Z	phase angle between the wave-induced force and the heaving motions (radians, degrees)
φ	two-dimensional velocity potential (m^2/s)
φ_{sd}	velocity potential for a two-dimensional source (m^2/s)
φ_s	velocity potential for a two-dimensional point source (m^2/s)
φ_ξ	two-dimensional velocity potential in eq. 11.57 (m^2/s)
$\varphi(Y,Z)$	spatial potential in eq. 11.175 (m^2/s)
Φ_S	velocity potential for a two-dimensional line source (m^2/s)
χ	rolling angular displacement measured from the y-axis (radians, degrees)
ω_e	circular frequency of encounter (rad/s)
$'$	indicating per unit length
α_F	phase angle between the wave-induced force and the wave in eq. 11.101 (radians, degrees)
α_M	phase angle between the wave-induced moment and the wave in eq. 11.106 (radians, degrees)
ε_θ	phase angle between the wave-induced moment and the pitching motions (radians, degrees)
ζ	$\varepsilon + i\in$ in Figure 11.7a
$\zeta(t)$	vertical displacement of a strip (m)
$\underline{\zeta}$	$i\zeta$ in eq. 11.49 and Figure 11.10a
θ	pitching angular deflection measured about the y-axis (radians, degrees)
Θ	trim angle in eq. 11.70 (radians, degrees)
ξ	distance from G to strip (m)
σ	yawing angular displacement measured about the z-axis (radians, degrees)

Subscripts

cir	property associated with a circular section
cr	critical damping
D	diffracted wave property
e	excitation
fs,FS	free-surface property
$gain$	gained buoyancy
h	hydrodynamic force
I	incident wave property
Lew	Lewis form property
$lost$	lost buoyancy
o	identifies a point in space
o	amplitude of a property
n	resonant condition value
r	radiation property
R	reflected wave property
S	property associated with a line source
v	viscous property
w	wave property
X,Y,Z	respective surging, swaying, and heaving properties
B_r	radiation damping moment coefficient in eq. 12.125 (N-m-s/rad)
B	breadth (in the x-direction) of a spread-footing structure at the soil line (m)
B_B	Bretschneider spectrum coefficient (1/s^4) in eq. 12.112
B_c	buoyant force of a mooring line (N)
c_S	apparent soil cohesion in eq. 12.22 (N/m^2)
wo	defined in eq. 11.101
wc	defined in eq. 11.101
ws	defined in eq. 11.101
wz	vertical component of the water particle motions
wZ	vertical wave-body interaction properties
W	total wave excitation
Z	property of vertical motions
ξ	property at a distance ξ from G
θ	pitching property
σ	yawing property
χ	rolling property
ω	wave-induced property

Chapter 12

a	soil strength constant in eq. 12.24 (N/m^2)
a_a	added mass (kg)
a	spar radius (m)
A_a	added-mass moment of inertia in eq. 12.124 (N-m-s^2/rad)
ALP	articulated-leg platform
A_B	Bretschneider spectrum coefficient (m^2/s^4) in eq. 12.112
A_y	cross-sectional area of the leg material (m^2)

b	soil strength constant in eq. 12.24 (N/m^3)
b_r	radiation damping coefficient
C_{sj}	effective soil damping constant for motions in the j (= x, z, or θ) directions (N-s/m or N-m-s/rad)
$C_{a,b}$	integration constants in eq. 12.7
$C_{c,d}$	constants in eq. 12.42
C_{dD}	drag coefficient based on spar diameter
$\mathbb{C}_{dDj}$	equivalent linear drag coefficient in eq. 12.134
d_0	draft of surface-piercing body (m)
d	moored draft of a tension-moored body (m)
d	embedment depth of a spread-footing structure (m)
d_1	thickness of the subsurface plastic zone (m)
D	spar diameter (m)
$D_{in,out}$	inner and outer diameters for a pile (m)
E_s	modulus of elasticity of the soil (N/m^2)
E	modulus of elasticity (Young's modulus) of the leg material (N/m^2)
E_{pile}	modulus of elasticity of the pile material (N/m^2)
E_s	modulus of elasticity of the mooring line material (N/m^2)
$f_n(\,)$	see eq. 12.123
FOT	flexible offshore tower
$F_{xo}(\omega)$	complex force amplitude (N)
$F_{x,z}$	shear and axial forces in a leg (N)
F_α	axial force in a diagonal member (N)
$F_{X,Z}$	horizontal and vertical reaction forces at the fixed eye-point of a mooring line (N)
F_d	pressure-drag force (N)
F_{Zs}	shear force at the interface of the elastic and plastic zones (N)
F_0	applied force at the soil-water interface (N)
G_s	shear modulus of the soil in eq. 12.50
$H(\omega)$	amplitude response function
$\mathbf{H}_0(\,)$	Struve function of zero order
I_Y	second moment of area with respect to the Y-axis
I_y	mass moment of inertia with respect to the *y*-axis ($N\text{-}m\text{-}s^2/rad$)
k_s	effective elastic modulus in eq. 12.23a (N/m^2)
K_j	effective soil spring constant for motions in the j (= x, z, or θ) directions (N/m or N-m/rad)
KC	Keulegan-Carpenter number in eqs. 9.45 and 12.129
K_s	effective spring constant of a mooring line (N/m)
K_{sp}	effective horizontal spring constant for mooring lines in parallel in eq. 12.16 (N/m)
K_{ss}	effective horizontal spring constant for mooring lines in series in eq. 12.17 (N/m)
K_{sX}	effective horizontal spring constant for a mooring lines in eq. 12.14 (N/m) (see Figure 12.6b)
$K_{X,Y}$	effective spring constants for redundant moorings in the X- and Y-directions (N/m)
L	length of a diagonal (m)
L_α	length of a cross-brace (m)

l	relaxed mooring line length (m)
m_n	lumped mass at the nth node (kg)
m_m	mass of a mooring buoy (kg)
M_{Zs}	bending moment at the interface of the elastic and plastic zones (N-m)
M_0	applied moment at the soil-water interface (N-m)
n	modal number
n_L	length scale factor
N_s	force coefficient in eq. 12.23a
N_s	number of tension mooring lines
N	node number at the platform
p	soil resistance in eq. 12.23 (N/m)
P,Q	see eq. 12.30
r	mooring line radius (m)
R_{pas}	Rankine passive earth coefficient in eq. 12.27
s	mooring line segment length (m)
$S(\,)$	wind-wave spectral density (m^2/s)
T	mooring line tension (N)
T_l	tension in a tension mooring line (N)
TLP	tension-leg platform
TRAP	tension-restrained articulated platform
V	velocity (m/s)
$\vee$	volume (m^3)
W_c	net weight (weight minus buoyancy) of a mooring line (N)
W'_c	net weight per unit line length (N/m)
W	weight of floating structure (N)
W_c	structural weight of a mooring line (N)
$X(\omega)$	complex surging amplitude (m)
X,Y,Z	slack mooring coordinates (m)
X,Y,Z	coordinates over the center of gravity of a moored structure (m)
X_x	sliding coordinate on the bed (m)
Z_z	heaving coordinate on the bed (m)
Z_s	soil coordinate (m)
α	cable angle from the horizontal direction in the vertical plane (radians, degrees)
α_θ	phase angle between the wave and wave-induced moment (radians, degrees)
α	angle of a diagonal member to the horizontal (radians, degrees)
β	cable angle to the X-direction in the horizontal plane (radians, degrees)
δ_α	elongation of a diagonal member (m)
ε_s	mooring line strain
ε_z	phase angle between the heaving motions and the wave-induced force (radians, degrees)
θ_y	rocking coordinate on the bed (radians, degrees)
Θ	angular displacement of a tension mooring line from the vertical (radians, degrees)
$\Lambda(ka)$	MacCamy-Fuchs amplitude function in eq. 9.72
ν_s	Poisson's ratio of the soil
yield	property at the elastic-plastic boundary

Θ	rocking property
ξ	property associated with horizontal mooring
1	property at the inflection point of the mooring line
ξ	horizontal length of a relaxed mooring line in Figure 12.8b (m)
ρ_s	mass density of the soil (kg/m^3)
σ	normal stress (N/m^2)
$\sigma(ka)$	MacCamy-Fuchs phase angle in eq. 9.73
τ_{ss}	mooring line normal stress (N/m^2)
τ_S	shear or soil strength in eq. 12.22 (N/m^2)
ϕ_s	friction angle in eq. 12.22 (radians, degrees)

Subscripts

a,b,m	properties as shackle a, b, m
avg	averaged value
B	property associated with buoyancy
B	associated with the Bretschneider wave spectrum
c	property of a mooring line or cable
C	property associated with the damping
d	nonlinear drag property
d	equivalent linear drag property
$\Im$	imaginary coefficient
MF	associated with the MacCamy-Fuchs equation
rms	root-mean-square value
$\Re$	real coefficient
s	soil property
s	mooring line property
v	viscous property
W	property associated with the incident waves
x,y,θ	sliding, heaving, and rocking properties
x,z	property in *x*- and *z*-directions
pile	pile property
yield	property at the movable eye-point of a mooring line

Appendices

a,b	lengths in Figure H1
C	constant in eq. E17
C	Cp
C_a	added-mass constant
C_A	added-mass moment-of-inertia constant
D	diameter of a sphere (m)
$E(\varepsilon,Q)$	elliptic integrals of the first kind
E'_k	kinetic energy per unit length (N-m/m)
$E(Q)$	$E(\pi/2,Q)$, complete elliptic integral of the first kind
$f(z)$	inverse Fourier transform
F	arbitrary function in eq. C3
$F(\varepsilon,Q)$	elliptic integrals of the second kind
$\mathscr{F}(z)$	Fourier transform

G	Green's function (1/m)
G	arbitrary function in eq. C3
$k_{a,b,c,d}$	Runga-Kutta constants in eq. B9
K	separation-of-variables constant (1/m)
$K(Q)$	$F(\pi/2,Q)$, complete elliptic integral of the second kind
M	three-dimensional source strength (m/s)
M	constant in eq. E17
n	index in Runga-Kutta method
N	separation of variables constant (1/m)
n	integer order of a Bessel function
P	source point
p,q	velocity potential values in eq. G2
Q	q/p
Q	parameter in eqs. G8 and G9
R	radius of the sphere (m)
R	radial position (m) (see Figure D1)
s	surface coordinate (m)
S	surface area (m^2)
T(t)	time variable (1/s) in eq. E14
$T(t)$	time variable (1/s) in eq. E3
U	fluid velocity (m/s)
V	volume (m^3)
w	complex velocity potential (m^2/s)
z	complex variable (m)
β	cylindrical angular coordinate (radians, degrees)
$\kappa_{a,b,c,d}$	Runga-Kutta constants in eq. B10
ν	order of a Bessel function
ϕ	two-dimensional velocity potential (m^2/s)
Φ	three-dimensional velocity potential (m^2/s)
Ω	solid angle (radians, degrees) in eq. D6

OCEAN ENGINEERING MECHANICS

1 Introduction

The field of *ocean engineering* was formally identified as such in the 1960s. Prior to that decade, civil, mechanical, and electrical engineers and naval architects concentrated on ocean-related technologies in rather narrow areas. Two books that helped define ocean engineering in the 1960s are those by Wiegel (1964) and edited by Myers, Holm, and McAllister (1969). These books are still often referenced today. The integrated field of ocean engineering primarily resulted from the discovery of massive oil deposits beneath the sea beds. This discovery led to the increased production of both fixed and floating offshore structures and ancillary systems designed to support extraction systems for the energy resource.

Some of the contemporary ocean engineering areas are listed in Table 1.1. The areas discussed in this book are identified by an asterisk (*) in the table. The primary focus of this book is on wave-induced forces and the subsequent effects of ocean structures.

A large number of the engineering problems that must be faced in the design of ocean engineering systems involve *water waves* in one form or another. The engineer's ability to deal with these problems depends on the extent of their knowledge of the physics of water waves. In this book, basic and intermediate analytical techniques used in water-wave hydromechanics are presented. Each chapter contains a number of worked examples designed to help the reader better understand the various wave-related phenomena. An excellent physical discussion of ocean waves is also found in the paperback book by Bascom (1964), which is available in book stores.

1.1 Generation of a Sea

As previously written, a majority of the problems encountered by ocean engineers are associated with ocean waves. Extremely large ocean waves can destroy the most sturdy of ocean structures by impact whereas moderate persistent waves can cause elastic fatigue of structural components. It is then appropriate to introduce the topic of wave generation early in this book.

Winds generate seas containing waves of various heights (H) and periods (T). Wind waves in a sea are statistical in nature, as discussed in Chapter 5. Wave heights and periods vary randomly from place to place and from time to time. The average

Table 1.1. *Areas of ocean technology*

Arctic Engineering	*Navigation*
Acoustical Communications	*Pollution and Environmental Control*
*Coastal Engineering**	*Powering*
Control of Marine Vehicles	*Resistance (Wave and Viscous)**
Diving	*Seafloor Mechanics**
Dredging	*Seakeeping**
*Energy Conversion**	*Sediment Transport**
*Foundations of Marine Structures**	*Sonar*
Harbor Design	*Structural Design**
*Hydromechanics**	*Structural Mechanics**
Instrumentation	*Undersea Optics*
Life Support	*Underwater Vehicles (Manned and Remote)*
Maneuvering	*Wave Mechanics**
Materials Science and Engineering	*Wave-Structure-Soil Interactions**
Mining	

values of these wave properties depend on the nature and duration of the generating winds. A slight, sustained breeze will create the smallest water waves called *capillary waves*. The name of these waves signifies the fact that the wave properties are strongly influenced by the surface tension of the water. Hurricane winds create high-energy waves, where the energy is distributed over a range of wave periods. The distribution of the wave energy is called the *wave spectrum*.

To gain an understanding of storm-generated seas, consider the wind-wave tank sketched in Figure 1.1. The wind field is over a rectangular harbor on a side of the large tank. The sea in the region from $x = 0$ to $x = F_{min}$ of the wind field is

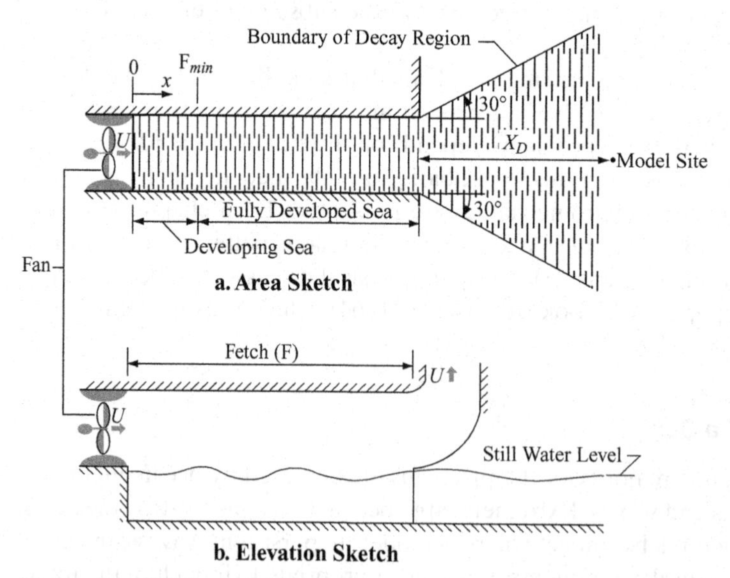

Figure 1.1. *Sketches of a Wind-Wave Flume Used in the Study of Wind-Generated Waves.* The wind generated by the fan on the left blows over the initially calm water surface. Air turbulence-induced pressure fluctuations on the water surface initially produce small waves. The waves then become subject to wind shear on the water surface, causing the waves to grow in both height and length. This process is described in Chapter 3 and is illustrated in Figure 3.1.

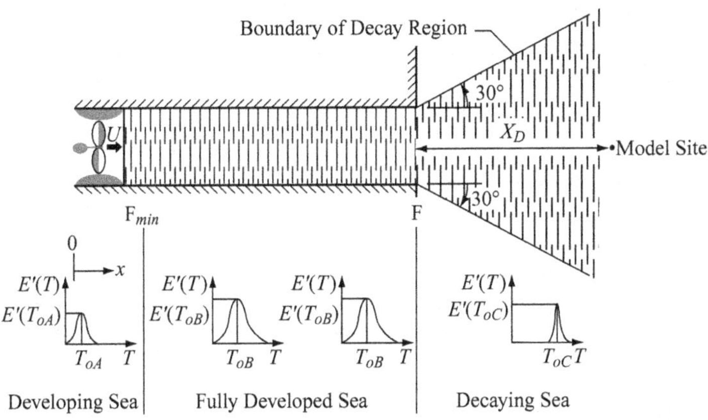

Figure 1.2. *Energy Spectra in a Wind-Generated Wave Field.* The peak energy value in the developing sea is less than that in the fully developed sea. That is, $E'(T_{oA}) < E'(T_{oB})$, where T_{oA} and T_{oB} are called the *modal periods*. The spectral density, $E'(T)$, is independent of both position and time within $F_{min} < x \leq F$. Also, the modal period in the developing sea is less than that in the fully developed sea, that is, $T_{oA} < T_{oB}$. Comparing the spectra (plots of wave energy versus period) in the fully developed sea and the decaying sea, we find $E'(T_{oB}) > E'(T_{oC})$ and, in addition, $T_{oB} < T_{oC}$. The latter condition is due to dispersion, discussed in Chapter 3.

said to be *developing.* This term means that the statistical averages of the wave properties (height and period) are increasing with x. The reason for the increase is that the wind energy absorbed by the sea is increasing with x. The distance F_{min} is called the *minimum fetch.* Over the region $F_{min} < x \leq F$ (where F is called the *storm fetch*), the statistical averages are both uniform over distance and constant in time, and the sea in this region is said to be *fully developed.* When the waves escape the wind field, they travel into a *decay region.* The length of this wind-free region is defined by either a land mass or a site of interest. This length is called the *decay length* and is represented by X_D. Due to the phenomenon called diffraction, the waves over the decay length expand into the quiet waters on either side of the decay region. According to the *Shore Protection Manual* [see U.S. Army (1984)], the *diffraction spread* has boundaries at approximately 30° to the wave direction, as sketched in Figure 1.1a. Because energy is transferred along the crest into the quiet waters, the *energy intensity* [$E'(T_{oC})$ – spectral density or the wave energy per crest width] decreases as x increases over $F < x < X_D$. The energy intensity is proportional to the square of the wave height (H^2); so the wave height also decreases as x increases in the decay region. The *spectra* (plots of the spectral density versus wave period) in the decay region along the x-axis resemble those in Figure 1.2. In this figure, the *modal period* (period of maximum energy intensity) increases with x while the overall energy intensity (represented by the area under the curve) decreases. The increasing modal period is due to the phenomenon called dispersion, where the longer waves (of greater period) outrun the shorter waves (of lesser period).

As wind waves travel in the decay region away from a storm, their energies are spread over increasingly wider crest widths. This energy spreading decreases the local energy intensity of the waves. The energy intensity decrease results in a reduction in the wave height. As a result, sites close to a storm will experience rather large waves of varying periods while distant sites will experience small waves of rather long periods in a rather narrow-period bandwidth. The prediction of a storm

Table 1.2. *Beaufort Wind-Force Scale (BWFS)*

BWFS	U (km/hr)	Wind/wave descriptions	ISSS value	H_{avg} (m)
0	0.00–1.85	Dead Calm/Mirror Surface	0.00	0.00
1	1.85–5.56	Light Air/Ripples	0.00	0.00
2	7.41–11.12	Light Breeze/Small Wavelets	1	0.00–0.30
3	12.97–18.53	Gentle Breeze/Large Wavelets	1–2	0.30–0.61
4	20.39–29.65	Moderate Breeze/Small Waves	2–3	0.61–1.22
5	31.51–38.92	Fresh Breeze/Moderate Waves	4–5	1.22–2.44
6	40.77–50.04	Strong Breeze/Large Waves	5–6	2.44–3.96
7	51.89–61.16	Moderate Gale/Small Breaks with Foam and Wind Steals	5–7	3.96⇒⇒
8	63.01–74.13	Fresh Gale/Moderate Breaks with Visible Spray	7–8	⇒⇒⇒
9	75.98–87.10	Strong Gale/Large Breaks with Dense Foam Streaks	8–9	⇒⇒6.10
10	88.96–101.93	Whole Gale/White Sea	9	6.10–9.14
11	103.78–116.76	Storm/Exceptionally High Waves	9	9.14–13.72
12	118.61–131.58	Hurricane/Exceptionally High Waves with Air Filled with Foam and Spray	9	>13.72

event is short term, that is, the time lapse between the prediction and the event is usually a matter of days. Statistically, the extreme heights of storm waves can be estimated using the methods presented in Chapter 5.

1.2 Wind Classification and Sea State

From the discussion in Section 1.1, we see that the energy content of a sea depends on the winds responsible for the waves. The energy of the sea is characterized by the heights and periods of the waves comprising the sea. As discussed by Bascom (1964), Bretschneider (1969), and others, storm winds and the corresponding seas each have scales that quantify their characters. These are the *Beaufort Wind-Force Scale* (BWFS) and the *International Sea State Scale* (ISSS). A seafarer might write in a ship log that the ship was in a storm with a "sea state 8" if the average wave height would be about 20 meters. An oceanographer might characterize the same storm as one having "Beaufort 11" winds, where the wind speed would be 60 knots per hour (112 kilometers per hour). The wind and sea scales are described in Table 1.2. In that table, U is the wind-speed range and H_{avg} is the range of average wave heights.

A discussion of wind-wave generation and its mathematical analysis are contained in the publication by Earle and Bishop (1984). For a historical perspective of wave analysis, the paper by Craik (2003) is recommended.

1.3 Ocean Engineering Literature

The literature for the various ocean engineering areas is abundant, both in books and journals. Many of the professional societies in the ocean-oriented countries have journals devoted to specific ocean-related areas. In addition, there are a number of independent journals that are more general in coverage. Some of the journals and their sponsoring organizations are listed in Table 1.3. Because advances in most

Table 1.3. *Ocean-oriented journals*

Applied Ocean Research (Computational Mechanics Publishers)
Atmosphere-Ocean (Canadian Meteorological and Oceanography Society)
Biological Oceanography (Taylor and Francis Company)
Bulletin of Marine Science (University of Miami)
China Ocean Engineering (Elsevier-Pergamon Press)
Coastal Engineering (Elsevier Science Publishers)
Coastal Engineering Journal (Japan Society of Civil Engineers)
Coastal Management (Taylor and Francis Company)
Corrosion (Elsevier-Pergamon Press)
Corrosion (National Association of Corrosive Engineers)
Deep-Sea Research (Elsevier-Pergamon Press)
Estuaries (Estuarine Research Federation)
Estuarine Coastal and Shelf Science (Academic Press)
Energy Conversion (Institute of Electrical and Electronic Engineers)
Engineering for the Maritime Environment (Institution of Mechanical Engineers)
The Journal of the Acoustical Society of America
Journal of Energy Resources (American Society of Mechanical Engineers)
Journal of Hydraulic Engineering (American Society of Civil Engineers)
Journal of Marine Research (Yale University Press)
Journal of Ocean Engineering (Institute of Electrical and Electronic Engineers)
Journal of Offshore Mechanics and Arctic Engineering (American Society of Mechanical Engineers)
Journal of Physical Oceanography (American Meteorological Society)
Journal of Ship Production (Society of Naval Architects and Marine Engineers)
Journal of Ship Research (Society of Naval Architects and Marine Engineers)
Journal of Sound and Vibration (Academic Press)
Journal of Vibrations and Acoustics (American Society of Mechanical Engineers)
Limnology and Oceanography (American Society of Limnology and Oceanography)
Marine Biology (Springer-Verlag Press)
Marine Geology (Elsevier Science Publishers)
Marine Geotechnology (Taylor and Francis Company)
Marine Pollution Journal (Pergamon Press)
Marine Research Bulletin (U.S. Office of Naval Research)
Marine Technology (Society of Naval Architects and Marine Engineers)
Marine Technology Society Journal Materials Science and Engineering (Elsevier Science Publishers)
Materials Research Bulletin (Pergamon Press)
Naval Engineers Journal (American Society of Naval Engineers)
Naval Research Reviews (U.S. Office of Naval Research)
Ocean Engineering (Elsevier Science)
Ocean Engineering International (Engineering Committee on Oceanic Resources, Memorial University of Newfoundland)
Ocean Industry (Gulf Publishing Company)
Ocean Science and Engineering (Marcel Dekker)
Ocean and Shoreline Management (Elsevier Science Publishers)
Oceanology (USSR Academy of Science)
Proceedings of the Institute of Marine Engineers
Progress in Oceanography (Pergamon Press)
Sedimentology (Blackwell Scientific Publishers)
Transactions of the Institute of Marine Engineers
Transactions of the Royal Institute of Naval Architects

of the ocean technology areas occur almost daily, the reader is advised to consult the journals of interest on a regular basis. Many of the journals listed in Table 1.3 can be accessed online.

There are a number of fine books covering various areas of ocean engineering. Many of those devoted to wave mechanics are referred to in the chapters that follow, and are listed in the References at the end of the book. For sources of information on specific topics, web searches are useful.

2 Review of Hydromechanics

The term *hydromechanics* normally refers to that part of fluid mechanics devoted to both the hydrostatics and hydrodynamics of incompressible flows. The term includes the effects of free surface at the *air-sea interface.* Although the focus of the discussion of hydrodynamic topics is on incompressible flows in this chapter, a discussion of hydrostatics includes the effects of the compressibility of seawater at great water depths. Because this is a review chapter, all aspects of hydrodynamics are not addressed. The reader is referred to the book by Robert A. Granger (1985) for an expanded coverage of the topics.

We begin our review of hydromechanics with a discussion of *hydrostatics.* Although this subject is basic to a course in fluid mechanics, hydrostatics is often neglected in favor of topics that are of more interest to the instructor. However, for the designer of deep-submergence vehicles, a thorough knowledge of the fundamentals of hydrostatics is required.

2.1 Hydrostatics

A discussion of hydrostatics must begin by paraphrasing Archimedes' Principle: *A body placed in a liquid loses an amount of weight equal to the weight of the liquid that it displaces.* From this simple observation, the hydrostatic equation can be derived. Consider the can buoy sketched in Figure 2.1, which displaces a volume ($\vee$) of water. In that sketch, W is the buoy weight, A is the cross-sectional area, and d is the buoy draft. From Archimedes' Principle, the mathematical expression for the static equilibrium of the body is

$$W = \gamma \vee = (\rho g) A d \tag{2.1}$$

The displaced water is referred to by naval architects as the *displacement* of the body. Assume that the body is in salt water, where the weight density (γ) of the water is approximately 10.1×10^3 N/m^3, and the mass density (ρ) is about 1.03×10^3 kg/m^3 for salt water. Also in eq. 2.1, g is the gravitational acceleration (9.81 m/s^2). The density values are those at sea level under standard atmospheric conditions. The product $\rho g d$ in eq. 2.1 is the *hydrostatic pressure* acting over the bottom of the buoy.

From its equilibrium position, let the buoy sketched in Figure 2.1 be given a small vertically downward displacement, $-\delta z$. The negative sign is due to the fact

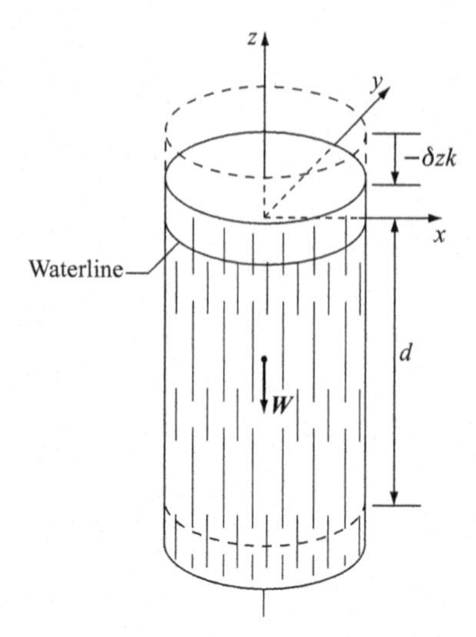

Figure 2.1. *Sketch of a Freely Floating Can Buoy.* The dashed lines show the position of the buoy in static equilibrium, whereas the solid lines show the buoy in a displaced condition. The origin of the coordinate system is in the still water plane.

that the positive direction of the z-coordinate is upward, with its origin on the free surface. The application of Archimedes' Principle to the new equilibrium position results in the following equation:

$$\gamma(\delta\vee) = -\rho g(\delta z)A = (\delta p)A \tag{2.2}$$

Here, the change in hydrostatic pressure (δp) acting on the bottom of the buoy is that which results from the additional displacement. The hydrostatic equation can now be obtained by rearranging eq. 2.2 and passing to the limit as δz approaches zero. The result is

$$\frac{dp}{dz} = -\rho g = -\gamma \tag{2.3}$$

Equation 2.3 is a total derivative because the hydrostatic pressure does not vary on planes parallel to the free surface.

According to King (1969), the weight density of salt water increases linearly with depth. As previously stated, at sea level the weight density is 10.1×10^3 N/m^3, and is denoted by γ_0. At a depth of approximately 9.15×10^3 m, the weight density is 10.46×10^3 N/m^3. The inclusion of this variation in the hydrostatic equation results in the following expression:

$$\frac{dp}{dz} \simeq -\gamma_0 + 0.0396z \tag{2.4}$$

in units of N/m^3. The integration of eq. 2.4 from sea level ($z = 0$) to an arbitrary depth yields

$$p \simeq -\gamma_0 z + 0.0198z^2 \tag{2.5}$$

in N/m^2. For a depth of 10^3 m, the difference in the incompressible and the compressible pressure values is less than 0.3%. Hence, for moderate depths the compressibility of seawater can be neglected.

There are a number of ocean engineering situations in which the hydrostatic pressure is the primary design factor. These include the determination of the wall thickness of a life capsule of a deep-submergence vehicle, the analysis of the compression or volume reduction and collapse depth of deep-submergence capsules at

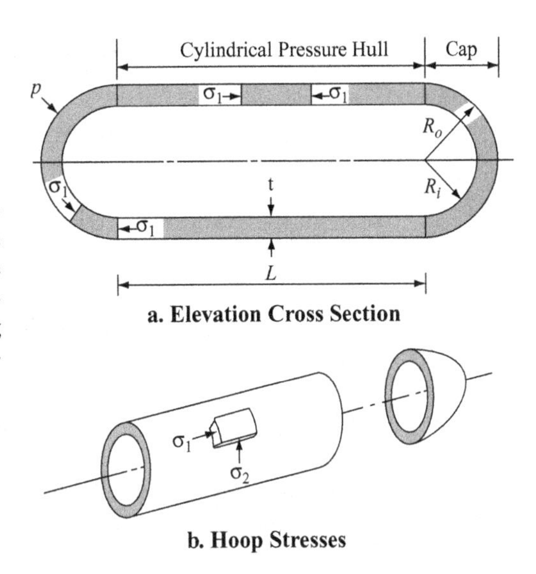

Figure 2.2. *Sketches of a Pressure Hull of a Deep-Submergence Vehicle (DSV).* Also illustrated are the hoop stresses and the ambient pressure. The configuration consists of the two most stable geometries of structures under pressure, those being the circular cylinder and the sphere. Here, hemispherical caps are shown.

extreme depths, and the determination of the conditions for static stability of floating bodies. An example of the determination of the collapse depth is found in the book by McCormick (1973).

EXAMPLE 2.1: PRESSURE HULL ANALYSIS Here we apply the hydrostatic equation (eq. 2.3 for incompressible water or eq. 2.4 for compressible water) to the design of a pressure hull or life capsule of a deep-submergence vehicle (DSV) to determine the weight-to-buoyancy ratio. Both the cost-effectiveness and the design efficiency of a DSV increase as this ratio decreases in value. The basic configuration of a pressure hull is either a spherical shell or a cylindrical shell with hemispherical ends. The most common of these pressure hulls are made of high-yield steel, titanium, or aluminum alloys, depending on both the mission and operational depth. Garvey (1990) recommends the use of organic-matrix composite pressure hulls because composites are positively buoyant. He states that if the 15-ton, forged-steel, two-person life capsule of the DSV *Trieste* was replaced by an equivalent composite capsule, the gasoline used for buoyancy would not be needed. The gasoline displaces about 125 tons of seawater.

The analysis of the effects of the hydrostatic pressure on pressure hulls requires the use of *thin-wall shell theory*, as stated by Garvey (1990). The term "thin-wall" simply means that the thickness of the hull is small when compared to the radius of the shell. Referring to the hull sketched in Figure 2.2, one can determine the weight and buoyancy for both the cylindrical and hemispherical components of the capsule. The strength and stability characteristics of the capsule are determined from the thin-wall theory, as discussed by Ross (1990) and others. From this theory, one obtains the relationships among the axial stress (σ_1), the hoop stress (σ_2), and the net pressure (p) for the cylindrical portion of the hull. For a wall thickness, t, these relationships are obtained as follows: The expression for the axial stress in the cylindrical wall is

$$\sigma_1 = pR_o/(2t) = pR_o/[2(R_o - R_i)] \tag{2.6}$$

where the radii R_i and R_o are those of the inner and outer surfaces of the shell wall, respectively. The hoop stress is

$$\sigma_2 = 2\sigma_1 \tag{2.7}$$

Referring to Figure 2.2, the weight of the cylindrical portion of the pressure hull is

$$W_c = \rho_m g \pi \left(R_o^2 - R_i^2\right) L \tag{2.8}$$

where ρ_m is the mass density of the hull material. The magnitude of the buoyant force of the cylindrical section is obtained from

$$F_{Bc} = \rho g \pi R_o^2 L \tag{2.9}$$

where ρ is the mass density of the water. The ratio of the weight to buoyancy is then

$$\frac{W_c}{B_c} = \frac{\rho_m}{\rho}\left(1 - \frac{R_i^2}{R_o^2}\right) \tag{2.10}$$

As previously stated, our objective is to minimize this ratio while maintaining structural stability. For the analysis of the structural stability, the axial normal stress can be expressed as

$$\sigma_1 = \sigma_u / \hat{S} \tag{2.11}$$

where the *ultimate strength* of the cylindrical wall, σ_u, is the ratio of the maximum test load before breaking and the initial cross-sectional area. The *safety factor*, $\hat{S}$, is a function of both the type of material and the configuration of the structure. For the cylinders, spheres, and connecting joints studied by Garvey (1990), the safety factor values vary from 1.5 to 2.2. Continuing by eliminating the radii R_o and R_i, eqs. 2.6, 2.10, and 2.11 can be combined to obtain the *weight-to-buoyancy ratio*, which is

$$\frac{W_c}{B_c} = 2\left(\frac{\rho_m}{\rho}\right)\left(\frac{p\hat{S}}{\sigma_u}\right)\left(1 - \frac{p\hat{S}}{2\sigma_u}\right) \tag{2.12}$$

In eq. 2.12, the term $p\hat{S}/\sigma_u$ (non-dimensional pressure) is negligible if $z < -10^3$ m. However, for extreme depths an increase in the value of this term results in a decrease of the weight-to-buoyancy ratio by up to 15% for both titanium and composite hulls.

For the hemispherical caps of the hull sketched in Figure 2.2, the tangential stress must be equal to the axial stress (σ_1) of the cylindrical section of the hull because this is a boundary condition at the joint of the sections. Hence, from the application of the thin-wall theory to the spherical hull, the tangential stress is given by eq. 2.6. The total weight of the hemispherical caps is

$$W_s = \rho_m g (4/3)\pi \left(R_o^3 - R_i^3\right) \tag{2.13}$$

The magnitude of the total buoyant force on the hemispherical caps is

$$B_s = \rho g (4/3)\pi R_o^3 \tag{2.14}$$

The combination of eqs. 2.6, 2.11, 2.13, and 2.14 yields the weight-to-buoyancy ratio for the hemispherical caps, which is

$$\frac{W_s}{B_s} = 1.5\frac{\rho_m}{\rho}\left(\frac{p\hat{S}}{\sigma_u}\right)\left[1 - \frac{1}{2}\frac{p\hat{S}}{\sigma_u} + \frac{1}{12}\left(\frac{p\hat{S}}{\sigma_u}\right)^2\right] \tag{2.15}$$

Garvey (1990) states that the value of the non-dimensional pressure, $p\hat{S}/\sigma_u$, is negligible for $z < -6{,}100$ m. However, for extreme depths ($z \geq -10{,}000$ m) an increase in the value of the non-dimensional pressure results in a decrease of the weight-to-buoyancy ratio of up to 10% for titanium hulls and 20% for composite hulls.

We see that the non-dimensional pressure ($p\hat{S}/\sigma_u$) is a critical element in the design of both the cylindrical and hemispherical hull components. To gain an idea of the values of the non-dimensional pressure, consider a titanium hull operating in the Mariana Trench at a depth of 11×10^3 m. For 64I4V titanium, according to Garvey (1990), the ultimate strength is 8.62×10^8 N/m^2 and the safety factor ($\hat{S}$) is 1.5 for both the cylindrical and hemispherical components of the hull. At the 11×10^3-m depth, the combination of these values with the pressure obtained from eq. 2.5 yields

$$p\hat{S}/\sigma_u = 0.198 \tag{2.16}$$

Neglecting compressibility of the water (the second term in eq. 2.5) yields a value of the non-dimensional pressure of 0.193. The respective compressible and incompressible values of the weight-to-buoyancy ratio of the cylindrical hull component, obtained from eq. 2.12, are 1.34 and 1.38. Hence, the inclusion of compressibility results in less than a 3% difference in the values of W_c/B_c.

We have devoted quite a bit of discussion to the pressure hull problem because it is normally treated lightly in the published literature. An excellent introductory discussion of small cylindrical pressure hulls is presented in the book by Dawson (1983), whereas Ross (1990) discusses the advanced topics.

2.2 Conservation of Mass

The conservation of mass applied to a fluid flow is expressed by the equation of continuity. In words, this equation expresses the fact that the internal time-rate of decrease of mass within a control volume equals the net efflux of mass through the surface area of the volume. The equation can be written in differential form as

$$-\frac{\partial \rho}{\partial t} = \nabla \cdot (\rho V) \tag{2.17}$$

where the del operator is defined in Cartesian coordinates by

$$\nabla \equiv \frac{\partial}{\partial x}\boldsymbol{i} + \frac{\partial}{\partial y}\boldsymbol{j} + \frac{\partial}{\partial z}\boldsymbol{k} \tag{2.18}$$

and the *fluid velocity vector* is

$$\boldsymbol{V} = u\boldsymbol{i} + v\boldsymbol{j} + w\boldsymbol{k} \tag{2.19}$$

In eqs. 2.18 and 2.19, the unit vectors of the Cartesian coordinates are $\boldsymbol{i}$, $\boldsymbol{j}$, and $\boldsymbol{k}$, referring to Figure 2.3 for their orientation. The units of eq. 2.17 are in terms of time-rate of mass flow per unit volume. For a *steady flow* (one that does not vary in time), eq. 2.17 reduces to

$$\nabla \cdot (\rho \boldsymbol{V}) = 0 \tag{2.20}$$

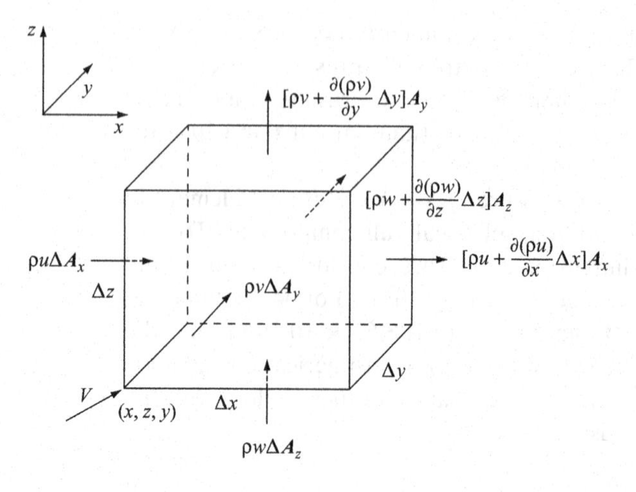

Figure 2.3. *Sketch of an Elemental Volume Illustrating the Conservation of Mass.*

When the fluid is incompressible, then eq. 2.17 becomes

$$\nabla \cdot \boldsymbol{V} = 0 \tag{2.21}$$

for both steady and unsteady flows. The conservation of mass can also be expressed in integral form. The *integral continuity equation* is

$$-\iiint_{\vee} \frac{\partial \rho}{\partial t} d\vee = \iint_{A} \rho \boldsymbol{V} \cdot d\boldsymbol{A} \tag{2.22}$$

where, referring to Figure 2.3, the elemental volume and surface area are, respectively,

$$d\vee = dxdydz \tag{2.23}$$

and

$$d\boldsymbol{A} = dydz\boldsymbol{i} + dxdz\boldsymbol{j} + dxdy\boldsymbol{k} \tag{2.24}$$

Note: The vector direction of an area element is outward from the fluid. For incompressible flows, eq. 2.22 is simplified to

$$\iint_{A} \boldsymbol{V} \cdot dA = 0 \tag{2.25}$$

Many incompressible flows can be treated as being either *uniform* (not varying in space) or *spatially averaged* over the flow area. If the control volume through which the flow passes has N entrances and exhausts, then eq. 2.25 can be expressed in terms of the spatially averaged velocities through the N flow areas. The resulting equation is

$$\sum_{j=1}^{N} \boldsymbol{V}_j \cdot \boldsymbol{A}_j = \sum_{j=1}^{N} V_j A_j \cos(\theta_j) = \sum_{j=1}^{N} Q_j = 0 \tag{2.26}$$

In eq. 2.26, θ_j is the angle between the jth velocity vector and the corresponding flow area, and Q_j is the *volume flow rate* through that area. The use of this equation is illustrated in the following example.

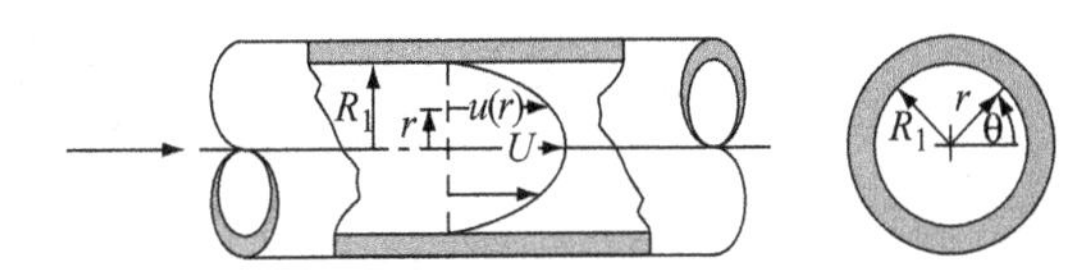

a. Notation for Laminar Flow in a Tube

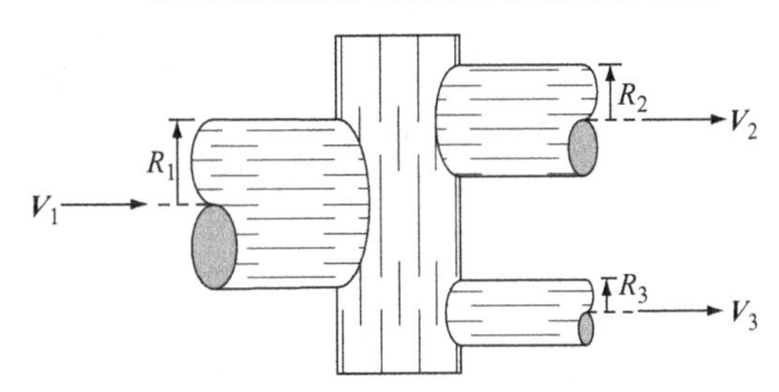

b. Flow through a Three-Branch Manifold

Figure 2.4. *Sketches of a Section of a Piping Manifold Illustrating the Conservation of Mass.*

EXAMPLE 2.2: FLOW THROUGH A MANIFOLD Consider the *laminar flow* (flow in "laminae" or "layers") through the manifold sketched in Figure 2.4a. From the book by Granger (1985) and others, the radial velocity distribution in a laminar pipe flow is parabolic, and is represented mathematically by

$$u = \frac{U}{R^2}(R^2 - r^2) \tag{2.27}$$

where, referring to Figure 2.4b, U is the maximum fluid velocity in the center of the pipe of radius R, and r is the radial distance from the centerline. The spatially averaged velocity in any of the component pipes is

$$V = \frac{1}{\pi R^2}\int_0^{2\pi}\int_0^{R} urdrd\theta = \frac{U}{2} \tag{2.28}$$

Now, the application of eq. 2.26 to the flow in the manifold results in the following:

$$\begin{aligned} &V_1\boldsymbol{i}\cdot\boldsymbol{A}_1 + V_2\boldsymbol{i}\cdot\boldsymbol{A}_2 + V_3 i\cdot\boldsymbol{A}_3 = \\ &V_1A_1\cos(180^\circ) + V_2A_2\cos(0^\circ) + V_3A_3(0^\circ) = 0 \end{aligned} \tag{2.29}$$

From eq. 2.29, we obtain the following relationship of the average velocities in the component pipes:

$$V_1A_1 = V_2A_2 + V_3A_3 \tag{2.30}$$

Because the terms in eqs. 2.29 and 2.30 are volume flow rates, eq. 2.30 can also be written as

$$Q_1 = Q_2 + Q_3 \tag{2.31}$$

where the units of Q are m^3/s in the International System and ft^3/s in the British system.

Laminar flow in pipes and tubes is referred to both as *Poiseuille flow* and *Hagen-Poiseuille flow*, as J. L. M. Poiseuille presented the theoretical analysis

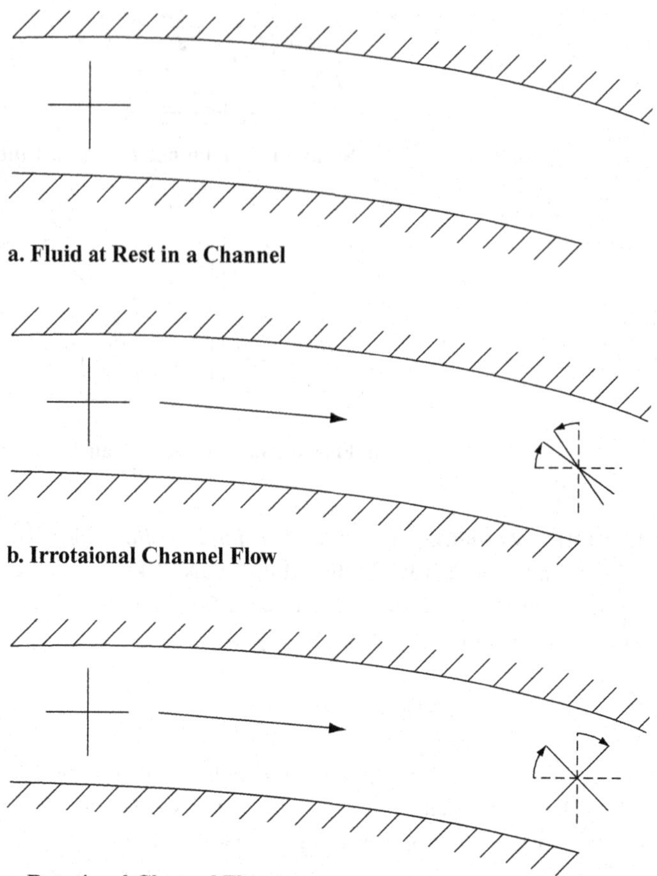

Figure 2.5. *Illustration of Irrotational and Rotational Flows.* In (a) the fluid is at rest, and the cross (+) is shown in its original position. In (b) the flow is irrotational because there is no net rotation of the cross members. That is, the positive rotation of one arm of the cross is equal in magnitude to the negative rotation of the other arm. In Figure 2.5c, the flow is rotational because the rotations of the arms are in the same direction.

of the flow in 1840 while, G. H. L. Hagen, working independently of Poiseuille, presented results of an experimental study of the flow in 1839. Both Poiseuille and Hagen found that the maximum velocity of eq. 2.28 is a linear function of the pressure gradient in the pipe.

2.3 Rotational and Irrotational Flows

The concept of the *rotationality* of a flow can be understood by considering the convection of a small cross (+) in a channel flow. In Figure 2.5a, the cross is shown at rest in the static fluid. In Figures 2.5b and 2.5c, the cross has migrated from its resting position through a portion of a bend in the channel. In Figure 2.5b, the cross has been deformed by having its component legs equally rotated in opposite directions. The net rotation about the center of the cross is zero, and the flow is said to be *irrotational.* In Figure 2.5c, the legs of the cross equally rotate in the clockwise direction, so that there is a net rotation about the center of the cross. The flow in this case is said to be *rotational*, and the positive direction of the rotation is counterclockwise.

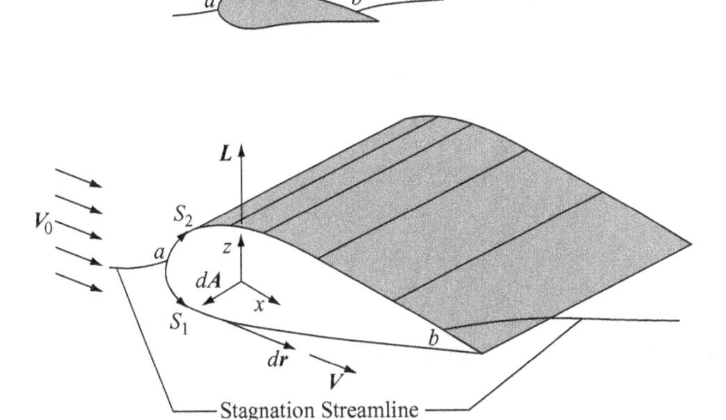

Figure 2.6. *Concept of Circulation about a Hydrofoil.* The two-dimensional flow about the hydrofoil is divided by the *stagnation streamline*, which includes the foil itself. When the line integrals of the product of the tangential flow velocity, V, and the surface direction vector, dr, are equal, there is no net circulation about the foil. The circulation, Γ, is related to the lift on the foil according to $L = \rho V\Gamma$. This relationship is called the *Kutta-Joukowski theorem.*

As is shown later in this chapter, the rotationality of a flow can be associated with flow losses.

A. Circulation

Associated with the rotationality of a flow is the concept of circulation. Referring to Figure 2.6, the *circulation* is defined as the line integral of the tangential velocity of a fluid about a closed path, S. In Figure 2.6, the closed path is on the surface of a hydrofoil. Mathematically, using the dot ($\cdot$) to represent a scalar product and the cross ($\times$) to represent a vector product, the circulation is defined by

$$\Gamma \equiv \oint_S \boldsymbol{V} \cdot d\boldsymbol{r} = \iint_A \nabla \times \boldsymbol{V} \cdot d\boldsymbol{A} \tag{2.32}$$

where $\boldsymbol{V}$ is the velocity vector defined in eq. 2.19, $d\boldsymbol{A}$ is the elemental vector area defined in eq. 2.24, and

$$d\boldsymbol{r} = dx\boldsymbol{i} + dy\boldsymbol{j} + dz\boldsymbol{k} \tag{2.33}$$

is the position vector. The relationship between the line and surface integrals in eq. 2.32 is the result of *Stokes' integral theorem*. See the book of Courant (1968) for a derivation of this theorem. Referring to Figure 2.6, the line integral in eq. 2.32 can be written as

$$\oint_S \boldsymbol{V} \cdot d\boldsymbol{r} = \int_a^b \boldsymbol{V} \cdot d\boldsymbol{r}|_{s_1} + \int_b^a \boldsymbol{V} \cdot d\boldsymbol{r}|_{s_2} \tag{2.34}$$

If the circulation is zero, then

$$\nabla \times \boldsymbol{V} = 0 \tag{2.35a}$$

in the surface integration of eq. 2.32, and the line integral is likewise equal to zero, resulting in

$$\int_a^b \boldsymbol{V} \cdot d\boldsymbol{r}|_{s_1} = -\int_b^a \boldsymbol{V} \cdot d\boldsymbol{r}|_{s_2} \tag{2.35b}$$

from eq. 2.34.

B. **The Velocity Potential**

In eqs. 2.32 and 2.33, the scalar product $V \cdot dr$ can be expressed as a scalar differential according to

$$\boldsymbol{V} \cdot d\boldsymbol{r} = d\phi \tag{2.36}$$

Furthermore, the scalar differential $d\phi$ in this expression can be expressed as

$$d\phi = \nabla\phi \cdot d\boldsymbol{r} \tag{2.37}$$

assuming that the function ϕ is continuous. Comparing the expressions in eqs. 2.36 and 2.37, we see that the velocity vector can be expressed in terms of the gradient of a continuous scalar function, that is,

$$\boldsymbol{V} = \nabla\phi \tag{2.38}$$

or, in component form,

$$u\boldsymbol{i} + v\boldsymbol{j} + w\boldsymbol{k} = \frac{\partial\phi}{\partial x}\boldsymbol{i} + \frac{\partial\phi}{\partial y}\boldsymbol{j} + \frac{\partial\phi}{\partial z}\boldsymbol{k} \tag{2.39}$$

The function ϕ is called the *velocity potential.* When the velocity vector in the surface integral of eq. 2.32 is replaced by the gradient of the velocity potential, the following mathematical identity results:

$$\nabla \times \nabla\phi \equiv 0 \tag{2.40}$$

From this result, we conclude that the velocity potential can be used to represent the fluid velocity if and only if the circulation is zero, that is, the flow must be irrotational. In other words, eq. 2.35 must be satisfied.

The advantage of using the velocity potential is that its use reduces the number of dependent variables from three (u, v, w) to one (ϕ). To illustrate, consider the equation of continuity for an incompressible flow, as expressed by eq. 2.21. A general solution of that equation cannot be obtained if the equation has more than one independent variable. If the flow can be considered to be irrotational, then the velocity vector can be represented by the velocity potential, as in eq. 2.38. The result obtained by combining eqs. 2.21 and 2.38 is

$$\nabla^2\phi = 0 \tag{2.41}$$

where the differential operator, called the Laplacian, is defined by

$$\nabla^2 \equiv \frac{\partial^2}{\partial x^2} + \frac{\partial^2}{\partial y^2} + \frac{\partial^2}{\partial z^2} \tag{2.42}$$

Equation 2.41 is called *Laplace's equation*, and is a basic equation of the linear wave theory discussed in Chapter 3. Laplace's equation does have a general solution

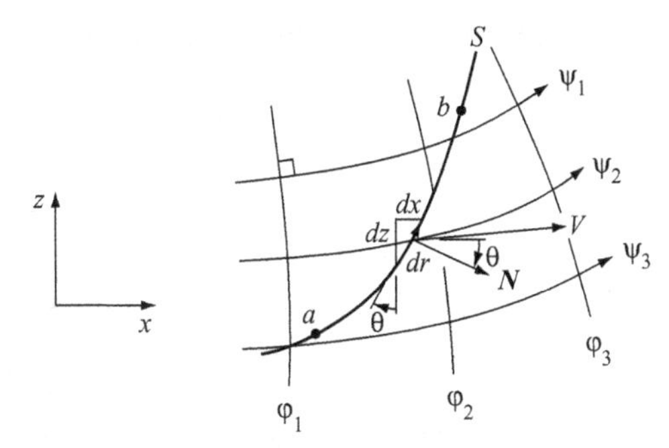

Figure 2.7. *Velocity Potential* (ϕ) *and Stream Function* (ψ) *Geometries.* The stream function can be used only in the analysis of two-dimensional flows, both irrotational and rotational. Because axially symmetric flows are two-dimensional, the stream function is a useful tool in the analysis of these flows, as discussed in Section 2.6. The lines representing the velocity potential and the stream function (*streamlines*) are orthogonal.

because it is a linear equation involving only one dependent variable. See Appendix E for the details.

C. The Stream Function

As discussed in the previous section, the velocity potential can be used in the analysis of three-dimensional irrotational flows. Consider the two-dimensional flow sketched in Figure 2.7. Our interest is in the *volume flux* (per unit depth into the page) of fluid across the line S. That flux is mathematically represented by

$$\int_a^b \boldsymbol{V}\cdot\boldsymbol{N}|d\boldsymbol{r}| = \int_a^b d\psi = \int_a^b \nabla\psi\cdot d\boldsymbol{r} \tag{2.43}$$

These relationships are similar to those in eqs. 2.36 and 2.37. That is, we have replaced the scalar product of the velocity (eq. 2.19) and the line element (eq. 2.33) by the scalar element $d\psi$.

The vectors in eq. 2.43 can be written in component form as

$$\boldsymbol{V} = u\boldsymbol{i} + w\boldsymbol{k} \tag{2.44}$$

$$d\boldsymbol{r} = dx\boldsymbol{i} + dz\boldsymbol{k} \tag{2.45}$$

and

$$\boldsymbol{N} = \cos(\theta)\boldsymbol{i} - \sin(\theta)\boldsymbol{k} = \frac{dz}{dr}\boldsymbol{i} - \frac{dx}{dr}\boldsymbol{k} \tag{2.46}$$

The scalar differentials in eq. 2.43 are

$$dr = |d\boldsymbol{r}| \tag{2.47}$$

and

$$d\psi = \frac{\partial\psi}{\partial x}dx + \frac{\partial\psi}{\partial z}dz \tag{2.48}$$

The combination of eqs. 2.43 through 2.48 yields

$$\int_a^b \boldsymbol{V}\cdot\boldsymbol{N}dr = \int_a^b u dz - \int_a^b w dx = \int_a^b \frac{\partial\psi}{\partial x}dx + \int_a^b \frac{\partial\psi}{\partial z}dz \tag{2.49}$$

From a comparison of the integrands of the last four integrals of eq. 2.49, the following relationships for the velocity components are

$$u = \frac{\partial \psi}{\partial z}, \; w = -\frac{\partial \psi}{\partial x} \tag{2.50}$$

The scalar function ψ is called the *stream function.*

The advantage of using the stream function is that its use reduces the number of dependent variables from two (u,w) to one (ψ). Furthermore, no assumption of irrotationality is required. Therefore, the stream function can be used in the analyses of both rotational and irrotational two-dimensional flows. When the flow is both two-dimensional and irrotational, then the Cartesian velocity components can be represented by either the velocity potential or the stream function. Mathematically, we can write the velocity components as

$$u = \frac{\partial \phi}{\partial x} = \frac{\partial \psi}{\partial z} \tag{2.51}$$

in the horizontal direction, and

$$w = \frac{\partial \phi}{\partial z} = -\frac{\partial \psi}{\partial x} \tag{2.52}$$

in the vertical direction. Equations 2.51 and 2.52 are called the *Cauchy-Riemann equations.*

The curves described by the potential function and the stream function can be found by determining the conditions for which the differentials of these functions are zero. That is, we seek the respective lines on which the functions are constant. First, consider the condition for which the differential of the velocity potential is zero,

$$d\phi = \frac{\partial \phi}{\partial x}dx + \frac{\partial \phi}{\partial z}dz = udx + wdz = 0 \tag{2.53}$$

From the last equality, the following is obtained:

$$\frac{u}{w} = -\left.\frac{dz}{dx}\right|_{\phi = constant} \tag{2.54}$$

Similarly, the condition for the constant stream function is

$$d\psi = \frac{\partial \psi}{\partial x}dx + \frac{\partial \psi}{\partial z}dz = -wdx + udz = 0 \tag{2.55}$$

from which we obtain

$$\frac{u}{w} = \left.\frac{dx}{dz}\right|_{\psi = constant} \tag{2.56}$$

Comparing the results of eqs. 2.54 and 2.56, we find that, for a two-dimensional irrotational flow, lines of constant ϕ and ψ are orthogonal, as illustrated in Figures 2.7 and 2.8. The lines of constant ψ are called *streamlines* and are tangent to the velocity vectors. In Figure 2.8, the boundaries of the flow are identified by the stream functions ψ_0 and ψ_4. In general, any streamline can be considered to be a boundary of the flow and, as such, no flow can cross streamlines.

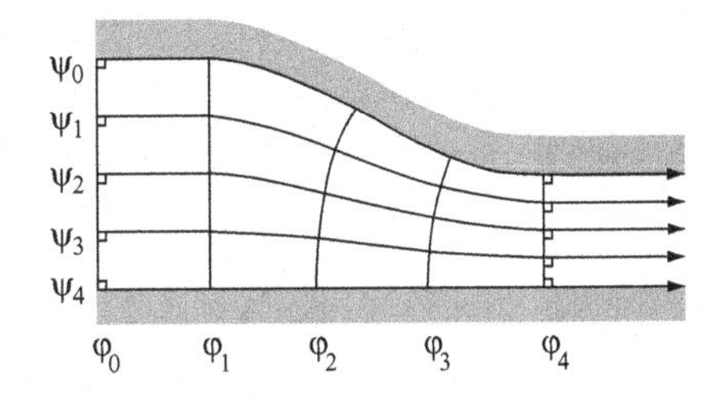

Figure 2.8. *Flow in a Converging Channel.* The boundaries of the flow are identified by constant steam function values. Hence, any streamline can be considered to be a flow boundary as there can be no flow across a streamline.

D. Superposition of Irrotational Flow Patterns

For an irrotational flow, the equation of continuity can be written in terms of the velocity potential, ϕ. The result is Laplace's equation, eq. 2.41. That equation is a linear, second-order, partial differential equation. Because it is linear, the various solutions of the equation can be combined to form other solutions by the *principle of superposition*. Physically, this means that irrotational flow patterns can be superimposed on each other to obtain other flow patterns. Several of the basic flow patterns are shown in Figure 2.9 along with the associated expressions for the velocity potential. The principle of superposition is illustrated in the following example.

EXAMPLE 2.3: TWO-DIMENSIONAL IRROTATIONAL FLOW ABOUT A CIRCULAR CYLINDER The geometric body that is most widely used in offshore ocean engineering structures is the circular cylinder. For most fixed and floating platforms, the circular cylinder is used for both the surface-piercing legs and the cross-bracing between the legs. To obtain the ideal flow pattern about a cylindrical cross-section, combine the velocity potentials for the uniform horizontal flow (Figure 2.9a) and the doublet (Figure 2.9d). The resulting velocity potential is

$$\begin{aligned} \phi &= \phi_{flow} + \phi_{doublet} \\ &= V_0 x + C\left(\frac{x}{x^2 + z^2}\right) \\ &= \left(V_0 r + \frac{C}{r}\right)\cos(\theta) \end{aligned} \tag{2.57}$$

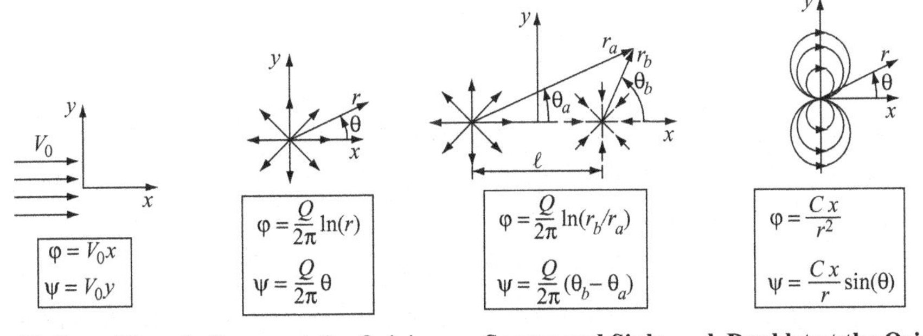

a. Uniform Flow **b. Source at the Origin** **c. Source and Sink** **d. Doublet at the Origin**

Figure 2.9. *Basic Two-Dimensional Flow Patterns.* These and other flow patterns can be combined to represent more complex flows by simply adding the velocity potentials or the stream functions representing the component flows. For example, the addition of the velocity potentials for the uniform flow in (a) and the doublet in (d) yields the velocity potential for the flow about a circular cylinder, as sketched in Figure 2.10.

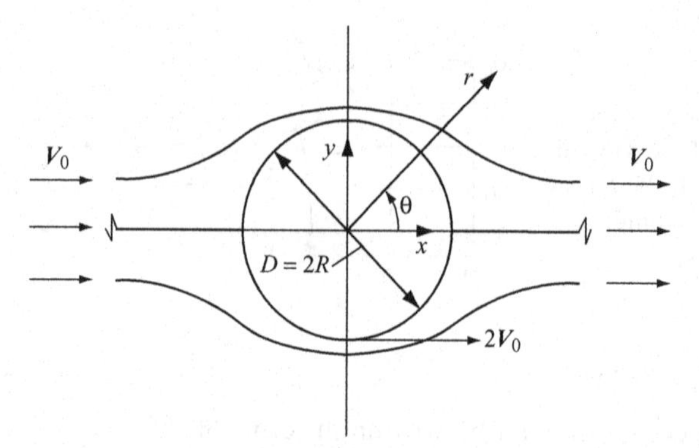

Figure 2.10. *Two-Dimensional Potential Flow Past a Circular Cylinder.* Low-Reynolds number (V_0D/ν, where ν is the *kinematic viscosity*) flows represented by potential or irrotational flow patterns are in good agreement with those experimentally observed. For example, see the photograph of Sadatoshi Taneda of the *quasi*-two-dimensional flow about a circular cylinder at a Reynolds number of approximately 0.16 in the book of Milton Van Dyke (1982).

and the flow pattern is sketched in Figure 2.10. It is advantageous to use the polar coordinate system because of the body geometry. The Cauchy-Riemann equations of eqs. 2.51 and 2.52 in polar coordinates are

$$V_r = \frac{\partial\phi}{\partial r} = \frac{1\partial\psi}{r\partial\theta} \tag{2.58}$$

and

$$V_\theta = \frac{1}{r}\frac{\partial\phi}{\partial\theta} = -\frac{\partial\psi}{\partial r} \tag{2.59}$$

On the surface of the cylinder, where the surface is intersected by the x-axis, the velocity must be zero. For this reason, the points of intersection are called *stagnation points*, and the streamline that is both coincident with the x-axis and defines the surface of the cylinder is called the *stagnation streamline.* On the cylinder, there can be no radial velocity component because flow cannot cross streamlines. The radial velocity component is obtained by combining eqs. 2.57 and 2.58. At the stagnation points, the radial velocity is

$$V_r|_{stagnation} = \frac{1}{R}\left(V_0 - \frac{C}{R^2}\right)(\pm 1) = 0 \tag{2.60}$$

Therefore, the constant in eqs. 2.57 and 2.60 is

$$C = V_0 R^2 \tag{2.61}$$

The velocity potential of eq. 2.57 is then

$$\phi = V_0 R\left(\frac{r}{R} + \frac{R}{r}\right)\cos(\theta) \tag{2.62}$$

The velocity adjacent to the top and bottom of the cylinder can be only in the angular direction. That is, at $r = R$ and $\theta = 90^\circ$ and 270°, the adjacent velocity is

$$V_\theta|_{90,270} = \mp 2V_0 \tag{2.63}$$

where the minus and plus signs refer to the top and bottom of the cylinder, respectively.

Returning to the Cauchy-Riemann relationships of eqs. 2.58 and 2.59, we can determine the stream function by combining those equations with the velocity potential of eq. 2.62 and integrating the resulting expressions to obtain

$$\psi = V_0 R\left(\frac{r}{R} - \frac{R}{r}\right)\sin(\theta) \tag{2.64}$$

Consider the conditions for which the value of the stream function is zero. From the results in eq. 2.64, we see that the $\psi = 0$ when $r = R$ and when $\theta = 0°$ and $180°$. From this observation, we can conclude that the zero-value of the stream function defines the stagnation streamline. This streamline, in turn, defines the body in the flow.

It is possible to determine the irrotational flow patterns that are far more complicated than those presented in Figure 2.9 by using conformal transformation of complex variables. There are a number of excellent books on the subject of flow representation using conformal transformations. The text by Granger (1985) gives an excellent introduction to the subject. For more advanced treatments of the subject, the books of Vallentine (1967) and Karamcheti (1966) are recommended.

2.4 Conservation of Momentum and Energy

Newton's second law of motion applied to fluid motion can be paraphrased as the time-rate of change of linear momentum of the fluid is equal to the sum of the external forces acting on the fluid. For an incompressible flow, this statement can be expressed mathematically by

$$m\frac{DV}{Dt} = m\left(\frac{\partial \boldsymbol{V}}{\partial t} + \boldsymbol{V}\cdot\nabla V\right) = \Sigma \boldsymbol{F} \tag{2.65}$$

where m is the mass of the fluid in a control volume, $\boldsymbol{V}$ is the velocity vector, and $\boldsymbol{F}$ is an external force acting on the mass in the control volume. In eq. 2.65, the notation DV/Dt is used to represent the total time derivative of the velocity. Also in the equation, there are two types of fluid acceleration. The first $(\partial \boldsymbol{V}/\partial t)$ is called the *local acceleration*, and exists only if there is a velocity variation in time. For example, the flow in Figure 2.9a has a local acceleration if the velocity $\boldsymbol{V}_0$ is time-dependent. The second type of acceleration $(\boldsymbol{V}\cdot\nabla\boldsymbol{V})$ is called the *convective acceleration*, and exists because of a change in flow geometry. For example, particles traveling along a streamline from the position of the ϕ_0-line to the ϕ_4-line in Figure 2.8 accelerate because of the change in the flow area between the two potential lines.

The external forces acting on the fluid are *body forces*, such as the gravitational force $(-\rho g\boldsymbol{k})$, and *surface forces*, which include those due to pressure gradient $(-\nabla p)$ and shear stress $(\mu\nabla^2 V)$. The reader is referred to the books by Schlichting (1979) and Granger (1985) for the derivations of the expressions for these forces. Applied to an incompressible flow, the expression in eq. 2.65 is

$$\rho\left(\frac{\partial \boldsymbol{V}}{\partial \boldsymbol{t}} + \boldsymbol{V}\cdot\nabla\boldsymbol{V}\right) = -\rho g\boldsymbol{k} - \nabla p + \mu\nabla^2\boldsymbol{V} \tag{2.66}$$

This expression is a form of the *Navier-Stokes equations*. This vector equation represents three equations, one in each of the component directions. Equation 2.66 is nonlinear because of the convective acceleration. As such, the Navier-Stokes equations have no general solution. However, there are numerous specific solutions that have engineering significance. These solutions can be found in the book by Schlichting (1979).

When the viscous forces are neglected $(\mu = 0)$, the flow is said to be *inviscid*. For an inviscid flow, the Navier-Stokes equations reduce to a form called

Euler's equations, that is,

$$\rho\left(\frac{\partial \boldsymbol{V}}{\partial t}+\boldsymbol{V}\cdot\nabla\boldsymbol{V}\right)=-\rho g\boldsymbol{k}-\nabla p \tag{2.67}$$

It should be noted that like the Navier-Stokes equations, Euler's equations are nonlinear because of the convective acceleration. Hence, there is no general solution to Euler's equations.

The nonlinear convection acceleration term in eqs. 2.66 and 2.67 can be replaced by the following vector identity:

$$\boldsymbol{V}\cdot\nabla\boldsymbol{V}=\nabla\left(\frac{V^2}{2}\right)-\boldsymbol{V}\times(\nabla\times\boldsymbol{V}) \tag{2.68}$$

When the flow is irrotational, then eq. 2.35 is satisfied, and the last term in eq. 2.68 vanishes. For an irrotational flow, the velocity potential, ϕ, can be introduced using the expression in eq. 2.38. The combination of eqs. 2.67, 2.68, and 2.35 yields the irrotational form of Euler's equations:

$$\rho\nabla\left(\frac{\partial\phi}{\partial t}+\frac{V^2}{2}+gz+\frac{p}{\rho}\right)=0 \tag{2.69}$$

As in eq. 2.37, the vector product of the gradient in eq. 2.69 and the direction vector of eq. 2.33 results in the differential of the scalars in the brackets of eq. 2.69. This differential scalar can be integrated to obtain

$$\rho\frac{\partial\phi}{\partial t}+\rho\frac{V^2}{2}+\rho gz+p=f(t) \tag{2.70}$$

which is *Bernoulli's equation*. The time function, $f(t)$, results from the spatial integration. The respective terms on the left side of Bernoulli's equation are the *unsteady kinetic energy*, *kinetic energy*, *potential energy*, and the *flow energy*, all per unit fluid volume. Bernoulli's equation is a mathematical expression of the conservation of energy for an irrotational flow. This equation is one of the basic equations of the wave theories of Airy and Stokes, both of which are discussed in the next chapter. The units of each term in eq. 2.70 are those of pressure. From left to right on the left side of the equation, the first two terms represent the *dynamic pressure* while the third term is the *hydrostatic pressure* and the last term is the *total pressure.* Note: The units of pressure are equivalent to those of energy per unit volume.

EXAMPLE 2.4: PRESSURE DISTRIBUTION ON A CYLINDER IN AN IRROTATIONAL FLOW

In Example 2.3, the velocity potential and stream function are presented for an irrotational flow about a two-dimensional circular cylinder, as sketched in Figure 2.10. Using the results of that example, we can determine the pressure distribution on the cross-section of a vertical cylinder in a uniform flow of velocity V_0. On the cylinder, there is no radial velocity component because the flow cannot cross a streamline, which in this case is the surface of the cylinder. Hence, only the angular velocity component exists. The mathematical expression for the velocity adjacent to the cylinder is

$$V_\theta\big|_{r=R}=\frac{1}{r}\frac{\partial\phi}{\partial\theta}\bigg|_{r=R}=-2V_0\sin(\theta) \tag{2.71}$$

from the results of eqs. 2.59 and 2.62. Because the flow is irrotational, the pressure distribution on the cylinder can be obtained from Bernoulli's equation,

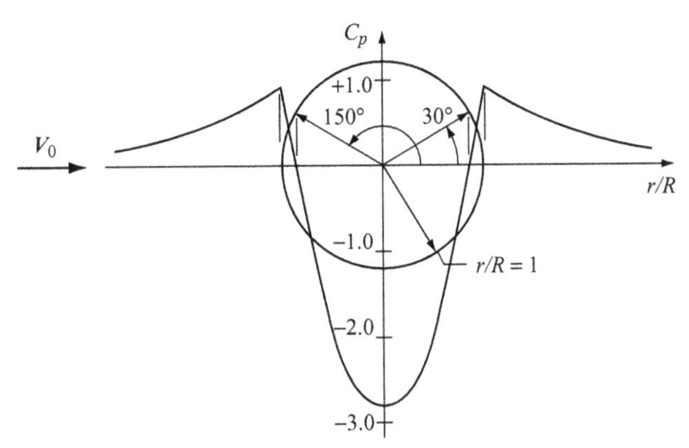

Figure 2.11. *Dynamic Pressure Coefficient Distribution on a Circular Cylinder in a Potential Flow.*

eq. 2.70. Assume that the flow is steady so that the time-derivative of the velocity potential is zero and the function $f(t)$ is a constant. When Bernoulli's equation is applied to the free surface or air-sea interface and well away from the cylinder, the total pressure (p) is zero-gauge (atmospheric), and the hydrostatic pressure (ρgz) is zero if the original of z is on the free surface. The resulting expression for the time function, $f(t)$, is $\rho(V_0^2/2)$. The expression for the total pressure at any submerged point on the cylinder is then

$$p|_{r=R} = \frac{\rho}{2}\left(V_0^2 - V_\theta^2\right) - \rho gz = \frac{\rho V_0^2}{2}[1 - 4\sin^2(\theta)] - \rho gz \tag{2.72}$$

From this result, the reader can see that the pressures at $\theta = 0°$ and 180° equal the free-stream dynamic pressure plus the hydrostatic pressure. The referred-to points on the cylinder are called *stagnation points.* The pressure at a stagnation point is called the *stagnation pressure.* In Figure 2.11, the dynamic pressure distribution at $z \leq 0$ is plotted in the non-dimensional form of the *pressure coefficient*, defined as

$$C_p = \frac{\frac{\rho}{2}\left(V_0^2 - V_\theta^2\right)}{\rho V_0^2/2} \tag{2.73}$$

where the parameters having the subscript "0" are those at $x = \pm\infty$. The stagnation pressure is that where $C_p = 1$. In that plot, we see that the dynamic pressure is zero-gauge at $\theta = \pm 30°$.

A phenomenon that is encountered in high-speed hydromechanics is that of cavitation. The term *cavitation* is the name given to low-pressure boiling. Fresh water boils if the ambient pressure at a point equals the vapor pressure of water,

$$p_v = -9.96 \times 10^4 \text{ N/m}^2 \text{ (absolute)} \tag{2.74}$$

From the results of Example 2.4, we see that the cavitation will first occur at the point $z = 0$, $\theta = \pm 90°$. The determination of the cavitation speed is illustrated in Example 2.5.

EXAMPLE 2.5: INCIPIENT CAVITATION ON A VERTICAL CIRCULAR CYLINDER The results of eq. 2.71 show that the maximum velocity on the circular cylinder occurs where $\theta = \pm 90°$. The two points of maximum velocity at the free surface are the points of minimum pressure, as can be seen from the results in eq. 2.72.

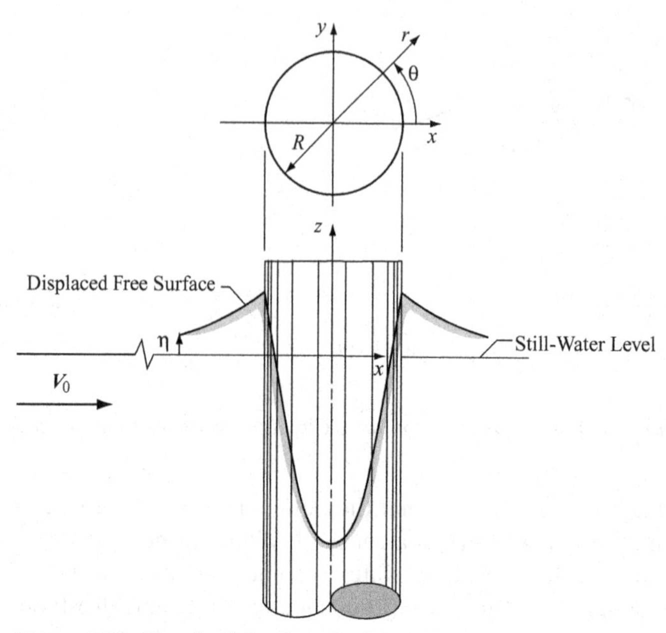

Figure 2.12. *Sketch of the Free-Surface Displacement about a Vertical Circular Cylinder in a Steady Flow, as Predicted by Irrotational Flow Theory.* The free-surface displacement, η, is measured from the still-water level.

From that equation, the expression for the minimum dynamic pressure is

$$p_{dynamic} = -1.5\rho V_0^2, \text{ at } r = R, \; \theta = \mp 90°, \; z = 0 \tag{2.75}$$

For the vapor pressure in eq. 2.74, the cavitation speed is

$$V_0 = \sqrt{-\frac{p_v}{1.5\rho}} = 8.15 \text{ m/s} \tag{2.76}$$

for fresh water, where the mass density is $\rho = 1{,}000 \text{ kg/m}^3$.

If the total pressure expression of eq. 2.72 is applied to the free surface where the static pressure is the ambient pressure (zero-gauge), then the deflection of the free surface about the cylinder can be determined. Referring to Figure 2.12 for notation, the free-surface deflection obtained from eq. 2.72 is

$$z|_{r=R,P=0} \equiv \eta = \frac{V_0^2}{2g}[1 - 4\sin^2\theta] \tag{2.77}$$

The results of eq. 2.77 are plotted in Figure 2.12. In this figure, the free-surface deflection is seen to be greatest on both the upstream and downstream stagnation lines of the cylinder while being lowest on the sides. The effects of viscosity cause the actual curvature of the free surface about the cylinder to be different from that sketched because energy is being transformed into heat by the action of viscosity. The effects of viscosity are discussed in the next section.

2.5 Viscous Flows

Real fluids have the ability to somewhat resist shear deformation. The property associated with that ability is called *viscosity*. Because of this property, the flow patterns near solid bodies are significantly altered from the inviscid patterns predicted

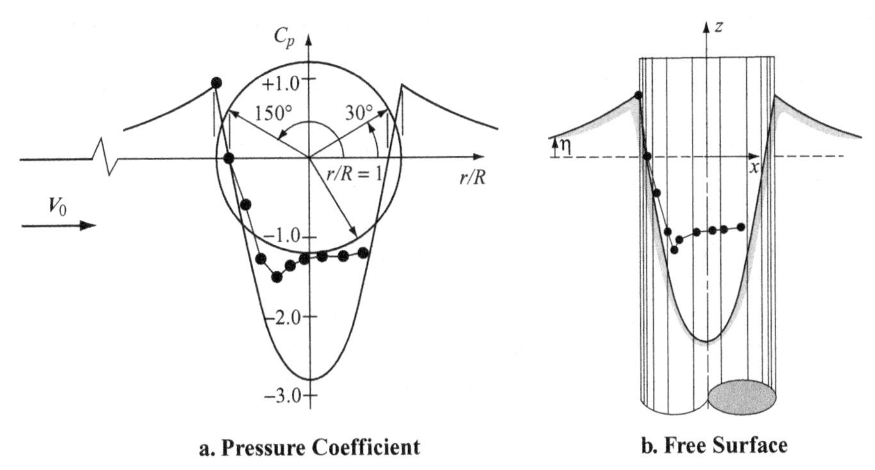

Figure 2.13. *Pressure-Coefficient and Free-Surface Displacements for Flow about a Vertical, Surface-Piercing Circular Cylinder in Steady Irrotational and Viscous Flows.* The measured data points are represented by •.

by the irrotational theory. Consider again the flow about a vertical circular cylinder. The irrotational flow pattern around a fully submerged cross section is sketched in Figure 2.10, whereas the free surface is sketched in Figure 2.12. The effects of viscosity on the fully submerged flow pattern are shown in Figure 2.13a, and the viscous effects on the free surface are illustrated in Figure 2.13b. Because of viscosity, the energy is lost from the flow and absorbed by the body in the form of heat.

Consider a fluid particle traveling along the upstream stagnation streamline. We know that the particle comes to a halt at the stagnation point. It is then divided, halves of the particles traveling to both sides of the cylinder. The property of viscosity causes the particles adjacent to a solid body to adhere to the surface. The particles adjacent to those adhering to the surface are slowed by attraction to the static particles, and the continuum must be preserved. The flow of the particles further away from the surface is slowed to a lesser extent. The net effect is a loss of kinetic energy in the region near the surface. Although the effects of viscosity theoretically extend an infinite distance from the surface, we can assume that there is a finite distance from the surface beyond which viscosity has no significant effect. Referring to Figure 2.14, the region in which viscous effects are evident is called the *boundary layer*, and the distance between the edge of the affected region and the surface is called the *boundary layer thickness*, denoted by δ.

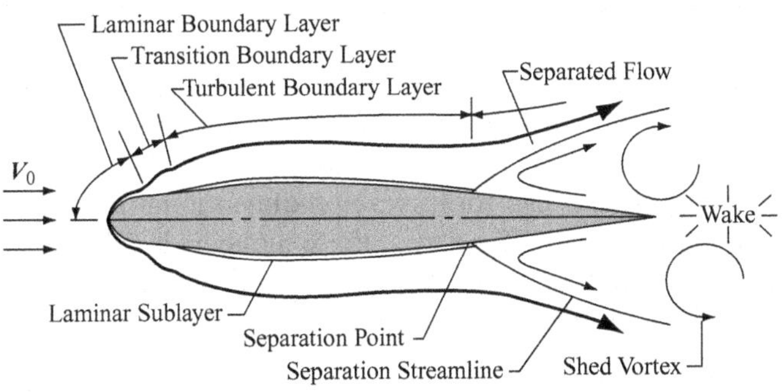

Figure 2.14. *Boundary Layers and Wake Created by a Viscous Flow Past a Two-Dimensional Symmetric Hydrofoil.* The scale of the sketch is exaggerated. A better understanding of the scale of such flows can be obtained from Photograph 156 in the book by Van Dyke (1982).

There are two forces that act on a particle in the flow near a solid. The first is the *inertial force*, and the second is the *viscous force*. The effects of these two forces are opposite in nature. The inertial force keeps the particle in motion, whereas the viscous force tries to stop the particle. One can imagine that these two forces are contesting with each other for control over the particle. When the viscous force is dominant, then the flow is slow and in well-ordered layers called *laminae*, and the flow is said to be *laminar*. Because molecular attraction causes the particles to adhere to the surface, viscous effects are strong near the body. Away from the body, the inertia becomes dominant. In the area close to the front of the body, the boundary layer is very thin. In this thin boundary layer, the viscosity is dominant and the flow is laminar. This viscous flow region is called the *laminar boundary layer*. As the particles in the boundary layer travel further downstream from the stagnation point, the effects of the inertia are experienced closer to the surface. The "contest" between the inertial and viscous forces on the particles result in sporadic transverse motions. The region in which these sporadic motions occur is called the *transitional boundary layer*. Because the transverse motions result in transverse momentum transfer between particles, the boundary layer thickness, δ, increases at a greater rate in the flow direction. The change in the curvature of the boundary layer causes more inertia to be absorbed and, eventually, the inertial effects are dominant over most of the boundary layer. This dominance causes random particle motions in both the transverse and streamwise directions that are superimposed on the mean flow. The nature of the flow is *turbulent*, and the affected region is called the *turbulent boundary layer*.

One must keep in mind that energy of the flow is being both lost to surface heating and redirected to the transverse direction in the turbulent boundary layer. For the flow over both the upper and lower surfaces of the two-dimensional body in Figure 2.14, there are points where the kinetic energies of the particles are zero. At these points, the particles momentarily stop and are subsequently pushed out into the adjacent flow by the particles that follow. Referring to the flow over the upper surface of the body in Figure 2.14, there is a streamline called the *separation streamline* along which the separated particles travel. Downstream of this streamline, some of the fluid particles travel in vortices in the wake of the body. The point on the body where the separation streamline begins is called the *point of separation*. The position of the point of separation is farther downstream in a turbulent boundary layer than in a laminar boundary layer because more inertia is absorbed due to the steeper turbulent boundary layer. The flow in the wake can be either laminar or turbulent. Furthermore, the vortices can remain in a relatively fixed position with respect to the body, or can be shed if the free-stream velocity (V_0) increases. *Vortex shedding* is a major area of concern to ocean engineers because the phenomenon causes periodic transverse and streamwise forces that, in turn, can cause unwanted body motions. One such situation is that of strumming of towing cables.

From what has been said concerning the relative effects of the inertial and viscous forces on particles near a body, we can form two non-dimensional numbers that represent the forces and fluid motions. The first of these numbers is called the *Reynolds number*, which results from the ratio of the inertial force and the viscous force. For both a circular cylinder and sphere, the Reynolds number is

$$R_{eD} = \rho VD/\mu = VD/\nu \tag{2.78}$$

where D is the body diameter, V is the free-stream (or towing) velocity, μ is the *dynamic viscosity*, ρ is the mass density of the fluid, and ν is called the *kinematic viscosity*. Physically, the dynamic viscosity is the ratio of the shear stress and the velocity gradient normal to the surface.

The second non-dimensional number is called the *drag coefficient*. This is the ratio of the drag force and the inertial force. It is defined mathematically by

$$C_{dD} = \frac{F_d}{\frac{1}{2}\rho V^2 A_d} \tag{2.79}$$

In eq. 2.79, F_d is the drag force and A_d is the projected area of the body. For a cylinder of length L normal to V, the projected area is DL whereas that of a sphere is $\pi D^2/2$.

In addition to the drag on circular cylinders, the *vortex shedding frequency* (f_V) can also be represented by a non-dimensional number called the Strouhal number. The *Strouhal number* is defined by the following equation:

$$S_{tD} = f_V D/V \tag{2.80}$$

There have been many experimental studies to determine the relationships of both the drag coefficient (C_d) and the Strouhal number (S_t) with the Reynolds number (R_e) for flows about circular cylinders. McCormick (1981) summarizes the results of many of the experimental studies, and his results are presented in Figure 2.15. The investigators responsible for those data are listed. Although not shown in Figure 2.15, the relationship between the two non-dimensional parameters is approximately linear on the log-log scale for Reynolds numbers less than approximately 4. For this Reynolds number range, the flow in the boundary layer is laminar and there is no wake. Between 4 and 40, a wake appears in which there are attached vortices. The flows in both the boundary layer and wake are laminar, and the drag is proportional to the velocity. For Reynolds numbers greater than 40, instabilities appear in the boundary layer and the wake begins to oscillate. At a Reynolds number of about 65, the vortices in the wake begin to shed. The flow in the shed vortices is laminar to a Reynolds number of about 800. As can be seen in Figure 2.15, the drag coefficient is approximately 1 in the Reynolds number range from approximately 800 to 6×10^3. Over this range, the flow in the boundary layer is laminar whereas that in the wake is turbulent. Above a Reynolds number of approximately 6×10^3, the flows in both the boundary layer and wake are mostly turbulent. As can be seen, the drag data for Reynolds numbers greater than 15×10^3 have much scatter. Furthermore, there are few data beyond 10^7 because experimental studies in the ultrahigh Reynolds regions are extremely difficult to perform.

The inverse of the Strouhal number is presented in Figure 2.15 because its behavior resembles that of the drag coefficient. The reader can see that there is much scatter in the data, with the maximum scatter occurring in the Reynolds number range from approximately 1.25×10^5 to 2.5×10^6. It is recommended that the lower curves of Figure 2.15 be used in engineering calculations because the data are concentrated near these curves.

A photographic album of numerous viscous flows is found in the book by Van Dyke (1982). This publication is highly recommended to the reader.

EXAMPLE 2.6: DRAG AND VORTEX SHEDDING FOR AN OTEC COLD-WATER PIPE
Ocean thermal energy conversion (OTEC) is one of the six ocean thermal energy conversion options. The other five energy sources are waves, ocean currents, salinity gradients, tides, and the biomass. The principle of OTEC is

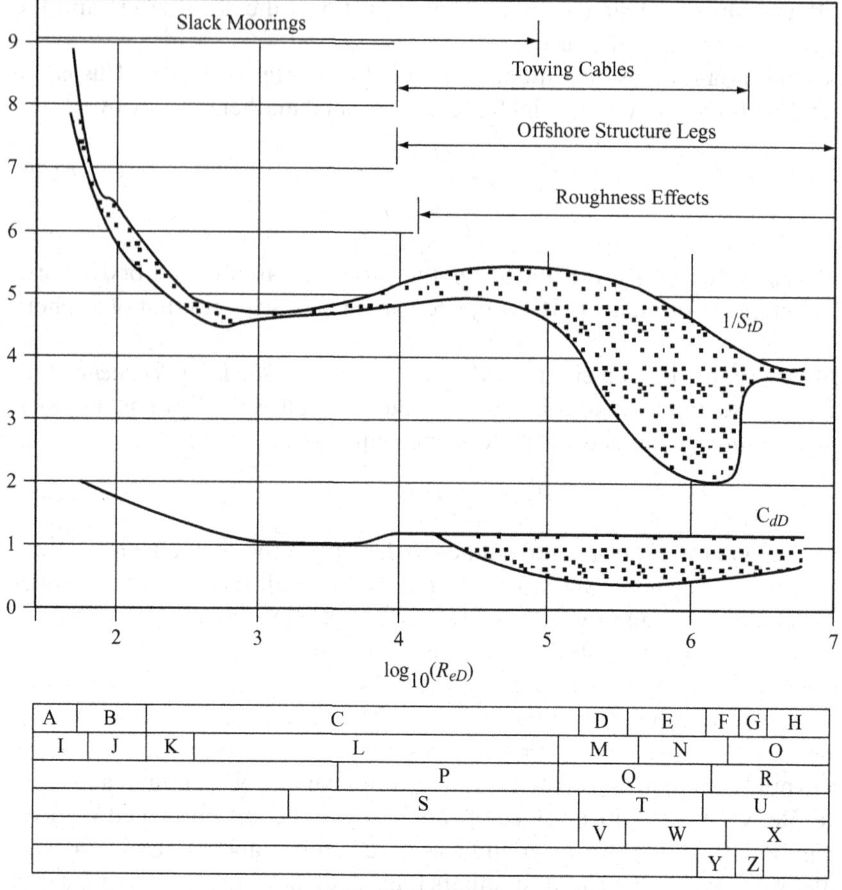

Figure 2.15. *Experimental Drag Coefficient and Strouhal Number as Functions of the Reynolds Number (Based on Diameter) for Circular Cylinders in Steady Flows.* The letters shown below the graph identify both the experimental observations and the regions studied by the experimenters, as listed in Table 2.1. The figure is after McCormick (1981).

as follows: Cold water from deep ocean water is upwelled through a large-diameter, cold-water pipe (CWP) to a condenser of a Rankine-type engine. Simultaneously, relatively warm surface water is passed through the evaporator of the engine. A working fluid such as ammonia is circulated in a closed system, absorbing heat from the evaporator and losing heat to the cold water in the condenser. When the working fluid absorbs the heat, it changes from a liquid phase to a gaseous phase, and the gas expands through a turbine. The working fluid changes from the gaseous phase to a liquid phase in the condenser as it loses heat to the upwelled cold water. The cycle then repeats itself. For those seeking more information on OTEC, the book by Avery and Wu (1994) is highly recommended. A simplistic OTEC floating power plant is sketched in Figure 2.16.

A hydrodynamic problem associated with an OTEC system concerns the vertical CWP. Because of the large diameter of the CWP, there are few or no reliable drag or vortex-shedding data as the Reynolds number values are too

Table 2.1. *Flow phenomena and data sources for Figure 2.15 (reference numbers are in parentheses)*

A	twin attached vortices (5)
B	laminar oscillatory wake and vortex shedding (5)
C	nonperiodic wake and transition to turbulence in the shear layer (5)
D	laminar "bubble" and turbulent wake (5 and 6)
E	questionable periodicity (3)
F	no periodicity observed (3)
G	small peaks in energy at various Strouhal number values (1)
H	peak in energy at a Strouhal number of approximately 0.3 (1)
I	unstable vortices (9)
J	laminar vortex shedding (9)
K	turbulence in free shear layer (9)
L	laminar boundary layer and turbulent vortex wake (9)
M	transition in the boundary layer (9)
N	questionable periodicity (3)
O	shear layer near separation is turbulent (2)
P	turbulent wake (8)
Q	vortex shedding frequency is the dominant frequency in spectrum (10)
R	turbulent boundary layer plus regular vortex shedding in the wake (3)
S	subcritical flow with the separation angle between 75° and 95° (6)
T	critical flow with the separation angle approximately 140° (6)
U	supercritical flow with the separation angle approximately 120° (7)
V	critical flow (8)
W	supercritical flow (8)
X	transcritical flow (4)
Y	wide band spectrum (10)
Z	narrow band spectrum (10)
Experimental References	
1	Sarpkaya (1979)
2	Jones, Cincotta, and Walker (1969)
3	Roshko (1961)
4	Miller (1976)
5	Morkovin (1964)
6	Achenbach (1971)
7	Achenbach (1968)
8	Schlichting (1979)
9	Reynolds (1974)
10	Every, King, and Weaver (1982)

large for practical experiments. To illustrate, consider an OTEC system having a CWP with a 30-m outside diameter (D) located in the Florida Current where the a subsurface current velocity (V_0) can exceed 2 m/s. Assuming that the kinematic viscosity (ν) of salt water is 1.2×10^{-6} m^2/s at 14°C, the Reynolds number (R_e) from eq. 2.78 is 5.0×10^7. From the results in Figure 2.15, we see that there are no drag coefficient (C_d) data or Strouhal number (S_t) data for this large Reynolds number value. Hence, a design dilemma exists because no practical experiment can be performed to supply the required data. To gain some idea of the magnitudes of the drag force and vortex-shedding frequency for the CWP, extrapolate the lower drag and vortex-shedding curves in Figure 2.15. The approximate results for the respective drag coefficient and the

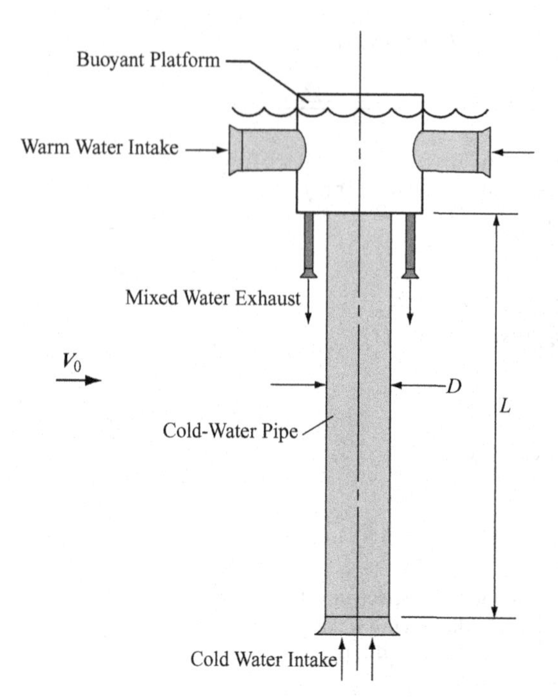

Figure 2.16. *Simplistic Sketch of One of the First Conceptual Designs of an Ocean Thermal Conversion (OTEC) System.* The conceptual design of the floating system was performed in the mid-1970s. More contemporary OTEC concepts are found in the book by Avery and Wu (1992). The operation of the sketched system is as follows: Deep cold water enters the bottom of the cold-water pipe (CWP) and flows up to a condenser in the platform. Near-surface warm water enters at the top of the platform and flows through an evaporator. The condenser and evaporator comprise a Rankine-type thermal engine. The diameter of the CWP varies from 32 to 40 m. When operating in strong currents, the large-diameter (D) CWP can pose a high-Reynolds number problem for both drag and vortex-shedding predictions.

inverse of the Strouhal number are 1.0 and 3.5, the latter yielding a Strouhal number of 0.286. For these values, the drag on a 300-m-long CWP is 1.85×10^7 N from eq. 2.79, and the vortex-shedding frequency is 0.0191 Hz from eq. 2.80. Concerning the vortex-shedding frequency, the beam type of vibrations of the CWP should be considered to determine if the fundamental or modal frequencies are near the vortex-shedding frequency. If the energy of the shed vortices is large, then there is the possibility of a near-resonance fatigue problem. For this hypothetical problem, the large drag would offer a challenge to the designer of the mooring system.

2.6 Hydrodynamics of Submerged Bodies

Naval architects have been aware that bodies of revolution are the best designs for high-speed fully submerged vehicles. The hydrodynamic analyses of the conceptual designs of these vehicles are usually based on the assumptions that the flows about a hull are both axially symmetric and irrotational. For this reason, the basics of ideal axially symmetric flows are presented herein. An excellent discussion of the flows past various bodies of revolution can be found in the book edited by Thwaites (1987).

In Section 2.3, it was shown that the stream function (ψ) can be used only in the analyses of two-dimensional flow situations. The advantage of using the stream function is that it can be used to represent both irrotational and rotational flows, the latter flow including the effects of viscosity. Axially symmetric flows are, by their nature, two-dimensional if the proper coordinate system is chosen. To illustrate, consider the flow about the sphere sketched in Figure 2.17. If we choose a Cartesian coordinate system, then the analysis of the flow cannot be two-dimensional because there are velocity components in all three coordinate directions in the neighborhood of the sphere.

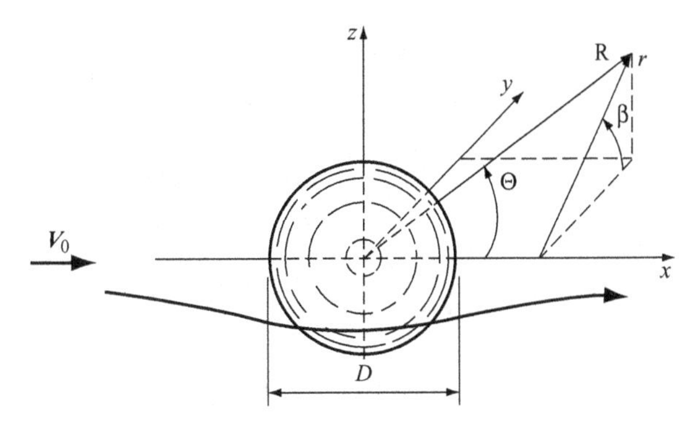

Figure 2.17 *Coordinate Systems Used in Axially Symmetric Flow Analyses.* To illustrate, the flow past a sphere is sketched. The cylindrical (x, r, β) and spherical (R, Θ, β) coordinate systems are shown with the Cartesian system (x, y, z).

A good choice of a coordinate system for the flow about a cylinder is the cylindrical system (x, r, β), which is also shown in Figure 2.17. The reader can see that for any value of r, the flow is independent of β. Hence, the flow is two-dimensional with respect to the cylindrical coordinate system. The stream function for this coordinate system is defined in terms of the volume rate of flow, Q, which can be mathematically defined as

$$Q = 2\pi(\Psi_b - \Psi_a) = 2\pi(\delta\Psi) \tag{2.81}$$

Here, the three-dimensional stream function (Ψ) is the *Stokes stream function.* From eq. 2.81, the volume rate of flow between the stream surfaces defined by the two functions (Ψ_a and Ψ_b) is known if the values of those stream functions are specified. The surfaces defined by specific stream functions are called *stream surfaces.* The Stokes stream function can also be represented in the spherical coordinate system $(\mathrm{R}, \Theta, \beta)$, also shown in Figure 2.17. Examples of flows in both the cylindrical and spherical coordinate systems are presented in Figure 2.18.

The velocity potential for irrotational axially symmetric flows is denoted by Φ. The relationships between the stream function and the velocity potential for irrotational flows are determined from the velocity components as follows: In cylindrical coordinates,

$$V_x = \frac{1}{r}\frac{\partial \Psi}{\partial r} = \frac{\partial \Phi}{\partial x} \tag{2.82}$$

and

$$V_r = -\frac{1}{r}\frac{\partial \Psi}{\partial x} = \frac{\partial \Phi}{\partial r} \tag{2.83}$$

For spherical coordinates,

$$V_\Theta = \frac{1}{\mathrm{R}\sin(\Theta)}\frac{\partial \Psi}{\partial \mathrm{R}} = \frac{1}{\mathrm{R}}\frac{\partial \Phi}{\partial \Theta} \tag{2.84}$$

and

$$V_\mathrm{R} = \frac{1}{\mathrm{R}^2\sin(\Theta)}\frac{\partial \Psi}{\partial \Theta} = \frac{\partial \Phi}{\partial \mathrm{R}} \tag{2.85}$$

There is one major difference between the relationships of the stream function and velocity potential in two-dimensional and axisymmetric flows. In irrotational two-dimensional flows, lines corresponding to constant values of ϕ and ψ are orthogonal whereas surfaces corresponding to constant values of Φ and Ψ are not always orthogonal.

Cylindrical: $\Phi = V_0 x$, $\Psi = 0.5 V_0 r^2$
Spherical: $\Phi = V_0 \mathrm{R}\cos(\Theta)$, $\Psi = 0.5 V_0 \mathrm{R}^2 \sin^2(\Theta)$

a. Uniform Flow

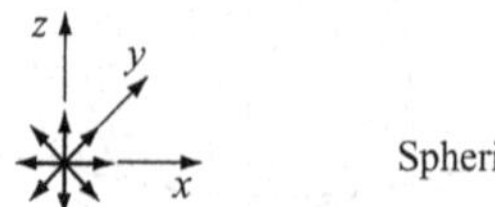

Spherical: $\Phi = -M/\mathrm{R}$, $\Psi = -M\cos(\Theta)$

b. Source at the Origin

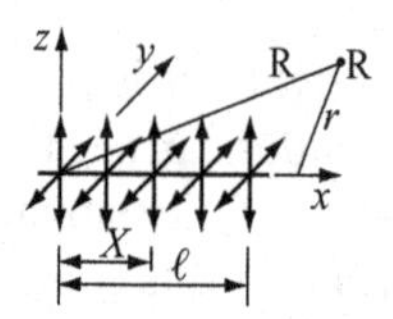

Cylindrical: $\Psi = \int_\ell \mathrm{M}_+(x)\,(x - X)[(x - X)^2 + r^2]^{-1/2} dX$

c. Line Source

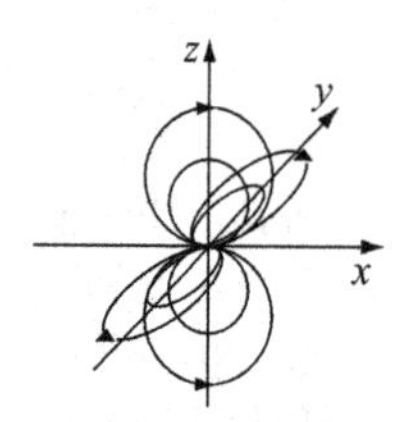

Spherical: $\Phi = (\mu/\mathrm{R}^2)\cos(\Theta)$, $\Psi = -(\mu/\mathrm{R})\sin^2(\Theta)$

d. Doublet at the Origin

Figure 2.18. *Basic Axially Symmetric Flow Patterns.* In (b) M is the strength of the source. For a sink, the negative sign (−) is replaced by a positive sign (+). In (c), $\mathrm{M}_+(x)$ is the stength of the line source. In (d), μ is the strength of the doublet.

EXAMPLE 2.7: FLOW ABOUT A SPHERE In Example 2.3, the velocity potential and stream function for the two-dimensional irrotational flow about a cylinder are derived by combining the respective expressions for a uniform free-stream flow and a doublet. Following the same procedure, we obtain the expression for the flow about a sphere of diameter D. Referring to Figure 2.19 for notation, the stream function for that flow is

$$\Psi = \frac{V_0}{2}\left[\mathrm{R}^2 - \frac{D^3}{8\mathrm{R}}\right]\sin^2(\Theta) \tag{2.86}$$

The strength of the doublet (μ in Figure 2.18a) is found by equating the stream function to zero on the sphere, that is, at $\mathrm{R} = D/2$. Adjacent to the sphere, there can only be an angular velocity component because there can not be flow across a stream surface (in this case, the sphere itself). The expression for the angular velocity component in spherical coordinates, found from the combination of eqs. 2.84 and 2.86, is

$$V_\Theta = -\frac{V_0}{2}\left[2 + \frac{D^3}{8\mathrm{R}^3}\right]\sin(\Theta) \tag{2.87}$$

The velocity adjacent to the sphere is then

$$V_\Theta|_{\mathrm{R}=\frac{D}{2}} = -\frac{3}{2} V_0 \sin(\Theta) \tag{2.88}$$

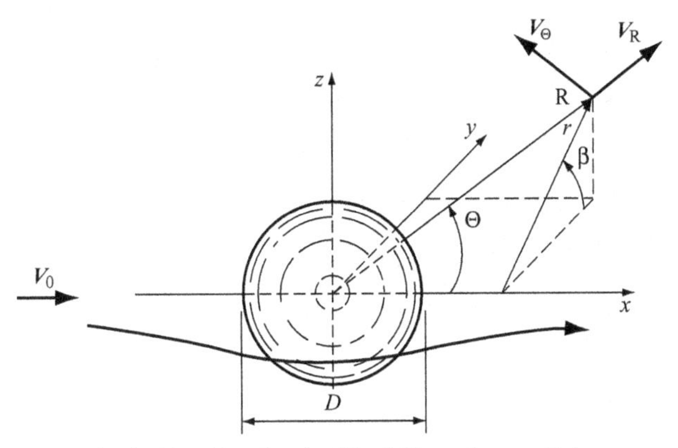

Figure 2.19. *Notation for the Ideal Flow about a Sphere.*

The pressure distribution on the sphere is found by applying Bernoulli's equation (eq. 2.70). The flow is assumed to be steady; hence,

$$\frac{\partial \Phi}{\partial t} = 0 \tag{2.89}$$

Furthermore, the total energy per unit volume on the stagnation streamline at $x = \pm\infty$ is

$$f(t) = \frac{1}{2}\rho V_0^2 + p_0 \tag{2.90}$$

in eq. 2.70. The resulting expression for the pressure on the sphere is then

$$p|_{R=D/2} = \frac{\rho V_0^2}{2}\left[1 - \frac{9}{4}\sin^2(\Theta)\right] + p_0 \tag{2.91}$$

The first term on the right side of eq. 2.91 is the difference in the dynamic pressures, whereas the last term in the equation is the ambient pressure. A plot of the dimensionless dynamic pressure difference on a horizontal plane bisecting the sphere is presented in Figure 2.20. Comparing the velocity expressions

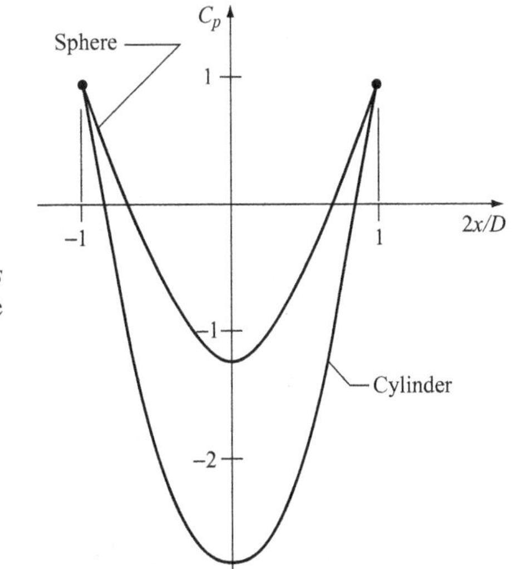

Figure 2.20. *Non-Dimensional Pressure Distributions over a Sphere and a Cylinder of Equal Diameters.* The pressure coefficient is defined in eq. 2.73.

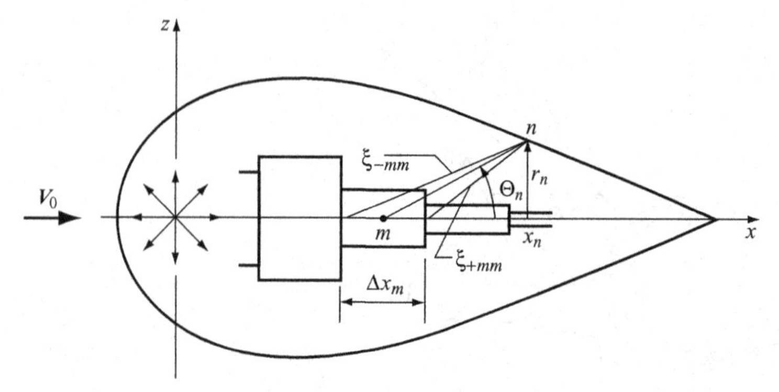

Figure 2.21. *Body of Revolution in a Uniform Flow.* The body shape is created by combining line sources of various strengths with a uniform flow that is parallel to the x-axis.

of eqs. 2.71 and 2.88, we see that the maximum velocity on the sphere is 75% that on the cylinder. Furthermore, the minimum dynamic pressure on the sphere (from eq. 2.91) is approximately 42% greater than that on the cylinder (eq. 2.72). These results illustrate the flow and pressure "relief" due to the three-dimensionality of the flow about the sphere.

By combinations of sources and sinks (of various strengths), a uniform flow can result in the flow about a more complicated body of revolution. Referring to the sketch in Figure 2.21, different body shapes are obtained by simply adjusting the strengths of the component sources and sinks. For a body with a desired (design) shape, the accuracies of both the theoretical body shape and the flow increase as the number of the sources and sinks increases and the distance between them decreases. The sources and sinks are collectively referred to as *line sources*, whereas the points on the body are called *body points*. It should be noted that the number of body points (actually rings of radii r_n) is equal to the combined number of sources and sinks. In Figure 2.21, the line sources are represented by blocks in the two-dimensional figure, and the distances between end points of block m to a body point n by ξ_{-mn} (from the left end of the block) and ξ_{+mn} (from the right end of the block). If M_m is the strength of the source, then the stream function at point n due to the line source is

$$\begin{aligned}\delta\Psi_{mn} &= -M_m(\xi_{-mn} - \xi_{+mn}) \\ &\simeq -M_m \delta x_m \cos(\Theta_n)\end{aligned} \tag{2.92}$$

where Θ_n is that shown in Figure 2.21. The approximation is valid if the elemental length δx is very small. Now combine all of the elemental stream functions to obtain

$$\begin{aligned}\Psi_n &= -\sum_{m=1}^{m_l} M_m(\xi_{-mn} - \xi_{+mn}) \\ &= -\sum_{m=1}^{m_l} M_m \delta x_m \cos(\Theta_n)\end{aligned} \tag{2.93}$$

The expression in eq. 2.93 is that of the stream function at point n. To obtain the Stokes stream function for a continuous line source, pass to the limit as δx

approaches zero in eq. 2.93. The result is

$$\Psi_n = -\int_0^l \mathrm{M}(x)\cos(\Theta_n)dx_n \tag{2.94}$$

The cosine term in eq. 2.94 can be replaced by

$$\cos(\Theta_n) = \frac{x - x_n}{\sqrt{(x - x_n)^2 + r_n^2}} \tag{2.95}$$

Hence, eq. 2.94 can be rewritten as

$$\Psi_n = -\int_0^l \mathrm{M}(x)\frac{x - x_n}{\sqrt{(x - x_n)^2 + r_n^2}}dx_n \tag{2.96}$$

An excellent discussion of the use of line sources and sinks is found in the book by Shames (1962).

We now combine the line source with a point source and a uniform flow to obtain the flow about a body similar to that sketched in Figure 2.21. As is the case for the flow about a sphere, the body shape can be obtained combining various stream functions and setting the resulting combination equal to zero. We choose to combine a line sink of length ℓ and strength M_- with a point source at the origin of strength M_+ and a uniform flow of velocity V_0 parallel to the x-axis. The resulting stream function is

$$\Psi = \int_0^l \mathrm{M}_-(x)\frac{x - x_n}{\sqrt{(x - x_n)^2 + r_n^2}}dx_n - M_+\cos(\Theta) + \frac{V_0}{2}r_n^2 \tag{2.97a}$$

and the body then is defined by $\Psi = 0$, or

$$\int_0^l \mathrm{M}_-(x)\frac{x - x_n}{\sqrt{(x - x_n)^2 - r_n^2}}dx_n = M_+\cos(\Theta) - \frac{V_0}{2}r_n^2 \tag{2.97b}$$

There are two methods available for applying the results in eq. 2.97b to axially symmetric flows about bodies of revolution. The first is to choose a body shape by specifying $r(x)$ on the body and solve for the line-source strength, $\mathrm{M}_-(x)$. In this case, eq. 2.97 is an integral equation of the first kind because the strength of the line source in the integrand is assumed to be the only unknown. The equation has no analytical solution, and must be solved numerically. The book by Dettman (1969) is recommended to those readers interested in the techniques in solving integral equations.

The second method of application of the results in eq. 2.97b is to specify the function describing the line source strength, $\mathrm{M}_-(x)$, and subsequently determine the body shape from the resulting $r(x)$ expression. An application of this method is given in the example that follows.

EXAMPLE 2.8: FLOW ABOUT A BODY OF REVOLUTION A body shape that is similar to the hulls of modern-day submarines can be obtained by combining a point source of strength M_+ at the origin, a line sink of uniform strength M_- between the origin and $x = \ell$, and a uniform flow of velocity V_0 in the positive x-direction.

The strength of a point source is related to the volume flow rate, Q_+, from the source by

$$M_+ = \frac{Q_+}{4\pi} \tag{2.98}$$

The relationship between the strength of the line sink of length l and the volume flow rate into the line sink, Q_-, is

$$M_- = -\frac{Q_-}{4\pi l} \tag{2.99}$$

To obtain a closed body shape, the magnitudes of the flow rates of the point source and line sink must be equal, that is, $Q_+ = Q_-$. This requirement produces the following relationship between the strengths of eqs. 2.98 and 2.99:

$$M_- = -\frac{M_+}{l} \tag{2.100}$$

Using the equality in eq. 2.100, the expression for the stream function describing the flow about the body of revolution is

$$\Psi = -M_+\left(\frac{x}{R_1} - \frac{R_1}{l} + \frac{R_2}{l}\right) + \frac{V_0}{2}r^2 \tag{2.101}$$

where, referring to Figure 2.22,

$$R_1 = \sqrt{x^2 + r^2} \tag{2.102}$$

and

$$R_2 = \sqrt{(x-l)^2 + r^2} \tag{2.103}$$

The expressions for the axial and radial velocity components are obtained by combining eq. 2.101 with eqs. 2.82 and 2.83, respectively. The results are

$$V_x = M_+\left(\frac{x}{R_1^3} + \frac{1}{lR_1} - \frac{1}{lR_2}\right) + V_0 \tag{2.104}$$

and

$$V_r = \frac{M_+}{r}\left(\frac{1}{R_1} - \frac{x^2}{R_1^3} - \frac{x}{lR_1} + \frac{x-l}{lR_2}\right) \tag{2.105}$$

The pressure distribution on the body is obtained by combining the velocity components of eqs. 2.104 and 2.105 with Bernoulli's equation, eq. 2.70. For steady flow, the resulting pressure distribution on a horizontal plane bisecting the body is obtained from

$$p = p_0 + \frac{\rho}{2}\left(V_0^2 - V_x^2 - V_r^2\right) \tag{2.106}$$

To illustrate, consider a body in a horizontal flow where $V_0 = 10$ m/s. A point source of strength $M_+ = 200$ m^3/s is at the origin, and a line sink of strength $M_- = -M_+/l$ extends from the origin to $x = l = 100$ m. The resulting body profile and pressure distribution in the horizontal plane bisecting the body are obtained numerically, and are shown in Figure 2.22. The body is approximately

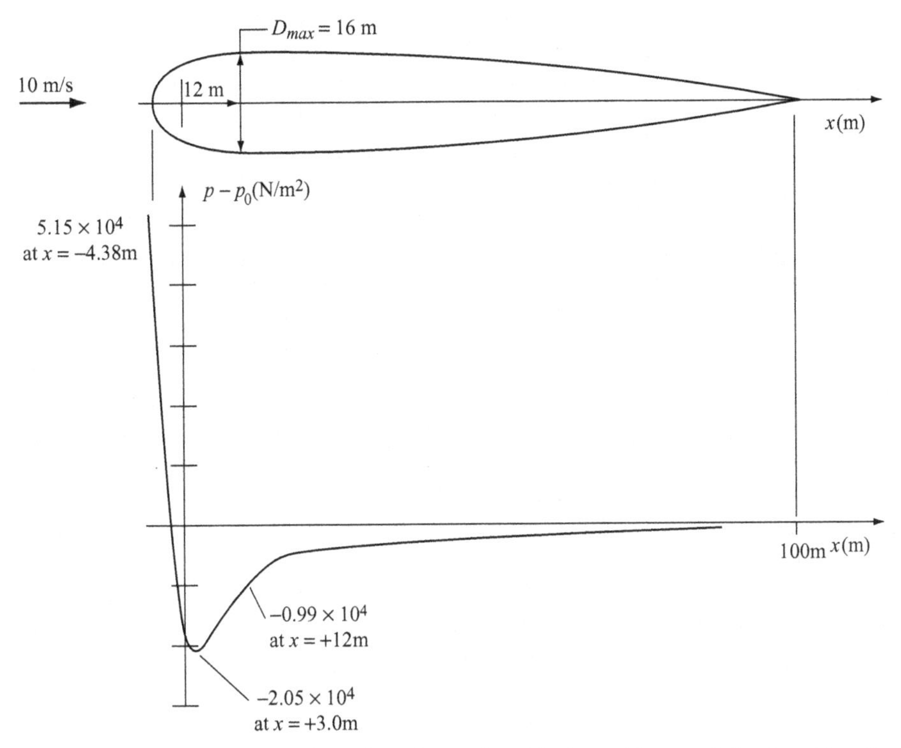

Figure 2.22. *Profile of a Body of Revolution in a Uniform Flow and the Resulting Pressure Distribution over the Body*. The values (in N/m^2) shown are rounded off. The reader should note that the minimum pressure does not occur at the maximum diameter of the body. This pressure phenomenon is discussed in the book edited by Thwaites (1987).

104 m in length with a maximum diameter of approximately 16 m located at $x = 12$ m. The maximum dynamic pressure is 51,500 N/m^2 at the bow and the minimum pressure is approximately $-20{,}500$ N/m^2, and occurs at about $x = 3$ m, or 9 m forward of the maximum body diameter.

The reader is encouraged to consult the book edited by Thwaites (1987) for a comprehensive discussion of flows past bodies of revolution. Many of the analytical and experimental studies of flows about bodies of revolution were performed in the early twentieth century, when airships were of interest to the aerodynamicists.

2.7 Scaling

The purpose of an experiment is to obtain performance data for a modeled system with the hope that the experimental data can be used to predict the performance of a prototype or full-scale system. The success in accomplishing this purpose depends on the scale of the experiment. For example, if we perform an experiment in a table-top wave tank to determine the properties of ocean waves, the surface tension of the water can distort the experimental data so that those data cannot be scaled up. That is, the effects of this small-scale phenomenon are negligible on the prototype scale but not on the model scale.

The first step in scaling an experiment is to perform a *dimensional analysis* to determine the non-dimensional parameter groupings that have the same values on both the model and prototype scales. In Section 2.5, three non-dimensional numbers were presented in the discussion of viscous flows. Those numbers were the Reynolds number (R_e) in eq. 2.78, the drag coefficient (C_d) in eq. 2.79, and the Strouhal number (S_t) in eq. 2.80. An excellent discussion of both physical modeling and dimensional analysis is found in the book by Chadwick and Morfett (1986).

A scaling analysis usually begins by specifying the ratio of the model length (L_{m}) and the prototype length (L_{p}). This is called the length scale factor,

$$n_L = \frac{L_{\mathrm{m}}}{L_{\mathrm{p}}} \tag{2.107}$$

and is the choice of the experimenter. To avoid small-scale effects, the value of n_L should be as large as possible for the experimental facility to be used. Note: The subscripts identifying the model and prototype are "m" and "p," respectively.

There are forces that cannot be scaled simultaneously. For example, the total drag on a ship is the sum of the viscous drag and wave drag. The former depends on viscosity of the liquid, and the latter depends on gravity. The problem is that these cannot be scaled simultaneously. For viscous drag, the non-dimensional number used in scaling is the Reynolds number based on ship length (L),

$$\boldsymbol{R}_{e_L} = VL/\upsilon \tag{2.108}$$

where V is the speed of the ship and υ is the kinematic viscosity. For the drag due to the waves produced by the motion of the ship, a non-dimensional number resulting from the ratio of the inertial force and the gravitational force is used. This parameter is called the *Froude number* based on ship length, and is defined as

$$\boldsymbol{F}_{r_L} = V/\sqrt{gL} \tag{2.109}$$

where g is the gravitational acceleration (9.81 m/s^2). Because we cannot Reynolds-scale and Froude-scale simultaneously, an experimental technique must be devised to exactly perform one type of scaling while approximating the other. On the model scale, Froude scaling is easily accomplished whereas Reynolds scaling is not. The reason for the difficulty in Reynolds scaling is as follows: On the model, the transition from laminar flow to turbulent flow in the boundary layer adjacent to the hull does not occur at the same relative position as on the prototype. This transition phenomenon is discussed in Section 2.5. To simulate the prototype transition on the model, *boundary-layer trips* are placed on the hull model, as shown in Figure 2.23. These trips are small projections from the hull that artificially produce turbulent flow downstream of the trips.

As previously stated, if the scale factor (n_L) is very small, then surface tension can significantly affect the experimental results. The relative effects of the surface tension is determined by the *Weber number*,

$$\boldsymbol{W}_e = \frac{\rho V^2 L}{\sigma_s} \tag{2.110}$$

which results from the ratio of the forces of inertia and surface tension (σ_s).

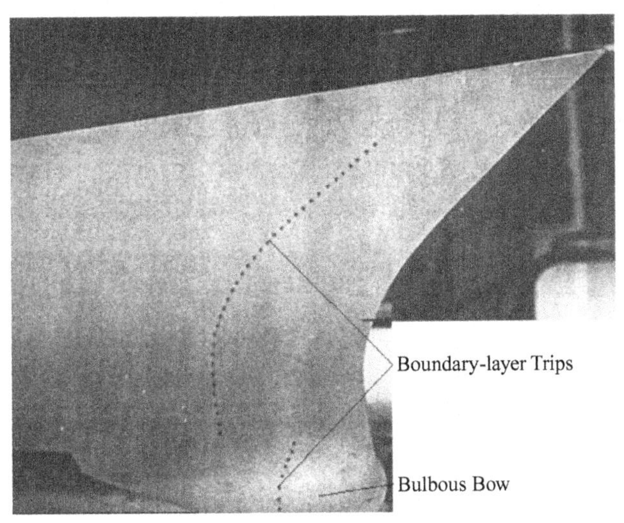

Figure 2.23. *Boundary-Layer Trips on a Ship Model.* The photograph of the bow of the ship model shows boundary-layer trips on both the hull and the bulbous bow. The bulbous bow is designed to reduce the wave drag on the ship.

In the ship-drag (or ship-resistance) study, the total drag on the ship is represented in non-dimensional form by the *drag coefficient*,

$$\boldsymbol{C_d} = \frac{F}{\frac{1}{2}\rho V^2 A} \tag{2.111}$$

where A is the relevant area and F is the total drag force on the ship. For a submarine that is well submerged and traveling at a low speed, the area in question would be the surface area because only viscosity is involved. At greater speeds, the forces due to the wake become dominant, and the area in eq. 2.110 is the maximum cross-sectional area (or projected area).

Using the results of eqs. 2.107, 2.108, 2.109, and 2.111, we can now determine the scale factors for velocity, time, force, pressure, and power. We assume that the model and prototype operate in the same medium (either fresh water or salt water), so that neither the mass density (ρ) nor the kinematic viscosity (ν) need be scaled. The scale factors obtained from Reynolds scaling and Froude scaling are as follows:

a. **Length Scale:** For both Reynolds scaling and Froude scaling, the length scale factor is n_L of eq. 2.107. Again, this is a choice of the experimenter and is normally based on the size of the experimental facility.

b. **Time Scales:**

1. ***Reynolds Scaling:*** We require the following equality:

$$\boldsymbol{R}_{e_{\mathrm{m}}} = \boldsymbol{R}_{e_{\mathrm{p}}} \tag{2.112}$$

From the definition of the Reynolds number in eq. 2.108, the ratio of the Reynolds numbers is

$$\frac{V_{\mathrm{m}} L_{\mathrm{m}}}{V_{\mathrm{p}} L_{\mathrm{p}}} = \frac{L_{\mathrm{m}}^2 / t_{\mathrm{m}}}{L_{\mathrm{p}}^2 / t_{\mathrm{p}}} = \frac{n_L^2}{n_t} = 1 \tag{2.113}$$

From the last equality of eq. 2.113, the *Reynolds time scale factor* is

$$n_t = n_L^2 \tag{2.114}$$

2. ***Froude-Scaling:*** Here we require the Froude number equality, that is,

$$\boldsymbol{F}_{\boldsymbol{r}_{\mathrm{m}}} = \boldsymbol{F}_{\boldsymbol{r}_{\mathrm{p}}} \tag{2.115}$$

which yields

$$\frac{V_{\mathrm{m}}\sqrt{L_{\mathrm{m}}}}{V_{\mathrm{p}}/\sqrt{L_{\mathrm{p}}}} = \frac{(L_{\mathrm{m}}/t_{\mathrm{m}})/\sqrt{L_{\mathrm{m}}}}{(L_{\mathrm{p}}t_{\mathrm{p}})/\sqrt{L_{\mathrm{p}}}} = \frac{\sqrt{n_L}}{n_t} = 1 \tag{2.116}$$

The *Froude time scale factor* is then

$$n_t = \sqrt{n_L} \tag{2.117}$$

c. *Velocity Scales:*

1. ***Reynolds Scaling:*** From eq. 2.112, we obtain

$$\frac{V_{\mathrm{m}}L_{\mathrm{m}}}{V_{\mathrm{p}}L_{\mathrm{p}}} = n_V n_L = 1 \tag{2.118}$$

From the last relationship, the *Reynolds velocity scale factor* is

$$n_V = n_L{}^{-1} \tag{2.119}$$

2. ***Froude Scaling:*** From eq. 2.116, we obtain

$$\frac{V_{\mathrm{m}}/\sqrt{L_{\mathrm{m}}}}{V_{\mathrm{p}}/\sqrt{L_{\mathrm{p}}}} = \frac{n_V}{\sqrt{n_L}} = 1 \tag{2.120}$$

The *Froude velocity scale factor* is then

$$n_V = \sqrt{n_L} \tag{2.121}$$

d. *Force Scales:* First, we must require drag coefficient equality, that is,

$$\boldsymbol{C}_{\boldsymbol{d}_{\mathrm{m}}} = \boldsymbol{C}_{\boldsymbol{d}_{\mathrm{p}}} \tag{2.122}$$

which, from eq. 2.111, yields

$$\frac{F_{\mathrm{m}}/V_{\mathrm{m}}^2 A_{\mathrm{m}}}{F_{\mathrm{p}}/V_{\mathrm{p}}^2 A_{\mathrm{p}}} = \frac{n_F}{(n_V n_L)^2} = 1 \tag{2.123}$$

By rearranging the last two terms of eq. 2.123, one obtains

$$n_F = n_V^2 n_L^2 \tag{2.124}$$

1. ***Reynolds Scaling:*** The combination of eqs. 2.119 and 2.124 yields the *Reynolds force scale factor*, which is

$$n_F = 1 \tag{2.125}$$

2. ***Froude Scaling:*** The *Froude force scale factor* is obtained from the combination of eqs. 2.121 and 2.124. The resulting relationship is

$$n_F = n_L^3 \tag{2.126}$$

e. ***Pressure Scales:*** Equation 2.123 can be rewritten such that the numerator and denominator are expressed as pressure coefficients. The result is

$$\frac{(F_\mathrm{m}/A_\mathrm{m})/V_\mathrm{m}^2}{(F_\mathrm{p}/A_\mathrm{p})/V_\mathrm{p}^2} = \frac{p_\mathrm{m}/V_\mathrm{m}^2}{p_\mathrm{p}/V_\mathrm{p}^2} = \frac{n_p}{n_\mathrm{V}^2} = 1 \tag{2.127}$$

From the last equality, one obtains

$$n_p = n_V^2 \tag{2.128}$$

1. ***Reynolds Scaling:*** The combination of the results of eqs. 2.119 and 2.128 yields the *Reynolds pressure scale factor*,

$$n_p = n_L^{-2} \tag{2.129}$$

2. ***Froude Scaling:*** The *Froude pressure scale factor* is obtained from the combination of eqs. 2.121 and 2.128. The result is

$$n_p = n_L \tag{2.130}$$

f. ***Power Scales:*** The power, P, of a flow is the product of the force and the velocity. Hence, we can define the *power coefficient* by

$$C_P = \frac{P}{FV} \tag{2.131}$$

For a model study, we require the following equality:

$$C_{P\mathrm{m}} = C_{P\mathrm{p}} \tag{2.132}$$

Equations 2.131 and 2.132 are combined to obtain

$$\frac{P_\mathrm{m}/F_\mathrm{m}V_\mathrm{m}}{P_\mathrm{p}/F_\mathrm{p}V_\mathrm{p}} = \frac{n_P}{n_F n_V} = 1 \tag{2.133}$$

from which, the *power scale factor* is

$$n_P = n_F n_V \tag{2.134}$$

1. ***Reynolds Scaling:*** By combining eqs. 2.119, 2.125, and 2.134, we obtain the *Reynolds power scale factor*,

$$n_P = n_L^{-1} \tag{2.135}$$

2. ***Froude Scaling:*** The *Froude power scale factor* is obtained by combining eqs. 2.121, 2.126, and 2.134. The resulting expression is

$$n_P = n_L^{7/2} \tag{2.136}$$

From these results, we see that Reynolds scaling and Froude scaling cannot be performed simultaneously if the model and prototype are studied in the same medium. The use of the scale factors presented in this section is illustrated in the following example.

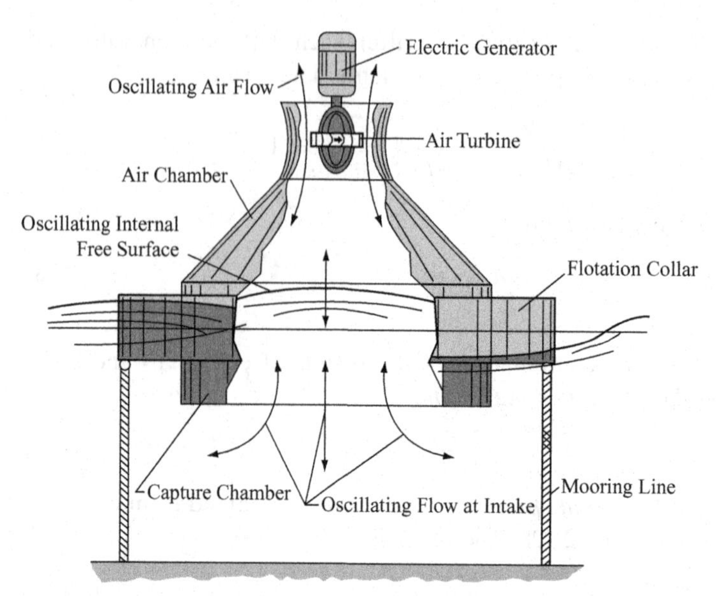

Figure 2.24. *Sketch of a Pneumatic, Oscillating Water Column Wave Energy Conversion System.* The internal water column is excited in a vertical motion by the passing waves. The spatially averaged internal water surface acts as the face of a pneumatic *quasi*-piston.

EXAMPLE 2.9: WAVE POWER CONVERSION Because of the dwindling petroleum resources, solar energy has received much attention since the early 1970s. Solar energy is transformed into other energy forms, a number of which are in the world's oceans. These solar-ocean energy options include the energies of biomass, currents, salinity gradients, tides (in part), and waves. Our interest in this example is in the exploitation of ocean waves. A historical treatment of wave energy conversion is found in the book by Ross (1979), whereas McCormick (1981, 2007) and Shaw (1982) present technical discussions of the subject.

Consider the pneumatic wave energy conversion system sketched in Figure 2.24. This system converts the energy of wave-excited motions of the internal water column into electrical energy by using a bi-directional air turbine. The water column acts as a piston, alternately exhausting and inhaling air through the air turbine. Our goal is to predict the prototype performance by experimentally studying a quarter-scale ($n_L = 1/4$) model of a prototype. The model is found to generate an average power of 1 kW (kilowatt) in a wave tank where the wave height (H_m) is 0.25 m and the wave period (T_m) is 3.0 s. Froude scaling is the logical choice for predicting the performance of the prototype as water waves exist because of gravity. The prototype waves corresponding to those of the quarter-scale model have a height of $H_p = H_m/n_L = 1.0$ m. Using the time-scale factor in eq. 2.116, the prototype wave period value is $T_p = T_m/n_t = T_m/n_L^{1/2} = 6.0$s. Finally, the power that can be expected from the prototype system, according to the Froude power-scale factor of eq. 2.136, is $P_p = P_m/n_P = P_m/n_L^{7/2} = 128$ kW. Note: In the contiguous United States, the average electrical power requirement for each citizen is 1 kW, so the ideal prototype should satisfy the electrical needs of 128 citizens.

2.8 Closing Remarks

The purpose of this chapter is to present a review of the basic topics in hydromechanics. From the author's experience, these topics are most applicable to problems involving ocean waves and wave-structure interactions. Many of the analytical techniques that are discussed in this chapter are used in the chapters that follow. The reader is encouraged to consult the references for thorough discussions of the various topics.

3 Linear Surface Waves

Ocean waves are caused by the motions of celestial bodies, seismic disturbances, moving bodies, and winds. The waves produced by these phenomena differ in size and character, and the consequences of each must be dealt with differently.

The gravitational attractions of both the moon and the sun cause the largest water waves, called the *tides*. The predictable *tidal wave* can be treated as a shallow-water wave because its length is much greater than the water depth. Extreme tides, called *spring tides*, occur when the attractive forces of both the moon and sun are aligned and in the same direction. These tides can cause flooding of lowlands if dikes or levees are not present. The tides can also be exploited by converting their energies into useable energy forms. This is normally accomplished by creating tidal barriers equipped with hydroturbines, taking advantage of the tidal-induced water level changes on opposite sides of the barriers. An excellent book on tidal energy conversion is that written by Charlier (1982).

Both sub-marine earthquakes and volcanic eruptions can produce a long, high-energy wave called a *tsunami* (the Japanese word for a tidal wave, although the wave referred to is not tidal in nature). This type of wave can pass a ship in the open ocean and not be noticed by the ship's crew because of the small wave height-to-wavelength ratio (called the *wave steepness*). As the tsunami approaches a land mass, the energy of the wave is transformed from mostly kinetic to mostly potential. This causes the wave steepness to increase significantly, and the resulting high wave can be devastating to coastal areas. An excellent discussion of the nature and consequences of tsunami is found in the book of Professor Robert L. Wiegel (1964). The book is a classic in the ocean engineering literature.

The speed of a ship traveling in a restricted waterway is normally limited because of the potential damage caused by *ship waves*. These waves travel in a group away from the line of motion of the ship. The energy of the ship-wave group depends on the speed and geometry of the ship. When ship waves travel into shallow water, they can become rather steep without breaking and, as a result, retain much of their energy. The steep waves can cause excessive motions of small moored vessels that, in turn, can result in mooring failures. Furthermore, when the steep ship waves break near the shore, erosion can occur.

Wind-generated waves and their consequences are the primary focus of this book. As discussed in the next section, winds produce waves that vary in length from the short capillary wave to the long swell. Wind waves can be classified as *linear waves* (having sinusoidal profiles), *nonlinear waves* (having nonsymmetric profiles with respect to the still-water level), and *breaking waves*. Each of these waves has a special significance in ocean engineering, as discussed herein. The wind scale and the sea state of wind waves are presented and discussed in Chapter 1.

3.1 Wind-Wave Generation

Let us first consider the effects of a slight breeze of velocity $\boldsymbol{U}_0$ over the free surface (air-sea interface) of calm water of an open ocean. As discussed in Section 2.5, because the air velocity is relatively small, the air flow is laminar. This laminar air flow simply drags the water particles on the free surface in the direction of the flow due to the viscosities of both the air and water. This air flow produces no wave, as illustrated in Figure 3.1a.

When the speed of the air increases such that the flow in the boundary layer adjacent to the free surface is turbulent, then the pressure fluctuations on the free surface beneath the turbulent air flow deform the free surface, and small waves are created. These waves are called *capillary waves*, and have a profile similar to that sketched in Figure 3.1b. In that figure, we see that the crest of the wave is broad, while the trough of the wave is narrow. The cause of this rather odd wave profile is the surface tension, which is the dominant force. Capillary waves travel in the direction of the air flow because of the shear stress on the free surface. For most engineering problems involving water waves, capillary waves are of little significance because of their low energy content.

As the air speed increases, the energy in the air turbulence increases as does that of the surface shear stress. The air flow can now be referred to as a wind rather than a breeze. The water converts the energies transferred to it by wind turbulence and shear stress into longer waves having sinusoidal profiles, as sketched in Figure 3.1c. These sinusoidal waves are called *linear waves* because they can be analyzed using linearized equations, as discussed in the next section. One of the linear properties of the waves is that of *superposition*, that is, linear waves of different heights (H) and lengths (λ) can be combined to form other wave profiles. As discussed in Chapter 5, the most basic analysis of random wave phenomena is that which exploits the property of superposition.

When the wind speed increases further, both the height of the wave and the wavelength increase due to the horizontal pressure gradient resulting from the separation of the air flow on the leeward side of the wave. The altered profile has a narrow crest and a broad trough, as sketched in Figure 3.1d. We note that the *mean-water level* (MWL) – the mean way between the crest and trough – is above the *still-water level* (SWL). The reason for this is that volumes of water above and below the SWL must be equal. Waves having a profile similar to that in Figure 3.1d are called *nonlinear waves* because their properties can no longer be well predicted using linearized equations. To analyze nonlinear waves, we use the expansion theory of Stokes (1847, 1880), presented in Chapter 4.

A further increase in the wind speed can produce *breaking waves*, as sketched in Figure 3.1e. A *break* is defined as the condition where the horizontal water-particle velocity at the crest equals the wave velocity ($\boldsymbol{c}$). The profile of a breaking wave has a pointed crest and a broad trough. After the wave breaks, the wind can shear off the crest, producing foamy white water that spills down the leeward side of the wave. This white-water spill is called a *white cap*. Breaking waves are discussed in Section 4.3.

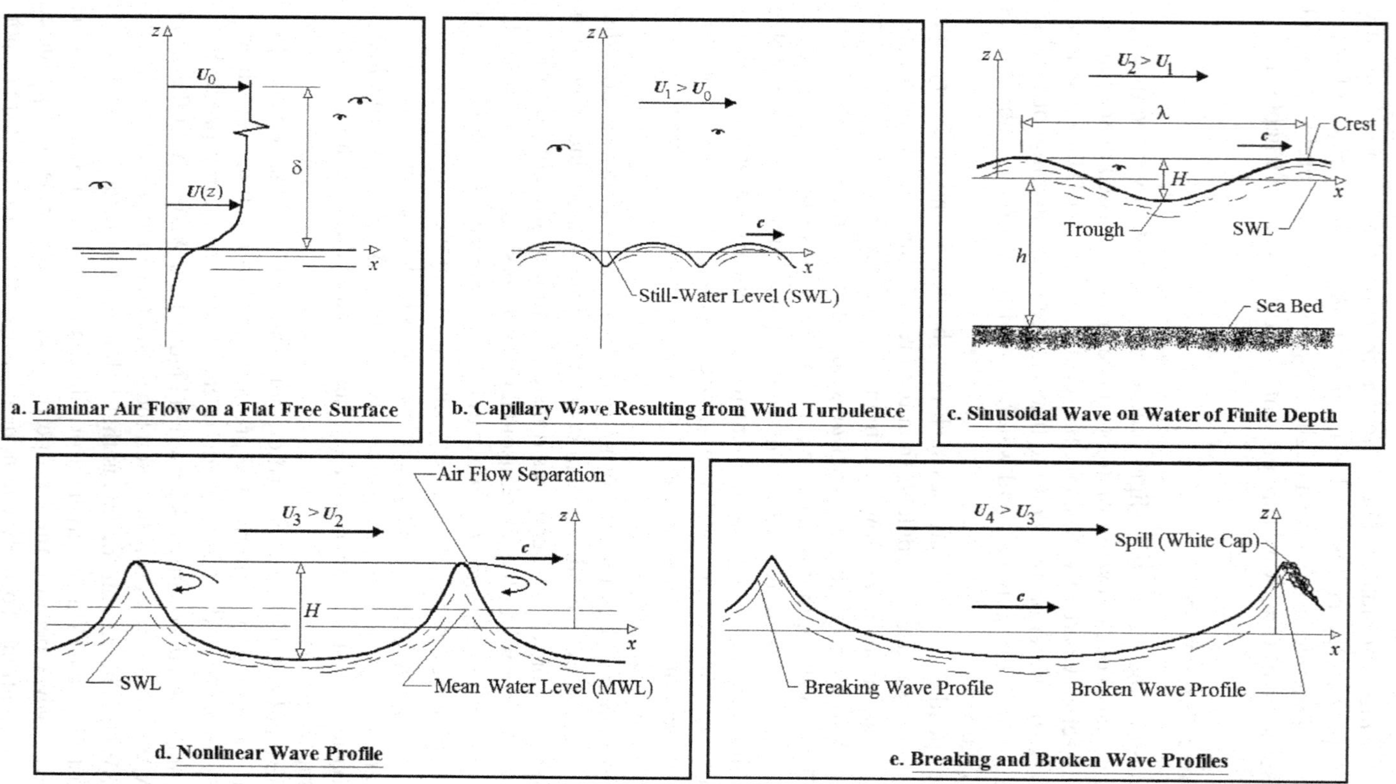

Figure 3.1. *Scenario of Wind-Wave Generation with Increasing Wind Velocity, U.* Some of the notation used in this chapter is defined in (c). The capillary wave profile in (c) is influenced by the surface tension. The separation of the air flow at the crests in (d) and the resulting wakes contribute to the crest profile. In (e), the breaking wave profile is sharp-crested. The broken wave has foam on the leeward side of the crest.

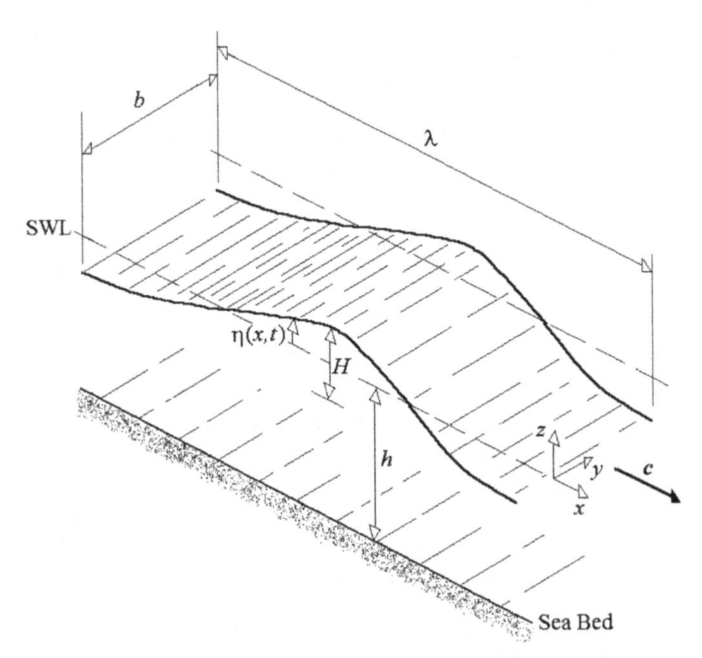

Figure 3.2. *Sketch and Notation for the Linear Wave Analysis.*

There are a number of hypotheses concerning the generation and growth of wind waves. These are well discussed and illustrated by St. Denis (1969). In the remainder of this chapter, the analysis of linear waves is presented. In Chapters 4 and 5, the respective nonlinear waves and random waves are discussed and analyzed.

3.2 Airy's Linear Wave Theory

The first meaningful analysis of surface waves was performed by George B. Airy in the middle of the nineteenth century; see Airy (1845). His analysis is known as either the *Airy wave theory* or the *linear wave theory*. The latter name results from the nature of Airy's analysis, which involves the solution of the linear equation of continuity for an irrotational flow (eq. 2.41) and the application of linearized boundary conditions. Although the theory is somewhat basic, the kinematic wave properties derived from the theory agree quite well with those actually observed, provided that the *wave steepness* (H/λ) is small. In this section, we outline the linear wave theory and discuss the predicted behavior of linear waves. Thorough discussions of the theory are found in the classical books by Kinsman (1984), Lamb (1945), Lighthill (1979), Phillips (1966), and Stoker (1957).

Before embarking on the analysis of linear waves, it is helpful to consider the physical properties of a traveling surface wave, as sketched in Figure 3.2. In this figure, we see that the origin of the Cartesian coordinate system is on the SWL, which is the calm-water position. The water depth, h, is measured from the sea floor to the SWL. The wave has a height, H (measured from the trough to crest), and a wavelength, λ (measured from crest to crest), and travels in the x-direction at a celerity or phase velocity, $\mathbf{c}$. The vertical *free-surface displacement*, $\eta(x, t)$, is measured from the SWL and, as indicated, is a function of both time, t, and distance, x. Note: Books that are directed at the civil engineering aspects of ocean engineering normally represent the water depth by d and the wavelength by L.

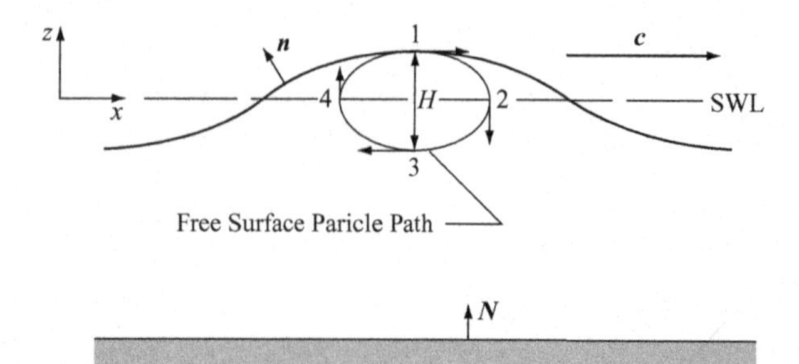

Figure 3.3. *Elliptical Path of a Surface Particle of a Linear Wave.* The numbers correspond to the passing of the wave crest (1), the nodes (2 and 4), and the trough (3). The unit vectors n and N are the outward normal vectors on the free surface and sea floor, respectively.

We begin our analysis by assuming that the flow beneath the free surface is irrotational, where irrotational flow is discussed in Section 2.3. By making this assumption, the velocity of the water particles can be represented by the velocity potential, ϕ, defined in eq. 2.38. The equation of continuity for an incompressible, irrotational flow is expressed by Laplace's equation, eq. 2.41. That equation is a specialized form of the wave equation, which is a second-order linear equation describing most physical wave phenomena, including light and sound. In our analysis, the general solution of Laplace's equation is obtained first, and the solution is then subjected to linearized boundary conditions. Those boundary conditions are the following:

a. ***Kinematic Free-Surface Condition.*** This boundary condition requires that the same water particles comprise the free surface at all times. Referring to Figure 3.2, this *kinematic free-surface condition* is mathematically represented by

$$\boldsymbol{V}|_{z=\eta} = \boldsymbol{V}_n \tag{3.1}$$

Physically, this equation expresses the condition that the velocity of a particle on the free surface must equal the velocity of the free surface itself. Because irrotational flow is assumed, eq. 3.1 can also be written as

$$\boldsymbol{V}|_{z=\eta} = \nabla\phi|_{z=\eta} = \boldsymbol{n}\frac{\partial\phi}{\partial n}\Big|_{z=\eta} \tag{3.2}$$

where $\boldsymbol{n}$ is the outward unit normal vector on the free surface, as sketched in Figure 3.3. Assume that the free-surface displacement, $\eta(x,t)$, is small compared to the wavelength, λ, so that the boundary condition described in eq. 3.2 can be approximated by

$$\boldsymbol{V}|_{z=\eta} \simeq \frac{\partial\eta}{\partial t}\boldsymbol{k} \simeq \frac{\partial\phi}{\partial z}\Big|_{z=0}\boldsymbol{k} \tag{3.3}$$

Note: The boundary condition in eq. 3.3 is mixed. That is, the left term is applied at $z = \eta$ while the right term is applied at $z \approx 0$. The space and time dependencies of the free-surface displacement are implied in the notation η. For the most part, this notation will be used in the remainder of this chapter.

b. ***Sea-Floor Condition.*** Here, it is required that the water particles adjacent to the floor (or *bed*) cannot cross that solid boundary. Mathematically, if the bed is at $z = -h$, then

$$\boldsymbol{V}\cdot\boldsymbol{N}|_{z=-h} = \frac{\partial\phi}{\partial N}\Big|_{z=-h} = 0 \tag{3.4}$$

where $\boldsymbol{N}$ is the outward normal unit vector at the sea floor, sketched in Figure 3.3.

c. ***Dynamic Free-Surface Condition.*** For this condition, the pressure on the free surface is zero (gauge) at any position, x, and any time, t. Because the flow is assumed to be irrotational, Bernoulli's equation (eq. 2.70) can be applied to the flow to obtain the pressure at any point in the water. Let us first examine Bernoulli's equation when applied to the free surface during a calm-water condition. Because there is no motion, the first two terms of eq. 2.70 equate to zero. On the free surface, the coordinate z is zero because the origin of the coordinate system is on the SWL. Furthermore, because the pressure on the free surface (here, the SWL) is zero, the time function, $f(t)$, in eq. 2.70 must also be equal to zero. Still water can then be considered to be a wave with a zero height and an infinite wavelength. Because $f(t) = 0$ for this special wave, the equality must hold for all waves.

Now consider a wave having a finite height and length, where the height is much smaller than the wavelength ($H \ll \lambda$), and apply Bernoulli's equation to the free surface. The result is

$$\left\{\frac{\partial\phi}{\partial t} + g\eta + \frac{1}{2}V^2\right\}\Big|_{z=\eta} = 0 \tag{3.5}$$

Equation 3.5 is nonlinear because of the velocity-squared term. For small values of the wave steepness, the nonlinear velocity term in eq. 3.5 can be neglected, that is, the magnitude of the term is of second order when compared to the other two terms. The linearized dynamic free-surface condition can be expressed mathematically by

$$\eta = -\frac{1}{g}\frac{\partial\phi}{\partial t}\Big|_{z=\eta} \tag{3.6}$$

Physically, the linearization of eq. 3.5 resulting in the expression of eq. 3.6 is based on the assumption that the vee-squared kinetic energy of the fluid particles is much less than the other mechanical energies of the fluid.

The reader might ask what threshold values of wave properties are required to justify the linearization of eq. 3.5. The following example, a comparison of small and large waves, is presented to answer that question.

EXAMPLE 3.1: LINEARIZATION Both in the laboratory and in the field, it has been observed that water particles travel in nearly circular paths if the ratio H/λ (the wave steepness) is small, and if the wave is not influenced by the bed. The condition on the bed is termed the *deep-water condition.* Such a circular path is a special case of the elliptical path sketched in Figure 3.3 for a particle on the free surface. As can be seen in that figure, the minor axis of the elliptical path is equal to the wave height. Therefore, when the wave is in deep water, the minor and major axes of the path are equal, and the wave height is then the diameter of a circular path. For the circular path, when two consecutive crests pass over a time period T, the particle travels a distance πH with an average speed $V = \pi H/T$.

Using this information, consider two waves having lengths of 100 m and periods of 8.0 sec but with respective heights of 10 m and 1 m. The particles on the steeper wave (for which $H = 10$ m) have a speed of 3.93 m/s whereas those on the less-steep wave (for which $H = 1$ m) have a speed of 0.393 m/s. With these values, compare the last two terms of eq. 3.5. For the 10-m wave, the maximum values of the respective terms are 49.0 $(\text{m/s})^2$ and 7.71 $(\text{m/s})^2$. For the 1-m wave, the values of these terms are, respectively, 4.90 $(\text{m/s})^2$ and 0.0771 $(\text{m/s})^2$. The differences in the values of these two terms is less than an order of magnitude for the 10-m wave but greater than an order of magnitude for the 1-m wave. Hence, the linearization is justified for the smaller wave but not for the larger wave. In general, the limits of applicability of the linearized wave theory depend on the wave height, period, and water depth. An excellent discussion of these limits is given by Le Méhauté (1969). In Chapter 4, the limits of the various wave theories are discussed.

The free-surface conditions of eqs. 3.3 and 3.6 are now combined by eliminating η to obtain the linearized free-surface condition, which is

$$\left\{\frac{1}{g}\frac{\partial^2\phi}{\partial t^2}+\frac{\partial\phi}{\partial z}\right\}\Big|_{z=\eta\simeq 0}=0 \tag{3.7}$$

Because the flow in the wave is assumed to be irrotational, the equation of continuity is expressed by Laplace's equation, eq. 2.41. That equation is

$$\nabla^2\phi=0 \tag{3.8}$$

where the Laplacian operator, ∇^2, is defined in eq. 2.42. Equation 3.8 is an elliptic partial differential equation and, as such, can be solved using a product solution. There are two forms of the product solution that can be used. The first is of the form

$$\phi=X(x)Z(z)T(t) \tag{3.9}$$

and results in a *standing wave*, as shown in the book by McCormick (1973) and others. The second product solution is for a *traveling wave*, and is

$$\phi=X(x\pm ct)Z(z)=X(\xi)Z(z) \tag{3.10}$$

The coordinate system for eq. 3.9 is fixed at a point, whereas that of eq. 3.10 moves with the wave. For the traveling wave, substitute the expressions of eq. 3.10 into that of eq. 3.8, and separate terms of the same variables to obtain

$$\frac{1}{X}\frac{d^2X}{d\xi^2}=-\frac{1}{Z}\frac{d^2Z}{dz^2}=-k^2 \tag{3.11}$$

where k is a constant. The negative sign arises because the free-surface profile is sinusoidal in the ξ-direction. An expression similar to that in eq. 3.11 is obtained from the combination of eqs. 3.9 and 3.8.

The general ξ- and z-solutions of the ordinary differential equations in eq. 3.11 are, respectively,

$$X(\xi)=C_\xi\sin(k\xi+\alpha) \tag{3.12}$$

and

$$Z(z)=C_z\cosh(kz+\beta) \tag{3.13}$$

where C_ξ, C_z, α, and β are arbitrary constants. Because the origins of the horizontal coordinates ξ and x are arbitrary, the constant α can be assigned a zero value without loss of generality. Equation 3.12 can then be written as

$$X(\xi)=C_\xi\sin(k\xi) \tag{3.14}$$

To determine the constant β, apply the sea-floor condition of eq. 3.4 to the velocity potential expression resulting from the combination of eqs. 3.10 and 3.13. Assuming that the sea floor is uniformly flat and horizontal, only the z-term is affected by the boundary condition. That is, the sea-floor condition results in the equation

$$\frac{dZ}{dz}\Big|_{z=-h} = kC_z \sinh(-kh + \beta) = 0 \tag{3.15}$$

from which $\beta = kh$. The expression in eq. 3.13 is then

$$Z(z) = C_z \cosh[k(x + h)] \tag{3.16}$$

The combination of the expressions of eqs. 3.14 and 3.16 with that of eq. 3.10 results in

$$\phi = C_\phi \cosh[k(z + h)] \sin(k\xi) \tag{3.17}$$

where $C_\phi = C_\xi C_z$.

Now consider the nature of the horizontal coordinate, ξ. From eq. 3.10, that coordinate is defined as

$$\xi = x \pm ct \tag{3.18}$$

The origin of the coordinate corresponds to

$$x = \mp ct \tag{3.19}$$

From the result in eq. 3.19, we see that the value of x decreases as the time, t, increases for the upper sign ($-$). Hence, the wave must travel in the negative x-direction at a celerity or phase speed, c. The waves corresponding to the upper signs in eqs. 3.18 and 3.19 are then called *left-running waves*. Following the same line of reasoning, the lower signs in those equations correspond to *right-running waves*. The respective coordinates for the right- and left-running waves are then

$$\xi^+ = (x - ct) \tag{3.20a}$$

and

$$\xi^- = (x + ct) \tag{3.20b}$$

Returning to the expression for the velocity potential in eq. 3.17, the arbitrariness of the coefficient C_ϕ in the equation can now be removed. To do this, combine the expressions in eq. 3.6 (the linearized free-surface condition) and eq. 3.17 by eliminating the velocity potential, ϕ. This combination yields the following expression for the free-surface displacement of a sinusoidal wave:

$$\eta^\pm = \frac{ckC_\phi^\pm}{g} \cosh(kh) \cos(k\xi^\pm) = \frac{H}{2} \cos(k\xi^\pm) \tag{3.21}$$

where the last equality results from our knowledge that the wave is sinusoidal in both time and space. In the right term of eq. 3.21, H is the wave height. Comparing the terms of the last equality of eq. 3.21, we obtain the expression for the coefficient C_ϕ, which is

$$C_\phi^\pm = \pm \frac{H}{2} \frac{g}{kc} \frac{1}{\cosh(kh)} \tag{3.22}$$

Now combine this expression and that of eq. 3.20 with that of eq. 3.17 to obtain the final expression for the velocity potential of a traveling wave, that is,

$$\phi^\pm = \pm \frac{H}{2} \frac{g}{kc} \frac{\cosh[k(z + h)]}{\cosh(kh)} \sin[k(x \mp ct)] \tag{3.23}$$

The velocity potential yields the velocity components of the particles in the irrotational flow beneath traveling waves. The velocity potential expression in eq. 3.23 is the primary result of the Airy wave theory. The other properties of the linear waves can now be derived.

3.3 Traveling or Progressive Waves

By convention, right-running waves are normally considered in two-dimensional wave mechanics problems. This convention is followed herein. Consider the right-running wave sketched in Figure 3.2. The free-surface displacement caused by the wave is

$$\eta = \frac{H}{2}\cos[k(x - ct)] \tag{3.24}$$

from the combination of equations 3.20a and 3.21. Because the free-surface profile is sinusoidal in both time and space, the maximum displacement from the SWL, or the *crest*, occurs when

$$k(x - ct) = 0, \ \pm 2\pi, \pm, \pm 4\pi, \ldots. \tag{3.25}$$

Consider first the case for which $t = 0$. The distance in the x-direction between two successive crests is the *wavelength*, λ. From eq. 3.25, we obtain

$$k = \frac{2\pi}{\lambda} \tag{3.26}$$

The wave parameter, k, is called the *wave number*.

Next, consider the expression of eq. 3.25 when $x = 0$. The time lapse between successive crests is the *wave period*, T. From the results in eq. 3.25, we obtain

$$kc = \frac{2\pi}{T} = 2\pi f \equiv \omega \tag{3.27}$$

In this equation, f is the *wave frequency* in units of Hertz (Hz), and ω is the *circular wave frequency* in units of radians per second. Combining eqs. 3.26 and 3.27 by eliminating the wave number yields the expression for the *celerity* or *phase velocity*,

$$c = \frac{\lambda}{T} \tag{3.28}$$

Returning to the velocity potential of eq. 3.23, the expression in that equation can now be written in terms of the circular wave frequency as

$$\phi = \frac{H}{2}\frac{g}{\omega}\frac{\cosh[k(z+h)]}{\cosh(kh)}\sin(kx - \omega t) \tag{3.29}$$

Combine this expression with that of eq. 3.7 by eliminating the velocity potential. After some simplification and the introduction of the expression in eq. 3.27, one obtains the following relationship for the circular wave frequency:

$$\omega = \sqrt{gk\tanh(kh)} \tag{3.30}$$

Waves are dispersive in that the celerity depends on the frequency and length of the wave and, in addition, on the water depth. Equations 3.26 and 3.30 can be

Table 3.1. *Linear theory parameters*

h/λ	kh	$\tanh(kh)$
$0.01 = 1/100$	0.06283..	0.06275..
$0.04 = 1/25$	0.25313..	0.24616..
$0.05 = 1/20$	0.31415..	0.30421..
$0.10 = 1/10$	0.62831..	0.55689..
$0.50 = 1/2$	3.14159..	0.99627..
$1.00 = 1/1$	6.28318..	0.99999..

combined to obtain the expression for the wavelength, which is a form of the dispersion equation, that is,

$$\lambda = \frac{2\pi}{k} = \frac{2\pi g}{\omega^2}\tanh(kh) = \frac{gT^2}{2\pi}\tanh\left(\frac{2\pi h}{\lambda}\right) = cT \tag{3.31}$$

The equation for the wavelength is transcendental because λ cannot be isolated. That is, the solution of λ cannot be obtained analytically, so the values of the wavelength must be determined using a numerical technique. One such technique is that called successive approximations or successive substitutions, as discussed by Carnahan (1969) and others. McCormick (1973) demonstrates the application of this numerical technique to wave mechanics problems. The significance of eq. 3.31 is illustrated in the following examples.

EXAMPLE 3.2: WAVELENGTH VARIATION WITH WATER DEPTH From a knowledge of hyperbolic functions, we know that $\tanh(kh)$ approaches unity as kh approaches infinity. This behavior of kh would be observed if either $h \to \infty$ or $\lambda \to 0$. At the other extreme, where the values of kh are very small, $\tanh(kh) \to kh$. Physically, this extreme corresponds to either relatively small water depths or relatively large wavelengths. The term "relative" is used because the behavior depends on the ratio h/λ. Some practical values of this ratio are presented in Table 3.1 along with the corresponding values of kh and $\tanh(kh)$ for the sake of illustration.

The results in Table 3.1 show that the difference in the values of kh and $\tanh(kh)$ is about 3.16% or less for $h/\lambda \leq 1/20$, whereas for $h/\lambda \geq 1/2$ the difference between $\tanh(kh)$ and unity is less than 0.4%. The significance of these approximation ranges is now discussed.

Because of the approximations presented in Example 3.2, it is a common practice in ocean engineering to partition the infinite range of h/λ into three regions, those being:

a. **Shallow Water:** $h/\lambda \leq 1/20$ so that $\tanh(kh) \approx kh$
b. **Intermediate Water:** $1/20 < h/\lambda < 1/2$ so that $\tanh(kh) = \tanh(2\pi h/\lambda)$
c. **Deep Water:** $h/\lambda \geq 1/2$ so that $\tanh(kh) \approx 1$

In intermediate water, the transcendental relationship of eq. 3.31 applies as it is written, and must be solved numerically. The numerical method of successive approximations, as applied to eq. 3.31, is demonstrated in Example 3.3. Note: Coastal engineers working in the field can use the shallow-water approximation to gain a fairly good idea of the length of waves in the coastal zone. Outside the coastal

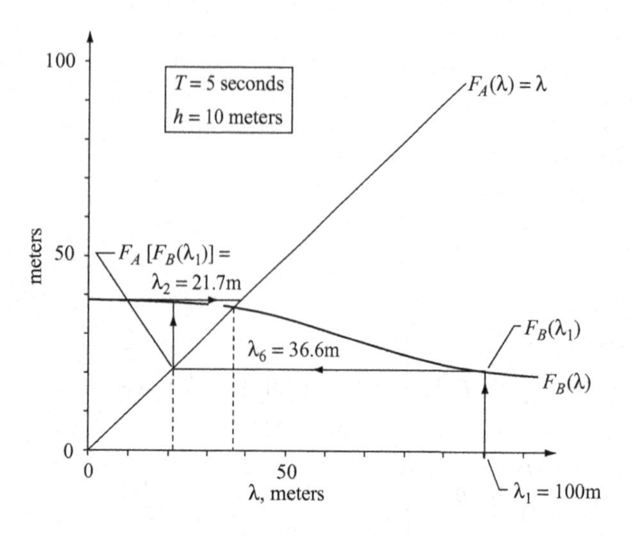

Figure 3.4. *Graphical Illustration of a Successive Approximation Solution for the Wavelength.*

zone, offshore engineers find the deep-water approximation to be very useful in field calculations of the wavelength.

EXAMPLE 3.3: WAVELENGTH SOLUTION BY SUCCESSIVE APPROXIMATIONS A wave having a period (T) of 5 sec travels in water that is 10 m deep (h = 10 m). The wavelength value determined from eq. 3.31 is obtained using the method of successive approximations. The method can best be illustrated as follows: Define two functions in the equation, those being

$$F_A(\lambda) = \lambda \tag{3.32}$$

and

$$F_B(\lambda) = \frac{gT^2}{2\pi} \tanh\left(\frac{2\pi h}{\lambda}\right) \tag{3.33}$$

Plot these functions as shown in Figure 3.4. Obviously, the solution of eq. 3.31 is obtained when

$$F_A(\lambda) = F_B(\lambda) \tag{3.34}$$

or where the lines in the figure cross. In fact, depending on the degree of accuracy desired, one could use the graphical solution. However, to increase the degree of accuracy we use successive approximations. To find the wavelength value numerically, assume a starting value of λ equal, say, $\lambda_1 = 10h = 100\,\text{m}$, as illustrated in Figure 3.4. The initial value of $F_A(\lambda)$ is then 100 m. If this is the solution, then $F_B(\lambda_1)$ should also have the value of λ_1. However, for this first approximation the value of the function is 21.7 m. To test the accuracy, let us define the *difference function* as

$$\Delta_F = |F_B(\lambda_i) - F_A(\lambda_i)| = |F_B(\lambda_i) - \lambda_i| \tag{3.35}$$

The value of this function for the first approximation (i = 1) is 78.3 m. For this example, let the desired value of Δ_F be 0.1 m. The second approximation ($i = 2$) is obtained by letting $F_A(\lambda_2) = \lambda_2 = 21.7\,\text{m}$ and evaluating $F_B(\lambda_2)$. The new value of that function is 38.8 m, and the value of the difference function is 17.1 m. By following this procedure, the desired accuracy is achieved in the sixth approximation ($i = 6$), and the value of λ_6 is 36.6 m. A flow chart for successive

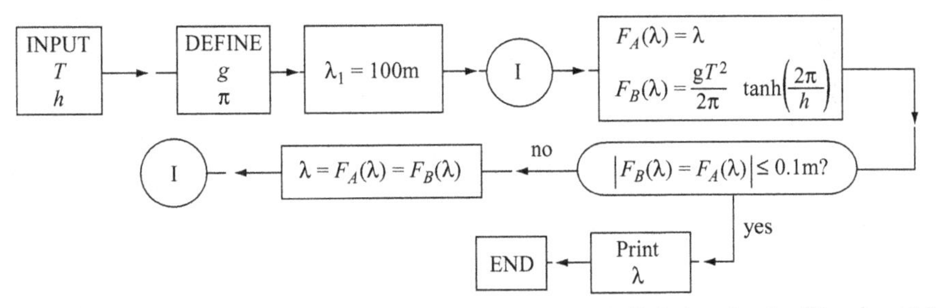

Figure 3.5. *Flow Chart for the Successive Approximation Solution for the Wavelength Value.*

approximation is in Figure 3.5. The only restrictions on the initial value of λ are that it must be both greater than zero and finite.

From the graph in Figure 3.4, we can see that the number of iterations required to achieve the desired accuracy is reduced as the value of λ_1 approaches the actual solution. However, from the results of this example it is apparent that the process rapidly converges on the solution.

Now apply the deep- and shallow-water approximations to eq. 3.31. First, for deep water the wavelength approximation is

$$\lambda \equiv \lambda_0 \simeq \frac{gT^2}{2\pi} \simeq c_0 T \tag{3.36}$$

where the subscript 0 is, by convention, used to identify deep-water properties. Physically, the result in eq. 3.36 shows that the deep water wavelength and celerity are functions of the wave period only, both increasing as the period increases.

The shallow-water approximation of eq. 3.31 is

$$\lambda \simeq \frac{gT^2}{2\pi} kh = \frac{gT^2}{2\pi}\frac{2\pi h}{\lambda} = \frac{gT^2 h}{\lambda} \simeq cT \tag{3.37}$$

from which

$$\lambda = \sqrt{gh}T \simeq cT \tag{3.38}$$

From this relationship, we see that the shallow-water wavelength is then a function of both depth and period, decreasing as both decrease. However, the shallow-water celerity is a function of depth and independent of period. From these results, we see that waves both shorten and slow down as they approach the shoreline. The approximations of both eqs. 3.36 and 3.38 are applied to the conditions in Example 3.3 in the following example.

EXAMPLE 3.4: DEEP- AND SHALLOW-WATER WAVELENGTH APPROXIMATIONS In Example 3.3, the wave period is 5 sec, and the water depth is 10 m. For this period value, the deep-water approximation in eq. 3.36 yields a wavelength value of 39.0 m. The shallow-water wavelength approximation from eq. 3.38 for the given period and water depth is 49.5 m. To an accuracy of 0.1 m, the numerical solution of eq. 3.31 yields 36.6 m. For the conditions in Example 3.3, the deep-water approximation is rather good as a first estimate whereas the shallow-water approximation is relatively poor.

A knowledge of traveling waves is needed for engineering analyses of both offshore and coastal structures. The linear wave theory for traveling waves discussed in this section is useful in the analysis of the motions of floating bodies, as discussed

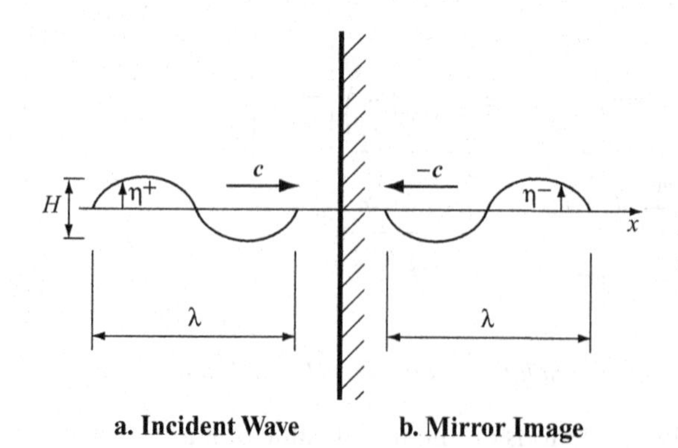

Figure 3.6. *Illustration of the Method of Images as Applied to Perfect Reflection.*

in Chapter 10. However, as waves travel into shallow water the steepness of the waves increases and the wave profiles become nonlinear. For nonlinear waves, the analytical methods are discussed in Chapters 4 and 6.

3.4 Standing Waves

In Section 3.2, it is stated that there are two forms of the product solution of the equation of continuity, expressed as Laplace's equation, eq. 2.41. The first is the standing wave solution of eq. 3.9 and the second is the traveling wave solution of eq. 3.10. The velocity potential for the traveling wave (both right- and left-running), as derived from the linear theory, is presented in eq. 3.23. We can use that equation to determine the velocity potential for a standing wave by taking advantage of the property of superposition, discussed in Section 2.3D, that is, the velocity potential of any two (or more) linear waves can be added together to obtain an additional wave pattern.

Referring to Figure 3.6, consider two traveling waves having the same heights and periods passing each other in the x-z plane, one wave being right-running and the other left-running. The vector celerities of the respective waves are $\boldsymbol{c}^{+}$ and $\boldsymbol{c}^{-}$. Because these vectors are equal in magnitude and opposite in direction, they cancel each other when the two wave patterns are superimposed. The resulting wave pattern is one having zero celerity, or a *standing wave.* By matching the signs in eq. 3.23, the respective velocity potentials (ϕ^{+} and ϕ^{-}) are obtained. Again, because of the superposition property, we can add these potentials to obtain the potential for a standing wave,

$$\Phi = \phi^{+} + \phi^{-} = \frac{-Hg}{\omega} \frac{\cosh[k(z+h)]}{\cosh(kh)} \cos(kx)\sin(\omega t) \tag{3.39}$$

In a similar manner, the free-surface displacements of the right-running wave, η^{+} of eq. 3.24, and the left-running wave (η^{-}) can be added to obtain the following:

$$\eta_S = \eta^{+} + \eta^{-} = H\cos(kx)\cos(\omega t) = \frac{\mathrm{H}}{2}\cos(kx)\cos(\omega t) \tag{3.40}$$

after using trigonometric identities and simplifying (see Figure 3.6). In eq. 3.40, we see that the standing wave height, H, is twice that of the traveling wave height, H,

$$\mathrm{H} = 2H \tag{3.41}$$

as illustrated in Figure 3.7.

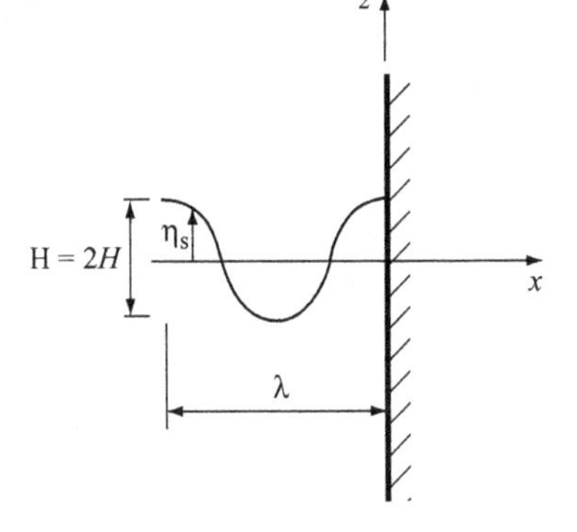

Figure 3.7. *Sketch of a Standing Wave Profile Resulting from a Perfect Reflection.*

In both eqs. 3.39 and 3.40, we note that the spacial (x) and temporal (t) functions are independent of each other. Because of this, the positions of the *nodes* (where $\eta = 0$) and the *antinodes* (positions of the crests and troughs) are fixed in space. Now take the origin of the spacial coordinate system to be as shown in Figure 3.7. The nodes occur when

$$kx = \frac{2\pi}{\lambda}x = \frac{\pi}{2}, \frac{3\pi}{2}, \ldots \tag{3.42}$$

So the nodal points are at

$$x = \frac{\lambda}{4}, \frac{3\lambda}{4}, \ldots \tag{3.43}$$

where λ is the wavelength of both the traveling and standing waves.

The velocity potential expression in eq. 3.39 can be used in the Cauchy-Riemann expressions of eqs. 2.51 and 2.52 to determine the *stream function* for the standing wave pattern. The result is

$$\Psi = \frac{Hg}{2\omega}\frac{\sinh[k(z+h)]}{\cosh(kh)}\sin(kx)\sin(\omega t) \tag{3.44}$$

As discussed in Section 2.3C, the stream function can be used for both two-dimensional rotational or irrotational flows to establish the flow streamlines. In fact, any constant value of Ψ (the notation for a standing wave stream function) will define a specific streamline. Furthermore, because no flow can cross a streamline, any streamline can be considered to be a flow boundary. In Example 2.3, we see that a zero value of the stream function defines the stagnation streamline and the surface of a circular cylinder in a uniform flow. Consider the streamlines corresponding to $\Psi = 0$ when $t > 0$. The zero value of the stream function occurs where

$$kx = 0, \pi, 2\pi\ldots. \tag{3.45}$$

and when $z = -h$, i.e., on the sea floor. The condition in eq. 3.45 corresponds to

$$x = 0, \frac{\lambda}{2}, \lambda, \ldots \tag{3.46}$$

as sketched in Figure 3.8. To illustrate, consider the following example.

EXAMPLE 3.5: STANDING WAVES AT A SEAWALL Traveling waves having a height of 0.5 m and a period of 5 sec perfectly reflect from a vertical seawall where the water depth is 2 m. So, $H = 0.5$ m, $T = 5$ sec, and $h = 2$ m. These values of period and water depth in eq. 3.31 yield a wavelength of approximately 20.9 m.

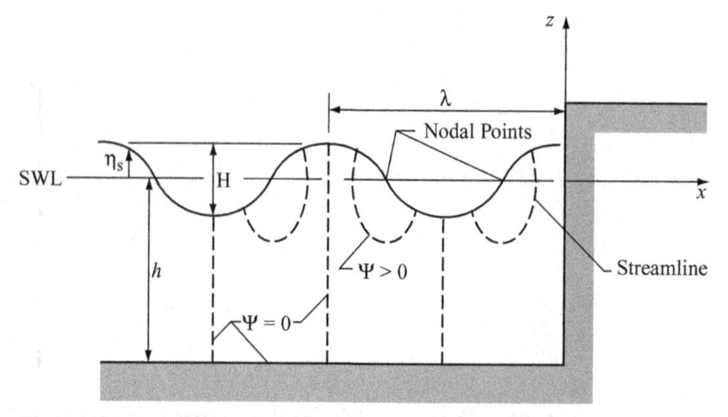

Figure 3.8. *Sketch of Streamlines within a Standing Wave.*

The water depth-to-wavelength ratio for the stated conditions is $h/\lambda = 0.096$, or approximately 1/10. From the discussion in Section 3.3, this value corresponds to intermediate water. After the wave reflects from the wall, the resulting standing wave height is twice that of the incident wave, so $\mathrm{H} = 1.0$ m according to eq. 3.41. The steepness of the incident traveling wave (H/λ) is 0.0239, whereas that of the standing wave (H/λ) is 0.0478. The resulting stream function from eq. 3.44 is

$$\Psi = 0.335 \sinh(0.601 + 0.301z) \sin(0.301x) \sin(1.26t) \tag{3.47}$$

From this equation, the geometries of the streamlines corresponding to constant values of Ψ are determined. The resulting streamlines are similar to those sketched in Figure 3.8, where the time is $t = \pi/2\omega$. In that figure, we see that the fluid is partitioned into flow cells, the cells being half a wavelength in width, or 10.45 m for the stated conditions. The boundaries of each cell correspond to $\Psi = 0$, as does the sea bed. Antinodes (crests and troughs) occur at the sides of the cells and nodes are at the center of the cells.

Because there can be no flow across streamlines, any liquid spilled into the water when standing waves are present should (theoretically) remain in the cell in which the liquid is poured, provided that there are no other currents in the cell. Contaminants should then be containable adjacent to sea walls.

From the discussion, we see that the reflection of traveling waves can result in a standing wave, the height of which is twice that of the incident traveling wave, as in eq. 3.41. However, the wavelength is the same for both the traveling and standing waves because the wavelength depends only on the water depth and wave period. One very important consequence of wave reflection from seawalls or quays is the doubling of the wave steepness, as illustrated in Example 3.5. That is, because the wavelength is unaffected by reflection, the wave steepnesses of the standing wave and the incident traveling wave are related by

$$\frac{\mathrm{H}}{\lambda} = 2\frac{H}{\lambda} \tag{3.48}$$

This condition follows directly from eq. 3.41. If a ship or a boat is moored to the seawall, the increased wave steepness can cause excessive motions of the vessel, with possible damage to the hull or the moorings. For this reason, when storms are forecast large ships are usually taken from their berths to sea, whereas the smaller boats are taken to more sheltered waters. Standing wave formation is a major concern in harbor design.

3.5 Water Particle Motions

In Section 2.3, the relationships between the velocity potential, stream function, and velocity components of the water particles in a two-dimensional irrotational flow are introduced as the Cauchy-Riemann equations, eqs. 2.51 and 2.52. In Section 3.2, Airy's theory yields the expression for the velocity potential. For right-running traveling waves, the velocity potential expression is in eq. 3.29, whereas for a standing wave, the potential is given by eq. 3.39.

To obtain the expressions for the two-dimensional water particle velocity components in traveling waves in irrotational flows, simply combine eqs. 2.51 and 2.52, respectively, with eq. 3.29 to obtain the following:

$$\begin{aligned} u &= \frac{Hgk\cosh[k(z+h)]}{2\omega\cosh(kh)}\cos(kx-\omega t) \\ &= \frac{H\omega\cosh[k(z+h)]}{2\sinh(kh)}\cos(kx-\omega t) \end{aligned} \tag{3.49}$$

in the horizontal direction, and

$$\begin{aligned} w &= \frac{Hgk\sinh[k(z+h)]}{2\omega\cosh(kh)}\sin(kx-\omega t) \\ &= \frac{H\omega\cosh[k(z+h)]}{2\sinh(kh)}\sin(kx-\omega t) \end{aligned} \tag{3.50}$$

in the vertical direction, where the last term in each equation results from eq. 3.31.

For standing waves, the expressions for the velocity components are obtained from the combinations of eqs. 2.51 and 2.52 with eq. 3.39. The respective results are

$$U = \frac{H\omega\cosh[k(z+h)]}{2\sinh(kh)}\sin(kx)\sin(\omega t) \tag{3.51}$$

in the horizontal direction, and

$$W = -\frac{\mathrm{H}\omega\sinh[k(z+h)]}{\sin(kh)}\cos(kx)\sin(\omega t) \tag{3.52}$$

in the vertical direction.

In the remainder of the present chapter, we concentrate on the motions of particles within right-running traveling waves. The horizontal and vertical particle displacements (ξ, ζ) about a fixed mean point (x_o, z_o) are found by replacing (x,z) by $(x_o+\xi, z_o+\zeta)$, expanding the results in Maclaurin series in ξ and ζ and, finally, integrating the respective velocity expressions over time. Referring to Figure 3.9, the resulting displacement expressions are, to the first order,

$$\xi = \int u dt|_{x_o,z_o} = -\frac{H\cosh[k(z_o+h)]}{2\sinh(kh)}\sin(kx_o-\omega t) \tag{3.53}$$

in the horizontal direction, and

$$\zeta = \int w dt|_{x_o,z_o} = \frac{H\sinh[k(z_o+h)]}{2\sinh(kh)}\cos(kx_o-\omega t) \tag{3.54}$$

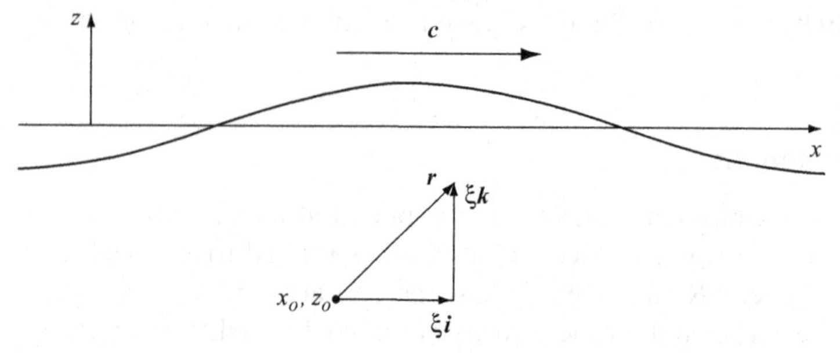

Figure 3.9. *Position Vector for a Particle Beneath a Traveling Wave.*

in the vertical direction. The subscript "o" identifies the mean position of the particle. The integration constants in eqs. 3.53 and 3.54 are neglected by adjustment of the time origin. The location vector of the particle from the mean position is

$$\begin{aligned} \boldsymbol{r} &= \xi \boldsymbol{i} + \zeta \boldsymbol{k} \\ &= \frac{H/2}{\sinh(kh)}\{-\cosh[k(z_o + h)]\sin(kx_o - \omega t)\boldsymbol{i} + \sinh[k(z_o + h)]\cos(kx_o - \omega t)\boldsymbol{k}\} \end{aligned} \tag{3.55}$$

Referring to the graphs of the hyperbolic functions in Figure 3.10, the expression in eq. 3.55 results in the following case.

CASE 1. DEEP WATER ($1/2 \leq h/\lambda < \infty$, or $\pi \leq kh < \infty$) For this range, $\sinh(kh) \approx \cosh(kh) \approx e^{kh}/2$ and $\tanh(kh) \approx 1$. With these approximations, the expression in eq. 3.55 is approximately

$$\mathbf{r} = \frac{H_0}{2} e^{k_0 z_o}[-\sin(k_0 x_o - \omega t)\boldsymbol{i} + \cos(k_0 x_o - \omega t)\boldsymbol{k}] \tag{3.56}$$

Here, the subscript "0" identifies deep-water properties and, again, "o" identifies the mean position of the particle. Because the values of z_o are negative beneath the SWL, the expression for $\boldsymbol{r}$ describes circular particle paths having diameters that decrease with depth. Plots of these paths are sketched in Figure 3.11. The maximum diameter corresponds to a mean position on the SWL, where $z_o = 0$.

CASE 2. INTERMEDIATE WATER ($1/20 < h/\lambda < 1/2$, or $2\pi/20 < kh < \pi$) For this range of kh, the equation for the location vector given in eq. 3.55 must be used as

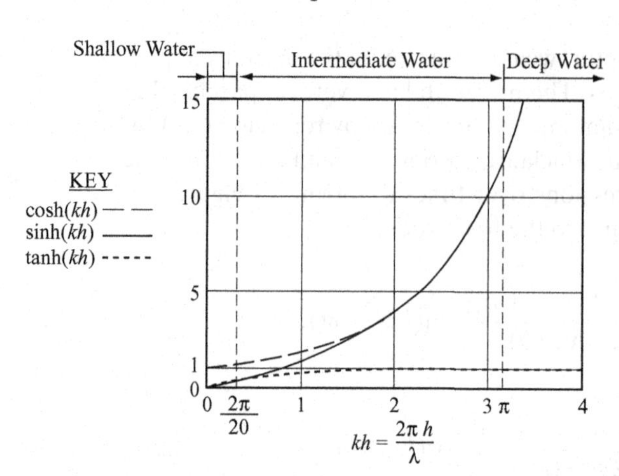

Figure 3.10. *Behaviors of Hyperbolic Functions.*

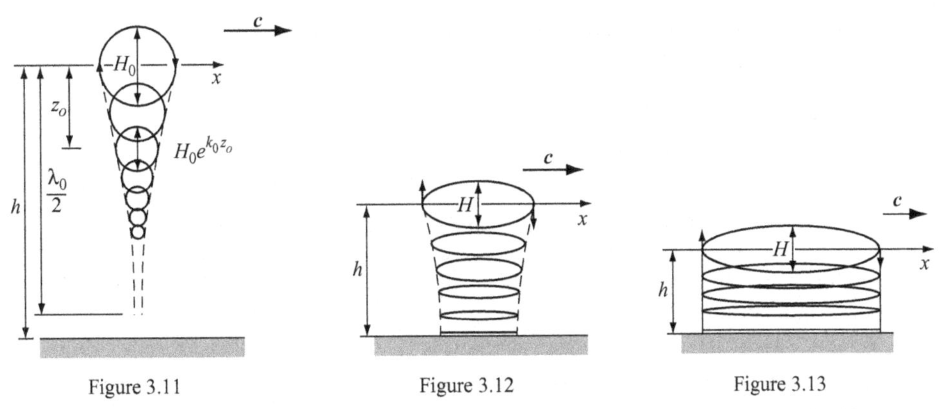

Figure 3.11 Figure 3.12 Figure 3.13

Figures 3.11, 3.12 and 3.13. *Particle Paths Predicted by Airy's Linear Wave Theory.* The deep-water paths in Figure 3.11 are circular with diameters that decrease exponentially with depth. In the figure, $z_o \leq 0$. In Figure 3.12, the intermediate water paths are elliptical with the major and minor axes decreasing with depth. The shallow-water paths in Figure 3.13 are elliptical with minor axes that decrease with depth and vertically uniform major axes.

stated. Plots resulting from the equation are sketched in Figure 3.12. In that figure, we see that the particle paths are ellipses with major and minor axes that decrease with depth. At $z_o = -h$, the minor axis is zero, and the particle motions are adjacent to the bed.

CASE 3. SHALLOW WATER ($0 < h/\lambda \leq 1/20$, or $0 < kh \leq 2\pi/20$) Again, referring to the plots of the hyperbolic functions in Figure 3.10, the shallow-water range of kh justifies the approximations $\sinh(kh) \approx \tanh(kh) \approx kh$ and $\cosh(kh) \approx 1$. With these approximations, eq. 3.55 simplifies to

$$\boldsymbol{r} = \frac{H}{2kh}\{-\sin(kx_o - \omega t)\boldsymbol{i} + [k(z_o + h)]\cos(kx_o - \omega t)\boldsymbol{k}\} \tag{3.57}$$

This equation also results in elliptic particle paths; however, in this case the major axes are uniform and the minor axes decrease with depth. These elliptic paths are sketched in Figure 3.13.

The particle paths described by eqs. 3.55 through 3.57 are good approximations of those actually observed when the wave steepness (H/λ) is small, or when the wave profile is approximately sinusoidal. As the steepness increases, or as the profile becomes nonlinear, there is a net convection of the particles in the direction of wave travel, as is discussed in Chapter 4.

3.6 The Wave Group

It is demonstrated in Section 2.3D that various flow patterns in irrotational flows can be superimposed upon each other to obtain other flow patterns. Because the linear wave theory is based on the assumption of irrotational flow, we can superimpose *regular* wave patterns (those having specified heights and periods) to obtain *irregular* wave patterns (those for which the heights and periods vary).

One such pattern is called the *wave group*, which is formed by superimposing two waves having the same heights and slightly different periods. When many waves of differing heights and periods are superimposed, the resulting pattern can appear

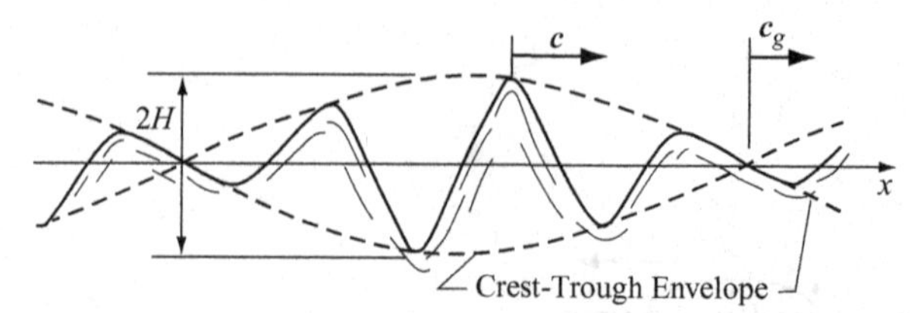

Figure 3.14. *Sketch of a Wave Group*. In deep and intermediate waters, the waves form at the left side of the group and travel to the right side, where they disappear because the celerity (c) is greater than the group velocity (c_g). In shallow water, the wave profile is uniform over the group length because the celerity and the group velocity are equal.

to be random in the time, as discussed in Chapter 5. In this section, our interest is in the simple wave group.

Consider two waves of height H and periods that differ by ΔT. The period difference results in small differences in both the period ($\Delta\omega$) and the wave number (Δk), as expected. The free-surface displacements of the two waves are described by

$$\eta_1 = \frac{H}{2}\cos(kx - \omega t) \tag{3.58}$$

and

$$\eta_2 = \frac{H}{2}\cos[(k + \Delta k)x - (\omega + \Delta\omega)t] \tag{3.59}$$

The superposition of these waves results in an irregular wave pattern described by

$$\eta = \eta_1 + \eta_2 = H\cos\left(\frac{\Delta k}{2}x - \frac{\Delta\omega}{2}t\right)\cos[(k + \Delta k)x - (\omega + \Delta\omega)t] \tag{3.60}$$

where the expression is obtained with the help of trigonometric identities. Because it is assumed that all of the incremental quantities are relatively small, that is, $k \gg \Delta k$ and $\omega \gg \Delta\omega$, the expression in eq. 3.60 can be approximated by

$$\eta \approx H\cos(kx - \omega t)\cos\left(\frac{\Delta k}{2}x - \frac{\Delta\omega}{2}t\right) = 2\eta_1\cos\left(\frac{\Delta k}{2}x - \frac{\Delta\omega}{2}t\right) \tag{3.61}$$

The wave pattern is called a *wave group*, and consists of a wave component, η_1, and an overriding function that modifies the wave height. An example of a wave group is sketched in Figure 3.14. The waves within the group are described by the expression in eq. 3.24, and have a celerity or phase velocity (c) obtained from eq. 3.31.

Now imagine that we are riding with the group such that the angle of the overriding cosine is held constant. The speed at which we must travel to obtain this effect is found by taking the time derivative of this constant angle to obtain

$$c_g = \frac{dx}{dt} = \lim_{\Delta k \to 0}\frac{\Delta\omega}{\Delta k} = \frac{d\omega}{dk} \tag{3.62}$$

This velocity is called the *group velocity*. From eq. 3.30, we find $\omega = \sqrt{[kg\tanh(kh)]}$. The combination of this expression with that of eq. 3.62 yields

$$c_g = \frac{c}{2}\left[1 + \frac{2kh}{\sinh(2kh)}\right] \tag{3.63}$$

To gain an idea of the significance of the expression in eq. 3.63, consider the group velocity at the extremes of the depth. In deep water, $\sinh(2kh) \to 0.5e^{2kh}$, and the

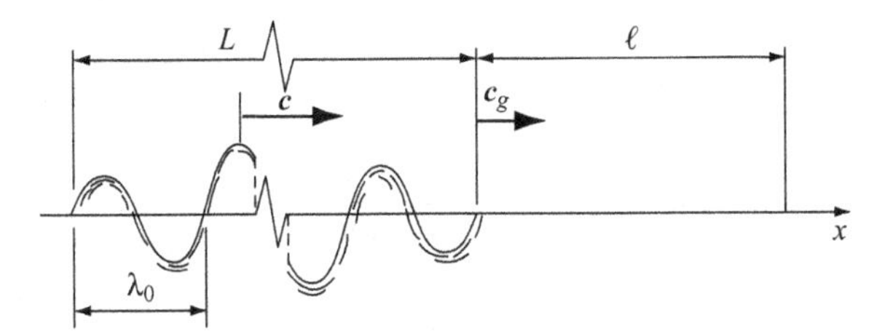

Figure 3.15. *Schematic Sketch of a Deep-Water Wave Group for Example 3.6.*

group velocity is approximately

$$c_{g0} = \lim_{h \to \infty} (c_g) = \frac{c_0}{2} \tag{3.64a}$$

where, as before, the subscript "0" refers to the deep-water conditions. Physically, the component waves of the group travel twice as fast as the group itself. Hence, in deep water, waves appear at the back of the group, travel to the front of the group, and then disappear.

In shallow water, $\sinh(2kh) \to 2kh$, and eq. 3.63 is approximately

$$c_g = \lim_{h \to 0} (c_g) = c \tag{3.64b}$$

From this result, we see that the component waves of the group remain in the same positions with respect to the front and back of the group in shallow water.

EXAMPLE 3.6: DEEP-WATER WAVE GROUP A 5-sec wave traveling in deep water is found to have a wavelength of approximately 39.0 m, according to the expression in eq. 3.36. The celerity of the wave is $c_0 = \lambda_0 / T = 7.8\,\text{m/s}$ and, from eq. 3.64, the group velocity is $c_{g0} = c_0/2 = 3.9\,\text{m/s}$. Consider a group composed of ten of these deep-water waves. We wish to determine how far the wave group travels when a component wave appears at the rear of the group and travels to the front of the group. In other words, our interest is in how far the group travels during the life of a component wave. Referring to the sketch in Figure 3.15, the length of the group is $L = 10\lambda_0 = 390\,\text{m}$, whereas the length of group travel is $\ell = c_{g0}t$. The component wave must travel the sum of the group length and the distance that the group travels, or $L + \ell = c_0 t$. By eliminating the time t, the length equations can be combined. The resulting value of the group travel distance is $\ell = L = 390\,\text{m}$. That is, the group travels a distance equal to its length, and the component waves travel twice the group length.

The engineering consequences of wave grouping are often debated. Most of the attention devoted to the subject has been in the area of structural design in the coastal zone. One of the advocates of including the effects of wave grouping is Per Bruun (1985), who states that the stability of rubble mound structures is sensitive to wave groups. However, J. W. van der Meer (1988) counters that wave groups play only a minor role in the stability problem. More recently, Medina, Fassardi, and Hudspeth (1990) present a comprehensive discussion on the wave group effects on the stability of rubble mound breakwaters. The reader should bear in mind that the wave height in the center of the group is twice that of the component waves, according to eq. 3.61. This can cause a problem in shallow water where, as discussed in Chapter 4, the wave breaks when the height is approximately 0.9 times the water depth. For this reason, the center waves can be quite high in the coastal zone before breaking. This is one argument for accounting for wave groupiness in the design of coastal structures.

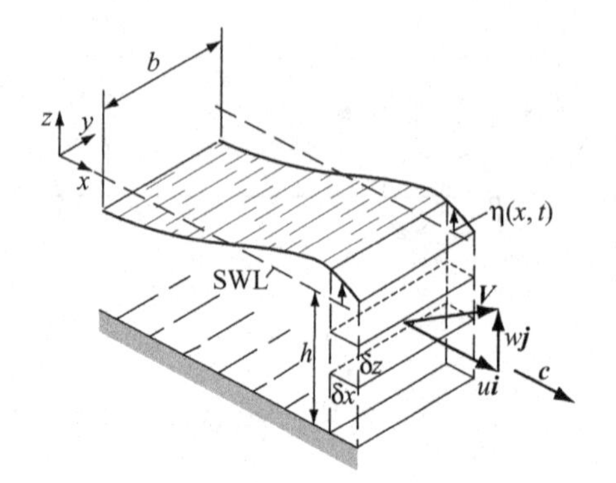

Figure 3.16. *Notation for the Wave Energy and Energy Flux Analyses.*

3.7 Wave Energy and Power

All physical waves (water, acoustical, optical, etc.) represent energy in transition. That is, energy is being transmitted by waves from one region to another. Free-surface water waves are produced by the motions of bodies, such as ships, earthquakes, and winds, as discussed previously in this chapter. In this section, we derive the expressions for both the average energy and the average energy flux or power of water waves.

We begin our analysis by considering the potential energy of an elemental mass of water displaced above the SWL, as sketched in Figure 3.16. The mass of that element is

$$\delta m = \rho\eta(\delta x)b$$

where b is the width of the wave crest under consideration, and the free-surface displacement (η) is positioned at the vertical centerline of the horizontally and vertically symmetric element. The length of the element, δx, is assumed to be small enough to allow us to both neglect the curvature of the free surface over this length and to assume that the center of mass is a distance $\eta/2$ above the SWL. The elemental potential energy of this elemental mass is then

$$\delta E_p = g(\delta m)\frac{\eta}{2} = \frac{1}{2}\rho g\eta^2(\delta x)b$$

The expression for the total potential energy of the wave is found by combining this expression with that of eq. 3.24 and integrating the resulting expression over one wavelength. The resulting expression for the *total potential energy* is then

$$E_p = \frac{\rho g H^2}{8}\int_0^\lambda \cos^2(kx - \omega t)dxb = \frac{\rho g H^2 \lambda b}{16} \tag{3.65}$$

Next, we consider the kinetic energy of the submerged elemental mass of water sketched in Figure 3.16. The elemental kinetic energy is

$$\delta E_k = \frac{1}{2}\rho(u^2 + w^2)b\delta x\delta z$$

where, for a linear wave, the respective particle velocity components (u, w) are obtained from eqs. 3.49 and 3.50. The combination of these equations with the δE_k

expression, and the subsequent integration over both the water depth and wavelength, results in the *total kinetic energy* of the wave:

$$E_k = \frac{1}{2}\rho \int_{-h}^{0}\int_{0}^{\lambda} (u^2 + w^2)dxdzb = \frac{\rho g H^2 \lambda b}{16} \tag{3.66}$$

Comparing the results of eqs. 3.65 and 3.66, we see that the total energy of a linear wave is equally divided between potential and kinetic energies. Hence, the *total energy* of the wave is the sum of the expressions in eqs. 3.65 and 3.66:

$$E = E_p + E_k = \frac{\rho g H^2 \lambda b}{8} \tag{3.67}$$

From this result, we see that a doubling of the wave height (H) results in a four-fold increase in the wave energy, whereas the energy is a linear function of the wavelength. The depth effects on the wave energy are demonstrated in the following example.

EXAMPLE 3.7: DEEP- AND SHALLOW-WATER WAVE ENERGY The combination of wave-energy expression of eq. 3.67 and the deep-water wavelength expression of eq. 3.36 results in

$$E_0 = \frac{\rho g^2 H^2 T^2 b}{16\pi} \tag{3.68}$$

Similarly, the combination of the shallow-water wavelength expression of eq. 3.38 and the energy expression of eq. 3.67 yields

$$E = \frac{\rho g^{3/2} H^2 \sqrt{h} T b}{8} \tag{3.69}$$

Comparing the results of eqs. 3.68 and 3.69, we see that the wave energy is proportional to the square of the wave period in deep water, but directly proportional to the period in shallow water.

To gain an idea of the magnitudes of the deep- and shallow-water energies, apply eqs. 3.68 and 3.69 to waves having a 1-m height and a 5-sec period. For the deep-water wave, the total energy per unit meter of crest (E_0/b) is 49,300 N-m/m from eq. 3.68. For the shallow-water wave, applying eq. 3.69 to a wave in 1 m of water yields an energy per unit meter of crest of 19,800 N-m/m. From these results, we see that waves of equal height and period in deep and shallow waters will have significantly different energies. This is true because the wavelength decreases with depth, as illustrated in Example 3.4.

As stated previously, a wave represents energy in transition. Of interest then is the rate of energy transmission in the direction of wave propagation, referred to as the energy flux. To obtain a mathematical expression for the energy flux, begin with the equation of energy conservation for an irrotational flow, Bernoulli's equation (eq. 2.70). When this equation is applied to a surface wave, two simplifications are made. First, as in the derivation of eq. 3.5, the time function $f(t)$ in eq. 2.70 is zero. Second, as in the derivation of the linearized free-surface condition of eq. 3.6, the nonlinear kinetic energy term ($\rho V^2/2$) is negligible for waves of very small steepness, that is, $H/\lambda \ll 1$. Bernoulli's equation applied to linear waves is then

$$\rho\frac{\partial\phi}{\partial t} + pgz + p = 0 \tag{3.70}$$

The first term represents both the unsteady kinetic energy per unit volume and the dynamic pressure, with the latter interpretation of interest here. The second term of eq. 3.70 represents both the potential energy per unit volume and the hydrostatic pressure. The third term in that equation is the flow energy per unit volume or the total pressure in the water. The only energy term in eq. 3.70 that is explicitly time-dependent is the kinetic energy term. Hence, the *energy flux*, which is the time-rate of change of energy per unit area normal to the flow direction, is simply the product of the dynamic pressure and the fluid velocity:

$$\rho\frac{\partial\phi}{\partial t}V = \rho\frac{\partial\phi}{\partial t}\nabla\phi \tag{3.71}$$

where the mathematical expression for the velocity potential, ϕ, is given in eq. 3.29. To obtain the *total energy flux* in the direction of wave travel, combine eqs. 3.71 and 3.29 and integrate the resulting expression over both the normal area (bh) and the wave period (T). The resulting expression can actually be considered to be the *wave power* in the direction of wave travel. Hence, we represent the resulting expression by the wave-power expression

$$\boldsymbol{P} = \rho b\frac{1}{T}\int_0^T\int_{-h}^0 \frac{\partial\phi}{\partial t}\nabla\phi dz dt = \frac{\rho g H^2 cb}{16}\left[\frac{2kh}{\sinh(2kh)} + 1\right]\boldsymbol{i} \tag{3.72a}$$

In this expression, we recognize the formula for the group velocity, c_g, given in eq. 3.63. Eq. 3.72 can then be rewritten in the more popular form,

$$\boldsymbol{P} = \frac{\rho g H^2 c_g b}{8}\boldsymbol{i} \tag{3.72b}$$

Physically, the wave power is convected in the direction of wave travel at the velocity of the wave group.

To obtain the deep-water approximation for the wave-power expression, combine eqs. 3.72b and 3.64 to obtain

$$\boldsymbol{P} = \frac{\rho g H^2 c_0 b}{16}\boldsymbol{i} = \frac{\rho g^2 H^2 T b}{32\pi}\boldsymbol{i} \tag{3.73}$$

Comparing this expression with the deep-water energy expression of eq. 3.68, we see that the energy is proportional to the square of the period, whereas the power is directly proportional to the period. In shallow water, the results in eqs. 3.72b, 3.64, and 3.38 are combined to obtain

$$\boldsymbol{P} = \frac{\rho g H^{3/2} H^2 \sqrt{h} b}{8}\boldsymbol{i} \tag{3.74}$$

In shallow water, the wave power is independent of the period.

EXAMPLE 3.8: DEEP- AND SHALLOW-WATER WAVE POWER To obtain an idea of the magnitude of the wave power, consider the 1-m, 5-sec wave of Example 3.7 in deep water. The value of the power per unit meter of crest obtained from eq. 3.73 is $\boldsymbol{P}/b$ = 4,930 watts/meter (W/m) or 4.93 kilowatts/meter (kW/m).

If the wave height increases to 2 m while the period is kept at 5 sec, then the power per unit length increases to 19.7 kW/m. That is, by doubling the wave height and maintaining the period, the power is increased by a factor of 4. When the shallow-water power expression of eq. 3.74 is applied to the 1-m, 5-sec wave in 1 m of water, as in Example 3.7, the resulting power per unit crest width is 3.96 kW/m, which is approximately 80% of the deep-water power associated with the same wave height and period.

EXAMPLE 3.9: WAVE POWER CONVERSION In Example 2.9, the scaling laws are applied to a pneumatic wave-energy conversion system. The system is sketched in Figure 2.24, and its operation is described in that example. Now, assume that the prototype system of Example 2.9 operates in 2 m of water ($h = 2.0$ m) where the wave height (H) is 1.0 m and the wave period (T) is 6 sec. From the example, the prototype power (P_p) is 128 kW. McCormick (2007) and others state that the maximum efficiency (ϵ) for the system sketched in Figure 2.24 is 50%; that is, a maximum of 50% of the incident wave energy or power can be converted to the energy or power of the oscillating water column. For this reason, 256 kW of incident wave power is required to obtain the desired prototype power under ideal conditions. Note that the efficiency applies to the hydraulics and pneumatics of the system and not to the turbine, generator, and other power take-off components. Our goal is to determine the width (b_{wc}) of the water column needed to produce the 128 kW of usable power, assuming that there are no point-absorbing effects, as described in Chapter 10.

First, the wave properties in 2 m of water must be determined for the 6-sec wave. The transcendental wavelength expression of eq. 3.31 yields a value of 25.6 m for λ. The approximate wavelength expression of eq. 3.38 yields a value of 26.6 m, a difference of less than 4%. Hence, we assume that the wave conditions are those corresponding to shallow water so that the wave-power expression of eq. 3.74 can be used. The equation that yields the water column width (b_{wc}) is then

$$P_p = 128{,}000\,\text{Watts} = \epsilon P = \epsilon \frac{\rho g^{3/2} H^2 \sqrt{h} b_{wc}}{8} \tag{3.75}$$

where $\epsilon = 0.5$, $\rho = 1{,}030$ kg/m^3, $g = 9.81$ m/s^2, $H = 1$ m, and $h = 2$ m. The width value is then 45.8 m.

The quarter-scale model of the prototype system yields a power of 1 kW from Example 2.9. The width of the model water column is 11.4 m. This model operates in a tank having a 0.5-m depth, where the respective wave height and period are 0.25 m and 3.0 sec.

The area of *wave-energy conversion* is receiving considerable attention as waves are a viable alternative energy resource for both island and isolated coastal communities. The reason for this interest can be understood by considering the following: The average power required by a citizen of the contiguous United States is 1 kW. The island communities that are candidates for wave-energy conversion have a requirement of about 100 W per person. The prototype of Examples 2.9 and 3.9 would then provide the average power for 1,280 island citizens. This converted wave power could be used directly in the production of either electricity or potable water.

3.8 Shoaling

A wave is said to "shoal" when the properties of the wave are affected by changes in the water depth. The same term is used to identify a rise in the sea bed that affects certain waves. The phenomenon of waves responding to changes in the water depth is called *shoaling*. The behaviors of both the wavelength and celerity with changing depth are mathematically described by the relationships in eq. 3.31, assuming that the waves are linear. In addition, the behavior of the group velocity with depth is described by the expression in eq. 3.63. We now focus our attention to the depth effects on the wave height.

To determine the relationship describing the depth effects on the wave height, we assume that the energy flux (wave power) is conserved as a wave passes from one water depth to another. To visualize, consider the sketches in Figure 3.17. In that figure, waves are shown to approach a shoreline over a flat, inclined bed. The *shoreline* is defined as the intersection of the SWL and the bed. An area sketch of the wave train is in Figure 3.17a, and an elevation sketch is in Figure 3.17b. The angle between the bed and the horizontal direction is assumed to be small so that there is no wave reflection.

As sketched in Figure 3.16, assume that the waves travel in a "channel" of width b having virtual side walls. These "walls" are seen as straight lines in Figure 3.17a, and are referred to as *orthogonals* because they are normal to the wave front. The flow between the orthogonals is assumed to be irrotational; therefore, there are no energy losses as the waves progress. With this assumption, the energy flux (wave power) is invariant with x. That is, wave powers passing through vertical areas at two offshore positions, say x_A and x_B, are equal. We can express this conservation of wave power mathematically by

$$\boldsymbol{P}_a = \boldsymbol{P}_b \tag{3.76a}$$

or, from the results in eq. 3.72b, we can write

$$\frac{\rho g H_A^2 c_{gA} b}{8}\boldsymbol{i} = \frac{\rho g H_B^2 c_{gB} b}{8}\boldsymbol{i} \tag{3.76b}$$

where c_g is the group velocity, expressed by eq. 3.63. From the results in eq. 3.76b, the relationship between the wave heights is found to be

$$H_B = H_A\sqrt{\frac{c_{gA}}{c_{bB}}} \tag{3.77}$$

Let position A be in deep water and position B be in the shoal. By convention, the deep-water wave properties are identified by the subscript "0," whereas the wave properties in the shoal are not subscripted. The expression in eq. 3.77 relating the deep-water wave height and that in the shoal can be written as

$$K_S \equiv \frac{H}{H_0} = \sqrt{\frac{c_{g0}}{c_g}} = \sqrt{\frac{\cosh^2(kh)}{\sinh(kh)\cosh(kh) + kh}} \tag{3.78}$$

where K_S is called the *shoaling coefficient*. The last term in eq. 3.78 is obtained from the combination of the expressions in eqs. 3.31, 3.36, 3.63, 3.64, and several hyperbolic function identities. The shoaling coefficient then describes the variation in the height of a wave as it passes from deep water into a shoaling region.

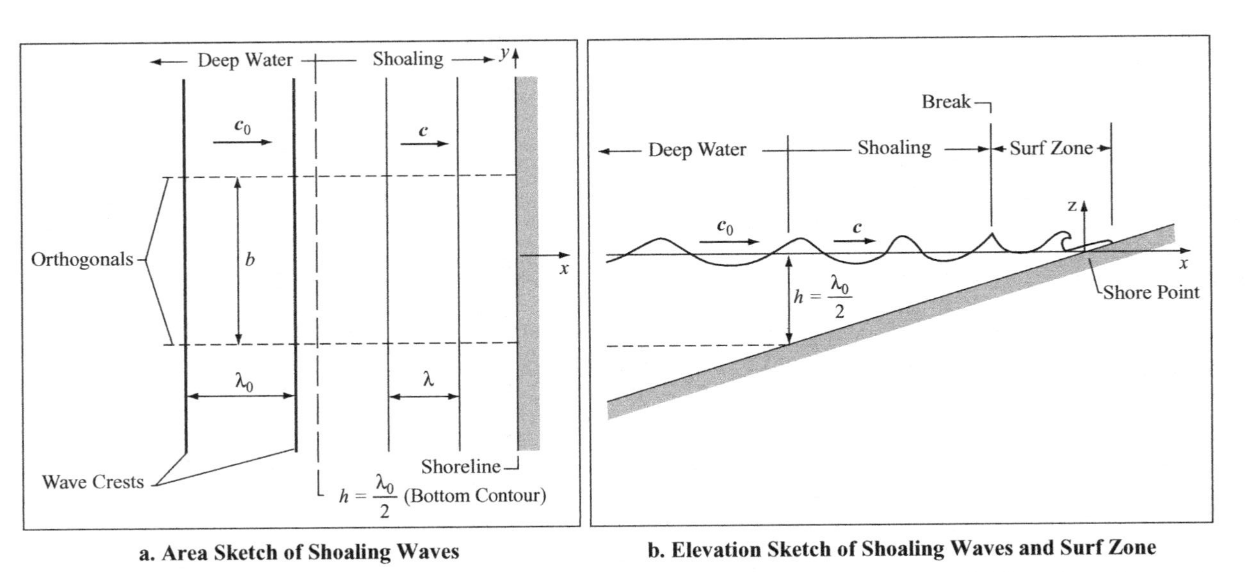

Figure 3.17. *Shoaling, Breaking, and Runup*. In (a) shoaling without breaking is illustrated. The waves shorten in length and slow down as the waves approach the shoreline in this idealized situation (predicted by Airy's theory). In (b) shoaling and real effects are shown. At the end of the shoaling zone, the waves break and travel beyond the beach as broken waves. The maximum height of the water on the beach is called the *runup*. These phenomena are discussed in some detail in Chapter 7.

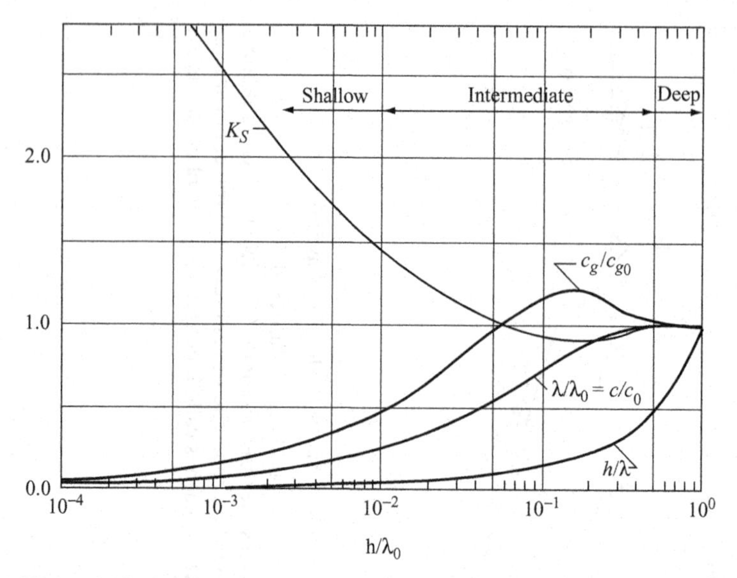

Figure 3.18. *Linear Shoaling Curves*. These curves apply to waves over flat, horizontal beds, that is, where the bed slope is zero. Effects of the bed slope can be seen in the experimental results shown in Figure 3.19.

It is also convenient to relate the other wave properties (wavelength, celerity, and group velocity) over a shoal to those in deep water. To obtain these relationships, first combine eqs. 3.31 and 3.36 to obtain the relationships for the wavelength and celerity. The resulting expression is

$$\frac{\lambda}{\lambda_0} = \frac{c}{c_0} = \tanh(kh) \tag{3.79}$$

To relate the group velocities over a shoal and in deep water, simply rearrange eq. 3.78. The resulting expression is

$$\frac{c_g}{c_{g0}} = \frac{\sinh(kh)\cosh(kh) + kh}{\cosh^2(kh)} = \frac{1}{K_S^2} \tag{3.80}$$

Curves resulting from eqs. 3.78, 3.79, and 3.80 are called the *shoaling curves*, and are presented in Figure 3.18. These curves represent the behaviors of the wavelength, celerity, group velocity, and wave heights with changing water depth, according to Airy's linear wave theory. In Chapter 4, it is shown that the wavelength, celerity, and group velocity curves of Figure 3.18 are also those that correspond to Stokes' second-order wave theory, discussed in Section 4.2.

As a test of the ability of the linear theory to predict wave heights in shoaling waters, let us apply the shoaling coefficient expression of eq. 3.78 to experimental data obtained for shoaling on beaches of small slopes. The experimental data are those of Brink-Kjær and Jonsson (1973), as reported by Svendsen and Jonsson (1976). These data and the theoretical curve resulting from eq. 3.78 are presented in Figure 3.19. The reader should note that the experimental data depend on both the beach slope and the deep-water wave steepness, whereas the theoretical curve from the linear theory is independent of both. A comparison of the theoretical and experimental results shows that the minimum values of the experimental data are up to 10% lower than the minimum theoretical value. Furthermore, the slopes of the

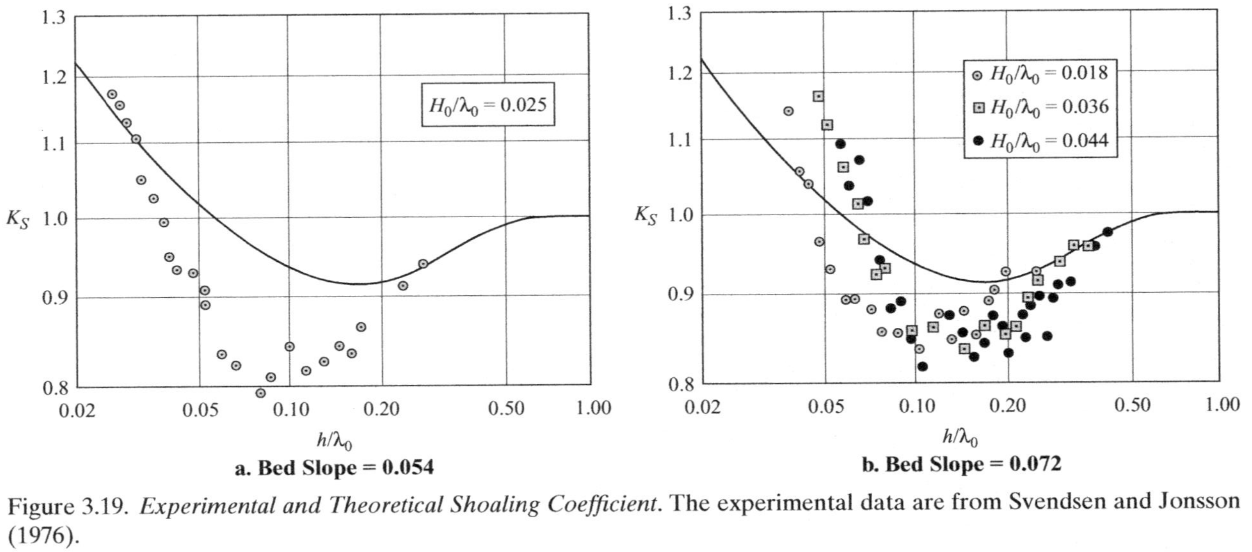

Figure 3.19. *Experimental and Theoretical Shoaling Coefficient*. The experimental data are from Svendsen and Jonsson (1976).

experimental data in shallow water are greater than that of the theoretical curve. Considering the limitations imposed by the assumption of linearity, Airy's theory does an acceptable job in predicting the qualitative behavior of shoaling waves.

3.9 Closing Remarks

The water wave theory of G. B. Airy (1845), called the linear wave theory, is the basic wave theory used by ocean engineers. This extraordinary theory was presented for the first time over a century-and-a-half ago. The range of validity of the linear theory is quite broad considering the simplicity of the theory. In the following chapters, we find many references to formulas of the linear theory that are developed and presented in this chapter.

One final note concerning the measurement of water waves: There are a number of experimental and field techniques used in the measurement of wave properties (including the direction of wave travel). This important subject requires far more discussion than can be afforded herein. For those readers interested in the wave measurement area, the book of Tucker and Pitt (2001) is recommended.

4 Nonlinear Surface Waves

The study of *nonlinear waves* began early in the nineteenth century. Results of the first studies of these waves were used to determine the *quasi*-static wave loads on ships and other tethered floating structures. Although the first analytical techniques were rather simplistic, these techniques were used well into the twentieth century. In the mid-nineteenth century, more mathematically sophisticated analytical methods were introduced. These methods are used by both physical oceanographers and ocean engineers in their respective predictions of real wave properties and the time-dependent loadings on offshore structures.

There are a number of theories that can be used to approximately predict the properties of nonlinear waves. Probably the earliest of these theories is that of Gerstner (1808). The theory of Gerstner is a geometric type, and is commonly referred to as the *trochoidal theory*. This name results from the predicted profile of the breaking wave (see Figure 3.1e) which, from Gerstner's theory, is a *trochoid*, having a cusp at the crest. The trochoidal theory is rotational in a hydrodynamic sense (see Section 2.3), with the rotational direction being opposite of that actually observed. Results of the theory are still used today by some structural naval architects to predict the extreme *quasi*-static wave loads on ship hulls in both sagging (crests at both the bow and stern, and the trough amidships) and hogging (troughs at both the bow and stern, and a crest amidships). A discussion of the trochoidal theory is found in the books of Wiegel (1964), Kinsman (1965), McCormick (1973), Horikawa (1978), and Sarpkaya and Isaacson (1981), whereas Bhattacharyya (1978) discusses the analyses of the wave loading on ships under sagging and hogging conditions.

An irrotational expansion theory was introduced in the mid-nineteenth century by Stokes (1847). That theory is based on the assumption that the wave properties can be represented by perturbation series. The desired accuracy of the theory simply depends on the number of terms maintained in each series. As is shown later in this chapter, the Stokes second-order theory yields excellent results for certain ranges of depth-wavelength ratio, that is, h/λ. De (1955) derived the generalized fifth-order theory, and expansions up to the fifth order have been found to yield good results in the prediction of wave loading on fixed structures. We shall confine our discussions to the second-order theory because of its relative simplicity. A discussion of the higher-order Stokes theories can be found in the book of Patel (1989), whereas

Dawson (1983) discusses the application of the Stokes fifth-order theory to wave loading situations.

In the latter part of the twentieth century, several nonlinear wave theories were developed to exploit the capabilities of high-speed computers. One such theory is the *stream-function theory* of Dean (1965), which is outlined in the books of Sarpkaya and Isaacson (1981) and Dean and Dalrymple (1984). Results of the theory are obtained numerically.

A nonlinear theory that can be used for wave profile prediction in the coastal zone is that of Korteweg and deVries (1895). That theory is called the *cnoidal theory*, the name of the theory chosen because of the use of the Jacobian elliptic cosine function, *cn*. The cnoidal theory is used to predict the properties of long waves in shallow water, that is, where the wavelength (λ) is much greater than the water depth (h). Useful results of the theory are presented in the *Shore Protection Manual*, published by the U. S. Army Corps of Engineers (see U.S. Army, 1984).

A limiting case of the cnoidal theory is the *solitary wave*, a wave first recognized by Russell (1838, 1844). See the discussion of Russell's observations in the book by Crapper (1984). Theoretically, the solitary wave is a shallow-water wave having an infinite length and, therefore, an infinite period. The passage of a solitary wave is then a one-time event. In addition, the free surface of the wave is (theoretically) entirely above the SWL. Although these wave properties are extraordinary, the analytical wave properties obtained from solitary wave theory do compare rather well with measured long-wave properties in shallow water. The analysis of the solitary wave was first satisfactorily performed by Boussinesq (1872), and later by Rayleigh (1876) and McCowan (1891). The solitary wave theory is discussed later in this chapter.

The analytical validity ranges of the various nonlinear wave theories are presented in Figure 4.1 with that of the linear theory. Versions of the figure appear in a technical report by Le Méhauté (1969), publications by Dean (1967, 1974), and elsewhere. Muga and Wilson (1970) also discuss the limitations of the validity curves. In Figure 4.1, we see that the validity range of Airy's linear wave theory, discussed in Chapter 3, decreases with the water depth, h. The linear theory has application to the long-period waves (swell) in both deep and intermediate waters. However, Stokes' theory has a larger validity range, extending to the breaking wave limit in deep water. In shallow water, the cnoidal theory has a wide range of validity, where the upper limit of that theory is the breaking line for the solitary theory. The Stokes, cnoidal, and solitary wave theories are discussed in depth in the following sections. Although the results of these theories are somewhat cumbersome, they can be easily programmed for either laptop computers or advanced pocket calculators for use by engineers working in the field.

4.1 Nonlinear Wave Properties

Before discussing the nonlinear wave theories, let us familiarize ourselves with the properties of nonlinear waves. The most striking difference between linear and nonlinear waves is the *wave profile*, as can be seen by comparing the sketches in Figures 3.1c and 3.1d. The profile of any monochromatic wave depends on the wave height (H), period (T), and water depth (h). In deep water, where we assume $h > \lambda_0/2$, the wave height and period can be combined into a single parameter called

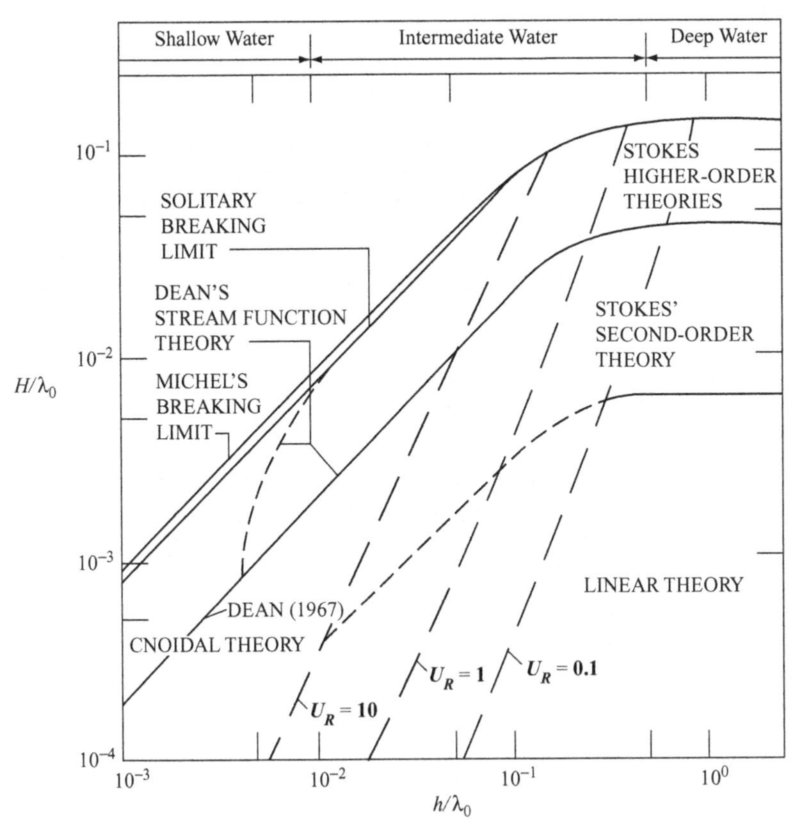

Figure 4.1. *Analytical Validity Ranges of the Linear and Nonlinear Wave Theories*. The breaking wave is the limiting case for each of the theories, as discussed in Section 4.7. The Ursell (1953) parameter, $U_R \equiv H\lambda^2/2h^3$, represents the ratio of the convective inertia and the local inertia, and is used to determine the regions of analytical validity. According to Le Méhauté (1969), the linear theory is valid for $U_R \ll 1$, whereas the cnoidal and solitary theories are mathematically valid for $U_R \gg 1$.

the *wave steepness*,

$$\frac{H_0}{\lambda_0} = \frac{2\pi H_0}{gT^2} \tag{4.1}$$

where the deep-water wavelength expression is in eq. 3.36. As the value of the wave steepness increases, the wave profile becomes more nonlinear, as can be seen in Figure 4.2. In that figure, experimentally observed deep-water wave profiles are presented in non-dimensional forms, where the ratio of the free-surface displacement to amplitude $(2\eta/H)$ is shown as a function of x/λ for several values of the wave steepness. The sinusoidal profile is also sketched in that figure for the sake of comparison. For the smallest value of H/λ, the profile is approximately sinusoidal or linear. As the value of the wave steepness increases, the crest of the wave becomes both narrow and high while the trough becomes both broad and shallow. As sketched in Figure 3.1d, the MWL, which is midway between the trough and the crest, is above the SWL for a nonlinear wave. For a linear wave, the MWL and SWL are coincident. Above a specific steepness value, the wave breaks. The conditions for breaking are discussed later in this chapter.

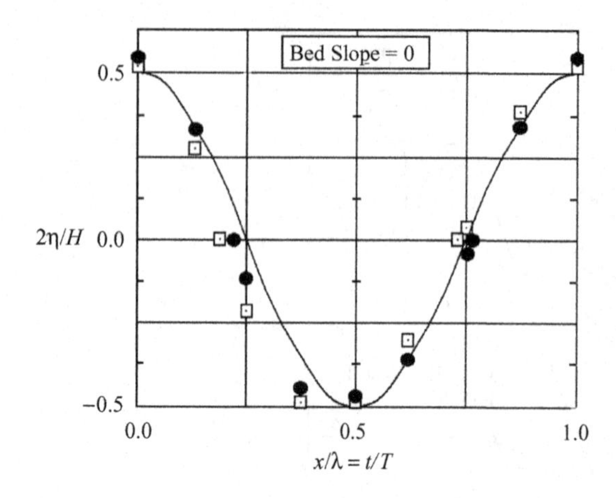

Figure 4.2. *Linear Free-Surface Profile and Experimental Free-Surface Displacements.* The data, obtained by Svendsen and Hansen (1978), represented by □, correspond to $H_0/\lambda_0 = 0.026$, $h/\lambda_0 = 2$, and $U_R = 0.016$. Those data represented by • correspond to $H_0/\lambda_0 = 0.032$, $h/\lambda_0 = 0.88$, and $U_R = 0.023$.

In shallow water, where we assume $h < \lambda/25$, the water depth has a strong effect on the wave profile. This is illustrated in Figure 4.3, where experimentally obtained profiles are presented in non-dimensional forms for various values of the ratio of the mean depth to wavelength (h/λ). The data were obtained by Svendsen and Hansen (1978) on a bed with a 1/35 slope. For this type of shallow-water wave, the wave-induced force on a structure has a dual nature. As a high, narrow crest passes the structure, the force in the horizontal direction is nearly impulsive, that is, a relatively large force is experienced over a short time duration. When the trough passes, the force is a *quasi*-steady hydrodynamic type.

In this chapter, the expansion theory of Stokes and the cnoidal theory are developed. To illustrate the use of the Stokes theory, the second-order theory is discussed. Also, the solitary theory is shown to be the shallow-water limiting case of the cnoidal theory.

4.2 Stokes' Wave Theory

The expansion theory of G. G. Stokes (1847) is based on two basic assumptions. First, as in the case of Airy's wave theory in Chapter 3, the flow beneath the free surface of the wave is assumed to be irrotational. This assumption allows the water

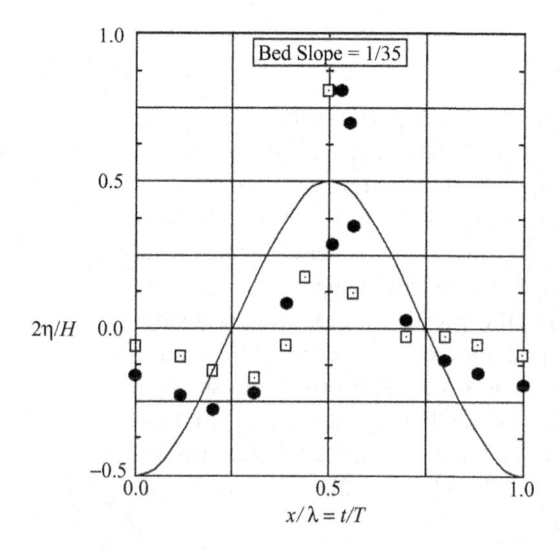

Figure 4.3. *Linear Profile and Measured Data in Shallow and Intermediate Waters.* The data, obtained by Svendsen and Hansen (1978), represented by □ correspond to $H_0/\lambda_0 = 0.01$, $h/\lambda_0 = 0.04$, and $U_R = 275$. Those data represented by • correspond to $H_0/\lambda_0 = 0.017$, $h/\lambda_0 = 0.065$, and $U_R = 68.5$.

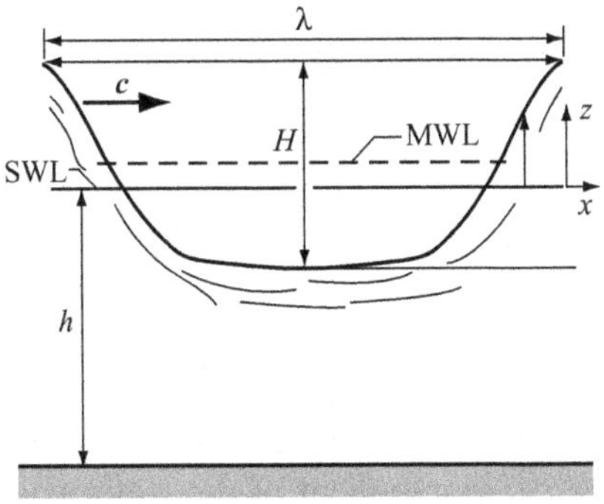

Figure 4.4. *Notation for the Stokes Wave Theory.*

particle velocity components to be represented by the velocity potential, φ, as in eqs. 2.38 and 2.39. The second assumption is that the wave properties can be represented by perturbation series. Physically, this assumption is equivalent to producing a nonlinear wave by superimposing linear waves upon each other. In Section 3.6, we superimposed two linear waves of equal height but of slightly differing periods to obtain the wave group. In the development of the Stokes theory, both the height and period of the component waves must be determined. In this section, Stokes' wave theory is described in general terms, and the second-order theory is demonstrated. When using Stokes wave theory, one must be conscious of the analytical validity ranges shown in Figure 4.1.

The nonlinear form of Bernoulli's equation applied to the free surface is presented in eq. 3.5. Physically, the form of that equation can be considered to represent the energy of the flow per unit water mass. Let us rewrite eq. 3.5 with one change, that is, replace the velocity term, V, with the gradient of the velocity potential, as in eq. 2.38. Referring to Figure 4.4 for notation, Bernoulli's equation applied to the free surface is then

$$\left\{\frac{\partial \varphi}{\partial t} + g\eta + \frac{1}{2}(\nabla\varphi)^2\right\}\Big|_{z=\eta} = 0 \tag{4.2}$$

where φ is the velocity potential, η is the free-surface displacement from the SWL, and g is the gravitational constant (9.81 m/s^2). This equation is the nonlinear and unsteady form of the dynamic free-surface condition. Instead of this form of the free-surface condition, a *quasi*-steady form is used in the present analysis. Following the method of Stokes, the unsteady term can be eliminated in Bernoulli's equation if the inertial frame of reference is changed to a relative frame fixed to the wave, as sketched in Figure 4.5. The two coordinate systems are related by

$$\begin{aligned} X &= x - ct \\ Z &= z \end{aligned} \tag{4.3}$$

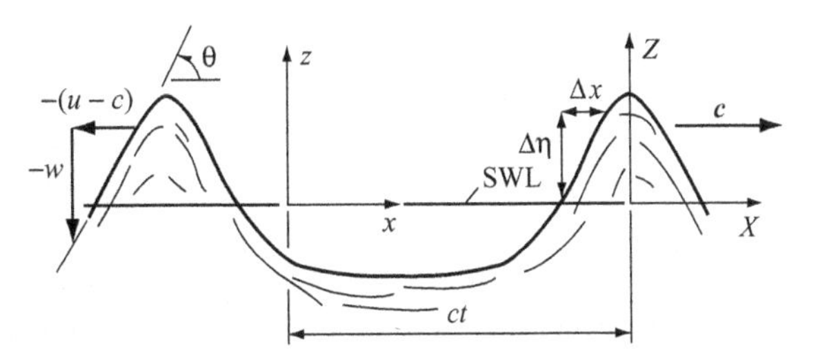

Figure 4.5. *Notation for a Nonlinear Wave of Fixed Form.*

From the first of these relationships, the horizontal velocity component in the relative or wave frame of reference is

$$\frac{dX}{dt} = \frac{dx}{dt} - c = u - c = \frac{\partial \varphi}{\partial x} - c \tag{4.4}$$

Using this result, the dynamic free-surface condition in the relative frame is

$$g\eta + \frac{1}{2}\left[\left(\frac{\partial \varphi}{\partial x} - c\right)^2 + \left(\frac{\partial \varphi}{\partial z}\right)^2\right]\Bigg|_{z=\eta} = \mathrm{K} \tag{4.5}$$

The kinematic free-surface condition is derived as follows: The vertical velocity component of the water particle motion can be written in terms of c, η, and φ by taking advantage of the geometric relationships shown in Figure 4.5. In that figure, the slope of the free surface at any point is

$$\tan(\theta) = \lim_{\Delta X \to 0} \frac{\Delta \eta}{\Delta X} = \frac{d\eta}{dX} = \frac{w}{u - c}\bigg|_{z=\eta} \tag{4.6}$$

where θ is the angle of the free surface with respect to the horizontal direction. From eq. 4.6, we obtain the following expression for the vertical velocity component on the free surface, which is the kinematic free-surface condition:

$$w = \frac{\partial \varphi}{\partial z}\bigg|_{z=\eta} = \frac{d\eta}{dX}\left\{\frac{\partial \varphi}{\partial x} - c\right\}\bigg|_{z=\eta} \tag{4.7}$$

Of the several approaches to Stokes' wave theory, we choose that outlined by Dean (1965) because of its directness. Following Dean (1965), we represent the respective velocity potential and the free-surface displacement by perturbation series as

$$\varphi = \varepsilon^1 \varphi_1 + \varepsilon^2 \varphi_2 + \cdots + \varepsilon^n \varphi_n + \cdots = \sum_{i=1}^{\infty} \varepsilon^i \varphi_i \tag{4.8}$$

and

$$\eta = \varepsilon^1 \eta_1 + \varepsilon^2 \eta_2 + \cdots + \varepsilon^n \eta_n + \cdots = \sum_{i=1}^{\infty} \varepsilon^i \eta_i \tag{4.9}$$

where ε is an arbitrary constant assumed to be $0 < \varepsilon \leq 1$ to ensure series convergence. Each higher-order term is then a perturbation to the sum of the lower-order terms. Each term of the series in eq. 4.8 is a velocity potential of a specific wave. Hence, each of these potentials must satisfy Laplace's equation, eq. 2.41,

$$\nabla^2 \varphi_i = 0 \tag{4.10}$$

The seafloor condition of eq. 3.4 must also be satisfied by each of the potential functions. Assuming a flat, horizontal bed at $z = -h$, this boundary condition is

$$\frac{\partial \varphi_i}{\partial z}\bigg|_{z=-h} = 0 \tag{4.11}$$

Returning to the perturbation series representations, there are two additional terms in eq. 4.5 to expand, those being the celerity, c, and the constant, K. There is one difference between the series of eqs. 4.8 and 4.9 and those for these two terms. The terms of eq. 4.5, Bernoulli's equation, are in units of energy per unit mass. Because the origin of the relative coordinate system is traveling away from the origin of the inertial system, there is a kinetic energy introduced. If no waves are present, both η and ϕ are equal to zero on the free surface. Hence, the imposed energy must be equal to K, and that energy per unit mass is

$$K = \frac{1}{2}c^2 \tag{4.12}$$

This means that there are no terms in the expansions for both c and K that represent the case of no waves. The perturbation series for these respective terms are then

$$c = C + \sum_{i=1}^{\infty} \varepsilon^i c_i \tag{4.13}$$

and

$$K = \mathrm{K} + \sum_{i=1}^{\infty} \varepsilon^i K_i \tag{4.14}$$

where C and K are the zeroth-order terms of the respective expansions.

Equations 4.5 and 4.7 represent the free-surface boundary conditions. The derivatives of the velocity potential, φ, in those equations are applied at $z = \eta$ or about the SWL at $z = 0$. Because of this, the derivatives of the velocity potential can be expanded in a Maclaurin series, resulting in the following expressions:

$$\left.\frac{\partial \varphi}{\partial x}\right|_{x=\eta} = \left.\frac{\partial \varphi}{\partial x}\right|_{z=0} + \left.\frac{1}{1!}\frac{\partial^2 \varphi}{\partial x \partial z}\right|_{z=0} \eta + \left.\frac{1}{2!}\frac{\partial^3 \varphi}{\partial x \partial z^2}\right|_{z=0} \eta^2 + \cdots \tag{4.15}$$

and

$$\left.\frac{\partial \varphi}{\partial z}\right|_{z=\eta} = \left.\frac{\partial \varphi}{\partial z}\right|_{z=0} + \left.\frac{1}{1!}\frac{\partial^2 \varphi}{\partial z^2}\right|_{z=0} \eta + \left.\frac{1}{2!}\frac{\partial^3 \varphi}{\partial z^3}\right|_{z=0} \eta^2 + \cdots \tag{4.16}$$

Equations 4.5 and 4.7 are now combined with both of the expansions of eqs. 4.15 and 4.16 and the perturbation series of eqs. 4.8, 4.9, 4.13, and 4.14. The resulting free-surface boundary conditions, the continuity equation of eq. 4.10, and the seafloor condition of eq. 4.11 comprise the set of equations of the *Stokes nth-order wave theory*. The equation sets for the various orders are obtained by equating the coefficients of ε, the perturbation constant. For the zeroth-, first-, and second-order theories, those equations are as follows.

a. Dynamic Free-Surface Condition:

$$\varepsilon_0\text{:} \quad \frac{C^2}{2} = \mathrm{K}_0 \tag{4.17}$$

$$\varepsilon_1\text{:} \quad g\eta_1 + Cc_1 - C\left.\frac{\partial \varphi_1}{\partial x}\right|_{z=0} = \mathrm{K}_1 \tag{4.18}$$

$$\varepsilon_2\text{:} \quad g\eta_2 + \frac{1}{2}\left.\left[\left(\frac{\partial \varphi_1}{\partial x}\right)^2 + \left(\frac{\partial \varphi_1}{\partial z}\right)^2\right]\right|_{z=0} - \left.\left[C\frac{\partial \varphi_2}{\partial x} + c_1\frac{\partial \varphi_1}{\partial x} + C\eta_1\frac{\partial^2 \varphi_1}{\partial x \partial z}\right]\right|_{z=0}$$
$$+ Cc_2 + \frac{c_1^2}{2} = \mathrm{K}_2 \tag{4.19}$$

b. Kinematic Free-Surface Condition:

$$\varepsilon_0: \quad 0 = 0 \tag{4.20}$$

$$\varepsilon_1: \quad -C\frac{d\eta_1}{dx} = \left.\frac{\partial \varphi_1}{\partial z}\right|_{z=0} \tag{4.21}$$

$$\varepsilon_2: \quad \frac{d\eta_1}{dx}\left.\frac{\partial \varphi_1}{\partial x}\right|_{z=0} - c_1\frac{d\eta_1}{dx} - C\frac{d\eta_2}{dx} = \left.\frac{\partial \varphi_2}{\partial z}\right|_{z=0} + \eta_1\left.\frac{\partial^2 \varphi_1}{\partial z^2}\right|_{z=0} \tag{4.22}$$

c. Equation of Continuity (Laplace's Equation):

$$\varepsilon_0: \quad 0 = 0 \tag{4.23}$$

$$\varepsilon_1: \quad \nabla^2 \varphi_1 = 0 \tag{4.24}$$

$$\varepsilon_2: \quad \nabla^2 \varphi_2 = 0 \tag{4.25}$$

d. Seafloor Condition:

$$\varepsilon_0: \quad 0 = 0 \tag{4.26}$$

$$\varepsilon_1: \quad \left.\frac{\partial \varphi_1}{\partial z}\right|_{z=-h} = 0 \tag{4.27}$$

$$\varepsilon_2: \quad \left.\frac{\partial \varphi_2}{\partial z}\right|_{z=-h} = 0 \tag{4.28}$$

In Section 3.2, we find that the general solution of Laplace's equation (eq. 3.8), subject to the seafloor condition of eq. 3.4, is eq. 3.17. Similarly, the respective general solutions of eqs. 4.24 and 4.25 are

$$\varphi_1 = C_{\varphi 1}\cosh[k_1(z+h)]\sin(k_1 X_1) \tag{4.29}$$

and

$$\varphi_2 = C_{\varphi 2}\cosh[k_2(z+h)]\sin(k_2 X_2) \tag{4.30}$$

where the respective wave numbers are

$$k_1 = \frac{2\pi}{\lambda_1} \tag{4.31}$$

and

$$k_2 = \frac{2\pi}{\lambda_2} \tag{4.32}$$

An inspection of eqs. 4.17, 4.20, 4.23, and 4.26 leads to the conclusion that the zeroth-order theory describes water having a flat surface. In the moving frame of reference, the water is flowing in the negative X-direction with a speed of C.

To derive the equations for the first-order theory, we begin by taking a derivative of eq. 4.18 with respect to x, and combining the resulting expression with that of eq. 4.21 by eliminating the derivative $d\eta_1/dX$. We note that the derivatives with respect to both X and x are equal from eq. 4.3. The result of the combination is an expression for the celerity in terms of both first-order wave number and water depth, that is,

$$C = \sqrt{\frac{g}{k_1}\tanh(k_1 h)} \tag{4.33}$$

The same form of the celerity from the linear theory is obtained by rearranging eqs. 3.30 and 3.31. Now, combine the expressions in eqs. 4.21, 4.29, and 4.33, and integrate the resulting expression from the crest at $X = 0$ to the trough at $\lambda_1/2$. This results in the expression for the parametric coefficient of the velocity potential expression of eq. 4.29. That expression is

$$C_{\varphi 1} = \frac{gH_1}{2k_1 C}\frac{1}{\cosh(k_1 h)} \tag{4.34}$$

where $H_1 = (\eta_{max} - \eta_{min})$ is the wave height resulting from the integration. The expression for the first-order velocity potential is then

$$\varphi_1 = \frac{gH_1}{2k_1 C}\frac{\cosh[k_1(z+h)]}{\cosh(k_1 h)}\sin(k_1 X) \tag{4.35}$$

from eq. 4.29. Now combine eqs. 4.33, 4.35, and 4.21, and integrate the resulting expression from the crest at the origin to X to obtain the expression for the first-order free-surface deflection. The result is

$$\eta_1 = \frac{H_1}{2}\sin(k_1 X) \tag{4.36}$$

Finally, for the first order, the combination of the expressions in eqs. 4.35, 4.36, and 4.18 produces a relationship that is not trivial only if

$$\mathrm{K}_1 = Cc_1 \tag{4.37}$$

Several observations are made concerning the results of the first-order theory. First, the expressions for both φ_1 and η_1 are identical in form to the corresponding velocity potential and free-surface displacement expressions obtained from the linear theory. Second, when the X-coordinate in eqs. 4.35 and 4.36 is replaced by the expression of eq. 4.3, and the product of the wave number and celerity in those equations is replaced by the circular frequency, the resulting expressions are identical with those of the linear theory. We conclude that the first-order Stokes theory is identical to the linear theory of Airy.

Following the introduction of the perturbation constant, ε, in eqs. 4.8 and 4.9, the statement is made to the effect that the value of this positive arbitrary constant should be less than or equal to unity to ensure convergence of the perturbation series. Because we have determined that the first-order theory is identical with that of Airy, the linear wave is that upon which we build to obtain higher-order (nonlinear) waves. For this reason, the value of the perturbation constant is

$$\varepsilon = 1 \tag{4.38}$$

Because the expression for η_1 in eq. 4.36 is linear, we can assume that the second-order free-surface displacement is also linear because of the form of the velocity potential in eq. 4.30. Physically, the assumption made in the Stokes theory is that large waves can be formed by simply superimposing linear waves. The second-order free-surface deflection can be written as

$$\eta_2 = \frac{H_2}{2}\cos(k_2 X) \tag{4.39}$$

where the wave height, H_2, is to be determined. Before continuing with the second-order theory, one assumption is made here. The second-order wave has some crests that coincide with those of the first-order theory. Mathematically, this assumption can be represented by the wavelength relationship,

$$\lambda_1 \equiv \lambda = n\lambda_2, \quad n = 1, 2, 3, \ldots. \tag{4.40}$$

In terms of the wave number, eq. 4.40 is equivalent to

$$k_1 \equiv k = 2\pi/\lambda = k_2/n, n = 1, 2, 3, \ldots. \tag{4.41}$$

In these equations, the integer, n, must be determined. To determine the value of this integer, combine the second-order kinematic free-surface boundary condition of eq. 4.22 with the expressions in eqs. 4.30, 4.35, 4.36, 4.37, 4.39, and 4.41. The resulting expression is not trivial only if

$$n = 2 \tag{4.42}$$

With this integer value in eq. 4.40, we see that the length of the perturbation wave is half that of the first order.

Take a derivative with respect to X of the expression in eq. 4.19, and combine the result with eq. 4.22 by eliminating $d\eta_2/dX$. In the resulting expression, replace φ_1, φ_2, and η_1 by the respective expressions in eqs. 4.35, 4.30, and 4.36. The result is not trivial only if

$$c_1 = 0 \tag{4.43}$$

In eq. 4.37, then,

$$\mathrm{K}_1 = 0 \tag{4.44}$$

In addition, the combination of the equations leading to the result of eq. 4.43 yields the expression for the parametric coefficient of the second-order velocity potential. That is, in eq. 4.30,

$$C_{\varphi 2} = \frac{3}{32}\frac{CkH_1^2}{\sinh^4(kh)} \tag{4.45}$$

The resulting second-order velocity potential expression is

$$\varphi_2 = \frac{3}{32}CkH_1^2\frac{\cosh[2k(h+z)]}{\sinh^4(kh)}\sin(2kX) \tag{4.46}$$

It is apparent that the results of the second-order theory can be expressed in terms of the first-order wave height. Hence, from this point forward, we shall omit the subscript "1" from that wave height and use the following wave height notation:

$$H = H_1 \tag{4.47}$$

The second-order wave height, H_2, is found from the kinematic free-surface expression of eq. 4.22. To obtain this height, first combine the free-surface displacement expression of eq. 4.36 and the velocity potential expressions of eqs. 4.35 and 4.46 with eq. 4.22. Then, integrate $d\eta_2$ from the trough to the crest of the second-order wave. The resulting wave height expression is then

$$H_2 = \eta_{2\max} - \eta_{2\min} = \frac{H^2k}{8}\frac{\cosh(kh)}{\sinh^3(kh)}[2 + \cosh(2kh)] \tag{4.48}$$

where, again, the subscript "1" has been omitted from the first-order wave properties in eq. 4.48.

The final two unknowns of the second-order theory are the constant, K_2, and the celerity, c_2. The relationship between the two can be found by applying eq. 4.18 to the special case of an infinitely long wave of zero height (the no-wave condition), where $\eta_1 = \eta_2 = \varphi_1 = \varphi_2 = 0$. The result is

$$K_2 = Cc_2 \tag{4.49}$$

With this result, the application of eq. 4.5 along with the first- and second-order expressions applied to the no-wave condition yields

$$c_2 = K_2 = 0 \tag{4.50}$$

The results obtained from Stokes' second-order theory applied to traveling waves are now summarized. The celerity or phase velocity expression to the second order is found by combining eqs. 4.13, 4.38, 4.33, 4.43, and 4.5. The resulting expression is

$$c = \frac{\lambda}{T} = \sqrt{\frac{g}{k}\tanh(kh)} \tag{4.51}$$

This expression can also be obtained from eqs. 3.30 and 3.31 of Airy's linear theory. Next, the velocity potential, obtained from the combination of eqs. 4.8 (to the second order), 4.38, 4.35, 4.46, and 4.3, is

$$\varphi = \frac{gH}{2kc}\frac{\cosh[k(z+h)]}{\cosh(kh)}\sin(kx-\omega t) + \frac{3}{32}ckH^2\frac{\cosh[2k(z+h)]}{\sinh^4(kh)}\sin[2(kx-\omega t)] \tag{4.52}$$

Finally, the free-surface displacement expression to the second order results from the combination of eqs. 4.9, 4.38, 4.36, 4.39, and 4.48. That expression is

$$\eta = \frac{H}{2}\cos(kx-\omega t) + \frac{kH^2}{16}\frac{\cosh(kh)}{\sinh^3(kh)}[2+\cosh(2kh)]\cos[2(kx-\omega t)] \tag{4.53}$$

This expression shows that the second-order free-surface displacement is simply the superposition of two linear waves having wave numbers that differ by a factor of 2, as illustrated in Figure 4.6. The free-surface profiles obtained from both Stokes' second-order wave theory and the linear wave theory are illustrated in the following example.

EXAMPLE 4.1: DEEP- AND SHALLOW-WATER FREE-SURFACE PROFILES The free-surface displacement expression resulting from Stokes' second-order theory is given in eq. 4.53. This expression can be simplified by using the respective approximations leading to eqs. 3.56 and 3.57 to obtain the free-surface profiles in deep and shallow waters. The resulting approximations are the following: For deep water, the free-surface displacement is

$$\eta_0 \simeq \frac{H_0}{2}\cos(k_0x-\omega t) + \frac{k_0H_0^2}{8}\cos[2(k_0x-\omega t)] \tag{4.54}$$

where, as in Chapter 3, the subscript "0" indicates deep water, and for shallow water,

$$\eta \simeq \frac{H}{2}\cos(kx-\omega t) + \frac{3}{16}\frac{H^2}{k^2h^3}\cos[2(kx-\omega t)] \tag{4.55}$$

Before applying these approximations, let us determine the wave height of the nonlinear wave. If we define that height as being the vertical distance

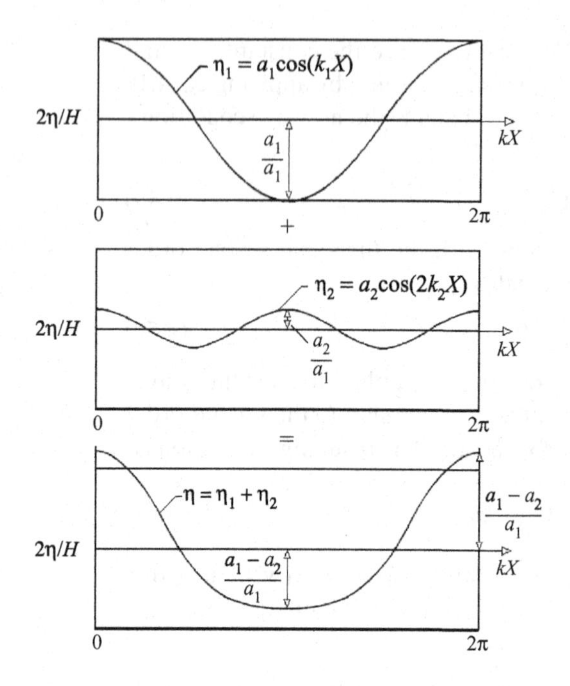

Figure 4.6. *Superposition of Linear Waves to form a Stokes Second-Order Wave.* From Figure 4.5, the relationship between the horizontal coordinates is $X = x - ct$.

between the crest of the wave and the displacement midway between crests, then we find that the height of the second-order wave is always equal to the height of the basic linear wave, H. This can be proven by representing the amplitudes of the two sine terms in eq. 4.53 by a_1 and a_2, where a_2 is less than a_1.

Returning to the approximate free-surface displacements, as represented by eqs. 4.54 and 4.55, let us examine the deep-water free-surface profiles under three conditions. First, referring to the validity ranges of the theories sketched in Figure 4.1, the free-surface profiles obtained using eq. 4.54 are shown in Figure 4.7 in non-dimensional form for two wave steepness conditions. Those conditions correspond to (a) the upper validity limit of the linear theory and (b) the upper limit of the second-order theory. In each case, the wave height is assumed to be equal to 1 m. For curves a and b the respective periods are then found to be 10.1 sec and 3.57 sec. As the period decreases with the wave height held constant, the wave steepness, H/λ, in deep water increases. Hence, as the steepness increases, the crest height above the SWL increases while the trough becomes broader and more shallow.

Next, we apply eq. 4.55 to the shallow-water conditions for two points on the shallow-water line in Figure 4.1. Those points correspond to ratios of (a) $H/\lambda_0 = 0.000377$ and (b) $H/\lambda_0 = 0.002010$, where $h\lambda_0 = 0.0100$, and the

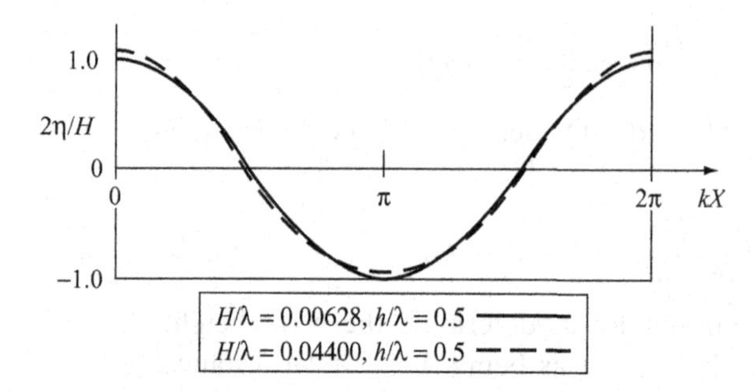

Figure 4.7. *Wave Profiles from Stokes' Second-Order Theory Applied to Deep Water.* The continuous line corresponds to the upper limit of the linear theory in Figure 4.1, whereas the dashed line corresponds to the upper limit of Stokes' second-order theory.

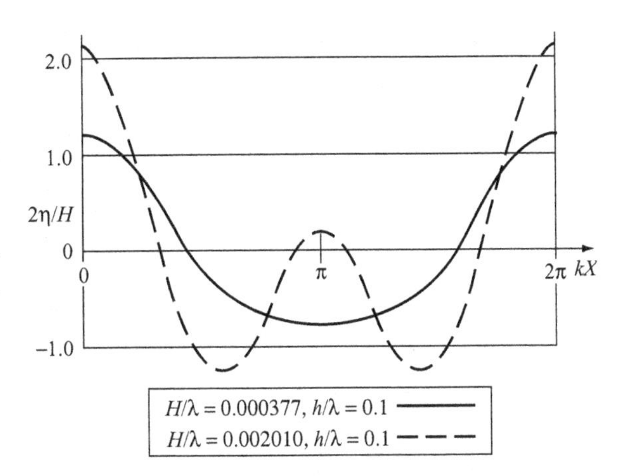

Figure 4.8. *Free-Surface Profiles Obtained from Stokes' Second-Order Theory.* The reader should check the validity of the Stokes' second-order theory for these two wave steepness values by consulting Figure 4.1.

resulting profiles are the curves in Figure 4.8. For both points studied in Figure 4.1, the wave height is 1.0 m. For the wave height ratio value of 0.000377, the period is 41.2 sec and the water depth is 26.5 m. The profile for this wave is nearly sinusoidal, as in Figure 4.7. For the larger wave height ratio value (0.002010), the profile is nonlinear. The curve corresponding to condition (b) clearly shows the wave corresponding to the second-order term of eq. 4.55. The reader is encouraged to apply the conditions for (b) to Figure 4.1 to check the validity of Stokes' second-order theory. In addition, the reader is referred to the book by Wiegel (1964) for comparisons of experimental and theoretical data for these shallow-water irregular waves.

Before discussing the particle motions, it is worth noting here that we now have a measure of the nonlinearity of water waves using the results of Stokes' second-order theory. That is the free-surface displacement expression of eq. 4.53. That displacement expression in non-dimensional form can be plotted as a function of the wave steepness. Mathematically, the relationship between the two non-dimensional parameters can be written as

$$\frac{2\eta_{\max}}{H} = 1 + f_\eta\left(\frac{H}{\lambda}\right) \tag{4.56}$$

For a linear wave, $f_\eta(H/\lambda)$ has a value of zero. As the wave profile becomes more nonlinear, the value of the function increases. This measure of the nonlinearity is illustrated in the following example, where results from eq. 4.56 are compared with experimental data.

EXAMPLE 4.2: FREE-SURFACE DISPLACEMENT IN DEEP WATER Let us apply eq. 4.56 to deep-water conditions, using the results of eq. 4.54 to define the function of that equation. The resulting deep-water expression is

$$\frac{2\eta_{\max}}{H_0} = 1 + \frac{\pi H_0}{2\lambda_0} \tag{4.57}$$

This equation is now applied to an experiment that was conducted in the 36.6-m wave and towing tank at the U.S. Naval Academy. That tank is 2.44 m wide and 1.52 m deep. For this experimental study, ten waves were created in two five-wave groups. The first five waves had a frequency of 0.7 Hz and heights that varied from 2.8 to 22.5 cm, whereas the second five had a frequency of 0.8 Hz

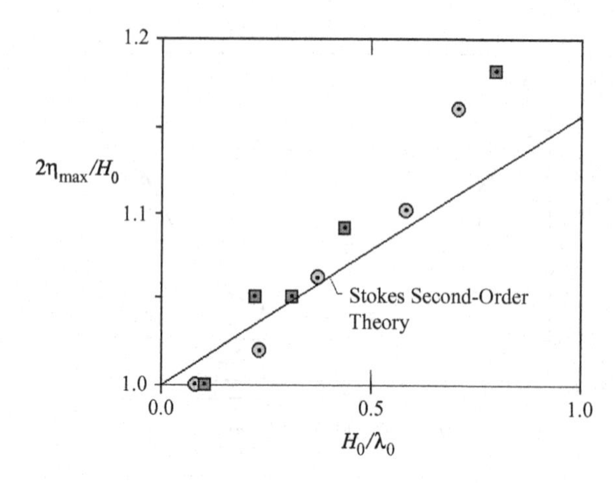

Figure 4.9. *Experimental and Theoretical Non-Dimensional Crest Height as Functions of the Deep-Water Wave Steepness.* The data represented by □ correspond to a wave frequency of 0.8 Hz whereas those represented by ∘ correspond to a wave frequency of 0.7 Hz. For both frequencies, the water depth is 1.52 m. The line results from eq. 4.57, Stokes second-order theory.

and heights that varied from 2.5 cm to 19.3 cm. The conditions for the 0.7-Hz wave were not exactly deep water, because, according to eq. 3.36, the deep-water wavelength is 3.19 m. Half of that value is 1.59 m, which is slightly larger that the 1.52-m tank depth. Using the non-dimensional form of eq. 4.57, the experimental data are presented in Figure 4.9, with the theoretical line resulting from eq. 4.57. As can be seen in that figure, the theoretical and experimental results agree quite well for wave steepness values below 1/17. Furthermore, the maximum difference between the experimental data and the value of $2\eta_{max}/H = 1.0$ (from the linear theory) is 10% for steepness values of 1/17 or less. Because of this good agreement, we can use the linear theory with some confidence for wave steepness values up to 1/17 in deep water.

4.3 Second-Order Particle Motions

The plots in Figure 4.9 of the non-dimensional free-surface displacement predicted by Stokes' second-order theory show that for a given water depth, the wave crests become high and narrow and the troughs become shallow and broad as the wave steepness, H/λ, increases. With this observation, attention is now directed to the motions of the water particles on and beneath the free surface. In general, the two-dimensional particle motions can be analyzed by first determining the particle velocity components, and then integrating these components over time, as is done for the linear motions in Section 3.5. Using the velocity potential expression in eq. 4.52, the second-order horizontal and vertical velocity components can be obtained from the results of eqs. 2.51 and 2.52, respectively. The results are the following: In the horizontal direction, the velocity component is

$$u = \frac{gH}{2c}\frac{\cosh[k(z+h)]}{\cosh(kh)}\cos(kx-\omega t) + \frac{3}{16}ck^2H^2\frac{\cosh[2k(z+h)]}{\sinh^4(kh)}\cos[2(kx-\omega t)] \tag{4.58}$$

and the vertical component is

$$w = \frac{gH}{2c}\frac{\sinh[k(z+h)]}{\cosh(kh)}\sin(kx-\omega t) + \frac{3}{16}ck^2H^2\frac{\sinh[2k(z+h)]}{\sinh^4(kh)}\sin[2(kx-\omega t)] \tag{4.59}$$

At first glance, it is difficult to determine the nature of the particle motions from eqs. 4.58 and 4.59. The first terms in these respective equations are the same as those in eqs. 3.49 and 3.50, which result from Airy's linear theory. Hence, as noted previously in this chapter, the second-order effects are simply superimposed on the linear motions of the water particles. As is done in Section 3.5, Stokes (1847) assumed that the paths of the particles could be obtained with respect to some fixed point (x_0, z_0) in the water. For the linear motions described in Chapter 3, this point is found to be a mean point about which the particles travel.

Following the method of Stokes (1847), we replace the spacial coordinates in eqs. 4.58 and 4.59 by

$$x = x_o + \xi \tag{4.60}$$

and

$$z = z_o + \zeta \tag{4.61}$$

where both ξ and ζ are functions of time, and (x_o, z_o) is a fixed reference point. Using a Maclaurin series, expand both the hyperbolic functions in ζ and the trigonometric functions in ξ in the resulting expressions for u and w. To the first order of ξ and ζ, the expression for the horizontal velocity component is

$$\begin{aligned} u &= \frac{d\xi}{dt} \\ &= \frac{gH}{2c\cosh(kh)}\left\{\cosh[k(z_o+h)]\cos(kx_o-\omega t) - \frac{k\xi}{1!}\cosh[k(z_o+h)]\sin(kx_o-\omega t)\right. \\ &\quad \left. + \frac{k\xi}{1!}\sinh[k(z_o+h)]\cos(kx_o-\omega t)\cdots\right\} \\ &\quad + \frac{3}{16}\frac{ck^2H^2}{\sinh^4(kh)}\left\{\cosh[2k(z_o+h)]\cos[2(kx_o-\omega t)]\right. \\ &\quad - \frac{2k\xi}{1!}\cosh[2k(z_o+z)]\sin[2(kx_o-\omega t)] \\ &\quad \left. + \frac{2k\xi}{1!}\sinh[2k(z_o+h)]\cos[2(kx_o-\omega t)]\cdots\right\} \end{aligned} \tag{4.62}$$

Again, following Stokes (1847), simply replace the variables ξ and ζ in eq. 4.62 by the linear expressions of eqs. 3.53 and 3.54, respectively. Actually, in his 1847 paper Stokes only used the linear series in eq. 4.62. We shall apply the method to both series. From an order-of-magnitude consideration, the three terms of the linear expansion and the first term of the second-order expansion shown in eq. 4.62 are retained. To the second order, the resulting approximate expression for the horizontal velocity component is

$$\begin{aligned} u &= \frac{d\xi}{dt} \\ &\simeq \frac{gH}{2c}\frac{\cosh[k(z_o+h)]}{\cosh(kh)}\cos(kx_o-\omega t) \\ &\quad + \frac{ck^2H^2}{8\sinh^2(kh)}\left\{\frac{3}{2}\frac{\cosh[2k(z_o+h)]}{\sinh^2(kh)} - 1\right\}\cos[2(kx_o-\omega t)] \\ &\quad + \frac{ck^2H^2}{8}\frac{\cosh[2k(z_o+h)]}{\sinh^2(kh)} \end{aligned} \tag{4.63}$$

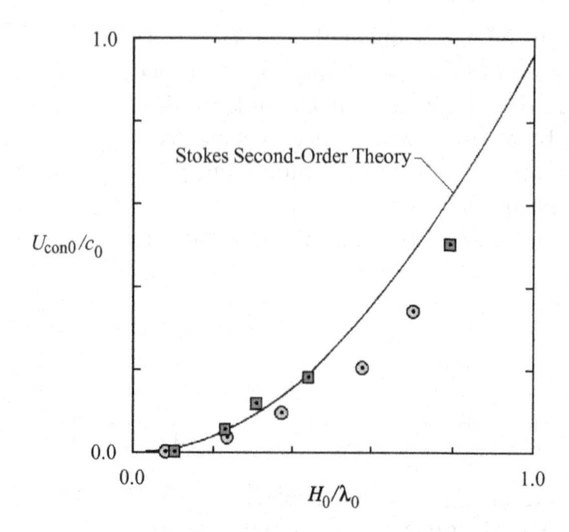

Figure 4.10. *Experimental and Theoretical Dimensionless Convection Velocity of the Surface Particles in Deep Water as Functions of the Deep-Water Wave Steepness.* The theoretical curve is obtained from eq. 4.66 at $z_0 = 0$. The frequencies and water depth are those for the data in Figure 4.9. The data represented by ○ correspond to a wave frequency of 0.7 Hz, and those represented by □ correspond to a wave frequency of 0.8 Hz.

Similarly, the approximate expression for the vertical velocity component of a water particle is

$$\begin{aligned} w &= \frac{d\xi}{dt} \\ &\simeq \frac{gH}{2c}\frac{\sinh[k(z_o+h)]}{\cosh(kh)}\sin(kx_o-\omega t) \\ &\quad + \frac{3}{16}ck^2H^2\frac{\sinh[2k(z_o+h)]}{\sinh^4(kh)}\sin[2(kx_o-\omega t)] \end{aligned} \tag{4.64}$$

to the same order of approximation as that of eq. 4.63.

4.4 Water Particle Convection

Comparing the approximate expressions of eqs. 4.62 and 4.63, we see that there is a constant term in the u-expression of eq. 4.62. No such constant is in eq. 4.63. Physically, the constant term in the horizontal velocity expression represents a constant velocity at which the water particles migrate in the direction of wave travel. This velocity is called the *convection velocity*, and is expressed by

$$U_{con} = \frac{ck^2H^2}{8}\frac{\cosh[2k(z_o+h)]}{\sinh^2(kh)} \tag{4.65}$$

Dimensionless results obtained from eq. 4.65 are presented in Figure 4.10 with experimental data. The approximate expressions for the deep- and shallow-water convection velocities are, respectively,

$$U_{con0} = \frac{c_0k_0^2H_0^2}{4}e^{2k_0z_o} = \frac{\omega^3}{4g}H_0^2e^{2k_0z_o} \tag{4.66}$$

and

$$U_{con} = \frac{cH^2}{8h^2} \tag{4.67}$$

For the case of intermediate water convection velocities, eq. 4.65 is used as shown. The deep-water, intermediate-water, and shallow-water particle convection is

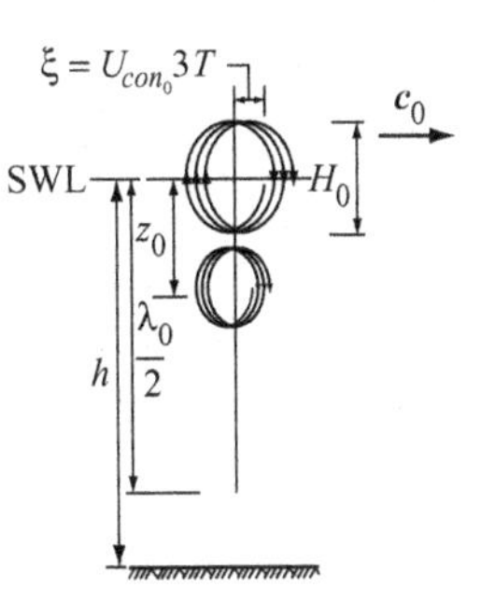

Figure 4.11. *Deep-Water, Second-Order Particle Paths in Deep Water (Not to Scale).* Compare these paths with those predicted by the linear theory in Figure 3.11.

illustrated in Figures 4.11 through 4.13. The significance of the convection velocity is demonstrated in the following example.

EXAMPLE 4.3: DEEP- AND SHALLOW-WATER PARTICLE CONVECTION VELOCITIES
Apply the approximations for the extremes of water depth (discussed in Section 3.5) to eq. 4.65. As in Chapter 3, the subscript "0" indicates deep-water properties in eq. 4.66, by convention. The results of eq. 4.66 show that the maximum migration for a given deep-water wave occurs on the free surface, as expected, and as illustrated in Figure 4.11. This corresponds to $z_o = 0$ because z_o is negative below the SWL. Furthermore, the convection velocity is seen to be a strong function of the circular wave frequency, ω, being proportional to the cube of the frequency. In shallow water, the convection velocity is seen to be independent of position, from the results of eq. 4.67. That is, the particles on the surface migrate as fast as those adjacent to the seafloor, as illustrated in Figure 4.13. The shallow-water celerity or phase velocity, c, is a function of depth only, from the results of eq. 3.38. The convection velocity in shallow water is then independent of the wave frequency, according to eq. 4.67. Also from the results of that equation, the convection velocity is seen to increase as the water depth decreases. This depth dependence is important to engineers working in the coastal zone, where much of coastal engineering deals with sediment transport by waves.

Apply the deep-water convection velocity expression of eq. 4.66 to waves having a 1-m height and a 4-sec period. For these waves, $H/\lambda_0 = 0.040$, which is near the upper bound of the validity range of the second-order theory in Figure 4.1. For this wave condition, the convection velocity of the particles on the free surface is 0.0988 m/s, according to eq. 4.66. Note: The deep-water celerity value is 6.24 m/s from eq. 3.36. The results in eq. 4.66 also show that the convection of water particles decreases rapidly with the mean depth position of the particle. Equation 4.66 is also applied to the conditions of Example 4.2 and plotted in Figure 4.12 in non-dimensional form.

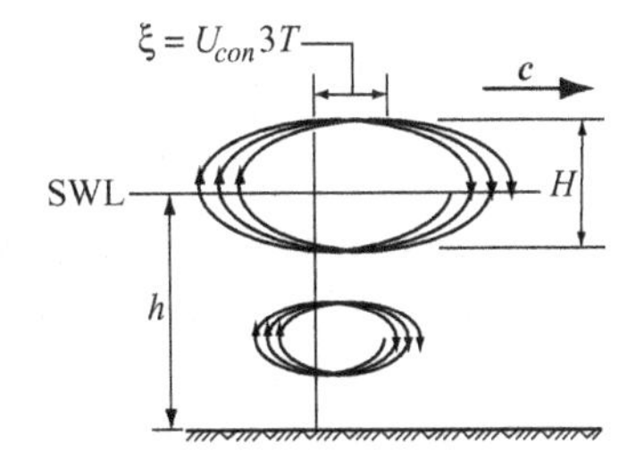

Figure 4.12. *Intermediate-Water, Second-Order Particle Paths in Deep Water (Not to Scale).* Compare these paths with those predicted by the linear theory in Figure 3.12.

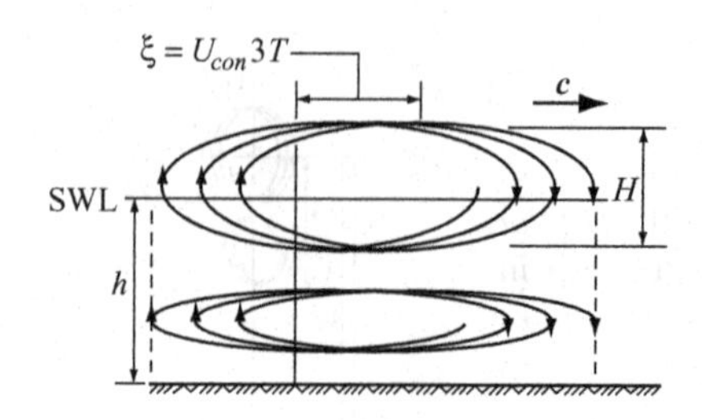

Figure 4.13. *Shallow-Water, Second-Order Particle Paths in Deep Water (Not to Scale)*. Compare these paths with those predicted by the linear theory in Figure 3.13.

Return now to the discussion of the time-dependent water particle motions. The displacement components of a particle about the position (x_0, z_0) is found by integrating eqs. 4.63 and 4.64 over time to obtain

$$\xi = -\frac{H}{2}\frac{\cosh[k(z_o+h)]}{\sinh(kh)}\sin(kx_o-\omega t)$$
$$-\frac{kH^2}{16}\frac{1}{\sinh^2(kh)}\left\{\frac{3}{2}\frac{\cosh[2k(z_o+h)]}{\sinh^2(kh)}-1\right\}\sin[2kx_o-\omega t)]+U_{con}t \qquad (4.68)$$

in the horizontal direction, and

$$\zeta = \frac{H}{2}\frac{\sinh[k(z_o+h)]}{\sinh(kh)}\cos(kx_o-\omega t)$$
$$+\frac{3}{32}kH^2\frac{\sinh[2k(z_o+h)]}{\sinh^4(kh)}\cos[2(kx_o-\omega t)] \qquad (4.69)$$

in the vertical direction. In eq. 4.68, we see that the first two terms are sinusoidal, whereas the last term is the *convection displacement*, which is linear in time. The terms in eq. 4.69 are both sinusoidal. Hence, (x_0, z_0) is not a mean position of a particle as it is in a linear wave in Section 3.5.

The deep- and shallow-water approximations of eqs. 4.68 and 4.69 are obtained by using the extreme-value approximations of the hyperbolic functions given in Section 3.5. The results are as follows:

CASE 1. DEEP WATER ($1/2 \le h/\lambda < \infty$, or $\pi \le kh < \infty$) In deep water, the horizontal particle displacement expression is approximated by

$$\xi_0 = -\frac{H_0}{2}e^{k_0 z_o}\sin(k_0 x_o-\omega t)+\frac{\omega^3}{4g}H_0^2 e^{2k_0 z_o}t \qquad (4.70)$$

where in the vertical direction, the approximate displacement expression is

$$\zeta_0 = \frac{H_0}{2}e^{k_0 z_o}\cos(k_0 x_o-\omega t) \qquad (4.71)$$

Particle paths predicted by eqs. 4.70 and 4.71 are presented in Figure 4.11.

CASE 2. INTERMEDIATE WATER ($1/20 \le h/\lambda < 1/2$, or $\pi/10 \le kh < \pi$) In this range, the respective expressions for ξ and ζ in eqs. 4.68 and 4.69 apply as written. Examples of the particle paths obtained by using these expressions are shown in Figure 4.12.

CASE 3. SHALLOW WATER ($0 < h/\lambda < 1/20$, or $0 < kh < \pi/10$) The shallow-water approximations of the expressions in eqs. 4.68 and 4.69 are, respectively,

$$\xi = -\frac{H}{2}\sin(kx_o - \omega t) + \frac{3}{32k}\frac{H^2}{h^2}\sin[2(kx_o - \omega t)] - \frac{cH^2}{8h^2}t \tag{4.72}$$

and

$$\zeta = \frac{H}{2}\frac{(z_o + h)}{h}\cos(kx_o - \omega t) + \frac{3}{16}\frac{1}{(kh)^2}\frac{H^2}{h^2}(z_o + h)\cos[2(kx_o - \omega t)] \tag{4.73}$$

Results obtained from eqs. 4.72 and 4.73 are sketched in Figure 4.13.

EXAMPLE 4.4: WAVE-INDUCED SPREADING OF A SURFACE SPILL As noted in Example 4.3, because the celerity decreases with decreasing water depth, the horizontal excursions of the water particles in shallow water increase as the wave approaches the shoreline. This poses a problem for ocean environmental engineers when contaminants, such as oil, are spilled at sea. In the absence of both winds and currents, the spills spread slowly in deep water when compared to their spread in shallow water, according to the results of eqs. 4.70 and 4.72, respectively.

To illustrate, consider the migration of a thin oil layer on the surface of the water. Let us determine the distances traveled by the oil, first on a deep-water swell (where the respective wave height and period are 0.5 m and 7 sec), and then on the swell in shallow water (where $h/\lambda = 1/25$). We shall determine the distance of travel over 70 sec in each case. Assuming that $x_o = t = 0$ initially, eq. 4.70 yields a convection distance of 0.323 m in deep water. In shallow water, the celerity is approximately

$$c = \frac{\lambda}{T} \simeq \sqrt{gh} = 25\frac{h}{T} \tag{4.74}$$

from eq. 3.38. Equation 4.74 can be solved to determine the water depth where the 7-sec wave becomes shallow. The result is

$$h = \frac{gT^2}{25^2} = 0.769 \text{ m} \tag{4.75}$$

The value of the celerity in shallow water is then 2.75 m/s. The wave height of the 7-sec wave at this water depth is obtained from the shallow-water approximation of the shoaling coefficient expression of eq. 3.78,

$$K_S = \frac{H}{H_0} \simeq \sqrt{\frac{1}{2kh}} = \sqrt{\frac{\lambda}{4\pi h}} = 1.41 \tag{4.76}$$

The wave height in 0.769 m of water is then 1.08 m. With these values for the celerity, water depth, and wave height, the last term of eq. 4.72 yields a distance of 47.6 m. From these convection distances, we see that the oil spill in deep water spreads slowly and, hence, is easy to contain. However, in shallow water the convection distance over the same time is more than two orders of magnitude greater than that in deep water.

This concludes the basic discussion of Stokes' second-order wave theory. The theory is used to determine breaking conditions later in this chapter. In the next section, both the cnoidal and solitary wave theories are presented.

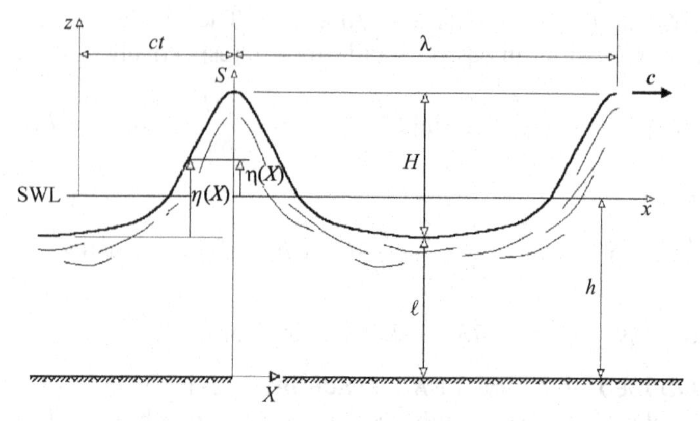

Figure 4.14. *Notation for the Cnoidal Wave Theory*. The wave in the moving coordinate system (X, S) is of permanent form.

4.5 Long Waves in Shallow Water

Lord Rayleigh (1876) extended the nonlinear method of Stokes by specifying the forms of the series expansions of the water particle velocity components (u, w). These expansions were then used by Korteweg and de Vries (1895) to develop a theory for long periodic waves in shallow water, that is, where the wavelength, λ, is much greater than the water depth, h. It is also assumed by Korteweg and de Vries that the free-surface displacement, η, is much less than the water depth. The waves resulting from the theory were called *cnoidal waves* by the investigators because of the presence of the Jacobian elliptic cosine function (cn) in the expression for the free-surface displacement. See Abramowitz and Stegun (1965) for a discussion of this function.

The *cnoidal wave theory* of Korteweg and de Vries (1895) has been the subject of many investigations since its conception. See the papers of Wiegel (1960) and Fenton (1979) and the book by Lighthill (1979), among others. The cnoidal theory bridges the theoretical gap between very long, shallow-water waves and moderately long waves in finite depths in that the respective extremes of the theory are identical with the solitary wave theory of Boussinesq (1872) and McCowan (1891) and the theory of Stokes (1847).

A. Cnoidal Wave Theory

Following the analysis of Lord Rayleigh (1876), we transform our (x, z) coordinate system with its origin on the SWL to the (X, S) on the seafloor. The two systems are related by $X = x - ct$ and $S = z + h$, as sketched in Figure 4.14. The origin of the coordinate system is then fixed below the crest of the wave, and the wave is of permanent form. This is a slight modification of the technique used in the derivation of Stokes' theory in Section 4.2. In terms of the transformed coordinate system, we represent the velocity components by rapidly convergent series of the forms

$$\mathrm{U} = u - c = S^0 f_0 + S^1 f_1 + S^2 f_2 + \cdots = \sum_{j=0}^{\infty} S^j f_j \tag{4.77}$$

and

$$\mathrm{W} = w = S^1 F_1 + S^2 F_2 + \cdots = \sum_{j=1}^{\infty} S^j F_j \tag{4.78}$$

where f_j and F_j are functions of X. Assuming a flat, horizontal bed, the seafloor condition where $\mathrm{W}(X, 0) = 0$ is satisfied. As is the case for any incompressible flow, the equation of continuity (eq. 2.21) must be satisfied. The combination eq. 2.21 (noting that $\partial(\ )/\partial X = \partial(\ )/\partial x$, etc. in this application) and the expressions in eqs. 4.77 and 4.78 yield the following relationship:

$$F_j = -\frac{1}{j}\frac{\partial f_{j-1}}{\partial X} \tag{4.79}$$

By assuming the irrotational flow condition of $\partial U/\partial S - \partial W/\partial X = 0$, we obtain

$$f_1 = 0 \tag{4.80}$$

and, following Korteweg and de Vries (1895),

$$f_j = \frac{1}{j}\frac{\partial F_{j-1}}{\partial X} = -\frac{1}{j(j-1)}\frac{\partial^2 f_{j-2}}{\partial X^2} \tag{4.81}$$

by incorporating eq. 4.79. The final forms of the velocity component expressions of eqs. 4.77 and 4.78 are, respectively,

$$\mathrm{U} = f_0 - \frac{S^2}{2!}\frac{\partial^2 f_0}{\partial X^2} + \frac{S^4}{4!}\frac{\partial^4 f_0}{\partial X^4} - \cdots \tag{4.82}$$

and

$$\mathrm{W} = -\frac{S}{1!}\frac{\partial f_0}{\partial X} + \frac{S^3}{3!}\frac{\partial^3 f_0}{\partial X^3} - \frac{S^5}{5!}\frac{\partial^5 f_0}{\partial X^5} + \cdots \tag{4.83}$$

The corresponding forms of the respective velocity potential and stream function are

$$\varphi = \int f_0 dX - \frac{S^2}{2!}\frac{\partial f_0}{\partial X} + \frac{S^4}{4!}\frac{\partial^3 f_0}{\partial X^3} - \cdots \tag{4.84}$$

and

$$\psi = S f_0 - \frac{S^3}{3!}\frac{\partial^2 f_0}{\partial X^2} + \frac{S^5}{5!}\frac{\partial^4 f_0}{\partial X^4} - \cdots \tag{4.85}$$

from the Cauchy-Riemann relationships of eqs. 2.51 and 2.52.

The expressions in eqs. 4.82 through 4.85 must satisfy the dynamic free-surface condition. The form of that condition used here is similar to that given in eq. 4.5, that is,

$$g S_\eta + \frac{1}{2}[\mathrm{U}^2 + \mathrm{W}^2]|_{S_\eta} = \mathrm{K} \tag{4.86}$$

In eq. 4.86, the position of the free surface of the water with respect to the seafloor is denoted as S_η. The velocity component expressions must also satisfy the kinematic free-surface condition. Assuming a small free-surface curvature due to the long-wave assumption, that condition can be expressed by

$$\mathrm{W}|_{S_\eta} \simeq \mathrm{U}|_{S_\eta}\frac{\partial S_\eta}{\partial X} \tag{4.87}$$

This expression is similar to the expression in eq. 4.7. In eqs. 4.82 through 4.85, the function f_0 has units of velocity. Korteweg and de Vries (1895) represent the function by an expression similar to

$$f_0 = -(q_0 + \Gamma) \tag{4.88}$$

where Γ (a free-surface function of X) is much less than q_0. With this representation of f_0, combine the expressions in eqs. 4.82 and 4.83 with the X-derivative of eq. 4.86. In the resulting expression, let

$$S_\eta = \ell + \eta \tag{4.89}$$

where ℓ is the height of the trough of the wave above the floor, and ε (small when compared to ℓ) is the elevation of the free surface above the trough. These are related to the still-water depth, h, and the free-surface displacement, η, in Figure 4.14. Note that the free-surface displacement from the SWL is η, whereas that measured from the trough is η.

Differentiate eq. 4.86 with respect to X and combine the results with eqs. 4.82, 4.83, 4.88, and 4.89. To the first order, the resulting expression is

$$q_0 \frac{\partial \Gamma}{\partial X} + g \frac{\partial \eta}{\partial X} = 0 \tag{4.90}$$

The combination of eqs. 4.82, 4.83, 4.88, and 4.89 with eq. 4.87 yields

$$q_0 \frac{\partial \eta}{\partial X} + \ell \frac{\partial \Gamma}{\partial X} = 0 \tag{4.91}$$

to the first order, where $\ell \gg \eta$. The origin of η is on a horizontal line through the trough, whereas the origin of η is on the SWL, as sketched in Figure 4.14. Equations 4.90 and 4.91 are satisfied if the first-order celerity is

$$q_0 = \sqrt{g\ell} \tag{4.92}$$

and the velocity function is

$$\Gamma = -\frac{q_0}{\ell}(\eta + \mathrm{A}) \tag{4.93}$$

where A is a constant of integration. Equations 4.92 and 4.93 are the first approximation of the Korteweg-de Vries long-wave theory.

The second approximation of the long-wave theory is based on the first by assuming

$$f_0 = -(q_0 + \Gamma + \delta\Gamma) = -\left[q_0 - \frac{q_0}{\ell}(\eta + \mathrm{A}) - \frac{q_0}{\ell}\mathrm{B}\right] \tag{4.94}$$

where $\delta\Gamma = -q_0 B/\ell$, and the function B is assumed to be small with respect to both η and A, that is, B is a perturbation function. Following the same procedure that leads to eqs. 4.90 and 4.91 while eliminating B, the following expression results from the inclusion of the second-order terms:

$$\frac{d}{dX}\left(\frac{1}{2}\eta^2 + \frac{2}{3}\mathrm{A}\eta + \frac{\ell^3}{9}\frac{d^2\eta}{dX^2}\right) = 0 \tag{4.95}$$

where small-valued terms have been neglected. The same procedure can be followed to determine the expression for B (that expression is presented later). Two integrations of eq. 4.95 yield

$$\frac{\ell^3}{3}\left(\frac{d\eta}{dX}\right)^2 + \eta^3 + 2A\eta^2 + 6C_1\eta + C_2 = 0 \tag{4.96}$$

where C_1 and C_2 are constants of integration, as is α is in eq. 4.93. To determine the C_2 in eq. 4.96, we simply apply the equation at a trough where $\eta = d\eta/dX = 0$. From this, we find $C_2 = 0$. With this value, the slope of the free surface is obtained directly from eq. 4.96, as

$$\frac{d\eta}{dX} = \pm\sqrt{-\frac{3}{\ell^3}\eta(\eta^2 + 2A\eta + 6C_1)} \tag{4.97}$$

which is known as the *Korteweg-de Vries equation.*

There are two values resulting from the quadratic portion of eq. 4.97 for which the slope of the free surface is zero. One of these values must correspond to the crest, where $\eta = H$, while the other is at $\eta = -C$, where C must be determined. The negative sign ($-$) in the latter is required to make the term $6C_1$ negative. Because ε is never negative, the expression in eq. 4.97 would be imaginary unless $6C_1 \leq \eta^2 + 2\alpha\eta$. For this reason, we represent the quadratic portion of eq. 4.97 by

$$\eta^2 + 2A\eta + 6C_1 = (\eta - H)(\eta + C) \tag{4.98}$$

where $2A = C - H$ and $6C_1 = -HC$. The differential equation of Korteweg and de Vries (1895), eq. 4.97, is now

$$\frac{d\eta}{dX} = \pm\sqrt{\frac{3}{\ell^3}\eta(H - \eta)(C + \eta)} \tag{4.99}$$

With some manipulation, the integration of eq. 4.99 results in an expression for the free-surface displacement involving the Jacobian elliptic (cosine) function, cn(). See Abramowitz and Stegun (1965) for a discussion of this function. The limits of integration are based on the assumption that the origin of the variable X corresponds to a wave crest, that is, $\eta = H$ at $X = 0$. The resulting free-surface expression is then

$$\eta = h - \ell + \eta = H\left[\mathrm{cn}^2\left(\sqrt{\frac{3(H+C)}{4\ell^3}}X\right)\right]; \quad M = \frac{H}{H+C} \tag{4.100}$$

where the wave height is $H = \eta_{\max}$, and M is the parameter of the function. For those interested, the integration of eq. 4.99 is performed by first replacing the variable η by v^2. The resulting integral is found on page 596 of the book edited by Abramowitz and Stegun (1965). A plot of η/H is presented in Figure 4.15 for three values of M, after Milne-Thomson (1950).

The wavelength, λ, can be determined from the results in eq. 4.100 by applying that equation to the trough located half a wavelength away from the origin of X. At this point, the value of cn^2 must be zero, and the corresponding value of the argument of the function is K, that is, K is the value of the argument of cn() at the first zero-crossing. Mathematically, $\mathrm{cn}^2(K) = 0$. In addition, K is also the complete

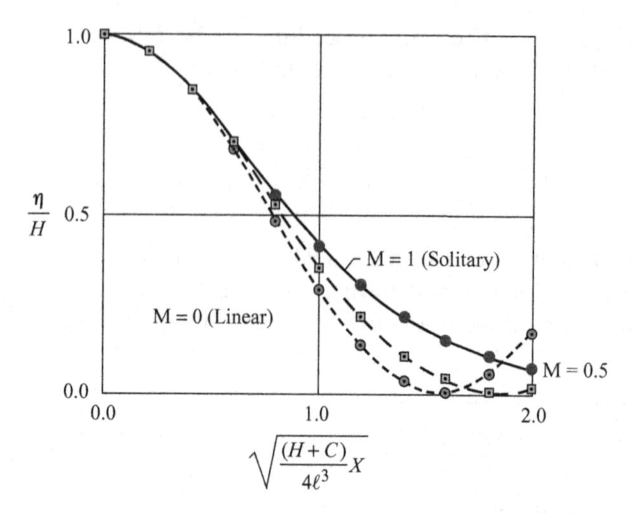

Figure 4.15. *Free-Surface Profiles Resulting from the Cnoidal Wave Theory.* These dimensionless profiles are obtained from eq. 4.100. As shown, the limiting values of the parameter M correspond to the linear wave theory (M = 0) and the solitary theory (M = 1).

elliptic function of the first kind. See Abramowitz and Stegun (1965) for a discussion of this function. The expression for the wavelength of the cnoidal wave is then

$$\lambda = 4\kappa\sqrt{\frac{\ell^3}{3(H+C)}} = 4\kappa\sqrt{\frac{M\ell^3}{3H}} \tag{4.101}$$

Again, consider the respective velocity component expressions of eqs. 4.77 and 4.78. The final expressions for these velocity components must first be determined. To do this, we follow the derivation of eq. 4.95 except, in this case, we retain the perturbation function B of eq. 4.94. The resulting expression for this function is

$$B = -\frac{\eta^2}{4\ell} + \frac{\ell^2}{3}\frac{d^2\eta}{dX^2} \tag{4.102}$$

where, from the resets of eq. 4.95,

$$\frac{d^2\eta}{dX^2} = -\frac{3}{2\ell^3}[3\eta^2 - 2(H-C)\eta - CH] \tag{4.103}$$

using the notation of eq. 4.98. With these results, the first-order velocity component expressions of eqs. 4.77 and 4.78 are, respectively,

$$U = u - c \simeq -\sqrt{g\ell}\left[1 - \frac{1}{2\ell}(C-H) - \frac{CH}{2\ell^2} - (\ell + H - C)\frac{\eta}{\ell^2}\right] \tag{4.104}$$

and, noting that $d\eta/dX$ is given in eq. 4.99,

$$W = w = -S\frac{d\eta}{dX}\sqrt{\frac{g}{\ell}} \tag{4.105}$$

The expression for the time-dependent horizontal velocity component of the water particle obtained from eq. 4.104 is

$$u = c - \sqrt{g\ell}\left[1 - \frac{(C-H)}{2\ell} - \frac{CH}{2\ell^2} - (\ell + H - C)\frac{\eta}{\ell^2}\right] \tag{4.106}$$

This velocity is assumed to be zero at the nodes. That is, at the end of a trough and the beginning of the next crest where the free surface crosses the SWL, the horizontal particle velocity is assumed to be zero. By applying this condition to the expression in eq. 4.110, the expression for the wave celerity is obtained. That

Figure 4.16. *Notation for Water Particle Displacement from the Reference Point* (x_0, S_0) *in a Cnoidal Wave.* The time-dependent displacement components are with respect to the inertial frame of reference (x, z) because the wave is of permanent form in the relative frame of reference (X, S).

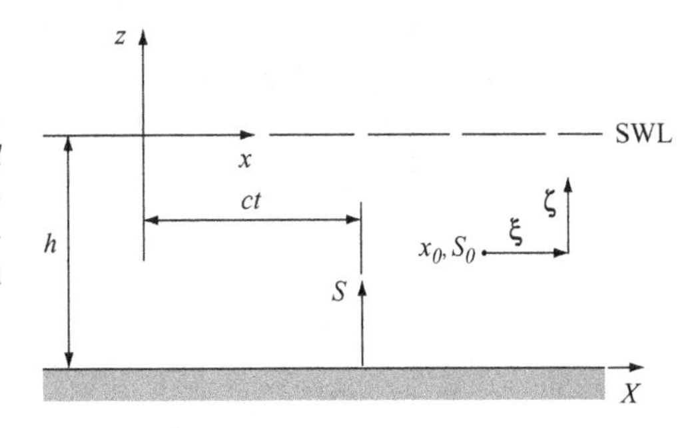

expression is

$$c = \sqrt{g\ell}\left[1 - \frac{(\mathrm{C} - H)}{2\ell} - \frac{CH}{2\ell^2} - (\ell + H - \mathrm{C})\frac{\eta_0}{\ell^2}\right] \tag{4.107}$$

where η_0 is the vertical distance between the trough and the SWL. The combination of eqs. 4.106 and 4.107 yields the following expression for the horizontal particle velocity:

$$u = \sqrt{\frac{g}{\ell}\frac{(\ell + H - C)}{\ell}}(\eta - \eta_0) \tag{4.108}$$

The expression for η_0 is found from that for the volume of water displaced by the free surface. That volume is

$$\mathrm{V}_\eta = \int_0^\lambda \eta dX = \left[(H + \mathrm{C})\frac{E(2K\,|\,\mathrm{M})}{2K} - \mathrm{C}\right]\lambda = \eta_0\lambda \tag{4.109}$$

where $E()$ is a complete elliptic integral of the second kind. From Abramowitz and Stegun (1965), we obtain the identity $E(2K) = 2E(K)$. The vertical offset of the trough from the SWL obtained from eq. 4.109 is then

$$\eta_0 = (H + \mathrm{C})\frac{E(K)}{K} - \mathrm{C} = h - \ell \tag{4.110}$$

Referring to Figure 4.16 for notation, the expressions for the components of the particle displacement vector from a reference point (x_0, S_0) are found by integrating the u- and w-expressions over time from $t = 0$ to some time t. The horizontal displacement is found by first replacing the relative coordinate X by $x_0 - ct$ in eq. 4.100. Without a loss in generality, the reference value of x_0 is set to zero. When $t = 0$, a wave crest is above $(0, S_0)$. Combine the resulting expression with that of eq. 4.110 in eq. 4.108, and integrate the result to obtain

$$\xi = \int_0^t udt = \sqrt{\frac{4gH}{3\mathrm{M}}\frac{(\ell + H - \mathrm{C})}{c}}\left\{E\left(\frac{2K}{T}t\right) - \left[1 - \mathrm{M}\left(1 - \frac{\eta_0}{H}\right)\right]\frac{2K}{T}t\right\} \tag{4.111}$$

where T is the wave period. From this equation, we can see that there is a *convection* in the x-direction over time. In a similar manner, the vertical displacement from the point $(0, S_0)$ is obtained from integrating eq. 4.105 over time, again, noting that a crest is initially over the point. The resulting expression is

$$\zeta = \int_0^t wdt = \sqrt{\frac{g}{\ell}}\frac{H}{c}\left[\mathrm{cn}^2\left(2K\frac{t}{T}\right) - 1\right]S_0 \tag{4.112}$$

One final condition is imposed on the fluid motion in the absolute coordinate system. That condition is that the net change in momentum of the fluid within the control volume beneath a single wave must be zero. Mathematically, that condition leads to the following relationship:

$$\int_0^{\lambda}\left(\int_0^{\ell+\eta} u ds\right) dX = ch\lambda - \sqrt{g\ell^3}\lambda\left[1 - \frac{C-H}{2\ell}\left(1 + \frac{\vee_\eta}{\lambda\ell}\right)\right] = 0 \tag{4.113}$$

The term $\vee_\eta$ for the two-dimensional flow represents the volume per unit crest width. From this result, the following expression for the celerity is obtained by neglecting terms with the coefficient $1/\ell^2$:

$$c = \frac{\lambda}{T} \simeq \frac{\sqrt{g\ell^3}}{h}\left[1 - \frac{(C-H)}{2\ell}\right]\left(1 + \frac{\vee_\eta}{\lambda\ell}\right) \tag{4.114}$$

In his discussion of the cnoidal theory, Lamb (1945) states that in the four basic equations of the theory (the two in eq. 4.100 and those in eqs. 4.101 and 4.110), there are six unknowns. Those unknowns are C, H, K, M, λ, and ℓ, and the water depth, h, is known. By specifying any two of the unknowns, the others can be determined. Theoretical physical oceanographers and applied mathematicians usually choose a value of M that in turn determines a value of K and $E(K)$. The values of the physical wave properties are then determined. This method is not satisfactory for design engineering because the design conditions at a site must be specified. That is, at a site the values of h, H, and T are known. In the sections that follow, we shall use more practical forms of the equations of the theory and illustrate their use.

B. Application of the Cnoidal Theory

Wiegel (1960) presents a practical method of using the cnoidal wave theory along with parametric plots that are useful to engineers. In what follows, Wiegel's work is summarized and an example is given.

The expression in eq. 4.110 is rearranged to obtain an expression for C. The resulting expression is then equated to the C-expression resulting from the second expression in eq. 4.100. The result is

$$C = \frac{(h-\ell)K - HE(K)}{E(K) - K} = \frac{H}{M}(1 - M) \tag{4.115}$$

From the last equality, we can obtain an expression for the vertical position of the wave trough with respect to the seafloor, that is,

$$\ell = h - \left(\frac{E(K)}{K} + M - 1\right)\frac{H}{M} \tag{4.116}$$

In addition, Wiegel (1960) assumes that $\ell \simeq h$ except in eq. 4.116. With this approximation applied to eq. 4.101, the following expression for the wavelength is obtained:

$$\lambda \simeq 4K\sqrt{\frac{Mh^3}{3H}} \tag{4.117}$$

Equation 4.117 can be rearranged in the form of the *Ursell number*, defined in the caption of Figure 4.1:

$$U_R \equiv \frac{H\lambda^2}{2h^3} = \frac{8}{3}K^2\mathrm{M} \tag{4.118}$$

Wiegel (1960) then derives approximate expressions for the celerity and period, which are, respectively,

$$c \simeq \sqrt{gh}\left[1 + \frac{H}{h\mathrm{M}}\left(\frac{1}{2} - \frac{E(K)}{K}\right)\right] \tag{4.119}$$

from eq. 4.114, and

$$T = \frac{\lambda}{c} \simeq \frac{4K\sqrt{\dfrac{\mathrm{M}h^2}{3gH}}}{1 + \dfrac{H}{h\mathrm{M}}\left(\dfrac{1}{2} - \dfrac{E(K)}{K}\right)} \tag{4.120}$$

using the results of eqs. 4.117 and 4.119. Again, we note that the values of the depth, wave height, and wave period at a site are known. Hence, the expression in eq. 4.120 is used to determine the value of the parameter M. This parametric value is then used to determine the complete elliptic integrals of the first and second kinds, that is, K(M) and E(M), respectively. Because the expression in eq. 4.120 is transcendental (the parameter cannot be separated), the value of the parameter must be obtained numerically. As in the solution of the wavelength in Example 3.3, the method of successive approximations can be used to obtain the numerical values.

Finally, the free-surface displacement from the SWL can be written in terms of both the absolute coordinate system and time by replacing the relative coordinate, X, by $x - ct$ in eq. 4.100. The result is

$$\eta = \ell - h + \eta = \ell - h + \mathrm{cn}^2\left[\frac{2K}{\lambda}(x - ct)\right]H \tag{4.121}$$

Again, we find in the equations of the theory the complete elliptic integrals of the first kind, K(M), and the second kind, E(M). Abramowitz and Stegun (1965) present polynomial approximations of these functions that can be used to facilitate the solution of M. Those respective approximations are the following:

$$K(\mathrm{M}) \equiv K = 1.38729 + 0.11197(1 - \mathrm{M}) + 0.07252(1 - \mathrm{M})^2$$
$$+ [0.5 + 0.12134(1 - \mathrm{M}) + 0.02887(1 - \mathrm{M})^2]\ln\left[\frac{1}{1 - \mathrm{M}}\right] + \delta_K(\mathrm{M}) \tag{4.122}$$

where $|\delta_K(\mathrm{M})| \leq 3 \times 10^{-5}$, and

$$E(\mathrm{K}) \equiv E \simeq 1 + 0.46301(1 - \mathrm{M}) + 0.10778(1 - \mathrm{M})^2$$
$$+ [0.24527(1 - \mathrm{M}) + 0.04124(1 - \mathrm{M})^2]\ln\left[\frac{1}{1 - \mathrm{M}}\right] + \delta_E(\mathrm{M}) \tag{4.123}$$

where $|\delta_E(\mathrm{M})| \leq 4 \times 10^{-5}$. See Figure 4.17 for plots of the functions in eqs. 4.122 and 4.123.

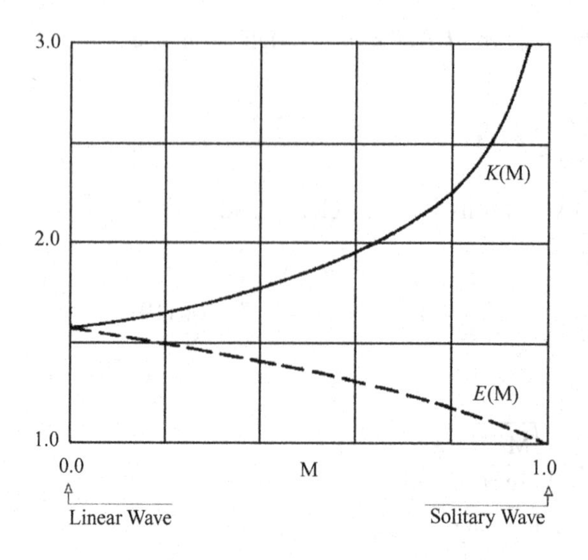

Figure 4.17. *Complete Elliptic Integrals of the First Kind (K) and the Second Kind (E) as Functions of the Parameter M.* The curves result from eqs. 4.122 and 4.123, respectively.

The method of solution using Wiegel's approach is as follows:

(a) Specify the conditions at the site, that is, h, H, and T.
(b) Determine the parameter M for the conditions in (a) by solving eqs. 4.120, 4.122, and 4.123 simultaneously using a numerical method of solution. See Example 3.3 for the method of successive approximations.
(c) Determine the values of the complete elliptic integrals, K and E(K), from eqs. 4.122 and 4.123, respectively.
(d) Determine the wavelength, λ, from eq. 4.117. Using this value, determine the Ursell number, U_R, from eq. 4.118. The value of the Ursell number can be used to check the validity of the application of the theory with the site conditions in Figure 4.1.
(e) Calculate the values of ℓ, c, and C from the respective expressions in eqs. 4.116, 4.119, and 4.115.
(f) Determine the expression for the free-surface displacement from eq. 4.121 as a function of time and space.
(g) Determine the particle velocity components (u, w) from eqs. 4.106 and 4.105 and displacement components (ξ, ζ) from eqs. 4.111 and 4.112.

To illustrate, the following example is presented.

EXAMPLE 4.5: APPLICATION OF THE CNOIDAL THEORY At a site on the coast of Southern California, a fixed wave staff in 10 m of water records an average wave height of 1.0 m and an average wave period of 13 sec. Using Wiegel's (1960) approximations of the cnoidal wave theory equations, we shall calculate the wave properties at the staff and compare our results with those obtained from the linear wave theory of Chapter 3.

For the site conditions of $h = 10$ m, $H = 1.0$ m, and $T = 13$ sec, the simultaneous solution of eqs. 4.120, 4.122, and 4.123 yields M $\simeq 0.7$. This value in eqs. 4.122 and 4.123, respectively, yields K $\simeq 2.076$ and $E[K(\mathrm{M})] \simeq 1.242$. Using these values, the wavelength from eq. 4.117 is $\lambda = 127$ m and the celerity from eq. 4.120 is $c = 9.76$ m/s. From linear theory, the wavelength and celerity values are 124 m and 9.54 m/s, respectively, using the results of eq. 3.31. The value of the deep-water wavelength according to linear theory is $\lambda_0 = 264$ m. From eqs. 4.116 and 4.115, respectively, we obtain $\ell = 9.57$ m and $C = 0.429$ m. The trough of the wave is $\eta_0 = h - \ell = 0.43$ m below the SWL. Note that for a linear wave having a 1.0-m height, the value of η_0 is 0.5 m. The maximum horizontal velocity of the fluid particle is $u_{max} = 0.877$ m/s, and the corresponding Ursell number for the conditions is $U_R = 8.06$. Referring to Figure 4.1, for $h/\lambda_0 = 0.0379$ and $H/\lambda_0 = 0.00379$ we see that the cnoidal theory applied to the stated site conditions lies in the validity range of the second-order Stokes theory.

Before concluding this section, it must be stated that only the first-order cnoidal theory is presented herein. Korteweg and de Vries (1895) present higher-order versions of the theory in their paper. These higher-order theories produce very interesting results, including asymmetric wave profiles. The original paper is most interesting, and is recommended to the reader.

In the next section, we apply the cnoidal wave theory to extremely shallow water. The approximations in this application yield the same results as the solitary wave theory, which is used extensively in coastal engineering problems. For a derivation of the solitary theory, see the paper by McCowan (1891).

C. The Solitary Wave

As mentioned in the introductory paragraphs of this chapter, J. Scott Russell (1838, 1845, 1881), a naval architect, reported that he was able to observe a "wave of translation" or a "solitary wave" while studying the motions of a towed barge in a canal. According to Russell (1881), his work on this type of wave was conducted in the years of 1832 and 1833. Russell states that his work was a result of an earlier observation of the phenomenon by a William Houston, a canal boat owner and operator on the Glasgow and Ardrossan Canal. This wave was observed to have unusual properties. First, it was a one-time event. That is, when a fast-moving, horse-drawn barge was suddenly accelerated (or brought to rest) in a canal, a single crest (or trough) resulted that was totally above (or below) the SWL. Second, the form of the free surface depended on both the volume of water displaced by the barge motion and the water depth.

According to McCowan (1891), "the first sound approximate theory of the wave was given by Boussinesq in 1871," and was able to predict the wave phenomena observed by Russell. However, McCowan (1891) is credited with a form of the solitary theory that is used even today. In this section, we shall use the shallow-water approximations of the first-order cnoidal theory of Korteweg and de Vries (1895) to obtain the solitary wave equations.

We begin our analysis by noting that a solitary wave is, by definition, one having both an infinite period and an infinite wavelength. The expression for the wavelength from the cnoidal theory is in eq. 4.101. For that expression to be infinite, the parameter M must be 1. This parametric value results in an infinite value of K from eq. 4.122, a value of unity for E from eq. 4.123, a zero value of C from eq. 4.100, and the equality $\ell = h$ from eq. 4.116. This latter result is the same as $\eta_0 = 0$ in eq. 4.110, indicating that the free surface is entirely above (or below) the SWL. Hence, by passing to the limit as $M \rightarrow 1$ in eq. 4.100, we obtain the following expression for the free-surface displacement of a solitary wave:

$$\lim_{M\rightarrow 1}(\eta) = \eta = (H)\lim_{M\rightarrow 1}\left\{\mathrm{cn}^2\left[\sqrt{\frac{3H}{4\ell^3}}(x-ct)\right]\right\} = (H)\mathrm{sech}^2\left[\sqrt{\frac{3H}{4h^3}}(x-ct)\right] \tag{4.124}$$

In addition to this result, the celerity expression obtained from eq. 4.107 is

$$c = \sqrt{gh}\left(1 + \frac{H}{2h}\right) \simeq \sqrt{g(h+H)} \tag{4.125}$$

The expressions in eqs. 4.124 and 4.125 are the same as those derived by Boussinesq (1872).

The particle velocity components are obtained from eq. 4.108 for u, and from eq. 4.105 for w. The resulting expressions are

$$u = \sqrt{\frac{g}{h}}\left(1 + \frac{H}{h}\right)\eta \tag{4.126}$$

and

$$w = -\sqrt{\frac{g}{h}}(z+h)\frac{d\eta}{dX} \tag{4.127}$$

respectively. The corresponding particle displacements from point (x_0, S_0) over time are

$$\xi = \int_0^t u dt = \sqrt{\frac{4}{3}H(h+H)}\tanh\left(\sqrt{\frac{3H}{4h^3}}ct\right) \tag{4.128}$$

and

$$\zeta = \int_0^t w dt = -\sqrt{\frac{H^2}{h(h+H)}}\left[\tanh^2\left(\sqrt{\frac{3H}{4h^3}}ct\right)\right]S_0 \tag{4.129}$$

from eqs. 4.111 and 4.112, respectively. From the results of eq. 4.128, according to both the cnoidal and solitary theories, the horizontal displacements of the particles are independent of their vertical positions. Also from eq. 4.128, the horizontal displacement from $x_0 = 0$ asymptotically approaches $\sqrt{[4H(h+H)/3]}$ as $t \to \infty$. Similarly, from eq. 4.129, the vertical downward displacement from S_0 approaches $\sqrt{[H^2/h(h+H)]}S_0$.

To obtain an idea of the accuracy of the displacement components described by eqs. 4.128 and 4.129 for a solitary wave, the experimental data of Daily and Stephan (1952) are used for comparison. The experimental and theoretical results are presented in Figure 4.18 for $S_0/h \simeq 0.36$, 0.98, and 1.35, where $H/h \simeq 0.51$ and $h =$ 7.9 cm. As can be seen in Figure 4.18, the agreement is excellent for the lower values of S_0/h, and is acceptable for the largest value. Daily and Stephan (1952) also present celerity data. Some of those data are presented in Figure 4.19 with the shallow-water linear expression of c obtained from eq. 3.38 and the exact and approximate celerity expressions of eq. 4.125.

Because of the scales used in that figure, the spread in the data is somewhat exaggerated. Actually, the two theoretical curves obtained from the solitary and the experimental data differ by no more than 5% for $H/h \leq 1.0$.

EXAMPLE 4.6: APPLICATION OF THE SOLITARY THEORY Nearer the shoreline than the site of Example 4.5, a wave staff is located in 2 m of water. It records a wave height of 1.0 m and a period of 15 sec. The equality of eq. 4.125 yields a celerity value of 5.54 m/s, whereas the shallow-water approximation of the linear theory yields 4.43 m/s. The maximum horizontal particle velocity at the crest of

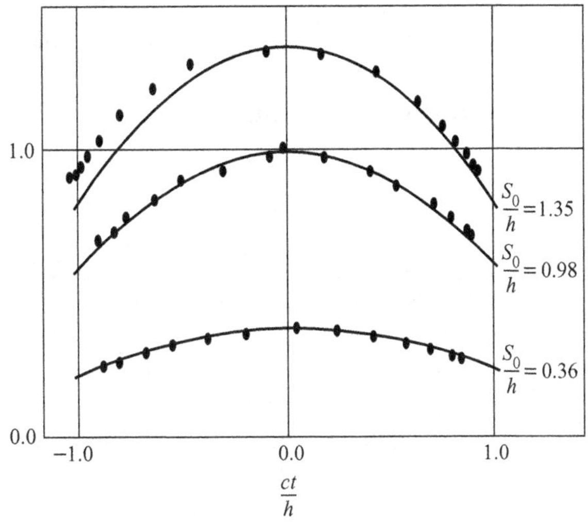

Figure 4.18. *Theoretical and Experimental Particle Paths Beneath a Solitary Wave.* The theoretical curves result from the application of eqs. 4.128 and 4.129, whereas the experimental data (•) are from Daily and Stephan (1952).

the wave is 3.32 m/s from eq. 4.126. Hence, the maximum particle velocity is approximately 60% of the celerity. The Ursell (1953) parameter,

$$U_R = \frac{H\lambda^2}{2h^3} = \frac{Hc^2T^2}{2h^3} \tag{4.130}$$

as defined in Figure 4.1, has a value of 4.32 for the conditions at this near-shore site. Note that the wavelength in this expression is replaced by the product of the celerity and period because, for a solitary wave, the wavelength is theoretically infinite. That product gives a wavelength of 83.1 m. Hence, the solitary theory yields approximate results for extremely shallow-water wave properties.

Munk (1949) applied the solitary theory to waves in the surf zone region. In his paper, an expression for an effective wavelength for the solitary wave is introduced. The Munk solitary wavelength expression is presented in the next section, which is

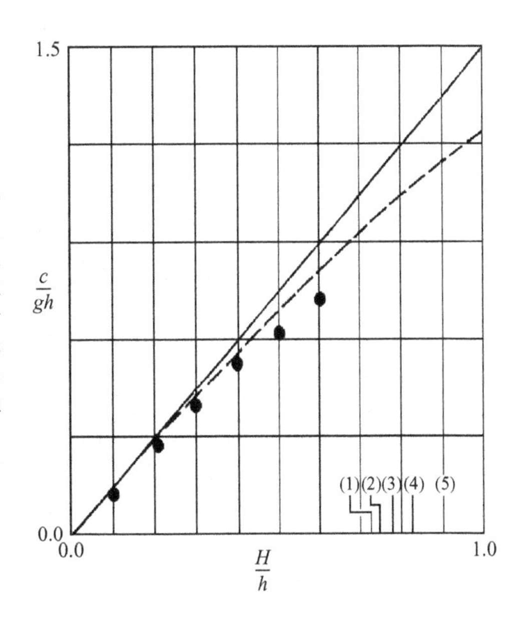

Figure 4.19. *Theoretical and Experimental Celerity Ratios for the Solitary Wave.* The solid line results from the equality in eq. 4.125, whereas the dashed line results from the approximate expression in that equation. The experimental data (•) are from Daily and Stephan (1952). The numbered values of *H/h* are the breaking limits of (1) Boussinesq (1872), (2) McCowan (1891), (3) McCowan (1894), (4) Lenau (1966), and (5) Miche (1944).

devoted to breaking waves. The reader should note that the limiting condition for the Stokes, cnoidal, and solitary wave theories is that of breaking.

4.6 Breaking Waves

Our interest is in the waves at the seaward boundary of the surf zone. The *surf zone* for a beach is defined as the region between the outermost break to the position of the uppermost runup. The *runup* is defined as the highest vertical position (above the SWL) attained by water on a beach. The surf zone is discussed in more detail in later chapters. Munk (1949) summarizes observations made by himself and others concerning the abilities of the various wave theories (Airy's linear theory, Stokes' expansion theory, and the solitary theory) to predict wave properties near the surf zone. His comments, based on comparisons of theoretical results and field and laboratory data, are as follows: (a) The wave height of a wave approaching a beach is considerably under-predicted by the linear theory. (b) The Stokes series converges "more and more slowly as one approaches the breaker points." (c) "As waves travel into water of depth less than, say, three times the wave height, the previously flat crests 'hump' into narrow crests separated by long flat troughs, and the character of these isolated crests scarcely depends on the distance (λ) between crests...," as is approximated by the solitary theory. From these observations, Munk (1949) concludes that the solitary theory is the most appropriate for waves near the surf zone.

For specific given wave period and water depth values, the highest wave is that which is breaking. The *breaking condition* is defined as that for which the horizontal particle velocity at the crest equals the celerity of the wave. Mathematically, the breaking condition is

$$u_{crest} \equiv u_c = c \tag{4.131}$$

As Munk (1949) states, wave mechanicians observed that when this condition is met, the wave crest is pointed, as sketched in Figure 3.1e. From this observation, the linear wave theory cannot be used to accurately determine the properties of a breaking wave because the free-surface profile of a linear wave is always sinusoidal, as sketched in Figure 3.1c. Hence, we must look to the nonlinear theories to describe breaking waves.

In the following sections, the analyses of breaking waves in deep water and in water of finite depth are presented. The breaking-wave analyses presented are those using the Stokes theory, Miche's formula, and the solitary theory.

A. Stokes' Deep-Water Analysis

In the collection of his papers, Stokes (1880) observes that the crest of a breaking wave is *pointed*, that is, there is a discontinuity in the first derivative of the profile at the break. The tangent lines of the weather (up-wave) and leeward (down-wave) sides of the free surface then meet at the crest and form an angle α_s, as sketched in Figure 4.20. To determine that angle, Stokes outlines a procedure which is as follows: Begin by placing the origin of a polar coordinate system at the breaking crest, as is done in Figure 4.20. Assuming irrotational flow, the equation of continuity is Laplace's equation (eq. 3.8), which for the polar coordinate system is

$$\frac{\partial^2 \varphi}{\partial R^2} + \frac{1}{R}\frac{\partial \varphi}{\partial R} + \frac{1}{R^2}\frac{\partial^2 \varphi}{\partial \alpha^2} = 0 \tag{4.132}$$

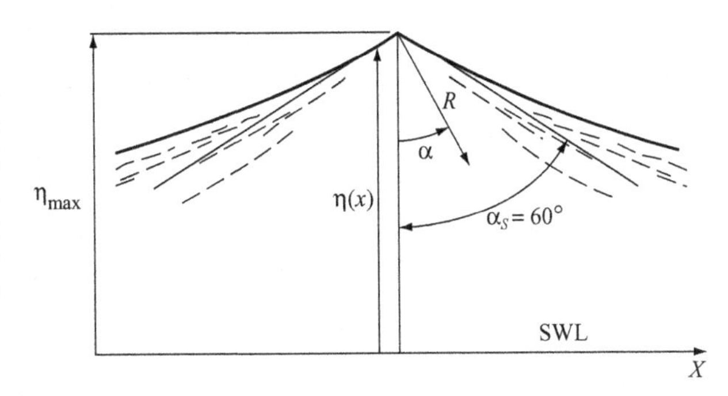

Figure 4.20. *Breaking Crest Profile Predicted by Stokes* (1880). This profile did not result from Stokes' original wave analysis, which was published in 1847. The value of the crest angle for the breaking wave was subsequently obtained by Michell (1893) and by Wilton (1913).

where φ is the velocity potential, R is the radial coordinate, and α is the angle from the negative vertical direction. A solution of eq. 4.132 for the velocity potential is

$$\varphi = C_n R^n \sin(n\alpha) \tag{4.133}$$

where C_n is a constant. The corresponding expression for the stream function is also required. To obtain this expression, the Cauchy-Riemann relationships of eqs. 2.58 and 2.59 are used. The expression for the stream function resulting from the combination of these equations and eq. 4.133 is

$$\psi = -C_n R^n \cos(n\alpha) \tag{4.134}$$

The velocity components obtained from this function are

$$v_R = \frac{\partial \varphi}{\partial \mathrm{R}} \simeq C_n R^{n-1} \sin(n\alpha) \tag{4.135}$$

and

$$v_\alpha = \frac{1}{R}\frac{\partial \varphi}{\partial \alpha} \simeq C_n n R^{n-1} \sin(n\alpha) \tag{4.136}$$

The expression for the velocity of water particles near the crest is then

$$V = \sqrt{v_R{}^2 + v_\alpha^2} = C_n n R^{n-1} \tag{4.137}$$

From this result, we see that the magnitude of the particle velocity is independent of the direction in the crest region, although the velocity components do vary with the angle.

The dynamic free-surface condition applied near the crest is expressed by

$$g\eta + \frac{1}{2}V^2 = K = \frac{c^2}{2} \tag{4.138}$$

which is similar to the expression in eq. 4.5. The relationship between K and c is given in eq. 4.12. At the crest, $V = 0$. Hence, the expression in eq. 4.138 becomes

$$\eta_{max} = \frac{c^2}{2g} \tag{4.139}$$

From the sketch in Figure 4.20, the free-surface displacement near the crest can be approximately represented by

$$\eta = \eta_{max} - R\cos(\alpha_s) = \frac{c^2}{2g} - R\cos(\alpha_s) \tag{4.140}$$

where the subscript s indicates at the free surface. At any point on the free surface near the crest, the combination of eqs. 4.137 through 4.140 yields

$$-gR\cos(\alpha_s) + \frac{1}{2}C_n^2 n^2 R^{2(n-1)} = 0 \tag{4.141}$$

This equation is valid if $n = 3/2$. In addition, for the fixed wave form, the stream function at the free surface must be equal to zero. By setting the expression in eq. 4.134 equal to zero and letting $n = 3/2$ and $\alpha = \alpha s$, the following is obtained:

$$\frac{3}{2}\alpha_s = \frac{\pi}{2} \tag{4.142}$$

The half-angle of the crest is then

$$\alpha_s = \pm\frac{\pi}{3}(\pm 60^\circ) \tag{4.143}$$

From this result, the tangents to the free surface at the crest are separated by an angle of 120°. This angular value was subsequently obtained by both Michell (1893) and Wilton (1913). McCowan (1894) theoretically demonstrates that the crest angle for a breaking wave in shallow water is also 120°.

The analyses of Stokes (1880) presented in this section and that presented in Section 4.2 are not the same, which should be obvious to the reader. The second-order theory of Stokes (1847) does not predict a pointed crest of a breaking wave when the breaking condition of eq. 4.131 is applied to the analysis. Higher-order approximations of Stokes do approach a sharp breaking crest. Penny and Price (1952) demonstrate this fact when applying the fifth-order theory to a deep-water breaking wave.

Values of the wave steepness $(H_0/\lambda_0)_b$ of a deep-water breaking wave were obtained by Michell (1893) and Wilton (1913). Michell's value, the more widely accepted, is

$$\left.\frac{H_0}{\lambda_0}\right|_b = 0.142455\cdots \simeq \frac{1}{7} \tag{4.144}$$

Later in this chapter, the profiles of deep-water breaking waves obtained from Stokes' (1847) second-order and fifth-order theories will be compared with that obtained using Michell's (1893) complex variable theory.

B. Miche's Formula: Breaking Waves in Waters of Finite Depth

To gain an insight into how the breaking wave steepness changes with water depth, we first apply the breaking condition of eq. 4.131 to the linear wave theory of Chapter 3. That is, by combining eq. 4.131 with the expressions for the celerity in eq. 4.33 and the horizontal component of the particle velocity of eq. 3.49 (the latter applied at a crest), the following relationship results:

$$c_b = \frac{H_b\omega}{2}\frac{1}{\tanh(k_b h_b)} = C_b\frac{H_b}{T}\frac{1}{\tanh(k_b h_b)} = \frac{\lambda_b}{T} \tag{4.145}$$

where, to generalize the formula, the proportionality constant C_b is introduced. The subscript "b" is used to indicate the breaking condition. The last equality of eq. 4.145 can be rearranged to obtain an expression for the breaking wave steepness, that is,

$$\frac{H_b}{\lambda_b} = \frac{1}{C_b}\tanh(k_b h_b) \tag{4.146a}$$

In deep water, the hyperbolic tangent approaches unity, and the constant must be equal to the inverse of the deep-water breaking wave steepness. Eq. 4.146a can then be written as

$$\frac{H_b}{\lambda_b} = \left.\frac{H_0}{\lambda_0}\right|_b \tanh(k_b h_b) \tag{4.146b}$$

Using the deep-water wave steepness value of Michell (1893) in eq. 4.144, the expression in eq. 4.146 can be approximated by

$$\frac{H_b}{\lambda_b} = \frac{1}{7}\tanh(k_b h_b) \tag{4.147}$$

This expression of the breaking steepness is called *Miche's formula*, and is an approximation of the expression found in his 1944 paper. Miche actually presents a coefficient value of 0.140 in his paper. The difference in this value and the 1/7 approximation is off by 2%. When applied to deep water, we see that the resulting approximation is identical with that of eq. 4.144. Hence, according to eq. 4.147 the wave steepness is maximum in deep water. In shallow water, eq. 4.147 is approximately

$$\frac{H_b}{\lambda_b} \simeq \frac{2\pi h_b}{7\lambda_b} \tag{4.148}$$

From this expression, we obtain the following shallow-water ratio of wave height-to-water depth:

$$\frac{H_b}{h_b} \simeq \frac{2\pi}{7} = 0.897\cdots \simeq 0.9 \tag{4.149}$$

This value has been found to agree reasonably well with experimental data for waves breaking over horizontal beds. In engineering practice, it is a good "field equation." Galvin (1972), a coastal engineer who has devoted much of his professional career to waves in the coastal zone, carries the approximation of eq. 4.149 further when he states "as a first approximation, the wave breaks when it reaches a depth equal to the height." The differences in observed and theoretically predicted breaking conditions and wave properties are discussed later.

EXAMPLE 4.7: THEORETICAL, DEEP-WATER BREAKING WAVE PROFILES In this example, the free-surface profiles obtained from Stokes' (1847) second-order theory and from Michell's (1893) theory are compared. First, non-dimensionalize the second-order expression of Stokes in eq. 4.54 by dividing through by the deep-water wavelength, λ_0. Then apply Michell's deep-water breaking steepness value of eq. 4.144 to the resulting expression. The result is

$$\left.\frac{\eta_0}{\lambda_0}\right|_b = \frac{1}{2}\left.\frac{H_0}{\lambda_0}\right|_b \cos(k_0 X) + \frac{\pi}{4}\left(\left.\frac{H_0}{\lambda_0}\right|_b\right)^2 \cos(2k_0 X) \tag{4.150}$$

where the coordinate X is defined by

$$X = x - ct \tag{4.151}$$

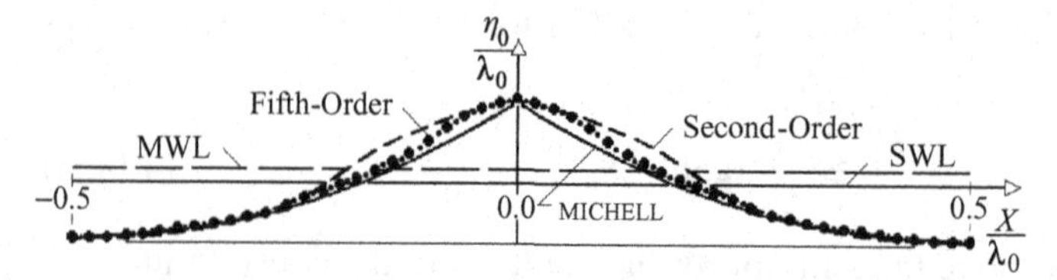

Figure 4.21. *Deep-Water Breaking-Wave Profiles Predicted by the Theories of Stokes* (1847) *and Michell* (1893). Only the Michell profile predicts a breaking crest angle of 120°. Michell's breaking-wave steepness value of $(H_0/\lambda_0)_b = 0.142$ is used for both of the Stokes profiles. Both of the Stokes profiles are horizontal at their crests.

as illustrated in Figure 4.16. Apply Michell's deep-water breaking steepness value, $H_0/\lambda_0|_b = 0.142$, to the expression in eq. 4.150. The result is plotted in Figure 4.21, along with results obtained from both the fifth-order theory of Stokes (1847) and Michell's (1893) theory. As previously mentioned, Michell's theory predicts a breaking crest angle of 120°, as first predicted by Stokes (1880) and later by Wilton (1913).

The reader can see that the results of Stokes' second-order and fifth-order theories both have zero slopes at the crests, whereas the Michell profile is pointed. The fifth-order theory, discussed by Penny and Price (1952), more closely approximates the Michell profile than does the second-order theory, as expected.

C. Breaking Solitary Waves

When the breaking condition of eq. 4.131 is applied to a solitary wave, the ratio of the breaking wave height-to-water depth value is somewhat different than that of eq. 4.149, as is shown in the following analysis. To apply the breaking condition ($u_{crest} = c$), equate the expression of eq. 4.126 applied at a wave crest (where $\eta = H_b$) to the celerity expression of eq. 4.125. The resulting wave height-to-water depth value is

$$\frac{H_b}{h_b} = 0.780776\cdots \simeq 0.781 \tag{4.152}$$

A comparison of this value with that of eq. 4.149 shows a difference of slightly more than 15%. Galvin (1972) points out that the various analyses of solitary waves from that of Boussinesq (1872) to Lenau's (1966) analysis have yielded values of this ratio varying from a low value of 0.73 to a high of 1.03. Although the value of 0.9 in eq. 4.149 is a good field value, McCowan's (1894) value of 0.78 has been adopted by the coastal engineering community for waves breaking in shallow water.

The expression from which the profile of a breaking solitary wave is determined is found by combining eqs. 4.124 and 4.149. The result of that combination, non-dimensionalized by dividing through by the breaking wave height, is

$$\frac{\eta_b}{H_b} = \text{sech}^2\left[\sqrt{\frac{3}{4}\left(\frac{H_b}{h_b}\right)^3}\frac{X}{H_b}\right] = \text{sech}^2\left(0.597\frac{X}{H_b}\right) \tag{4.153}$$

The free-surface profile obtained from this expression is presented in Figure 4.22, with the solitary wave profiles corresponding to $H/h = 0.1$ and 0.5. From the results

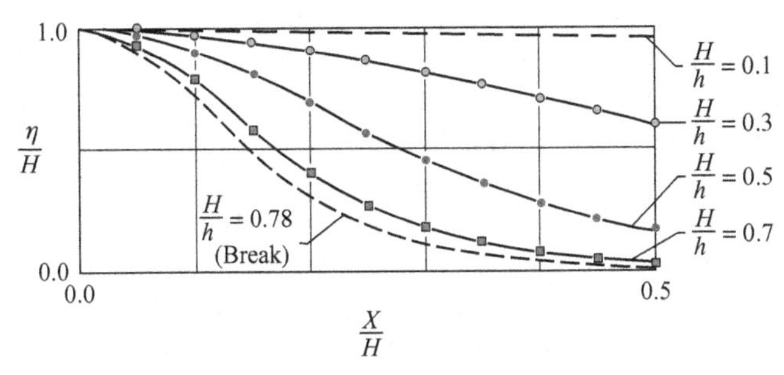

Figure 4.22. *Solitary Wave Profiles*. The profiles are predicted by eq. 4.124 for four non-breaking waves ($H/h < 0.78$) and one breaking wave ($H/h = 0.78$) from Michell's equation, eq. 4.153. The reader should note that all of the profiles are horizontal at the crest.

in the figure, one sees that there are dramatic changes in the profiles as the wave height-to-depth ratio is increased.

The expression for the breaking height-to-depth ratio in eq. 4.152 shows no dependence on the wave properties prior to the break. To make the solitary theory more practical, Munk (1949) derives a relationship between the breaking height and the deep-water wave steepness by using a modified solitary theory. The modification is accomplished by introducing an effective wave period that is based on both the displaced water volume of a solitary wave and the energy-flux conservation of the wave.

The displaced volume of water of per crest width of a solitary wave between points $\pm x$ on either side of the crest (at time $t = 0$), is found from the integration of eq. 4.124. The result is

$$Q' = \int_{-x}^{x} \eta dx = 2\int_{0}^{x} \eta dx = 4h^2\sqrt{\frac{1}{3}\left(\frac{H}{h} + \frac{\eta}{h}\right)} \tag{4.154}$$

where the free-surface displacement, $\eta(x)$, is given in eq. 4.124. We note that the bed is assumed to be horizontal so that the wave profile is symmetric about the crest. The displaced volume over the entire solitary wave is obtained from eq. 4.154 by passing to the limit as $x \to \infty$. This volume is

$$Q = 4h^2\sqrt{\frac{H}{3h}} \tag{4.155}$$

Following Munk (1949), we find that $Q'/Q \simeq 0.98$ over the length defined by $-2.5 \leq x/h \leq 2.5$ when $H/h = 0.5$. That is, approximately 98% of the displaced volume is found to be in the near-region of the crest.

According to Munk (1949), the energy of a solitary wave is approximately divided equally between potential energy and kinetic energy. The energy of the solitary wave per crest width over the length defined by $\pm x$ is

$$\mathrm{E}' = \frac{4}{3}\rho g h^3 \frac{H}{h}\left(2 + \frac{\eta}{H}\right)\sqrt{\frac{1}{3}\left(\frac{H}{h} - \frac{\eta}{h}\right)} \tag{4.156}$$

The total energy per crest length of a solitary wave is found by passing to the limit as $x \to \infty$, where $\eta \to 0$. The resulting total energy expression is

$$E' = \lim_{X\to\infty} \mathrm{E}' = \frac{8}{3}\rho g h^3\sqrt{\frac{1}{3}\left(\frac{H}{h}\right)^3} \tag{4.157}$$

Munk (1949) finds that $E'/E' \simeq 0.98$ over the length defined by $-2.1 \leq x/h \leq 2.1$ when $H/h = 0.5$. As is the case for the displaced volume, Munk concludes that most of the energy of the solitary wave is in the near-region of the crest.

The conclusions concerning the concentrations of both the displaced volume and the energy of a solitary wave lead Munk (1949) to conclude that an effective wavelength (λ_{eff}) can be introduced for the solitary wave. Munk states that "the assumption of a single solitary wave is fulfilled to a high degree of accuracy if the actual wave length (λ) of the waves exceeds the *effective* wave length…" Applying this wavelength concept to a shallow-water breaking wave, the *effective wavelength* is

$$\lambda_{eff}|_b = \sqrt{gh_b}T_{eff}|_b \tag{4.158}$$

where the *effective wave period* is

$$T_{eff}|_b = 1.17\sqrt{h_b} = 1.32\sqrt{H_b} \tag{4.159}$$

This period expression is a limiting case of the effective-period expression derived by Bagnold (1947), which in turn was based on the theory of McCowan (1894). These effective properties are used simply to establish lower limits for the use of the solitary theory.

Following Munk (1949), the breaking solitary wave height expression is found by assuming that the energy flux is conserved between orthogonals as the wave purely shoals, as is discussed in Section 3.7 for linear waves. Assuming that the wave period is invariant with position, the conservation of the energy flux per crest width is expressed by

$$c_{g_0}E'_0 = \frac{c_0}{2}E'_0 = \frac{c_0}{2}\left\{\frac{1}{8}\rho g H_0^2\lambda_0\right\} = c_b E'_b \tag{4.160}$$

where the expression for the energy at the shallow-water break is found by applying the energy expression of eq. 4.157 at h_b. The resulting breaking wave height expression is

$$H_b = \frac{H_0}{3.3\left(\dfrac{H_0}{\lambda_0}\right)^{1/3}} \tag{4.161}$$

Results obtained from this equation are plotted in Figure 4.23 in non-dimensional form. Munk (1949) plots this curve with field data, and shows that the curve bisects the scatter of the measured data over the range of the deep-water steepness shown in Figure 4.23. For $H_0/\lambda_0 > 0.02$, Munk shows that the linear theory of Airy yields satisfactory results.

EXAMPLE 4.8: BREAKING HEIGHT OF A SHOALING SOLITARY WAVE A deep-water wave having a wave height of 1 m and a period of 8 sec shoals without refraction on a straight, parallel-contoured beach. From both the linear and Stokes second-order theories, the deep-water wavelength is approximately $\lambda_0 \simeq g\,T^2/2\pi \simeq 100$ m. The deep-water wave steepness is then $H_0/\lambda_0 \simeq 0.01$. From eq. 4.161, the breaking wave height is approximately 1.99 m. From eq. 4.152, this wave breaks in a water depth of approximately 2.52 m.

A final note: For the free-surface displacement expression in eq. 4.124, the value of the derivative of η with respect to X is zero when applied to a crest. Because this

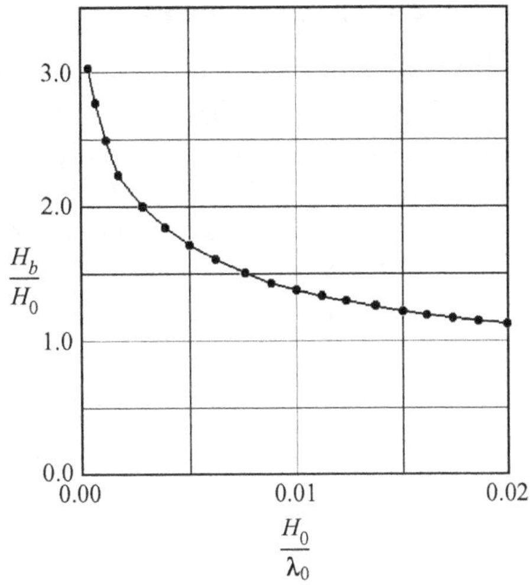

Figure 4.23. *Breaking Wave Height Ratio as a Function of the Deep-Water Steepness for a Solitary Wave.* From eq. 4.161.

is true for any value of the height-to-depth ratios up to and including the breaking value, the angle for a breaking solitary wave crest angle is also 0°.

4.7 Summary

In this chapter, three classical nonlinear theories are outlined and discussed. Although these theories were formulated in the nineteenth and early twentieth centuries, they are used today by a number of oceanographers and ocean engineers because of both their elegance and applicability under specific conditions. The analytical accomplishments of Stokes, Rayleigh, Korteweg, de Vries, Michell, and the other renaissance wave mechanicians of that time period are truly extraordinary.

There have been numerous studies of nonlinear waves in the twentieth century, the list of these being far too large to present here. By following the thought processes of the early wave analysts, modern investigators have both extended and expanded the early theories. Some of the more recent studies include the following:

a. *Stokes Wave Theory:* Bhattacharyya (1995) discusses two methods used in deriving the fifth-order theory.
b. *Cnoidal Wave Theory:* Yamaguchi (1992) reviews the studies of cnoidal theory dating back to 1960. The application of the Korteweg-de Vries (1895) equation to transient, axisymmetric waves is presented by Khangaonkar and Le Méhauté (1991).
c. *Solitary Wave Theory:* Most of the mid- to late-twentieth-century works on solitary waves have focused on shoaling. For example, Grilli, Subramanya, Svendsen, and Veeramony (1994) discuss the shoaling of solitary waves on plane beaches.
d. *Breaking Waves:* Longuet-Higgins, Cokelet, and Fox (1976) present a short but informative discussion of steep waves, including breaking waves. Broeze (1992) applies the "panel method" in determining the properties of breaking waves.

Although these references are few, it would be of benefit to the reader to consult these works for both their discussions and their references.

4.8 Closing Remarks

Most of the waves observed in the world's oceans are nonlinear. The reader will find the results in Figure 4.1 useful in determining the appropriate theory to be used in dealing with engineering problems involving nonlinear water waves. Fortunately, in the conceptual design phase of many engineering projects, the linear theory presented in Chapter 3 is satisfactory in predicting the behaviors of wave-induced forces and motions of ocean structures.

5 Random Seas

Let us begin by stating the term *random* is synonymous with *unpredictable.* A truly random phenomenon should by definition defy mathematical analysis. One might question then if *random waves* can be analyzed. By making certain assumptions, waves that are random in the time domain can be shown to have rather predictable properties in the frequency domain. In this chapter, a brief history of random wave analysis is first presented. This is followed by discussions and illustrations of the various statistical methods of wave analysis.

5.1 Introduction

We can safely assume that seafarers down through the ages have been aware of the randomness or unpredictability of ocean waves. The occurrence of "rogue waves" has been documented again and again. These are extremely high waves that occur in the open ocean without warning. The earliest attempts to deal mathematically with random waves were confined to averaging observed wave heights and periods. The data were obtained by visual means in a laboratory setting (as by Weber and Weber, 1825, according to St. Denis, 1969), in lakes, onboard a ship (as by Abercromby, 1888), or in coastal waters. When log-keeping came into being, wave height and period observations were recorded along with wind speeds. The more sophisticated mariners also recorded wavelength estimations obtained by comparing the wavelengths with lengths of their vessels. The accuracy of the wave height estimations from visual observations from ships is discussed by Cornish (1910). He reports that the estimated wave height observations at the time of his writing were about 90% accurate provided that the observer was well practiced. More examples of wave height and wavelength estimations made during open-ocean voyages are given by Cornish (1934). In addition, an excellent historical perspective on wave measurement and analysis is presented by St. Denis (1969).

There are many instruments that have been created to measure the properties of waves. The earliest laboratory devices were either vertical staffs or tank walls with painted height scales. These were later replaced by capacitance probes and digital wave staffs. At-sea wave studies are now conducted by using such devices as digital wave staffs, floats equipped with accelerometers, and submerged pressure gauges. All these systems are designed for studies of the wave climates at specific sites. Wider perspectives of ocean waves are obtained from both air and space crafts equipped with both photographic and microwave imagery systems. For a discussion of these techniques, the reader is referred to the collection of papers resulting from the symposium on *Measuring Ocean Waves from Space* at the Johns Hopkins University's Applied Physics Laboratory in 1986 (see Beal, 1987). Excellent discussions of the various wave measurement techniques are found in the books by Tucker (1991) and Young (1999). Also in those references are clear and concise discussions of the various statistical wave analysis methods.

Statistical averaging of wave heights and periods continued into the twentieth century. Early in that century, oceanographers and applied mathematicians began to consider the probabilities of occurrence of specific properties of ocean waves. Many of the probabilistic studies were based on the work of John W. Strutt. In 1880, Strutt (then Baron Rayleigh, and later Lord Rayleigh) developed the well-known Rayleigh probability distribution by considering the superposition of "a large number of vibrations of the same pitch and of random phase" (see Strutt, 1880). Strutt's (Rayleigh's) analysis was analogous to superimposing linear water waves of the same height and frequency but randomly differing in phase. Approximately 70 years later, random phasing was used in the analysis of ocean waves by Pierson (1952), Longuet-Higgins (1952), and others, and in the analysis of ship motions by St. Denis and Pierson (1953).

In the mid-twentieth century, the method of time-series analysis was introduced to physical oceanography. The work of Taylor (1921) (in which the correlation technique was used in the analyses of diffusion and turbulence), Rice (1944) (in the study of noise), and Tuckey (1949) (in spectral analysis) inspired a number of investigators to apply time-series techniques in the analysis of random ocean waves. In one year, papers by Darbyshire (1952), Longuet-Higgins (1952), Neumann (1952), Pierson (1952), and Putz (1952) were published, all containing analyses of ocean wave spectra. According to Ewing (1971), the first detailed measurements of waves were made by Barber and Ursell in 1945 (see Barber and Ursell, 1948, for details of their study). Those investigators were able to estimate wave spectra. Later, empirical wave spectral density formulas of Phillips (1958), Pierson and Moskowitz (1964), Hasselman et al. (1973), and others appeared, each accounting for certain physical conditions. The first analytical wave spectral density expression is attributed to Bretschneider (1959). The works of both Pierson and Moskowitz (1964) and Bretschneider (1959) are discussed in Section 5.7 of this book. The reader is also referred to the book edited by Le Méhauté and Hanes (1990) and that written by Sorensen (1993) for excellent discussions of these and other spectral representations. At the end of the twentieth century, McCormick (1998b, 1999) applied the probability formula of Weibull (1951) to water-wave spectral analysis with some success. This method is also described and discussed herein.

In this chapter, the statistical analysis of ocean waves is developed. First, the wave height (H) and period (T) values are determined from a wave trace, and a bivariate (H-T) nomograph is constructed. Based on this data sample, various statistical averages, the probability, and the probability density are defined. Also, the wave spectral density for discrete data is defined and applied to the H-T data. The statistical functions are then applied to continuous data. Finally, the directionality of random or irregular seas is discussed. All discussions contained herein are introductory in nature. The reader is encouraged to consult the references for more expansive discussions on wave measurement and analysis.

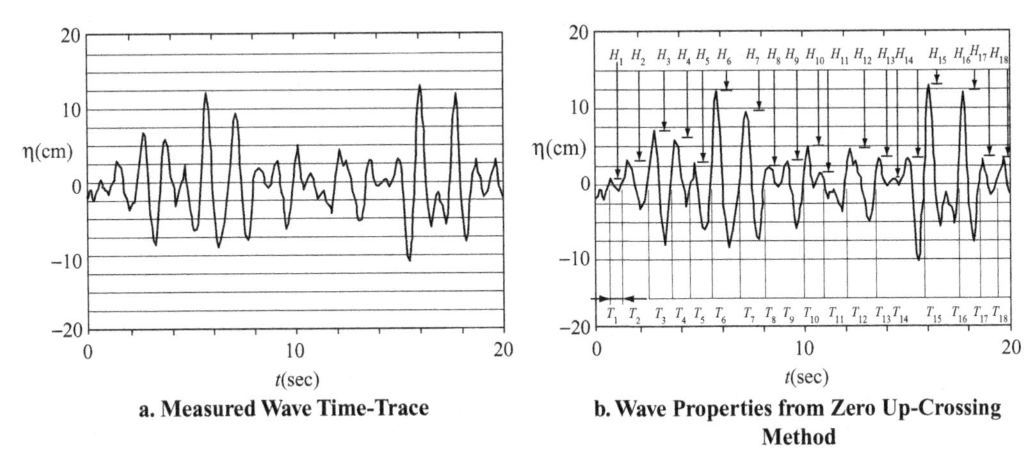

a. Measured Wave Time-Trace **b. Wave Properties from Zero Up-Crossing Method**

Figure 5.1. *Measured Wave Properties.* In (a) the trace was obtained in a wave-tank simulation of a wind-generated sea. In (b) the zero up-crossing method of wave identification is applied. As can be seen, some waves are not identified by the zero up-crossing method because they are totally above or below the horizontal axis. For large data sets, these missed waves are considered to be statistically insignificant.

5.2 Statistical Analysis of Measured Waves

The properties of waves in the open ocean are typically determined by analyzing signals from floating wave gauges. These signals are either recorded onboard the floats or telemetered to ships, aircraft, or shore facilities for analysis. Such signals might be transmitted for twenty minutes of each hour, twenty hours each day for several years. A signal from an at-sea measurement system would be similar to that shown in Figure 5.1a, which shows the displacement (η) of the free surface from the SWL as a function of time (t). The trace shown in Figure 5.1a was obtained in a laboratory simulation of a wind-generated sea. The wave pattern was produced by a wave maker creating a number of *regular waves* (having specific heights and periods but differing phases) over a period of time. In other words, these waves were physically superimposed on each other to produce an irregular wave pattern. Although the trace appears to be random in time, there is a definite energy distribution in the frequency domain of the input signal to the wave maker. Because the nature of a wind-generated sea is somewhat well defined in the frequency (or period) domain, the seas are often referred to as *irregular seas* instead of random seas.

There are several methods of identification of waves in an irregular trace such as that shown in Figure 5.1a. One technique that is widely used by those performing "hand and eye" analyses of wave traces is the *zero up-crossing method*. This method is illustrated in Figure 5.1b, where the waves in the trace in Figure 5.1a are identified. The method is as follows: Each time the wave trace crosses the time axis with a positive slope (that is, a *zero up-crossing*), assume that one wave has passed and another is just appearing. The zero up-crossing then is at the "front" of the trailing wave and the "back" of the leading wave. The *wave period* is defined as the time between consecutive up-crossings, whereas the *wave height* is defined as the vertical distance between the maximum displacement and the minimum displacement of the trace between consecutive up-crossings. The reader might question the accuracy of the method as small waves that are entirely above or below the time axis are missed

Table 5.1. *Example of wave data obtained from at-sea measurements*

H(m) – measured	T(sec) – measured	H(m) – rounded off	T(sec) – rounded off
0.56	3.12	0.5	3.0
1.27	6.36	1.5	6.5
0.78	5.77	1.0	6.0
....	....	....	....

by the method. Statistically speaking, there are millions of waves that pass a point in the open ocean each year. The sample is so large that the number of small waves missed do not significantly affect the statistics.

The tabulated wave data would resemble the wave height and period data shown in Table 5.1. Those data are taken with some specified accuracy. The question might be raised as to how accurate we should be in these measurements. That decision is strictly up to the individual reducing the data. To illustrate, the data in Table 5.1 first appear with an accuracy of two places past the decimal point for both the heights and periods. If this accuracy is maintained, the amount of data to be analyzed is somewhat overwhelming. Instead of maintaining this accuracy, round off the data to 0.5 m for the wave heights and 1.0 sec for the wave period, as illustrated in the table. Those rounded-off data are used to create a nomograph, such as that in Figure 5.2. The 100 data points in the figure are used to illustrate the derivations and applications of the statistical wave properties discussed herein. The total sample of wave heights and periods in Figure 5.2 is

$$N = \sum_{i=1}^{i=11} n_{T_i} = \sum_{j=1}^{j=7} n_{H_j} = 100 \tag{5.1}$$

In eq. 5.1, i and j are indices of the wave periods and heights, respectively. To illustrate, the $j = 4$ wave height is 2 m, and there are 11 waves with this rounded-off wave height value, that is, $n_{H4} = 11$.

We can now define some of the basic statistical properties of the wave data. First, the *cumulative frequency of occurrence* of the wave height sample in Figure 5.2 is defined as

$$\mathrm{P}(H \leq H_J) = \sum_{j=1}^{J} \frac{n_{H_j}}{N} \tag{5.2}$$

This is simply the percentage or fraction of the sample wave heights that are less than or equal to the wave height, H_J. For an infinite sample, the cumulative frequency of occurrence becomes the *cumulative probability of occurrence*, which is the probability that any wave height, H, will be less than or equal to the wave height, H_J. Mathematically, this probability is defined as

$$P(H \leq H_J) = \lim_{N \to \infty} \mathrm{P}(H \leq H_J) = \lim_{N \to \infty} \sum_{j=1}^{J} \frac{n_{H_j}}{N} \tag{5.3}$$

From this, a deterministic sea is one for which

$$P(H \leq \infty) = 1 \tag{5.4}$$

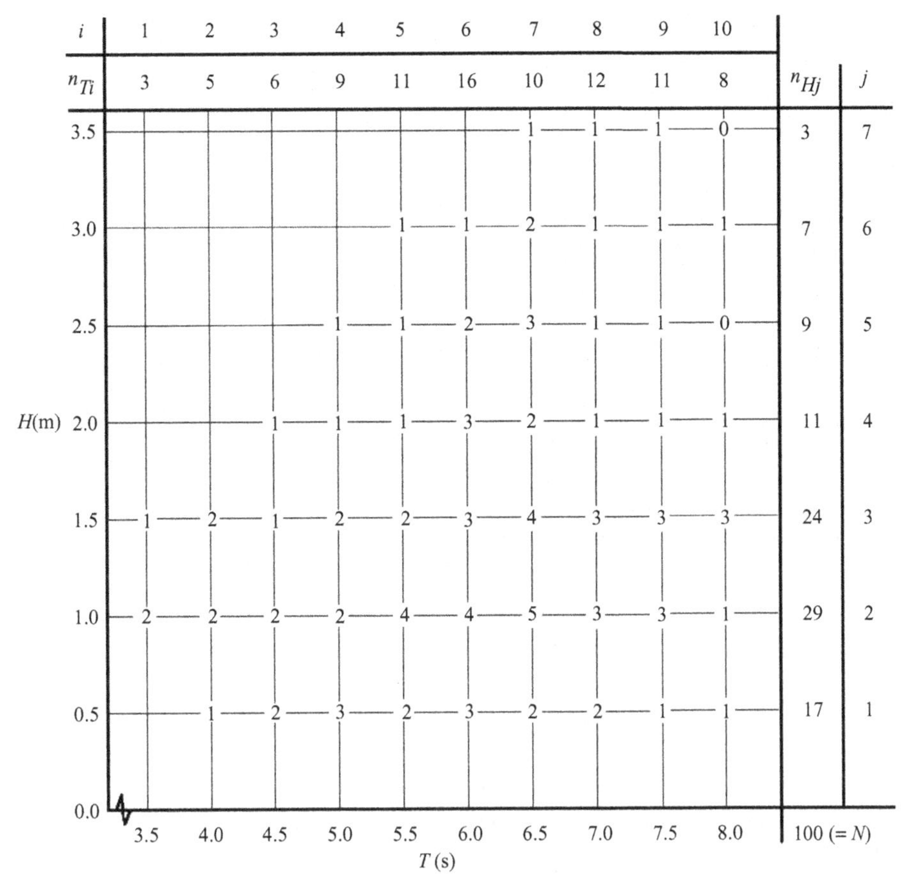

Figure 5.2. *Wave Height – Period Nomograph.* In this figure, i and j are the indices of the respective period (T) and wave height (H), and n_{Ti} and n_{Hj} are the numbers of observation for the respective period and wave height values. The total sample is N.

because all wave heights must be less than infinite. At the other extreme, an impossible sea is one for which

$$P(H \leq 0) = 0 \tag{5.5}$$

because no wave heights can be less than zero, and a zero wave height is undefined. For a finite number of data points, as in Figure 5.2, the cumulative frequency of occurrence can be considered to be an approximate cumulative probability of occurrence.

EXAMPLE 5.1: CUMULATIVE PROBABILITY OF OCCURRENCE The probability that any measured height in Figure 5.2 is less than or equal to 2 m ($J = 4$, $H_4 = 2$ m) is approximately

$$P(H \leq 2\,\text{m}) \simeq \text{P}(H \leq 2\,\text{m}) = \frac{(n_{H_1} + n_{H_2} + n_{H_3} + n_{H_4})}{N} = 0.81$$

The term "approximately" is used because an infinite sample is required to have a true probability. The probability is approximately equal to the cumulative frequency of occurrence of eq. 5.2 for this finite sample. Following this example, a chart of the probability as a function of wave height is constructed (see Figure 5.3).

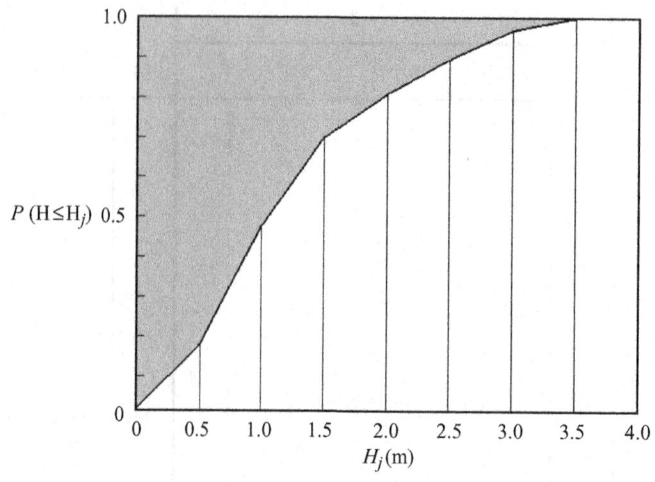

Figure 5.3. *Cumulative Frequency of Occurrence for the Data in Figure 5.2.* This can be considered to be an approximate probability of occurrence for the wave data.

We define the *probability density function* as the slope of the probability curve. For continuous wave height data, the probability density function is

$$p(H_J) = \frac{d[P(H \leq H_J)]}{dH} = \lim_{\Delta H \to 0} \frac{\Delta[P(H \leq H_J)]}{\Delta H} \tag{5.6}$$

Using the last term without passing to the limit, we can obtain an approximate probability density function for a discrete data set, such as that in Figure 5.2. The expression of that approximation is

$$p(H_J) \simeq \sum_{j=1}^{J} \frac{n_{H_j} - n_{H_{j-1}}}{N(\Delta H)} = \frac{(n_{H_J}/N)}{\Delta H} \tag{5.7}$$

The *most-probable wave height, H_{mp}*, is that wave height corresponding to the maximum value of the probability density. This wave height is then determined from

$$\frac{d[p(H_J)]}{dH}\Big|_{H_{mp}} = 0 \tag{5.8}$$

if the wave height distribution is continuous. As is demonstrated in Sections 5.5 and 5.6, the most probable value of a random variable, from either experimental studies or field measurements, can be used to determine the form of the probability density function.

EXAMPLE 5.2: PROBABILITY DENSITY AND MOST-PROBABLE WAVE HEIGHT As in Example 5.1, we focus our attention on the 2-m wave height $(J = 4)$ in Figure 5.2. The value of the probability density, from eq. 5.7, is

$$p(H_4) = p(2\,\text{m}) = \frac{(n_{H_4}/N)}{\Delta H} = \frac{(11/100)}{0.5} = 0.22\,\text{m}^{-1}$$

The *probability distribution* (the plot of the probability density) for the wave height data in Figure 5.2 is presented in Figure 5.4. From the condition of eq. 5.8, the most probable wave height in Figure 5.4 is $H_{mp} = 1.0$ m.

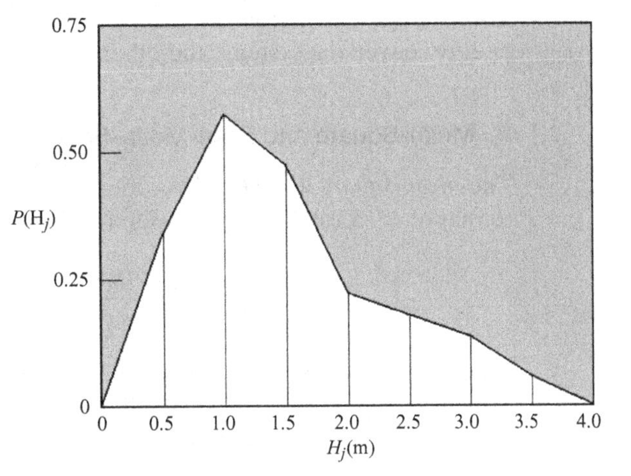

Figure 5.4. *Probability Density for the Wave Height Data in Figure 5.2.*

In addition to the cumulative probability of occurrence and the probability density, other statistical properties of use in random wave analysis are listed in the following.

A. Average Wave Period and Wave Height

The *average wave period* over the period range $T_i < T \leq T_I$ is

$$T_{iI} = \frac{\sum_i^I \frac{n_{T_i} T_i}{N}}{\sum_i^I \frac{n_{T_i}}{N}} = \frac{\sum_i^I p(T_i) T_i \Delta T}{\sum_i^I p(T_i) \Delta T} \tag{5.9a}$$

where the expression for $p(T_i)$ is obtained by simply replacing the wave height (H) by the period (T) and the indices (j, J) by (i, I) in eq. 5.7. When the expression in eq. 5.9a is applied to the total sample (N) of wave periods, then the expression for the average wave period is obtained, that is,

$$T_{avg} = \frac{\sum_{i=1}^{i_{max}} \frac{n_{T_i} T_i}{N}}{1} = \frac{\sum_{i=1}^{i_{max}} p(T_i) T_i \Delta T}{1} \tag{5.9b}$$

In a similar manner, the *average wave height* over the range $H_j < H \leq H_J$ is

$$H_{jJ} = \frac{\sum_j^J \frac{n_{H_j} H_j}{N}}{\sum_j^J \frac{n_{H_j}}{N}} = \frac{\sum_j^J p(H_j) H_j \Delta H}{\sum_j^J p(H_j) \Delta H} \tag{5.10a}$$

The average wave height of the entire sample in Figure 5.2 is then

$$H_{avg} = \frac{\sum_{j=1}^{j_{max}} \frac{n_{H_j} H_j}{N}}{1} = \frac{\sum_{j=1}^{j_{max}} p(H_j) H_j \Delta H}{1} \tag{5.10b}$$

The average expressions of the wave properties are used in the design of wave-energy conversion systems and other dynamic ocean systems.

B. Mean-Square and Root-Mean-Square Wave Heights

The *mean-square wave height* is obtained from an averaging formula that is similar to that of eq. 5.10b. The expression for this wave property is

$$\overline{H^2} = \sum_{j=1}^{j_{max}} \frac{n_{H_j} H_j^2}{N} = \sum_{j=1}^{j_{max}} p(H_j) H_j^2 \Delta H \tag{5.11}$$

where the unity value in the denominator (the value of the probability of occurrence over the entire sample) is assumed here and henceforth. The square root of this is called the *root-mean-square wave height*, that is,

$$H_{rms} = \sqrt{\overline{H^2}} \tag{5.12}$$

The root-mean-square wave height is considered to be a measured wave property. It is extensively used in many statistical formulas including various expressions for the probability density function for continuous wave data.

C. Variance of the Wave Heights

The *variance* of the wave heights is a measure of the spread of the data about the average wave height, and is mathematically defined by

$$\overline{(H - H_{avg})^2} = \sum_{j=1}^{j_{max}} \frac{n_{H_j}(H_j - H_{avg})^2}{N} = \sum_{j=-1}^{j_{max}} p(H_j)(H_j - H_{avg})^2 \Delta H \tag{5.13}$$

In the following example, the averages of the wave height and period and the mean-square, root-mean-square, and variance of the wave heights are demonstrated.

EXAMPLE 5.3: AVERAGE WAVE PERIOD AND STATISTICAL WAVE HEIGHTS For the data in Figure 5.2, the values of the respective average, mean-square, root-mean-square wave heights, and the wave height variance are

$$\begin{aligned} H_{avg} &= 1.50\ \text{m} \\ \overline{H_2} &= 2.87\ \text{m}^2 \\ H_{rms} &= 1.69\ \text{m} \\ \overline{(H - \overline{H})^2} &= 0.64\ \text{m}^2 \end{aligned}$$

These values are obtained from eqs. 5.10b, 5.11, 5.12, and 5.13, respectively. The average wave period obtained from eq. 5.9b is $T_{avg} = 6.13$ sec.

D. Significant Wave Height and Period

As previously stated in this chapter, most of the open-ocean wave height data prior to the mid-twentieth century resulted from visual observations. Although Cornish (1910) estimated the accuracy of such observations to be about 90%, Sverdrup and Munk (1947) had different thoughts on the matter. According to Sverdrup and Munk, the visually observed "average" wave heights more closely correspond to the

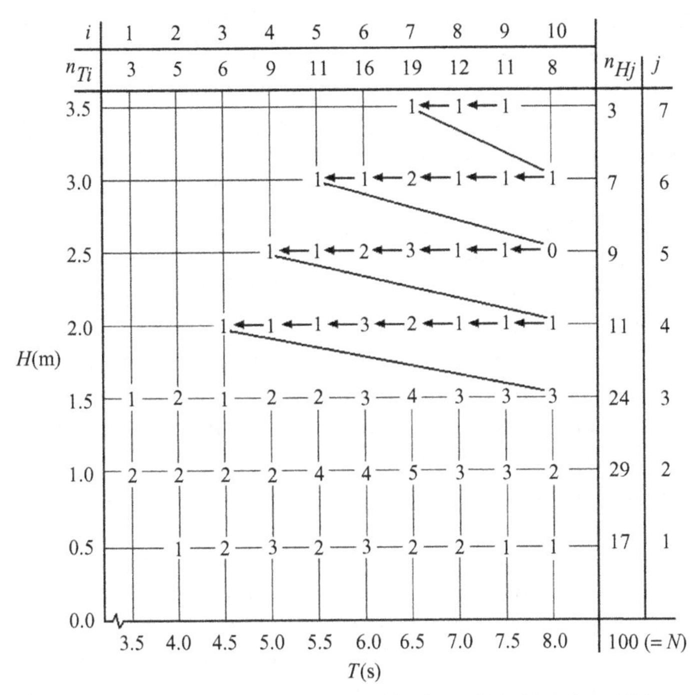

Figure 5.5. *Method for Determining the One-Third Highest Waves.* By convention, the count begins in the upper right corner of the nomograph. For the 100-wave sample, the highest one-third of the sample is approximately 33 waves. Although there are 24 1.5-m waves, only 3 on the right side are used in the count by our convention.

average height of the one-third highest waves. They reason that the visual observer is selective in that one tends to neglect the smaller waves. Again, because so much of the wave height data available at the time of their writing resulted from visual observations, Sverdrup and Munk suggested that the average height of the one-third highest waves be a standard of measure, and called that wave height the *significant wave height.* The average period of the one-third highest waves is called the *significant wave period.*

Consider again the wave height and period data of Figure 5.2. To determine the significant wave height and significant wave period, we must first identify the highest one-third waves in our data set. Because the sample (N) for the data in Figure 5.2 is 100, the highest 33 waves are used. By convention, our count of the highest waves begins with the data point in the upper right corner of the nomograph, as illustrated in Figure 5.5. The indices corresponding to the data are $i = 9$ and $j = 7$. We count from right to left, and terminate our count after counting 33 waves, that is, we terminate the count for index values of $i = 10$ and $j = 3$. The significant wave height and period values for the data of Figures 5.2 and 5.5 are given in Example 5.4.

EXAMPLE 5.4: SIGNIFICANT WAVE PROPERTIES The highest one-third wave heights for the data in Figure 5.2 are identified in Figure 5.5. The average height of these waves is the significant wave height, whereas the average period corresponding to the one-third highest wave is the significant wave period. The respective approximate values for these are

$$H_s = 2.44\,\text{m}, \qquad T_s = 6.58\,\text{sec}$$

The term "approximate" is used here as 33/100 is approximately 1/3. By comparing our results with the average wave height and period values of Example 5.3, we see that the significant wave height is about 1.63 times the average height, whereas the average period is about 1.07 times the average period. For these data then, the significant and average wave periods are approximately equal.

To avoid "hand and eye" measurements of wave data, as described in this section, we seek similarities between our measured data and well-established statistical properties of continuous wave data. In other words, after collecting as much wave data as we can, we try to fit established formulas and curves to the data to generalize our results. For example, if the sea is assumed to be ergodic (having statistical properties that are invariant in time and space), then the statistically averaged wave periods and heights at a specific site in the ocean should be the same as those at other sites. When we take into account the generation of these waves by the wind, then we must qualify the ergodic hypothesis by specifying both the wind speed and duration of the storm generating the seas. That is, for a given wind condition (speed and duration) at a point, the statistical averages of the wave heights and periods will be the same as those measured at some other time and place in the ocean having the same wind conditions. Wave properties that are invariant in time are said to be *statistically stationary,* whereas those that are invariant in space are said to be *statistically homogeneous.* We assume that the waves are ergodic in our discussions.

5.3 Continuous Probability Distributions

As stated in the previous section, it is not practical to measure wave properties at every site of interest in the open ocean. For this reason, we try to match our finite statistical wave height and period data with some widely used expressions for either the probability of occurrence or the probability density. These established expressions are continuous functions of the property in question. This means that the sample (N) of random data is now considered to be infinite. With this assumption, we can express eqs. 5.9, 5.10, 5.11, and 5.13 as integrals by passing to the limit as $N \to \infty$. The respective results are shown in the following.

A. Average Wave Period and Wave Height

For continuous wave data, the summations in eqs. 5.9a and 5.9b evolve into integral expressions of the average wave period and mean wave period, respectively. The mathematical expression for the average wave period for the period range of $T_i < T \leq T_I$ is

$$T_{iI} = \frac{\int_{T_i}^{T_I} p(T)TdT}{\int_{T_i}^{T_I} p(T)dT} \tag{5.14a}$$

where $p(T)$ is the period probability density function. When the lower limits of the integrals of eq. 5.14a are 0 and the upper limits are ∞, the resulting expression is that of the average wave period, that is,

$$T_{avg} = \int_0^\infty p(T)TdT \tag{5.14b}$$

Similarly, the average wave height over the wave height range of $H_j < H \leq H_J$ is

$$T_{jJ} = \frac{\int_{H_j}^{H_J} p(H)HdH}{\int_{H_j}^{H_J} p(h)dH} \tag{5.15a}$$

where $p(H)$ is the probability density of the wave heights. Over the semi-infinite range $0 < H \leq \infty$, eq. 5.15a becomes

$$H_{avg} = \int_0^\infty p(H)HdH \tag{5.15b}$$

which is the expression for the mean wave height.

The mathematical expressions of the probability density functions in eqs. 5.14 and 5.15 must be specified. Several of these expressions are introduced and discussed in the next three sections.

B. Mean-Square Wave Height

The result of passing to the limit of the sum defining the mean-square wave height in eq. 5.11 is

$$\overline{H^2} = \int_0^\infty p(H)H^2dH \tag{5.16}$$

C. Variance of the Wave Heights

The variance for a continuous probability distribution of wave heights from eq. 5.13 is

$$\overline{(H - H_{avg})^2} = \int_0^\infty p(H)(H - H_{avg})^2dH \tag{5.17}$$

where, again, the expression for the wave height probability function, $p(H)$, must be specified. This is done later in this chapter.

D. Significant Wave Height

Using the average wave height expression in eq. 5.15a, the significant wave height for an infinite sample of wave heights can be determined by performing the integration from the smallest wave height of the highest one-third heights. Represent this wave height by $H_{33\%}$. The expression for the significant wave height is then

$$H_s = \frac{\int_{H_{33\%}}^{\infty} p(H) H dH}{\int_{H_{33\%}}^{\infty} p(H) dH} \tag{5.18}$$

It is shown later in this chapter how one determines the value of $H_{33\%}$ for a data sample.

Our task is to find suitable probability distributions for the wave height data. Of special interest to oceanographers and ocean engineers is the Rayleigh probability distribution for the wave heights. The use of this distribution is of particular value in determining wave loads of fixed and floating structures in the open ocean.

5.4 Rayleigh Probability Distribution of Wave Heights

The reader is again reminded that the probability distribution is a plot of the probability density function. There are a number of probability density functions that have resulted from statistical analyses of various phenomena. As previously mentioned, the Rayleigh probability density function (see Strutt, 1880) resulted from a study of the random vibrations created by superimposing linear, monochromatic vibrations having the same frequency but differing in phase, the phase angle being randomly chosen. In this section, the probability density of Lord Rayleigh (J. W. Strutt) is used to represent the statistical distributions of wave heights in random seas. The Rayleigh distribution applied to ocean waves is discussed in the classical paper by Longuet-Higgins (1952).

For wave heights measured in the open ocean, the plot of the probability density function (probability distribution) was found to resemble the *Rayleigh probability distribution*. The expression of the Rayleigh probability density function for wave heights is

$$p(H) = \frac{2H}{H_{rms}^2} e^{-\left(\frac{H}{H_{rms}}\right)^2} \tag{5.19}$$

where H_{rms} is the root-mean-square wave height, defined by eq. 5.12. Note that the value of H_{rms} is considered to be a measured quantity.

As written in eq. 5.6, the probability density function is the derivative of the probability. Hence, the expression for the *Rayleigh probability of occurrence* can be obtained by integrating eq. 5.19 between two wave height limits. That is, the probability that a measured wave height (H) will be between $H = H_j$ and $H = H_J$ is obtained by integrating eq. 5.19 between these limits. The resulting expression for the probability of occurrence is

$$P(H_j < H \leq H_J) = \int_{H_j}^{H_J} p(H) dH = e^{-\left(\frac{H_j}{H_{rms}}\right)^2} - e^{-\left(\frac{H_J}{H_{rms}}\right)^2} \tag{5.20}$$

The application of the Rayleigh probability relationships is demonstrated in the following example.

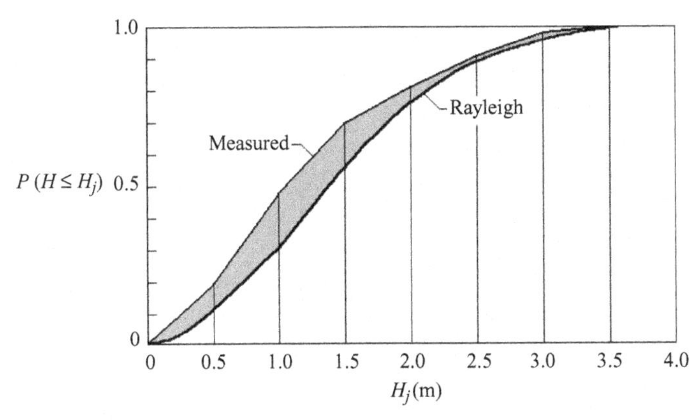

Figure 5.6. *Cummulative Frequency of Occurrence for the Data in Figure 5.2.* The Rayleigh probability formula in eq. 5.20 is used to obtain the theoretical curve.

EXAMPLE 5.5: RAYLEIGH PROBABILITY DENSITY AND PROBABILITY FUNCTIONS To apply the probability density expression of eq. 5.19 and the expression for the probability of occurrence of eq. 5.20 to the data in Figure 5.2, all that is needed is the value for the root-mean-square wave height, the value of which is $H_{rms} = 1.69$ m from Example 5.3. The expression for the Rayleigh probability density function is then

$$p(H) = 0.700He^{-0.350H^2}, \qquad \text{m}^{-1}$$

The corresponding Rayleigh cumulative probability of occurrence (the probability that H lies between 0 and H_J) is

$$P(0 < H \leq H_J) = \int_0^{H_J} p(H)dH = 1 - e^{-0.350H_J^2}$$

This is also called the *probability of non-exceedance* by some wave analysts. Results obtained from the application of this probability expression to the data in Figure 5.3 are presented in Figure 5.6. The probability distribution, $p(H)$, is presented in Figure 5.7 with the values found in Figure 5.4.

A. Average Wave Height

The combination of the Rayleigh probability function of eq. 5.19 and the generic equation for the average wave height of eq. 5.15a yields the following expression for the average wave height in $H_j < H \leq H_J$:

$$H_{jJ} = \frac{H_j e^{-\left(\frac{H_j}{H_{rms}}\right)^2} - H_J e^{-\left(\frac{H_J}{H_{rms}}\right)^2} + H_{rms}\frac{\sqrt{\pi}}{2}\left[\text{erf}\left(\frac{H_J}{H_{rms}}\right) - \text{erf}\left(\frac{H_j}{H_{rms}}\right)\right]}{P(H_j < H \leq H_J)} \qquad (5.21a)$$

where erf( ) is called the error function. Values of this function can be obtained from the book edited by Abramowitz and Stegun (1965). From that reference, a rather accurate approximate formula is found, and is presented following eq. 5.34. In the denominator of eq. 5.21a is the probability of occurrence, the expression for which is in eq. 5.20.

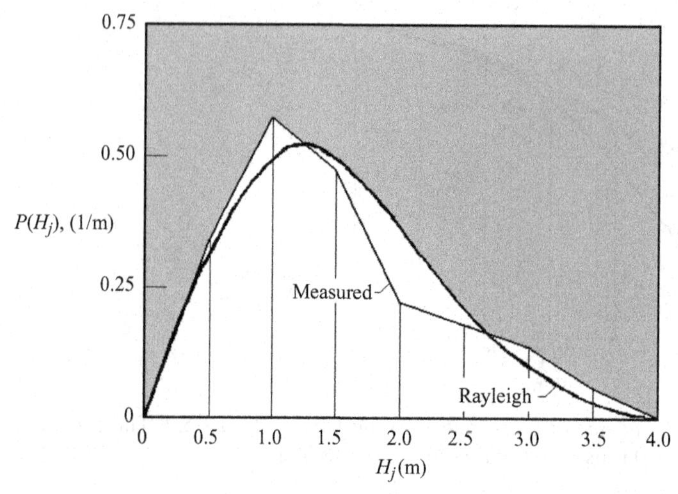

Figure 5.7. *Probability Density for the Data in Figure 5.2.* The Rayleigh curve is obtained from eq. 5.19.

When the expression in eq. 5.21a is applied to the entire range of wave heights, that is, from $H_j = 0$ to $H_J = \infty$, the expression for the average wave height corresponding to a Rayleigh wave height distribution results, that is,

$$H_{avg} = \frac{\sqrt{\pi}}{2} H_{rms} \tag{5.21b}$$

As before, the root-mean-square wave height is normally considered to be a measured quantity in these equations.

B. Probability of Exceedance

The probability that a measured wave height will be greater than, say, H_K is called the *probability of exceedance.* From eq. 5.20, this probability is found by integrating from H_K to ∞. That probability is then expressed by

$$P(H_K < H \leq \infty) = e^{-\left(\frac{H_K}{H_{rms}}\right)^2} = f \tag{5.22}$$

where f is the fraction of waves having heights greater than H_K. One application of the formula is in the determination of the significant wave height, discussed in Section 5.2D. From the expression in eq. 5.22, we can determine the value of H_K for the percentage of waves in question. The expression for that wave height is

$$H_K = H_{rms}\sqrt{\ln\left(\frac{1}{f}\right)} \tag{5.23}$$

C. Significant Wave Height

The *significant wave height* is defined in Section 5.2D as the average height of the one-third highest waves, and is mathematically represented by eq. 5.1 for a continuous probability distribution of wave heights. To determine the value of the significant wave height, we need to know the smallest wave height value of the highest one-third waves, which is also the lower limit of the integral in eq. 5.18. This lower limit is obtained from eq. 5.23, where the fraction is $f = 1/3$ and the subscript is $K =$ 33%. The result is $H_{33\%} = 1.048H_{rms}$. Replace the lower limits of the integrals of eq. 5.18 to obtain the expression for the significant wave height for a Rayleigh wave height probability distribution. The result is

$$H_s = 1.416 H_{rms} \tag{5.24}$$

D. Extreme Wave Height

Equation 5.23 can be used to determine the *extreme wave height* of a sample of wave heights assumed to have a Rayleigh distribution. Suppose that we determine (from measurements) that there are N_M waves passing a site in the open ocean in M years. For that sample, we wish to determine the value of the height of the highest wave. In other words, our interest is in the height of the $f = 1/N_M$ wave because there is only one wave that can be the highest. That extreme (maximum) wave height is obtained from eq. 5.23 by using this frequency value, that is,

$$H_{\max} = H_{rms}\sqrt{\ln(N_M)} = \frac{2}{\sqrt{\pi}} H_{avg}\sqrt{\ln(N_M)} \tag{5.25}$$

Here, the results of eq. 5.21b have been incorporated. For the data in Figure 5.2, $N_M = 100$ and $H_{rms} = 1.69$ m from Example 5.3. These values in eq. 5.25 yield $H_{\max} = 3.63$ m. This extreme value is close to the maximum wave height in the Figure 5.2 data, which is 3.5 m.

EXAMPLE 5.6: SIGNIFICANT AND EXTREME WAVE HEIGHTS At a site in the open ocean, a wave staff is placed to determine the statistical wave properties. Measurements are taken for 20 min each hour, 20 hours each day for 365 days. In eq. 5.25, then, $M = 1$. The average measured wave properties are those found in Example 5.3. From that example, we find

$$H_{avg} = 1.5\,\text{m}, \quad H_{rms} = 1.69\,\text{m}, \quad T_{avg} = 6.13\,\text{sec}$$

The significant wave height expression of eq. 5.24 then yields $H_s = 2.39$ m. To determine the extreme wave height, the total sample of waves passing the wave staff in 1 year must be determined. To make that determination, simply divide the total number of seconds in 1 year (365 days × 24 hours × 3600 sec = 3.1536×10^7 sec) by the average wave period. The result is $N_1 \simeq 5.14 \times 10^6$. Substitute the wave property values and the N_1 value into eq. 5.25 to obtain $H_{\max} = 6.63$ m.

The reader can see that much information can be obtained if the wave height probability distribution resembles that predicted by a Rayleigh probability density function. This distribution is most valid in the open ocean. When in a coastal area, the wave height distribution might be somewhat different than that predicted by eq. 5.19. In that situation, a more general probability density function is desirable. One such distribution is that of Weibull (1951), which is discussed in the next section.

5.5 Weibull Probability Distribution of Wave Heights

The Weibull probability density function is a parametric relationship that has wide application. In fact, most of the commonly used probability density functions can be obtained from Weibull's equation by the proper choice of the parameters in that equation. In his 1951 paper, W. Weibull states that any cumulative probability of occurrence of a variable (say, the wave height, H) being less than or equal to a value (say, H_J.) can be represented by

$$P(0 < H \leq H_J) = 1 - e^{-F(H_J)} \tag{5.26}$$

where $F(H_J)$ is a function that is non-negative, non-decreasing, and vanishing at some value, say, A. Weibull (1951) states that the "most simple function" satisfying the conditions is

$$F(H_J) = \frac{(H_J - \mathrm{A})^{\mathrm{m}}}{\mathrm{B}^{\mathrm{m}}} \tag{5.27}$$

With this relationship, the expression in eq. 5.26 is a *three-parameter Weibull probability,* where the parameters are A, B, and m. It must be noted that these expressions have no theoretical basis. The probability density function corresponding to the probability expression in eq. 5.26 (coupled with the function in eq. 5.27) is found by simply taking the derivative of the probability with respect to H_J. The result is

$$p(H_J) = \mathrm{m}\frac{(H_J - \mathrm{A})^{\mathrm{m}-1}}{\mathrm{B}^{\mathrm{m}}} e^{-\frac{(H_J-\mathrm{A})^{\mathrm{m}}}{\mathrm{B}^{\mathrm{m}}}} \tag{5.28}$$

By comparing the expressions in eqs. 5.19 and 5.28, we see that the Rayleigh probability density function corresponds to the parametric values of $\mathrm{m} = 2$ and $\mathrm{A} = 0$, where $\beta = H_{rms}$.

The average value of the variable is obtained from the combination of eqs. 5.15b and 5.28. The resulting expression is

$$H_{avg} = \mathrm{B}\Gamma\left(\frac{\mathrm{m}+1}{\mathrm{m}}\right) + \mathrm{A} \tag{5.29}$$

where $\Gamma(\,)$ is a gamma function, the values of which can be found in the book edited by Abramowitz and Stegun (1965) for various values of the parameter m.

By combining eqs. 5.16 and 5.28, we obtain both the mean-square and root-mean-square of the variable. The mean-square expression is

$$\begin{aligned}\overline{H^2} &= \mathrm{B}^2\Gamma\left(\frac{\mathrm{m}+2}{\mathrm{m}}\right) + 2\mathrm{AB}\Gamma\left(\frac{\mathrm{m}+1}{\mathrm{m}}\right) + \mathrm{A}^2 \\ &= H_{rms}^2\end{aligned} \tag{5.30}$$

In general, the applications of the Weibull equation to the wave height and period result in two-parameter expressions because the probability of occurrence for the zero values of H and T is zero. For the free-surface displacement (η), the three-parameter Weibull equation applies because this variable can have zero and negative values.

Excellent discussions of the application of the Weibull probability distribution function to ocean waves can be found in the texts by Tucker (1991) and Gren (1992). Forristall (1978) applied a two-parameter ($\mathrm{A} = 0$) Weibull probability density function to storm wave data, as reported by Tucker (1991). Because the root-mean-square wave height is normally a measured property, the value of m can be obtained from eq. 5.30 for a two-parameter expression. This is illustrated in the following example.

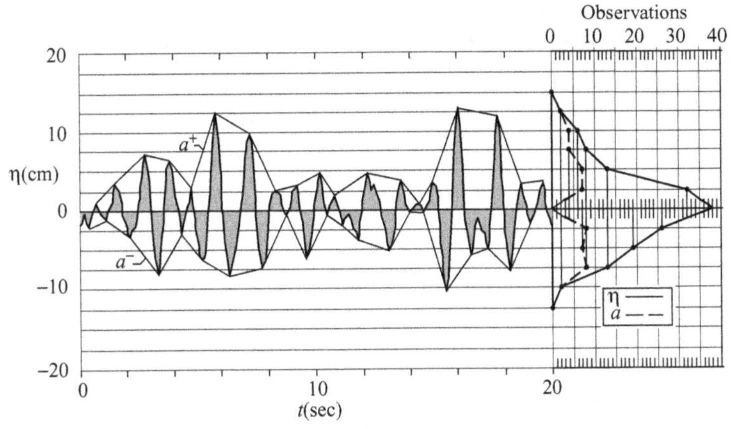

Figure 5.8. *Amplitude (a^+, a^-) and Displacement (η) Distributions for the Wave Trace in Figure 5.1.*

EXAMPLE 5.7: WEIBULL PROBABILITY DENSITY AND PROBABILITY FUNCTIONS
Apply the two-parameter (A = 0) Weibull probability and probability function to the wave height data of Figure 5.2. The combination of the expression of eqs. 5.12 and 5.30 yields the expression for the root-mean-square wave height. That expression, combined with the value of H_{rms} from Example 5.3, yields

$$H_{rms} = \sqrt{\overline{H^2}} = \mathrm{B}\sqrt{\Gamma\left(\frac{\mathrm{m}+2}{\mathrm{m}}\right)} = 1.69\,\mathrm{m}$$

In a similar manner, combine the average wave height value of Example 5.3 with the two-parameter expression of eq. 5.29 to obtain

$$H_{avg} = \mathrm{B}\Gamma\left(\frac{\mathrm{m}+1}{\mathrm{m}}\right) = 1.50\,\mathrm{m}$$

The simultaneous solution of these equations, with the help of the gamma function tables in Abramowitz and Stegun (1965), yields $m \simeq 2.0$ and $\mathrm{B} \simeq 1.69$ m. These values correspond to the Rayleigh values. Hence, for the data in Figure 5.2, the Weibull probability density function and probability function are the same as those in Example 5.5.

The equations of Weibull can be used to analyze both experimental and at-sea data. Within the past several decades, the Weibull equations have found increasing favor with physical oceanographers and ocean engineers.

5.6 The Gaussian-Rayleigh Sea

Let us again examine the experimental time trace of the free-surface displacement (η) in Figure 5.1a. We are now interested in the distributions of both the time-dependent free-surface displacement and the time-dependent positive and negative maxima of the free-surface displacement. These are illustrated in Figure 5.8. In that figure, we see that the trace in Figure 5.1a is enveloped by dashed lines drawn through the positive maxima (denoted by a^+) and negative minima (denoted by a^-). Following Longuet-Higgins (1952), these maxima and minima are referred to collectively as *amplitudes*, and the curves are called *amplitude curves.* In addition,

Table 5.2. *Experimental data for probability distributions*

η (cm)	Observations	a^+ (cm)	Observations	a^- (cm)	Observations
−10.0	2	–	–	−10.0	2
−7.5	8	–	–	−7.5	8
−5.0	19	–	–	−5.0	7
−2.5	26	–	–	−2.5	8
0.0	38	0.0	0	0.0	0
+2.5	32	+2.5	7	–	–
+5.0	13	+5.0	7	–	–
+7.5	8	+7.5	4	–	–
+10.0	6	+10.0	4	–	–
+12.5	2	+2.5	2	–	–

there are lines passing through these curves that are parallel to the time axis. These parallel lines intersect the solid $\eta(t)$ curve and the dashed $a^+(t)$ and $a^-(t)$ curves. To the right of these curves are data plots of the number of intersections. To illustrate, for the parallel line corresponding to an η-value of 1 cm, there are 13 intersections with the η-curve and 7 intersections with the a^+-curve. The data gathered in this manner are presented in Table 5.2. The data plots at the right side of Figure 5.8 are used to determine the probability distributions of the time-dependent data. These data are also presented in Figure 5.9 along with theoretical probability distributions. To approximate the probability density of the free-surface displacement, the *Gaussian probability distribution* is shown, whereas Rayleigh probability distributions are used to approximate the probability densities of the amplitude data.

Assuming that the displacement of the free surface is a Gaussian process, the following probability density function is used:

$$p(\eta) = \frac{1}{\eta_{rms}\sqrt{2\pi}} e^{-\frac{1}{2}\left(\frac{\eta}{\eta_{rms}}\right)^2} \tag{5.31}$$

where η_{rms} is considered to be a measured quantity. For the time segment shown in Figure 5.8, the root-mean-square free-surface displacement is

$$\eta_{rms} = \sqrt{\sum_{j=1}^{j=10} \frac{n_j \eta_j^2}{N_\eta}} \tag{5.32}$$

the value of which is 4.54 cm for the data in Table 5.2. In both Table 5.2 and eq. 5.32, n_j is the number of intersection corresponding to the value of η_j. The total number (sample) of these intersections is N_η. For the data in Table 5.2, that number is 154. The Gaussian distribution of the free-surface displacement resulting from the combination of the last two equations is presented in Figure 5.9a.

By replacing the wave height terms in eq. 5.19 and the free-surface displacement in eq. 5.32 by the amplitudes (a^+ or a^-), one obtains the expression for the Rayleigh probability densities of the free-surface maxima and minima (amplitudes). That is, for either the maxima or minima, we can write

$$p(a) = \frac{2a}{a_{rms}^2} e^{-\left(\frac{a}{a_{rms}}\right)^2} \tag{5.33}$$

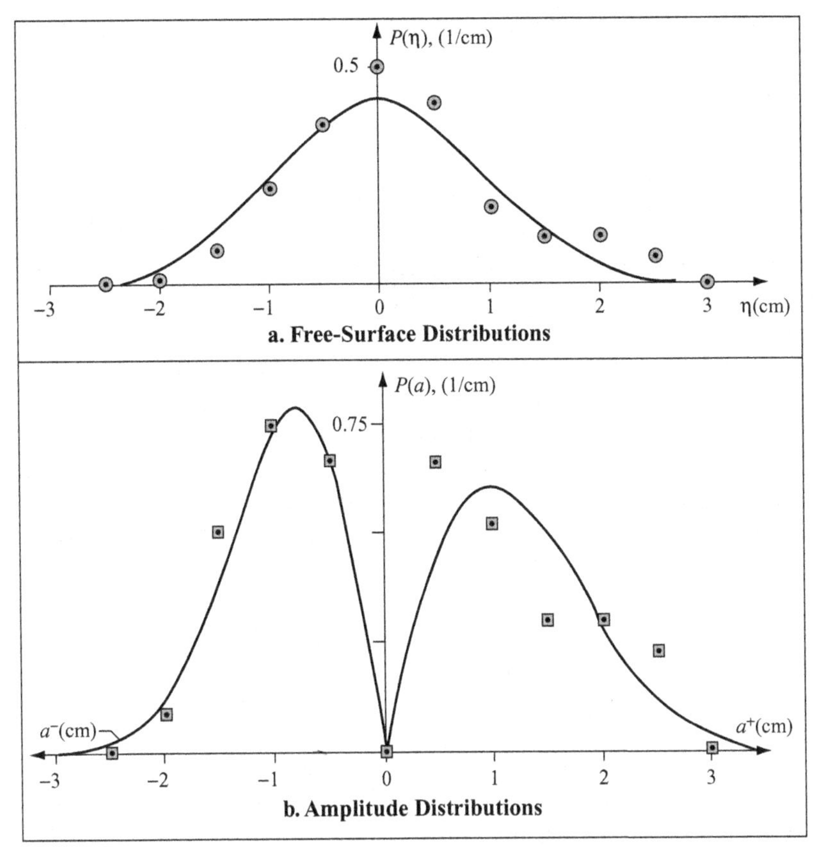

Figure 5.9. *Probability Distributions of the Free-Surface Displacement and Wave Amplitudes.* In (a), the Gaussian distribution obtained from eq. 5.31 is presented with the discrete free-surface data. In (b), results from the Rayleigh formula in eq. 5.33 are presented with the discrete amplitude values.

where the respective root-mean-square values of the maximum and minimum amplitudes are $a^{+}_{rms} \simeq 6.94$ cm and $a^{-}_{rms} \simeq 5.93$ cm. The Rayleigh amplitude distributions, obtained by substituting these values into eq. 5.33, are presented in Figure 5.9b. In Section 5.3, various formulas resulting from the assumption of a Rayleigh distribution of wave heights are presented. Of particular interest are the probability of occurrence of eq. 5.20, the averaging formula of eq. 5.21, and the extreme value expression of eq. 5.25. These formulas can be applied to the amplitude curves, again by replacing the wave height by the amplitude, as in eq. 5.33.

For a Gaussian distribution of the time-dependent free surface, the average value is zero. The various probabilities can be obtained by integrating the expression in eq. 5.31 over any two values. The expression for the probability of occurrence is then

$$P(\eta_j < \eta \leq \eta_J) = \int_{\eta_j}^{\eta_J} p(\eta)d\eta = \frac{1}{2}\left[\mathrm{erf}\left(\frac{2}{\sqrt{2}}\frac{\eta_J}{\eta_{rms}}\right) - \mathrm{erf}\left(\frac{1}{\sqrt{2}}\frac{\eta_j}{\eta_{rms}}\right)\right] \tag{5.34}$$

In eq. 5.34, the terms on the right involve error functions, defined as

$$\operatorname{erf}(Z) = \frac{2}{\sqrt{\pi}} \int_0^Z e^{-\xi^2} d\zeta$$

$$\simeq 1 - \frac{1}{(1 + 0.278393Z + 0.230389Z^2 + 0.000972Z^3 + 0.078108Z^4)^4}$$

The polynomial approximation of the error function is obtained from Abramowitz and Stegun (1965), and is valid for $0 \leq Z < \infty$, with an error that is less than 0.0005. Using that approximation, we see that $\operatorname{erf}(0) = 0$ and $\operatorname{erf}(\infty) = 1$. In addition, the error function is an odd function, so $\operatorname{erf}(-Z) = -\operatorname{erf}(Z)$. Hence, the probability that $-\infty < \eta \leq \infty$ is unity, as expected.

The agreement between the measured values and the theoretical curves in Figure 5.9 is good, but not perfect. For perfect agreement, the wave system would need to be Gaussian-Rayleigh or simply Gaussian. As demonstrated by Longuet-Higgins (1952) and discussed by Buckley et al. (1984) and many others, a Gaussian sea results from the superposition of linear waves. If there are nonlinear components, as in Figures 5.1a, then the distribution of the time-dependent free surface will be non-Gaussian. For the time-dependent amplitude curves, the probability distributions in Figure 5.9 would be better represented by two-parameter Weibull (1951) probability distribution functions. The method of evaluation of the parameters is the same as that in Example 5.7.

Let us now determine a relationship between the root-mean-square free-surface displacement and the root-mean-square wave height. To accomplish this, assume that the sea is composed of a large number (N) of superimposed linear waves. The sea is then Gaussian and is referred to as an irregular sea. In this sea, the free-surface of the nth wave is

$$\eta_n = \frac{H_n}{2} \cos(k_n x - \omega_n t)$$

from eq. 3.24. The spacial average of the square of this component wave over the surface area $\lambda_n b$ can be expressed as

$$\langle \eta^2 \rangle_n = \frac{H_n^2}{8}$$

where the brackets $\langle\rangle$ indicate a spatial average. For N waves, the average of this expression over the entire ensemble is the variance of the free-surface displacement, that is,

$$\overline{\eta^2} \equiv \eta_{rms}^2 = \sum_{n=1}^{N} \frac{\langle \eta^2 \rangle_n}{N} = \sum_{n=1}^{N} \frac{H_n^2}{8N} = \frac{\overline{H^2}}{8} = \frac{H_{rms}^2}{8} \tag{5.35a}$$

From this expression, we obtain the root-mean-square, which is

$$\eta_{rms} = \frac{H_{rms}}{2\sqrt{2}} \simeq 0.3536 H_{rms} \tag{5.35b}$$

where the value of H_{rms} is known from measurements.

EXAMPLE 5.8: ROOT-MEAN-SQUARE WAVE PROPERTIES It has been demonstrated that the data in Table 5.2 are not quite Gaussian, and therefore some nonlinear wave components are present. We shall determine if the Gaussian assumption leading to eq. 5.35 is satisfactory for these data. Using the amplitude (a^+ and a^-) data in Table 5.2, the root-mean-square value of the wave height is found to be

$$H_{rms} = \sqrt{\sum_{j=1}^{j=N} \frac{n_j 2(a_j^2)}{N}} = \sqrt{141.35} \simeq 11.9\,\text{cm}$$

where, for a given amplitude value, $H^{\pm} = 2a^{\pm}$. There are 49 measured wave heights in Table 5.2, so the sample or ensemble (N_a) equals 49.

The application of eq. 5.32 to the data in Table 5.2 yields $\eta_{rms} = 4.54$ cm, which is approximately 0.382 H_{rms}. A comparison of this result with the expression in eq. 5.35b shows a difference in the two expressions of approximately 7%. Hence, for the data in Table 5.2, the Gaussian assumption leading to the expression in eq. 5.35 is satisfactory.

5.7 Wave Spectral Density

This section is devoted to the relationship between the energies of both wind-generated seas and long-term seas and the statistical properties of the wave fields. *Long-term seas* are those for which data are recorded and averaged over time periods of months or years. Wind-generated seas have received far more attention than long-term seas. The reason is that extreme wave heights in wind-generated seas are used in the probabilistic design of ocean structures for survivability. The phenomena associated with wind-wave generation is rather complex. In Chapters 3 and 4, the waves are treated as two-dimensional phenomena, and the heights of the monochromatic waves are assumed to be independent of the wave period. This is not the case in a *wind-generated sea.* The wind eddies that convect in the boundary layer adjacent to the free surface each have energies that are associated with their convection velocities. Pressure variations on the free surface result from the passage of these eddies. Convecting pressure pulses can deform the compliant free surface, thereby creating the first waves, called *ripples.* There is a transverse curvature of the first waves created that is partially the result of the increasing transverse curvature of the rotational axes of the convecting wind eddies, as illustrated in Figure 5.10. The interaction of such eddies and compliant surfaces is discussed by McCormick and Mouring (1995). The pressure-induced deformations of the free surface then have definite geometrical patterns. These geometrical patterns change due to wind-wave interactions in which the wind stress acts on the free surface, and wind wakes are formed on the leeward side of the waves, as illustrated in Figure 3.1d. The generation of wind-generated seas is also discussed in Section 1.1. The classification of wind-generated seas is discussed in Section 1.2.

The literature for both wind-generated seas and long-term seas is extensive. The first definitive analysis of wind-generated waves is attributed to Phillips (1957). Discussions of this analytical method are found in the books by Kinsman (1965) and Phillips (1966). More recent analyses and discussion of wind-generated waves are given by Donelan (1990), in the book edited by Le Méhauté and Hanes (1990), and by Tucker and Pitt (2001). The subject of wind stress on waves is discussed by Wu (1980). The statistics of long-term seas are discussed by Isaacson and MacKenzie

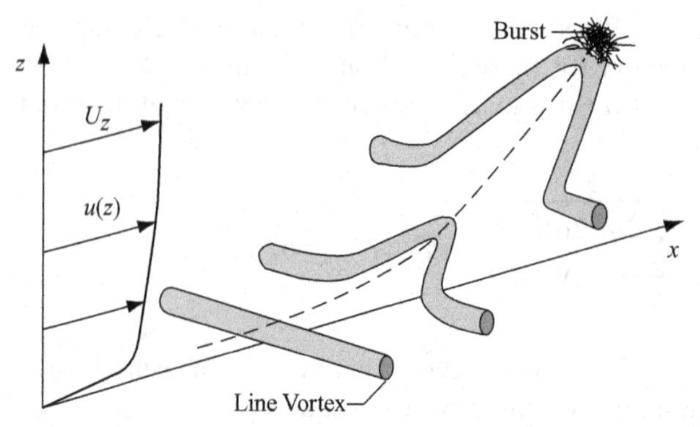

Figure 5.10. *Sketch of Boundary-Layer Vortex Distortion and Migration.* The vortices form at various heights above the calm-water surface. Because of the distortion in the line vortex, additional energy is absorbed at points along the line, and the vortex grows (strengthens) at these points. As the vortex grows, it is subjected to greater wind speeds and lower wind pressures. The changing wind properties further distort the vortex and cause it to grow and migrate in both the horizontal and vertical directions over the calm-water surface. The time-scale for the phenomenon depends on the height of the initial vortex formation because that scale depends on the vertical velocity distribution of the wind.

(1981), Hogben (1990), and McCormick (1998b, 1999). Wind-wave spectra are analyzed in Section 5.8, whereas long-term spectra are analyzed in Section 5.9.

In the following subsections, general discussions of measured wave spectra and empirical spectral expressions are presented.

A. Point Spectra from Discrete Wave Data

To analyze the energy content of a random sea from data obtained from point measurements, we first assume that the sea is composed of a large number of linear waves that are superimposed upon each other. Because of this assumption, let us refer to this assumed sea as irregular rather than random. For a component wave of an irregular sea, eq. 3.67 is used to obtain the energy per free-surface area (called the *energy intensity,* $\underline{E}$) per unit weight of water, that is,

$$\underline{E} \equiv \frac{(E/\lambda b)}{\rho g} = \frac{\mathrm{E}}{\rho g} = \frac{H^2}{8} \tag{5.36}$$

where λb is the surface area and ρg is the weight-density of the water.

The expression in eq. 5.36 can be applied to the data in Figure 5.2 to determine the average energy intensity distribution over the wave periods. For a period T_i in that figure, we can write

$$\mathrm{E}(T_i) = \rho g \sum_{j_{(i)\min}}^{j_{(i)\max}} \frac{n_{j_{(i)}} H_{j_{(i)}}^2}{8N} = \rho g S(T_i)\Delta T \tag{5.37}$$

where ΔT is the period increment (equal to 0.5 sec in Figure 5.2), $j_{(i)}$ is the wave height index corresponding to the ith period, and N is the total sample (equal to 100 for the Figure 5.2 data). To illustrate, for $i = 2$ in Figure 5.2, $T_2 = 4.0$ sec, whereas $j_{(2)\min} = 1$ and $j_{(2)\max} = 3$. The function $S(T_i)$ is called the *wave spectral density*. The purpose of this function is to show the period dependence of the average energy intensity of a random sea. From eq. 5.37, the mathematical expression of the spectral density for discrete data is

$$S(T_i) = \sum_{j(i)\min}^{j(i)\max} \frac{n_{j(i)} H_{j(i)}^2}{8N\Delta T} \tag{5.38}$$

The average energy intensity of all of the waves in the sample is simply the sum of the terms in eq. 5.37 over all of the measured periods, that is,

$$\sum_{i=1}^{i_{\max}} \mathrm{E}(T_i) = \rho g \sum_{i=1}^{i_{\max}} \sum_{j(i)\min}^{j(i)\max} \frac{n_{j(i)} H_{j(i)}^2}{8N} = \rho g \sum_{j=1}^{j_{\max}} \frac{n_j H_j^2}{8N}$$
$$= \rho g \sum_{i=1}^{i_{\max}} S(T_i)\Delta T \tag{5.39}$$

Note that the double summation of the second term is equivalent to the single summation of the third term as both sum over the entire sample. We can also rewrite the summation of the third term in terms of the mean-square wave height and the probability density function using the results of eq. 5.11. The result is

$$\overline{H^2} = H_{rms}^2 = \sum_{j=1}^{j_{\max}} p(H_j) H_j^2 \Delta H = 8 \sum_{i=1}^{i_{\max}} S(T_i)\Delta T \tag{5.40}$$

The applications of the spectral density formula of eq. 5.38 and the mean-square wave height expression of eq. 5.40 are illustrated in the following example by using the data in Figure 5.2.

EXAMPLE 5.9: WAVE SPECTRAL DENSITY FROM DISCRETE WAVE DATA The application of the spectral density expression of eq. 5.38 to the data in Figure 5.2 yields the following results:

$$S(T_1) = S(3.5\,\text{sec}) = [2(1.0\,\text{m})^2 + 1(1.5\,\text{m})^2]/[8(100)0.5\,\text{sec}] \simeq 0.0106\,\text{m}^2/\text{s}$$
$$S(T_2) = S(4.0\,\text{s}) \simeq 0.0169\,\text{m}^2/\text{s}$$
$$S(T_3) = S(4.5\,\text{s}) \simeq 0.0219\,\text{m}^2/\text{s}$$
$$S(T_4) = S(5.0\,\text{s}) \simeq 0.0438\,\text{m}^2/\text{s}$$
$$S(T_5) = S(5.5\,\text{s}) \simeq 0.0706\,\text{m}^2/\text{s}$$
$$S(T_6) = S(6.0\,\text{s}) \simeq 0.1125\,\text{m}^2/\text{s}$$
$$S(T_7) = S(6.5\,\text{s}) \simeq 0.1788\,\text{m}^2/\text{s}$$
$$S(T_8) = S(7.0\,\text{s}) \simeq 0.1044\,\text{m}^2/\text{s}$$
$$S(T_9) = S(7.5\,\text{s}) \simeq 0.1038\,\text{m}^2/\text{s}$$
$$S(T_{10}) = S(8.0\,\text{s}) \simeq 0.0550\,\text{m}^2/\text{s}$$

When these values are multiplied by eight times the period increment ($\Delta T = 0.5$ sec) and summed as in eq. 5.40, the result is 2.87 m^2. This value is the same as that of the mean-square wave height in Example 5.3, as expected.

The values of the spectral density are shown in Figure 5.11 with their corresponding period values. The plot of the $S(T)$ is called the *wave spectrum*. This plot represents the distribution of the energy content of the sea over a range of measured wave periods. The most energetic waves in this sample occur at a period of 6.5 sec. This period is called the *modal period*. The modal period is one to avoid in

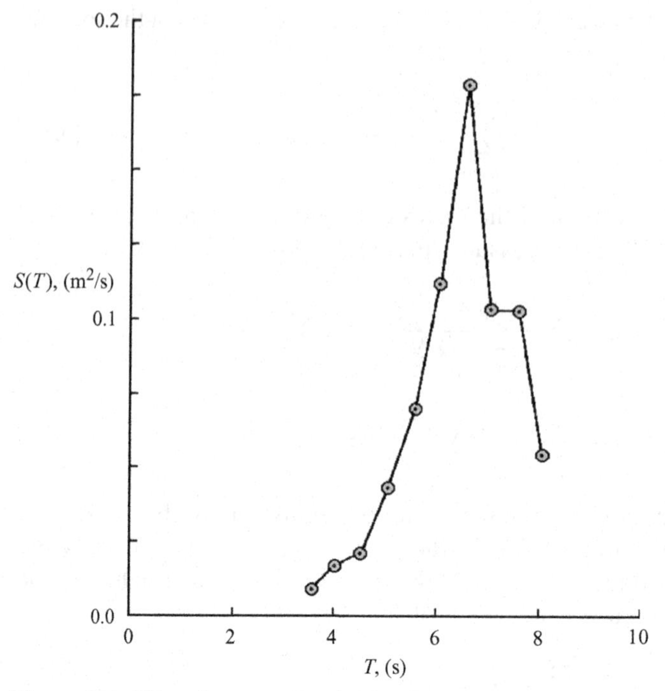

Figure 5.11. *Wave Spectrum for the Data in Figure 5.2 and Example 5.9.*

the engineering design of a compliant offshore structure. On the other hand, the design period for an ocean wave energy conversion system would be the modal period.

B. Empirical Expressions of the Point Spectral Density

An enormous amount of wave data now exist. These data have resulted from point measurements at various open-ocean and confined-water sites. Because of the abundance of the data, a number of empirical expressions for the wave spectral density have been formulated. Some of the works leading to these empirical expressions are referred to in Section 5.1.

An empirical expression of the spectral density is a continuous functional relationship. To obtain such a function, the expression for the mean-square wave height of eq. 5.40 is transformed by first passing to the limit as $\Delta T \to 0$ and $\Delta H \to 0$, and then integrating the resulting expression over the wave period from $T = 0$ to $T = \infty$. The result is

$$\overline{H^2} = H_{rms}^2 = \int_0^\infty p(H)H^2 dH = 8\overline{\eta^2} = 8\int_0^\infty S(T)dT \tag{5.41a}$$

where the results of eq. 5.35a have been incorporated. A more general form of this equation is

$$\overline{H^2} = H_{rms}^2 = C\int_0^\infty S(T)dT \tag{5.41b}$$

where the constant, C, is determined from field data. The integral is sometimes referred to as the *zero moment* because many physical oceanographers and ocean engineers refer to

$$\int_0^\infty T^m S(T)dT$$

as the *mth moment*. We shall avoid referring to these expressions as "moments" as the term could cause confusion in later chapters dealing with physical moments on fixed and floating structures.

For *fully developed seas* (seas for which the statistical properties are both uniform over the free surface and constant in time), the generic equation used by a number of investigators is

$$S(T) = AT^m e^{-BT^n} \tag{5.42}$$

where the coefficients A and B depend on the statistical wave properties. In turn, these properties depend on those associated with the wind for wind-generated seas. The wind properties are the nominal wind speed (U), the fetch (F), and the duration (t_d). The *fetch* is the length of the sea over which the winds blow, and the *duration* is the time-life of the wind event. In the empirical formulas, m and n result from curve-fitting of the data. Four wind-wave spectra that are represented by eq. 5.42 are the following: Neumann (1952), for which $m = 4$ and $n = 2$; Bretschneider (1959), for which $m = 3$ and $n = 4$; Pierson and Moskowitz (1964), for which $m = 3$ and $n = 4$; and the long-term formula of McCormick (1998b), for which $m = 6$ and $n = 7$. The choice of the exponent m is rather critical in representing the condition of the sea, as discussed by Pierson and Moskowitz (1964). The value of the exponent must be such that it is neither small enough to result in prematurely breaking-wave conditions (Phillips, 1958) nor large enough to introduce viscous dissipation (Hamada, 1964, as reported by Pierson and Moskowitz, 1964).

The spectral density expression of eq. 5.42 can be combined with eq. 5.41b to obtain a relationship for the mean-square and root-mean-square wave heights. The result is

$$\overline{H^2} = H_{rms}^2 = C\frac{A}{nB^{\frac{m+1}{n}}}\Gamma\left(\frac{m+1}{n}\right) \tag{5.43}$$

5.8 Wind-Wave Spectra

Because the formulas of Bretschneider (1959) and of Pierson and Moskowitz (1964) have the same exponent values, that version of eq. 5.42 is used herein for the sake of illustration. The Bretschneider and Pierson-Moskowitz spectra have been widely used since their original appearances in the literature. The International Ship Structures Committee (ISSC) has modified the Bretschneider formula, whereas the International Towing Tank Committee (ITTC) has modified the Pierson-Moskowitz formula. For discussions of these respective modifications, see ISSC (1967) and ITTC (1972). An additional modification of the Bretschneider spectrum was performed by Mitsuyasu (1970), according to Goda (1985, 1990). Goda refers to the modified spectrum as the *Bretschneider-Mitsuyasu spectrum*. The 1990 paper by Goda appears as an appendix in the book edited by Pilarczyk (1990). Based on measurements in the German Bight of the North Sea, Hasselmann et al. (1973) modified the Pierson-Moskowitz formula to account for limited fetch. The resulting spectral formula is called the *JONSWAP spectrum,* the acronym derived from the Joint North Sea Wave Project. Hogben (1990) and Komen et al. (1994) give excellent discussions of the various wave spectral density formulas.

The Bretschneider (1959) and Pierson-Moskowitz (1964) spectra expressions have the following form in the wave period domain:

$$S(T) = AT^3 e^{-BT^4} \tag{5.44a}$$

We can also write the spectral expressions in the wave frequency domain ($f = 1/T$), where $S(T)\mathrm{dT} = -S(f)df$. This relationship results from the fact that the energy of the wave field over the frequency domain must be equal to that over the period domain. The negative sign results from the derivative df/dT. Solving for the spectral density in the frequency domain, we then obtain

$$S(f) = -S(T)\frac{dT}{df} = T^2 S(T) = \frac{A}{f^5}e^{-\frac{B}{f^4}} \tag{5.44b}$$

The preference in this book is to work with the period-domain spectra.

It must be noted here that the wave spectral density functions of Bretschneider (1959) and of Pierson and Moskowitz (1964) are defined differently, that difference being in the value of C in eq. 5.41b. Bretschneider (1959, 1963) specifies $C = 1$, whereas Pierson and Moskowitz (1964) specify $C = 1/8$, as in eq. 5.39.

The coefficients A and B are directly related to the wave properties. The expressions for these coefficients distinguish the various wave spectral density formulas. Bretschneider (1959, 1963) obtains expressions for the coefficients by assuming Rayleigh probability distributions of the wave heights and wavelengths. The Pierson and Moskowitz (1964) coefficients are based the similarity hypothesis of Kitaigorodskii (1962) and on at-sea measurements of the wave properties. Before discussing the specific spectral expressions, we shall determine relationships for the various statistical periods.

A. Statistical Wave Periods

Up to this point, most of the discussion of random waves has been focused on statistical wave heights. With the introduction of the concept of the wave spectral density, much information concerning the statistical wave periods can be obtained. In the analyses that follow, the expressions in eq. 5.44 are used. The peak value of the period-based spectral density of eq. 5.44 is found by setting the derivative dS/dT equal to zero. The period that yields that peak value is called the *modal period*, and is expressed as

$$T_o = \left(\frac{3}{4B}\right)^{1/4} \tag{5.45}$$

This formula is important to the design engineer because it represents the wave period for which the energy of a wind-generated sea is a maximum. Sarpkaya and Isaacson (1981) note that the modal period is not that for which the frequency-based spectral density [$S(f)$ in eq. 5.44b] is a maximum. They refer to that period as the *peak period*, T_p. This period is determined by setting the derivative of the expression in eq. 5.44b equal to zero. The result is

$$T_p = \frac{1}{f_p} = \left(\frac{5}{4B}\right)^{1/4} = \left(\frac{5}{3}\right)^{1/4} T_o \tag{5.46}$$

The last equality is obtained by solving eq. 5.45 for B and combining the result with the third term in eq. 5.46.

The statistical averages of the wave period can be obtained by using the spectral expression in eq. 5.44a. The average period is found from

$$T_{avg} = \frac{\int_0^\infty TS(T)dT}{\int_0^\infty S(T)dT} = \frac{\Gamma\left(\frac{1}{4}\right)}{4B^{1/4}} \simeq \frac{3.6256}{4B^{1/4}} \tag{5.47}$$

where, again, the gamma function is discussed both by Abramowitz and Stegun (1965). In a similar manner, the expressions for the mean-square period and the root-mean-square period are obtained. The results are

$$\overline{T^2} = T_{rms}^2 = \frac{\int_0^\infty T^2 S(T)dT}{\int_0^\infty S(T)dT} = \frac{\Gamma\left(\frac{3}{2}\right)}{\sqrt{B}} = \frac{1}{2}\sqrt{\frac{\pi}{B}} \tag{5.48}$$

By eliminating B in the combination of eqs. 5.47 and 5.48, we obtain the relationship between the average period and the root-mean-square period, which is

$$T_{avg} \simeq 0.96282 T_{rms} \tag{5.49}$$

The *significant wave period*, T_s, is defined as the average period of the highest one-third waves. According to Sarpkaya and Isaacson (1981), Bretschneider (1977) proposed the following relationship between the significant period and the peak period of eq. 5.46:

$$T_s = \left(\frac{4}{5}\right)^{\frac{1}{4}} T_p \tag{5.50}$$

With this relationship, we can now relate the significant, peak, modal, mean, and root-mean-square wave periods as

$$T_s \simeq 0.946 T_p \simeq 1.075 T_o \simeq 1.104 T_{avg} \simeq 1.063 T_{rms} \tag{5.51}$$

The reader will find these relationships to be of value when dealing with the various spectral density formulas that are used in ocean engineering studies. The reader will notice that the significant wave period and the other periods in eq. 5.51 do not vary by more than 10%. When the concept of the significant wave period was first introduced, it was common practice to equate the significant period to the mean period and other periods of interest in engineering studies.

EXAMPLE 5.10: STATISTICAL WAVE PERIODS In Example 5.3, we find that the average period for the data in Figure 5.2 is 6.13 sec. The average period for the highest 33 waves of those data is approximately the significant wave period, T_s. Referring to the data in Figure 5.5, the value of the significant wave period is approximately 6.58 sec, which is 1.073 times the average period value. This value differs by less than 3% from the coefficient value of 1.104 in eq. 5.51.

B. The Bretschneider Spectrum

Following Bretschneider (1959, 1963), assume that the wave heights have a Rayleigh probability distribution, as expressed in eq. 5.19. By doing so, we can relate average wave height and the significant wave height to the root-mean-square wave height, as in eq. 5.21b and eq. 5.24, respectively. These relationships are

$$H_{avg} = \frac{\sqrt{\pi}}{2} H_{rms} \simeq 0.886 H_{rms} \tag{5.52}$$

and

$$H_s \simeq 1.416 H_{rms} \tag{5.53}$$

Bretschneider (1959, 1963) applies his analysis to deep-water waves, where the wavelength-period relationships are

$$\lambda_0 \simeq \frac{gT^2}{2\pi}, \quad \frac{d\lambda_0}{dT} \simeq \frac{gT}{\pi} \tag{5.54}$$

from eq. 3.36. The subscript "0" signifies deep-water conditions. Assume that the wavelengths also have a Rayleigh probability distribution in deep water. The probability density function for the wavelengths is

$$p(\lambda_0) = \frac{2\lambda_0}{\lambda_{0rms}^2} e^{-\left(\frac{\lambda_0}{\lambda_{0rms}}\right)^2} \tag{5.55}$$

which is similar to the wave height expression in eq. 5.19. The root-mean-square wavelength in eq. 5.55 is related to the average wavelength, the mean-square, and root-mean-square periods, according to

$$\lambda_{0rms} = \frac{2}{\sqrt{\pi}} \lambda_{0avg} = \frac{g}{\pi^{3/2}} T_{rms}^2 \tag{5.56}$$

We now require that $p(T)dT = p(\lambda_0)\, d\lambda_0$. By combining this relationship with the results of eqs. 5.54 through 5.56, we obtain the probability density function for the deep-water wave periods, which is

$$p(T) = p(\lambda_0)\frac{d\lambda_0}{dT} = \pi \frac{T^3}{T_{rms}^4} e^{-\frac{\pi}{4}\left(\frac{T}{T_{rms}}\right)^4} \simeq 2.70 \frac{T^3}{T_{avg}^4} e^{-0.675\left(\frac{T}{T_{avg}}\right)^4} \tag{5.57}$$

where the results of eq. 5.49 have been used to obtain the last approximation. With the probability density function for the wave period, we can rewrite the expression for the average wave period of eq. 5.47 as

$$T_{avg} = \frac{\int_0^\infty TS_B(T)dT}{\int_0^\infty S_B(T)dT} = \frac{\int_0^\infty TS_B(T)dT}{H_{rms}^2} = \int_0^\infty Tp(T)dT \tag{5.58}$$

The subscript "B" identifies the spectral density function as that of Bretschneider (1959, 1963). From the second equality, we see that $C = 1$ in eq. 5.41b for the Bretschneider spectral density function. Hence, the integral of the Bretschneider spectral function is eight times that of eq. 5.41a, by definition. Bretschneider defined the spectrum in this manner. From the last equality in eq. 5.58, the relationship between the spectral density and the probability density function is obtained. By replacing the probability density function of the integrand of the last integral in eq. 5.58 by the expression in eq. 5.57, the *Bretschneider spectral density formula* is obtained. The result is

$$S_B(T) = H_{rms}^2 p(T) = 1.27 H_{avg}^2 p(T) = 3.437 \frac{H_{avg}^2}{T_{avg}^4} T^3 e^{-0.675\left(\frac{T}{T_{avg}}\right)^4} \tag{5.59}$$

A comparison of the spectral density expressions in eqs. 5.44a and 5.59 shows that the expression for the A coefficient is

$$A_B = 3.437 \frac{H_{avg}^2}{T_{avg}^4} \tag{5.60}$$

while that of the B coefficient is

$$B_B = 0.75 \left(\frac{1}{T_0^4}\right) \simeq 0.675 \left(\frac{1}{T_{avg}^4}\right) \tag{5.61}$$

As stated in Section 5.7B, when the spectral density expression of Bretschneider is integrated over $0 < T < \infty$, the result is the mean-square of the wave height, that is,

$$\int_0^\infty S_B(T)dT = \frac{A_B}{4B_B} \simeq 1.27 H_{avg}^2 = H_{rms}^2 \int_0^\infty p(T)dT = H_{rms}^2 \tag{5.62}$$

C. The Pierson-Moskowitz Wind-Wave Spectrum

The wave spectral density expression of Pierson and Moskowitz (1964) is based on the assumption of Kitaigorodskii (1962) that the wave spectrum is a function of four variables. Those are the fetch (F), gravitational acceleration (g), frictional velocity of the wind (U_+), and the frequency (f). Pierson and Moskowitz (1964) choose to use the wind speed ($U_{19.5}$) measured by a manometer at $z = 19.5$ m in their formula in place of the frictional velocity. Their spectral formula contains constants that satisfy the measured data of Moskowitz (1964) in open-ocean, fully developed seas. The fetch is then assumed to be infinite. The *Pierson-Moskowitz spectral density function* is

$$S_{PM} = \frac{0.00810}{(2\pi)^4} g^2 T^3 e^{-\frac{0.74}{(2\pi)^4}\left(\frac{g}{U_{19.5}} T\right)^4} \tag{5.63}$$

where the subscript PM refers to this particular spectral density formula. The constant in eq. 5.41b is $C = 1/8$. A comparison of the expressions of eqs. 5.43 and 5.63 results in the following expressions for the Pierson-Moskowitz coefficients:

$$A_{PM} = \frac{0.00810 g^2}{(2\pi)^4} \tag{5.64}$$

and

$$B_{PM} = \frac{0.74}{(2\pi)^4} \left(\frac{g}{U_{19.5}}\right)^4 \tag{5.65}$$

The integral of the Pierson-Moskowitz wave spectral density expression [$S_{PM}(T)$ in eq. 5.63] is equal to the variance of the free-surface displacement, η.

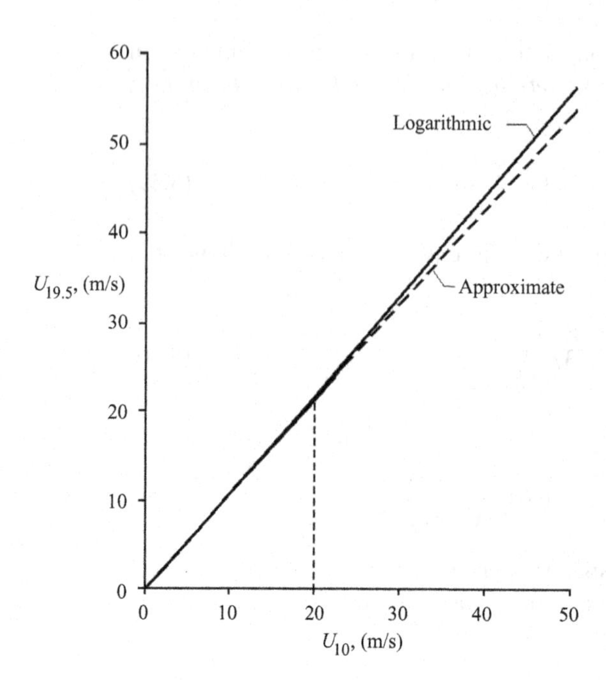

Figure 5.12. *Relationship Between the Average Wind Speeds at $z = 19.5$ m and $z = 10$ m.* The logarithmic and approximate curves result from eq. 5.67b. The vertical dashed line is the upper limit of the JONSWAP data, as reported by Carter (1982).

For a Gaussian sea, the average free-surface displacement is zero, and the variance is simply equal to the mean-square of the free-surface displacement. By using the results in eq. 5.35a, the integral of the Pierson-Moskowitz spectrum is

$$\int_0^\infty S_{PM}(T)dT = \frac{A_{PM}}{4B_{PM}} \simeq 2.74x10^{-3}\frac{(U_{19.5})^4}{g^2} = \frac{H_{rms}^2}{8} = \overline{\eta^2} \tag{5.66}$$

Compare this expression with the discrete data expression of eq. 5.40. From the comparison, we see that the Pierson-Moskowitz spectrum and the discrete-data spectrum have similar definitions. However, a comparison of the expressions in eqs. 5.40 and 5.62 shows that the integral of the Bretschneider spectrum is eight times that of the Pierson-Moskowitz spectrum, and also approximately eight times the sum of the discrete-data spectrum.

Following Pierson (1964), the relationship between the wind speed U_z (measured at any z) and that measured at $z = 10$ m is given by the following form of logarithmic law:

$$U_z = U_{10}\left[1 + \frac{\sqrt{0.80 + 0.114U_{10}}}{0.4 \times 10^{3/2}}\ln\left(\frac{z}{10}\right)\right] \tag{5.67a}$$

The reference height of 10 m is that used by Bretschneider (1959) and others. The relationship between the nominal wind speeds at $z = 19.5$ m and $z = 10$ m is then

$$U_{19.5} = U_{10}\left[1 + 0.0528\sqrt{0.80 + 0.114U_{10}}\right] \simeq 1.075U_{10} \tag{5.67b}$$

The approximation is used by Carter (1982) in an analysis of the JONSWAP data, and is valid to approximately $U_{10} \simeq 30$ m/s. The approximate expression is also used later in this chapter. Curves obtained from the exact and approximate expressions in eq. 5.67b are presented in Figure 5.12. The approximation introduces an error of approximately 6.7% for $U_{10} = 50$ m/s. The logarithmic expression is presented

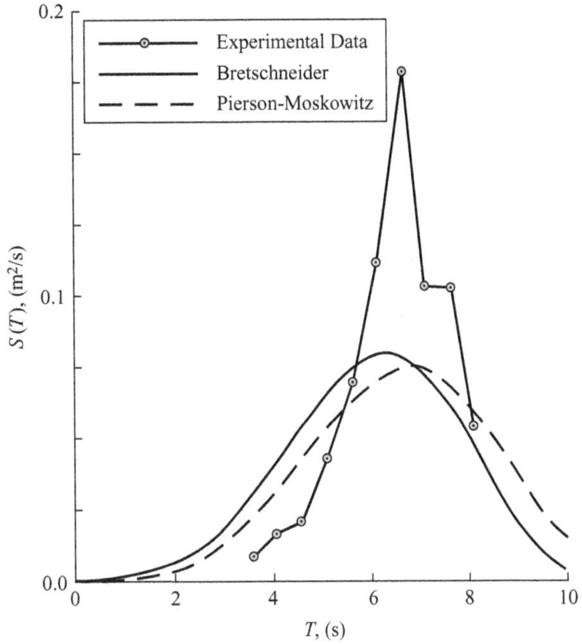

Figure 5.13. *Comparison of the Bretschneider, Pierson-Moskowitz Spectra and the Long-Term Spectral Data of Figure 5.11.*

because it appears in the paper of Moskowitz (1964). However, Kinsman (1965) expresses doubt about the validity of the logarithmic law for wind over water.

EXAMPLE 5.11: COMPARISON OF OPEN-OCEAN WIND-WAVE SPECTRA The wave spectral density values corresponding to the data in Table 5.2 are plotted in Figure 5.13. Let us see how well the spectrum of these values compares with the spectra predicted by the Bretschneider formula (eq. 5.59) and the Pierson-Moskowitz formula (eq. 5.63). From the results of eqs. 5.62 and 5.66, the two empirical spectral formulas are related by

$$\frac{A_B}{4B_B} = \frac{2A_{PM}}{B_{PM}} = H_{rms}^2 \tag{5.68}$$

To apply these two formulas to the data, we must use the experimental value of the mean-square wave height, which is 2.87 m^2 from Examples 5.3 and 5.8. First, the wind speed in eq. 5.66 can be determined. That value is approximately 10.6 m/s. Combine this value with the Pierson-Moskowitz spectral density expression of eq. 5.63. Next, the mean wave height value of 1.50 m and the mean period value of 6.13 sec (both values from Example 5.3) are combined with the Bretschneider spectral density expression of eq. 5.59, and the resulting expression is multiplied by eight. The results are presented in Figure 5.13. In that figure, we see that the discrete spectrum is sharply peaked at a 6.5-sec period. The empirical spectra have peak values of less than half of the discrete data peak value. The peak values of the empirical spectra occur at the modal period (T_0), defined by eq. 5.45. For the Pierson-Moskowitz spectrum, the modal period value is 6.81 sec. For the Bretschneider spectrum, the value is 6.29 sec. The reader must keep in mind that comparisons of the spectra in this example are for the purpose of illustration. The wave height and period data in Figure 5.2 are representative of long-term seas but not of wind-generated seas.

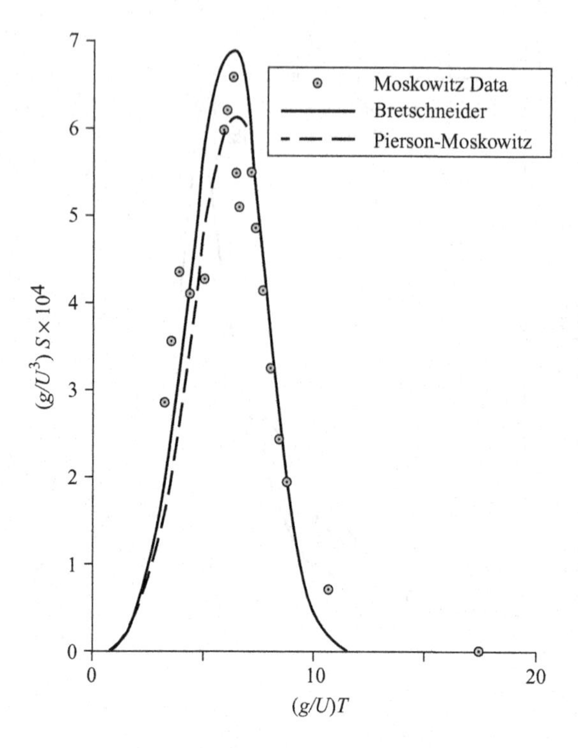

Figure 5.14. *Non-Dimensional Bretschneider and Pierson-Moskowitz Spectra and the Observed Wind-Wave Data of Moskowitz* (1964).

For open-ocean wind-generated seas, spectral results obtained by using the Bretschneider and Pierson-Moskowitz formulas compare well with measured spectra. This fact is demonstrated by the results in Figure 5.14, where the spectral data of Moskowitz (1964) and the results of eq. 5.59 (subject to the condition of eq. 5.68) and eq. 5.63 are presented. The reader should note that the parameters of the Pierson-Moskowitz spectral density formula (A_{PM} and B_{PM}) are based on the Moskowitz (1964) data. The good agreement of the Bretschneider spectral results and the Moskowitz data indicates that the statistics of open-ocean wind-generated seas are Rayleigh in nature.

D. The JONSWAP Spectra – the Fetch-Limited Sea

The discussions to this point have been concerned with open-ocean wind-generated seas, that is, where the fetch is assumed to be infinite. If our engineering interest is in a site that is near the coast of a land mass, then the wave statistics can be significantly altered by the limited fetch. To establish fetch-limited statistics, the Joint North Sea Wave Project (JONSWAP) was conducted in 1969, and reported by Hasselmann et al. (1973). As discussed by Komen et al. (1994), Hasselmann et al. (1973) modified the form of the Pierson-Moskowitz spectral representation (eq. 5.63) for an open-ocean fully developed sea to obtain a four-parameter spectral representation of a fetch-limited sea. The result is the five-parameter *JONSWAP wave spectral density* formula, which is

$$S_{JON}(T) = A_{JON} T^3 e^{-B_{JON} T^4} \gamma^{e^{-\frac{1}{2\sigma_{a,b}^2}\left(\frac{T_p}{T}-1\right)^2}} \tag{5.69}$$

where

$$A_{JON} = \frac{0.076}{(2\pi)^4} g^2 \left(\frac{Fg}{U_{10}^2} \right)^{-0.22} \tag{5.70}$$

and

$$B_{JON} = \frac{1.25}{T_p^4} \tag{5.71}$$

In the spectral formula, the five parameters are A_{JON}, T_p, γ, σ_a, and σ_b. The first two are called the *scale parameters*, and the last three are the *shape parameters*. The term γ is called the *peak enhancement parameter*, whereas σ_a and σ_b modify the width of the spectrum and thereby also enhance the peak. As reported by Hasselmann et al. (1976), the mean values of the shape parameters are $\gamma = 3.3$ and $\sigma_a = 0.07$, which apply to the high-period side of the peak period, T_p, in eq. 5.46 and $\sigma_b = 0.09$, which applies to the low-period side.

The mathematical form of the JONSWAP spectral formula of eq. 5.69 presents a problem in obtaining explicit relationships. Relationships among the root-mean-square wave height, peak spectral period, nominal wind speed, and fetch are obtained using numerical techniques. These relationships are presented in the next section.

For more in-depth discussions of the JONSWAP spectral density, the reader is urged to consult the publications of Goda (1985), Hogben (1990), Tucker (1991), Komen et al. (1994), and McCormick (1999).

E. Wave Property Relationships from Empirical Wind-Wave Spectra

In the following subsections, the empirical spectra for three situations are discussed. Those are the fully developed open-ocean seas, the fetch-limited seas, and the duration-limited seas.

(1) Fully Developed, Open-Ocean Seas

For wind-generated seas in the open ocean, results of the formula of Pierson and Moskowitz (1964) can be used to obtain the relationships between average wave properties (wave height and period) and the nominal wind speed. Following Bretschneider (1959), who assumes that the Rayleigh probability density function is valid for both the wave heights and wavelengths, the average wave heights and periods can be related to the nominal velocities ($U_{19.5}$ and U_{10}). The wind-speed approximation of eq. 5.67b is used to relate the nominal speeds at $z = 10$ m and $z = 19.5$ m. The resulting wave height and period expressions are as follows.

Root-Mean-Square Wave Height: The expression for the root-mean-square wave height can be obtained directly from eq. 5.66. The result is

$$H_{rms} \simeq 0.148 \frac{U_{19.5}^2}{g} \simeq 0.171 \frac{U_{10}^2}{g} \tag{5.72}$$

Average Wave Height: To obtain the mean wave height expression, combine the expressions of eq. 5.72 and the Rayleigh result of eq. 5.21b. This combination

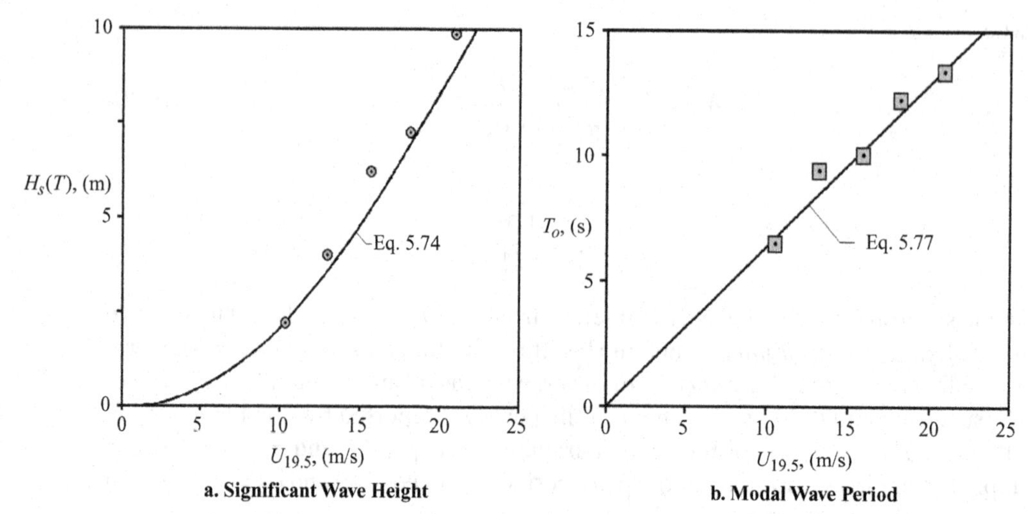

Figure 5.15. *Comparison of Moskowitz* (1964) *Data and Analytical Curves.* The significant wave height (H_S) and the modal period (T_o) are shown as functions of the wind speed measured at 19.5 m above the SWL.

results in

$$H_{avg} = \frac{\sqrt{\pi}}{2} H_{rms} \simeq 0.131 \frac{U_{19.5}^2}{g} \simeq 0.152 \frac{U_{10}^2}{g} \tag{5.73}$$

Significant Wave Height: Again, assuming that the Rayleigh distribution of wave heights is valid, eqs. 5.24 and 5.72 can be combined to obtain the expression for the significant wave height. The resulting expression is

$$H_s \simeq 1.416 H_{rms} \simeq 0.210 \frac{U_{19.5}^2}{g} \simeq 0.243 \frac{U_{10}^2}{g} \tag{5.74}$$

Results of this expression are presented in Figure 5.15a with the observed data reported by Moskowitz (1964).

Average Period: To obtain the relationship between the average wave period and the nominal velocity, combine eqs. 5.47 and 5.65 by eliminating the B_{PM} coefficient. The result is

$$T_{avg} \simeq 6.14 \frac{U_{19.5}}{g} \simeq 6.60 \frac{U_{10}}{g} \tag{5.75}$$

Root-Mean-Square Period: From the combination of eqs. 5.48 and 5.65, the expression for the root-mean-square wave height is

$$T_{rms} \simeq 6.38 \frac{U_{19.5}}{g} \simeq 6.86 \frac{U_{10}}{g} \tag{5.76}$$

Modal Period: Equations 5.45, 5.46, and 5.65 are combined to obtain the expressions for the modal and peak spectral periods, which is

$$T_o \simeq 0.88 T_p \simeq 6.30 \frac{U_{19.5}}{g} \simeq 6.77 \frac{U_{10}}{g} \tag{5.77}$$

Results obtained from eq. 5.77 are presented in Figure 5.15, with the modal period values corresponding to the Moskowitz (1964) data.

The reader can see that the results obtained from the empirical equations and the Moskowitz (1964) data in Figure 5.15b compare quite well. These good comparisons reinforce the conclusion that the probability distributions of the wave heights and the wavelengths for a fully developed, open-ocean sea are well represented by the Rayleigh probability density function, as assumed by Bretschneider (1959).

(2) Fetch-Limited Seas

Further analyses of the data leading to the JONSWAP spectral formula of eq. 5.69 yield averaged heights and periods for *fetch-limited seas*, that is, the fetch length (F) over which the wind blows is limited by the presence of a land mass located upwind from the site in question. Two such analyses are those of Hasselman et al. (1976) and Carter (1982). The Carter results are presented here.

The Carter (1982) equations are presented in terms of the fetch (F) in kilometers, wind duration (t_D) in hours, and the nominal wind speed (U_{10}) in meters per second. The wave properties can be determined by either the fetch or the duration, depending on value of the *critical duration*, defined by

$$t_{DF} \equiv 2.315\left(\frac{U_{10}}{g}\right)\left(g\frac{F}{U_{10}^2}\right)^{0.7} \tag{5.78}$$

For the fetch-limited sea, Carter (1982) states that

$$t_D > t_{DF} = 1.167\frac{F^{0.7}}{U_{10}^{0.4}} \tag{5.79}$$

where, again, the fetch is in kilometers, the duration in hours, and the wind speed in meters per second. The gravitational constant (g) used is 9.81 m/s^2. For this condition, the following expressions apply:

Significant Wave Height:

$$H_s \simeq 0.0511\left(\frac{U_{10}^2}{g}\right)\left(\frac{gF}{U_{10}^2}\right)^{0.5} \simeq 0.0163U_{10}\sqrt{F} \tag{5.80}$$

Peak Spectral Period:

$$T_p \simeq 2.80\left(\frac{U_{10}}{g}\right)\left(\frac{gF}{U_{10}^2}\right)^{0.3} \simeq 0.566U_{10}^{0.4}F^{0.3} \tag{5.81}$$

Average Wave Period:

$$T_{avg} \simeq 0.857T_p \simeq 2.40\left(\frac{U_{10}}{g}\right)\left(\frac{gF}{U_{10}^2}\right)^{0.3} \simeq 0.485U_{10}^{0.4}F^{0.3} \tag{5.82}$$

where the relationship between the two periods is obtained from eq. 5.51. Rather than using the mean period as in eq. 5.82, Carter (1982) presents an expression for the zero up-crossing period (T_z), which he states is approximately equal to $0.710T_p$.

(3) Duration-Limited Seas

For the duration-limited sea, the condition reported by Carter (1982) is

$$t_D \le t_{DF} = 1.167\frac{F^{0.7}}{U_{10}^{0.4}} \tag{5.83}$$

For this condition, the following expressions apply:

Significant Wave Height:

$$H_s \simeq 0.0280\left(\frac{U_{10}^2}{g}\right)\left(\frac{gt_D}{U_{10}}\right)^{5/7} \simeq 0.0146U_{10}^{9/7}t_D^{5/7} \tag{5.84}$$

Peak Spectral Period:

$$T_p \simeq 1.99\left(\frac{U_{10}}{g}\right)\left(\frac{gt_D}{U_{10}}\right)^{3/7} \simeq 0.540U_{10}^{4/7}t_D^{3/7} \tag{5.85}$$

Average Wave Period:

$$T_{avg} \simeq 0.857T_p \simeq 1.70\left(\frac{U_{10}}{g}\right)\left(\frac{gt_D}{U_{10}}\right)^{3/7} \simeq 0.461U_{10}^{4/7}t_D^{3/7} \tag{5.86}$$

By using the results of Sections (1), (2), and (3), the limiting conditions of the JONSWAP relationships can be determined. That is, by assuming that the JONSWAP and Pierson-Moskowitz are identical at limiting values of either the fetch or duration for a given wind speed, we can obtain the limiting expressions for F and t_D for a developing sea. We shall do this by equating the various significant wave height expressions presented in the subsections. For the fetch-limited case, the Pierson-Moskowitz expression of eq. 5.74 and the JONSWAP expression of eq. 5.80 are equated to obtain the minimum fetch for a fully developed sea (F_{fds}). The result is

$$F_{fds} = 2.32U_{10}^2 \tag{5.87}$$

in kilometers. Hence, when $F \ge F_{fds}$, the sea is fully developed. For the duration-limited case, the minimum duration for a fully developed sea is obtained by equating the expressions in eqs. 5.74 and 5.84 and solving for the duration. The result is

$$t_{Dfds} = 2.10U_{10} \tag{5.88}$$

The analysis of the situation described in Example 5.12 illustrates the use of the equations in hours. So when $t_D \ge t_{Dfds}$, the sea is fully developed. Following the example of Carter (1982), a decision tree for the selection of the appropriate expressions is given in Table 5.3.

EXAMPLE 5.12: FETCH-LIMITED AND DURATION-LIMITED SEAS Prior to the commencement of a beach nourishment project near Ocean City, Maryland, on the Atlantic coast of the contiguous United States, offshore operations at several borrow sites are terminated at sea states above certain values. *Borrow sites* are large sand bars that are the sand resources. The reader should consult the papers by Grosskopf and Kraus (1994) and Dean (2003) for excellent discussions of beach nourishment. For this particular operation, the maximum possible sea states at the offshore sites are based on a 5-hour storm having a nominal wind

Table 5.3. *Decision tree for wave height and period equations*

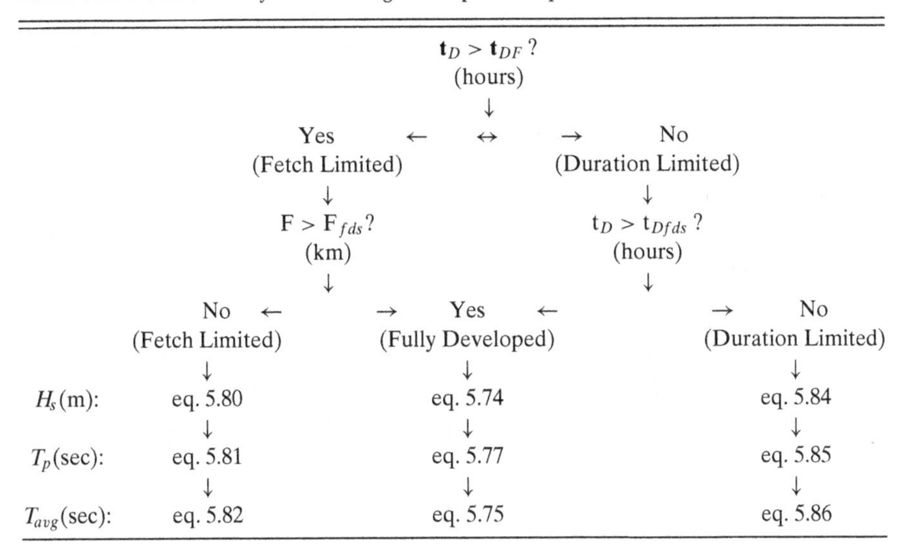

speed of 15 m/s. Hence, $U_{10} = 15$ m/s and $t_D = 5$ hours. The borrow sites are located 1 km, 5 km, and 10 km east of the Ocean City shoreline. Assume all sites are in deep water. For winds out of the west, the JONSWAP equations are to be used, and for winds out of the east, the Pierson-Moskowitz equations apply.

For the winds out of the west, we must first determine the value of the critical duration, defined in eq. 5.78 for each borrow site. Applying the expression in that equation to the various sites results in the following critical duration values:

$$t_{DF} \simeq 0.395\mathrm{F}^{0.7} \simeq \begin{array}{l} 0.40\,\text{hour}(\mathrm{F} = 1\,\text{km}) \\ 1.22\,\text{hour}(\mathrm{F} = 5\,\text{km}) \\ 1.98\,\text{hour}(\mathrm{F} = 10\,\text{km}) \end{array}$$

Because our 5-hour duration is greater than these values, as in eq. 5.79, the seas are fetch-limited at each site. Following the decision tree in Table 5.3, we must now determine if the sea is fully developed. From eq. 5.87, the minimum fetch for a fully developed sea is 522 km for our 15 m/s wind speed. So at each site, the sea is developing when the winds are from the west. The significant wave height values at each site are determined from eq. 5.80. Those values are

$$H_s \simeq 0.244\sqrt{\mathrm{F}} \simeq \begin{array}{l} 0.244\ \text{m}(\mathrm{F} = 1\ \text{km}) \\ 0.546\ \text{m}(\mathrm{F} = 5\ \text{km}) \\ 0.772\ \text{m}(\mathrm{F} = 10\ \text{km}) \end{array}$$

The application of eq. 5.82 to each site yields the following average period values:

$$T_{avg} \simeq 1.137\mathrm{F}^{0.3} \simeq \begin{array}{l} 1.137\,\text{sec}(\mathrm{F} = 1\,\text{km}) \\ 1.843\,\text{sec}(\mathrm{F} = 5\,\text{km}) \\ 2.269\,\text{sec}(\mathrm{F} = 10\,\text{km}) \end{array}$$

When the winds are out of the east at Ocean City, the seas will be fully developed if the fetch exceeds 522 km, from eq. 5.87. Assume that this is the case for the 10 km borrow site. However, note that the duration does not exceed 31.5 hours – the condition of eq. 5.88 for a duration-limited sea. From Table 5.3, for the fetch-limited sea where the fetch is the minimum for a fully developed sea, we can use the Pierson-Moskowitz results of eqs. 5.74 and 5.75 to obtain the respective values of the significant wave height and average period. Those results are

$$H_s \simeq 5.57\,\text{m and } T_{avg} \simeq 10.09\,\text{sec}$$

Comparing the statistical wave properties resulting from the two wind conditions, we conclude that operations can continue at any of the sites for the 15 m/s wind out of the west, but should terminate long before the 5-hour duration when the winds are out of the east.

In lieu of a decision tree, such as that presented in Table 5.3, the U.S. Army (1984) *Shore Protection Manual* contains curves that result from the JONSWAP equations. The reader should consult that reference to see if the curves are more to their liking.

F. Directional Properties of Wind-Generated Waves

The expressions for the spectral density up to this point are those that describe *point spectra*, that is, the expressions describe the energy of the wave climate at a single point in the ocean without consideration of the directions of travel of the component waves. If three or more closely spaced wave gauges are simultaneously used to measure the wave properties, then the *directionality* of the sea can be determined, that is, the energy distribution as a function of direction can be determined by correlating the signals from the gauges. The directionality of the waves can be important in a number of engineering applications. For example, McCormick (1978) takes into account the directionality of the sea in determining the wave energy resource for arrays of wave energy conversion devices.

Wind waves are directional due to the nature of their creation. This fact is demonstrated by the following simplistic analysis of the waves created by pressure fluctuations on the free surface of the water caused by the passage of eddies in the wind's turbulent boundary layer, adjacent to the surface. Consider such a boundary layer adjacent to the free surface of (initially) calm water. The randomly occurring eddies, convecting in the nominal direction of the wind, are at various heights above the free surface. Because of the velocity distribution in the boundary layer, as illustrated in Figure 5.16, the convection velocity increases with distance above the surface. In addition, the energies associated with the various eddies also depend on their vertical position.

Let us consider one of these eddies. As described by McCormick and Mouring (1995) and McCormick, Bhattacharyya, and Mouring (1997) and others, the eddies first appear as rectilinear line vortices, as sketched in Figure 5.10. Certain segments of the line vortex acquire energy more rapidly than other segments. As the energy of a segment increases, the vortex grows and subsequently migrates away from the free surface. The time scale of the events just described depends on the elevation of the formation of the vortex in question. The evidence of the events on the water surface is in the form of pressure fluctuations. The magnitudes of the pressure fluctuations depend on the position of the vortex and, hence, on the speed at which the vortex is convected.

Consider two surface-pressure pulses on the calm-water surface that are convecting at velocities u_1 and u_2, where $u_2 > u_1$, as illustrated in Figure 5.17. The pulses

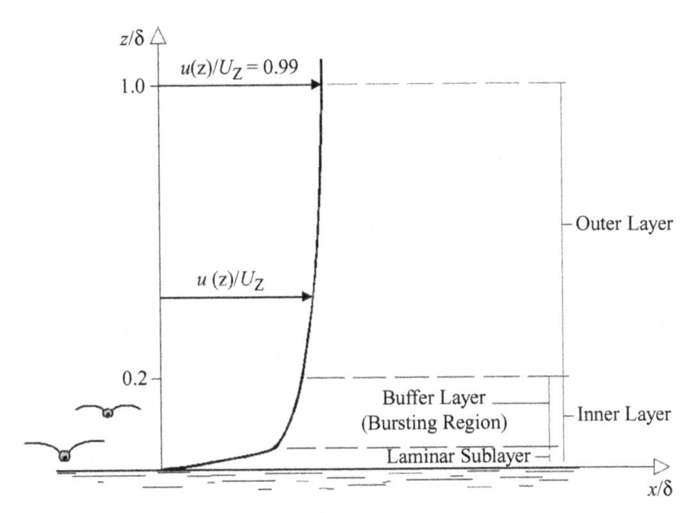

Figure 5.16. *Sketch of the Variation of Wind Speed in a Turbulent Boundary Layer over Calm Water.* As illustrated in Figure 5.10, the bursting phenomena originates in the buffer region. The outer region is characterized by large-scale eddies. The boundary layer thickness (δ) is actually time-dependent because the turbulent eddies cause a somewhat wavy interface with the external air flow. The reader is referred to the excellent discussion of the various boundary layer phenomena found in the book by Granger (1995).

have respective periods of T_1 and T_2. The effect of each pressure pulse on the water surface is to create a small radial wave. The expression for the free-surface deflection of the deep-water, radial wave originating at some point x_a on the line of action is

$$\begin{aligned}\eta_a(r_a, t) &= -\frac{\omega}{g} A_o J_1(k_0 r_a - \omega t) \\ \underset{(r_a \to \infty)}{\to} & -\frac{\omega}{g} A_0 \sqrt{\frac{2}{\pi(k_0 r_a - \omega t)}} \cos\left(k_0 r_a - \omega t - \frac{\pi}{4}\right)\end{aligned} \tag{5.89}$$

The expression in eq. 5.89 results from the solution of Laplace's equation (eq. 3.8) in the cylindrical coordinate system, as derived by Lamb (1945), McCormick (1973),

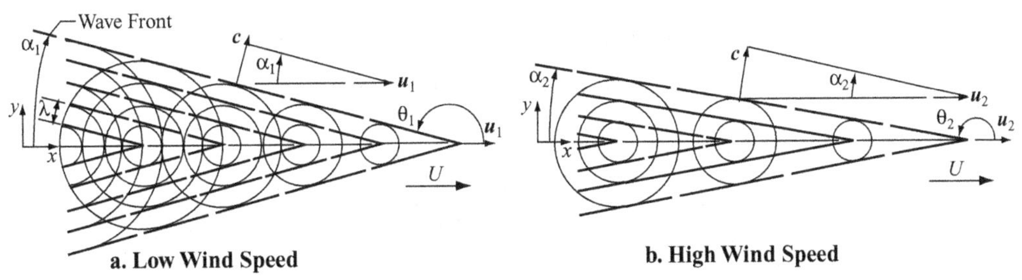

Figure 5.17. *Area Sketch of Two Convecting (Wind) Pressure Pulses on a Calm-Water Surface.* Each pulse has the same period; however, the pulse associated with the speed $u_1 (< u_2)$ is closer to the water surface. The angle, α, is analogous to the Mach angle in supersonic aerodynamics. The actual wave patterns formed by continual traveling pressure disturbances are known as *Kelvin wave patterns* because of the original work on the phenomenon by Lord Kelvin (Sir W. Thomson). See Lamb (1945) for a discussion of waves generated by traveling impulses, and Sorensen (1973, 1993) for the related topic of ship-generated waves.

and others. In eq. 5.89, A_o is the amplitude of the wave at its origin and r_a is the radial coordinate from the origin of the wave. In the far field (where $r_a \to \infty$), the radial wave fronts are approximately sinusoidal, having a deep-water wavelength corresponding to the period of the pulse, as in eq. 5.54. The initial wave amplitude (A_o) on the x-axis, or line of action, depends on the magnitude of the pressure pulse. As the waves travel outward from their point of origin on the line of action, their amplitudes decrease because the energy supplied by the pulse is distributed over an expanding circumferential wave front. After a number of periods, the outermost wave fronts approximately coalesce to form rectilinear wave fronts, as sketched in Figure 5.17. The *wave angle* θ (the angle between the direction of wave travel of the rectilinear wave and the line of action) is a function of both the wave celerity, c, and the convection speed, u. The expression for the wave angle is

$$\theta = \cos^{-1}\left(\frac{c}{u}\right) = \frac{\pi}{2} - \alpha \tag{5.90}$$

from the geometries in Figure 5.17. This simplified analysis of events is similar to that in the production of supersonic waves, where the rectilinear wave front becomes a shock wave, the velocity ratio in eq. 5.90 is called the Mach number, and the angle α is called the Mach angle. After the creation of a radial wave, the surface wind-shear stress acts to enhance the growth of wave segments in $|\theta| < \pi/2$. Although this description of wind-wave generation is simplistic, it should give the reader an idea of the nature of the phenomenon. The reader is referred to the works of Phillips (1957, 1958, 1966, and 1985) for the mathematical physics of wind-wave generation.

Many papers have been written that are either devoted to the subject of directionality of wind-generated seas or include extensive discussions of the subject. Included are those by Arthur (1949), Pierson (1955), Longuet-Higgins, Cartwright, and Smith (1963), Cartwright (1963), Mitsuyasu et al. (1975), Le Méhauté (1982), Ewing (1986), and Niedzwecki and Whatley (1991), among many others. The book edited by Wiegel (1982) is devoted to engineering aspects of directional spectra. In addition, excellent discussions of the subject are found in the books of Goda (1985) and Tucker (1991).

One of the most popular mathematical expressions describing the directionality of wind-wave spectra is that presented by Cartwright (1963), which in turn was based on the analysis of Longuet-Higgins, Cartwright, and Smith (1963). The standard method of presenting the directional wave data is to do so in terms of true north. The angle, Θ, is measured from true north in a clockwise direction. In terms of this angle, the Cartwright formula for the *directional spectral density* function is

$$S(T,\Theta) = S(T)G(T,\Theta) \tag{5.91}$$

where the *spreading function* is given as

$$G(T,\Theta) \equiv G(s)\left|\cos\left(\frac{\Theta - \Theta_0}{2}\right)\right|^{2s} \tag{5.92}$$

In eq. 5.92, s is called the *spreading parameter*, and Θ_0 is the *mean direction* of the energy of the sea. Both s and Θ_0 are functions of the wave period. The normalizing factor given by Cartwright (1963) has the following form:

$$G(s) = \frac{2^{2s-1}}{\pi}\frac{\Gamma^2(s+1)}{\Gamma(2s+1)} \tag{5.93}$$

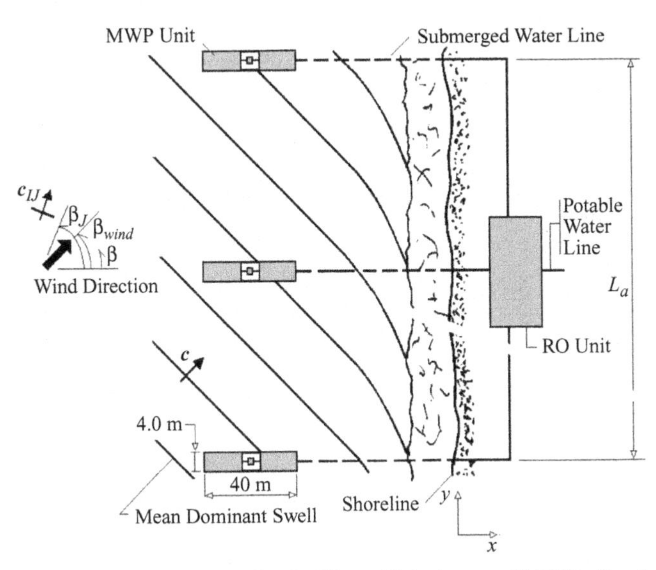

Figure 5.18. *Area Sketch of a Three-Unit Array of MWPs Deployed in a Wind-Generated Sea.* Each unit is a hinged-barge system consisting of three barges, as sketched in Figure 5.19. The wave-induced motions of the forward and after barges excite water pumps on the inertial central barge. These pumps in turn force pressurized water through submerged hoses leading to either a reverse-osmosis (RO) desalination system, as illustrated, or an electrical generator, depending on the purpose of the system.

Because the point spectral density function, $S(T)$, must result from an integration of eq. 5.91 over all angles, the following condition must be true:

$$\int_{-\pi}^{\pi} G(T, \Theta) d\Theta = 1 \tag{5.94}$$

There have been a number of at-sea studies resulting in modifications to both the spreading parameter and the normalizing function. One of the most popular is that of Mitsuyasu et al. (1975). From the data reported in that study, the authors recommend the following expressions for the spreading parameter:

$$s = 11.5 \frac{(2\pi U_{10})_{\mu}}{gT} \tag{5.95}$$

where the exponent values are $\mu = -2.5$, for $T < T_p$, and $\mu = 5$, for $T \geq T_p$. The use of the directional wave spectrum is illustrated in Example 5.13, based on the work of McCormick (1978).

EXAMPLE 5.13: WAVE POWER RESOURCE DETERMINATION FOR A DIRECTIONAL SEA
One of the first expressions for the directional spectral density for open-ocean waves was based on the Arthur (1949) "cosine observation." Referring to the sketch in Figure 5.18, that expression is

$$S(T, \beta) = S(T)\, G_0(\beta) = S(T) \frac{2}{\pi} \cos^2(\beta - \beta_{wind}), \quad -\frac{\pi}{2} < \beta - \beta_{wind} < \frac{\pi}{2} \tag{5.96}$$

from Pierson (1955). Equation 5.96 is based on the assumption that the mean direction of the waves is the same as the nominal direction of the wind. We

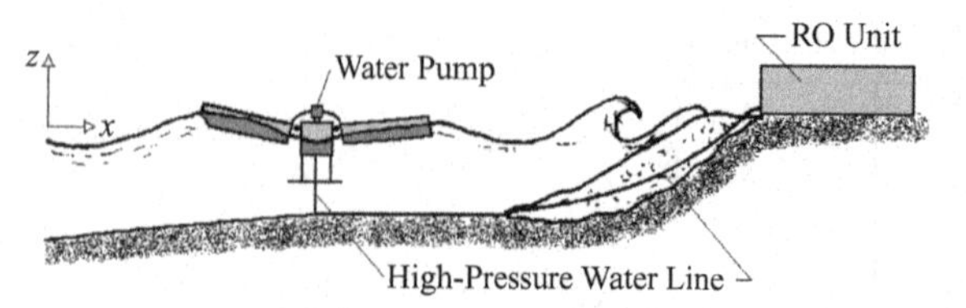

a. Elevation Sketch of the MWP System

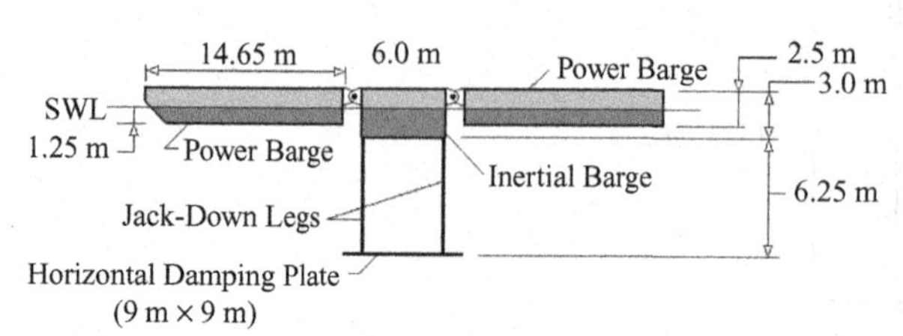

b. Dimensions of the MWP

Figure 5.19. *Elevation Sketches of the McCabe Wave Pump*. The dimensions shown are those of the prototype that was deployed in the Shannon estuary from 1996 to 2003. The system is designed to supply potable water or electricity, depending on the needs at the site.

shall use this expression to illustrate the method of determining the wave power resource for an array of wave energy conversion systems.

Floating wave energy converters appear to be best if oriented in line arrays. That is, the most cost-effective orientation of such systems is the line array consisting of a number of wave energy conversion modules, as opposed to large single units. To illustrate, a line array of modules of a system called the McCabe wave pump (MWP) is sketched in Figure 5.18. The MWP, sketched in Figure 5.19, is a three-barge wave energy converter that is designed to produce either electricity or potable water for isolated coastal communities or inhabited islands. See the paper by McCormick and Kim (1997) for the details of the MWP operation. Our attention is focused on an offshore array operating in deep-water waves. The waves at the site in question have the same properties as those in Examples 5.3 and 5.11, namely, a 1.50-m average wave height and an average wave period of 6.13 sec. Referring to Figure 5.18, the wind direction is $\beta_{wind} = 45°$ to the onshore direction.

To determine the wave power resource in a random sea, we must first determine the relationship between the average wave power and the directional spectral density. This is accomplished by using a method similar to that used in the derivation of the average energy intensity expression of eq. 5.39. For a monochromatic wave, the energy intensity and energy flux (or power) can be related by using the results of eqs. 3.72b and 5.36. For deep-water waves, the result is

$$\frac{E_0}{\lambda_0 b} = \rho g \frac{H_0^2}{8} = \frac{|P_0|}{c_{g0} b} \simeq 4\pi \frac{|P_0|}{gTb} \tag{5.97}$$

where b is the crest width and c_{g0} is the deep-water group velocity ($\simeq gT/4\pi$). Apply this expression to the Ith wave traveling in the Jth direction in a random sea, as illustrated in Figure 5.18. As in the process leading to eq. 5.39, the power of this wave is found to be

$$\begin{aligned} \boldsymbol{P}_{IJ} &\simeq \frac{\rho g^2}{4\pi} S(T_{IJ}, \beta_J) T_{IJ} b_J \left[\cos(\beta_J)\boldsymbol{i} + \sin(\beta_J)\boldsymbol{j}\right] \Delta T \Delta\beta \\ &= \boldsymbol{P}_{IJx} + \boldsymbol{P}_{IJy} \end{aligned} \tag{5.98}$$

In this expression, the power vector has two components. Those are the onshore component ($\boldsymbol{P}_{IJx}$), which is that available to the array of wave energy converters, and the alongshore component ($\boldsymbol{P}_{IJy}$). Also in eq. 5.98 are the respective onshore and alongshore unit vectors ($\boldsymbol{i}, \boldsymbol{j}$). The crest width ($b_J$), not shown in the figure, is related to the array length (L_a) by

$$b_J = L_a \cos(\beta_J) \tag{5.99}$$

After combining the expressions in eqs. 5.98 and 5.99, we pass to the limits as both ΔT and $\Delta\beta$ approach zero, and integrate the resulting power component expressions. The resulting expression for the onshore power component (that available to the line array) is

$$\boldsymbol{P}_x \simeq \frac{\rho g^2 L_a}{4\pi} \int_{\beta_{wind}-\frac{\pi}{2}}^{\beta_{wind}+\frac{\pi}{2}} \int_{0}^{\infty} S(T,\beta) T \cos^2(\beta) dT d\beta \boldsymbol{i} \tag{5.100}$$

Combine the expression in eq. 5.100 with both the directional spectral formula of eq. 5.96 and the generic point spectral expression of eq. 5.44a, and divide the result by the array length (L_a). The resulting expression is that for the onshore power per unit array length:

$$\frac{\overline{P_x}}{L_a} = \frac{\rho g^2}{64\pi} \frac{A}{B^{5/4}} \Gamma\left(\frac{5}{4}\right) [2 + \cos(2\beta_{wind})] \tag{5.101}$$

where the gamma function value is approximately 0.9064, from Abramowitz and Stegun (1965).

For a 1.50-m mean wave height and a 6.13-sec mean period, the corresponding wind speed at $z = 19.5$ m is 10.6 m/s from Example 5.11. We use the Pierson-Moskowitz spectral coefficients in our analysis, which are $A_{PM} \simeq 5.00 \times 10^{-4} \text{m}^2/\text{s}^4$ and $B_{PM} \simeq 3.48 \times 10^{-4} \text{s}^{-4}$ from eqs. 5.64 and 5.65, respectively. Substitute these values and the wind angle of 45° into the expression in eq. 5.101 to obtain approximately 10,400 W/m, or 10.4 kW/m. Note that for each citizen living in the contiguous United States, an average electrical power of 1 kW is required. Hence, at this site the power available to each meter of the line array is enough for 10.4 U.S. citizens if that power can be converted to electricity at 100% efficiency. The actual conversion efficiency of the MWP is about 75%. The busbar power supplied by a MWP module with a 4-m beam (B) operating in this sea is $P_x B/L_a = 31.2$ kW. If the sea is a head sea ($\beta_{wind} = 0$), then the busbar power would be approximately 46.8 kW. For the three-unit system sketched in Figure 5.18, the busbar power is about 140 kW in a head sea.

One note concerning the exploitation of ocean waves for the production of either electricity or potable water: The design of a wave energy conversion system should be based on long-term wave statistics rather than short-term wind-wave statistics. Short-term wind wave statistics are normally associated with storms. Long-term wave statistics are those usually determined on a monthly or yearly basis. Long-term statistics are used also in the design of control systems for wave-energy conversion systems. Long-term wave statistics are discussed in the next chapter.

5.9 Long-Term Wave Statistics

In the designs of ocean structures and systems, there are two major considerations. The first of these is survivability, and the second is performance. Most attention is given to survivability because, without it, performance is of little consequence. The survivability of an ocean system is based on extreme wave events, particularly the maximum wave height that would be encountered in wind-generated seas over the life of the system. Based on a Rayleigh probability density function for the wave heights in a random sea, the expression for the maximum wave height is given in eq. 5.25, and is discussed in Example 5.6. For example, for a structure located at the position in the ocean for which the wave data in Figure 5.2 apply, the (statistically) maximum wave height encountered in 100 years is 6.64 m, assuming yearly samples of 6,000,000 waves. To survive the passage of this wave, the designer must determine the force associated with the wave, multiply by some safety factor, and design accordingly.

The performance of an offshore system, such as the MWP wave energy conversion system (sketched in Figures 5.18 and 5.19), depends on *long-term wave statistics*, which are based on wave properties averaged over months or years. These statistics are discussed by Hogben (1990) and others. McCormick (1998a) demonstrates that the use of wind-wave formulas to predict the performance of an ocean system can lead to erroneous results. Performance analyses of wave energy converters should be done using long-term statistical formulas. To this end, McCormick (1998b) derives such formulas based on the two-parameter Weibull (1951) probability function, discussed in Section 5.5. McCormick obtains expressions for the probability density functions for the wave heights and periods and, in addition, an expression for the spectral density function by applying the derived formulas to the open-ocean observations reported by Mollison (1982). Again, the data reported by Mollison (1982) are for Station Porcupine in the Atlantic Ocean, several hundred kilometers west of Ireland. Based on the Weibull (1951) two-parameter probability density formula of eq. 5.28 (where A = 0), the expression for the wave height probability density function for a long-term, open-ocean sea is

$$p(H) = \frac{\mathrm{m}_H}{H_{rms}^{\mathrm{m}_H}} \Gamma_2^{0.5\mathrm{m}_H} H^{\mathrm{m}_H - 1} e^{-\Gamma_2^{0.5\mathrm{m}_H} \left(\frac{H}{H_{rms}}\right)^{\mathrm{m}_H}} \tag{5.102a}$$

where

$$\Gamma_2 \equiv \Gamma\left(\frac{\mathrm{m}_H + 2}{\mathrm{m}_H}\right)$$

Let us apply this formula to the deep-water wave data of Mollison (1982). The value of H_{rms} is approximately 1.025 m for the data reported by Mollison (1982), whereas the most-probable wave height (the wave height satisfying the condition of eq. 5.8) is 0.625 m. The value of the shape factor ($\mathrm{m}_H = 1.80$) is found by curve-fitting. The resulting probability density expression is

$$p(H) \simeq 1.80 H^{0.80} e^{-1.001 H^{1.80}} \tag{5.102b}$$

Plots obtained from the expression in eq. 5.102b, the Rayleigh formula of eq. 5.19, and the probability density values for the observed wave heights are presented in Figure 5.20. The wave height probability distributions presented in that figure show that both the long-term distribution and the Rayleigh distribution under-predict the

Figure 5.20. *Long-Term and Rayleigh Probability Distributions for Observed Data.*

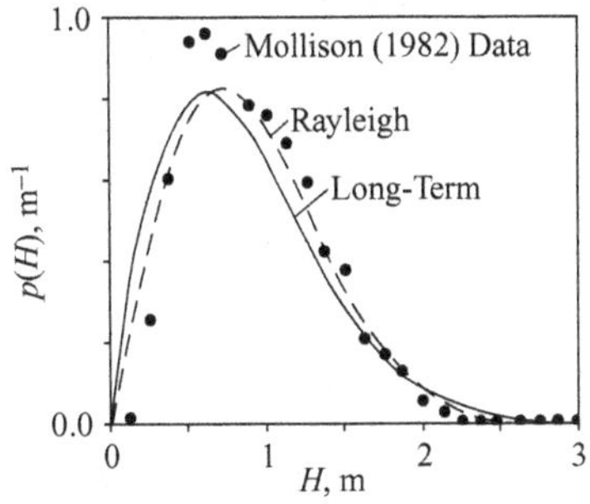

peak value occurring at the most-probable wave height by about 15%. However, the shapes of both empirical curves are approximately the same. The Rayleigh curve more closely fits the at-sea data than the long-term curve. We can conclude from the results in Figure 5.20 that the long-term probability distribution of wave heights in the open ocean is approximately Rayleigh.

McCormick (1998b) also presents a formula for the period probability density function, which is

$$p(T) = \frac{\mathrm{m}_T}{T_{avg}^{\mathrm{m}_T}} \Gamma_1^{\mathrm{m}_T} T^{\mathrm{m}_T - 1} e^{-\Gamma_1^{\mathrm{m}_T}\left(\frac{T}{T_{avg}}\right)^{\mathrm{m}_T}} \tag{5.103a}$$

Here,

$$\Gamma_1 \equiv \Gamma\left(\frac{\mathrm{m}_T + 1}{\mathrm{m}_T}\right)$$

Applying this expression to the Mollison (1982) data yields

$$p(T) \simeq 6.01 \times 10^{-7} T^6 e^{-8.56\times 10^{-8} T^7} \tag{5.103b}$$

where the average wave period (T_{avg}) value for the data reported by Mollison (1982) is 9.56 sec. The value of the shape factor (m_T in eq. 5.28) is 7. Also, for the Mollison data, $T_{\mathrm{rms}} = 9.69$ sec. Results obtained from the empirical formula of eq. 5.103b, the Bretschneider-Rayleigh expression of eq. 5.57, and at-sea probability density values for the Mollison data are presented in Figure 5.21. In that figure, we see an excellent agreement between the long-term empirical distribution and the Mollison data. The Bretschneider-Rayleigh curve under-predicts the peak probability density value by about 40%, and has a much broader base. Hence, the empirical expression of eq. 5.103a is recommended for long-term seas.

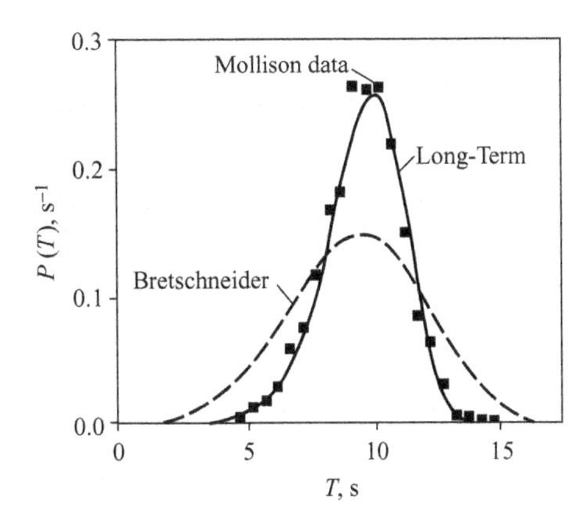

Figure 5.21. *Long-Term and Bretschneider Period Probability Distributions for the Observed Data.*

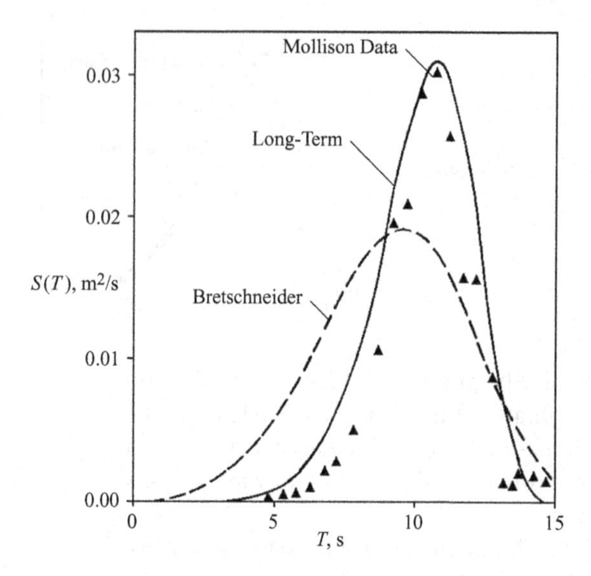

Figure 5.22. *Long-Term and Bretschneider Wave Spectra for the Mollison* (1982) *Observed Data.*

Based on the modal period (T_o), McCormick (1998b) obtains the following expression for the long-term spectral density function:

$$S_{LT}(T) = A_{LT}T^{m_T-1}e^{-B_{LT}T^{m_T}} = \frac{(m_T-1)}{T_o^{m_T}}\frac{H_{rms}^2}{8}T^{m_T-1}e^{-\left(\frac{m_T-1}{m_T}\right)\left(\frac{T}{T_o}\right)^{m_T}} \tag{5.104a}$$

where the subscript LT refers to "long term." The application of the long-term spectral formula to the Mollison (1982) data results in

$$S_{LT}(T) \simeq 4.75 \times 10^{-8}T^6e^{-5.17\times10^{-8}T^7} \tag{5.104b}$$

Here, for the Mollison (1982) observations T_o and H_{rms} are approximately 10.75 sec and 1.025 m, respectively. Results obtained from eq. 5.104b are presented in Figure 5.22, with results obtained using the Bretschneider spectral formula of eq. 5.59 (divided by eight) and spectral density values corresponding to the observed data of Mollison (1982). The division of the expression in eq. 5.59 by eight is dictated by the relationship between the Bretschneider and Pierson-Moskowitz formulas (see eq. 5.66). The long-term spectral formula of eq. 5.104 is seen to agree well with the observed data. Again, the Bretschneider formula is based on the assumption of a Rayleigh probability distribution of wavelengths, as in eq. 5.56. The results presented in Figure 5.22 then lead us to conclude that the long-term probability distribution of wavelengths is not Rayleigh in nature.

5.10 Wave Spectra in Waters of Finite Depth

The discussions of the various wave spectral density formulas in Sections 5.8 are based on the assumption that the water is infinitely deep. That is, the formulas of Bretschneider (1959, 1963), Pierson and Moskowitz (1964), Hasselmann et al. (1973), and McCormick (1998b) predict deep-water spectra. These respective formulas are presented in eqs. 5.59, 5.63, 5.69, and 5.104a. The spectral formulas are all proportional to the statistical wave heights raised to the second power. Assume that this statement also applies in waters of finite depth. Furthermore, assume that the constant coefficients and period-dependent coefficients in the spectral formulas do not change with water depth. With these assumptions, the ratio of the finite-depth

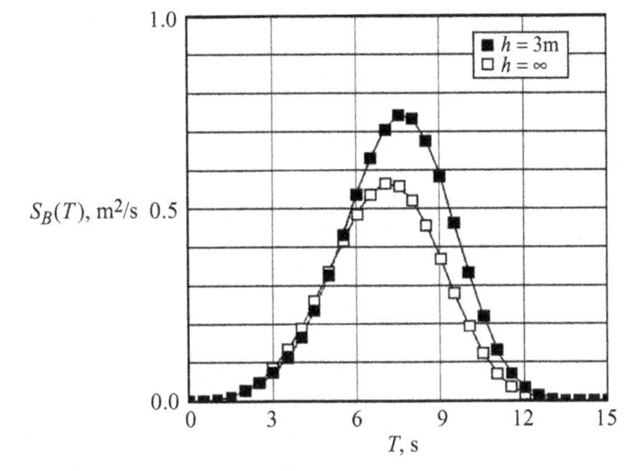

Figure 5.23. *Bretschneider Spectra in Deep Water and Where h = 3 m.* The conditions for these spectra are given in Example 5.14.

and the infinite-depth spectral formulas is

$$\frac{S_J(T)|_h}{S_{J0}(T)} = \left(\frac{H_{rms}|_h}{H_{0rms}}\right)^2 = K_S^2 = \frac{\cosh^2(kh)}{\sinh(kh)\cosh(kh) + kh} \tag{5.105}$$

Here, as is the custom, the subscript "0" refers to the deep-water condition and the subscript J is generic. As an example of the latter, for the Bretschneider spectral formula, $J = B$. Also in eq. 5.105 is K_S, the shoaling coefficient defined in eq. 3.78. By rearranging the terms in eq. 5.105, we obtain the expression for the wave spectral density in water of finite depth, which is

$$S_J(T)|_h = \left[\frac{\cosh^2(kh)}{\sinh(kh)\cosh(kh) + kh}\right] S_{J0}(T) \tag{5.106}$$

The shoaling coefficient expression in eq. 5.105 is that resulting from the linear wave theory. As discussed in Section 3.8, this expression is somewhat limited in that it is independent of both the deep-water wave steepness (H_0/λ_0) and the slope of the bed. From the results in Figure 3.19, this is not the case in actuality.

EXAMPLE 5.14: BRETSCHNEIDER SPECTRA IN DEEP WATER AND WATERS OF FINITE DEPTH In this example, our task is to compare the Bretschneider spectra in deep water and in a water depth of 3 m. The measured deep-water average wave properties are $H_{avg0} = 1.50$ m and $T_{avg} = 7.00$ sec. For the Bretschneider spectrum corresponding to these average values, the modal period is $T_o \simeq 7.19$ sec. The deep-water Bretschneider formula for this application is

$$S_{B0}(T) = 0.00322T^3 e^{-0.000281T^4}$$

The deep-water spectrum obtained from this equation and that resulting from the substitution of the equation into eq. 5.106 are presented in Figure 5.23. We note that the modal period is greater where $h = 3$ m and the peak spectral density value is also larger.

The results in Figure 5.23 are somewhat misleading. It would appear that the energy in the waters of finite depth is greater than that of the corresponding waves in the open ocean because the area under the spectrum is greater. This is not the case. The energy intensity is greater where the depth is finite. That is, from eqs. 5.36 and 5.37, we see that the spectral density can be interpreted as the energy per unit water weight per unit surface area. The surface area in eq. 5.36 is $b\lambda$. In the shoaling process, the wavelength, λ, decreases as the depth (h) decreases. Thus, the energy per unit area (energy intensity) increases.

5.11 Closing Remarks

One of the more complicated areas of wave mechanics is that dealing with wave statistics. With this in mind, the early sections of this chapter are written assuming that the reader has only an elementary knowledge of statistics. Applications of the statistical techniques described in this chapter are applied to ocean engineering situations in Chapters 9 and 10.

The cited references (presented in the References at the end of the book) are those that the author believes are significant in the evolution of water-wave statistical analyses. The reader is encouraged to consult these references to obtain a thorough understanding of the topics discussed herein.

6 Wave Modification and Transformation

In Chapters 3 through 5, respectively, linear, nonlinear, and random waves are introduced, analyzed, and discussed. These waves are assumed to be affected by two physical boundaries, those being a flat, horizontal seafloor (or sea bed) and the free surface (at the air-water interface). In the present chapter, other boundaries are considered. These include vertical and sloping walls and sloping beds. The presence of these boundaries can cause the waves to be both modified (affecting the wave properties) and transformed (affecting the wave energy or energy flux). Specifically, the presence of boundaries causes waves to reflect, shoal, refract, and diffract. These wave phenomena and some of their engineering ramifications are discussed in the present chapter. An excellent "working document" covering the coastal engineering aspects of wave reflection, shoaling, refraction, and diffraction is the *Shore Protection Manual* of the Coastal Engineering Research Center (CERC) of the U.S. Army Corps of Engineers (see U.S. Army, 1984). A more recent CERC publication is the *Automated Coastal Engineering System* (User's Guide and Technical Reference), which is a computer-based document designed to assist coastal engineers in the predicting the behavior of waves (see Leenknecht, Szuwalski, and Sherlock, 1992). There are many other works available devoted to each wave phenomenon, the number being too large to individually reference in this chapter. For this reason, those works that are referred to in this chapter are those which are either encompassing or describe basic analyses, experiments, or prototype studies.

In Section 3.4, an introductory discussion of the phenomenon of reflection is presented. In that introductory discussion, it is shown that if left-running and right-running linear waves of equal heights and periods are superimposed upon each other, then a standing wave results, the height of which is twice that of either component traveling wave. The superposition of the waves then gives the same pattern as that resulting from perfect reflection from a vertical wall. Imperfect reflection results when part of the energy of the incident traveling wave is absorbed in by a barrier. In this chapter, perfect and imperfect wave reflections of monochromatic waves are discussed. The waves discussed include those that approach the reflecting barrier both directly and obliquely.

Also in Chapter 3 (Section 3.8), the phenomenon of shoaling is introduced. Shoaling is the process whereby the wave properties are affected by changing water depth. In Figure 3.19, results obtained from the application of the linear wave theory (based on the assumption of a horizontal bed) are presented with measured sloping-bed data. From the results in Figure 3.19, we see that the linear theoretical results agree qualitatively with the measured wave behavior but not quantitatively. The quantitative disagreement results from the linear theory's inability to account for either bed slope or deep-water wave steepness. In this chapter, we introduce the long-wave equation to improve our ability to analyze the shoaling process. The analysis is applied to shoaling monochromatic linear waves.

Two additional wave phenomena that result from the presence of physical barriers are refraction and diffraction. As discussed herein, refraction is the phenomenon whereby the direction of the wave front is changed due to changing water depths. Historically, the first analysis of water-wave refraction was done in a manner similar to that used in the analysis of optical wave refraction, leading to the application of Snell's law to water waves. Snell's law is attributed to the Dutch mathematician Willebrod Snell von Rayen, who lived from 1591 to 1626. In this chapter, the analysis based on Snell's law is introduced and discussed.

The phenomenon called diffraction involves the crestwise transfer of wave energy into regions of lower wave energy. These regions include shadow zones (regions hidden from direct wave action) and the neighborhoods of energy-absorbing bodies (either fixed or floating). The classical diffraction theory is first presented and discussed. That theory is then applied to diffraction caused by vertical thin barriers. The phenomenon of diffraction caused by energy-absorbing bodies is discussed and analyzed in Chapters 9 and 10.

An equation derived by J. C. W. Berkhoff (1972, 1976), called the mild-slope equation, has been found to predict all of the wave phenomena mentioned in the previous paragraphs, that is, shoaling, reflection, refraction, and diffraction. This equation is of the form of the Helmholtz equation, discussed in the book by Matthews and Walker (1970) and others. The mild-slope equation must be solved numerically because the coefficients of the equation change with water depth. Because of its versatility, much attention has been devoted to the equation since its introduction by Berkhoff. The mild-slope equation is derived in this chapter, and applications to both pure shoaling and shoaling with refraction are presented.

The material in this chapter leads to the analyses of the real effects in the coastal zone, presented in Chapter 7. The discussions in that chapter concentrate on empirical analyses (that is, those based on observations). From the empirical formulas, dimensionless parameters such as the surf similarity parameter result. Wave phenomena in the coastal zone have been identified with specific ranges of the surf similarity parameter. These ranges have been determined from both experimental and field studies.

6.1 Wave Reflection from Vertical Barriers

The term *wave reflection* refers to the reflection of the energy flux or power of the waves. See Section 3.7 for a discussion of the energy flux, and eq. 3.72 for a mathematical expression of that wave property resulting from the linear wave theory. The most general case involving wave reflection is that for which the energy flux of component waves of a directional sea (discussed in Section 5.8F) are partially reflected and partially absorbed by a structure. For example, see Yokoki, Isobe, and Watanabe (1992) and Dickson, Herbers, and Thornton (1995). In this section, discussions of perfect and imperfect reflections of monochromatic waves are presented, where the incident waves approach reflecting structures either directly or obliquely.

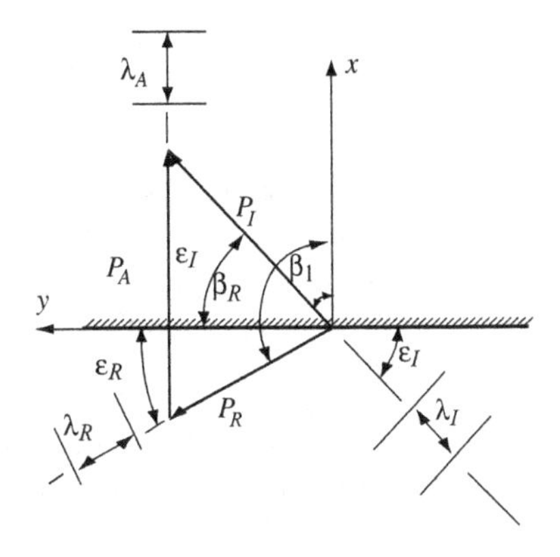

Figure 6.1. *Area Diagram of Incident Waves Partially Reflected and Partially Absorbed by a Vertical Structure.*

The general case of wave reflection is first considered. That is, the assumption is made that some structure, having the horizontal y-axis along its face, causes the wave energy flux to be partially reflected and partially absorbed. Consider the vector diagram in Figure 6.1. In that sketch, $\boldsymbol{P_I}$, $\boldsymbol{P_R}$, and $\boldsymbol{P_A}$ represent the respective energy fluxes of the incident, reflected, and absorbed waves. The "absorbed" wave is either an actual wave or an equivalent wave having an energy flux equal to that absorbed. The relationship involving these fluxes is

$$\boldsymbol{P_I} = \boldsymbol{P_R} + \boldsymbol{P_A} \tag{6.1}$$

As sketched in Figure 6.1a, the vector $\boldsymbol{P_A}$ is normal to the face of the structure in the positive x-direction. The energy flux vectors in eq. 6.1 can be expressed in terms of the component wave properties by using the linear wave expression in eq. 3.72. The resulting vector expression is

$$\frac{\rho g H_I^2 \boldsymbol{c_{gI}} b}{8} = \frac{\rho g H_R^2 \boldsymbol{c_{gR}} b}{8} + \frac{\rho g H_A^2 \boldsymbol{c_{gA}} b}{8} \tag{6.2}$$

where the vector $\boldsymbol{c_g}$ is the convection velocity vector of eq. 3.63, and b is the crest width. Referring to the diagram in Figure 6.2, this expression can be both simplified and written in terms of its respective onshore (in the x-direction) and alongshore (in the y-direction) components as

$$H_I^2|\boldsymbol{c_{gI}}|\cos(\beta_I) = H_R^2|\boldsymbol{c_{gR}}|\cos(\beta_R) + H_A^2|\boldsymbol{c_{gA}}| \tag{6.3a}$$

or

$$H_I^2 c_{gI}\sin(\varepsilon_I) = -H_R^2 c_{gR}\sin(\varepsilon_R) + H_A^2 c_{gA} \tag{6.3b}$$

in the onshore direction, and in the alongshore direction,

$$H_I^2|\boldsymbol{c_{gI}}|\sin(\beta_I) = H_R^2|\boldsymbol{c_{gR}}|\sin(\beta_R) \tag{6.4a}$$

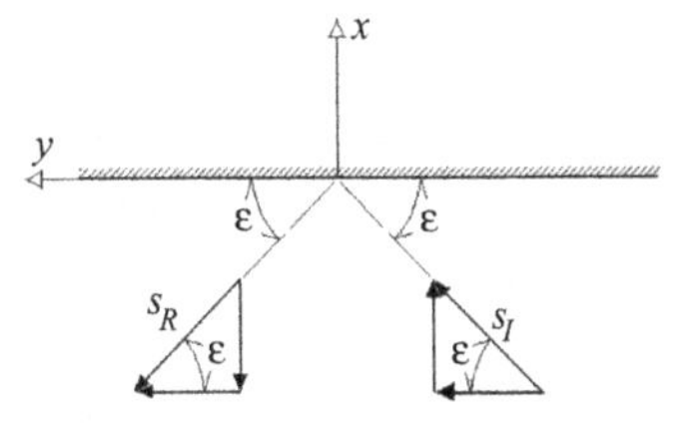

Figure 6.2. *Directional Vectors for Perfect Reflection from a Vertical, Flat Barrier.*

or

$$H_I^2 c_{gI} \cos(\varepsilon_I) = H_R^2 c_{gR} \cos(\varepsilon_R) \tag{6.4b}$$

where the angles ε and β are the directed angles of the component waves in Figure 6.1, positive in the clockwise direction. From the results in eq. 6.4, we see that the alongshore component of the energy flux is conserved.

Before applying these results, four basic assumptions are made. These are the following:

(a) The periods of the component waves represented in eq. 6.1 are equal.
(b) The bed in the region of the face, both for $x < 0$ and $x > 0$, is at a uniform depth.
(c) The face of the structure is vertical and pierces the free surface.
(d) The incident, reflected, and absorbed waves are all in phase at the face of the structure.

Assumptions (a) and (b) lead to the following relationships:

$$\lambda_I = \lambda_R = \lambda_A = \lambda \tag{6.5}$$

$$|\boldsymbol{c_I}| = |\boldsymbol{c_R}| = |\boldsymbol{c_A}| = c \tag{6.6}$$

and

$$|\boldsymbol{c_{gI}}| = |\boldsymbol{c_{gR}}| = |\boldsymbol{c_{gA}}| = c_g \tag{6.7}$$

A. Perfect Reflection of Linear, Monochromatic Waves

By definition, *perfect reflection* is the case when all of the incident energy flux is reflected back into the incident wave field. Also by definition, the absorbed energy flux vector, $\boldsymbol{P_A}$, in eq. 6.1, is zero; consequently, so are the last terms in both eqs. 6.3 and 6.4. For perfect reflection, eqs. 6.3 and 6.4 become

$$H_I^2 \sin(\varepsilon_I) = -H_R^2 \sin(\varepsilon_R) \tag{6.8}$$

and

$$H_I^2 \cos(\varepsilon_I) = H_R^2 \cos(\varepsilon_R) \tag{6.9}$$

respectively. By squaring and adding these equations, the wave height relationship is found to be

$$H_I = H_R = H \tag{6.10}$$

When this result is combined with the expression in eq. 6.8, the following relationship between the directed angles is obtained:

$$\varepsilon_I = -\varepsilon_R \tag{6.11a}$$

The results of eq. 6.11a show that for perfect reflection, the magnitudes of the angles of incidence and reflection are equal. In the remainder of this subsection, that angular magnitude is denoted as ε, that is,

$$|\varepsilon_I| = |-\varepsilon_R| = \varepsilon \tag{6.11b}$$

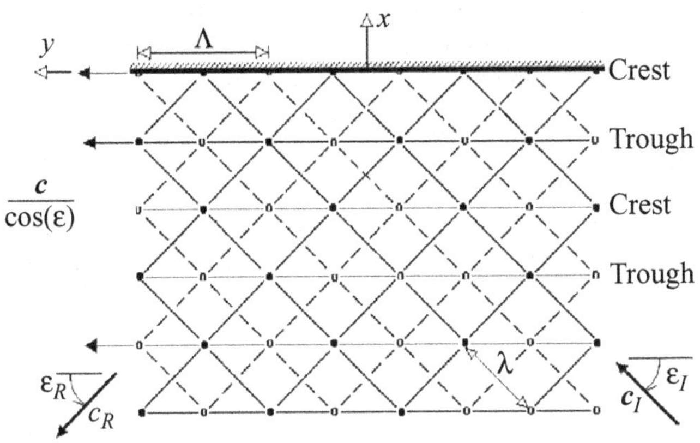

Figure 6.3. *Area Diagram of Perfect, Oblique Reflection (for an Incident Wave Angle of 45°).* Note that for nonlinear waves, the incident wave angle of 45° is a limiting case for the pattern shown. Wiegel (1964) discusses the oblique reflection of solitary waves from vertical walls, where $\varepsilon_I > 45°$. For this case, a near-wall wave front forms that is normal to the wall, and travels along the wall. Away from the wall, a wave pattern exists that is similar to that shown. Wiegel calls this reflection phenomenon the Mach-stem effect because of its similarity to the pattern of reflected shock waves at corners.

The free-surface displacements of the respective incident and perfectly reflected waves are mathematically represented by

$$\eta_I = \frac{H}{2}\cos(ks_I - \omega t) \tag{6.12}$$

and

$$\eta_R = \frac{H}{2}\cos(ks_R - \omega t) \tag{6.13}$$

Referring to the sketch in Figure 6.3, the magnitudes of the wave direction vectors ($\boldsymbol{s}_I$ and $\boldsymbol{s}_R$) can be written in terms of the onshore coordinate (x) and alongshore coordinate (y) as

$$s_I = x\sin(\varepsilon) + y\cos(\varepsilon) \tag{6.14}$$

and

$$s_R = -x\sin(\varepsilon) + y\cos(\varepsilon) \tag{6.15}$$

The free-surface displacement of the resulting wave is now obtained by, first, replacing wave coordinates in eqs. 6.12 and 6.13 by the respective expressions in eqs. 6.14 and 6.15, and then adding the results. The resulting expression for the free-surface displacement is

$$\begin{aligned}\eta &= \eta_I + \eta_R \\ &= \frac{H}{2}\cos\{k[x\sin(\varepsilon) + y\cos(\varepsilon)] - \omega t\} + \frac{H}{2}\cos\{k[-x\sin(\varepsilon) + y\cos(\varepsilon)] - \omega t\} \quad (6.16) \\ &= \frac{\mathrm{H}}{2}\cos[kx\sin(\varepsilon)]\cos[ky\cos(\varepsilon) - \omega t]\end{aligned}$$

where the height, $\mathrm{H} = 2H$, is that of the wave system, as in eq. 3.41. For direct (normal) reflection, $\varepsilon = \pi/2$. This angular value reduces eq. 6.16 to eq. 3.40.

The nodes (points of zero displacement) of the wave pattern corresponding to the expression in eq. 6.16 are found by equating that expression to zero and solving for the conditions on x and y. The resulting nodal lines have coordinates of

$$x_0 = -\frac{(2n-1)\lambda}{4\sin(\varepsilon)}, \quad y_0 = \left[\frac{t}{T} \pm \frac{(2n-1)}{4}\right]\frac{\lambda}{\cos(\varepsilon)}, \quad \text{where } n = 1, 2, \ldots. \tag{6.17}$$

The antinodes (points of either maximum or minimum displacements) are found by letting the derivatives of η with respect to x and y equal zero. The resulting coordinate expressions are

$$x_\pm = -\frac{n\lambda}{2\sin(\varepsilon)}, \quad y_\pm = \left[\frac{t}{T} \pm \frac{n}{2}\right]\frac{\lambda}{\cos(\varepsilon)}, \quad \text{where } n = 1, 2, \ldots. \tag{6.18}$$

The results in eqs. 6.17 and 6.18 show that the wave pattern is fixed with respect to the x-direction but moves in the y-direction. The speed at which these lines travel in the y-direction can be found by, first, letting the angle of the last cosine function in eq. 6.16 be equal to a constant, that is,

$$ky\cos(\varepsilon) - \omega t = \text{constant} \tag{6.19}$$

This corresponds to the free-surface displacement that we would see if we rode in a boat parallel to the barrier in the positive y-direction. The speed at which the boat would travel to maintain the condition in eq. 6.19 is found by taking the time-derivative of the expression in eq. 6.19. The resulting expression for the alongshore speed is then

$$\frac{dy}{dt} = \frac{\omega}{k}\frac{1}{\cos(\varepsilon)} = \frac{\lambda}{T}\frac{1}{\cos(\varepsilon)} = \frac{\Lambda}{T} = \frac{c}{\cos(\varepsilon)} \tag{6.20}$$

where Λ is the alongshore distance between two successive crests at the barrier, as illustrated in Figure 6.3. The expression in eq. 6.18 is infinite when the angle of incidence (ε) is $\pi/2$. Also in that figure, several wave conditions are illustrated. We see that the alongshore distance, Λ, increases with ε.

EXAMPLE 6.1: PERFECT OBLIQUE REFLECTION Consider the incident waves in Figures 6.1 and 6.2 to be approaching the reflecting barrier with an angle of incidence of 45°. From the results presented in eqs. 6.17 and 6.18, we can determine the wave pattern in the region $x \leq 0$. At time $t = 0$, the dimensionless coordinates of the nodal lines, from eq. 6.17, are

$$\frac{x_0}{\lambda} \simeq -0.354(2n-1), \quad \frac{y_0}{\lambda} \simeq \pm 0.354(2n-1), \quad \text{where } n = 1, 2, \ldots.$$

whereas the dimensionless coordinates for the antinodal lines, from eq. 6.18, are

$$\frac{x_\pm}{\lambda} \simeq -0.707n, \quad \frac{y_\pm}{\lambda} \simeq \pm 0.707n, \quad \text{where } n = 1, 2, \ldots.$$

The pattern is sketched in Figure 6.3. From the results of eq. 6.20, that pattern travels in the alongshore direction with a speed of

$$\frac{dy}{dt} \simeq 1.41c = 1.41\frac{\lambda}{T} \simeq \frac{\Lambda}{T}$$

The distance between successive crests at the barrier is then $\Lambda \simeq 1.41\lambda$.

B. Imperfect Reflection of Direct, Monochromatic, Linear Waves – Healy's Formula

In Section 3.4, the wave pattern resulting from the superposition of two waves of equal heights and periods traveling in opposite directions is analyzed. The result of the superposition is a standing wave having a height (H) that is twice the wave height (H) of either the incident or reflected traveling wave. The periods of the incident, reflected, and standing waves are equal in that case. Now, we shall assume that some of the energy is lost to the barrier when the incident wave approaches at an angle of 90° ($\epsilon = \pi/2$ in eq. 6.11b). Using the assumptions (a), (b), (c), and (d) stated prior to Section 6.1, the expression in eq. 6.4 vanishes, whereas the expression in eq. 6.3 reduces to

$$H_I^2 = H_R^2 + H_A^2 \tag{6.21}$$

Note that the negative sign in eq. 6.3 vanishes because of the relationship between the directed incident and reflected angles in eq. 6.11a.

It is common practice to write an expression that is equivalent to that in eq. 6.21 in terms of non-dimensional coefficients. The equivalent expression is

$$1 = K_R^2 + K_A^2 \tag{6.22}$$

which is obtained by dividing eq. 6.21 by the incident wave height. In eq. 6.22 are the *reflection coefficient*, defined by

$$K_R = \frac{H_R}{H_I} \tag{6.23}$$

and the *absorption coefficient*, defined by

$$K_A = \frac{H_A}{H_I} \tag{6.24}$$

Some writers, such as Wang and Ren (1992), define these coefficients as ratios of energies. Under our assumptions, the energy-based reflection and absorption coefficients are then equal to the squares of those in eqs. 6.23 and 6.24, respectively. The reader should also note that the absorption coefficient is sometimes referred to as the transmission coefficient. In the present analysis, the portion of the incident energy flux that is not reflected is assumed to be totally absorbed within the structure, with no transmission leeward of the structure. For this reason, the term absorption coefficient is used.

Now, let us examine the expression for the free-surface displacement of the wave pattern. That expression is

$$\begin{aligned}\eta &= \eta_I + \eta_R \\ &= \frac{H_I}{2}\cos(kx - \omega t) + \frac{H_R}{2}\cos(-kx - \omega t) \\ &= a_I\cos(kx - \omega t) + a_R\cos(-kx - \omega t) \\ &= (a_I + a_R)\cos(kx)\cos(\omega t) + (a_I - a_R)\sin(kx)\sin(\omega t)\end{aligned} \tag{6.25}$$

where $a_I (= H_I/2)$ and $a_R (= H_R/2)$ are the respective incident and reflected wave amplitudes. From the last equality, the maximum free-surface displacement from the SWL is

$$\eta_{\max} = a_I + a_R \tag{6.26}$$

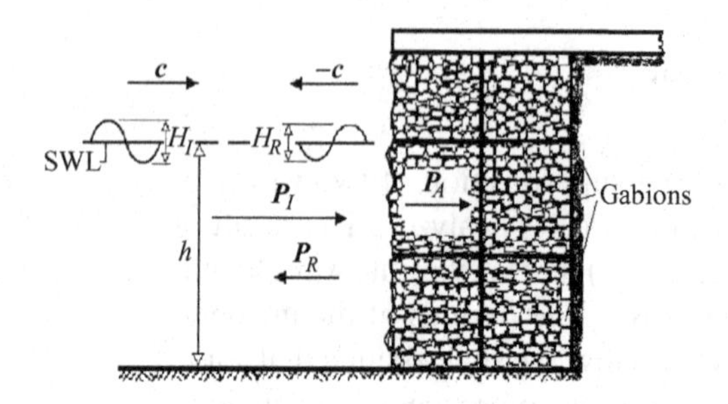

Figure 6.4. *Incident, Reflected, and Absorbed Energy Flux Vectors at a Gabion-Faced Barrier.*

and the minimum displacement is

$$\eta_{\min} = a_I - a_R \tag{6.27}$$

By solving eqs. 6.26 and 6.27 simultaneously, we find

$$a_I = \frac{\eta_{\max} + \eta_{\min}}{2} \tag{6.28}$$

and

$$a_R = \frac{\eta_{\max} - \eta_{\min}}{2} \tag{6.29}$$

Hence, the *reflection coefficient* for the case of partial reflection from a vertical barrier is

$$K_R \equiv \frac{H_R}{H_I} = \frac{a_R}{a_I} = \frac{\eta_{\max} - \eta_{\min}}{\eta_{\max} + \eta_{\min}} \tag{6.30}$$

The expression in eq. 6.30 was derived by Healy (1953), and is known as *Healy's formula.* The use of Healy's formula is illustrated in the following example.

EXAMPLE 6.2: DIRECT PARTIAL REFLECTION OF LINEAR WAVES In Example 3.5, waves of 0.5-m heights and 5-sec periods are perfectly reflected from a seawall in 2 m of water. The waves approach the seawall directly, so $\varepsilon = \pi/2$ in Figure 6.2. Because the reflection is direct and perfect, the height of the standing wave is 1.0 m and the wavelength (unaffected by the reflection) is about 20.9 m. A sketch of a standing wave is presented in Figure 3.8. Because of the relatively large value of the wave steepness (H/λ) of the standing wave, the docking of boats at the seawall under these wave conditions is somewhat precarious. To reduce the height of the wave pattern, a system of gabions is used, as illustrated in Figure 6.4. *Gabions* are cubic wire baskets that are filled with stones.

In Figure 6.5, the profiles of the free-surface displacements corresponding to the maximum and minimum wave components are shown. Theses profiles are obtained from eq. 6.25, using the results in eqs. 6.26 and 6.27. At the face of the modified structure in Figure 6.5, the maximum displacement (η_{max}) is measured to be 0.75 m. At a distance $x = -\lambda/4 - 5.23$ m (seaward) from the face of the structure, the maximum displacement is $\eta_{min} = 0.25$ m. The subscript *min* (for minimum) is used here because that value is the minimum value of the maximum free-surface displacement, as can be seen in Figure 6.5. Note that in Figure 3.8, we see that a node ($\eta = 0$) exists at $x = -\lambda/4$ when the incident wave is perfectly reflected. The resulting reflection coefficient value is 0.5 from eq. 6.30, and the corresponding *absorption coefficient* value is approximately 0.867 from eq. 6.22. The squares of the coefficients are proportional to the ratios of the respective energies and the incident wave energy. Then for this partially reflecting seawall, 0.25% of the incident energy is reflected and 0.75% is

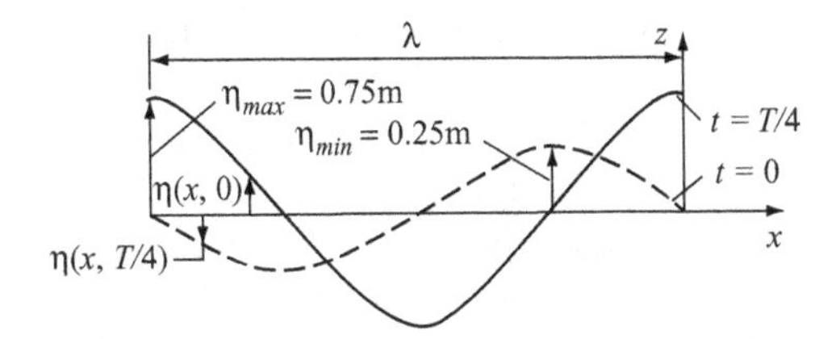

Figure 6.5. *Partially Reflected Wave Profiles at Times $t = 0$ and $t = T/4$.*

absorbed. The gabion system is then quite effective in reducing the wave reflection.

C. Reflection from a Vertical Porous Barrier

As stated in Example 6.2, the gabion system sketched in Figure 6.4 consists of cubical wire baskets filled with stones. Because of both the internal porosity and frictional effects, wave energy is both absorbed and dissipated within the structure. If the structure stands alone without a backing seawall, as sketched in Figure 6.6, then it is possible that some of the wave energy will be transmitted to the "quiet waters" in the lee of the structure. This type of structure is called a *breakwater*.

The analyses of the effects on waves of vertical porous structures include those of Sollitt and Cross (1972), Madsen (1974, 1983), Liu, Yoon, and Dalrymple (1986), and Wang and Ren (1992). Results of the two-dimensional analysis of Madsen (1974) are used here to illustrate how the porosity and internal friction relate to the reflection coefficient. As sketched in Figure 6.6, the rectangular, porous breakwater has a width B in the direction of wave travel. In his analysis of the internal flow caused by the waves, Madsen (1974) makes the following assumptions:

(a) The *porosity* (N), defined as the void volume-to-total volume ratio, is uniform throughout the structure.
(b) The structure is surface-piercing.
(c) The sea bed is flat and horizontal in $-\infty < x < +\infty$.
(d) The incident, reflected, "absorbed," and transmitted waves are all long, shallow-water linear waves, where $\eta \ll d$ and $\eta \ll \lambda$.

Referring to the sketch in Figure 6.6 for notation, Madsen (1974) finds that the reflection coefficient for the rectangular, porous breakwater is

$$K_R = \frac{H_R}{H_I} = \frac{1}{1 + \dfrac{2\mathrm{N}}{kBf_\mu}} = \frac{kBf_\mu}{kBf_\mu + 2\mathrm{N}} \tag{6.31}$$

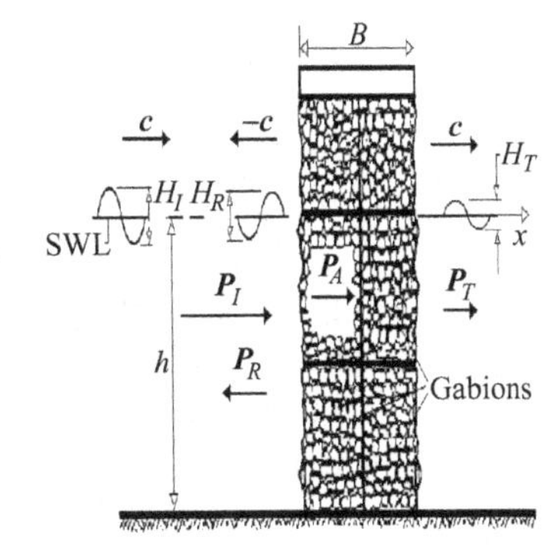

Figure 6.6. *Sketch of Incident, Reflected, "Absorbed," and Transmitted Waves at a Vertical Gabion Barrier.*

where H_R is the height of the reflected wave, N is the porosity, k is the wave number for all of the waves, B is the breadth of the breakwater, and f_μ is the linearized friction factor. The subscripts I, R, A, and T indicate properties of the incident, reflected, absorbed, and transmitted waves, respectively. The value of f_μ can be found analytically by using one of the various methods of Sollitt and Cross (1972), Madsen (1974, 1983), and others. The porosity values for selected stones are presented in Table 7–13 in the *Shore Protection Manual* (see U.S. Army, 1984). Corresponding to the reflection coefficient expression of eq. 6.31 is the *transmission coefficient*,

$$K_T = \frac{H_T}{H_I} = \frac{1}{1+\dfrac{kBf_\mu}{2\mathrm{N}}} = \frac{2\mathrm{N}}{2\mathrm{N}+kBf_\mu} \tag{6.32}$$

where H_T is the height of the transmitted wave.

The absorbed wave energy in this case is dissipated within the structure. The conservation of the energy flux for the system is expressed by

$$P_I = P_R + P_A + P_T \tag{6.33}$$

In eq. 6.33, the energy flux per crest width of any component wave is obtained from the shallow-water expression in eq. 3.74. The resulting expression is

$$\frac{P_j}{b} = \frac{\rho g^{3/2}(K_j H_I)^2\sqrt{h}}{8} \tag{6.34}$$

where b is the crest width and the subscript j identifies the wave in question. For the incident wave, $j = I$ and $K_I \equiv 1$. The absorbed, or dissipated, energy flux within the structure is obtained by combining the results of eqs. 6.31 through 6.34. The result is

$$\begin{aligned}\frac{P_A}{b} &= \frac{\rho g^{3/2}H_I^2\sqrt{h}}{8}\left(1-K_R^2-K_T^2\right)\\ &= \frac{\rho g^{3/2}H_I^2\sqrt{h}}{8}\left[\frac{4\mathrm{N}kBf_\mu}{(2\mathrm{N}+kBf_\mu)^2}\right]\end{aligned} \tag{6.35}$$

The expressions in eqs. 6.31, 6.32, and 6.35 have some physical limitations. The most apparent of these concerns the structural breadth, B. As the width increases without limit, the reflection coefficient of eq. 6.31 approaches unity. This result implies that a porous structure of semi-infinite width will reflect all of the wave energy, which is not true. For practical values of B (when compared to the wavelength), Madsen (1974) shows rather good agreement with experimental results.

The following example deals with experimental methods for determining the values of both the porosity and the linearized friction factor.

EXAMPLE 6.3: PARTIAL REFLECTION FROM A POROUS BREAKWATER A large-scale model of a gabion breakwater is to be studied in a wave tank. The breadth (B) of the structure, sketched in Figure 6.6, is 5 m, the water depth (h) is 2 m, and the tank breadth (W) is 2 m. The structure spans the tank, so the experiment is effectively two-dimensional. The goal of this study is to determine the effectiveness of the breakwater in protecting the quiet-water area leeward of the structure from the waves described in Example 6.2. Those waves are 0.5 m in height (H_I) and 5 sec in period (T). In that example, we find that the wavelength (λ) in 2 m of water is about 20.9 m.

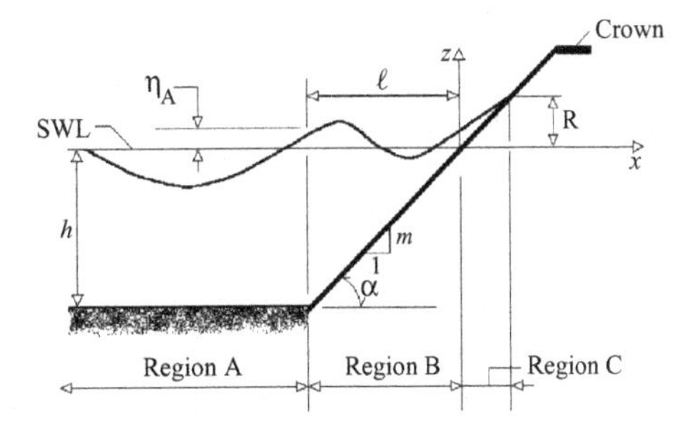

Figure 6.7. *Notation for Waves on Inclined Barriers.* The notation for the slope of the barrier face is $m \equiv \tan(\alpha)$.

First, we determine the volume of the gabions by placing them in still water and measuring the volume of water that they displace. That displaced volume is found to be approximately 10 m^3. The displaced volume divided by the total wetted volume of the structure ($5\,\text{m} \times 2\,\text{m} \times 2\,\text{m} = 20\,\text{m}^3$) is the value of the porosity; hence, $\text{N} = 0.5$.

In Example 6.2, the reflection coefficient (K_R) is found to be 0.5. We can now rearrange the expression in eq. 6.31 to obtain an expression for the linearized friction factor. That expression is

$$f_\mu = \frac{2\text{N}K_R}{kB(1 - K_R)} \tag{6.36}$$

where, for this study, $\text{N} = 0.5$, $K_R = 0.5$, $k = 2\pi/\lambda \simeq 0.301\,\text{m}^{-1}$, and $W = 5\,\text{m}$. The resulting value of the friction factor is approximately 3.33. This value in eq. 6.32 yields a transmission coefficient value of about 0.166. The percentage of the incident energy flux transmitted to the quiet water is proportional to ${K_T}^2$. Only 2.76% of the incident energy flux is transmitted. We can conclude then that the breakwater is an effective protective barrier for the wave conditions of this experiment.

The reader is encouraged to consult the references for more detailed discussions of partial reflection from porous structures.

6.2 Reflection from Inclined Barriers – The Long-Wave Equations

In the discussion of wave reflection from inclined barriers, it is important to keep in mind that some of the wave energy will be lost when breaking occurs on the structure. This is not the case for linear waves reflecting from vertical or near-vertical walls, as discussed in the previous section. For unbroken waves, there is an uprush of water on the barrier. The maximum height of the uprush is called the runup, *R*, as sketched in Figure 6.7. The ability to predict the value of the runup on near-vertical structures is needed to design for the prevention of *overtopping*. This phenomenon occurs when the runup is greater than the height of the crown of the barrier.

Most of the inclined barriers of interest are in the coastal zone, where the barriers are beaches, nearshore structures, or shoreline structures. High-energy, long-period waves in the coastal zone can be very damaging, causing either beach erosion or damage to the artificial structures. There is a class of long-period waves called long waves. Waves in this class lend themselves to rather simplistic mathematical analyses. *Long waves* are defined as those for which the wavelength is much greater than both the water depth and wave height. They are then shallow-water waves by nature. The experimental data of Murota and Yamada, as presented by Shuto

(1972), indicate that for long waves, near-perfect reflection from inclined barriers occurs for a barrier-face angle (α) of between approximately 45° and 90°. The reflection coefficient (K_R in eq. 6.23) is approximately equal to 1.0 for this slope range. For face angles less than 45°, the experimental results are somewhat mixed. Breaking may also occur on or seaward of the structure having small angles of inclination. Tsai, Wang, and Lin (1998) and others show that occurrence of breaking on an inclined barrier of a given wall slope (m) depends on the deep-water wave steepness and the depth-to-deep-water wavelength ratio.

There are a number of analytical methods available for predicting the behavior of waves on inclined barriers, such as seawalls and beaches. Many of these methods require the use of numerical techniques to obtain applicable results. In this section, we discuss both a *quasi*-closed-form method and a numerical method, both of which are based on the long-wave assumptions ($\lambda \gg h$ and $\lambda \gg H$). The long-wave equations are developed in the following subsection.

A. The Long-Wave Equations

Begin the analysis by assuming that two traveling linear waves are present well away from the barrier (in the far field). Those are an incident right-running wave (η_I) and a reflected left-running wave (η_R). Following Dean (1964), the resulting free surface can be mathematically represented by

$$\begin{aligned} \eta &= \eta_I + \eta_R \\ &= E_{I1}(x)\cos(\omega t + \sigma_I) + E_{I2}(x)\sin(\omega t + \sigma_I) \\ &= E_{R3}(x)\cos(\omega t + \sigma_R) - E_{R4}(x)\sin(\omega t + \sigma_R) \end{aligned} \tag{6.37}$$

where the amplitude functions, E(x), and the phase angles, σ, are determined from the boundary conditions. The boundary values for E(x) are defined later in this chapter.

Referring to the sketch in Figure 6.7, linear monochromatic waves are assumed to exist in Region A. In this region, the free-surface displacement is η_A, and the E-functions in eq. 6.37 are simply wave amplitudes of the component waves. The E-functions for the respective incident and reflected waves can be written in terms of wave heights as $E_{I1} = E_{I2} = H_{IA}/2$ and $E_{R1} = E_{R2} = H_{RA}/2$. The incident wave height (H_{IA}) is assumed to be known, and the height (H_{RA}) of the reflected wave is to be determined. Without loss in generality, we can assume that the phase angle of the incident wave in Region A is zero, and that of the reflected wave, σ_A, is to be determined. The expression for the free-surface displacement in Region A is then

$$\begin{aligned} \eta_A = \eta_{IA} + \eta_{RA} = {} & \frac{H_{IA}}{2}[\cos(k_A x)\cos(\omega t) + \sin(k_A x)\sin(\omega t)] \\ & + \frac{H_{RA}}{2}[\cos(k_A x)\cos(\omega t + \sigma_A) - \sin(k_A x)\sin(\omega t + \sigma_A)] \end{aligned} \tag{6.38}$$

In the analysis of the long waves encountering the inclined barrier, assume that there is no energy lost from the system due to either friction or percolation into the barrier. Furthermore, assume that waves travel in a channel of uniform width, b, and in the direction normal to the barrier waterline, that is, the wave properties do not vary in the y-direction (between the channel walls).

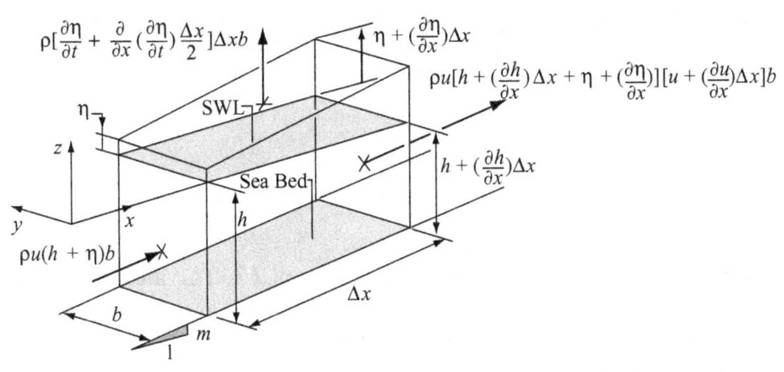

Figure 6.8. *Mass Flow through a Volume Element beneath the Free Surface.*

Now consider the wave-traveling element over the inclined barrier sketched in Figure 6.8. We can mathematically express the conservation of mass for the incompressible flow in the control volume in that figure by

$$\rho u(h+\eta)b$$
$$= \left[\rho\left(h+\frac{\partial h}{\partial x}\Delta x+\eta+\frac{\partial \eta}{\partial x}\Delta x\right)\left(u+\frac{\partial u}{\partial x}\Delta x\right)+\rho\left(\frac{\partial \eta}{\partial t}+\frac{\partial^2 \eta}{\partial x \partial t}\frac{\Delta x}{2}\right)\Delta x\right]b$$
$$\simeq \left[\rho u(h+\eta)+\rho h\frac{\partial u}{\partial x}\Delta x+\rho u\frac{\partial h}{\partial x}\Delta x+\rho\eta\frac{\partial u}{\partial x}\Delta x+\rho\frac{\partial \eta}{\partial t}\Delta x\right]b \tag{6.39a}$$

where ρ is the mass density of water. In the approximation in eq. 6.39a, the terms containing $(\Delta x)^2$ are assumed to be of second order and negligible. Because of this assumption, the expression in eq. 6.39a reduces to the following form of the continuity equation:

$$h\frac{\partial u}{\partial x}+u\frac{\partial h}{\partial x}+\eta\frac{\partial u}{\partial x}+\frac{\partial \eta}{\partial t}\simeq 0 \tag{6.39b}$$

The conservation of linear momentum of the water in the control volume is expressed by Euler's equation (eq. 2.67). In component form, the application of that equation to the water element in Figure 6.8 is

$$\rho\frac{\partial u}{\partial t}+\rho u\frac{\partial u}{\partial x}=-\frac{\partial p}{\partial x} \tag{6.40a}$$

and

$$\rho\frac{\partial w}{\partial t}+\rho w\frac{\partial w}{\partial z}=-\rho g-\frac{\partial p}{\partial z} \tag{6.40b}$$

The assumption of long waves allows for the simplification of these component equations. The first simplification is due to the fact that the spacial variations in the particle velocity components are of second order when compared to the other terms in the expressions of eqs. 6.40a and 6.40b. Secondly, long waves are by nature shallow-water waves. From the results in eq. 3.57, we see that the time variation of vertical displacement of a water particle is much smaller than that of the horizontal components in shallow water. Hence, the corresponding vertical velocity component and its derivatives in time and space are relatively small compared to the gravitational and pressure terms in eq. 6.40b. Neglecting these small terms and the second-order terms in the component equations of eq. 6.40 results in

$$\rho \frac{\partial u}{\partial t} \simeq -\frac{\partial p}{\partial x} \tag{6.41a}$$

and

$$0 = -\rho g - \frac{\partial p}{\partial z} \tag{6.41b}$$

The latter equation of these is the hydrostatic equation, eq. 2.3. The integration of the second equation from the free surface ($z = \eta$, where $p = 0$) to any depth position yields the pressure at any depth. The expression for the pressure is

$$p = \rho g(\eta - z) \tag{6.42}$$

The expressions in eqs. 6.39, 6.41a, and 6.42 can be combined to form a single equation by eliminating both the horizontal velocity term, u, and the pressure, p. The resulting expression is

$$-gh\frac{\partial^2 \eta}{\partial x^2} - g\frac{\partial \eta}{\partial x}\frac{\partial h}{\partial x} + \frac{\partial^2 \eta}{\partial t^2} \simeq 0 \tag{6.43}$$

where the depth is $h = h(x)$ in the most general case.

In the following two subsections, eq. 6.43 is solved for the two limiting cases of wave reflection. First, the case of perfectly reflecting barriers is analyzed. This is followed by an analysis of waves on a nonreflecting barrier.

B. Perfect Reflection from an Inclined Barrier

For an inclined barrier with a flat face of slope m, as sketched in Figure 6.7, the depth of the water at any position over the inclined barrier is expressed by

$$h = \tan(\alpha)(\ell - x) = m(\ell - x) \tag{6.44}$$

where the origin of the coordinate system is over the toe of the barrier at a distance ℓ from the intersection of the SWL and the structure (the barrier waterline). Equation 6.44 yields a positive water depth seaward of the origin. For the inclined flat barrier, the combination of eqs. 6.43 and 6.44 yields

$$-gm(\ell - x)\frac{\partial^2 \eta}{\partial x^2} + gm\frac{\partial \eta}{\partial x} + \frac{\partial^2 \eta}{\partial t^2} \simeq 0 \tag{6.45}$$

To solve eq. 6.43, we can assume the product solution,

$$\eta = \mathrm{X}(x)\mathrm{T}(t) \tag{6.46}$$

as is done in the solution of Laplace's equation, eq. 3.8. From the expression in eq. 6.38, we see that the time component of the separated equation is

$$\frac{d^2\mathrm{T}}{dt^2} = -\omega^2\mathrm{T} \tag{6.47}$$

where ω is the circular wave frequency. The solution of this second-order linear differential equation is a linear combination of sine and cosine functions of ωt. The combination of the expressions in eqs. 6.45, 6.46, and 6.47 results in the differential equation for $X(x)$ of a long wave on an inclined barrier. That equation is

$$-(\ell - x)\frac{d^2X}{dx^2} + \frac{dX}{dx} - \frac{\omega^2}{gm}X = 0 \tag{6.48}$$

This equation is in a form of the Bessel equation in x. See Chapter 9 of the book edited by Abramowitz and Stegun (1965). The solution of eq. 6.45 is then written in terms of Bessel functions of the first kind, $J_n(\ )$, and the second kind, $Y_n(\ )$, both of zero order ($n = 0$). See Appendix A for a summary of the properties of these functions. Following Dean (1964), the solution of eq. 6.45 for the free-surface displacement in Region B ($0 \le x \le \ell$) is

$$\begin{aligned}\eta_B &= A_I J_0(2K\sqrt{\ell - x})\cos(\omega t + \sigma_{I_{IB}}) + A_I Y_0(2K\sqrt{\ell - x})\sin(\omega t + \sigma_{IB}) \\ &\quad + A_R J_0(2K\sqrt{\ell - x})\cos(\omega t + \sigma_{RB}) - A_R Y_0(2K\sqrt{\ell - x})\sin(\omega t + \sigma_{RB})\end{aligned} \tag{6.49}$$

where

$$K \equiv \frac{\omega}{\sqrt{gm}} = k_A\sqrt{\frac{h_A}{m}} = k_A\sqrt{\ell} \tag{6.50}$$

The amplitude coefficients, A_I and A_R, are to be determined, as are the phase angles, σ_I and σ_R. Note that the argument of the Bessel functions is a real function seaward of the shoreline and an imaginary function landward of the shoreline. Landward of the shoreline ($x > \ell$) in Region C of Figure 6.7, the Bessel functions $J_0(\ )$ and $Y_0(\ )$ are replaced by functions of modified Bessel functions $I_0(\ )$ and $K_0(\ )$. The relationships for the zero-order Bessel functions are

$$J_0(is) = I_0(s), \quad Y_0(is) = i\,I_0(s) - \frac{2}{\pi}K_0(s) \tag{6.51}$$

and s is either a variable or a parameter. If s is a dependent variable, then the spacial derivatives of the Bessel functions of zero order are

$$\frac{dJ_0(s)}{dx} = -J_1(s)\frac{ds}{dx}, \quad \frac{dY_0(s)}{dx} = -Y_1(s)\frac{ds}{dx} \tag{6.52}$$

and

$$\frac{dI_0(s)}{dx} = I_1(s)\frac{ds}{dx}, \quad \frac{dK_0(s)}{dx} = -K_1(s)\frac{ds}{dx} \tag{6.53}$$

where the subscript "1" identifies Bessel functions of the first order. In our application, the variable s is $2K\sqrt{(\ell - x)}$, where K is defined in eq. 6.50. The modified Bessel functions are used landward of the origin, where $x > \ell$. Again, the properties of Bessel functions are summarized in Appendix A. From Appendix A and eq. 6.51, we see that both the J_0-function and I_0-function equal 1.0 at the origin ($x = \ell$), whereas the Y_0-function equals $-\infty$ and the K_0-function equals $+\infty$ at the origin.

Equation 6.49, which applies from the toe of the barrier ($x = 0$) to the maximum runup ($x = \ell + R/m$), has four unknowns. Those are the amplitudes, A_I and A_R, and the phase angles, σ_{IB} and σ_{RB}. Recall that the reflected wave height, H_{RA}, and phase angle, σ_A, in eq. 6.38 are also unknowns. Six boundary conditions are required to determine these six unknowns.

For large wall angles, assume that there is no energy lost due to breaking. The energy of the reflected wave is therefore equal to that of the incident wave. Because the wavelengths of both waves must be equal, the wave height of the reflected wave must be equal to that of the incident wave ($H_{RA} = H_{IA}$). Following Lamb (1945), we also require that the free-surface displacement be finite at the origin, which is at the barrier waterline. Because the absolute values of Bessel functions of the second kind are infinite at the origin, this requirement is met if A_I and A_R are equal in eq. 6.49, as are the phase angles σ_{IB} and σ_{RB}. In region B then, let the amplitude coefficient be B_B and the phase angle be σ_B. Equation 6.49 is now

$$\eta_B = 2B_B J_0(2K\sqrt{\ell - x})\cos(\omega t + \sigma_B) \tag{6.54}$$

Because the first-order Bessel function (J_0) becomes the modified Bessel function (I_0) in Region C, the free-surface displacement in this region is mathematically represented by

$$\eta_C = 2B_B I_0(2K\sqrt{x - \ell})\cos(\omega t + \sigma_B) \tag{6.55}$$

where, again, $I_0[2K\sqrt{(\ell - x)}]$ is a modified Bessel function of the first kind, zero order, as defined in eq. 6.51. Referring to the plots of the Bessel functions in Appendix A, the reader can see that the coefficient of the trigonometric function in the expression in eq. 6.54 plots as a "wavy" curve in Region B, whereas that in the expression in eq. 6.55 plots as a diverging curve. Physically, the water motion landward of the origin is that of an oscillating flume, as illustrated in Figure 6.11 and discussed in Example 6.4. The maximum values of η_B and η_C at the origin occur when the cosine terms in eqs. 6.54 and 6.55 are both equal to unity. Because the free surface at the origin is continuous, the amplitude coefficients must be equal, as must be the phase angles.

To determine the remaining unknowns (B_B, σ_A, and σ_B), the boundary conditions at the toe ($x = 0$) are used. At the toe of the barrier, we require the free-surface displacements of Regions A and B to be equal. Mathematically,

$$\eta_A|_{x=0} = \eta_B|_{x=0} \tag{6.56}$$

where, for a perfectly reflected wave, eq. 6.38 applied at the toe is

$$\begin{aligned}\eta_A|_{x=0} &= (\eta_{IA} + \eta_{RA})|_{x=0} \\ &= H_A \cos\left(\frac{\sigma_A}{2}\right)\cos\left(\omega t + \frac{\sigma_A}{2}\right)\end{aligned} \tag{6.57}$$

after some trigonometric manipulations. The expression in eq. 6.54 at the toe is

$$\eta_B|_{x=0} = 2B_B J_0(2K\sqrt{\ell})\cos(\omega t + \sigma_B) \tag{6.58}$$

where the wave number at the toe for the shallow-water wave is

$$k_A \equiv k|_{xA} = \frac{2\pi}{\sqrt{gh_A}T} = \frac{2\pi}{\lambda_A} \tag{6.59}$$

Also at the toe of the barrier, we require the equality of the slopes of the free surfaces, that is,

$$\frac{\partial \eta_A}{\partial x}|_{x=0} = \frac{\partial \eta_B}{\partial x}|_{x=0} \tag{6.60}$$

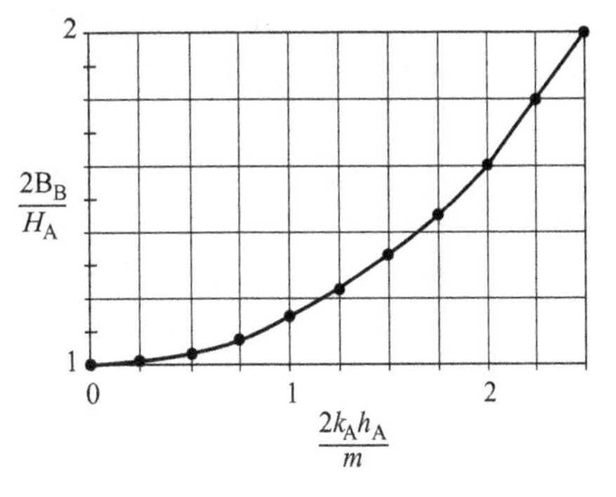

Figure 6.9. *Amplitude-Function Ratio Versus Non-Dimensional Water Depth.* In this figure, the bed slope is $m = \tan(\alpha)$, as sketched in Figure 6.7.

By applying these boundary conditions and separating the coefficients of $\cos(\omega t)$ and $\sin(\omega t)$, the desired unknown relationships are obtained. Those are the amplitude function,

$$\mathrm{B_B} = \frac{H_\mathrm{A}}{2}\frac{1}{\sqrt{J_0^2(2\mathrm{K}\sqrt{\ell}) + J_1^2(2\mathrm{K}\sqrt{\ell})}} \tag{6.61}$$

and the phase angle,

$$\begin{aligned}\sigma_\mathrm{A} &= -2\tan^{-1}\left[\frac{J_1(2\mathrm{K}\sqrt{\ell})}{J_0(2\mathrm{K}\sqrt{\ell})}\right] \\ &= 2\sigma_\mathrm{B}\end{aligned} \tag{6.62}$$

The dimensionless amplitude coefficient, $2\mathrm{B_B}/H_\mathrm{A}$, and the phase angle, σ_A, are presented as functions of $2k_\mathrm{A}h_\mathrm{A}/m$ in Figures 6.9 and 6.10, respectively. From the results in these figures, the reader can see that B_B approaches $H_\mathrm{A}/2$ and σ_A approaches zero as the slope, m, approaches infinity (the slope of a vertical wall).

The free-surface expressions in eqs. 6.54 and 6.55 are actually equivalent because a positive value of $(\ell - x)$ results in a real number in eq. 6.54, whereas a negative value results in an imaginary number. The latter results in the modified Bessel function in eq. 6.55. Because of this, we can replace these two equations by the following single equation:

$$\eta_\mathrm{BC} = \frac{H_\mathrm{A}\, J_0(2\mathrm{K}\sqrt{\ell - x})}{\sqrt{J_{0\mathrm{A}}^2 + J_{1\mathrm{A}}^2}}\cos\left[\omega t - \tan^{-1}\left(\frac{J_{1\mathrm{A}}}{J_{0\mathrm{A}}}\right)\right] \tag{6.63}$$

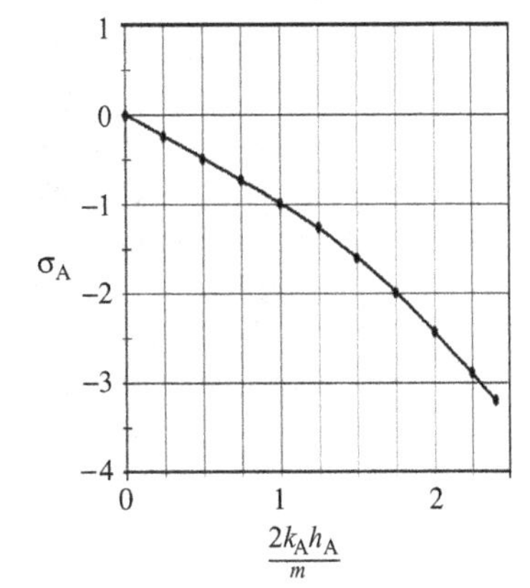

Figure 6.10. *Phase-Angle Variation with Non-Dimensional Water Depth.*

where the results of eqs. 6.61 and 6.62 have been incorporated. Also in eq. 6.63 are the following Bessel function symbols:

$$J_{0\mathrm{A}} \equiv J_0(2\mathrm{K}\sqrt{\ell}) = J_0\left(2k_{\mathrm{A}}\frac{h_{\mathrm{A}}}{m}\right) = J_0(2k_{\mathrm{A}}\ell) \tag{6.64}$$

and

$$J_{1\mathrm{A}} \equiv J_1(2\mathrm{K}\sqrt{\ell}) = J_1\left(2k_{\mathrm{A}}\frac{h_{\mathrm{A}}}{m}\right) = J_1(2k_{\mathrm{A}}\ell) \tag{6.65}$$

The runup, R, is the maximum value of η_{BC} at the intersection with the surface of the barrier (see the sketch in Figure 6.7). In other words, it is the maximum height of the uprush of water on the surface. The value of R is found from eq. 6.63 by replacing both η_{BC} by R and x by $\ell + \mathrm{R}/m$ while letting the angle of the cosine term equal zero. The resulting expression is

$$\eta_{\mathrm{BC}}|_{x=\ell+\frac{\mathrm{R}}{m}} = \mathrm{R} = \frac{H_{\mathrm{A}} I_0\left(2\mathrm{K}\sqrt{\frac{\mathrm{R}}{m}}\right)}{\sqrt{J_{0\mathrm{A}}^2 + J_{1\mathrm{A}}^2}} \tag{6.66}$$

This equation is transcendental because R cannot be isolated. The solution of R can be obtained using a numerical technique such as the method of successive approximations, the method used in the determination of intermediate wavelength values in Section 3.3. Again, our design goal is to ensure that the crown of the barrier is greater than the maximum (design) value of R to prevent overtopping, that is, the phenomenon of water passing over the barrier.

The behaviors of the waves on relatively steep inclined barriers is illustrated in the following two examples.

EXAMPLE 6.4: TOTALLY REFLECTED WAVES ON AN INCLINED BARRIER An inclined seawall is to be designed to protect a roadway. Up to this time, the roadway has been protected from both waves and high water by a vertical seawall. The purpose of the design is to see if inclining the seawall at an angle of 45° (for which $m = 1$ in Figure 6.7) will increase the seawall's effectiveness. The water depth (h_{A}) in the region is 1.5 m, and is uniform throughout Region A. The design wave is one for which the period (T) is 10.0 sec and the height (H_{A}) is 0.5 m. The deep-water wavelength corresponding to the 10-sec period is $\lambda_0 = 156$ m from eq. 3.36. To use the equations that are derived in this section, we must first ensure that the design wave is a shallow-water wave. To make this determination, use the method of successive approximations to determine the wavelength (λ_{A}) at the toe of the structure (see Section 3.3). The resulting wavelength is approximately 38.0 m, a value approximately predicted by the shallow-water expression in eq. 3.38. The depth-to-wavelength ratio is $0.0395 \cong 1/25.3$, which satisfies the shallow-water condition in Section 3.5, that is, $h_{\mathrm{A}}/\lambda_{\mathrm{A}} < 1/20$.

The behavior of the free surface over the seawall is determined from the expression in eq. 6.63. The application of that expression to the seawall in question results in

$$\eta_{\mathrm{BC}}|_{x\leq\ell} = 0.515 J_0(0.401\sqrt{\ell - x})\cos(0.628t - 0.248), \quad x \leq \ell$$

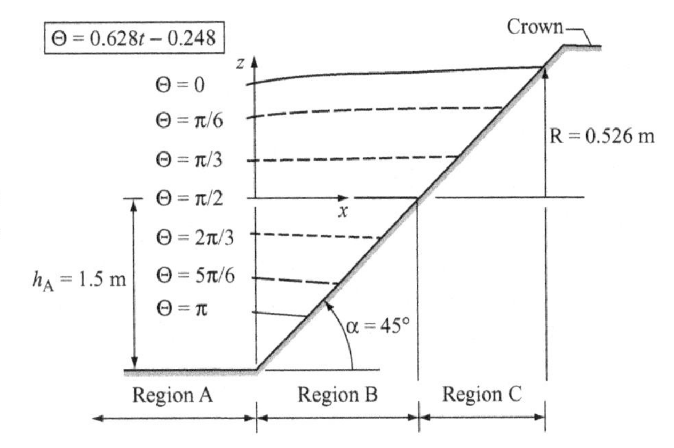

Figure 6.11. *Free-Surface Profiles at a Flat Barrier Inclined at 45°*. Hence, $m = 1$.

and

$$\eta_{BC}|_{x>\ell} = 0.515\, I_0(0.401\sqrt{x-\ell})\cos(0.628t - 0.248), \quad x > \ell$$

To obtain the free-surface profile landward of the toe over one period, simply allow the angle in the cosine term to take on values of 0, $\pi/4$, $\pi/2$, $3\pi/2$, π, and so on. The results are sketched in Figure 6.11. The intersection of the profile corresponding to the zero angle and the seawall is 0.526 m above the SWL. This then is the value of the runup, R. If the seawall is vertical, then the water would rise to a height equal to the incident wave height, H_A, or 0.5 m. We conclude that a decrease in slope causes an increase in the runup, at least in the range of $1 \leq m \leq \infty$, and our goal of preventing overtopping is less attainable by decreasing the face slope of the seawall.

One final comment concerning the runup on inclined barriers: Saville (1956) conducted a series of wave tank experiments in which he studied the effects of face slope (m), deep-water wave steepness (H_0/λ_0), and relative toe depth (h_A/H_0) on the relative runup (R/H_0). He found that for a given face angle, the relative runup varies significantly with the deep-water wave steepness, increasing as the steepness decreases. His results are also presented in the *Shore Protection Manual* (see U.S. Army, 1984).

C. Nonreflecting Beaches

We now focus our attention on beaches, which can be considered to be inclined barriers with small slopes. In this subsection, the slope (m) of the beach is assumed to be so small that there is no wave reflection. This is an approximation because all beaches reflect part of the incident energy back to sea. The breaking phenomenon, introduced in Chapter 3 and discussed further in Section 6.7, also reduces the reflected energy. When waves break on a beach of unsaturated sand, then the water (and its energy) percolates into the sand. Goda (1970) presents a breaking index that allows the coastal engineer to predict the breaking conditions on beaches of small slope. Based on the data of Goda and others, McCormick and Cerquetti (2003) present empirical formulas for the use of wave analysts.

In the analysis of waves on nonreflecting beaches, the long-wave equations of Section 6.2A are used. Referring to the sketch in Figure 6.12, the length (ℓ) of the

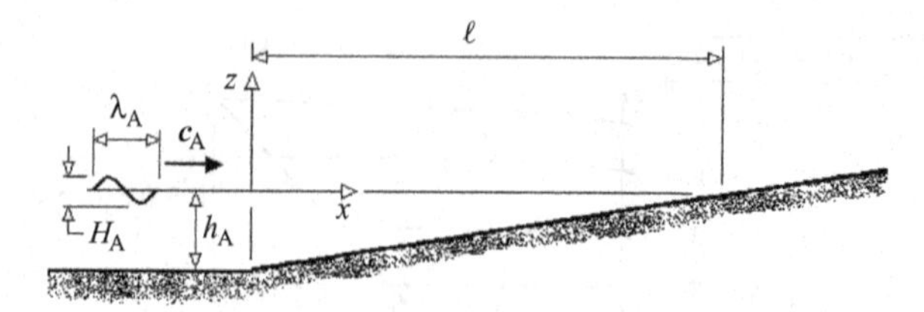

Figure 6.12. *Notation for Waves on a Nonreflecting Beach.*

beach, from the toe to the shoreline, is assumed to be much greater than the length (λ_A) of the incident wave. The free-surface displacement of this wave approaching the toe of the beach is

$$\eta_A = \frac{H_A}{2}\cos(k_A x - \omega t) \qquad (6.67)$$

where the origin of the x-axis is over the toe of the bed, as in Section 6.2B. As the wave travels over the sloping bed of the beach, the free-surface displacement is represented by

$$\eta_B = B_{B1} J_0(2K\sqrt{\ell - x})\cos(\omega t + \sigma_B) + B_{2B} Y_0(2K\sqrt{\ell - x})\sin(\omega t + \sigma_B) \qquad (6.68)$$

where B_{B1} and B_{B2} are amplitude coefficients.

At the toe of the beach, where $x = 0$, the boundary conditions of eqs. 6.56 and 6.60 apply. Physically, these respective conditions dictate that the magnitudes and slopes of the free surfaces of the incident and transmitted waves must be equal. From the application of the first toe condition, the following expression for the amplitude coefficient is obtained:

$$B_{B1} = \frac{H_A}{2}\frac{\sqrt{Y_{0A}^2 + Y_{1A}^2}}{(J_{1A}Y_{0A} - J_{0A}Y_{1A})} = \frac{Y_{1A}}{J_{0A}}B_{B2} = -\frac{Y_{0A}}{J_{1A}}B_{B2} \qquad (6.69)$$

where J_{0A} and J_{1A} are the abbreviated notation for the Bessel functions of the first kind, first and second order, respectively, and Y_{0A} and Y_{1A} are the abbreviated Bessel functions of the second kind, all applied at $x = 0$ (see eqs. 6.64 and 6.65). The phase angle in eq. 6.68 is

$$\sigma_B = \tan^{-1}\left(\frac{Y_{0A}}{Y_{1A}}\right) = -\tan^{-1}\left(\frac{J_{1A}}{J_{0A}}\right) \qquad (6.70)$$

from the results in eq. 6.62. The combination of eqs. 6.68, 6.69, and 6.70 yields the following expression for the free-surface displacement of the wave traveling on a nonreflecting beach:

$$\eta_B = \frac{H_A}{2}\frac{\sqrt{Y_{0A}^2 + Y_{1A}^2}}{(J_{0A}Y_{1A} - J_{1A}Y_{0A})}\{J_0(2K\sqrt{\ell - x})\cos(\omega t + \sigma_B) + Y_0(2K\sqrt{\ell - x})\sin(\omega t + \sigma_B)\} \qquad (6.71)$$

where K is defined in eq. 6.50.

EXAMPLE 6.5: SHOALING ON A NONREFLECTING BEACH The incident waves in Example 6.4 approach a beach having a 1/1000 (rise over run) slope, that is, $m = 0.001$. From Example 6.4, the toe of the beach is in 1.5 m of water ($h_A = 1.5$ m), where the incident wave has a height (H_A) of 0.5 m and a period (T) of 10 sec. In that example, we find that the deep-water wavelength (λ_0) is 156 m. The wavelength (λ_A) of the shallow-water wave is approximately 38.4 m, so the toe of the beach is about 39 wavelengths from the shoreline. Our interest is in the maximum free-surface displacement at a site halfway between the toe and the shoreline, that is, at $x = \ell/2 = 750$ m.

The small slope results in large arguments of Bessel functions. For large arguments (s), the Bessel functions can be approximated by

$$J_0(s) \simeq -Y_1(s) \simeq \sqrt{\frac{2}{\pi s}}\cos\left(s - \frac{\pi}{4}\right) + \mathrm{O}\left[\frac{1}{s}\right] \tag{6.72}$$

and

$$J_1(s) \simeq Y_0(s) \simeq \sqrt{\frac{2}{\pi s}}\sin\left(s - \frac{\pi}{4}\right) + \mathrm{O}\left[\frac{1}{s}\right] \tag{6.73}$$

from Abramowitz and Stegun (1965). The notation $\mathrm{O}[1/s]$ in these expressions signifies the order of magnitude of the remaining terms. The approximations are valid for $s \geq 15$. For the $m = 0.001$ beach, $2\mathrm{K}\sqrt{\ell} \simeq 491.384$, and the values of the Bessel Functions are $J_{0\mathrm{A}} \simeq -Y_{1\mathrm{A}} \simeq 0.031411$ and $J_{1\mathrm{A}} \simeq Y_{0\mathrm{A}} \simeq 0.017576$. At the site, where $x = \ell/2$, we have $2\mathrm{K}\sqrt{(\ell/2)} \simeq 347.461$. For this value, the Bessel functions are $J_0(\ell/2) \simeq -Y_1 \simeq 0.019402$ and $J_1(\ell/2) \simeq Y_0(\ell/2) \simeq 0.038154$. The numbers of terms retained to the right of the decimal points might seem spurious; however, they are required for the accuracy of the trigonometric functions. The angles of these functions are in radians.

After a few trigonometric manipulations, the approximation of the expression in eq. 6.71 is

$$\begin{aligned}\eta_B &= \frac{H_A}{2}\left(\frac{\ell}{\ell - x}\right)^{\frac{1}{4}}\cos\left(2\mathrm{K}\sqrt{\ell - x} - \frac{\pi}{4} + \omega t + \sigma_B\right) \\ &= \frac{H_B}{2}\cos[2\mathrm{K}(\sqrt{\ell - x} - \sqrt{\ell}) + \omega t] = \frac{H_B}{2}\cos[2\mathrm{K}(\sqrt{\ell} - \sqrt{\ell - x}) - \omega t]\end{aligned} \tag{6.74}$$

From the definition of the shoaling coefficient (K_S) in eq. 3.78, we find the following relationships for the wave heights:

$$\frac{H_B}{H_A} = \frac{H_B}{H_0}\frac{H_0}{H_A} = \frac{K_{SB}}{K_{SA}} = \left(\frac{h_A}{h_B}\right)^{1/4} \tag{6.75}$$

By applying the shoaling coefficient expression to the shallow-water conditions at A and over the beach, we find that the relationship in eq. 6.75 can also be obtained using Airy's linear theory discussed in Chapter 3. The linear theory can also be used to determine the celerity c_B and the wavelength λ_B. To prove this statement, simply "ride" with the crest of the wave so that the angle of the cosine in the last term in eq. 6.74 appears to be constant. The time-derivative of the constant angle leads to the same shallow-water celerity expression as the linear theory expression in eq. 3.38.

From the results in Example 6.5, we see that for nonreflecting beaches of small slope, Airy's linear theory yields the same wave height, celerity, and wavelength expressions as the theory outlined in this section for $2\mathrm{K}\sqrt{(\ell - x)} \geq 15$. For values less than 15, the expressions in eqs. 6.70 and 6.71 must be used.

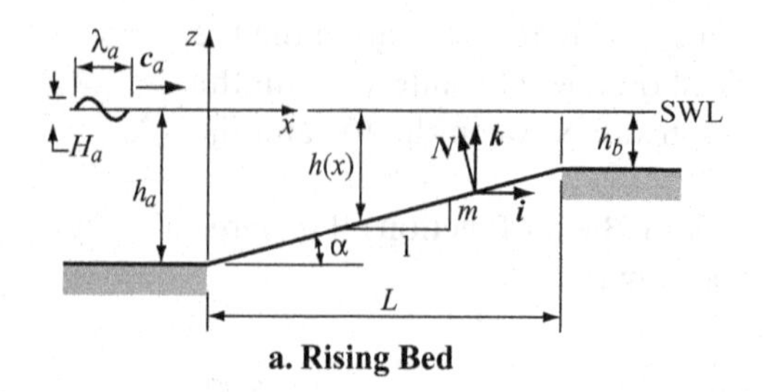

a. Rising Bed

b. Step Approximation

Figure 6.13. *Step Approximation for a Rising Sea Bed.*

D. Reflection from a Bed of Intermediate Slope

Up to this point, our study of wave reflection has been focused on totally reflecting vertical or near-vertical surfaces and on nonreflecting beaches. For both, the surfaces are assumed to be flat and surface-piercing. We now introduce an approximate method for determining the partial reflection from beds of arbitrary slope that may or may not be surface piercing.

As done by Rey, Belzons, and Guazzelli (1992), Twu and Liu (1999), and others, assume that the bed sketched in Figure 6.13a can be approximately represented by a series of small steps as in Figure 6.13b. The height (δh) and length (δx) of each step are related to each other by the local bed slope, $m(x)$, according to

$$\frac{\delta h}{\delta x} = -m(x) \tag{6.76}$$

The analysis can also be applied to down-sloping beds by simply changing the sign of m in eq. 6.76. The steps are assumed to reflect only the portion of the wave that is incident upon them, that is, a step reflects the energy in a vertical distance δh from the base of the step. After the wave passes over the nth step, the wave is assumed to have fully adjusted to the new water depth, which is $(h_n - \delta h)$ for the negatively sloping bed. The energy of the reflected mini-wave is then subtracted from that of the total of the incident wave, and the difference is the energy transmitted to the next step.

In the following derivation, the reflective bed surface is assumed to be flat, having a constant slope of $-m$, and extending from a horizontal bed of depth h_a to a horizontal bed of depth h_b, as illustrated in Figure 6.13. The sloping bed is approximated by N steps of equal size, although the equality of step size is not a requirement. If the horizontal length of the bed is L, then

$$\frac{h_a - h_b}{N} = m\frac{L}{N} \tag{6.77a}$$

Also, for any depth $h_a > h(x) > h_b$, we can write

$$\frac{h_a - h(x)}{n} = m\frac{x}{n} \tag{6.77b}$$

where n is the number of the step and $1 \leq n \leq N$. As is done by Twu and Liu (1999), we assume that the wave passing over any step is linear and that the sloping bed ends at the shoreline. The latter assumption then requires that $h_b = 0$. Twu and Liu further assume that the total energy of the reflected mini-wave is proportional to its kinetic energy. Mathematically, the time-averaged energy per unit volume of the mini-wave reflected from the nth step is

$$\begin{aligned} e_n &\propto \frac{1}{T}\int_0^T \frac{\rho}{2}\left(u_n^2 + w_n^2\right)\Big|_{h=h_n} dt \\ &\propto \cosh^2[k_n(z+h_n)] + \sinh^2[k_n(z+h_n)] = \cosh[2k_n(z+h_n)] \end{aligned} \tag{6.78}$$

where u_b and w_n are the respective horizontal and vertical velocity components of the fluid particles. By using the energy relationships of the reflected mini-wave, both over the step and over the depth h_a, we can write the expression for the wave height of the reflected mini-wave over h_a in terms of the incident wave height H_a. The result is

$$H_n = \frac{H_a}{\Sigma_n}\sqrt{\cosh^2[k_n(z+h_a)]} \tag{6.79}$$

where

$$\Sigma_N \equiv \sum_{n=0}^{N-1}\sqrt{\cosh\left(2nk_a\frac{h_a}{N}\right)} \tag{6.80}$$

One must keep in mind that the reflected mini-waves, when passing over the bed of depth h_a, have the same wavelength (and wave number) as the incident wave. The properties that differ in the incident and reflected waves are the wave height and the phase between the waves. These mini-waves coalesce to form the one reflected wave over h_a. The free-surface displacement of that wave is

$$\begin{aligned} \eta_R &= \frac{H_a}{2\Sigma_N}\left\{\sum_{n=0}^{N-1}\sqrt{\cosh\left(2nk_a\frac{h_a}{N}\right)}\cos\left(k_a x + \omega t - nk_a\frac{2L}{N}\right)\right\} \\ &= \frac{H_a}{2\Sigma_N}\left\{\sum_{n=0}^{N-1}\sqrt{\cosh\left(2nk_a\frac{h_a}{N}\right)}\left[\cos\left(k_a x + \omega t\right)\cos\left(nk_a\frac{2L}{N}\right)\right.\right. \\ &\quad \left.\left. + \sin(k_a x + \omega t)\sin\left(nk_a\frac{2L}{N}\right)\right]\right\} \\ &= \frac{H_R}{2}\cos(k_a x + \omega t - \sigma_R) = \frac{H_R}{2}[\cos(k_a x + \omega t)\cos(\sigma_R) \\ &\quad + \sin(k_a x + \omega t)\sin(\sigma_R)] \end{aligned} \tag{6.81}$$

In eq. 6.81, the wave height H_R is that of the reflected wave and σ_R is the phase angle between the incident wave and the reflected wave. In the first equality of eq. 6.81, the phase angle $nk_a(2L/N)$ is based on the total distance that the mini-wave (reflected by the nth step) must travel from the toe of the bed to the step and back to the toe. Because of the coefficients of cosine and sine terms of the second and fourth

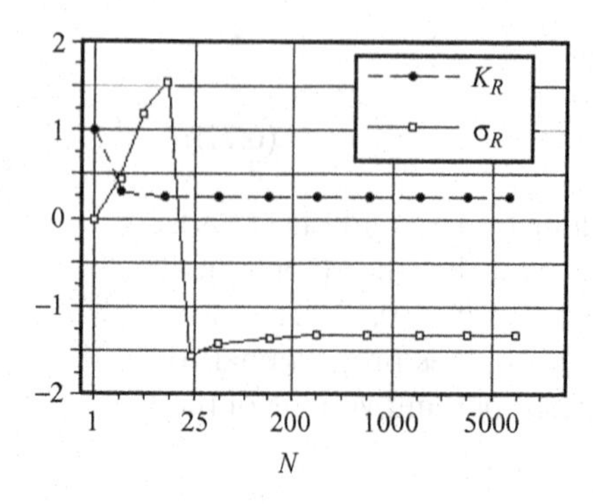

Figure 6.14. *Convergence for the Reflection Coefficient (K_R) and the Phase Angle (σ_R) for Bed Slope (m) of 0.5 and a Wave Period (T) of 7 sec.*

equalities, eq. 6.81 is actually a system of two equations with two unknowns, those unknowns being H_R and σ_R. Hence, these two wave properties can be determined. The resulting expressions for these respective properties are

$$H_R = \frac{H_a}{2\Sigma_N} \times \sqrt{\left\{\sum_{n=0}^{N-1} \sqrt{\cosh\left(2nk_a\frac{h_a}{N}\right)} \cos\left(nk_a\frac{2L}{N}\right)\right\}^2 + \left\{\sum_{n=0}^{N-1} \sqrt{\cosh\left(2nk_a\frac{h_a}{N}\right)} \sin\left(nk_a\frac{2L}{N}\right)\right\}^2} = H_a K_R \tag{6.82}$$

were K_R is the reflection coefficient, and,

$$\sigma_R = \arctan \frac{\left\{\sum_{n=0}^{N-1} \sqrt{\cosh\left(2nk_a\frac{h_a}{N}\right)} \sin\left(nk_a\frac{2L}{N}\right)\right\}}{\left\{\sum_{n=0}^{N-1} \sqrt{\cosh\left(2nk_a\frac{h_a}{N}\right)} \cos\left(nk_a\frac{2L}{N}\right)\right\}} \tag{6.83}$$

Obviously, the accuracy of the step method depends on the number N of steps chosen. Twu and Liu (1999) show that for some incident wave-steepness and bed-slope combinations, a good reflection coefficient convergence (in eq. 6.82) is obtained by using as few as seven steps. However, we found that the convergences for the phase angle in eq. 6.83 requires more steps. This is illustrated in the following example.

EXAMPLE 6.6: CONVERGENCE OF THE REFLECTION COEFFICIENT AND PHASE ANGLE
A 7-sec deep-water wave approaches a bed having a slope (m) of 0.5. The value of kh corresponding to the first shoaling contour (where $h_0 = \lambda_0/2$) is the deep-water value, that is, $k_0h_0 = \pi$. This value is that of k_ah_a in eqs. 6.82 and 6.83. From those respective equations, the convergences of K_R and σ_R are shown in Figure 6.14. From the results in Figure 6.14, the reader can see that the reflection coefficient value is approximately 0.24, with a satisfactory convergence obtained with 30 steps. For the phase angle, the value of approximately −1.32 radians is obtained with 50 steps.

Although the expressions in eqs. 6.82 and 6.83 appear to be somewhat "messy," they are quite straightforward, and are not difficult to program for a computer.

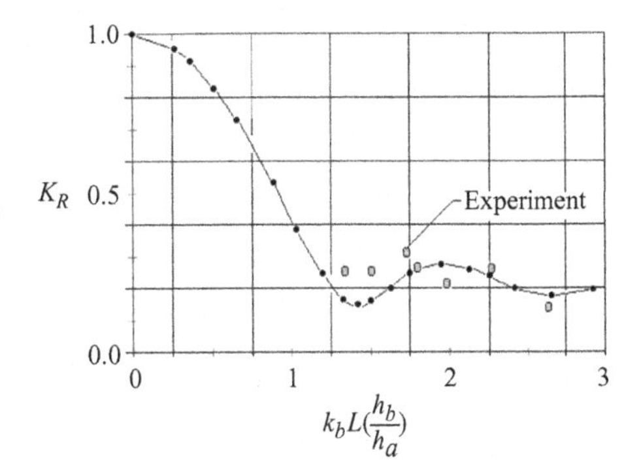

Figure 6.15. *Reflection Coefficients from Eq. 6.82 and Experiments of Bourodimus and Ippen (1966).*

EXAMPLE 6.7: REFLECTION FROM A BED TRANSITION In this example, we apply the analysis of Twu and Liu (1999) to a bed having a rising transition in the direction of wave travel, as sketched in Figure 6.13. In that sketch, we see that the bed has a depth of h_a before the rise and h_b following the rise. The relationships of these limiting water depths, the length (L) and the slope (m) of the transition, are found in eq. 6.77a. The values used in this example are those studied experimentally by Bourodimus and Ippen (1966). The water depths in that experiment are varied, as are the wave periods. The fixed experimental parametric values are $L = 2.44$ m and $m = 0.125$. Values of the reflection coefficient obtained from eq. 6.82 and the Bourodimus-Ippen experiment are presented in Figure 6.15. The parameter $k_b L(h_b/h_a)$ is that used by Bourodimus and Ippen (1966). Comparing the theoretical and experimental results in Figure 6.15, we see that the agreement between the predicted and observed values is good for the water depths and periods of the experiment. Hence, the expansion method of Twu and Liu (1999) can be used to predict the values of the reflection coefficient, not only from beaches but from transitions of the bed.

We now consider waves approaching shoals and beaches at angles greater than zero. The direction of these waves is altered by the changing depth of the bed – a process called refraction.

6.3 Refraction without Reflection – Snell's Law

Refraction occurs when a wave approaches the bottom contours at some nonzero angle. To understand the phenomenon, consider the case of linear deep-water waves obliquely approaching a shoreline over a bed having straight and parallel contours, sketched in Figure 6.16. The case of pure shoaling over a similar bed is sketched in Figure 3.17. For the case of refraction, the deep-water wave crests are at an angle β_0 to the first shoaling contour, that contour being at a depth (h_0) approximately equal to $\lambda_0/2$, assuming the linear wave theory of Chapter 3. Assume that a portion of the wave is traveling in a fictitious channel having vertical walls (called orthogonals) separated by a distance b along the crest. In deep water, this distance is b_0. In the analysis, assume that the energy flux (P) between orthogonals does not vary as the wave approaches the shoreline. From this assumption, the energy flux between orthogonals at any point in the shoal is equal to the energy flux in deep water. Mathematically, this condition is

$$P_0 b_0 = \frac{\rho g H_0^2 c_{g_0} b_0}{8} = P_b = \frac{\rho g H^2 c_g b}{8} \tag{6.84}$$

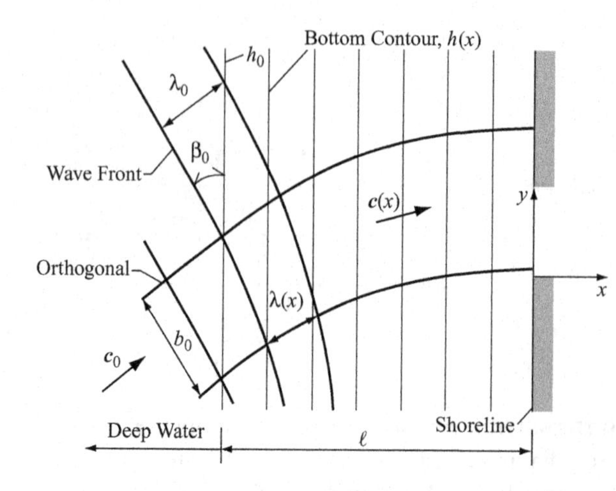

Figure 6.16. *Area Sketch of Refracting Waves on a Beach Having Straight and Parallel Bottom Contours.* The wave fronts are curved because the celerity values change along the wave fronts due to the decreasing water depth – the process called refraction.

From this equation, we obtain the following relationship between the deep-water wave height and that over any bed contour, which is

$$\frac{H}{H_0} = \sqrt{\frac{c_{g0}}{c_g}}\sqrt{\frac{b_0}{b}} \equiv K_S K_r \tag{6.85}$$

where K_S is the shoaling coefficient and K_r is called the *refraction coefficient.* In this section, the expression for the shoaling coefficient is that in eq. 3.78. To determine the values of the refraction coefficient, assume that the first shoaling contour is a nonreflecting step, as sketched in Figure 6.17. Hence, we can construct two triangles having a common hypotenuse of length Y_0, called the alongshore distance, over the deep-water contour. From geometric considerations, we can write

$$\frac{b_0}{b} = \frac{Y_0\cos(\beta_0)}{Y_0\cos(\beta)} = \frac{\cos(\beta_0)}{\cos(\beta)} = K_r^2 \tag{6.86}$$

and

$$\frac{Y_0}{Y_0} = 1 = \frac{\lambda_0/\sin(\beta_0)}{\lambda/\sin(\beta)} = \frac{c_0/\sin(\beta_0)}{c/\sin(\beta)} \tag{6.87}$$

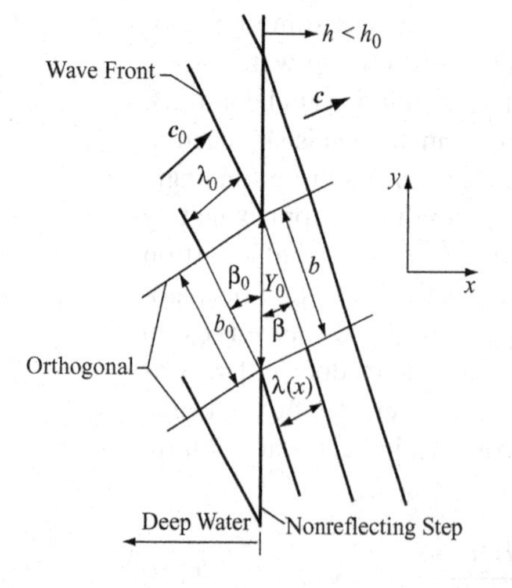

Figure 6.17. *Wave Refraction over a Nonreflecting Step.* The waves are assumed to be instantly bent as the waves pass over the step.

From eq. 6.86, we obtain the following expression for the refraction coefficient:

$$K_r = \sqrt{\frac{\cos(\beta_0)}{\cos(\beta)}} \tag{6.88}$$

The relationship between the wavelength and celerity with the wave angle is obtained from eq. 6.87, which is

$$\frac{\lambda_0}{\lambda} = \frac{c_0}{c} = \frac{\sin(\beta_0)}{\sin(\beta)} \tag{6.89}$$

This relationship is called *Snell's law*, as in the fields of optics and acoustics. The expression in eq. 6.89 is somewhat misleading. It appears that the wavelength ratio and the celerity ratio are functions of the wave angle. This is not the case because the wavelength and celerity are only functions of the water depth and wave period, as derived in Chapters 3 and 4. Snell's law then shows the dependence of the wave angle on the wavelength (or celerity).

The wave height relationship in eq. 6.85 can now be written as

$$\frac{H}{H_0} = \sqrt{\frac{c_{g0}\cos(\beta_0)}{c_g\cos(\beta)}} = \sqrt{\frac{\cos(0)}{\cos(\beta)}}\sqrt{\frac{c_{g0}}{c_g}}\sqrt{\frac{\cos(\beta_0)}{\cos(0)}} = \frac{H'}{H'}\frac{H'}{H'_0}\frac{H'_0}{H_0} \tag{6.90}$$

These relationships need some explanation. We shall see in Chapter 7 that most of the empirical formulas available to predict breaking and runup on beaches result from experiments that are without refraction. It is common practice to use the prime (′) to indicate pure shoaling, as is done in eq. 6.90. Unfortunately, it is also common practice to use the prime to represent the spatial derivatives, as in Chapter 9. In this chapter, the primes represent pure shoaling. Because the celerity and, therefore, the group velocity are independent of the wave angle, the middle terms of the last equality of eq. 6.90 must be equal; that is, the equivalent purely shoaling wave height ratio is

$$\frac{H'}{H'_0} = \sqrt{\frac{c_{g0}}{c_g}} \tag{6.91}$$

The last terms of the last equality of eq. 6.90 can be also be equated. From this equality, we obtain the equivalent deep-water wave height expression, which is

$$H'_0 = H_0\sqrt{\cos(\beta_0)} \tag{6.92}$$

as $\cos(0) = 1$. The importance of this last relationship will be demonstrated later in this chapter.

EXAMPLE 6.8: SHOALING AND REFRACTION ON A STRAIGHT, PARALLEL CONTOURED BEACH Deep-water 1.0-m, 7-sec waves approach a beach having a bed slope (m) of 0.05. On Monday, the waves approach directly ($\beta_0 = 0°$ in Figures 6.16 and 6.17), whereas on Tuesday, waves having the same wave height and period approach at a deep-water angle of 45°. To determine how refraction affects the wave height, we compare the wave heights at a water depth of 3 m on the two days. The onshore distance traveled by the incident waves from the h_0 ($= \lambda_0/2$) contour to the site contour is $L = (h_0 - h)/m \simeq 70$ m. The deep-water wavelength and celerity values are obtained from eq. 3.36, and the group velocity value for the deep-water waves is obtained from eq. 3.64. Those respective values are $\lambda_0 \simeq 76.5$ m, $c_0 \simeq 10.9$ m/s, and $c_{g0} \simeq 5.45$ m/s. From eq. 3.31, the respective wavelength and celerity values at the site are $\lambda \simeq 36.4$ m and $c \simeq 5.20$ m. The group velocity at the site is $c_g \simeq 4.79$ m/s from eq. 3.63. As expressed by the expression in eq. 6.89, Snell's law yields a wave angle at the site of $\beta \simeq 19.7°$. The shoaling coefficient and refraction coefficient values are, respectively, $K_S \simeq 1.0682$ (from eq. 3.78) and $K_r \simeq 0.8665$ (from eq. 6.88). These values applied to eq. 6.85 yield a wave height value of $H \simeq 0.926$ m. For $K_r = 1$ (pure shoaling), $H' \simeq 1.0682$ m.

The results of Example 6.8 show that the effect of depth-induced refraction is to retard the growth of the wave height. From eq. 6.88, we see that the refraction coefficient is equal to or less than one. The results in Figures 3.18 and 3.19 show that the shoaling coefficient can be equal to, less than, or greater than one, depending on the ratio of the values of the water depth at the site and the wavelength.

Assuming that the incident wave properties (height, period, and wave angle) and water depth values are known, the procedure in solving refraction problems is as follows:

(a) Determine the incident wavelength, celerity, and group velocity.
(b) Determine the wavelength, celerity, and group velocity at the site.
(c) Using the results in (a) and (b) determine the wave angle at the site using Snell's law (eq. 6.89).
(d) Determine the value of the shoaling coefficient from eq. 3.78.
(e) Using the site wave angle value from (c) determine the refraction coefficient from eq. 6.88.
(f) Determine the wave height at the site from eq. 6.85.

More will be said of refraction later in this chapter in the discussion of the mild-slope equation.

6.4 Diffraction

When waves travel past a body, a region called the *shadow zone* is created in the lee (down-wave) of the body. The water particles in this region are shielded from the incident waves approaching the body on the weather (up-wave) side. To illustrate, consider the situation sketched in Figure 6.18. In that figure, linear waves are incident upon a large-diameter circular cylinder. The term "large" here means that the diameter (D) of the cylinder is of the order of magnitude of the incident wavelength (λ). In the natural processes, nature avoids energy voids. Because the shadow zone is a region of no direct wave energy, energy is transferred into the shadow zone by the process called *diffraction*. We can define diffraction then as the transfer of wave energy in a direction that is parallel to the wave front and into regions of lower wave energy.

A classic of water-wave diffraction theories is that of Penny and Price (1944, 1952). That theory was extended and applied to engineering situations by Putnam and Arthur (1948), Blue and Johnson (1949), Johnson (1952), and Wiegel (1962). These studies are discussed and summarized in the book by Wiegel (1964). Graphical results of both Johnson (1952) and Wiegel (1962) are found in the *Shore Protection Manual* of the CERC of the U.S. Army Corps of Engineers (see U.S. Army, 1984). The diffraction theory of Sommerfeld (1896), which is applied to

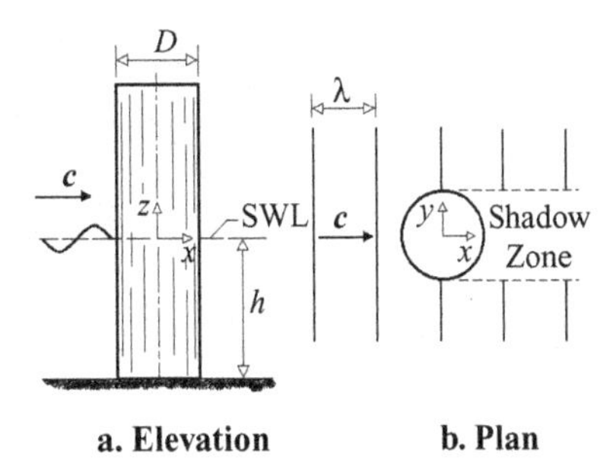

Figure 6.18. *Waves Incident on a Large-Diameter Vertical Cylinder.* The shadow zone is the region into which wave energy is transferred by the process called diffraction.

Fresnel-Kirchhoff diffraction of light waves, can be considered to be a foundation for the twentieth-century water-wave diffraction theories. Penny and Price (1944, 1952) modified that theory for application to water waves incident on both fully reflecting and totally absorbing semi-infinite breakwaters and, in addition, to waves passing through a gap separating the heads (ends) of a pair of semi-infinite breakwaters.

The Penny-Price diffraction analyses are presented here following an introductory discussion of diffraction phenomenon. The introductory discussion of diffraction is rather lengthy because it is intended to describe to the reader how the equations evolve. For application of the diffraction theory to breakwaters, the book by Wiegel (1964) is recommended. In addition, the reader is also encouraged to consult the book by Elmore and Heald (1985) for a thorough coverage of the mathematical physics of the diffraction of other wave forms, including light and sound waves.

A. Huygens' Principle

The diffraction phenomenon occurs in all wave forms, such as acoustic and optical waves. The earliest analytical works in wave diffraction were in the field of optics. One of these works was by the Dutch physicist Christian Huygens, who in 1678 derived a geometrical theory based on the assumption that light is a wave phenomenon, according to Halliday and Resnick (1978). Shortly before this, Robert Hooke had proposed that light was a traveling wave phenomenon, as noted by Elmore and Heald (1985). Referring to the sketch in Figure 6.19, an important concept attributed to Huygens is that the wave front at a time $t + \Delta t$ is the tangential surface to hemispherical wavelets originating on the front at time t. This is called *Huygens' principle*, and is used today as a basic principle in the analyses of both refraction and diffraction of different types of waves. The books by Sommerfeld (1954), Halliday and Resnick (1978), and Elmore and Heald (1985) are recommended for excellent discussions on the diffraction phenomena and Huygens' principle. A variation of Huygens' principle is used in the analysis of waves on the

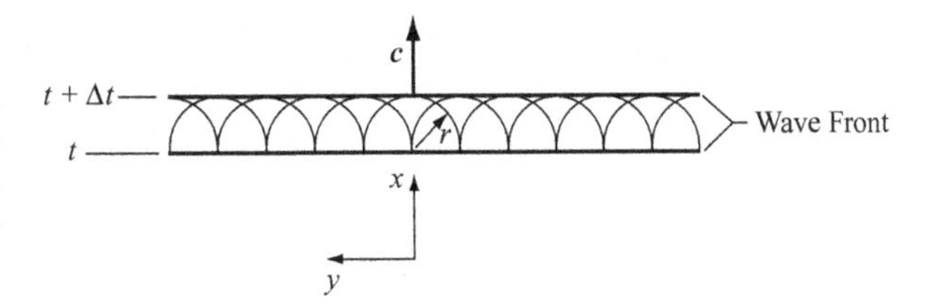

Figure 6.19. *Advancing Wave Fronts According to Huygens' Principle.* Each point on the wave front at time t is the source of a hemispherical wavelet. The wavelets coalesce at time $t + \Delta t$, forming the new wave front.

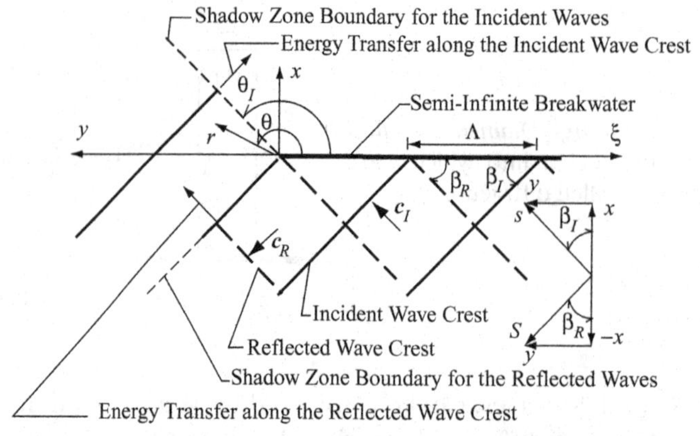

Figure 6.20. Wave Reflection and Diffraction at a Semi-Infinite Breakwater in Waters of Uniform Depth. Note that there are two shadow zones, one for the incident waves and one for the reflected waves. The case of perfect reflection is sketched.

free surface of a liquid. In that case, Huygens' wavelets are assumed to be semi-cylindrical rather than hemispherical. This variation of Huygens' principle is used herein.

B. Basic Equations and Boundary Conditions in the Analysis of Diffraction

Consider incident linear water waves approaching a thin, semi-infinite breakwater, as sketched in Figure 6.20. The water depth (h) is assumed to be uniform in the neighborhood of the breakwater. Let the waves be traveling in the direction of the s-coordinate, where the velocity potential (similar to that in eq. 3.23) is

$$\begin{aligned}\Phi_I = \varphi_I e^{-i\omega t} &= \frac{H_I}{2}\frac{g}{kc}\frac{\cosh[k(z+h)]}{\cosh(kh)}\sin(ks - \omega t)\\ &= \frac{H_I}{2}\frac{g}{kc}\frac{\cosh[k(z+h)]}{\cosh(kh)}\sin[k[x\cos(\beta_I) + y\sin(\beta_I)] - \omega t] \qquad (6.93)\\ &= Z(z)\Re\{ie^{ik[x\cos(\beta_I)+y\sin(\beta_I)]-i\omega t}\}\end{aligned}$$

where the subscript I identifies the incident wave properties. When the wave clears the head (end) of the breakwater, the process of diffraction begins. The velocity potential is altered due to diffraction, and can be written in complex form as

$$\Phi_D = \varphi_D e^{-i\omega t} = \frac{H_I}{2}\frac{g}{kc}\frac{\cosh[k(z+h)]}{\cosh(kh)}\Re\{(x, y)e^{-i\omega t}\} \qquad (6.94)$$

where the subscript D is used to identify the diffracted wave properties. The notation φ is used to represent the time-independent part of the velocity potential. The spacial function, $F(x,y)$, is to be determined. The velocity potential in eq. 6.94 must satisfy Laplace's equation, eq. 3.8, which is the mathematical expression for the equation of continuity for an incompressible, irrotational flow. When the expressions in eqs. 3.8 and 6.94 are combined, the result leads to

$$\nabla^2\varphi_D = \frac{\partial^2\varphi_D}{\partial x^2} + \frac{\partial^2\varphi_D}{\partial y^2} + k^2\varphi_D = 0 \qquad (6.95a)$$

The latter equality of this equation is called the *Helmholtz equation*. An alternative form of eq. 6.95a is obtained by using polar coordinates. The resulting expression is

$$\frac{\partial^2 \varphi_D}{\partial r^2} + \frac{1}{r}\frac{\partial \varphi_D}{\partial r} + \frac{1}{r^2}\frac{\partial^2 \varphi_D}{\partial \theta^2} + k^2 \varphi_D = 0 \tag{6.95b}$$

Our task is to solve either of these equations, subject to boundary conditions.

Referring to Figure 6.20, the boundary conditions are both on the breakwater and well away from that structure (for practical purposes, at an infinite distance away from the structure). We can treat the breakwater as being either rigid (being totally reflective) or compliant (being totally absorbent). The boundary condition for the rigid (reflecting) breakwater is that the normal velocity component of the water adjacent to the wall equals zero. Mathematically, this condition is

$$\left.\frac{\partial \varphi_D}{\partial x}\right|_{y \le 0}^{x_- = 0} = 0, \quad \left.\frac{1 \partial \varphi_D}{r \partial \theta}\right|_{\theta = 2\pi}^{r > 0} = 0 \tag{6.96}$$

In the first of these relationships, the notation x_- is used to indicate the weather side (upwave) of the zero-thickness breakwater. For the compliant (absorbent) breakwater, the structure somehow responds to the wave such that the pressure along the structure is independent of time. From the linearized form of Bernoulli's equation – eq. 2.70, where $f(t) = 0$ – the dynamic pressure is zero. In terms of the velocity potential, this condition is mathematically expressed by

$$\rho \left[\left.\frac{\partial \Phi_D}{\partial t}\right|_{y \le 0}^{x_- = 0} \right] e^{-i\omega t} = -i\omega \left[\varphi_D \big|_{y \le 0}^{x_- = 0} \right] e^{-i\omega t} = 0 \tag{6.97a}$$

and

$$\rho \left[\left.\frac{\partial \Phi_D}{\partial t}\right|_{\theta = 2\pi}^{r > 0} \right] e^{-i\omega t} = -i\omega \left[\varphi_D \big|_{\theta = 2\pi}^{r > 0} \right] e_{-i\omega t} = 0 \tag{6.97b}$$

Note that these conditions avoid the trivial condition of $\omega = 0$.

Let us obtain a general solution of eq. 6.95b. To do this, assume a product solution of the form

$$\varphi_D = \frac{\cosh[k(z+h)]}{\cosh(kh)} \mathrm{P}(r)\Theta(\theta) = \frac{\cosh[k(z+h)]}{\cosh(kh)} F(r, \theta) \tag{6.98}$$

where the z-function is due to the seafloor condition, as in eq. 6.94. The combination of eq. 6.98 and eq. 6.95b leads to the following radial component expressions:

$$\begin{gathered} \mathrm{P}_a(r) = J_0(kr) \\ \mathrm{P}_b(r) = Y_0(kr) \\ \mathrm{P}_c(r) = H_0^{(1)}(kr) \equiv J_0(kr) + iY_0(kr) \\ \mathrm{P}_d(r) = H_0^{(2)}(kr) \equiv J_0(kr) - iY_0(kr) \end{gathered} \tag{6.99a}$$

where the functions on the right sides of the equations are zero-order Bessel functions. See Appendix A for a summary of the properties of these functions, and

Abramowitz and Stegun (1965). The following approximations of the Bessel functions are valid if kr is very large:

$$\begin{aligned}
P_a(r) &= J_0(kr) \to \sqrt{\frac{2}{\pi kr}} \cos\left(kr - \frac{\pi}{4}\right) \quad \text{as } kr \to \infty \\
P_b(r) &= Y_0(kr) \to \sqrt{\frac{2}{\pi kr}} \sin\left(kr - \frac{\pi}{4}\right) \quad \text{as } kr \to \infty \\
P_c(r) &= H_0^{(1)}(kr) \equiv J_0(kr) + i\,Y_0(kr) \to \sqrt{\frac{2}{\pi kr}} e^{i\left(kr - \frac{\pi}{4}\right)} \quad \text{as } kr \to \infty \\
P_d(r) &= H_0^{(2)}(kr) \equiv J_0(kr) - i\,Y_0(kr) \to \sqrt{\frac{2}{\pi kr}} e^{i\left(kr - \frac{\pi}{4}\right)} \quad \text{as } kr \to \infty
\end{aligned} \tag{6.99b}$$

For very small values of kr, the following approximations of the Bessel functions of the first and second kind are valid:

$$\begin{aligned}
P_a(r) &= J_0(kr) \to 1,\ kr \ll 1 \\
P_b(r) &= Y_0(kr) \to \frac{2}{\pi} \ln(kr),\ kr \ll 1
\end{aligned} \tag{6.99c}$$

In our analysis, derivatives of these Bessel functions are required. For any of the zero-order Bessel functions, say $B_0(kr)$, the derivative is

$$\frac{dB_0(kr)}{dr} = -kB_1(kr) \tag{6.99d}$$

The far-field approximations in eq. 6.99b show that the solutions represent wave forms having amplitudes that decrease in the radial direction. Each possible solution satisfies the Sommerfeld radiation condition, the derivation of which can be found in the book of Sommerfeld (1949) for both spherical and cylindrical waves. The book by Brebbia and Walker (1979) also has a rather extensive discussion of the radiation condition. For cylindrical water waves, the Sommerfeld radiation condition is

$$\lim_{r\to\infty}\left[\sqrt{r}\left(\frac{\partial\varphi}{\partial r} - ik\varphi\right)\right] = 0 \tag{6.100}$$

C. Modified Huygens-Fresnel Principle

According to Elmore and Heald (1985), Augustus J. Fresnel in 1818 combined Huygens' principle (assuming hemispherical wavelets) with Thomas Young's principle of interference to account for the phenomenon of diffraction. The result was a basic analysis of the phenomenon. The *Huygens-Fresnel principle* is used here to give the reader a better understanding of how diffraction of water waves occurs. There is one major difference in the analysis of water-wave diffraction, which is that the wavelets in the water-wave applications are assumed to be cylindrical in form rather than hemispherical.

Following Lindsay (1960), begin by considering three points in Figure 6.21. Those points are P_S (a source of semicylindrical waves outside a harbor entrance), point P_S' (on the y-axis in the entrance gap separating the heads of the breakwaters), and point P_O (within the harbor). The harbor has a surface area, A. The small

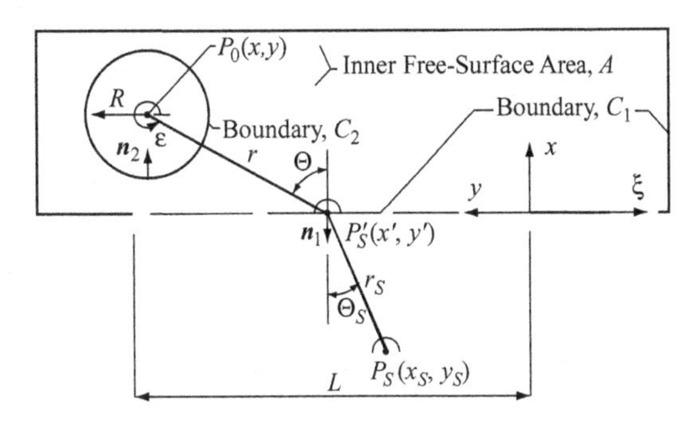

Figure 6.21. *Semicylindrical Wavelets Transmitting Energy into a Harbor Formed by Totally Absorbent Boundaries.*

circle, C_2, surrounding P_O defines a region that is outside of the area A. This small-radius circle is a mathematical artifice required in the analysis. The entrance gap ($x = 0, 0 \leq y \leq L$) is considered to be a part of the boundary of the enclosure. As is done by Elmore and Heald (1985), we refer to point P_S as a source point of semi-cylindrical wavelets at time $t = 0$, P_S' as a secondary source point of semicylindrical wavelets in the gap between the heads of the breakwater at time $t = t'$, and P_O as an observer point, where wavelets from P_S' arrive at time $t = t'$. Because our attention is given to the free surface, we can use the linearized dynamic free-surface condition of eq. 3.6 to represent the free-surface deflection at P_O in terms of the velocity potential. The result is

$$\eta_{P_o} = -\frac{1}{g}\frac{\partial \Phi_{P_o}}{\partial t}\Big|_{z=0} = \frac{\omega}{g}F(r, \Theta)e^{-i\omega t} \tag{6.101}$$

where r is the directed distance from the observation point to the secondary source point and $\omega = 2\pi/T$ is the circular wave frequency. The spacial function $F(r,\Theta)$ is similar to the function F(x,y) in eq. 6.98. Again, the water is assumed to be of uniform depth. Our goal is to determine the free-surface deflection (η_O) at the observation point (P_O) that results from the wavelets produced at all of the secondary source points (P_S') across the gap.

The function $F(r,\Theta)$ must satisfy the Helmholtz equation, similar to that in eq. 6.95, that is,

$$\nabla^2 F + k^2 F = 0 \tag{6.102}$$

To help us determine $F(r,\Theta)$, we introduce the function $G(r,\Theta)$, which is to be determined. We require this function to also satisfy the Helmholtz equation, so we can write

$$\nabla^2 G + k^2 G = 0 \tag{6.103}$$

The solutions of the Helmholtz equation are given in eq. 6.99. Hence, $G(r,\Theta)$ can either be one of these solutions or a sum of the solutions. The two-dimensional Green's theorem of eq. C8 in Appendix C is now applied to $F(r,\Theta)$ and $G(r,\Theta)$. The result of this application is

$$\iint_A [G\nabla^2 F - F\nabla^2 G]dA = \iint_A [-Gk^2 F + Fk^2 G]dA \equiv 0$$
$$= \int_C \left[G\frac{\partial F}{\partial n} - F\frac{\partial G}{\partial n}\right]dC \tag{6.104}$$

where eqs. 6.102 and 6.103 are incorporated. The area of the water surface bounded by the borders C_1 and C_2 is A. Now, noting that r is in the opposite direction of the outward normal coordinate n_2 on the circle C_2, perform the line integrations over C_1 and C_2 to obtain

$$\int_{C_1} \left[G\frac{\partial F}{\partial n_1} - F\frac{\partial G}{\partial n_1} \right] dC + \int_0^{2\pi} \left[G\left(-\frac{\partial F}{\partial r} \right) - F\left(-\frac{\partial G}{\partial r} \right) \right] r\,d\theta|_{r=R} = 0 \quad (6.105)$$

where the second integral is over the C_2 contour. We choose $G(r,\Theta)$ to be equal to $H_0^{(1)}(kr)$, where the Hankel function is defined in eq. 6.99a. This function has a singularity at the observation point, P_O. We are not concerned about the singularity because the point is within the circle and, therefore, not in area A. The chosen radial function is a form of Green's function, discussed in Section D2 of Appendix D. The function applied to the internal boundary C_2 is

$$\begin{aligned} G(r,\theta)|_{r=R} &= G(R) = H_0^{(1)}(kr)|_{r=R} \\ &= [J_0(kr) + iY_0(kr)]|_{r=R} \rightarrow \left[1 + i\frac{2}{\pi}\ln(kR) \right], \quad kR \ll 1 \end{aligned} \quad (6.106)$$

where, again, the radius (R) of the circle is very small, that is, much smaller than the wavelength (λ). From eq, 6.99d, the derivative of this function with respect to the outward normal direction on C_2 is

$$\begin{aligned} \frac{dG}{dn_2} &= -\frac{dG}{dR} = kH_1^{(1)}(kR) \\ &= k[J_1(kR) + iY_1(kR)] \rightarrow k\left[\frac{kR}{2} + i\frac{2}{\pi kR} \right], \quad kR \ll 1 \end{aligned} \quad (6.107)$$

In the integration over C_2 (the second integral in eq. 6.105), combine eqs. 6.105 through 6.107 and pass to the limit as $R \rightarrow 0$. By this process, one sees that the singularity at the observation point is excluded. The resulting expression is

$$\int_{C_1} \left[G\frac{\partial F}{\partial n_1} - F\frac{\partial G}{\partial n_1} \right] dC + i4F_{P_o} = 0 \quad (6.108)$$

By combining eqs. 6.106 and 6.107 with this expression, and noting that $r\cos(\Theta)$ is in the direction of n_1 over the gap (the only region of C_1 that contains the secondary source points), the value of $F(r,\Theta)$ at the observation point is found to be

$$F_{P_o} = \frac{i}{4}\int_{C_1} \left[H_0^{(1)}(kr)\frac{\partial F}{\partial n_1} + FkH_1^{(1)}(kr)\cos(\theta) \right] dC \quad (6.109)$$

Physically, this expression represents the contributions of all of the wavelets generated over the active part of boundary C_1, which is the gap. For the situation in Figure 6.21, the wavelets are produced by all of the secondary source points across the gap. So, the expression in eq. 6.109 applied to the harbor sketched in Figure 6.21 is

$$F_{P_o} = i\frac{1}{4}\int_0^L \left[H_0^{(1)}(kr)\frac{\partial F}{\partial n_1} + FkH_1^{(1)}(kr)\cos(\theta) \right] dy' \quad (6.110)$$

which is mathematically analogous to Huygens' principle.

As stated previously, the source point (P_S) emits semicylindrical sinusoidal wavelets. The source point is assumed to be many wavelengths away from the gap.

The value of the function $F(r_S,\Theta_S)$ at the gap is then assumed to be of the form

$$F(r_s,\theta_s) \equiv F_{P_S'} = \frac{B_F}{\sqrt{r_S}} e^{ikr_S} \tag{6.111}$$

where B_F is a dimensional constant. Note that the angle θ_S is not present in this equation. The reason is that the wavelets are radiated uniformly between two points. The normal derivative of the function at the gap involves θ_S because the direction of the derivative depends on the orientation of the gap. Because $r_S\cos(\Theta_S)$ is in the opposite direction of the outward normal vector, $\boldsymbol{n}_1$, the normal derivative of the expression in eq. 6.111 is

$$\begin{aligned}\frac{dF_{P_S'}}{dn_1} &= -\frac{dF}{dr_S}\cos(\Theta_S)\\ &= -B_F\left(-\frac{1}{2r_S} + ik\right)\frac{1}{\sqrt{r_S}}e^{ikr_S}\cos(\Theta_S) = B_F\left(\frac{1}{2r_S} - ik\right)F_{P_S'}\cos(\Theta_S)\end{aligned} \tag{6.112}$$

Similarly, assuming that P_O is many wavelengths from the gap, $G(r,\Theta)$ in eq. 6.106 and its derivative in eq. 6.107 at the gap are

$$G_{P_S'} = H_0^{(1)}(kr) \to B_G\sqrt{\frac{1}{r}}e^{irk}, \quad kr \gg 1 \tag{6.113}$$

and

$$\begin{aligned}\frac{dG_{P_S'}}{dn_1} &= \frac{dG}{dr}\cos(\Theta) = -kH_1^{(1)}(kr)\cos(\Theta)\\ &\to -B_G\sqrt{\frac{1}{r}}\left(\frac{1}{2r} - ik\right)e^{ikr}\cos(\Theta), \quad kr \gg 1\end{aligned} \tag{6.114}$$

Note that the phase angle $\pi/4$ in the limiting functions of eq. 6.99 is neglected in eqs. 6.113 and 6.114 because it can be absorbed into a complex coefficient B_G. Combine the expressions in eqs. 6.111 through 6.114 with that in eq. 6.110. The resulting expression for large values of kr and kr_S is

$$\begin{aligned}F_{P_o} &= i\frac{B_{FG}}{4}\int_0^L \left[\left(\frac{1}{2r_S} - ik\right)\cos(\Theta_S) + \left(\frac{1}{2} - ik\right)\cos(\Theta)\right]\frac{e^{ik(r+r_S)}}{\sqrt{rr_S}}dy'\\ &\simeq \frac{B_{FG}k}{2}\int_0^L \left[\frac{\cos(\Theta_S)+\cos(\Theta)}{2}\right]\frac{e^{ik(r+r_S)}}{\sqrt{rr_S}}dy'\end{aligned} \tag{6.115}$$

where $B_{FG} = B_F B_G$. The approximation in eq. 6.115 is called the *Fresnell-Kirchhoff diffraction formula*, and is based on the assumption that both P_S and P_O are far (many wavelengths) from the gap. The derivation of the formula is due to the German physicist Gustav Kirchhoff (the co-inventor of the spectroscope). The formula is difficult to work with; however, the derivation is physically enlightening and, for this reason, it is presented here.

Assuming that the bed is uniform in Figure 6.21, the free-surface displacement at the observation point is obtained by combining the expressions in eqs. 6.101 and 6.115. The result is

$$\eta_{P_o} = -\frac{1}{g}\frac{\partial \Phi_{P_o}}{\partial t}\bigg|_{z=0} \simeq B\frac{\omega k}{2g}\int_0^L \left[\frac{\cos(\Theta_S)+\cos(\Theta)}{2}\right]\frac{e^{i[k(r+r_S)-\omega t]}}{\sqrt{rr_S}}dy' \tag{6.116}$$

In summary, the physical reasoning leading to the expression in eq. 6.116 is the following: The free-surface displacement at the observation point leeward of the boundary results from Huygens-type semicylindrical wavelets originating at and distributed across the gap, and arriving simultaneously at the observation point. The integral equation is basic to the derivation of the Sommerfeld (1896) diffraction theory of light, upon which the Penny-Price (1952) diffraction analysis of water waves is based.

There are a few geometric manipulations and approximations, due to Fresnel, that must be performed before we arrive at the Penny-Price analysis. These are based on the assumptions that, first, the aperture width (L) is small compared to both r and r_S, and second, both the source point (P_S) and observation point (P_O) are relatively close to the centerline of the gap. The first of these assumptions requires the gap to be approximately a slit in the vertical breakwater. With these assumptions, the cosines of the angles Θ_S and Θ are

$$\cos(\Theta_S) = \frac{|x_S|}{r_S} \simeq \cos(\Theta) = \frac{x}{r} \tag{6.117}$$

The separation distances of the points can be represented by

$$r_S = \sqrt{x_S^2 + (y' - y_S)^2} = |x_S|\sqrt{1 + \frac{(y' - y_S)^2}{x_S^2}} \simeq |x_S|\left[1 + \frac{1}{2}\left(\frac{y'}{x_S}\right)^2\right] \tag{6.118}$$

and

$$r = \sqrt{x^2 + (y - y')^2} = x\sqrt{1 + \frac{(y - y')^2}{x^2}} \simeq x\left[1 + \frac{1}{2}\left(\frac{y'}{x}\right)^2\right] \tag{6.119}$$

The approximations in eqs. 6.118 and 6.119 are used in the exponent of the integrand of eq. 6.116. In the coefficient of the exponential term of the integrand, we can assume $r_S \simeq x_S$ and $r_O \simeq x_O$ to obtain

$$\frac{1}{2}\left[\frac{\cos(\Theta_S) + \cos(\Theta)}{\sqrt{r_S r}}\right] = \frac{1}{2}\left(\frac{|x_S|}{r_S} + \frac{x}{r_O}\right)\frac{1}{\sqrt{r_S r}} \simeq \frac{1}{\sqrt{|x_S| x}} \tag{6.120}$$

Again, the last approximation is based on the assumption that the source point and the observation point are near the centerline of the gap. With the relationships in eq. 6.117 and the approximation in eqs. 6.120, the integral equation in eq. 6.116 is approximately

$$\eta_{P_O} \simeq B\frac{\omega k}{2g}\frac{e^{-i\omega t}e^{k(x+|x_S|)}}{\sqrt{x|x_S|}}\int_0^L e^{i\frac{\pi}{\lambda}\left(\frac{1}{x}+\frac{1}{|x_S|}\right)(y')^2}dy' \tag{6.121}$$

In the integral of eq. 6.121, let

$$u = y'\sqrt{\frac{2}{\lambda}\left(\frac{1}{x} + \frac{1}{|x_S|}\right)} \tag{6.122}$$

The expression in eq. 6.121 becomes

$$\eta_{P_O} \simeq B\frac{\omega k}{2g}e^{-i\omega t}e^{i\pi(x+|x_S|)}\sqrt{\frac{\lambda}{2(x + |x_S|)}}\int_0^{u_L} e^{i\frac{\pi}{2}u^2}du \tag{6.123}$$

where the upper limit of the integral is the variable (introduced in eq. 6.122) evaluated at $y' = L$.

By using Euler's identity, the integral in eq. 6.123 can be written as

$$\int_0^{u_L} e^{i\frac{\pi}{2}u^2} du = \int_0^{u_L} \cos\left(\frac{\pi}{2}u^2\right) + i\int_0^{u_L} \sin\left(\frac{\pi}{2}u^2\right) du$$
$$\equiv C(u_L) + iS(u_L) = -C(-u_L) - iS(-u_L) \tag{6.124}$$

The integrals $C(u_L)$ and $S(u_L)$ are called *Fresnel integrals*. See Abramowitz and Stegun (1965) for the details of these integrals. From the last equality, we see that both $C(\sigma)$ and $S(\sigma)$ are asymmetric about $\sigma = 0$, according to Abramowitz and Stegun (1965), that is, $C(-\sigma) = -C(\sigma)$ and $S(-\sigma) = -S(\sigma)$. Furthermore, $C(\infty) = S(\infty) = 1/2$, so $C(-\infty) = -S(\infty) = -1/2$. As done by McCormick and Kraemer (2002), we can simplify the diffraction analysis by using the following approximations of the Fresnel integrals for positive values of u_L:

$$C(u_L) = -C(-u_L) \simeq \frac{1}{2} + \frac{(1 + 0.926u_L)\sin\left(\frac{\pi}{2}u_L^2\right)}{2 + 1.792u_L + 3.104u_L^2} - \frac{\cos\left(\frac{\pi}{2}u_L^2\right)}{2 + 4.142u_L + 3.492u_L^2 + 6.670u_L^3} + \mathrm{E}(u_L) \tag{6.125}$$

and

$$S(u_L) = -S(-u_L) \simeq \frac{1}{2} - \frac{(1 + 0.926u_L)\cos\left(\frac{\pi}{2}u_L^2\right)}{2 + 1.792u_L + 3.104u_L^2} - \frac{\sin\left(\frac{\pi}{2}u_L^2\right)}{2 + 4.142u_L + 3.492u_L^2 + 6.670u_L^3} + \mathrm{E}(u_L) \tag{6.126}$$

where $0 \le u_L \le \infty$. The remainder for both equations is $E(u_L) \le 0.002$. The limiting conditions for the Fresnel integrals, $C(0) = S(0) = 0$ and $C(\infty) = S(\infty) = 1/2$, are obtained from eqs. 6.125 and 6.126. Because the Fresnel integrals must be evaluated by using numerical methods, the reader will find these approximations to be most useful for engineering applications. The expressions in eqs. 6.125 and 6.126 can be evaluated using hand calculators by engineers in the field.

To summarize, the free-surface displacement at the observation point is obtained from the expression in eq. 6.123. This expression is based on the assumptions that both the gap width (L) and the off-center positions of both P_S and P_O are much less than either x or x_S in Figure 6.21. This is called the *Fresnel diffraction* for a narrow gap or slit. Fresnel extended this analysis to diffraction of light by a single straight edge. The Sommerfeld (1896) diffraction theory is based on the straight-edge diffraction theory of Fresnel. Subsequently, as previously mentioned, Penny and Price (1944, 1952) modified the Sommerfeld diffraction theory in their analysis of water-wave diffraction. A summary of the Sommerfeld (1896) diffraction theory is found in the book by Bateman (1964). The application of the Sommerfeld theory to water waves is found in the book by Stoker (1957).

D. Diffraction Analyses of Water Waves

As stated previously, the first major work on the application of diffraction theory to water waves is that of Penny and Price (1944, 1952). The Penny-Price diffraction analysis is in three parts, with only the first two parts presented here. Those are the parts devoted to the diffraction of water waves having their crests parallel to a semi-infinite breakwater, and to the diffraction analysis of waves approaching the semi-infinite breakwater at an angle, as is the case in Figure 6.20. The analysis of waves passing through a gap presented here is that of Williams and Crull (1993), which is an extension of the analysis in Section 6.4.C.

Penny and Price (1944, 1952) begin their analysis of the wave field about a semi-infinite breakwater, such as that sketched in Figure 6.20, with an expression for the velocity potential that is similar to that in eq. 6.94. To present the analysis as by Penny and Price, we use the coordinates (ξ,x) where, as shown in Figure 6.20, ξ is the coordinate that is coincident with the breakwater and x is the normal coordinate to the structure. The origins of both coordinates are at the head of the structure. The potential function is

$$\Phi = A\cosh[k(z+h)]\Re[F(x,\xi)e^{-iwt}] = A\cosh[k(z+h)]\Re[F(r,\theta)e^{-iwt}] \quad (6.127)$$

where $F(\xi\ x)$ is a complex function that is to be determined. Referring to Figure 6.20, the angle θ is that used in the derivation of Penny and Price (1944, 1952). The velocity potential in eq. 6.127 applies to the entire wave field, and must satisfy the continuity expression of the forms in eq. 6.95b. The combination of eqs. 6.127 and 6.96a results in the Helmholtz equation,

$$\nabla^2 F = \frac{\partial^2 F}{\partial \xi^2} + \frac{\partial^2 F}{\partial x^2} + k^2 F = 0 \quad (6.128a)$$

The combination of eqs. 6.127 and 6.96b yields the continuity equation in cylindrical form:

$$\frac{\partial^2 F}{\partial r^2} + \frac{1}{r}\frac{\partial F}{\partial r} + \frac{1}{r^2}\frac{\partial^2 F}{\partial \theta^2} + k^2 F = 0 \quad (6.128b)$$

The velocity potential must satisfy the boundary conditions at the barrier or seawall. For the rigid breakwater (totally reflective), the boundary conditions of eq. 6.96 are

$$\left.\frac{\partial F}{\partial \xi}\right|_{\xi\geq 0}^{x_-=0} = 0, \quad \left.\frac{1}{r}\frac{\partial F}{\partial \theta}\right|_{\theta=2\pi}^{r>0} = 0 \quad (6.129)$$

where x_- applies to the weather side (up-wave) of the zero-thickness breakwater. For the compliant breakwater (totally absorbent), eq. 6.97 applies. For this condition, the resulting expressions are

$$F\Big|_{\xi\geq 0}^{x_-=0} = 0, \quad F\Big|_{\theta=2\pi}^{r>0} = 0 \quad (6.130)$$

Equations 6.127 through 6.130 are basic to analyses presented in the following two subsections.

(1) Diffraction of Waves Directly Incident upon a Semi-Infinite Breakwater

In the diffraction analysis of water waves directly incident upon a semi-infinite breakwater, Penny and Price (1944, 1952) assume that the form of $F(\xi,\,x)$ is that

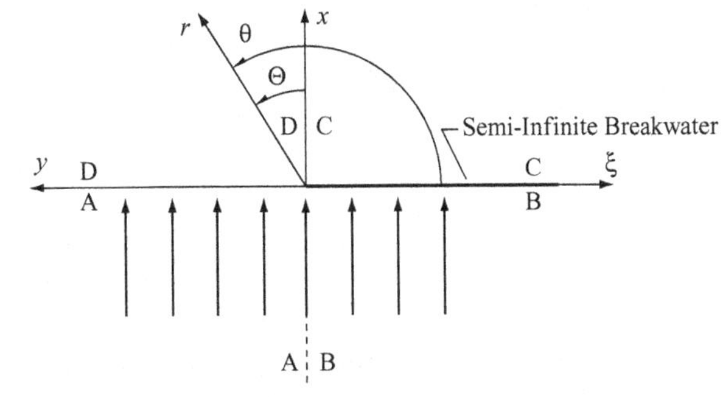

Figure 6.22. *Ray Diagram for Waves Directly Incident upon a Rigid Breakwater of Infinitely Small Thickness.*

derived by Sommerfeld (1896) for the diffraction of light waves by a sharp-edged, semi-infinite barrier. As mentioned previously, the Sommerfeld diffraction theory is based on the Fresnel diffraction theory, which is outlined in the previous section. The ray diagram in Figure 6.22 is useful in this analysis. The rays are parallel to the celerity or phase velocity vector of the incident wave. The breakwater is assumed to be rigid, so reflection of the incident wave occurs. Four regions are shown in Figure 6.22. Those are:

(a) *Region A:* In this region are the incident waves and the diffracted reflected waves.
(b) *Region B:* This region is occupied by the incident and reflected waves.
(c) *Region C:* This region is the shadow zone, occupied only by the diffracted incident waves.
(d) *Region D:* Incident waves primarily occupy this region.

The diffracted reflected waves are also in Region D and, to a lesser extent, in Region C. If the breakwater is compliant, such that it absorbs all of the wave energy incident upon it, then there is no reflection, and Regions A, B, and D are occupied only by incident waves.

From Penny and Price (1944, 1952), the application of the Sommerfeld expression to both the rigid breakwater and a compliant breakwater results in the following expression for $F(\xi, x)$:

$$
\begin{aligned}
F(\xi, x)\Big|_{\substack{rigid \\ Compliant}}
&= \frac{(1+i)}{2}\left[e^{-ikx}\int_{-\infty}^{\sigma} e^{-i\frac{\pi}{2}u^2}du \pm e^{ikx}\int_{-\infty}^{\sigma'} e^{-i\frac{\pi}{2}u^2}du\right] \\
&= \frac{(1+i)}{2}\left[e^{-ikx}\left(\int_{-\infty}^{0} e^{-i\frac{\pi}{2}u^2}du + \int_{0}^{\sigma} e^{-i\frac{\pi}{2}u^2}du\right) \pm e^{-ikx}\left(\int_{-\infty}^{0} e^{-i\frac{\pi}{2}u^2}du + \int_{0}^{\sigma'} e^{-i\frac{\pi}{2}u^2}du\right)\right] \\
&= \frac{(1+i)}{2}\left[e^{-ikx}\left(\frac{(1-i)}{2} + \int_{0}^{\sigma} e^{-i\frac{\pi}{2}u^2}du\right) \pm e^{ikx}\left(\frac{(1-i)}{2} + \int_{0}^{\sigma'} e^{-i\frac{\pi}{2}u^2}du\right)\right] \\
&= \frac{(1+i)}{2}\left[e^{-ikx}\left(\left\langle\frac{1}{2} + C(\sigma)\right\rangle - i\left\langle\frac{1}{2} + S(\sigma)\right\rangle\right) \pm e^{ikx}\left(\left\langle\frac{1}{2} + C(\sigma')\right\rangle - i\left\langle\frac{1}{2} + S(\sigma')\right\rangle\right)\right]
\end{aligned}
\tag{6.131}
$$

Table 6.1. *Penny-Price quadrant signs*

Quadrants in Figure 6.22	Signs of σ and σ'
A	+, −
B	+, +
C	−, −
D	+, −

where $C(\sigma)$ and $S(\sigma)$ are Fresnel integrals, as defined in eq. 6.124, and approximated in eqs. 6.125 and 6.126, respectively. The real and imaginary parts of the rigid breakwater function $F(\xi,x)|_{rigid}$ are

$$\begin{aligned} F_{\Re}|_{rigid} &= \frac{1}{2}\{[2 + C(\sigma) + C(\sigma') + S(\sigma) + S(\sigma')]\cos(kx)\} \\ &\quad + [C(\sigma) - C(\sigma') - S(\sigma) + S(\sigma')]\sin(kx)\} \\ &= \frac{1}{2}[A\cos(kx) + B\sin(kx)] \end{aligned} \tag{6.132}$$

and

$$\begin{aligned} F_{\Im}|_{rigid} &= \frac{1}{2}\{[C(\sigma) + C(\sigma') - S(\sigma) - S(\sigma')]\cos(kx) \\ &\quad - [C(\sigma) - C(\sigma') + S(\sigma) - S(\sigma')]\sin(kx)\} \\ &= \frac{1}{2}[D\cos(kx) - E\sin(kx)] \end{aligned} \tag{6.133}$$

The Sommerfeld expression for the respective real and imaginary parts of $F(\xi, x)$ that satisfy the boundary conditions for waves directly incident on a compliant or absorbent breakwater are

$$\begin{aligned} F_{\Re}|_{compliant} &= \frac{1}{2}\{[C(\sigma) - C(\sigma') + S(\sigma) - S(\sigma')]\cos(kx) \\ &\quad + [C(\sigma) + C(\sigma') - S(\sigma) - S(\sigma')]\sin(kx)\} \\ &= \frac{1}{2}[E\cos(kx) + D\sin(kx)] \end{aligned} \tag{6.134}$$

and

$$\begin{aligned} F_{\Re}|_{compliant} &= \frac{1}{2}\{[C(\sigma) - C(\sigma') - S(\sigma) + S(\sigma')]\cos(kx) \\ &\quad - [2 + C(\sigma) - C(\sigma') + S(\sigma) + S(\sigma')]\sin(kx)]\} \\ &= \frac{1}{2}[B\cos(kx) - A\sin(kx)] \end{aligned} \tag{6.135}$$

The upper limits of the integrals of eq. 6.131 are

$$\sigma = \pm 2\sqrt{\frac{r - x}{\lambda}}, \quad \sigma' = \pm 2\sqrt{\frac{r + x}{\lambda}} \tag{6.136}$$

where λ is the wavelength, which is uniform everywhere because the water depth (h) is assumed to be uniform. The radial term in eq. 6.136 and in Figure 6.20 is $r = \sqrt{(\xi^2 + x^2)}$. According to Penny and Price (1944, 1952), the positive (+) and negative (−) signs of the terms in eq. 6.136 are as follows:

Our interest is in the ratio of the wave heights of the diffracted and the incident waves. This ratio is called the *diffraction coefficient*. To obtain the expression for the diffraction coefficient, we assume that the waves are linear and that the free-surface displacements are represented by the linear relationship,

$$\eta = -\frac{1}{g}\frac{\partial \Phi}{\partial t}\Big|_{z=\eta\simeq 0} \tag{6.137}$$

The velocity potentials for the incident and diffracted waves are given in eqs. 6.93 and 6.94, respectively. For the present application, $y = -\xi$ and the reference point on the incident wave can be at the head of the breakwater, where $x = \xi = 0$. Following this procedure, the respective expressions for the diffraction coefficients for the respective rigid and compliant breakwaters are

$$\begin{aligned} K_D|_{rigid} &\equiv \frac{H_D}{H_I}\Big|_{rigid} = \frac{|\eta_D|}{\eta_I|}\Big|_{rigid} \\ &= |F(\xi, x)|_{rigid}| = \sqrt{\left(F_{\Re}{}^2|_{rigid} + F_{\Im}{}^2|_{rigid}\right)} \\ &= \frac{1}{2}\sqrt{(A^2 + D^2)\cos^2(kx) + 2(AB - DE)\cos(kx)\sin(kx) + (B^2 + E^2)\sin^2(kx)} \end{aligned} \tag{6.138}$$

and

$$\begin{aligned} K_D|_{compliant} &\equiv \frac{H_D}{H_I}\Big|_{compliant} = \frac{|\eta_D|}{|\eta_I|}\Big|_{compliant} \\ &= |F(\xi, x)|_{compliant}| = \sqrt{\left(F_{\Re}^2|_{compliant} + F_{\Im}^2|_{compliant}\right)} \\ &= \frac{1}{2}\sqrt{(E^2 + B^2)\cos^2(kx) - 2(AB - DE)\cos(kx)\sin(kx) + (A^2 + D^2)\sin^2(kx)} \end{aligned} \tag{6.139}$$

where A and B are defined in eq. 6.132, and D and E are defined in eq. 6.133. Also of interest are the phase angles between the diffracted waves and the incident wave passing the head of the breakwater, which are, respectively,

$$\varepsilon|_{rigid} = \tan^{-1}\left(\frac{F_{\Im}}{F_{\Re}}\right)\Big|_{rigid} \tag{6.140a}$$

and

$$\varepsilon|_{compliant} = \tan^{-1}\left(\frac{F_{\Im}}{F_{\Re}}\right)\Big|_{compliant} \tag{6.140b}$$

These expressions are of use in determining the planar geometries of the wave fronts. Results obtained using the expression for the diffraction coefficient in eq. 6.138 for a rigid breakwater are presented in Figure 6.23, where the distances are non-dimensionalized by dividing by the wavelength (λ), which is uniform throughout the region of uniform water depth (h). Although Penny and Price (1944, 1952) present plots of lines of constant values of K_D for waves directly approaching a rigid breakwater, we choose here to present the plot presented in the *Shore Protection Manual* of the CERC of the U.S. Army Corps of Engineers (U.S. Army, 1984; see Figure 6.24).

The diffraction diagram in Figure 6.24 is based on the assumption that the breakwater is rigid and totally reflective. The diffraction diagrams for totally

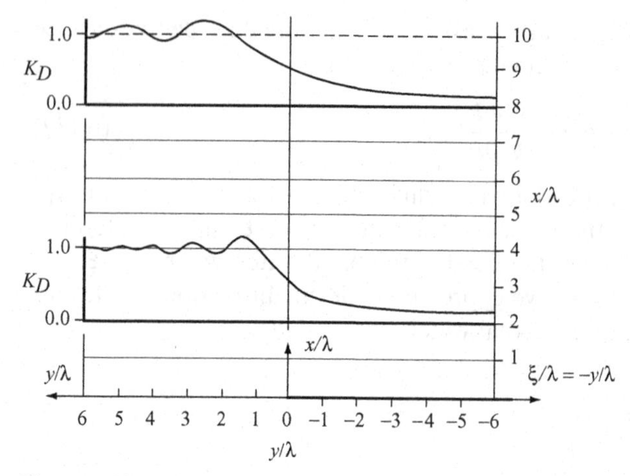

Figure 6.23. *Diffraction Coefficient versus the Alongshore Position at* $x/\lambda = 2$ *and* $x/\lambda = 8$.

absorbent breakwaters are somewhat different. This fact is illustrated in Figure 6.25, where the diffraction coefficients for both rigid and absorbent breakwaters are plotted as a function of y/λ, where $x/\lambda = 2$. Hence, caution must be taken in using diagrams to determine wave heights in the shadow zone, such as that in Figure 6.24. It should also be noted that Penny and Price (1952) write that "the conditions (K_D values) on the lee side of the breakwater at distances greater than about 2λ are nearly the same whether the breakwater is of the rigid or cushion (absorbent) type." From the results in Figure 6.25, this does not appear to be the case.

The results in Figure 6.25 were obtained by using the approximate polynomial representations of the Fresnel integrals in eqs. 6.125 and 6.126. A further demonstration of the use of these approximations is presented in Example 6.9, where we

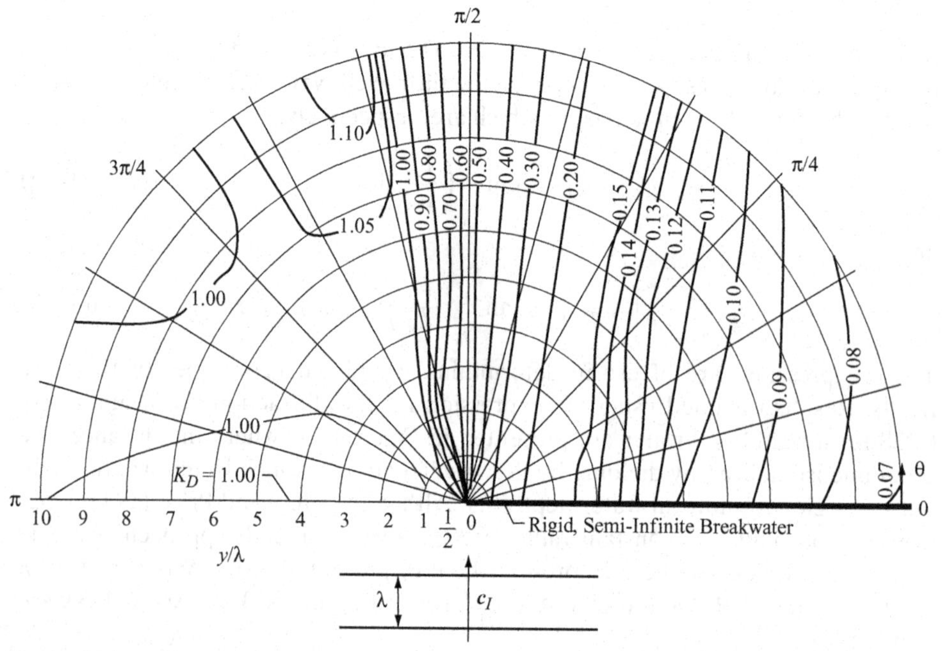

Figure 6.24. *Diffraction Coefficient Values for* $\theta_I = \pi/2$ *(after Wiegel, 1964).*

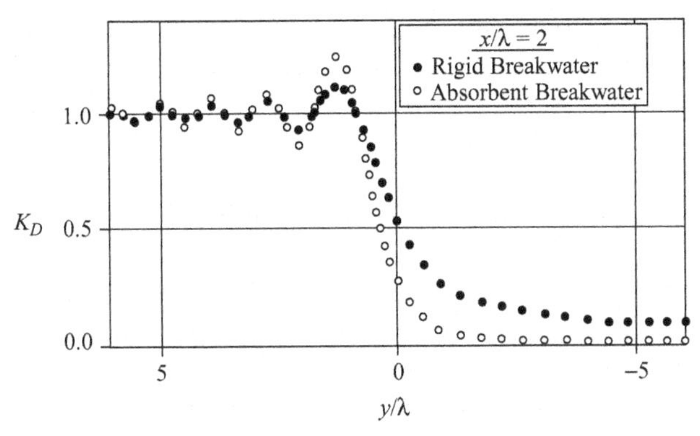

Figure 6.25. *Diffraction Coefficients for Rigid and Absorbent Breakwaters along a Line Parallel to the Leeward Face of the Breakwater.*

determine the values of the diffraction coefficient along the leeward side of both semi-infinite rigid and compliant breakwaters.

EXAMPLE 6.9: DIFFRACTION COEFFICIENTS ALONG THE LEEWARD SIDES OF RIGID AND COMPLIANT BREAKWATERS We are to design a breakwater to create a harbor, as sketched in the insert of Figure 6.26. The beaches protected by the breakwater are perfect absorbers. Hence, the reflection coefficient (K_R) values for the beaches are zero. The shoreline structure to the right of the breakwater in Figure 6.26 has the same reflection coefficient value at the breakwater. That is, for the rigid breakwater, the shoreline structure has a reflection coefficient value of one, whereas for the perfectly absorbent breakwater, the reflection coefficient of the shoreline structure is zero. Of particular interest is the value of the diffraction coefficient at 10 wavelengths from the head of the breakwater on the leeward side of the structure. From the results presented in Figure 6.26, we see that the diffraction coefficient value for the rigid breakwater is approximately 0.0705 and that of the absorbent breakwater is 0.005. The absorbent breakwater then provides more protection in the shadow zone of the harbor. However, from the results in Figure 6.25 the water to the left of the absorbent breakwater will

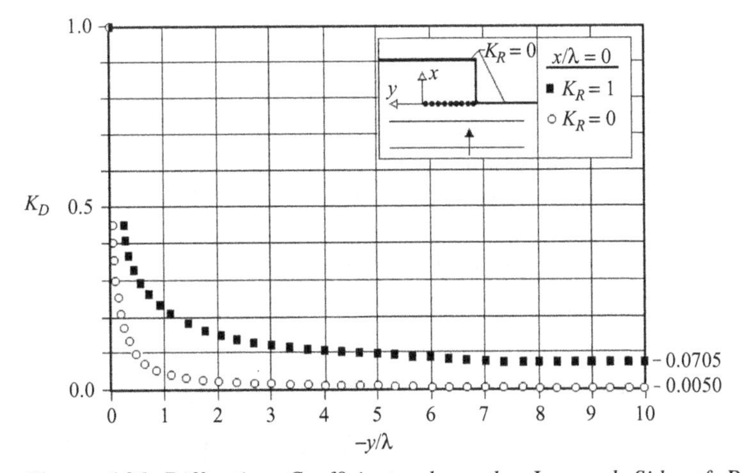

Figure 6.26. *Diffraction Coefficients along the Leeward Side of Rigid and Compliant Breakwaters.*

be a little rough. The reader must remember that the polynomial representations of the Fresnel integrals in eqs. 6.125 and 6.126 are valid for $0 \le \sigma, \sigma' \le \infty$. Along the leeward side of the breakwater, both $\sigma < 0$ and $\sigma' < 0$, so the Fresnel integrals values are absolute values.

(2) Diffraction of Waves Obliquely Incident upon a Semi-Infinite Breakwater

Our interest now is directed at the situation sketched in Figure 6.20, where the incident waves approach the breakwater at an angle $\beta_I \neq 0$ or $\theta_I \neq \pi/2$; that is, where the wave crests are oblique to the structure. Again, the subscript I identifies the incident wave properties. For this analysis, it is convenient to use the polar coordinate system (r, θ). To convert from the rectilinear coordinates to the polar coordinates, use the wave coordinates, s and S, shown in Figure 6.20. That is, for the incident wave,

$$s = x\sin(\theta_I) - y\cos(\theta_I) = r\sin(\theta)\sin(\theta_I) + r\cos(\theta)\cos(\theta_I) = r\cos(\theta - \theta_I) \tag{6.141}$$

and for the reflected wave,

$$S = -x\sin(\theta_I) - y\cos(\theta_I) = -r\sin(\theta)\sin(\theta_I) + r\cos(\theta)\cos(\theta_I) = r\cos(\theta + \theta_I) \tag{6.142}$$

The modified expression in eq. 6.131 is then

$$\begin{aligned}
F(r,\theta)|_{\substack{rigid\\compliant}} &= \frac{(1+i)}{2}\left[e^{-iks}\int_{-\infty}^{\sigma} e^{-i\frac{\pi}{2}u^2}du \pm e^{-ikS}\int_{-\infty}^{\sigma'} e^{-i\frac{\pi}{2}u^2}du\right] \\
&= \frac{(1+i)}{2}\left[e^{-ikr}\left(\left\langle\frac{1}{2}+C(\sigma)\right\rangle - i\left\langle\frac{1}{2}+S(\sigma)\right\rangle\right) \pm e^{-iks}\left(\left\langle\frac{1}{2}+C(\sigma')\right\rangle - i\left\langle\frac{1}{2}+S(\sigma')\right\rangle\right)\right] \\
&= \frac{(1+i)}{2}\left[e^{-ikr\cos(\theta-\theta_I)}\left(\left\langle\frac{1}{2}+C(\sigma)\right\rangle - i\left\langle\frac{1}{2}+S(\sigma)\right\rangle\right)\right. \\
&\left.\pm e^{-ikr\cos(\theta+\theta_I)}\left(\left\langle\frac{1}{2}+C(\sigma')\right\rangle - i\left\langle\frac{1}{2}+S(\sigma')\right\rangle\right)\right]
\end{aligned} \tag{6.143}$$

Expressions similar to those in eqs. 6.132 through 6.135, respectively, are

$$\begin{aligned}
F_{\Re}|_{rigid} &= \frac{1}{2}\{[1 + C(\sigma) + S(\sigma)]\cos(ks) \\
&+ [C(\sigma) - S(\sigma)]\sin(ks) + [1 + C(\sigma') + S(\sigma')]\cos(kS) \\
&+ [C(\sigma') - S(\sigma')]\sin(kS)\} \\
&= \frac{1}{2}\{\mathrm{A}\cos(ks) + \mathrm{B}\sin(ks) + \mathrm{M}\cos(kS) + \mathrm{N}\sin(kS)\} \\
&= \frac{1}{2}\{\mathrm{A}\cos[kr\cos(\theta - \theta_I)] + \mathrm{B}\sin[kr\cos(\theta - \theta_I)] \\
&+ \mathrm{M}\cos[kr\cos(\theta + \theta_I)] + \mathrm{N}\sin[kr\cos(\theta + \theta_I)]\}
\end{aligned} \tag{6.144}$$

$$\begin{aligned}
F_{\Im}|_{rigid} &= \frac{1}{2}\{\mathrm{B}\cos(ks) - \mathrm{A}\sin(ks) + \mathrm{N}\cos(kS) - \mathrm{M}\sin(kS)\} \\
&= \frac{1}{2}\{\mathrm{B}\cos[kr\cos(\theta - \theta_I)] - \mathrm{A}\sin[kr\cos(\theta - \theta_I)] \\
&+ \mathrm{N}\cos[kr\cos(\theta + \theta_I)] - \mathrm{M}\sin[kr\cos(\theta + \theta_I)]\}
\end{aligned} \tag{6.145}$$

$$
\begin{aligned}
F_{\Re}|_{compliant} &= \frac{1}{2}\{\mathrm{A}\cos(ks) + \mathrm{B}\sin(ks) - \mathrm{M}\cos(kS) - \mathrm{N}\sin(kS)\} \\
&= \frac{1}{2}\{\mathrm{A}\cos[kr\cos(\theta - \theta_I)] - \mathrm{B}\sin[kr\cos(\theta - \theta_I)] \\
&\quad - \mathrm{M}\cos[kr\cos(\theta + \theta_I)] - \mathrm{N}\sin[kr\cos(\theta + \theta_I)]\} \qquad (6.146)
\end{aligned}
$$

and

$$
\begin{aligned}
F_{\Im}|_{compliant} &= \frac{1}{2}\{\mathrm{B}\cos(ks) - \mathrm{A}\sin(ks) + \mathrm{N}\cos(kS) - \mathrm{M}\sin(kS)\} \\
&= \frac{1}{2}\{\mathrm{B}\cos[kr\cos(\theta - \theta_I)] - \mathrm{A}\sin[kr\cos(\theta - \theta_I)] \\
&\quad - \mathrm{N}\cos[kr\cos(\theta + \theta_I)] + \mathrm{M}\sin[kr\cos(\theta + \theta_I)]\} \qquad (6.147)
\end{aligned}
$$

where the coefficients A, B, M, and N are defined in eq. 6.144. The limits of integration in eq. 6.143 are

$$
\sigma = \pm 2\sqrt{\frac{r - s}{\lambda}} = \pm 2\sqrt{\frac{kr}{\pi}}\sin\left(\frac{\theta - \theta_I}{2}\right), \quad \sigma' = \pm 2\sqrt{\frac{r - S}{\lambda}} = \pm 2\sqrt{\frac{kr}{\pi}}\sin\left(\frac{\theta + \theta_I}{2}\right) \qquad (6.148)
$$

where λ is the uniform wavelength.

The respective diffraction coefficients for the rigid (totally reflecting) and compliant (totally absorbing) breakwaters are

$$
\begin{aligned}
K_D|_{rigid} &\equiv \frac{H_D}{H_I}|_{rigid} = \frac{|\eta_D|}{|\eta_I|}|_{rigid} \\
&= |F(r,\theta)|_{rigid}| = \sqrt{(F_r^2|_{rigid} + F_i^2|_{rigid})} \qquad (6.149)
\end{aligned}
$$

and

$$
\begin{aligned}
K_D|_{compliant} &\equiv \frac{H_D}{H_I}|_{compliant} = \frac{|\eta_D|}{|\eta_I|}|_{compliant} \\
&= |F(r,\theta)|_{compliant}| = \sqrt{(F_{\Re}^2|_{compliant} + F_{\Im}^2|_{compliant})} \qquad (6.150)
\end{aligned}
$$

The phase angles between the waves leeward of the y-axis and the incident waves at $y = 0$ are similar to those in eq. 6.140. We should note that by giving the phase angle a constant value, the solution of r, as a function of θ in eq. 6.140, yields the plan of the wave crest pattern. As noted by Penny and Price (1944, 1952), the crests are approximately circular in the shadow zone.

From the *Shore Protection Manual* of the CERC of the U.S. Army Corps of Engineers (U.S. Army, 1984), the source of Figure 6.24, we present two cases of oblique reflection. Those are for incident waves approaching a perfectly reflecting semi-infinite breakwater at angles of $\theta_I = 135°$ ($3\pi/4$ radians) in Figure 6.27, and $\theta_I = 45°$ ($\pi/4$ radians) in Figure 6.28. These figures result from the work of Wiegel (1962). Also see Wiegel (1964).

EXAMPLE 6.10: DIFFRACTION COEFFICIENTS ALONG THE LEEWARD SIDES OF A RIGID BREAKWATER In Example 6.9, we determined the values of the diffraction coefficients adjacent to the leeward faces of totally reflecting (rigid) and totally absorbent (compliant) breakwaters for waves approaching the breakwaters directly, that is, for $\theta_I = 90°$. The results of the study are presented in Figure 6.26. We now direct our attention to waves approaching a rigid breakwater at both $\theta_I = 135°$ and $\theta_I = 45°$. For the entire wave fields associated

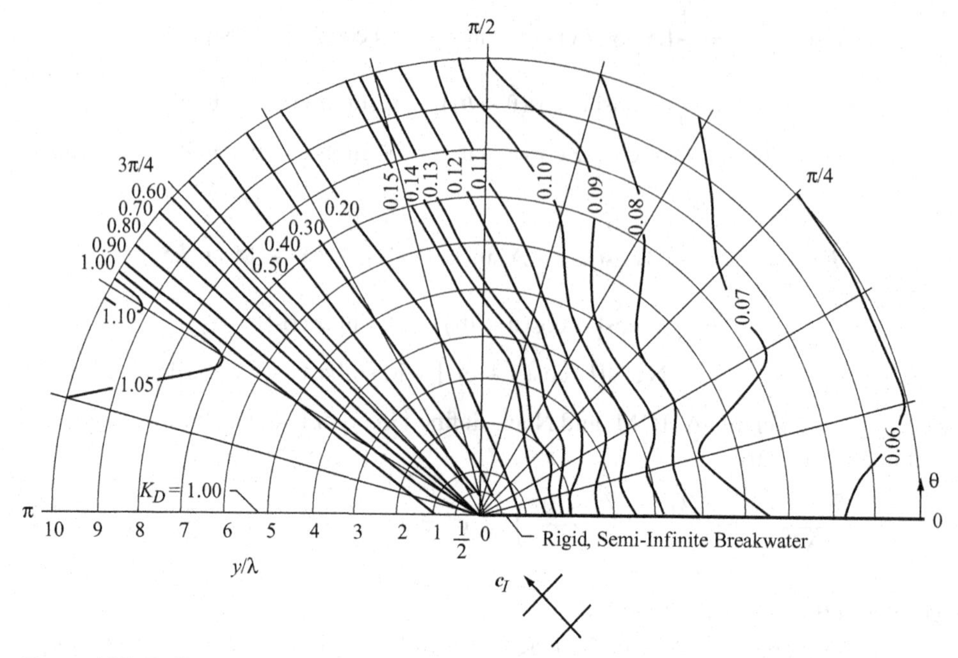

Figure 6.27. *Diffraction Coefficient Values for* $\theta_I = 3\pi/4$ *(after U.S. Army, 1984).*

with these incident wave conditions, the diffraction coefficients are presented in Figures 6.27 and 6.28, respectively. As is the case for Figure 6.24, these curves result from the analyses of Wiegel (1962) and the corresponding presentations in the *Shore Protection Manual* (see U.S. Army, 1984). Our interest is in the values of the diffraction coefficient along the leeward side of the

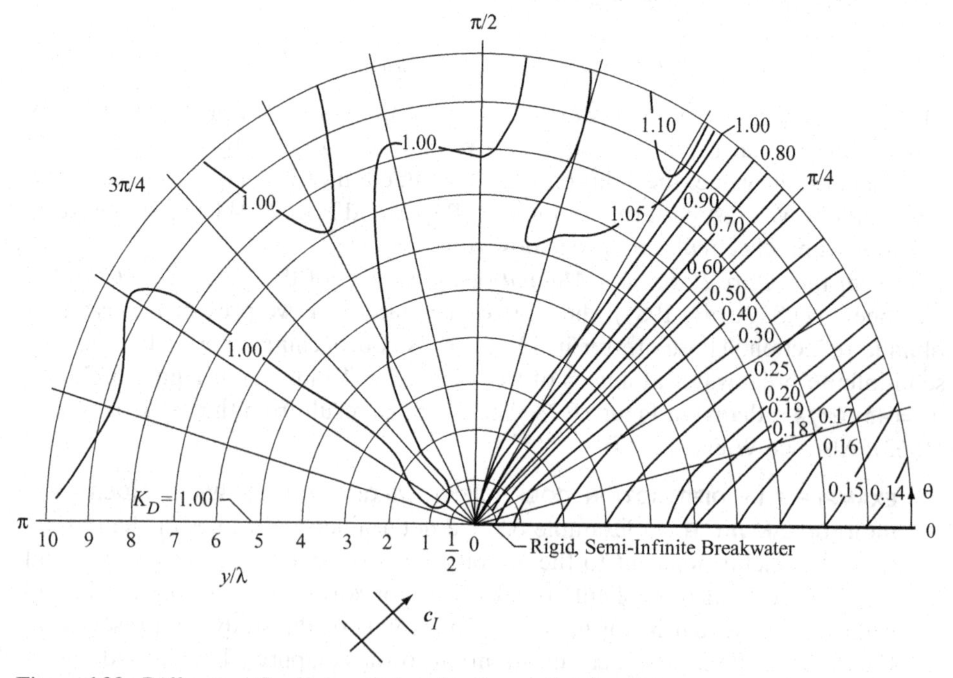

Figure 6.28. *Diffraction Coefficient Values for* $\theta_I = \pi/4$ *(after U.S. Army, 1984).*

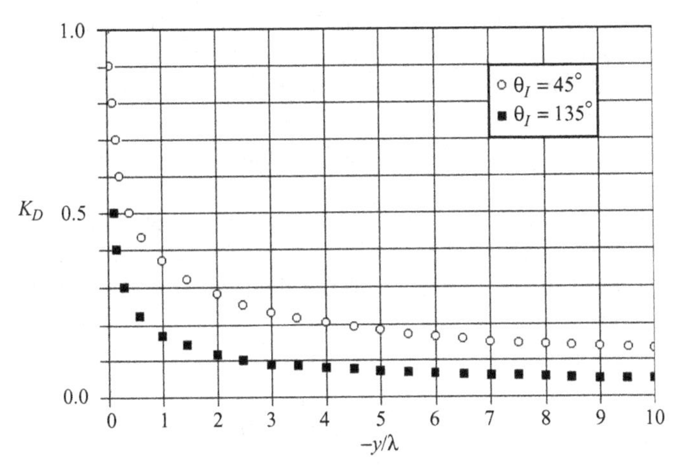

Figure 6.29. *Diffraction Coefficient versus Distance from the Head of a Semi-Infinite Rigid Breakwater along the Leeward Side.*

breakwater for the two oblique wave conditions. In eqs. 6.144, 6.145, 6.148, and 6.149, the angle θ is zero because our interest is in the leeward water adjacent to the structure. As in Example 6.9, the Fresnel integrals are determined from the polynomial representations in eqs. 6.125 and 6.126, using the absolute values of both σ and σ'. Following this, $C(-\sigma) = -C(\sigma)$ and $S(-\sigma) = -S(\sigma)$ are used in eqs. 6.144, 6.145, 6.148, and 6.149. The results are presented in Figure 6.29, where K_D is presented as a function of the non-dimensional distance (y/λ) from the head of the breakwater on the leeward side. At 10 wavelengths from the head, the diffraction coefficient values for the respective $\theta_I = 135°$ and $\theta_I = 45°$ conditions are 0.054 and 0.133. The ratios of the wave energy values at the site and at the head are proportional to $K_D{}^2$ at $y = 10$. The respective energy ratio values are then 0.0027 and 0.0173, which are relatively small.

(3) Diffraction of Waves by a System of Detached Breakwaters

There are several methods for determining the diffraction patterns caused by a system of detached breakwaters, as sketched in Figure 6.30. Two of these are the eigenfunction method of Dalrymple and Martin (1990) and the Green's function method of Achenbach and Li (1986), which is applied to water waves by Williams and Crull (1993). The Achenbach-Li scattering analysis is directed at acoustical, elastodynamic, and electromagnetic waves. The Dalrymple-Martin analysis involves an eigenfunction expansion of the velocity potential, whereas the Williams-Crull analysis uses a Chebyshev expansion to represent the change in the velocity potential

Figure 6.30. *Area Sketch of Waves Obliquely Incident on an In-Line System of an Infinite Number of Thin, Detached Breakwaters.*

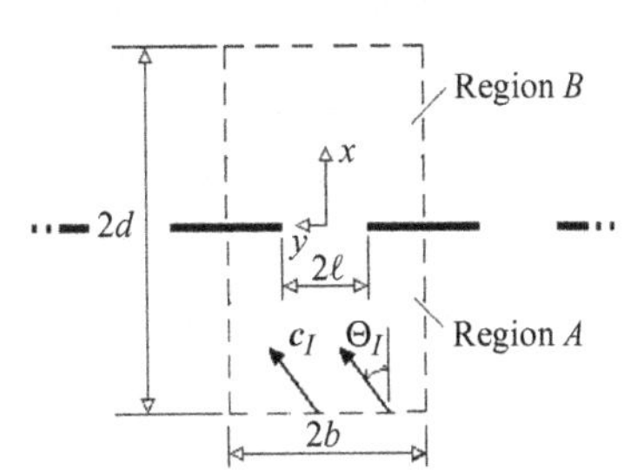

across the individual units of the system. Abul-Azm and Williams (1997) extend the Dalrymple-Martin analysis to determine the wave fields caused by waves obliquely incident on single breakwaters of finite length, gaps separating semi-infinite breakwaters, and an infinite linear array of finite breakwaters, as sketched in Figure 6.30. The analysis of Abul-Azm and Williams (1997), based on that of Dalrymple and Martin (1990), is presented herein.

In the following derivation, we use the term *scattered waves* when referring collectively to the reflected and diffracted waves of the wave field. The potential function (a real function, as indicated by $\Re$) representing the field composed of both incident and scattered waves is

$$\begin{aligned}\Phi &= \Re\varphi e^{-i\omega t} = \Re(\varphi_I + \varphi_S)e^{-i\omega t} \\ &= -i\frac{H_I}{2}\frac{g}{\omega}\frac{\cosh[k_I(z+h)]}{\cosh(k_I h)}\Re\{[e^{ik_I[x\cos(\Theta_I)+y\sin(\Theta_I)]} + F(x,y)]e^{-i\omega t}\} \\ &\equiv -i\,Z(z)\Re\{[e^{ik_I[x\cos(\Theta_I)+y\sin(\Theta_I)]} + F(x,y)]e^{-i\omega t}\}\end{aligned} \tag{6.151}$$

where the subscript I refers to properties of the incident waves. Note that in the function $Z(z)$ introduced in eq. 6.93, kc has been replaced by ω, the circular wave frequency. The function $F(x,y)$, which is to be determined, satisfies the Helmholtz equation, eq. 6.102. Because the wave patterns associated with the breakwater units in Figure 6.30 are repetitive in the y-direction, we can concentrate our analysis on any single unit. The unit of interest is that centered in the dashed-line rectangle in the figure. We refer to this rectangular area as a *cell.* The sides of the cell bisect breakwater units, and the top and the bottom of the cell are at large distances from the units, that is, d in Figure 6.30 is much greater than 2ℓ (gap width), $2b$ (length between unit centers), and the incident wavelength, λ_I. The cell of area $2bd$ is divided into an up-wave region, A, and a down-wave region, B.

The breakwaters are assumed to be rigid and, therefore, reflect perfectly. The boundary condition is mathematically represented by the expressions in eq. 6.129. In the notation of Figure 6.30, that condition is

$$\left.\frac{\partial\varphi}{\partial x}\right|_{x=0-} = \left.\frac{\partial\varphi}{\partial x}\right|_{x=0+} = 0, \quad \ell \le |y| \le b \tag{6.152}$$

which applies on both sides of a unit of negligible thickness. The notation $x = 0-$ identifies the weather (up-wave) side, and $x = 0+$ identifies the leeward (down-wave) side. On the weather side, the combination of the velocity potential in eq. 6.151 and the first term in eq. 6.152 results in

$$\left.\frac{\partial\varphi_S}{\partial x}\right|_{x=0} = -\left.\frac{\partial\varphi_I}{\partial x}\right|_{x=0} = -Z(z)k_I\cos(\Theta_I)e^{ik_I\sin(\Theta_I)y}, \quad \ell \le |y| \le b \tag{6.153}$$

The diffracted component waves of the scattered field must also satisfy the Sommerfeld radiation condition in eq. 6.100 at large values of $|x|$. Because the cell represented by the rectangular envelope in Figure 6.30 is the same for each unit, the wave field is repetitive in the y-direction. The repetitive nature of the wave field allows us to write

$$\varphi(x,b) = \varphi(x,-b), \quad -d \le x \le +d \tag{6.154}$$

and

$$\left.\frac{\partial\varphi}{\partial y}\right|_{y=-b} = \left.\frac{\partial\varphi}{\partial y}\right|_{y=+b}, \quad -d \le x \le +d \tag{6.155}$$

In Figure 6.30, the wave field within the cell is broken down into two regions, those being the up-wave region, A, and the down-wave region, B. These two regions meet at the y-axis, where the velocity potentials satisfy the following conditions:

$$\varphi^{(A)}(0, y) = \varphi^{(B)}(0, y), \quad 0 < |y| < \ell \tag{6.156}$$

and

$$\left.\frac{\partial \varphi^{(A)}}{\partial x}\right|_{x=0} = \left.\frac{\partial \varphi^{(B)}}{\partial x}\right|_{x=0}, \quad 0 < |y| < \ell \tag{6.157}$$

Following Abul-Azm and Williams (1997), assume that the time-independent velocity potential for the up-wave region (Region A) and the down-wave region (Region B) can be represented by

$$\varphi^{(A)} = -i\,Z(z)\Re\left\{e^{ik_I[x\cos(\Theta_I)+y\sin(\Theta_I)]} + \sum_{m=-\infty}^{\infty} \mathrm{A}_m e^{-i(\mu_m x - \upsilon_m y)}\right\} \tag{6.158}$$

and

$$\varphi^{(B)} = -i\,Z(z)\Re\left\{e^{ik_I[x\cos(\Theta_I)+y\sin(\Theta_I)]} - \sum_{m=-\infty}^{\infty} \mathrm{B}_m e^{-i(\mu_m x - \upsilon_m y)}\right\} \tag{6.159}$$

In these equations, the first term represents the incident waves, and the second term represents the reflected and diffracted wave modes. The complex coefficients A_m and B_m are to be determined. By combining the boundary conditions of eqs. 6.156 and 6.157 with the expressions of eqs. 6.158 and 6.159, we find that A_m and B_m are equal in magnitude. So, we can confine our attention to either Region A or Region B in the remainder of the analysis. Also, in eqs. 6.158 and 5.159 are the following:

$$\upsilon_m = k_I \sin(\Theta_I) + \frac{m\pi}{b} \tag{6.160}$$

and

$$\mu_m = \sqrt{k_I^2 + \upsilon_m^2} \tag{6.161}$$

The reflected and transmitted waves propagate in the respective $-x$ and $+x$ directions, so μ_m must be a real number. To ensure this, the radicand in eq. 6.161 must be positive. The corresponding condition on the wave number is

$$k \geq |\upsilon_m| = \left|k_I \sin(\Theta_I) + \frac{m\pi}{b}\right| \tag{6.162}$$

This inequality is always satisfied for $m = 0$. Physically, this means that there will always be one propagating wave mode. The other modes are those for evanescent standing waves, exponentially decaying in the x-direction in Region B of Figure 6.30.

Again, because A_m and B_m are equal in magnitude, we can write the boundary condition of eq. 6.156 as

$$E(y) = k_I \cos(\Theta_I) e^{ik_I y \sin(\Theta_I)} - \sum_{m=-\infty}^{\infty} \frac{\mu_m}{k_I} \mathrm{A}_m e^{i\upsilon_m y} = 0, \quad \ell \leq |y| \leq b \tag{6.163}$$

where we introduce $E(y)$ as a boundary-condition function. This function represents the boundary conditions at $x = 0$ for all values of y in $0 \leq |y| \leq b$, including the boundary condition in eq. 6.153. After dividing the expression in eq. 6.153 by k_I, the boundary-condition function applied to the up-wave side of a unit is

$$\varphi^{(A)}(0, y) - \varphi^{(B)}(0, y) = 0 \Rightarrow E(y) \equiv \sum_{m=-\infty}^{\infty} \mathrm{A}_m e^{i\upsilon_m y} = 0, \quad 0 < |y| < \ell \qquad (6.164)$$

Now, truncate the series in eqs. 6.163 and 6.164 at a finite index value $|M|$. That is, the series is approximated by choosing the respective upper and lower limits of the summation at $m = -M$ and $m = +M$. The goal is to minimize the error introduced by truncating the series in eqs. 6.163 and 6.164 with the appropriate choice of A_m. With the introduction of the function $E(y)$, we can use Legendre's least-squares method for determining the values of A_m, as done by Dalrymple and Martin (1990) and Abul-Azm and Williams (1997). This method is outlined in the book of Korn and Korn (2000) and is described in detail in books on numerical analysis. Using the truncated-series approximation, we minimize

$$S_{mn} = \begin{cases} 2\left(1 - \dfrac{\ell}{b}\right) = 2 - Q_{mn}, & n = m \\[2ex] -2\dfrac{\sin\left[(n-m)\pi\dfrac{\ell}{b}\right]}{(n-m)\pi} = -Q_{mn}, & n \neq m \end{cases} \qquad (6.165)$$

with respect to A_m by solving

$$\int_{-b}^{b} E^* \frac{\partial E(y)}{\partial A_m} dy = 0, \quad m = \pm 1, \pm 2, \ldots, \pm M \qquad (6.166)$$

where E^* is the complex conjugate of $E(y)$ and, therefore, a function of y.

The combination of the (truncated) expressions in eqs. 6.163 and 6.164 with that in eq. 6.166 yields

$$\int_{-b}^{b} |E(y)|^2 dy = \left\{ \int_{-b}^{b} |E(y)|^2 dy \right\} \Big|_{minimum} \qquad (6.167)$$

where the parametric terms are

$$\sum_{m=-M}^{M} \mathrm{A}_m^* \left(Q_{mn} + \frac{\mu_m^* \mu_n}{k_I^2} S_{mn} \right) \simeq \frac{\mu_n}{k_I} S_{n0} \cos(\Theta_I), \quad n = 0, \pm 1. \pm 2, \ldots, \pm M \qquad (6.168)$$

and

$$Q_{mn} = \begin{cases} 2\dfrac{\ell}{b}, & n = m \\[2ex] 2\dfrac{\sin\left[(n-m)\pi\dfrac{\ell}{b}\right]}{(n-m)\pi}, & n \neq m \end{cases} \qquad (6.169)$$

Equation 6.167 represents $2M + 1$ equations with the same number of unknowns (A_m^* being unknown). Abul-Azm and Williams (1997) find that satisfactory convergence is obtained for $M = 75$.

Recall that $m = 0$ corresponds to a propagating wave mode because, for this value, the condition of eq. 6.162 is satisfied. The complex coefficient A_0 represents

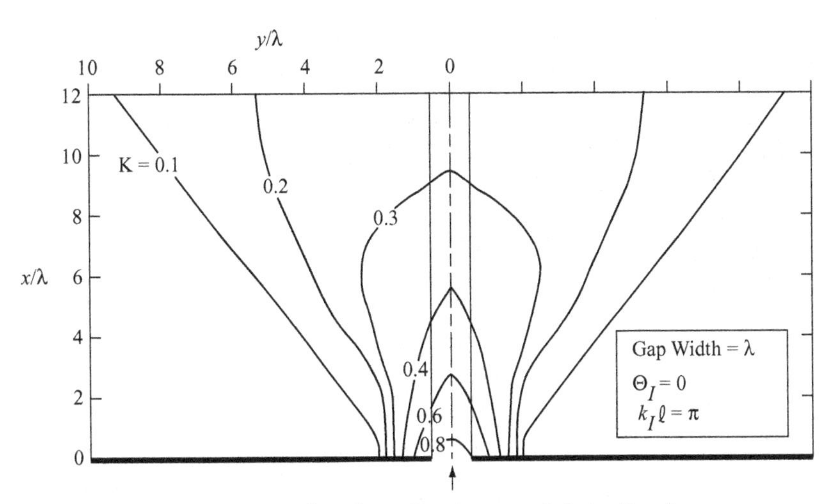

Figure 6.31. *Predicted* K*–Values for a Gap between Infinite Breakwaters.*

the zeroth-order reflection coefficient for the wave system, that is,

$$K_R|_{rigid} = \frac{H_R}{H_I} = |A_0| \tag{6.170}$$

Furthermore, Abul-Azm and Williams (1997) define *diffraction parameter* (which includes the diffraction coefficient) as the total free-surface elevation divided by the incident wave amplitude, that is,

$$\left.\begin{matrix} K^{(A)}|_{rigid} \\ K^{(B)}|_{rigid} \end{matrix}\right\} = |e^{ik_I[x\cos(\Theta_I)+y\sin(\Theta_I)]} \pm \sum_{m=-\infty}^{\infty} A_m e^{-i(\mu_m x - \upsilon_m y)}| \tag{6.171}$$

The superscripts of the parameter refer to Region A and Region B. Abul-Azm and Williams (1997) state that the changes in the values of diffraction parameter (K) for $M > 75$ are less than 1%. Plots of the diffraction parameter are presented in Figure 6.31 for a gap between (approximate) semi-infinite breakwaters, and in Figure 6.32 for an infinite number of finite detached breakwaters. These plots are similar to those presented by Abul-Azm and Williams (1997).

EXAMPLE 6.11: WAVES DIRECTLY INCIDENT ON A GAP BETWEEN SEMI-INFINITE BREAKWATERS In this example, we discuss the conditions used by Abul-Azm and Williams (1997) in obtaining the results in Figure 6.32. In the application, $\Theta_I = 0$, as expected, that is, the waves are directly incident upon the gap between the breakwaters. We also use the parametric value in Figure 6.31, that being $k\ell = \pi$. One would expect $b = \infty$ because the breakwaters are semi-infinite in length. However, this yields a rather trivial result for two reasons. First, from the expressions in eqs. 6.168 and 6.169, we see that for a gap problem, $Q_{mn} = 0$ for any combination of m and n values. Second, $S_{mn} = 0$ for $m \neq n$, but $S_{mn} = 2$ for $m = n$. Equation 6.167 then yields $A_0^* = 1$, and $A_m^* = 0$ for $m \neq 0$. From the results in eq. 6.170, we see that we have perfect reflection for the condition $b = \infty$.

To obtain the plots of the diffraction parameter in Figure 6.31, Abul-Azm and Williams (1997) assumed $kb = 184.8$, so that $\ell/b = 0.017$. Those investigators compare their results with those obtained by Johnson (1952), not shown

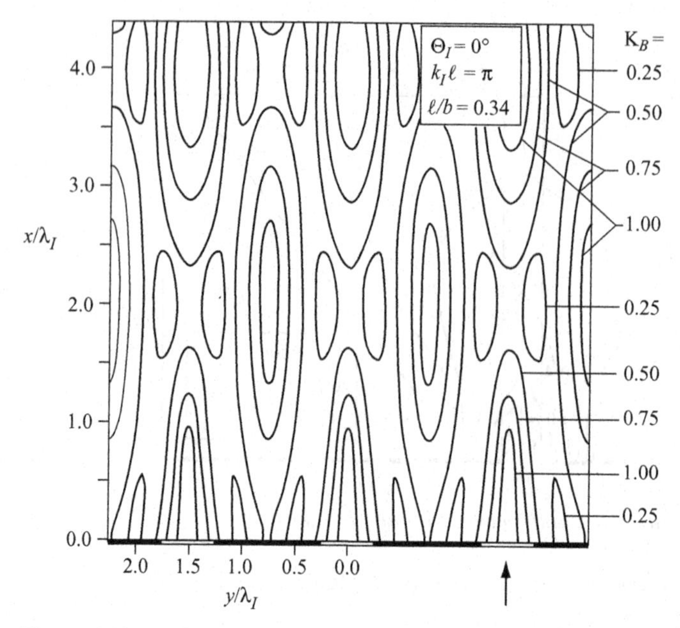

Figure 6.32. *Diffraction Parameter Values for Waves Directly Incident upon Detached Finite Breakwaters Separated by a Gap of One Incident Wavelength.*

in Figure 6.31, and find that the comparison is satisfactory. They attribute differences near the gap to the basic assumption made in their analysis, that is, because of the finite value of b, the analysis is still that of segmented breakwaters of finite length, although the lengths of each unit are relatively large. Abul-Azm and Williams (1997) show that the agreement with the curves of Johnson (1952) improves as $2\ell/\lambda$ increases for a given value of b.

Note that the diffraction parameter ($K^{(B)}$) and the diffraction coefficient of eqs. 6.149 and 6.150 are related by

$$K^{(B)}|_{rigid} = 1 + K_D \tag{6.172}$$

The diffraction coefficient is plotted in Figures 6.23 through 6.29, and the diffraction parameter is plotted in Figures 6.31 and 6.32.

In this section, several methods of diffraction analysis are presented. For engineering practice, the results presented in the *Shore Protection Manual* (SPM) of the U.S. Army Corps of Engineers (U.S. Army, 1984) are recommended. The SPM results are obtained from the works of Blue and Johnson (1949), Johnson (1952), and Wiegel (1962), which, in turn, are related to the work of Penny and Price (1944, 1952).

6.5 The Mild-Slope Equation

In Chapter 3, the linear theory is applied to waves passing over a bed having straight and parallel bottom contours, as sketched in Figure 6.16. The wave system in that analysis is required to conserve the energy flux between orthogonals as the waves approach the shoreline. That is, the expressions of the energy flux in eqs. 3.72a and 3.72b have fixed values for wave systems traveling from deep water to the shoreline. The analysis leads to the expression for the shoaling coefficient (K_S – a wave

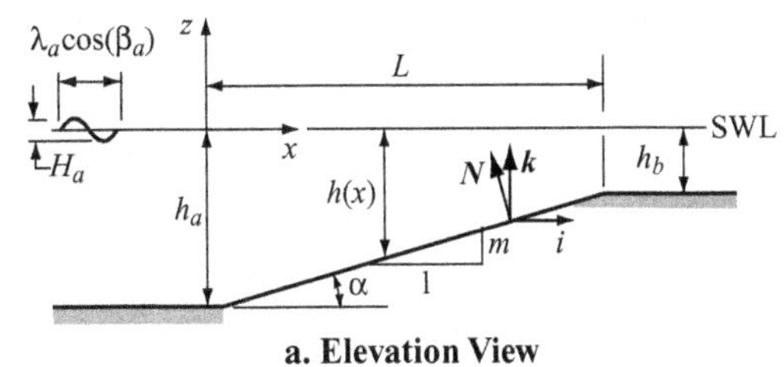

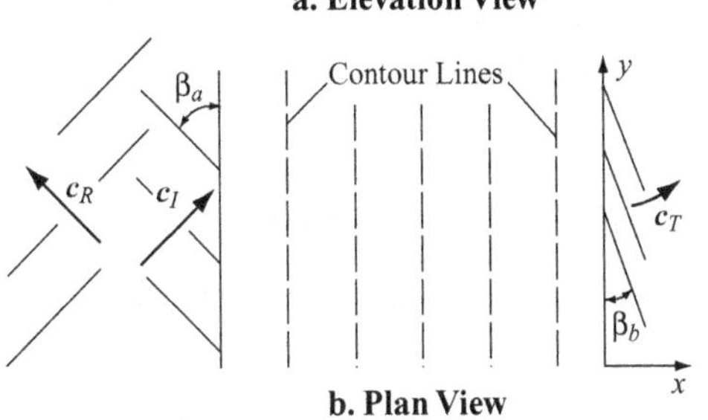

Figure 6.33. *Waves Passing over a Rectilinear Bed Rise.*

height ratio) in eq. 3.78. There are no terms in that equation that account for either the deep-water wave steepness or the slope of the bed. In Figure 3.19, the shoaling coefficient is represented by a single curve resulting from the linear analysis. However, the experimental data in that figure show dependencies on both the wave steepness and the bed slope.

In the following sections, the analysis of Berkhoff (1972, 1976), which accounts for the bed slope, is presented. That analysis leads to the mild-slope equation, which includes the effects of reflection, refraction, and diffraction.

A. Derivation of the Mild-Slope Equation

The analysis of Berkhoff (1972, 1976) leading to the *mild-slope equation* is presented herein. As the name of the equation implies, the analysis is directed at shoaling over moderately sloping beds. The analysis also includes the effects of both refraction and diffraction. Since its introduction in 1972, the mild-slope equation has been the subject of many studies. Two of the studies are those of Booij (1983), who studies the accuracy of the mild-slope equation, and of Chandrasekera and Cheung (1997). The latter includes the effects of both bottom curvature and the slope-squared terms that were neglected in the original analysis. In this section, the mild-slope equation is derived in its general form. The equation is then applied to the case of pure shoaling, where the bottom contours are both straight and parallel. One note concerning the equation: The mild-slope equation cannot be solved analytically, so we must rely on numerical techniques to obtain results. Copeland (1985) presents an alternative method to that of Berkhoff (1972, 1976).

Referring to the sketch Figure 6.33, consider linear waves approaching a rise in the sea bed at an angle β_a to the toe of the rise. As sketched in Figure 6.33a, reflection from the bed is assumed to occur. The bed has a moderate (mild) slope, and is comprised of straight and parallel bottom contours. The rise can either terminate at the shoreline, as is sketched in Figure 6.12, or at some offshore contour. The flow within the waves is assumed to be irrotational, so that a velocity potential can be used to mathematically represent the flow. Written in complex notation, the general form of the velocity potential for the shoaling wave is

$$\begin{aligned}\Phi &= \Re\{\varphi(x, y, z)Z[h(x), z]e^{-i\omega t}\} \\ &= \Re\{\varphi(x, y, z)Z[h(x), z][\cos(\omega t) - i\sin(\omega t)]\}\end{aligned} \tag{6.173}$$

where $\Re$ indicates the real part of the mathematical expression, and i is the imaginary number, $\sqrt{-1}$. For the sake of brevity, we shall omit $\Re$ in the remainder of this analysis; however, its presence is assumed. The dimensional term $\varphi(x, y, z)$ is a weak function of z, as is explained later. The potential function in eq. 6.173 must satisfy the equation of continuity in the form of Laplace's equation,

$$\nabla^2\Phi = 0 \tag{6.174}$$

and is subject to both the linearized free-surface condition,

$$\left\{\frac{\partial\Phi}{\partial z} - \frac{\omega^2}{g}\Phi\right\}\bigg|_{z=0} = \left\{Z\frac{\partial\varphi}{\partial z} + \varphi\frac{\partial Z}{\partial z} - \frac{\omega^2}{g}\varphi Z\right\}\bigg|_{z=0} = 0 \tag{6.175}$$

and the seafloor or bed condition,

$$(\nabla\Phi \cdot N)|_{z=-h} = 0 \tag{6.176}$$

In eq. 6.176, the outward unit normal vector on the bed is

$$\boldsymbol{N} = -\sin(\alpha)\boldsymbol{i} + \cos(\alpha)\boldsymbol{k} \tag{6.177}$$

where $\boldsymbol{i}$ and $\boldsymbol{k}$ are the unit vectors in the respective x- and z-directions, and β is the angle between the crest and the contour. The combination of the expressions in eqs. 6.176 and 6.177 results in

$$\left[-\frac{\partial\Phi}{\partial x}\sin(\alpha) + \frac{\partial\Phi}{\partial z}\cos(\alpha)\right]\bigg|_{z=-h} = 0 \tag{6.178}$$

Divide this expression by $\cos(\alpha)$, and note that

$$\tan(\alpha) \equiv m = \frac{\partial h}{\partial x} \tag{6.179}$$

where m is the *slope* of the inclined bed. The slope m can be either positive or negative, depending on the bed geometry. The combination of eqs. 6.178 and 6.179 yields the following inclined bed condition:

$$\left[\frac{\partial\Phi}{\partial z} + m\frac{\partial\Phi}{\partial x}\right]\bigg|_{z=-h} = \left[Z\frac{\partial\varphi}{\partial z} + \varphi\frac{\partial Z}{\partial z} + mZ\frac{\partial\varphi}{\partial x} + m\varphi\frac{\partial Z}{\partial x}\right]\bigg|_{z=-h} e^{-i\omega t} = 0 \tag{6.180}$$

Divide the second equality by φZ, noting that $\varphi|_{z=-h} = \varphi(x, y)|_{z=-h}$ and $Z|_{z=-h} = Z(x)|_{z=-h}$. The resulting expression in eq. 6.180 can then be separated into two equations, one for φ and the other for Z. The solution of the latter, subject to the conditions that there is no flow across the bed and that $Z(x, 0) = 1$ (imposed by Berkoff, 1972, 1976), is

$$Z(x, z) = \frac{\cosh\{k(x)[z + h(x)]\}}{\cosh[k(x)h(x)]} \tag{6.181}$$

In this expression, the wave number, $k = k(x)$, must satisfy the dispersion relationship,

$$\frac{\omega^2}{g} = k(x)\tanh[k(x)h(x)] \tag{6.182}$$

From this point on, the wave number and water depth will be represented by k and h, respectively. Their dependence on x is implied.

When the expressions in eqs. 6.181 and 6.182 are combined with the linearized free-surface condition of eq. 6.175, the result leads to the following:

$$\left[\frac{1\partial\varphi}{\varphi\partial z}\right]\Big|_{z=0} = 0 \tag{6.183}$$

Return now to the equation of continuity of eq. 6.174. The equation is satisfied if

$$\begin{aligned}\nabla^2(\varphi Z) &= \Delta^2(\varphi Z) + \frac{\partial^2(\varphi Z)}{\partial Z^2}\\ &= Z\Delta^2(\varphi) + Z\frac{\partial^2\varphi}{\partial z^2} + 2\frac{\partial Z}{\partial x}i\cdot\Delta(\varphi) + 2\frac{\partial Z\partial\varphi}{\partial z\partial z} + \varphi\left(\frac{\partial^2 Z}{\partial x^2} + \frac{\partial^2 Z}{\partial z^2}\right) = 0\end{aligned} \tag{6.184}$$

where the two-dimensional vector operator is defined by

$$\Delta(\;) \equiv \frac{\partial(\;)}{\partial x}\boldsymbol{i} + \frac{\partial(\;)}{\partial y}\boldsymbol{j} \tag{6.185}$$

Berkhoff (1972, 1974) uses $Z(x,z)$ as a weighting function, multiplying the expression in eq. 6.184 by that function, and integrating the resulting expression over the water depth. The result is

$$\int_{-h}^{0}\left\{Z^2\Delta^2(\varphi I) + Z^2\frac{\partial^2\varphi}{\partial z^2} + 2Z\left(\frac{\partial Z\partial\varphi}{\partial x\partial x} + \frac{\partial Z\partial\varphi}{\partial z\partial z}\right) + Z\varphi\left(\frac{\partial^2 Z}{\partial x^2} + \frac{\partial^2 Z}{\partial z^2}\right)\right\}dz = 0 \tag{6.186a}$$

which can be partially integrated to obtain

$$\begin{aligned}&\int_{-h}^{0}\left\{Z^2\Delta^2(\varphi) + 2Z\frac{\partial Z\partial\varphi}{\partial x\partial x} + Z\varphi\frac{\partial^2 Z}{\partial z^2}\right\}dz + m\left[Z^2\frac{\partial\varphi}{\partial x}\right]\Big|_{z=-h}\Big\}\\ &\quad + \int_{-h}^{0} Z\varphi\frac{\partial^2 Z}{\partial x^2}dz + m\left[Z\varphi\frac{\partial Z}{\partial x}\right]\Big|_{z=-h} = 0\end{aligned} \tag{6.186b}$$

Chandrasekera and Cheung (1997) retain all of the terms in eq. 6.186 in their analysis. The inclusion of the terms alters the results slightly, as discussed by Lee (1999). Berkhoff (1972) neglects the last two terms of this equation, assuming that they are of second order because of the gradually varying bed. The last two terms are also neglected in the analysis herein.

Following Berkhoff (1972, 1974), the dimensional function of the velocity potential in eq. 6.174 is represented by

$$\varphi(x,y,z) = \varphi_0(x,y) + z^1\varphi_1(x,y) + z^2\varphi_2(x,y) + \cdots = \sum_{n=0}^{\infty} z^n\varphi_n(x,y) \tag{6.187}$$

Berkhoff (1972, 1974) assumes that $\phi(x,y,z)$ is a weak function of z. That is, for $n \geq 1$ in the expansion, the terms are of second order both individually and collectively when compared to the first term in the expansion, so only the first term is retained. The result is

$$\varphi(x, y, z) \simeq \varphi_0(x, y) \tag{6.188}$$

Note that the subscript "*0*" is italicized here because the nonitalicized "0" is used to identify deep-water properties and functions. The combination of the expression in eq. 6.188 with the integral expression in eq. 6.186b results in

$$\begin{aligned}
&\int_{-h}^{0} Z^2 dz \Delta^2(\varphi_0) + \int_{-h}^{0} \frac{\partial (Z^2)}{\partial x} dz \frac{\partial \varphi_0}{\partial x} + \int_{-h}^{0} Z \frac{\partial^2 Z}{\partial x^2} dz \varphi_0 + \int_{-h}^{0} Z \frac{\partial^2 Z}{\partial z^2} dz \varphi_0 \\
&\simeq \int_{-h}^{0} Z^2 dz \Delta^2(\varphi_0) + \frac{\partial}{\partial x}\left[\int_{-h}^{0} Z^2 dz\right] \frac{\partial \varphi_0}{\partial x} + [0] + k^2 \int_{-h}^{0} Z^2 dz \varphi_0 \simeq 0
\end{aligned} \tag{6.189}$$

In eq. 6.189, the third integral of the first line can be shown to be dependent on m^2, which is of second order for beds of small slope, and is neglected on the second line. The fourth integral of the second line is a result of the combination of eq. 6.181 and the fourth integral of the first line. The second line of eq. 6.189 is the differential-integral form of the *mild-slope equation.* We shall introduce a notation similar to that used by Booij (1983) to represent the integrals in the second line of that equation, which is

$$G(x) \equiv \int_{-h}^{0} Z^2 dz = \frac{1}{4k}\left\{\frac{\sinh(2kh) + 2kh}{\cosh^2(kh)}\right\} = \frac{cc_g}{g} \tag{6.190}$$

The second equality in this equation results from the expression in eq. 3.63. With this notation, the differential form of the *mild-slope equation* is

$$G\Delta^2(\varphi_0) + \frac{\partial G \partial \varphi_0}{\partial x\, \partial x} + k^2 G\varphi_0 = \frac{\partial}{\partial x}\left(G\frac{\partial \varphi_0}{\partial x}\right) + G\frac{\partial^2 \varphi_0}{\partial y^2} + k^2 G\varphi_0 = 0 \tag{6.191}$$

Again, the onshore coordinate is x, and the alongshore coordinate is y, as sketched in Figure 6.33b. In the remainder of this section, the mild-slope equation will be applied to waves passing over a sea bed for which the contours are both straight and parallel. For applications to other bottom topographies, the references should be consulted.

B. Application to a Straight and Parallel Contoured Bed

Referring to Figure 6.33, the assumption of straight and parallel bottom contours allows us to represent the function in eq. 6.188 by

$$\varphi_0(x, y) = Q(x)Y(y) = Q(x)e^{ik_y y} = [Q_{\Re}(x) + iQ_{\Im}(x)]e^{ik_y y} \tag{6.192}$$

where $Q_{\Re}(x)$ and $Q_{\Im}(x)$ are the respective real and imaginary components and k_y is the component of the wave number in the y-direction. Note that the function $Y(y)$ results directly from the application of the separation-of-variables technique to eq. 6.191. The wave number relationships are

$$k_y = k_a \sin(\beta_a) = k(x)\sin[\beta(x)] \tag{6.193}$$

and

$$k_x(x) = k(x)\cos[\beta(x)] \tag{6.194}$$

where $\beta(x)$ is the angle between the contour and the wave crest. Physically, the wave component in the y-direction has both a uniform wave height and wavelength (and wave number) along the contour. The wave angle then does not vary along a specific contour. The behavior of k_y is also discussed in the next section, where the refraction analysis using Snell's law is discussed. The combination of eqs. 6.191 and 6.192 yields the following specialized form of the mild-slope equation:

$$G\frac{d^2Q}{dx^2} + \frac{dGdQ}{dxdx} + (k^2 - k_y^2)GQ = 0 \tag{6.195}$$

The expression for G is in eq. 6.190. The derivative of that function is

$$\begin{aligned}\frac{dG(x)}{dx} &= \frac{dG}{d(kh)}\frac{d(kh)}{dx} = \frac{dG}{d(kh)}\left\{\frac{km}{\left[1+\frac{k_0h}{\sinh(kh)}\right]}\right\} \\ &= \frac{m}{2}\frac{1}{\cosh^2(kh)}\left\{\frac{2\sinh(2kh) - 4kh\sinh^2(kh)}{\sinh(2kh)+2kh} - 1\right\}\end{aligned} \tag{6.196}$$

Equation 6.195 must be solved numerically. For the case of waves passing over a rising bed having straight and parallel contours, as sketched in Figure 6.33, the mild-slope equation can be solved using the Runga-Kutta numerical method, which is presented in the text by Adey and Brebbia (1983) along with other numerical techniques. Applications of the equation to more complicated bottom topography require the use of the finite-element method (FEM) as, for example, by Chandrasekera and Cheung (1997). See the book by Zienkiewicz and Taylor (2000) for the basics of this numerical technique.

Before presenting results obtained by using the mild-slope equation, several comments concerning the nature of the function Q must be made. First, from eq. 6.192, one sees that Q is complex. Hence, both the real and imaginary parts of the function must satisfy the mild-slope equation. Second, Q for the wave field in $x \leq 0$ must represent both incident and reflected waves.

We note that the mild-slope equation is a second-order, ordinary differential equation. Because Q is a complex function of x, there are two second-order equations to be solved – one corresponding to the real function, and the second corresponding to the imaginary function. Each equation requires two boundary conditions, either at the toe, where $x = 0$, or at $x = L$. Both the real and imaginary parts of Q are found to be "wavy" functions of x. At the toe and for $x < 0$, the velocity potential for the complete wave pattern is the sum of the potentials for both the incident waves and the reflected waves. At $x = L$, the potential represents the transmitted wave.

Use the results obtained from the linear theory of Chapter 3 to obtain the Q-values at both the toe or the crest of the rise. Specifically, we assume that the waves are linear in the region having the uniform water depth, h_a, and beyond the rise, where the depth is h_b.

The expression for the free-surface displacement is found by combining the linearized dynamic free-surface condition of eq. 3.6 with the expression for the velocity potential of eq. 6.173. That potential can be written as

$$\begin{aligned}\Phi &= \Re\{Q(x)Z(h,z)e^{i(k_y y-\omega t)}\} = \Re\{Q(x)Z(h,z)e^{i[k_a \sin(\beta_a)y-\omega t]}\}\\ &= \Re\{Q(x)Z(h,z)[\cos(k_y y-\omega t)+i\sin(k_y y-\omega t)]\}\\ &= \Re\{[Q_\Re(x)+iQ_\Im(x)]Z(h,z)[\cos(k_y y-\omega t)+i\sin(k_y y-\omega t)]\}\\ &= Z(h,z)\{Q_\Re(x)\cos[k_a\sin(\beta_a)y-\omega t]-Q_\Im(x)\sin[k_a\sin(\beta_a)y-\omega t]\}\end{aligned} \quad (6.197)$$

where, again, $\Re$ refers to the real component of the equation. The subscripts, $\Re$ and $\Im$, identify the respective real and imaginary functions, and the wave number relationship in the first line of the equation is from the results in eq. 6.193.

We can express the free-surface displacement at any position in terms of $Q_\Re$ and $Q_\Im$ as

$$\begin{aligned}\eta &= -\frac{1}{g}\Re\left\{\frac{\partial\Phi}{\partial t}\right\}|_{z\simeq 0} = \Re\left\{i\frac{\omega}{g}Q(x)e^{i(k_y y-\omega t)}\right\} = \Re\left\{\eta_x(x)e^{i(k_y y-\omega t)}\right\}\\ &= -\frac{\omega}{g}\{Q_\Re\sin[k_a\sin(\beta_a)y-\omega t]+Q_\Im\cos[k_a\sin(\beta_a)y-\omega t]\}\\ &= \eta_\Re+\eta_\Im\end{aligned} \quad (6.198)$$

where $Z_a(0)=1$ from eq. 6.181. This expression allows us to relate the real and imaginary parts of Q to the physical (linear) wave. In eq. 6.198, we see that $Q_\Re$ and $Q_\Im$ are out of phase by an angle $\pi/2$ because of the nature of real and imaginary components. Booij (1983) writes that "these ($Q_\Re$ and $Q_\Im$) can be interpreted as positions of the free surface at two times, a quarter of a period apart" when multiplied by the appropriate constants.

The relationship in eq. 6.198 allows us to write the specialized form of the mild-slope equation in eq. 6.195 in terms of the free-surface displacement, η. The result is

$$G\frac{d^2\eta}{dx^2}+\frac{dGd\eta}{dx\,dx}+(k^2-k_y^2)G\eta=0 \quad (6.199)$$

Copeland (1985), Girolamo, Kostense, and Dingemans (1988), and others use this form of the equation in their studies. We shall also use eq. 6.199 because the boundary conditions required for the solution are easily applied to the free-surface displacement.

At the toe of the rise and upwave of the rise (where $h=h_a$ and $x\le 0$), the free-surface displacement is the sum of those due to the incident wave (η_I) and the reflected wave (η_R). In this region, the free-surface displacement is then

$$\begin{aligned}\eta_a &= \eta_I+\eta_R\\ &= \frac{H_I}{2}\cos[k_a\cos(\beta_a)x+k_a\sin(\beta_a)-\omega t-\sigma_I]\\ &\quad+\frac{H_R}{2}\cos[-k_a\cos(\beta_a)x+k_a\sin(\beta_a)-\omega t+\sigma_R]\end{aligned} \quad (6.200)$$

where σ_I and σ_R are the respective phase angles. The incident wave leads the wave transmitted to the rise and by the angle σ_I while the reflected wave lags the transmitted wave by σ_R. As is the case for both optical and acoustical waves, the magnitudes of the reflected wave angle and the incident wave angles are equal, that is, $|\beta_R|=|\beta_I|=|\beta_a|$ (see Figure 6.33).

The expression for the free-surface displacement at the crest of the rise and downwave of that crest (where $h = h_b$ and $x \geq L$) is

$$\eta_b = \frac{H_T}{2}\cos[k_b\cos(\beta_b)x + k_b\sin(\beta_b) - \omega t + \sigma_T] \tag{6.201}$$

In eqs. 6.200 and 6.201, the wave numbers $k_a(=2\pi/\lambda_a)$ and $k_b(=2\pi/\lambda_b)$ are known from the dispersion equation, eq. 3.31. There are only transmitted waves both at the crest or the rise and downwave ($x > L$) of this crest. Those waves lag the waves on the rise by an angle σ_T.

In both the forms of eqs. 6.195 and 6.199, the mild-slope equation is a second-order differential equation. Because of this, a specific solution requires two boundary conditions. In solving the mild-slope equation of eq. 6.199, we specify the free-surface displacement and its derivative with respect to x. These conditions can be applied either at the toe of the rise or at the rise crest. Because there is only a transmitted wave both at and downwave of the rise crest of the rise, we find that $H_b = H_T$ and $\sigma_b = \sigma_T$. In the application of the mild-slope equation to breaking waves, Girolamo, Kostense, and Dingemans (1988) assume the property values of the transmitted waves (H_T and σ_T) well downwave of the crest rise. This may seem awkward to the reader as the incident wave properties are normally assumed. However, for this boundary-value problem this solution method is valid. Without loss in generality, we can dictate that a crest of the wave occurs at either the toe or the crest. At the toe, $\eta(0) = H_I/2$ and $\partial\eta/\partial x|_{x=0} = 0$. These conditions are also those used in Example 6.7.

It is important to remember that the $\eta(x)$-curve resulting from the solution of eq. 6.199 does not change by changing the crest position of the rise. That is, for given values of the slope (m) and the water depth (h_a), the behavior of $\eta(x)$ is determined. Hence, the curve includes the effects of the cumulative reflectivity of the rise as x increases. At $x = 0$, $\eta(0)$ must then equal the incident wave displacement η_I. The gross reflection from the rise is quantitatively represented by the reflection coefficient ($K_R \equiv H_R/H_I$). For the bed rise in question, we simply subtract the energy flux of the mild-slope transmitted wave (eq. 3.72) from the energy flux of the ideal (nonreflected) transmitted wave.

To demonstrate the application of the energy flux relationship of eq. 3.72, consider a segment of a crest (wave front) of a wave approaching a submerged step between two orthogonals (fictitious vertical walls, as in Figure 3.17a), illustrated in Figure 6.34. Also in that figure are wave fronts of the reflected and transmitted wave systems. The width, b_I, of the channel formed by the pair or orthogonals is determined by the incident wavelength (λ_I) and the crest angle (β_I) between the wave front and the step, as can be seen in the figure. Note that the alongshore step length is defined by the orthogonals and is common to the incident, reflected, and transmitted wave systems. As a result, the channel widths are related by

$$\frac{b_I}{\cos(\beta_I)} = \frac{b_R}{\cos(\beta_R)} = \frac{b_T}{\cos(\beta_T)} \tag{6.202}$$

Because the magnitudes of the incident and reflected wave angles are equal, we can write $|\beta_I| = |\beta_R| = |\beta_a|$ and $b_I = b_R = b_a$. Also from Figure 6.34, the relationships among the wavelengths at the step are

$$\frac{\lambda_I}{\sin(\beta_I)} = \frac{\lambda_R}{\sin(\beta_R)} = \frac{\lambda_T}{\sin(\beta_T)} \tag{6.203}$$

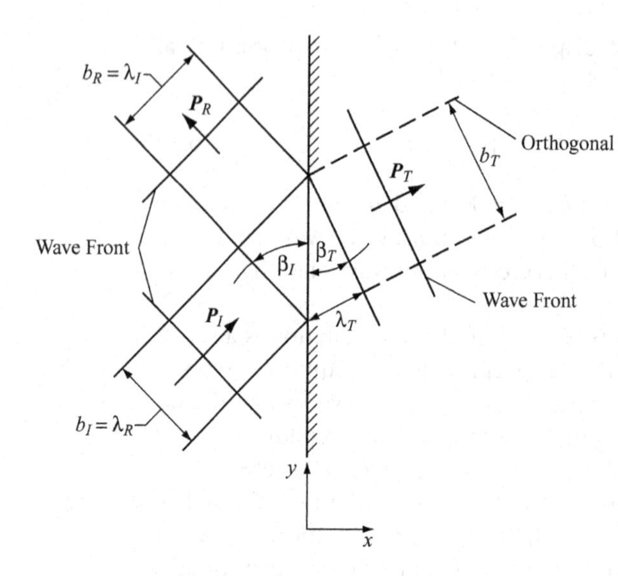

Figure 6.34. *Incident, Reflected, and Transmitted Waves at a Bed Step.*

The transmitted wave front angle is $\beta_T = \beta_b$.

Assuming the conservation of the energy flux, the magnitude of the energy flux (equal to that of the incident wave) is assumed to be unchanged after the reflection and transmission occur. Mathematically, this is expressed by

$$|\boldsymbol{P_I}| = |\boldsymbol{P_R}| + |\boldsymbol{P_T}| \tag{6.204}$$

So, the energy flux of the reflected wave is obtained from

$$|\boldsymbol{P_R}| = |\boldsymbol{P_I}| - |\boldsymbol{P_T}| \tag{6.205}$$

For the ideal case of no reflection from the step, the incident and transmitted energy fluxes are equal, as discussed in Section 6.3. That is,

$$|\boldsymbol{P_I}| = |\boldsymbol{P'_T}| \tag{6.206}$$

where the prime (′) is used to indicate the nonreflecting or pure shoaling assumption. In terms of the linear wave properties, eq. 6.206 can be expressed as

$$\frac{\rho g H_1^2 c_{ga} b_a}{8} = \frac{\rho g (H'_T)^2 c_{gb} b_b}{8} \tag{6.207}$$

where ρ is the mass density of the water, g is the gravitational constant, and c_g is the group velocity defined in eq. 3.63. Combine the expressions in eqs. 6.205 and 6.206 to obtain

$$|\boldsymbol{P_R}| = |\boldsymbol{P'_T}| - |\boldsymbol{P_T}| \tag{6.208}$$

This energy flux expression, expressed in terms of the wave properties, is

$$\begin{aligned}\frac{\rho g H_R^2 c_{ga} b_a}{8} &= \frac{\rho g (H'_T)^2 c_{gb} b_b}{8} - \frac{\rho g H_T^2 c_{gb} b_b}{8} \\ &= \frac{\rho g H_I^2 c_{ga} b_a}{8} - \frac{\rho g H_T^2 c_{gb} b_b}{8}\end{aligned} \tag{6.209}$$

In this expression, H'_T is the wave height without reflection. Because both the reflecting and nonreflecting wave systems have the same initial values, we divide the reflection and transmission terms of eq. 6.209 by $\rho g H_I^2 c_{ga} b_a / 8$, and take the square root of the resulting expression. The result is

$$\frac{H_R}{H_I} = \sqrt{\left(\frac{H_T'}{H_I}\right)^2 \frac{c_{gb}b_b}{c_{ga}b_a} - \left(\frac{H_T}{H_I}\right)^2 \frac{c_{gb}b_b}{c_{ga}b_a}} = \sqrt{1 - \left(\frac{H_T}{H_I}\right)^2 \frac{c_{gb}\cos_b}{c_{ga}\cos_a}} \tag{6.210}$$

In terms of coefficient notation, the last equality of this expression yields

$$\frac{H_R}{H_I} \equiv K_R = \sqrt{1 - \frac{K_T^2}{K_S^2 K_r^2}} \tag{6.211}$$

which relates the reflection coefficient (K_R) to the shoaling coefficient [$K_S \equiv \surd(c_{ga}/c_{gb})$, discussed in Chapter 3], the refraction coefficient [$K_r \equiv \surd(b_a/b_b) = \surd\{\cos(\beta_a)/\cos(\beta_b)\}$ from the results in eq. 6.88], and the transmission coefficient ($K_T = H_b/H_a$).

The application of the mild-slope equation is demonstrated in the following example, where the equation is applied to the pure shoaling (shoaling without refraction) of waves. That is, the waves directly approach a straight, parallel-contoured rise in the sea bed, hence, $\beta_a = \beta_b = 0$.

EXAMPLE 6.12: APPLICATION OF THE MILD SLOPE TO PURE SHOALING Consider a system of 1-m, 7-sec linear waves in deep water approaching a rising bed having straight, parallel, and evenly spaced bottom contours. The waves approach the beach directly, so the separation distance (b) between orthogonals is uniform throughout the region, and the wave angles (β) between the crest lines (wave fronts) and the h-contours are equal to zero in Figure 6.33b. The water depth at the crest of the bed rise is $h_b = 0.06\lambda_0 = 4.6$ m, and the slope of the rise is $m = 1/3$, which is the upper practical limit of the mild-slope equation. The mild-slope equation in eq. 6.195 is solved for both the real and imaginary components of Q, that is, $Q_\Re$ and $Q_\Im$, respectively. In addition, because we know that these components are out of phase by $\pi/4$, let the origin of the horizontal coordinate be where the following conditions occur:

$$\begin{aligned} Q_{\Re a} &= 0, \\ Q_{\Im a} &= -\frac{gH_I}{2\omega} \simeq -5.46\frac{m^2}{s}, \\ \left.\frac{\partial Q_\Re}{\partial x}\right|_{x=0} &= \frac{gK_a H_I}{2\omega} \simeq 0.449\frac{m}{s} \quad \text{and} \\ \left.\frac{\partial Q_\Im}{\partial x}\right|_{x=0} &= 0 \end{aligned}$$

These conditions result from the application of the expression in eq. 6.198 to the toe, when the incident wave first arrives. Because both y and t have infinite ranges, we assume that both are equal to zero in eq. 6.198 and in the spacial derivative of the equation. The conditions here are then the boundary conditions used in the solution of the mild-slope equation, eq. 6.195, when that equation is solved for $Q_\Re$ and $Q_\Im$ for waves approaching the rise directly ($\beta_a = 0$ and $k_y = 0$). Plots of $Q_\Re$ and $Q_\Im$ are presented in Figure 6.35. The mild-slope equation is solved using the Runga-Kutta numerical integration, which is presented in Appendix B of this book.

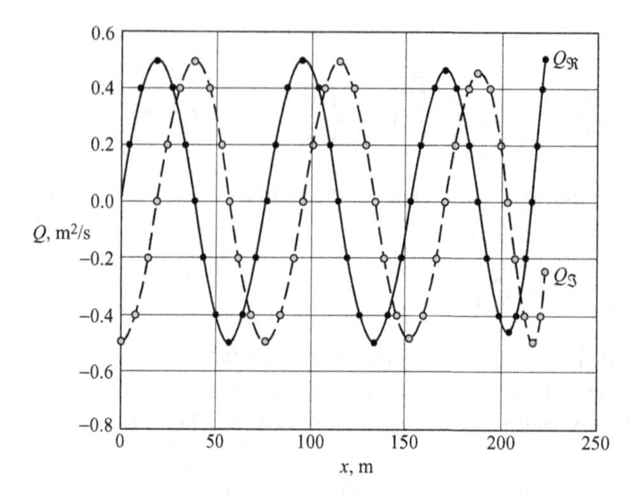

Figure 6.35. *Variation of the Real and Imaginary Potentials over a Bed Slope of 1/3.*

Our interest is in the behavior of the free surface over the bed rise, as predicted by the expression in eq. 6.198, and in the reflection and transmission coefficients resulting from the rising bed. The free-surface profile, η, corresponding to $Q_{\Im}$ is presented in non-dimensional form in Figure 6.36 as a function of x/L. In Figure 6.36, we note that the wave height first decreases and then rises as the wave shoals. This behavior is observed for nonreflecting shoaling waves, as discussed in Chapter 3. For comparison, in Figure 6.37 results obtained from the linear shoaling coefficient expression of eq. 3.78 are plotted with H/H_I values obtained from the mild-slope analysis as functions of h/λ_0. From the plots, we see that the linear and mild-slope analyses give approximately the same values for the larger h/λ_0 values. As h/λ_0 decreases the curves separate, where the lower-valued curve is that predicted by the mild-slope analysis. The separation of the curves is due to the cumulative effects of wave reflection from the rise. The reflection coefficient of eq. 6.211 is approximately 0.1365 for the given slope and rise-crest depth.

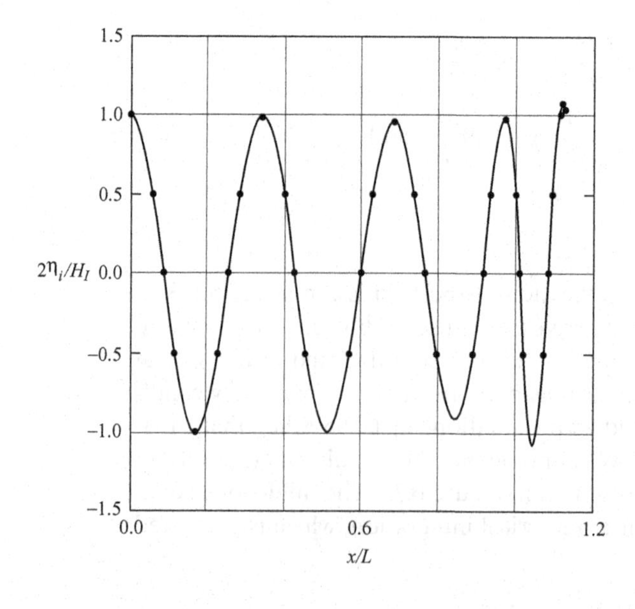

Figure 6.36. *Dimensionless Free-Surface Profile over a Rise Having a Slope of 1/3.*

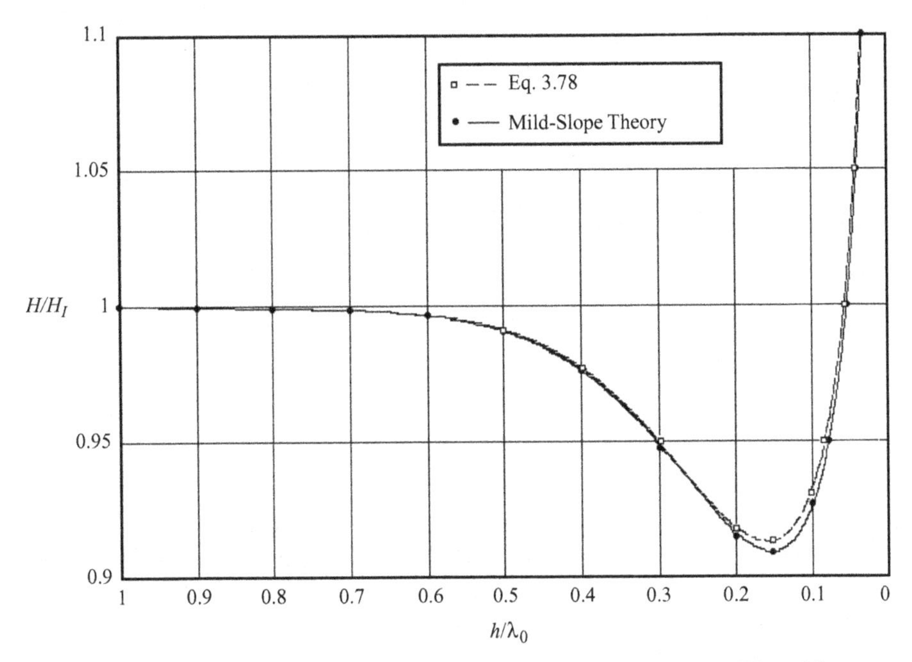

Figure 6.37. *Comparison of the Shoaling Coefficients from the Linear Wave Theory and the Mild-Slope Equation for a Bed Rise Having a Slope of 1/3.*

In summary, the mild-slope equation is both versatile and useful. It allows us to predict the wave profile, reflection coefficient, and transmission coefficients over beds having moderate changes in elevation. The application of the equation is not limited to beds having rectilinear elevations. One of the favorite applications of wave analysts is the bed having a circular mound, such as analyzed by Chandrasekera and Cheung (1997) and others. The comparison of the theoretical and experimental results for this bed geometry is quite good. Hence, we can use the mild-slope equation with confidence.

6.6 Closing Remarks

In this chapter, some of the available theoretical and numerical methods used in the analyses of wave modification and wave transformation are presented. Modification refers to a change in the wave properties, whereas transformation refers to the change in the wave energy and energy flux. The methods presented in this chapter are those that the author believes to be the best suited to the analyses and solutions of ocean engineering problems. The readers are encouraged to consult the references to obtain the details of the particular methods of interest.

7 Waves in the Coastal Zone

The field of coastal engineering has many facets. The reader is referred to the handbooks edited by Herbich (1999) and by Kim (2009) for discussions of most of the coastal engineering areas. In this chapter, the focus is on the coastal zone, where most of the attention of coastal engineers is focused on the effects of breaking and broken waves. The phenomenon of breaking is nonlinear in nature, as discussed in Section 4.6. The nonlinear behavior of breaking waves can be approximately predicted by theoretical analyses. The theoretical expressions for breaking waves presented in Section 4.6 are based on two assumptions. First, the water depth is assumed to be uniform, and second, the wave profile of the breaking wave is symmetric about a vertical plane containing the crest line. Even with these modeling constraints, the theoretical analyses have been found to have value in conceptual engineering design applications. Dean (1974) presents a detailed discussion of the limitations of the various wave theories when applied to waves at or near breaking.

After a shoaling wave breaks, energy losses occur, and the resulting behavior of the wave depends on phenomena that cannot be completely mathematically modeled. The behavior of the wave prior to the break is also affected by bed friction and, if the bed is porous, by percolation. These cause energy losses and complicate our ability to theoretically predict the behavior of the wave. Because of this, empirical formulas based on both experimental and field data have been developed. These formulas can be used to predict the type of the breaking wave (spilling, plunging, collapsing, and surging), the runup (resulting from the uprush on the beach), and the alongshore or longshore transport of the water particles within the surf zone.

In this chapter, the nature of the breaking wave is first discussed. This discussion is based on the extensive laboratory and field observations of Galvin (1968, 1972) and others. The empirical formulas for the breaker height index of McCormick and Cerquetti (2002), based on the works of Goda (1970a, 1970b), and the breaker depth index of Weggel (1972) are then presented. The empirical equations are used in the determination of the breaking height and breaking depth. The concept of surf similarity (Iribarren and Nogales, 1949; Battjes, 1974), resulting from a dimensional analysis of waves in the surf zone, is also introduced. From similarity considerations, the non-dimensional parameter called both the *surf similarity parameter* and the *Iribarren number* results. Specific ranges of this parameter have been shown to correspond well to the types of breaks, the ratio of reflected-to-incident wave

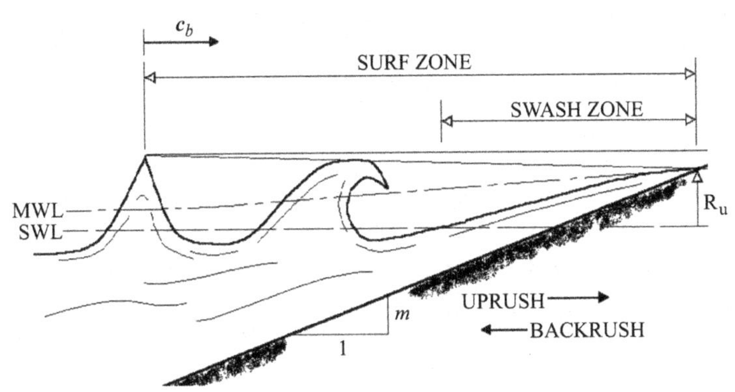

Figure 7.1. *Sketch of the Profiles of Breaking and Broken Regular Waves in the Surf Zone.* The reader should note that the maximum height of the water above the SWL decreases as the broken wave travels landward. The MWL is shown to be above the SWL. This phenomenon is called *wave set-up*. The *runup*, R_u, is the height above the SWL attained by the landward uprush of water. The seaward backrush of the water to the sea following the uprush is due to gravity. The swash zone is the region landward of the break where the water motions are flume-like. See U.S. Army (1984).

energies, runup, and the ratio of the breaking height-to-breaking depth for purely shoaling waves. It is then demonstrated how the results for the purely shoaling waves can be included in the analysis of refracting waves. Finally, the concept of radiation stress (Longuet-Higgins, 1970a, 1970b, 1972; Longuet-Higgins and Steward, 1960, 1961, 1962, 1964) is discussed. This concept leads to methods to determine the hydrodynamic phenomena within the surf zone, including set-up, set-down, and longshore transport.

The discussions in this chapter are somewhat limited. For more complete coverage of both coastal processes and coastal engineering, see the books by Dean and Dalrymple (2002), Massel (1989), Sorensen (1997), and Herbich (1999). The handbooks edited by Herbich (1999) and Kim (2009) contain extensive outlines of the *Automated Coastal Engineering System* (ACES) of the U.S. Army Corps of Engineers. The ACES is an updated and electronic version of the *Shore Protection Manual* (see U.S. Army, 1984).

7.1 Coastal Zone Phenomena

There are three major regions of the coastal zone that are of interest. These are called the littoral zone, the surf zone, and the swash zone. The *littoral zone* is the region extending from the first (outermost) wave-induced motions of sediment on the bed to the innermost extent of the uprush. Sand is the sediment of choice herein. The littoral zone is the region of interest in coastal engineering, which is primarily devoted to the prevention of either unwanted erosion (the loss of sand and other littoral materials) or unwanted accretion (the gain of littoral materials). Within the littoral zone is the *surf zone*, which is the region extending from the outermost break to the innermost extent of the uprush, as sketched in Figure 7.1. The *swash zone* is the region within the surf zone, landward of the break, where the water travels up the beach with a flume-like flow called the *uprush*. Some of the uprushed water can percolate into the sand, whereas the remainder of the water returns to sea in the gravitationally induced backrush.

As classified by Galvin (1968, 1972), there are four possible types of breaking waves in the coastal zone. Profiles of these breaking waves are sketched in Figure 7.2. From Galvin (1972), the descriptions of the types of breaking waves are as follows:

Spilling Break (Figure 7.2a): "Foam, bubbles, and turbulent water appear at the wave crest and eventually cover the front face of the wave. Spilling starts at the crest when a small tongue of water moves forward faster than the wave form as a whole. In its final stages, the spilling wave evolves into a bore or an undulatory bore." (We see in this description that the mathematical definition of the breaking wave in eq. 4.131 is physically realized.)

Plunging Break (Figure 7.2b): "The whole front face of the wave steepens until vertical; the crest curls over the front face and falls into the base of the wave; and a large sheet-like splash arises from the point where the crest touches down."

Collapsing Break (Figure 7.2c): "The lower part of the front face of the wave steepens until vertical, and this front face curls over as an abbreviated plunging wave. The point where the front face begins to curl over is landward of, and lower than, the point of maximum elevation of the wave."

Surging Break (Figure 7.2d): "The front face and crest of the wave remains relatively smooth and the wave slides up the beach with only minor production of foam and bubbles. Resembles a standing wave."

The sketches in Figure 7.2 are modeled after those of Galvin (1972). C. J. Galvin has devoted much of his professional life to the study of waves in the coastal zone. Although his referenced works are based on studies conducted several decades ago, both in the field and in large wave tanks, his observations are still considered to be most authoritative.

In Figure 7.1, two waves in the surf zone are sketched. Actually, there can be several waves, depending on the type of break. For spilling waves over beds of small slope, one can observe a number of waves, according to both Galvin (1972) and Battjes (1974).

7.2 Empirical Analyses of Breaking Waves on Beaches

In this section, both experimental data and empirical formulas for purely shoaling waves on beaches of uniform slope are presented and discussed. Of particular interest are the results of the experimental studies of Goda (1970a, 1970b) and the empirical equation of Weggel (1972), both of which appear in the *Shore Protection Manual* (U.S. Army, 1984), and are widely used by the coastal engineering community. Also presented is the empirical formula of McCormick and Cerquetti (2002), which is based on the Goda (1970a) data, and the formula of Komar and Gaughan (1972) for the breaking wave index over flat, horizontal beds. An excellent discussion of many of the empirical and theoretical formulas for the breaking indices H_b'/H_0' and h_b/H_0' is found in the paper of Wang and Du (1993). The prime ($'$) is used to indicate pure shoaling (without refraction). McCormick and Cerquetti (2002) find that the averaged data presented by Goda (1970a) for the breaking height index (H_b'/H_0')

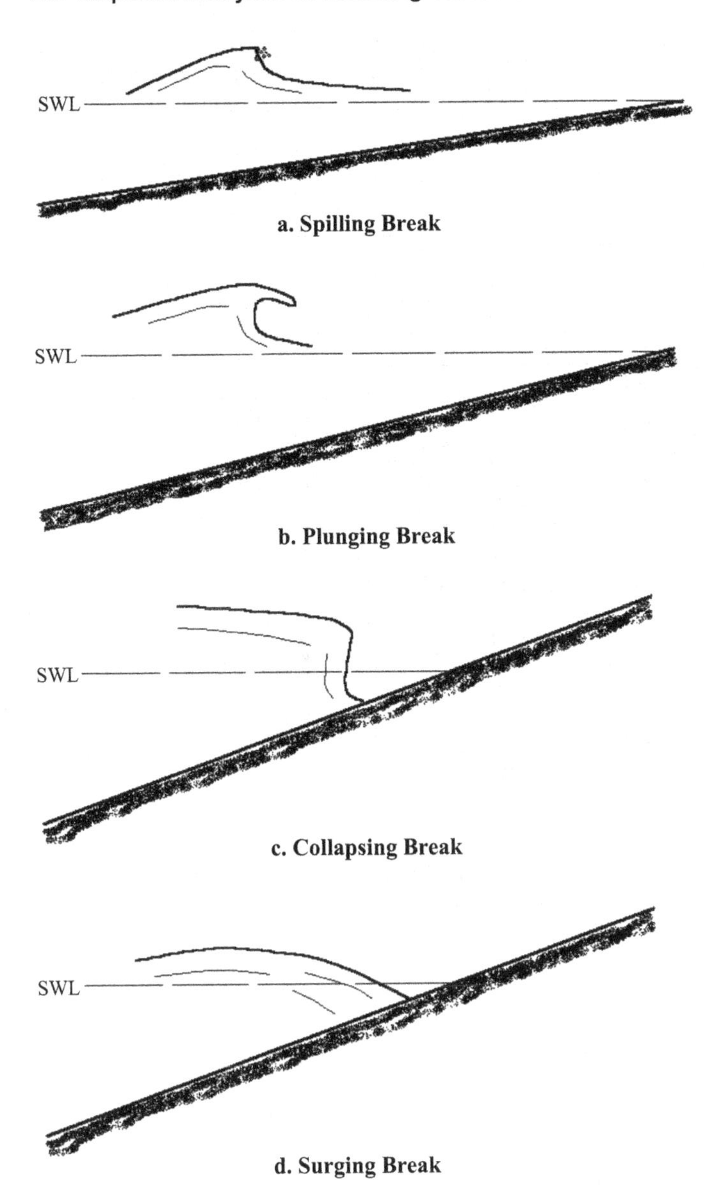

Figure 7.2. *Types of Breaking Waves*. See Galvin (1972).

for a purely shoaled wave over beds of uniform slope (m) can be approximated by

$$\frac{H_b'}{H_0'} = \{1 + \mathrm{A}\tanh[\mathrm{B}(m - 0.02)]\}\left\{\frac{\cosh^2\left(11.88\dfrac{H_0'}{\lambda_0}\right)}{0.5\sinh\left[2\left(11.88\dfrac{H_0'}{\lambda_0}\right)\right] + \left(11.88\dfrac{H_0'}{\lambda_0}\right)}\right\}^{0.265} \tag{7.1}$$

where

$$\mathrm{A} = 0.236 - 1.641\left(\frac{H_0'}{\lambda_0}\right) \tag{7.2}$$

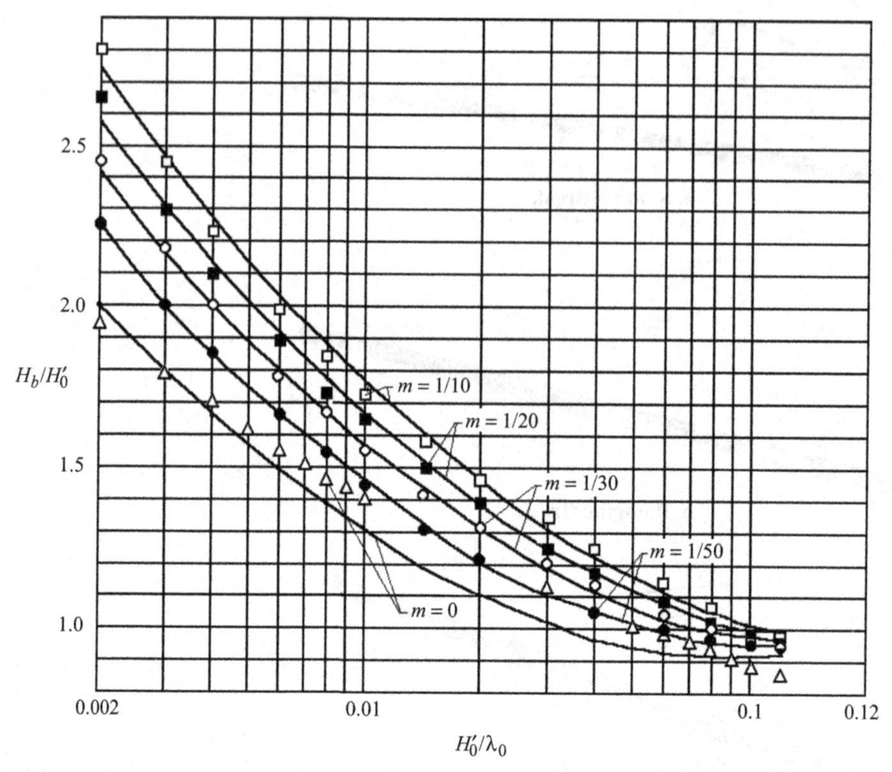

Figure 7.3. *Breaking Height Index Curves.* The solid lines are obtained from eq. 7.1 due to McCormick and Cerquetti (2002), and the data points for $m > 0$ are the averaged values reported by Goda (1970a) and presented in Table 7.1. The $m = 0$ values result from the expression in eq. 7.4 due to Komar and Gaughan (1972).

and

$$B = 25.48 + 25.41\left(\frac{H_0'}{\lambda_0}\right) \tag{7.3}$$

The expression in eq. 7.1 can be applied to breaking waves over both flat, horizontal beds and beds of uniform slope. Results obtained from eq. 7.1 along with the averaged numerical values of Goda (1970a) for $m > 0$ and the empirical data of Komar and Gaughan (1972) for $m = 0$ are presented in Figure 7.3. The empirical expression of Komar and Gaughan (1972) is

$$H_b' = \frac{0.56 H_0'}{\left(\dfrac{H_0'}{\lambda_0}\right)^{1/5}} \tag{7.4}$$

which is based on both the Airy's linear wave theory, discussed in Chapter 3, and observed data. Specific values of Goda (1970a) and those from eq. 7.1 are presented in Table 7.1. Again, the primes (′) indicate that the properties are for purely shoaling waves. As noted in the *Shore Protection Manual* (U.S. Army, 1984), the Goda values in Figure 7.1 and Table 7.1 are based on somewhat scattered data. The results from eq. 7.1, when applied to the case of a flat, horizontal bed ($m = 0$), agree to within 10% with the results from empirical expression of Komar and Gaughan

Table 7.1. *Breaking indices (H_b'/H_0') from Goda (1970a) and eq. 7.1*

$m =$	|	1/50	|	1/30	|	1/20	|	1/10
H_0'/λ_0	Goda	eq. 7.1	Goda	eq. 7.1	Goda	eq. 7.1	Goda	eq. 7.1
0.002	2.28	2.242	2.45	2.413	2.65	2.578	2.80	2.746
0.003	2.01	2.014	2.19	2.166	2.29	2.314	2.44	2.464
0.004	1.85	1.866	2.00	2.007	2.10	2.142	2.23	2.280
0.006	1.68	1.677	1.79	1.801	1.88	1.922	1.98	2.044
0.008	1.54	1.555	1.64	1.669	1.72	1.779	1.84	1.890
0.010	1.45	1.467	1.55	1.573	1.64	1.675	1.71	1.779
0.015	1.31	1.321	1.41	1.414	1.50	1.503	1.58	1.592
0.020	1.22	1.229	1.32	1.313	1.39	1.393	1.48	1.472
0.030	–	1.117	1.21	1.188	1.26	1.255	1.35	1.320
0.040	1.05	1.052	1.14	1.113	1.18	1.170	1.26	1.226
0.060	1.01	0.985	1.05	1.031	1.09	1.075	1.15	1.117
0.080	0.98	0.959	1.00	0.994	1.02	1.027	1.07	1.057
0.100	0.94	0.952	0.96	0.977	0.97	1.000	0.98	1.019
0.120	0.94	0.955	0.94	0.969	0.95	0.981	0.96	0.992

(1972) in eq. 7.4. Other empirical expressions for the breaker height index are compared and discussed by McCormick and Cerquetti (2002).

From either eq. 7.1 or the curves in Figure 7.3, the breaking height index for purely shoaled waves on impermeable beds of uniform slope is obtained. With this information, we can use the empirical formula of Weggel (1972) to obtain the breaking water depth. The Weggel formula is

$$\frac{h_b}{H_b'} = \frac{1}{b - \dfrac{aH_b'}{gT^2}} = \frac{1}{b - \dfrac{1}{2\pi}\dfrac{H_b'}{\lambda_0}} \tag{7.5}$$

where the slope-dependant coefficients a and b are, respectively,

$$a = 43.75(1 - e^{-19m}) \tag{7.6}$$

and

$$b = \frac{1.56}{1 + e^{-19.5m}} \tag{7.7}$$

Results of eq. 7.5 are presented in Figure 7.4, where the breaker depth index is as a function of the deep-water steepness (H_b'/λ_0). Note that the breaker depth index value for waves over a flat, horizontal bed is independent of H_b'/λ_0. The use of breaker index equations is illustrated in the following examples.

EXAMPLE 7.1: BREAKING WAVE PROPERTIES OVER A FLAT, HORIZONTAL BED In Chapter 4, the Munk (1949) expression for a solitary breaking wave is given in eq. 4.161. That expression is

$$H_b' = \frac{H_0'}{3.3\left(\dfrac{H_0'}{\lambda_0}\right)^{1/3}}$$

The bed slope, m, does not appear in this expression. Here, we apply the Munk expression, the McCormick-Cerquetti expression in eq. 7.1, and the Komar-Gaughan expression in eq. 7.4 to 8-sec deep-water waves traveling over a flat,

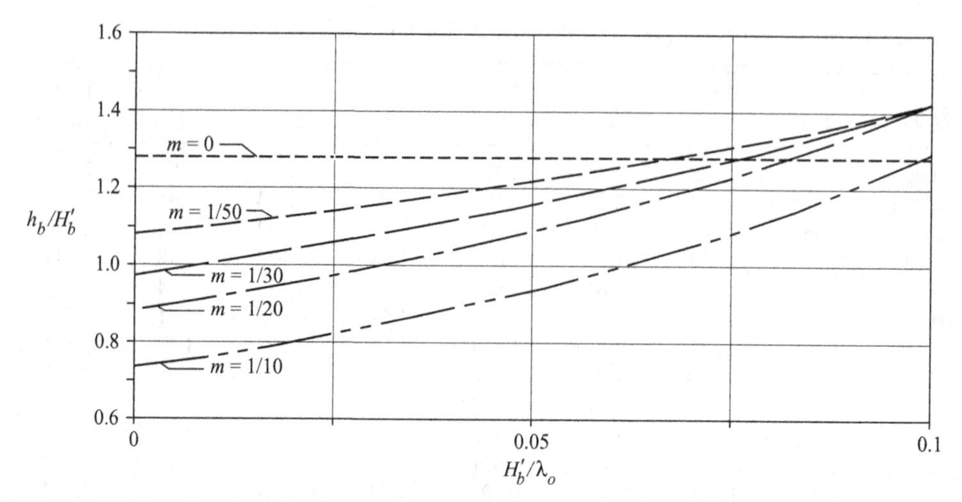

Figure 7.4. *Breaking Depth Index Curves.* The curves result from the empirical expression in eq. 7.5 due to Weggel (1972).

horizontal bed, and compare the resulting breaking height values. The wave heights are 0.2 m, 1 m, and 10 m, respectively. Because the deep-water wavelength for this period is $\lambda_0 \simeq gT^2/2\pi \simeq 100$ m, the deep-water wave steepness values for these waves are 0.002, 0.01, and 0.1, respectively. In Figure 7.1, these respective values approximately correspond to the left, center, and right conditions in that figure. The results are as follows:

H_b'/λ_0

H_0'/λ_0	McCormick-Cerquetti, Eq. 7.1	Komar-Gaughan, Eq. 7.4	Munk, Eq. 4.161
0.002	2.000	1.941	2.405
0.01	1.310	1.407	1.406
0.1	0.918	0.888	0.653

When compared to the results obtained from eq. 7.1, the Munk (1949) equation is seen to over-predict the breaker index at the lowest deep-water wave steepness value while under-predicting at the highest steepness value. Values obtained from eq. 7.1 agree with the Komar-Gaughan results to within 8%.

EXAMPLE 7.2: BREAKING WAVE PROPERTIES ON A BEACH OF UNIFORM SLOPE – EMPIRICAL EQUATIONS The 1-m, 8-sec deep-water waves in Example 7.1 approach a beach having a uniform slope (m) of 1/50. On Monday, the waves approach directly, whereas on Tuesday, the waves approach at a deep-water wave angle (β_0) of 30°. Using the expressions in eqs. 7.1 and 7.5, we are to determine the breaking height and breaking depth on each day. On Monday, the waves are not refracted; so, the equations can be applied directly. On Tuesday, Snell's law of eq. 6.89 must be incorporated to account for wave refraction.

As for the deep-water wave in Example 7.1, $H_0'/\lambda_0 \simeq 0.01$ for Monday's waves. From eq. 7.1, the breaking height is approximately 1.47 m. For the deep-water wavelength of $\lambda_0 \simeq gT^2/2\pi \simeq 100$ m, one finds $H_b'/\lambda_0 = 0.00147$. Using this value and the slope of 1/50 in eq. 7.5, we obtain $h_b/H_b' \simeq 1.08$. The breaking depth is then $h_b \simeq 1.59$ m. For the 1/50 slope, Monday's waves are found to break 79.3 m from the shoreline.

For Tuesday's waves, we must include refraction with the expressions in eqs. 7.1 and 7.5. First, the equivalent nonrefracting deep-water wave height must be determined. From eq. 6.90, that wave height is $H_0' = H_0\sqrt{[\cos(\beta_0)]} \simeq 0.931\,\text{m}$. Using this value, the equivalent deep-water steepness is $H_0'/\lambda_0 \simeq 0.00931$, and from eq. 7.1, $H_b'/H_0' \simeq 1.50$. The equivalent breaking wave height is $H_b' \simeq 1.40\,\text{m}$. In the Weggel (1972) expression of eq. 7.5, $H_b'/\lambda_0 \simeq 0.0140$, which yields $h_b/H_b' \simeq 1.11$. The water depth at the break is then $h_b \simeq 1.55\,\text{m}$, which is 77.5 m offshore from the shoreline. The wavelength at the breaking depth is $\lambda_b \simeq 30.7\,\text{m}$ from eq. 3.31, and the breaking wave angle is $\beta_b \simeq 8.83^{\text{o}}$ from eq. 6.89. Finally, the breaking wave height is found to be

$$H_b = \frac{H_b H_b' H_0'}{H_b' H_0' H_0} H_0 = \sqrt{\frac{\cos(0)}{\cos(\beta_b)}} \left(\frac{H_b'}{H_0'}\right) \sqrt{\frac{\cos(\beta_0)}{\cos(0)}} H_0$$
$$\simeq (1.01)\,(1.50)\,(0.931)\,1\,\text{m} \simeq 1.40\,\text{m} \tag{7.8}$$

The effects of refraction cause the Tuesday waves to break 1.8 m closer to shore than the Monday waves, and the breaking wave height of the Tuesday waves is 0.07 m smaller than that of the Monday waves. From these results, one can conclude that refraction delays the break of the wave and, in addition, retards the growth of the breaking wave height.

7.3 Surf Similarity

In Section 2.7, applications of dimensional analysis are presented to determine model-to-prototype scaling relationships. The non-dimensional groupings can be found by dimensional analysis or from experimental and field observations. Our focus here is the latter. For waves in the coastal zone, Table 7.2 is useful in the dimensional analysis. In that table, force is used as a primary dimension. Note that when sediment transport in the coastal zone is of interest, the friction coefficient on the bed and the properties of the sediment must be included. This topic is not discussed herein.

The spatial coordinates (x, y, z), the running time (t), and the particle velocity components (u, v, w) are not included in Table 7.2 because the dimensional analysis is directed at wave parameters rather than independent and dependent variables.

For the present, our attention is focused on the kinematic properties of purely shoaling waves traveling over a flat bed having a uniform slope, m. Therefore, there will be no β-dependence. The water depth-to-offshore distance ratio, h/x, is equal to the bed slope, m; hence, the bed slope is included in the analysis rather than h and x. The analysis can be further simplified by the findings in Chapter 3, where the kinematic wave properties over a horizontal bed are shown to depend on the wave steepness, H/λ, and the depth-to-wavelength ratio, h/λ.

Iribarren and Nogales (1949) and Battjes (1974) discuss the non-dimensional parameter known as both the *Iribarren number* and the *surf similarity parameter*, defined as

$$\xi_I \equiv \frac{m}{\sqrt{H_I'/\lambda_0}} \tag{7.9}$$

where H_I' is the height of a nonrefracting incident wave. The parameter in eq. 7.9 is referred to as the *surf similarity parameter* in the remainder of this chapter. For

Table 7.2. *Physical quantities and dimensions*

Physical quantity	Symbol	Dimensions
Length (wave height, wavelength, water depth, distance from shoreline)	H, λ, h, x_b	L
Time (period)	T	T
Mass	m	F-T^2-L^{-1}
Mass density	ρ	F-T^2-L^{-4}
Velocity (celerity, group, and particle velocities)	c, c_g, and u, v, w	L-T^{-1}
Gravitational acceleration	g	L-T^{-2}
Force	F	F
Pressure	p	F-L^{-2}
Dynamic viscosity	μ	F-T-L^{-2}
Bed slope	m	F^0-T^0-L^0
Wave angle	β	F^0-T^0-L^0*

* The dimensions of the angle are in radians or degrees.

a flat beach of uniform slope extending from the shoreline to deep water, the wave height in eq. 7.9 is the deep-water wave height, H_0', that is, the subscript I can be replaced by 0. In the general form in eq. 7.9, the incident wave height could be over a flat horizontal bed seaward of a foreshore of slope m.

A. Breaking Waves

As noted by Battjes (1974), Galvin (1968) uses an "offshore parameter," defined as $H_0'/(\lambda_0 m^2) = 1/(\xi_0)^2$, and an "inshore parameter," defined as $H_b'/(gT^2 m) = m/[2\pi(\xi_b)^2]$, in his classification of breaking waves into breaker types, where the subscript b identifies the wave properties at the break point. Galvin's observations are presented in Table 7.3. Again, the profiles of the breaker types are sketched in Figure 7.2.

Battjes (1974) presents "an expression for the condition at which the transition occurs between non-breaking and breaking waves approaching a slope which is plane in the neighborhood of the still-water line" of Iribarren and Nogales (1949). This expression is based on shallow-water trochoidal wave theory (for example, see McCormick, 1973), from which the breaking condition occurs when the breaking wave amplitude $(H_b'/2)$ equals the water depth (h_b). The reader should note that the trochoidal theory predicts a cusp-shaped crest for a breaking wave, the existence of which is physically impossible. Iribarren and Nogales (1949) assume that the breaking depth is a quarter-wavelength $(\lambda_b/4)$ from the shoreline, where the breaking wavelength is approximated by the shallow-water expression of the linear (and Stokes' second-order) theory. From eq. 3.38, $\lambda_b \simeq \sqrt{(gh_b)}T$. The resulting

Table 7.3. *Breaker type ξ-ranges from Galvin's (1968) observations*

Breaker type	ξ_0 range	ξ_b range (Battjes, 1974)
Surging or *collapsing*	$\xi_0 > 3.3$	$\xi_b > 2.0$
Plunging	$0.5 < \xi_0 < 3.3$	$0.4 < \xi_b < 2.0$
Spilling	$\xi_0 < 0.5$	$\xi_b < 0.4$

expression for the surf similarity parameter at the break is

$$\xi_b \equiv \frac{m}{\sqrt{H_b'/\lambda_0}} \simeq 2.3 \tag{7.10}$$

The reader can see that this value occurs in range for surging or collapsing breaks in Table 7.3.

For breaking waves on a beach of uniform slope extending into deep water, the following formula has been used to approximate the mean data presented by Battjes (1974):

$$\frac{H_b'}{h_b} = 0.781 + 0.2\xi_0, \quad \text{where} \quad \xi_0 \leq 2.0 \tag{7.11}$$

From the results presented in Table 7.3, we see that this expression applies to all spilling waves and some plunging waves. Note that the constant value in eq. 7.11 is that obtained from the solitary theory in eq. 4.152.

B. Wave Reflection

For waves approaching the shore directly ($\beta_0 = 0$), the energy flux (wave power) that is not lost in the breaking process will be reflected back to sea. Seaward of the line of breakers, the incident and reflected wave heights are $H_I'$and H_R', respectively. The ratio of these two wave heights is called the *reflection coefficient*. As defined in Chapter 6, the reflection coefficient is

$$K_R \equiv \frac{H_R'}{H_I'} \tag{7.12}$$

The ratio of the reflected-to-incident energy flux is proportional to K_R^2. Battjes (1974) approximates the upper limit of the reflection coefficient for widely scattered data by

$$\begin{aligned} K_R &= 0.1\xi^2, \quad \text{where} \quad 0.1\xi^2 < 1.0 \\ &= 1.0, \quad \text{where} \quad 0.1\xi^2 > 1.0 \end{aligned} \tag{7.13}$$

A continuous equation that well approximates the upper bound of the reflection coefficient data is

$$K_R = \tanh^{3.31}(0.527\xi), \quad \text{where} \quad 0.1\xi^2 \leq 1.2 \tag{7.14}$$

Results obtained from eqs. 7.13 and 7.14 are presented in Figure 7.5.

C. Runup

Landward of a shoaling wave break, the fluid motions are flume-like over the beach, traveling landward in the uprush and seaward in the backrush. The uprush and backrush are paramount in the transport of sand. Depending on the slope of the bed and the breaking wave conditions, the beach experiences either erosion or accretion. As discussed in Chapter 6, the maximum height above the SWL attained by a broken wave on a beach is called the *runup*, $\mathrm{R_u}$. The runup is illustrated in Figure 7.1. For waves on a beach extending into deep water, the runup for a purely shoaling wave is obtained from the empirical formula of Hunt (1959). The *Hunt formula* is

$$\mathrm{R_u} = C_p H_0' \xi_0, \quad \text{where} \quad 0.1 \leq \xi_0 \leq 2.3 \tag{7.15}$$

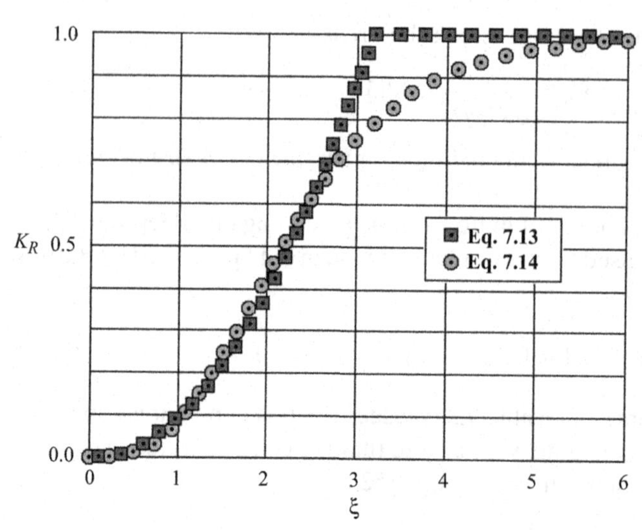

Figure 7.5. *Upper Bounds of the Reflection Coefficients versus the Surf Similarity Parameter.*

where C_p is called the *porosity factor*, defined as the volume of the void space and the total volume of the solid and void components. For a smooth beach, $C_p \simeq 1.0$. See Hughes (2004) for an extensive analysis of this condition. From the results in Table 7.3, we see that Hunt's formula applies to all spilling waves and some plunging waves.

EXAMPLE 7.3: BREAKING WAVE PROPERTIES ON A BEACH OF UNIFORM SLOPE – SURF SIMILARITY Consider the conditions in Example 7.2, where 1-m, 8-sec deep-water waves directly approach a 1/50 beach. For this condition, the deep-water wave steepness is $H_0'/\lambda_0 \simeq 0.01$. The value of the surf similarity parameter of eq. 7.9 (where $H_I' = H_0'$) is $\xi_0 = m/\sqrt{(H_0'/\lambda_0)} \simeq 0.2$. From Table 7.3, a spilling break will occur. Because $\xi_0 \leq 2.0$, the expression in eq. 7.11 can be used to determine the ratio of the breaking height to breaking depth. The result is $H_b'/h_b = 0.821$. In Example 7.2, using the Weggel (1972) expression in eq. 7.5, we find that $h_b/H_b' = 1.08$, or $H_b'/h_b = 0.926$. The difference in the values is slightly larger than 10%. The reader should note that the expressions in eqs. 7.5 and 7.11 both yield $H_b'/h_b|_{m=0} \simeq 0.78$ when applied to a flat, horizontal bed.

7.4 Surf Zone Hydromechanics – Radiation Stress

Coastal engineers are concerned with the wave- and current-induced motions of sand. The wave-induced sand motions occur primarily in the surf zone. The *surf zone* is the region between the outermost break and the landward extent of the uprush. The sand is primarily transported by broken waves in the alongshore or longshore direction; however, some time-averaged transport is both landward and seaward. To assess the need for shore protection, the coastal engineer needs to fully understand both the nature and magnitude of the longshore transport. To this end, Longuet-Higgins and Stewart (1960, 1961, 1962, 1964) and Longuet-Higgins (1970a, 1970b, 1972) introduced and developed the concept of radiation stress. In this section, the physical consequences of radiation stress are described and discussed.

Referring to Figure 7.6, a goal of this section is to determine the time-averaged and depth-averaged longshore velocity distribution, $V_\ell(x)$, and the phenomena of

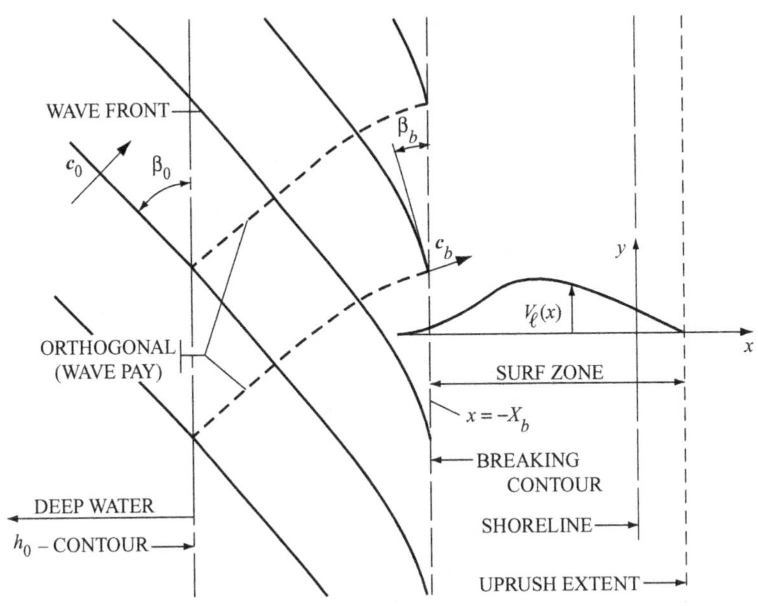

Figure 7.6. *Area Sketch of Waves Shoaling on a Parallel-Contoured Beach, Illustrating the Time-Averaged and Depth-Averaged Longshore Velocity* $[V_\ell(x)]$ *Distribution.*

wave set-up and set-down. Wave *set-up* is the rise of the MWL above the SWL. It results from wave-induced water-mass transport toward the shoreline. Wave *set-down* is the drop in the MWL below the SWL. Set-up can also be caused by wind stress on the free surface of the water. This phenomenon, known as *wind set-up*, is not discussed herein. The reader is referred to the *Shore Protection Manual* (see U.S. Army, 1984).

A. Radiation Stress

(1) Radiation Stress in the Direction of Wave Travel

As stated by Longuet-Higgins (1970a, 1970b, 1972), the wave set-up depends on the energy flux, and the longshore transport depends on the momentum flux. These combined mechanisms lead to the concept of radiation stress. From Longuet-Higgins and Stewart (1964), *radiation stress* is defined as "the excess flow of momentum due to the presence of the waves," or in other words, the excess momentum flux.

Radiation stress can be introduced on a theoretical level, as by Longuet-Higgins and Stewart (1962), or on a practical (heuristic) level, as by Longuet-Higgins and Stewart (1964). The latter is readily applicable to engineering situations and, for that reason, the heuristic approach is taken in this section.

Consider linear waves traveling over a flat, horizontal bed. As sketched in Figure 7.7, the coordinate X is in the direction of wave travel, and the crest-wise coordinate is Y. Referring to the coordinate box in Figure 7.7, the relationships between these coordinates and the inertial onshore (x) and alongshore (y) coordinates are

$$\begin{aligned} X &= x\cos(\beta) + y\sin(\beta) \\ Y &= -x\sin(\beta) + y\cos(\beta) \end{aligned} \tag{7.16}$$

Our attention is first directed at both the momentum flux and energy flux across a vertical Y-Z plane. These are functions of the respective wave-induced horizontal

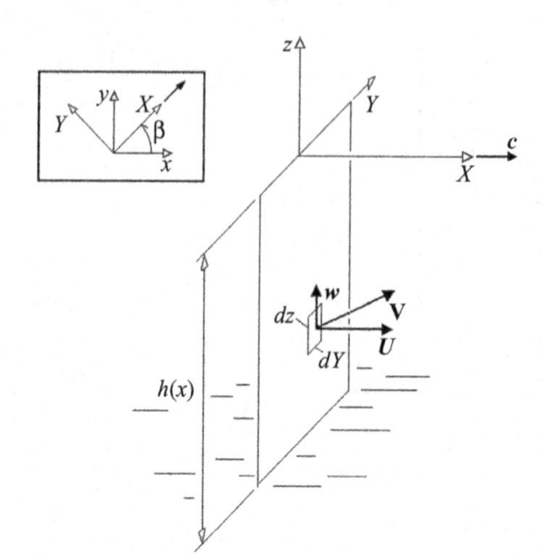

Figure 7.7. *Wave-Induced Water Particle Velocity Components on a Vertical Plane Normal to the Wave Direction.*

and vertical velocity components, which for linear waves are

$$U = \frac{\partial \varphi}{\partial X} = \frac{u}{\cos(\beta)} = \frac{H\omega}{2}\frac{\cosh[k(z+h)]}{\sinh(kh)}\cos(kX-\omega t) \tag{7.17}$$

where u is the horizontal landward velocity component, and

$$W = w = \frac{\partial \varphi}{\partial z} = \frac{H\omega}{2}\frac{\cosh[k(z+h)]}{\sinh(kh)}\sin(kX-\omega t) \tag{7.18}$$

from eqs. 3.49 and 3.50, respectively. In these equations, the velocity potential is

$$\varphi = \frac{H}{2}c\frac{\cosh[k(z+h)]}{\sinh(kh)}\sin(kX-\omega t) \tag{7.19}$$

There is no transverse (along-crest) velocity component. The expression for the velocity potential results from the combination of eqs. 3.29 and 3.30. Referring to the sketch in Figure 7.7, there are two components of the momentum flux (through the normal, vertical plane) – a horizontal component and a vertical combination of eqs. 3.29 and 3.30. According to Longuet-Higgins and Stewart (1964), the principal component of the radiation stress is the difference in the mean of the wave-induced horizontal momentum flux and the mean flux in the absence of waves. That radiation stress component is

$$\begin{aligned}
S_{XX} &\equiv \overline{\int_{-h}^{\eta(t)} (\rho U^2 + p)dz} - \int_{-h}^{0} p_0 dz \\
&= \overline{\int_{-h}^{\eta(t)} \rho U^2 dz} + \overline{\int_{-h}^{0} (p - p_0)dz} + \overline{\int_{0}^{\eta(t)} p dz} \\
&\simeq \int_{-h}^{0} \rho \overline{U^2} dz + \int_{-h}^{0} (\overline{p} - p_0)dz + \overline{\int_{0}^{\eta(t)} p dz} \\
&= S_{XX}^{(1)} + S_{XX}^{(2)} + S_{XX}^{(3)}
\end{aligned} \tag{7.20}$$

where the overline represents the time average and p_0 is the hydrostatic pressure. In this equation, $S_{XX}^{(1)}$ can be interpreted as the integral of a Reynolds stress over the vertical plane in Figure 7.7. See Granger (1995) and other books covering the subject of fluid mechanics for in-depth discussions of the Reynolds stress. Continuing, $S_{XX}^{(2)}$ results from the change in the mean pressure. According to Longuet-Higgins and Stewart (1964), "the mean flux of vertical momentum across any horizontal plane... must be just sufficient to support the weight of water above it." The result is that the integrand of the $S_{XX}^{(2)}$ integral in eq. 7.20 can be written as

$$\bar{p} - p_0 = -\rho\overline{W^2} \tag{7.21}$$

which is interpreted as a *Reynolds stress*. Finally, $S_{XX}^{(3)}$ in eq. 7.20 is approximately equal to the time-averaged potential energy per surface area. For linear waves, the principal component of the radiation stress is then

$$S_{XX} = \left[S_{XX}^{(1)} + S_{XX}^{(2)}\right] + S_{XX}^{(3)} = \left[\frac{\rho g H^2}{8}\frac{2kh}{\sinh(2kh)}\right] + \frac{\rho g H^2}{16} \tag{7.22}$$

(2) Radiation Stress Transverse to the Wave Travel

To obtain the expression for the transverse or along-crest component of the radiation stress, consider the momentum flux through the vertical X-Z plane. As stated previously, there is no flow across this plane because $V = 0$. Following the same line of reasoning that leads to the expressions for the principal component of radiation stress in eqs. 7.20 and 7.22, we find

$$\begin{aligned} S_{YY} &\equiv \int_{-h}^{0} (\bar{p} - p_0)dz + \overline{\int_{0}^{\eta(t)} p dz} \\ &= \int_{-h}^{0} (-\rho\overline{W^2})dz + \frac{1}{2}\rho g\overline{\eta^2} \\ &= S_{YY}^{(2)} + S_{YY}^{(3)} \\ &= \frac{\rho g H^2}{8}\left[\frac{kh}{\sinh(2kh)}\right] \end{aligned} \tag{7.23}$$

We note that because $V = 0$,

$$S_{XY} = S_{YX} = \overline{\int_{-h}^{\eta(t)} \rho UV dz} = 0 \tag{7.24}$$

(3) Radiation Stress Matrix

The radiation stress can now be represented by the following stress matrix (stress tensor):

$$[\mathbf{S}] \equiv \begin{bmatrix} S_{XX} & 0 \\ 0 & S_{YY} \end{bmatrix} \tag{7.25}$$

where the nonzero components are obtained from eqs. 7.22 and 7.23.

EXAMPLE 7.4: COMPARISON OF RADIATION STRESS IN DEEP AND SHALLOW WATERS
In this example, we determine the forms of the elements of the stress tensor of eq. 7.25 in both deep water and in shallow water. In deep water, where $h > \lambda_0/2$, only the principal diagonal stress $S_{XX} = S^{(3)}_{XX}\ (\simeq \rho g H_0^2/16)$ is in the matrix of eq. 7.25. In shallow water, where $h < \lambda/20$, both diagonal terms of the matrix appear. These respective radiation stress components are $S_{XX}(\simeq 3\rho g H^2/16)$ and $S_{YY}(\simeq \rho g H^2/16)$.

The value of the radiation stress in coastal engineering is primarily in the prediction of the wave set-up and the longshore current. These phenomena are discussed in later sections. To determine the set-up and longshore current expressions, we must first transform the radiation stress matrix from the wave coordinate system (X, Y, Z) to the inertial coordinate system (x, y, z).

(4) Transformation of the Radiation Stress Matrix

The combination of eqs. 7.22, 7.23, and 7.25 results in

$$[\mathbf{S}] \equiv \begin{bmatrix} S_{XX} & S_{XY} \\ S_{YX} & S_{YY} \end{bmatrix} = \frac{\rho g H^2}{8} \begin{bmatrix} \dfrac{2kh}{\sinh(2kh)} + \dfrac{1}{2} & 0 \\ 0 & \dfrac{kh}{\sinh(2kh)} \end{bmatrix} \tag{7.26}$$

where, again, X is the coordinate in the wave propagation direction and Y is the along-crest coordinate, as in Figure 7.7. The coefficient of the last matrix is the wave energy per surface area, that is, $E/\lambda b$ from eq. 3.67. Near the break, assume that shallow-water conditions exist, so that the expression in eq. 7.26 is approximated by

$$[\mathbf{S}]_{shallow} \simeq \frac{\rho g H^2}{8} \begin{bmatrix} 3/2 & 0 \\ 0 & 1/2 \end{bmatrix} \tag{7.27}$$

In deep water, the radiation stress matrix is

$$[\mathbf{S}]_{deep} \simeq \frac{\rho g H^2}{8} \begin{bmatrix} 1/2 & 0 \\ 0 & 0 \end{bmatrix} \tag{7.28}$$

as found in Example 7.4. Our goal is to define the matrix

$$[\mathbf{S}] = \begin{bmatrix} S_{XX} & S_{XY} \\ S_{YX} & S_{YY} \end{bmatrix} \tag{7.29}$$

that is equivalent to $[\mathbf{S}]$ in eq. 7.26. This is accomplished by considering the stresses on a vertical rectangular water column sketched in Figure 7.8. From the equilibrium of the radiation stress forces, we find the equivalent stresses. The normal component, needed to determine the set-up or set-down, is

$$\begin{aligned} s_{XX} &= S_{XX}\cos^2(\beta) + S_{YY}\sin^2(\beta) \\ &= \frac{E}{\lambda b}\left(\frac{1}{2} + \frac{2kh}{\sinh(2kh)}\right)\cos^2(\beta) + \frac{E}{\lambda b}\frac{kh}{\sinh(2kh)}\sin^2(\beta) \end{aligned} \tag{7.30}$$

where the energy intensity is $E/\lambda b = \rho g H^2/8$, and the lower-case s is used to represent the equivalent stress components. The diagonal component, needed for the

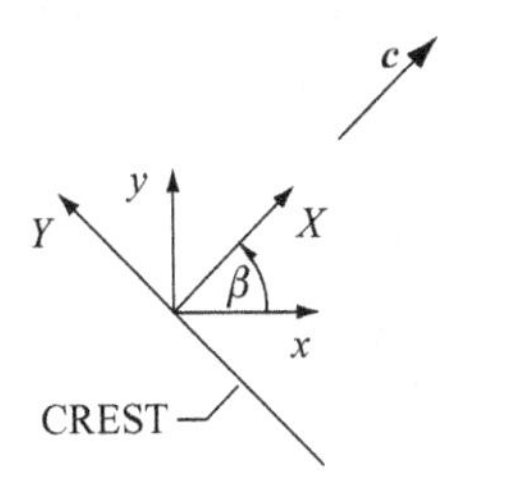

a. **Horizontal Plane Coordinates**

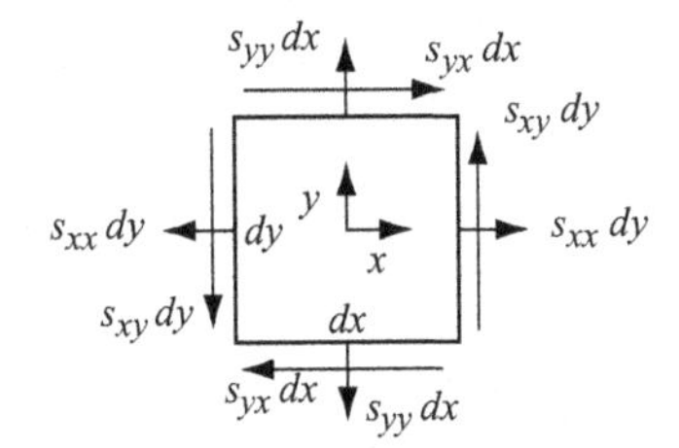

b. **Equivalent Radiation Stress**

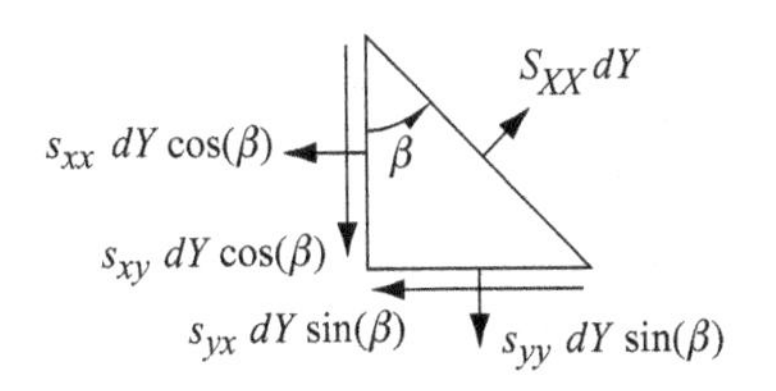

c. **Equilibrium for** ***dY***

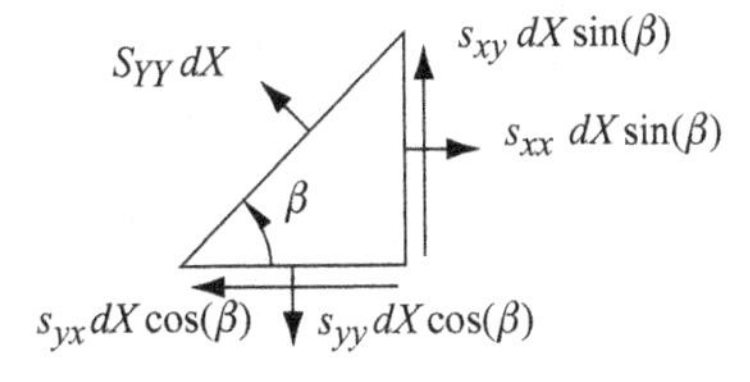

d. **Equilibrium for** ***dX***

Figure 7.8. *Radiation Stress Forces and Equivalent Radiation Stress Forces.* Both the radiation stresses, S, and the equivalent radiation stresses, s, are assumed to be vertically averaged over the faces of the vertical elemental water column.

determination of the longshore velocity, is

$$s_{xy}\,(=s_{yx}) = (S_{XX} - S_{YY})\cos(\beta)\sin(\beta) = \frac{\rho g H^2}{8}\left(\frac{1}{2} + \frac{kh}{\sinh(2kh)}\right)\cos(\beta)\sin(\beta)$$
$$= \frac{E}{\lambda b}\left(\frac{1}{2} + \frac{kh}{\sinh(2kh)}\right)\cos(\beta)\sin(\beta) = \frac{E}{\lambda b}\left(\frac{c_g}{c}\right)\cos(\beta)\sin(\beta) \quad (7.31)$$

The diagonal component can also be expressed in an integral form (similar to that in eq. 7.24) as

$$s_{xy} = \overline{\int_{-h}^{\eta(t)} \rho uv dz} \quad (7.32)$$

Again, the overline is used to denote time averaging. Physically, the integral expression in eq. 7.32 represents the momentum flux (parallel to the shoreline) across a shoreline-parallel vertical plane, as illustrated in Figure 7.9. The units of the momentum flux are in terms of force per unit length.

Following Longuet-Higgins (1970a), and referring to Figure 7.8, Snell's law of eq. 6.89 allows us to write

$$\frac{b}{\cos(\beta)} = \frac{b_0}{\cos(\beta_0)} = \frac{\lambda}{\sin(\beta)} = \frac{\lambda_0}{\sin(\beta_0)} \quad (7.33)$$

where b and b_0 are the respective separation distances between two orthogonals in water of finite depth and deep water. The radiation stress component in eq. 7.31 can now be written as

$$s_{xy} = \frac{E}{\lambda_0 b_0}\left(\frac{1}{2} + \frac{kh}{\sinh(2kh)}\right)\cos(\beta_0)\sin(\beta_0) = \frac{E}{\lambda_0 b_0}\left(\frac{c_g}{c}\right)\cos(\beta_0)\sin(\beta_0) \quad (7.34)$$

The expressions in eqs. 7.30 and 7.34 are used in the determination of the expressions for the wave set-up and longshore velocity, respectively.

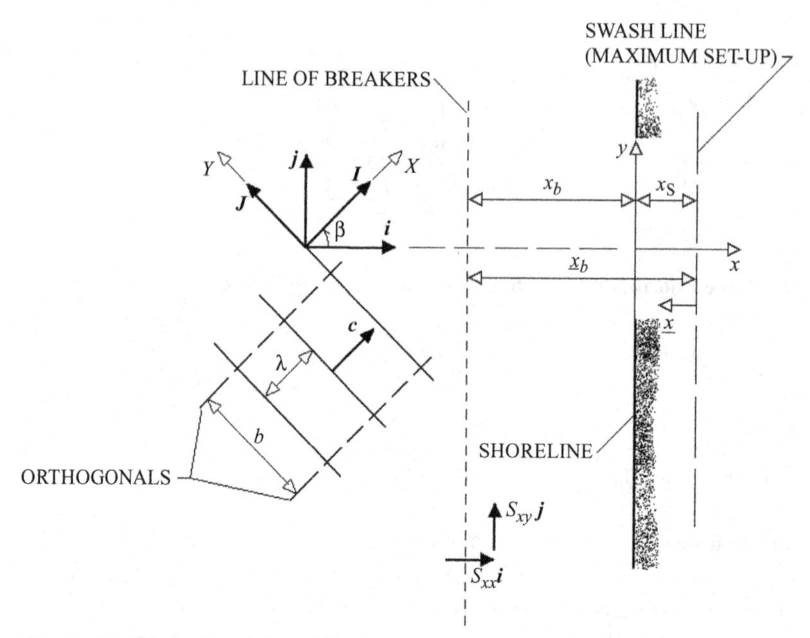

Figure 7.9. *Wave, Inertial, and Swash-Line Coordinates and Associated Unit Vectors.*

EXAMPLE 7.5: NORMAL AND DIAGONAL RADIATION STRESS COMPONENTS FOR A BREAKING WAVE In Example 7.2, a 1-m, 8-sec deep-water wave approaches a beach having a uniform slope (m) of 1/50. In deep water, the crest of the wave is at an angle (β_0) of 30° to the shoreline. In that example, the expressions in eqs. 7.1 and 7.5 are combined with 6.89 (Snell's law) to determine the breaking wave properties. The results are the following: $h_b = 1.55\,\mathrm{m}$, $X_b = 77.5\,\mathrm{m}$, $\lambda_b = 30.7\,\mathrm{m}$, and $\beta_b = 8.83°$. Note that $h_b/\lambda_b \simeq 1/20$, so shallow-water approximations can be used. At the break, the energy intensity (energy per surface area) is $E_b/\lambda_b b_b = \rho g H_b^2/8 = 2{,}480\,\mathrm{N\text{-}m/m^2}$. The values of the normal and diagonal radiation stress components in eqs. 7.30 and 7.31 are, resepectively, $s_{xxb} = 3{,}650\,\mathrm{N/m}$ and $s_{xyb} = 377\,\mathrm{N/m}$. The values of these components in deep water are $s_{xx0} = 474\,\mathrm{N/m}$ and $s_{xy0} = 547\,\mathrm{N/m}$. Comparing the values, we see that the normal component increases as the wave approaches the shoreline while the diagonal component decreases.

B. Wave Set-Up and Set-Down

When waves approach the shoreline over a flat bed of slope m, the radiation stress components will change as the wave approaches the shoreline, as demonstrated in Examples 7.4 and 7.5. Longuet-Higgins and Stewart (1964) write that "changes in radiation stress lead to changes in the level of the mean surface." The phenomenon that they describe is called *wave set-up* when the MWL rises above the SWL, and *set-down* when the MWL is below the SWL. We shall collectively refer to the phenomenon in the following derivation as the *set-up*. To analyze set-up, consider a wave approaching the shoreline directly, as sketched in Figure 7.10. Assume that the wave condition is that of shallow water and that the beach slope is small enough so that the time averages of both u^2 and uw are of second order. The time-averaged

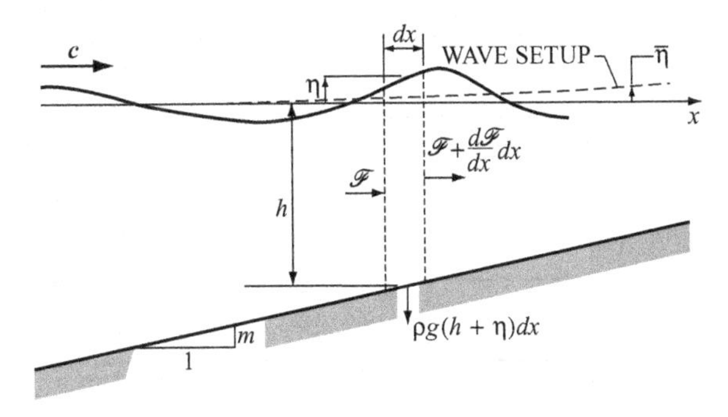

Figure 7.10. *Momentum Balance for Waves in Shallow Water.* The wave set-up is represented by the time-averaged free-surface elevation.

momentum flux ($\mathscr{F}$) entering the region of length dx is

$$\mathscr{F} \equiv s_{xx} + \int_{-h}^{\bar{\eta}(t)} \rho g(\bar{\eta} - z)dz = s_{xx} + \frac{1}{2}\rho g(\bar{\eta} + h)^2 \tag{7.35}$$

where the time average of the free-surface elevation, η, represents the wave set-up. The momentum exiting in the region has increased by

$$\frac{d\mathscr{F}}{dx}dx = \frac{d}{dx}\left[s_{xx} + \frac{1}{2}\rho g(\bar{\eta} + h)^2\right]dx = \frac{ds_{xx}}{dx} + \rho g(\bar{\eta} + h)\left(\frac{d\bar{\eta}}{dx} + \frac{dh}{dx}\right)dx \tag{7.36}$$

Because the bottom is not horizontal, there is an additional horizontal momentum flux due to the hydrostatic pressure on the bottom (shown in Figure 7.10). This additional momentum flux is

$$\frac{d\mathscr{F}}{dx}dx = \rho g\,(\bar{\eta} + h)\,\frac{dh}{dx}dx \tag{7.37}$$

The momentum flux expressions in eqs. 7.36 and 7.37 must be equal. Thus, the momentum flux balance yields

$$\frac{ds_{xx}}{dx}dx + \rho g(\bar{\eta} + h)\frac{d\bar{\eta}}{dx}dx \simeq \frac{ds_{xx}}{dx}dx + \rho gh\frac{d\bar{\eta}}{dx}dx = 0 \tag{7.38}$$

The approximation in eq. 7.38 is based on the assumption that the wave set-up is much less than the water depth. From eq. 7.38, we obtain the differential equation for the wave set-up, which is

$$\begin{aligned}\frac{d\bar{\eta}}{dx} &= -\frac{1}{\rho gh}\frac{ds_{xx}}{dx} = -\frac{1}{\rho gh}\frac{d}{dx}[S_{XX}\cos^2(\beta) + S_{YY}\sin^2(\beta)] \\ &= -\frac{1}{\rho gh}\frac{d}{dx}\left[\frac{E}{\lambda h}\left(\frac{1}{2} + \frac{2kh}{\sinh(2kh)}\right)\cos^2(\beta) + \frac{E}{\lambda h}\frac{kh}{\sinh(2kh)}\sin^2(\beta)\right]\end{aligned} \tag{7.39}$$

The expression for the radiation stress in eq. 7.30 has been included in eq. 7.39. The wave set-up is obtained by integrating eq. 7.39.

Consider the case where the waves directly approach a shoreline over a straight, parallel-contoured bed of uniform slope. In Figure 7.9, the wave angle is $\beta = 0°$. The expression in eq. 7.39 can be integrated to obtain

$$\begin{aligned}\bar{\eta}|_{\beta=0} &= -\int \frac{1}{\rho gh}ds_{xx} = -k\frac{H'^2}{8}\frac{1}{\sinh(2kh)} \\ &= -\frac{k_0 H_0'^2/16}{\tanh^2(kh)[\sinh(kh)\cosh(kh) + kh]}\end{aligned} \tag{7.40}$$

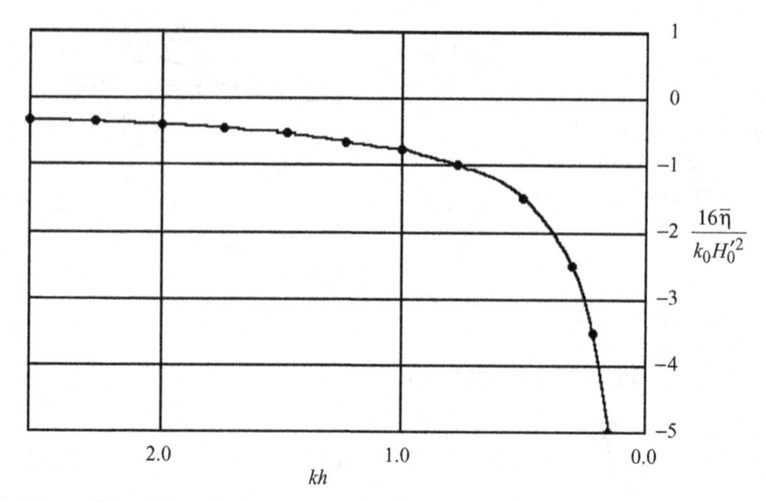

Figure 7.11. *Wave Set-Down Seaward of the Surf Zone for* $\beta = 0$, *from eq. 7.40.* The entire curve is below the SWL.

The prime (′) is used to indicate that the shoaling occurs without refraction. Details of the integration in eq. 7.40 are presented by Longuet-Higgins and Stewart (1962). We note that the bed slope (m) does not appear in the equation. This is as expected because the derivation of the expression in eq. 7.40 is based on the linear theory (discussed in Chapter 3), which assumes that the bed is flat at all points considered. The last equality in eq. 7.40 results from replacing H by K_S H_0, where K_S is the shoaling coefficient of eq. 3.78. For waves approaching the shoreline seaward of the break, results obtained from eq. 7.40 are presented in non-dimensional form in Figure 7.11. For nonbreaking waves, the set-up is always negative and, hence, instead of a set-up we have a set-down. The reason for the set-down is that the normal component of the radiation stress is increasing as the wave approaches the shore, as is demonstrated in Example 7.5.

In the surf zone, the radiation stress decreases, causing ds_{xx}/dx to be negative and the derivative of the set-up to be positive in eq. 7.30. The result is that there is a set-up within the surf zone. As noted by Longuet-Higgins and Stewart (1963, 1964), the behavior of the radiation stress in the surf zone must be predicted by using empirical expressions. To this end, we first determine the breaking height and breaking depth by using the empirical expressions in Sections 7.2 or 7.3. The former is recommended because of the relatively good accuracy of eqs. 7.1 and 7.5. For waves directly approaching the shoreline over a bed of uniform slope, eq. 7.1 is used to determine the breaking wave height, and eq. 7.5 is used in combination with the results obtained from eq. 7.1 to determine the corresponding water depth.

A definitive experimental study of wave set-up and set-down was performed by Bowen, Inman, and Simmons (1968). Prior to the publication resulting from that study, results of the laboratory study of Saville (1961) were used to obtain empirical coefficients for set-up in the surf zone. The Saville study was primarily concerned with overtopping of seawalls. Results obtained by Bowen, Inman, and Simmons (1968) for $m = 0.082 \simeq 1/12$ are presented in Figure 7.12. In that figure, a set-down is seen to extend from deep water to the plunge point. Landward of the plunge point (in the swash zone of Figure 7.1), the gradient of the radiation stress is positive, eventually resulting in a set-up. For the data shown in the figure, the respective wave period (T), deep-water wave height (H_0'), and breaking wave height (H_b') are

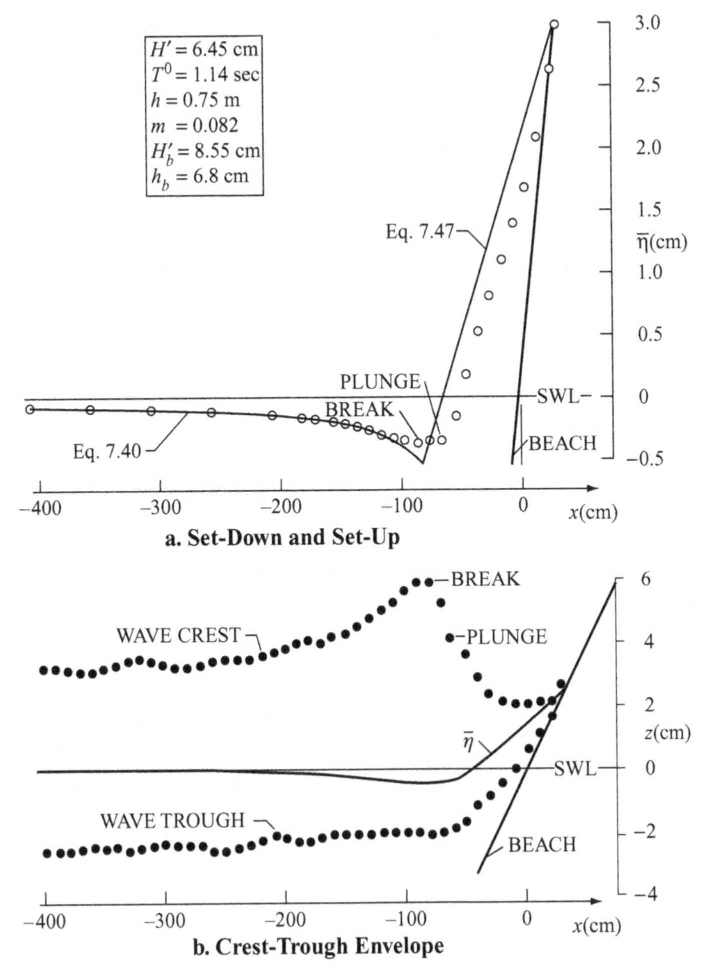

Figure 7.12. *Experimental, Theoretical, and Empirical Wave Set-Down and Set-Up and Crest-Trough Envelope.* The experimental data (set-down/set-up ∘ and crest-trough •) are from Bowen, Inman, and Simmons (1968).

1.14 sec, 0.0645 m, and 0.0855 m. According to Longuet-Higgins and Stewart (1964), the observed spacial derivative of the set-up is proportional to the spacial derivative of the water depth. However, the experimental results of Bowen, Inman, and Simmons (1968) show that there is a "residual wave height" at the beach, causing the behavior of the set-up in the swash zone to be slightly nonlinear with respect to x. This nonlinear behavior can be seen in Figure 7.12a, where the larger values of the set-up are shown to deviate from the rectilinear behavior of the lower set-up values in the swash. Other results presented by Bowen (1969a) show that the nonlinear behavior becomes more or less apparent, depending on both the wave conditions and the beach slope. According to both Bowen, Inman, and Simmons (1968) and Longuet-Higgins (1972), the linear approximation for the behavior of the set-up in the swash zone is in good qualitative and quantitative agreement with experimental data.

From Bowen, Inman, and Simmons (1968), the expression for the gradient of the set-up in the surf zone is

$$\frac{d\overline{\eta_s}}{dx} \simeq -\mathrm{K}\frac{dh}{dx} = \mathrm{K}m \tag{7.41}$$

where K is a proportionality constant, the subscript "s" designates surf-zone conditions, and the water depth at any position is $h = -mx$. Upon integration, the expression for the set-up is then found to be

$$\overline{\eta_s} = Kmx + B = -Kh + B \tag{7.42}$$

where B is a constant of integration.

To determine both K and B in eq. 7.42, two conditions for the set-up are required. The first of these can be determined by applying the shallow-water approximations to eq. 7.40 at the breaking point. The result is

$$\overline{\eta_b} \simeq -\frac{1}{16}\frac{(H_b')^2}{h_b} \tag{7.43}$$

where the breaking depth is obtained from eq. 7.5. The second condition involves the behavior of the wave height. According to Bowen, Inman, and Simmons (1968), "Inside the break point the wave energy decreases shoreward, leading to a decrease in the radiation stress. Using similarity arguments, we can assume that the height of the broken wave, or bore, remains an approximately constant proportion of the mean water depth." The mathematical expression of the latter observation is

$$H_s = \gamma(\overline{\eta_s} + h_s) \equiv \gamma \underline{h}_s \tag{7.44}$$

where γ is an experimental proportionality constant, and $\underline{h}_s$ is the actual water depth in the surf zone. The values of γ reported by Bowen, Inman, and Simmons (1968) range from 0.90 to 1.28. However, Longuet-Higgins (1972) recommends a value of 0.82. Again, the subscript "s" in eq. 7.44 indicates a property in the surf zone. By letting $\beta = 0$ and assuming shallow-water conditions, the normal component of the radiation stress in eq. 7.30 is

$$s_{xxs} = \frac{3}{16}\rho g H_s^2 = \frac{3}{16}\rho g \gamma^2 \underline{h}_s^2 \tag{7.45}$$

Combine this expression with the equality in eq. 7.38 to obtain the gradient of the set-up in the surf zone. Note that the approximation in eq. 7.38 is based on the assumption that the set-down is much less than the water depth seaward of the breaking point. This assumption cannot be made for the set-up in the surf zone. The combination of eqs. 7.45 and 7.38 results in

$$\frac{d\overline{\eta_s}}{dx} = -\frac{1}{\rho g(h+\overline{\eta_s})}\frac{ds_{xxs}}{dx} = -\frac{3}{8}\gamma^2\frac{d\underline{h}_s}{dx} = -\frac{3}{8}\gamma^2\left(\frac{d\overline{\eta_s}}{dx} + \frac{dh}{dx}\right) \tag{7.46a}$$

where, for the bed of uniform slope, $dh/dx = -m$. From the expression in eq. 7.46a, we obtain

$$\frac{d\overline{\eta_s}}{dx}\left(= -\frac{d\overline{\eta_s}}{d\underline{x}}\right) = \frac{\frac{3}{8}m\gamma^2}{1+\frac{3}{8}\gamma^2} = Km \tag{7.46b}$$

In this equation, the swash-line coordinate ($\underline{x}$) is introduced. Referring to Figure 7.9, this coordinate originates at the position of the maximum set-up. The swash-line coordinate is used later in this chapter to determine the longshore velocity expression.

By both applying the condition in eq. 7.43 to eq. 7.42 and using the last equality in eq. 7.46b, the following expression for the set-up in the surf zone is obtained:

$$\overline{\eta_s} = \frac{(3/8)\gamma^2}{1+(3/8)\gamma^2}(h_b - h) - \frac{1}{16}\frac{H_b^2}{h_b}$$
$$= \frac{(3/8)\gamma^2 m}{1+(3/8)\gamma^2}(x_b + x) - \frac{1}{16}\frac{H_b^2}{h_b} \tag{7.47a}$$

where $x = -x_b = -h_b/m$ is the location of the break, as sketched in Figure 7.9. An alternative expression for the set-up can be obtained from eq. 7.42 by applying the boundary condition $\underline{h}(x_S) = 0$, where $x = x_S$ is the position of maximum set-up. The resulting expression is

$$\overline{\eta_s} = \underline{h}_s - h_s$$
$$= \frac{m}{\left(1+\frac{3}{8}\gamma^2\right)}(x_s - x) - h_s = \frac{m}{\left(1+\frac{3}{8}\gamma^2\right)}(x_s - x) + mx$$
$$= \frac{m}{\left(1+\frac{3}{8}\gamma^2\right)}\underline{x} + m(x_s - \underline{x}) \tag{7.47b}$$

where, again, $\underline{x}$ is the swash-line coordinate.

By applying eq. 7.47a at $x = x_S$ (where $\underline{x} = 0$), the position of maximum set-up is found, as demonstrated in Example 7.7.

EXAMPLE 7.6: COMPARISON OF THEORETICAL AND EXPERIMENTAL WAVE HEIGHT, WATER DEPTH, AND SET-DOWN AT A BREAK AND RUNUP In this example, we compare the experimental data of Bowen, Inman, and Simmons (1968) in Figure 7.12 with results obtained from eq. 7.1 for the breaking wave height, eq. 7.5 for the breaking water depth, and eq. 7.43 for the set-down at the break. The experimental values of the deep-water wave height and wave period are, respectively, 6.45 cm and 1.14 sec. The slope (m) of the beach is approximately 0.082. For these data, eq. 7.1 predicts a breaking height of 8.21 cm. The observed breaking height value is 8.55 cm. Equation 7.5 predicts a breaking depth of 7.73 cm, whereas the observed value is 6.8 cm. The minimum observed set-down is –0.32 cm, and that predicted by eq. 7.43 is –0.55 cm. Hunt's formula in eq. 7.15 predicts a runup value of 2.97 cm, where the porosity factor is $C_p = 1$ for the smooth, impermeable bed. This factor is the ratio of the void volume to the total (void + material) volume. From eq. 7.9, the value of the deep-water surf similarity parameter is $\xi_0 = m/\sqrt{(H_0/\lambda_0)} = 1.45$, where the deep-water wavelength (λ_0) is approximately 2.03 m. The observed runup is 3.25 cm.

The predicted breaking wave height, water depth, and runup values are in relatively good agreement with the observed values. However, the agreement between the minimum set-down values is not quite as good. In Figure 7.12a, we see that the theoretical set-down values continue to decrease with kh. However, the experimental values level off at the break point and then increase landward of the plunge. As the wave approaches the break, the nonlinear phenomena associated with the break become pronounced, but are not accounted for by the theory.

EXAMPLE 7.7: SET-UP IN THE SURF ZONE As in Example 7.6, our interest is in predicting the behavior observed by Bowen, Inman, and Simmons (1968) leading to Figure 7.12. The conditions in that figure are the following: $m = 0.082$, $H_0 = 6.45$ cm, $T = 1.14$ sec. The proportionality constant in eq. 7.44 reported by Bowen and his colleagues is $\gamma = 1.15$. In Example 7.6, we find that the breaking wave height is $H_b = 8.21$ cm from eq. 7.1 and the breaking depth is $h_b = 7.73$ cm from eq. 7.5. In units of centimeters, these values applied to the expression in eq. 7.47 result in the following expression for the surf-zone set-up in Figure 7.12:

$$\overline{\eta_s} \simeq 0.0272(94.3 + x) - 0.545$$

The position of the maximum set-up value ($x = x_S$ or $x = 0$) can be found in terms of the breaking conditions by solving eq. 7.47a. The resulting expression is

$$x_S = \frac{3}{8}\gamma^2 x_b - \frac{\left(1 + \frac{3}{8}\gamma^2\right)}{m} \frac{1}{16} \frac{H_b'^2}{h_b} \tag{7.48}$$

For the parametric values, the onshore position of the maximum set-up is $x_S = 37$ cm from eq. 7.48. The predicted value of the maximum set-up is approximately 3.02 cm and the observed value is 2.95 cm. The set-up curve is presented in Figure 7.12 with the experimental results.

The nonlinearity of the experimental set-up data in the surf zone is evident in Figure 7.12. The linear assumption leading to eqs. 7.47 and 7.48 yields a maximum set-up value that agrees with that observed, at least for that particular experiment. In a study of longshore transport and wave decay in the surf zone, Miller (1987) examines the results obtained by assuming that the wave height variation in the surf zone varies exponentially with x, that is,

$$H_s = H_b e^{-\kappa(x + X_b)} \tag{7.49}$$

where κ is a parametric constant. Miller (1987) compares results obtained from this equation with both the observed data of Horikawa and Kuo (1966) and with the linear assumption. This comparison shows the linear expression in eq. 7.44 is satisfactory as a first approximation for the wave height variation in the surf zone.

C. Longshore Velocity

One of the focus areas of coastal engineering is beach stability. The goal is to preserve the balance between erosion (the loss of material) and accretion (the gain of material) along the sea coast. The material referred to is collectively called *littoral drift*. For most of the beaches of the contiguous United States, the littoral drift is primarily quartz and feldspar or, simply, beach sand. The longshore transport of material is primarily due to the wave-induced littoral currents in the surf zone. Based on the concept of radiation stress of Longuet-Higgins and Stewart (1960, 1961, 1962, 1964) and Longuet-Higgins (1970a, 1970b, 1972), the mechanics of the longshore littoral currents can be developed. The primary references for the discussion of the phenomenon herein are Longuet-Higgins (1972) and Bowen (1969b).

Longshore currents are time-dependent because the currents are caused by refracting waves. However, in the analysis of the currents our interest is in their time-averaged values where, as sketched in Figure 7.6, the longshore velocity is represented by $V_\ell(x)$. Longuet-Higgins (1972) states that the *quasi*-steady longshore currents result from a balance of forces, those being the driving force due to the gradient of the radiation stress tensor ([**S**] in eq. 7.25), friction, and the hydrostatic force due to the wave set-up. As shown in the derivation of eq. 6.84, the energy flux is conserved between orthogonals seaward of the break. In the shoaling region, the energy flux can then be written as

$$P = \frac{\rho g H^2 \lambda b}{8}\left(\frac{c_g}{\lambda}\right) = \frac{\rho g H_0^2 \lambda_0 b_0}{8}\left(\frac{c_{g0}}{\lambda_0}\right) = E\frac{c_g}{\lambda} = E_0\frac{c_{g0}}{\lambda_0} \tag{7.50}$$

In the second equality, replace b and b_0 by $Y\cos(\beta)$ and $Y_0\cos(\beta_0)$, respectively, where Y_0 is the longshore distance between orthogonals, as sketched in Figure 6.17. From the resulting relationship, we obtain

$$P = \mathrm{P}_x Y_0 \equiv \frac{\rho g H^2}{8} c_g Y_0 \cos(\beta) = \frac{\rho g H_0^2}{8} c_{g0} Y_0 \cos(\beta_0) \tag{7.51}$$

The energy flux seaward of the line of breakers is then, theoretically, uniform as no energy dissipation is included in the analysis. Because Y_0 is constant, $\mathrm{P}_x = P/Y_0$ must also be constant. By replacing E/λ in eq. 7.34 by its equivalent from the second equality of eq. 7.50, the relationship between the momentum flux and the onshore component of the energy flux is obtained. That relationship is

$$s_{xy} = \mathrm{P}_x \frac{\sin(\beta)}{c} = \mathrm{P}_x \frac{\sin(\beta_0)}{c_0} = \frac{\rho g H_0^2}{16} \cos(\beta_0)\sin(\beta_0) \tag{7.52}$$

where the second equality results from Snell's law, expressed by eq. 6.89. The normal component of the radiation stress (s_{xy}) is then independent of water depth seaward of the line of breakers because P_x is independent of depth.

Near the line of breakers, the velocity gradient adjacent to the bed becomes large if the fluid is viscous, and the bottom friction affects the wave energy. In the surf zone, bottom friction, turbulence resulting from breaking, and bed percolation all contribute to the dissipation of the wave energy. Here, we assume that the sea bed is both smooth and impermeable. According to Longuet-Higgins (1972), the spacial rate of change of the onshore component of the energy flux must equal the local rate of dissipation. We can then write

$$\frac{\partial \mathrm{P}_x}{\partial x} = \begin{cases} 0, & (-\infty \le x < -x_b, \quad \infty > \underline{x} > x_b + x_S) \\ -\mathrm{D}, & (-x_b \le x < x_S, \quad x_b + x_S > \underline{x} > 0) \end{cases} \tag{7.53}$$

where D is the local rate of energy dissipation, x_b is the seaward distance of the break from the shoreline, x_S is the distance from the shoreline to the maximum set-up, and $\underline{x}$ is the swash-line coordinate that is positive in the seaward direction. To determine the effects of the loss mechanisms in the surf zone, consider an elemental rectangular water column having a volume of *h-dx-dy*. An area sketch of the column is presented in Figure 7.13. The seaward side is shown at the line of breakers, although the water column can be at any point in the surf zone. On the element, the net radiation stress forces, the effective bed stress force, and the lateral mixing

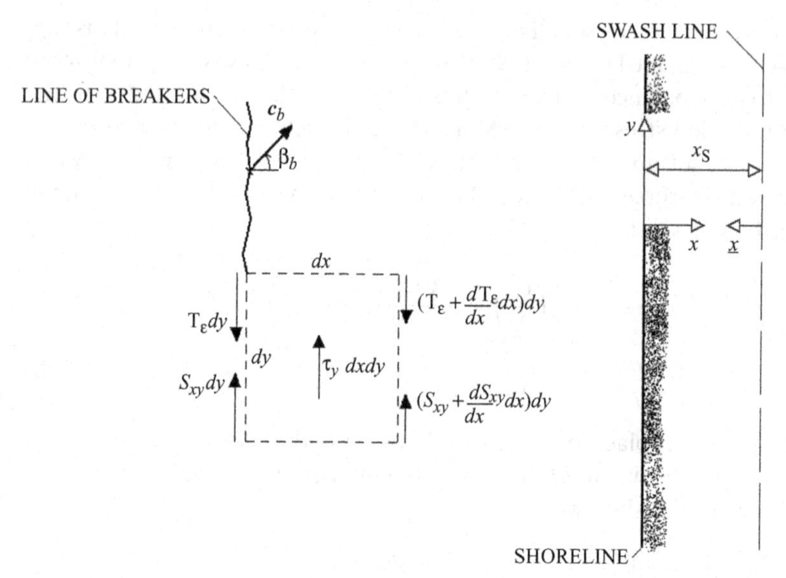

Figure 7.13. *Equilibrium of Radiation Stress Forces, Lateral Mixing Forces, and Bed Frictional Forces in the Surf Zone.*

forces due to turbulence (all averaged over time) are in equilibrium. The equilibrium relationship of the net respective forces is

$$\sum F_y = (\tau_{xys} + \tau_y - \tau_\varepsilon)dxdy$$
$$= \frac{ds_{xys}}{dx}dxdy + \tau_y dxdy - \frac{dT_\varepsilon}{dx}dxdy = 0 \qquad (7.54)$$

where τ_{xys} is the radiation stress, τ_y is the bed shear stress, and the function T_ε depends on the eddy viscosity. One might question the directions of the forces in eq. 7.54 shown in Figure 7.13. According to Longuet-Higgins (1970a), "when the orbital velocity is onshore, the direction of the bottom stress is inclined more in the positive y-direction (if v is positive); when the orbital velocity is offshore, the bottom stress, now almost in the opposite direction, is again more toward the positive y-direction." Hence, the forces due to the diagonal radiation stress component and bottom stress are in the same direction. Our goal is to solve the differential equation in eq. 7.54 for the time-averaged and depth-averaged alongshore or longshore velocity, $V_\ell(x)$.

The first term in eq. 7.54 is the force resulting from the diagonal component of the radiation stress. From the results in eq. 7.31, this radiation stress component applied to the surf zone is

$$s_{xys} = \frac{1}{8}\rho g H_s^2 \sin(\beta_s)\cos(\beta_s) \simeq \frac{1}{8}\rho g \gamma^2 \underline{h}_s^2 \sqrt{g\underline{h}_s}\frac{\sin(\beta_0)}{c_0}\cos(\beta_b) \qquad (7.55)$$

The last approximation results from two assumptions: First, the wave angle within the surf zone is small, as assumed by Longuet-Higgins in his authored and co-authored referenced papers, and is approximately equal to that at the break. The second assumption is made that Snell's law (eq. 6.89) is approximately valid in the surf zone.

In the remainder of this discussion, the derivation of the longshore velocity will be in terms of the swash-line coordinate, $\underline{x}$, as by Longuet-Higgins (1970b) and others. The radiation stress, which is the spacial derivative of the approximate expression in eq. 7.55, is

$$\tau_{xys} \equiv \frac{ds_{xys}}{d\underline{x}} = \frac{5}{16}\rho g^{3/2}\gamma^2 \frac{\sin(\beta_0)}{c_0}\cos(\beta_b)\underline{h}_s^{3/2}\frac{d\underline{h}_s}{d\underline{x}}$$
$$= \frac{5}{16}\rho g^{3/2}\gamma^2 \frac{\sin(\beta_b)}{\sqrt{g\underline{h}_b}}\cos(\beta_b)\underline{h}_s^{3/2}\frac{m}{\left(1+\frac{3}{8}\gamma^2\right)} \tag{7.56}$$
$$= \frac{5}{16}\rho g^{3/2}\gamma^2 \frac{\sin(\beta_b)}{\sqrt{g\underline{h}_b}}\cos(\beta_b)\left[\frac{m}{\left(1+\frac{3}{8}\gamma^2\right)}\right]^{5/2}\underline{x}^{3/2} \equiv C_{xy}\underline{x}^{3/2}$$

Also, from eq. 7.46b the parametric constant is

$$\mathrm{K} = \frac{\frac{3}{8}\gamma^2}{1+\frac{3}{8}\gamma^2} \tag{7.57}$$

where γ is the proportionality constant in eq. 7.44. In Figure 7.13, the momentum flux is decreasing in the x-direction and increasing in the $\underline{x}$-direction. That is, the maximum momentum flux must be at the break, and decreases to the swash line.

The second term in eq. 7.54 involves the time-averaged shear stress (τ_y) on the bed. There are several expressions available for τ_y that vary in complexity. As is done in the derivation herein, simplifications of the shear stress expression are based on two assumptions: The first is that the magnitude of the averaged longshore velocity within the surf zone is assumed to be much less than the maximum orbital velocity, which, according to Longuet-Higgins (1972) is $U_{max} = \alpha\sqrt{(gh_s)} \simeq 0.41\sqrt{(g\underline{h}_s)}$. Second, as previously stated, the wave angle within the surf zone is small. Liu and Dalrymple (1978) demonstrate the effects of the inclusions of both large longshore velocities and large wave angles while neglecting the effect of turbulent mixing.

The respective particle velocity components in the wave coordinate directions (X, Y, z in Figure 7.9) are U, V, w, whereas the components in the inertial coordinate directions (x, y, z) are u, v, w. Following Longuet-Higgins (1972), we can expect v to be relatively small if the wave angle at the break (β_b) is small. However, Liu and Dalrymple (1978) show that this is not always the case, based on both field and experimental data.

In Example 7.5, we find that the breaking wave angle is 8.83° for a deep-water wave angle of 30° over a bed having a 1/50 slope. If v is approximately an order of magnitude less than U_{max}, then the mean bed shear stress in the longshore direction can be mathematically represented by

$$\tau_y(\underline{x}) = \frac{1}{4\pi}\rho f U_{\max} V_\ell(\underline{x})[1+\sin^2(\beta_s)]$$
$$\simeq \frac{1}{8\pi}\rho f_\mu \gamma\sqrt{g\underline{h}_s}V_\ell(\underline{x}) = \frac{1}{8\pi}\rho f_\mu\gamma\sqrt{g}\left[\frac{m}{\left(1+\frac{3}{8}\gamma^2 a\right)}\right]^{1/2}\underline{x}^{1/2}V_\ell(\underline{x})$$
$$\equiv C_y\underline{x}^{1/2}V_\ell(\underline{x}) \tag{7.58}$$

according to Liu and Dalrymple (1978). In this expression, f_μ is a friction factor (the value of which depends on the smoothness of the bed) and $V_\ell(x)$ is the time-averaged longshore velocity. The approximation in eq. 7.58 follows from the assumption concerning the smallness of the breaking angle (β_b), as previously discussed. Also in eq. 7.58 is the maximum orbital velocity of the water particles, $U_{max} \simeq (H/2\underline{h})\surd(g\underline{h})$, where $H = \gamma\underline{h}$. The expression for $U_{\max}$ is obtained from the shallow-water approximation of the horizontal particle velocity expression in eq. 3.49.

The third term in eq. 7.54 involves the eddy viscosity resulting from the turbulent mixing within the broken wave. Following Longuet-Higgins (1970b), let the eddy viscosity function be represented by

$$\mu_\varepsilon = N_\varepsilon \rho \underline{x}\sqrt{g\underline{h}_s} = N_\varepsilon \rho \sqrt{g} \sqrt{\frac{m}{\left(1 + \frac{3}{8}\gamma^2\right)}}\underline{x}^{3/2} \tag{7.59}$$

where, from eq. 7.44, $\underline{h}_s$ is the actual water depth in the surf zone and $\underline{x}$ is the swash-line coordinate, as sketched in Figures 7.9 and 7.13. The expression for $\underline{h}_s(x)$ is found following eq. 7.46b. N_ε is a experimental parametric constant. The effect of the turbulent mixing due to the breaking waves is mathematically expressed by the function

$$T_\varepsilon = \mu_\varepsilon \underline{h}_s \frac{dV_\ell(\underline{x})}{d\underline{x}} = N_\varepsilon \rho \sqrt{g} \left[\frac{m}{\left(1 + \frac{3}{8}\gamma^2\right)}\right]^{3/2} \underline{x}^{5/2} \frac{dV_\ell(\underline{x})}{d\underline{x}} \tag{7.60}$$

The effective shear stress due to turbulent mixing is then

$$\begin{aligned}\tau_\varepsilon = \frac{dT_\varepsilon}{d\underline{x}} &= N_\varepsilon \rho \sqrt{g} \left[\frac{m}{\left(1 + \frac{3}{8}\gamma^2\right)}\right]^{3/2} \left[\frac{5}{2}\underline{x}^{3/2}\frac{dV_\ell}{d\underline{x}} + \underline{x}^{5/2}\frac{d^2 V_\ell}{d\underline{x}^2}\right] \\ &= C_\varepsilon \frac{5}{2}\underline{x}^{3/2}\frac{dV_\ell}{d\underline{x}} + C_\varepsilon \underline{x}^{5/2}\frac{d^2 V_\ell}{d\underline{x}^2}\end{aligned} \tag{7.61}$$

There have been many studies devoted to turbulent mixing in the surf zone since the mid-twentieth century. For example, Bowen (1969b) assumes that the eddy viscosity is uniform throughout the surf zone. By making this assumption, Bowen is able to solve the time-averaged Navier-Stokes equations (eq. 2.66) applied to shallow water. An excellent summary of the findings of these studies is presented by Longo, Petti, and Losada (2002).

The expression in eq. 7.54 can now be rewritten using the relationships in eqs. 7.56, 7.58, and 7.61, and rearranged to obtain

$$C_\varepsilon \underline{x}^{5/2}\frac{d^2 V_\ell}{d\underline{x}^2} + C_\varepsilon \frac{5}{2}\underline{x}^{3/2}\frac{dV_\ell}{d\underline{x}} - C_y \underline{x}^{1/2} V_\ell = -C_{xy}\underline{x}^{3/2} \tag{7.62}$$

We should note that this equation applies only to the surf zone, where $x_b > x \geq 0$.

In the following subsections, the relative effects of the shear stress on the bed and the lateral mixing due to the turbulence are examined. First, the lateral mixing, represented by the first two terms in eq. 7.62, is neglected. Then, the bed shear, represented by the third term, is assumed to be negligible.

(1) Negligible Lateral Mixing

If the assumption is made that the bed friction force is much larger than the apparent friction force due to lateral mixing, as assumed by Liu and Dalrymple (1978), then $C_\varepsilon \simeq 0$ in eq. 7.62, and the longshore velocity is found to behave according to

$$V_\ell(\underline{x})\,|_{C_\varepsilon=0} = \frac{C_{xy}}{C_y}\underline{x} = \frac{5}{2}\frac{\pi\gamma\sqrt{g}}{f\sqrt{\underline{h}_b}}\left[\frac{m}{\left(1+\frac{3}{8}\gamma^2\right)}\right]^2 \sin(\beta_b)\cos(\beta_b)\underline{x}$$

$$= \frac{5}{2}\frac{\pi\gamma g}{f}\left[\frac{m}{\left(1+\frac{3}{8}\gamma^2\right)}\right]^2 \frac{\sin(\beta_0)}{c_0}\cos(\beta_b)\underline{x} \equiv V_b\frac{\underline{x}}{\underline{x}_b} \equiv V_b\chi \tag{7.63}$$

According to this equation, the velocity distribution increases linearly with swash-line coordinate, maximizing at the line of breakers. This behavior is not supported by experimental or field observations, so we can conclude that the lateral mixing effects are not negligible. The last line of eq. 7.63 contains two identities, $V_b \equiv V_\ell(x_b)$ and $\chi \equiv x/x_b$. Longuet-Higgins (1970b, 1972) uses these identities in non-dimensionalizing the expression in eq. 7.62, as is done later in this chapter.

(2) Negligible Bed Friction

If the lateral mixing force is assumed to be much greater than the average bed friction, then $C_y \simeq 0$ in eq. 7.62. The differential equation for this condition is

$$C_\varepsilon \underline{x}^{5/2}\frac{d^2V_\ell}{d\underline{x}^2} + C_\varepsilon\frac{5}{2}\underline{x}^{3/2}\frac{dV_\ell}{d\underline{x}} = -C_{xy}\underline{x}^{3/2} \tag{7.64}$$

The solution of this equation, subject to the condition $V_\ell(x=0)=0$ at the swash line, is

$$V_\ell\,|_{C_y=0} = -\frac{2}{5}\frac{C_{xy}}{C_\varepsilon}\underline{x} \tag{7.65}$$

As is the case in eq. 7.63, this linear distribution of the longshore velocity is not supported by physical observations. Hence, the bed friction effects cannot be considered to be negligible.

(3) Combined Bed Friction and Lateral Mixing Effects

Equation 7.62 is a linear, second-order, nonhomogeneous differential equation. Following Longuet-Higgins (1970b, 1972), the differential equation can be transformed into a non-dimensional equation by dividing the equation by $C_y V_b \sqrt{x_b}$, where the reference velocity (V_b) is defined in eq. 7.63. The transformed equation can now be written as

$$\frac{C_\varepsilon}{C_y}\frac{d}{d\chi}\left[\chi^{5/2}\frac{d(V_\ell/V_b)}{d\chi}\right] - \chi^{1/2}(V_\ell/V_b) = -\frac{C_{xy}}{C_y}\frac{\underline{x}_b}{V_b}\chi^{3/2}$$

$$= \frac{C_\varepsilon}{C_y}\frac{d}{d\chi}\left[\chi^{5/2}\frac{d\underline{V}_s}{d\chi}\right] - \chi^{1/2}\underline{V}_s = -\chi^{3/2} \tag{7.66}$$

where $\underline{V}_s = V_\ell/V_b$ and $\chi \equiv \underline{x}/\underline{x}_b$. Longuet-Higgins (1970b, 1972) uses the velocity resulting from the assumption of negligible momentum transfer at the break (V_b)

as a reference velocity. Equation 7.66 is a second-order Cauchy-Euler type of equation. Following Zill (1986), the solution of this equation, subject to the boundary condition $\underline{V}_s(\chi = 0) = 0$ where $V_\ell(\underline{x} = 0) = 0$, is

$$\underline{V}_s \equiv \frac{V_\ell}{V_b} = \mathrm{E}_1 \chi^{-\frac{3}{4}+\sqrt{\frac{9}{16}+\frac{C_y}{C_\varepsilon}}} + \frac{1}{1 - \frac{5}{2}\frac{C_\varepsilon}{C_y}} \chi \tag{7.67}$$

where $0 \leq \chi < 1$ and $C_\varepsilon / C_y \neq 2/5$. When $C_\varepsilon / C_y = 2/5$, the form of the original differential equation in eq. 7.62 changes, and the solution involves the natural log of χ. In eq. 7.67, the first term on the right side of the equal sign results from the homogeneous equation, whereas the second term is the particular solution of the nonhomogeneous equation. The integration constant E_1 is determined by applying the boundary conditions at the break, that is, where $\chi \equiv 1$. These boundary conditions are not apparent, so one must be "manufactured," so to speak. In deep water, where $\underline{x} = \infty$, we know that there is no net longshore velocity, so we can write $V_\ell(\infty) = 0$. However, deep water is well outside the surf zone, and eq. 7.66 does not actually apply in deep water. Because $dS_{xy}/dx = 0$ seaward of the surf zone from eq. 7.52, we can let the coefficient $C_{xy} = 0$. The resulting equation is then valid outside the surf zone as both the lateral stress and bed stress are approximately equal to zero. Under these assumptions and conditions, eq. 7.66 is

$$\frac{C_\varepsilon}{C_y} \frac{d}{d\chi} \left[\chi^{5/2} \frac{d\underline{V}_s}{d\chi} \right] - \chi^{1/2} \underline{V}_s = 0 \tag{7.68}$$

This equation is then the homogeneous part of eq. 7.66. The solution of eq. 7.68, subject to $V_\ell(\infty) = 0$, is

$$\underline{V} = \frac{V_\ell}{V_b} = \mathrm{E}_2 \chi^{-\frac{3}{4}-\sqrt{\frac{9}{16}+\frac{C_y}{C_\varepsilon}}} \tag{7.69}$$

where $\sim \infty > \chi > 1$. Note that the subscript "s" is missing from the velocity ratio $\underline{V}$ because that subscript identifies surf-zone phenomena. At the break, the velocity expressions in eqs. 7.67 and 7.69 must be equal, and the same can be said of their spacial derivatives, that is,

$$\underline{V}|_{\chi=1} = \underline{V}_s|_{\chi=1} \tag{7.70a}$$

and

$$\frac{d\underline{V}}{d\chi}\Big|_{\chi=1} = \frac{d\underline{V}_s}{d\chi}\Big|_{\chi=1} \tag{7.70b}$$

These requirements result in two equations with two unknowns, E_1 and E_2. The results are

$$\mathrm{E}_1 = \frac{1}{2\left(\frac{5}{2}\frac{C_\varepsilon}{C_y} - 1\right)} \left[\frac{7}{4\sqrt{\frac{9}{16} + \frac{C_y}{C_\varepsilon}}} + 1 \right] \tag{7.71}$$

and

$$\mathrm{E}_2 = \frac{1}{2\left(\frac{5}{2}\frac{C_\varepsilon}{C_y} - 1\right)} \left[\frac{7}{4\sqrt{\frac{9}{16} + \frac{C_y}{C_\varepsilon}}} - 1 \right] \tag{7.72}$$

Summarizing our results, the longshore velocity can be written as follows:
In the surf zone, where $1 < \chi (\equiv \underline{x}/\underline{x}_b) \geq 0$,

$$V_s(\chi) \equiv \frac{V_\ell}{V_b} = \frac{1}{2\left(\frac{5}{2}\frac{C_\varepsilon}{C_y} - 1\right)} \left[\frac{7}{4\sqrt{\frac{9}{16} + \frac{C_y}{C_\varepsilon}}} + 1 \right] \chi^{-\frac{3}{4} + \sqrt{\frac{9}{16} + \frac{C_y}{C_\varepsilon}}} + \frac{1}{1 - \frac{5}{2}\frac{C_\varepsilon}{C_y}} \chi \tag{7.73a}$$

Seaward of the surf zone, where $\infty > \chi \geq 1$,

$$\underline{V}(\chi) = \frac{1}{2\left(\frac{5}{2}\frac{C_\varepsilon}{C_y} - 1\right)} \left[\frac{7}{4\sqrt{\frac{9}{16} + \frac{C_y}{C_\varepsilon}}} - 1 \right] \chi^{-\frac{3}{4} - \sqrt{\frac{9}{16} + \frac{C_y}{C_\varepsilon}}} \tag{7.73b}$$

Results obtained from eqs. 7.73a and 7.73b are presented in Figure 7.14. In that figure, an inflection point for $\underline{V}(\chi)$ is in the surf zone where the longshore velocity is maximum. In eqs. 7.73a and 7.73b, the respective C_{xy}, C_y, and C_ε are obtained from eqs. 7.56, 7.58, and 7.61. The velocity at the break (V_b) is defined in eq. 7.63, and γ from eq. 7.44 has a value of about 0.82, according to Longuet-Higgins (1970b). To find the expressions for the maximum longshore velocity and its position in the surf zone, we equate the derivative of the expression in eq. 7.73a to zero.

The corresponding expression for position of the maximum longshore velocity is

$$\chi_{(\max)} = \left(\frac{2\sqrt{\frac{9}{16} + \frac{C_y}{C_\varepsilon}}}{\frac{C_y}{C_\varepsilon} + \sqrt{\frac{9}{16} + \frac{C_y}{C_\varepsilon}} - \frac{3}{4}} \right)^{\frac{1}{\sqrt{\frac{9}{16} + \frac{C_y}{C_\varepsilon}} - \frac{7}{4}}} \tag{7.74}$$

Knowledge of the position of the maximum longshore velocity is important in the design of groins, which are structures projecting from shore into the surf zone. Groins are designed to trap the sand being transported alongshore.

The expression for the maximum value of the longshore velocity is

$$\underline{V}_{\max} = \left[\frac{\frac{C_y}{C_\varepsilon}}{\frac{C_y}{C_\varepsilon} - \frac{3}{4} + \sqrt{\frac{9}{16} + \frac{C_y}{C_\varepsilon}}} \right] \chi_{(\max)} \tag{7.75}$$

Results obtained from eqs. 7.74 and 7.76 are presented in Figure 7.14. Refaat, Tsuchiya, and Kawata (1990) show excellent agreement between the Longuet-Higgins model for the longshore velocity in eq. 7.73 and experimental results on a smooth beach at a deep-water wave angle of $\beta_0 = 45°$.

The reader should remember that the profile of the longshore velocity depends on the conditions in the surf zone. We have considered the relative effects of both the shear stress on the bed and the apparent shear stress due to turbulent mixing.

Other considerations involve the porosity of the bed and the air content of the water. Concerning the former: Cobble beaches absorb more water than beaches of

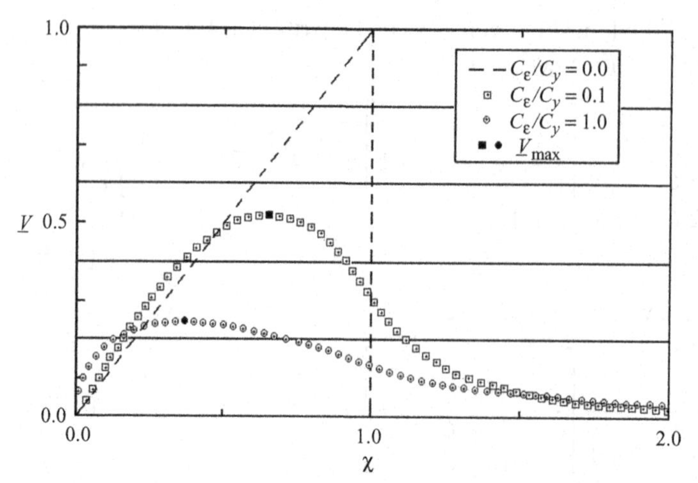

Figure 7.14. *Non-Dimensional Longshore Velocity Distribution.* The non-dimensional velocity is $\underline{V} = V_\ell / V_b$, where V_b is the velocity at the liner of breakers, from eq. 7.63. That velocity expression results from the assumption that no turbulent-mixing losses are present, that is, $C_\varepsilon = 0$. The non-dimensional spacial coordinate is $\chi = \underline{x}/x_b$, where $\underline{x}$ is the swash-line coordinate in Figure 7.13. The maximum non-dimensional velocity (V_{max}) values are obtained from eq. 7.75.

fine sand. Also, dry sand absorbs more water than wet sand. The water absorbed has both energy and momentum, so the longshore velocity is directly affected by absorption or percolation. The air content in the surf zone depends on the type of break (spilling, plunging, surging, and collapsing, illustrated in Figure 7.2). The spilling break has foam at the crest, which expands as the wave travels through the surf zone, so spilling waves can be assumed to be mildly aerated. Plunging waves will trap air as they curl and are highly aerated. The surging and collapsing waves will have the least air content of breaking waves. The presence of air causes the flow to be heterogeneous, and both the mass density and viscosity differ from those of a homogeneous flow.

EXAMPLE 7.8: MAXIMUM LONGSHORE VELOCITY Consider, again, the conditions in Example 7.2, where 1-m, 8-sec deep-water waves approach a beach at an angle (β_0) of 30° to the shoreline. In deep water, the wavelength is $\lambda_0 \simeq 100\,\text{m}$ and the corresponding wave number is $k_0 \simeq 0.0628\,\text{m}^{-1}$. The beach has a uniform slope (m) of 1/50. The following breaking conditions are found: $h_b = 1.55$ m, $\underline{x}_b = 77.5$ m, $\lambda_b = 30.7$ m, and $\beta_b = 8.83°$. It is noted that $h_b/\lambda_b \simeq 1/20$, so shallow-water approximations can be used. From measurements, the parameter C_ε/C_y is found to be approximately 0.1. The non-dimensional longshore velocity distribution for this parametric value is shown in Figure 7.14. From eq. 7.74, the maximum longshore velocity occurs at $\chi_{(max)} \simeq 0.65$, or $x_{(max)} \simeq 50.4$ m. The maximum value of the time-averaged non-dimensional longshore velocity is approximately 0.517 from eq. 7.75. The expression for the maximum velocity is

$$V_\ell(\underline{x}_{(max)}) = V_b \underline{V}_{max} = \frac{5}{2}\frac{\pi\gamma\sqrt{g}}{f\sqrt{\underline{h}_b}}\left[\frac{m}{\left(1+\frac{3}{8}\gamma^2\right)}\right]^2 \sin(\beta_b)\cos(\beta_b)\underline{x}_b\underline{V}_{max} \tag{7.76}$$

where, as previously mentioned, $\gamma \simeq 0.82$ according to Longuet-Higgins (1970b). To determine the value of $\underline{h}_b$, we must first determine the value of the set-down at the break from eq. 7.40. Assuming shallow water at the break, the approximate expression for the set-down is

$$\bar{\eta}_b \simeq -\frac{k_0 H_0^2 \cos(\beta_0)}{32(k_b h_b)^3} \tag{7.77}$$

the value of which is approximately −0.0534 m. Note: The expression in eq. 7.40 is for the nonrefracting condition, that is, $\beta_0 = 0°$. The expression can be applied to refracting waves by assuming that the wave heights in the expression are equivalent wave heights. That is, for the deep-water wave, the equivalent wave height is $H_0' = H_0\sqrt{\cos(\beta_0)}$, as in eqs. 6.90 and 7.8. The actual water depth at the break is found to be $\underline{h}_b \simeq 1.50$ m. Also needed in eq. 7.76 is the value of the friction factor (f) which, in turn, depends on the equivalent surface roughness of the bed, κ_e. According to Liu and Dalrymple (1978), the empirical expression for the friction factor is

$$f_\mu \simeq 1.41\left(\frac{4\pi\kappa_e}{\gamma\sqrt{g\underline{h}_b}T}\right)^{2/3} \tag{7.78}$$

This is based on the works of Jonsson (1966) and Kamphius (1975). Rather than assuming a value of κ_e, we use $f_\mu = 0.04$, the value given by Longuet-Higgins (1970b). By using this f_μ-value in eq. 7.76, the maximum longshore velocity value is approximately 0.638 m/s.

Refaat, Tsuchiya, and Kawata (1990) present experimental data showing rather good agreement with the predicted longshore velocity distribution in the surf zone. The expressions in eqs. 7.73 through 7.75 can then be used with some confidence in the planning phase of shore protection projects.

D. Average Longshore Volume Flow Rate

A knowledge of the average longshore velocity distribution helps the engineer in the design of a groin field to trap sand. In addition, the average longshore velocity is related to the longshore volume flow rate. The amount of sand being transported will be some fraction of the average total volume being transported alongshore in the surf zone. The volume rate of water flow is found by integrating eq. 7.73a over the average longshore flow area in the surf zone. The resulting expression for the *longshore volume flow rate* is

$$Q_{\ell s} = \int_0^{\underline{x}_b} V_\ell(\underline{x})\underline{h}(\underline{x})d\underline{x} = \frac{C_{xy}}{C_y}\left[\frac{E_1}{\left(\sqrt{\frac{9}{16}+\frac{C_y}{C_\varepsilon}}+\frac{9}{4}\right)} + \frac{1}{3\left(1-\frac{5}{2}\frac{C_\varepsilon}{C_y}\right)}\right]\frac{m}{\left(1+\frac{3}{8}\gamma^2\right)}\underline{x}_b^3$$

$$= \frac{\frac{5}{4}\pi\frac{\sqrt{g\gamma}}{f_\mu}\sin(2\beta_b)}{\left(\frac{5}{2}\frac{C_\varepsilon}{C_y}-1\right)}\left[\frac{\left(\frac{7}{4}+\sqrt{\frac{9}{16}+\frac{C_y}{C_\varepsilon}}\right)}{\left(\frac{9}{8}+2\frac{C_y}{C_\varepsilon}+\frac{9}{2}\sqrt{\frac{9}{16}+\frac{C_y}{C_\varepsilon}}\right)} - \frac{1}{3}\right]\left[\frac{m}{\left(1+\frac{3}{8}\gamma^2\right)}\right]^{9/2}\underline{x}_b^{5/2} \tag{7.79}$$

where, again, C_{xy}, C_y, and C_ε are obtained from eqs. 7.56, 7.58, and 7.61, respectively, whereas V_b is defined in eq. 7.63 and γ from eq. 7.44 has a value of about 0.82, according to Longuet-Higgins (1970b). The reader should note that the volume rate of flow seaward of the line of breakers might be significant. However, most of the sediment transported alongshore is in the surf zone.

EXAMPLE 7.9: LONGSHORE VOLUME TRANSPORT RATE The conditions in Examples 7.2 and 7.8 are the following: In deep water, $T = 8$ sec, $\beta_0 = 30°$, $\lambda_0 \simeq 100$ m, and $k_0 \simeq 0.0628\,\text{m}^{-1}$. On the uniform 1/50 beach, we find $h_b = 1.55$ m, $x_b = 77.5$ m, $\lambda_b = 30.7$ m, $\beta_b = 8.83°$, and $h_b/\lambda_b \simeq 1/20$, so the shallow-water approximations are permitted. Also, the breaking wave height is about 1.27 m from eq. 7.44, where, again, we assume $\gamma \simeq 0.82$. In Example 7.8, the parameter C_ε/C_y is about 0.1. With these values applied to eq. 7.79, the volume flow rate of water is approximately 7×10^{-3} m^3/s. The corresponding yearly volume flow rate is approximately 220,300 m^3/year.

The longshore volume flux or volume transport rate obtained from eq. 7.79 is for pure water. However, coastal engineers are interested in the mass transport rate of sediment because the stability (the balance between erosion and accretion) of a beach depends on this transport rate. Empirical formulas for the longshore sediment transport rate are important in the planning analyses in coastal engineering. These formulas are for the volume flow rate of the sediment within the longshore slurry. One of the most widely used of these formulas is that presented in the *Shore Protection Manual* of the U.S. Army (1984). Also, see the technical report issued by the U.S. Army Corps of Engineers (U.S. Army, 1990). Using that formula, the *sediment volume transport rate* is

$$Q_{sed} = \frac{K_{sed}}{(\rho_{sed} - \rho)C_p} \frac{\rho H_{sb}^2 c_{gb}}{16} \sin(2\beta_b) \tag{7.80}$$

where $C_p(\simeq 0.6$ for sand – the sediment of choice) is 1 minus the porosity $(\forall_{void}/\forall_{total})$ for the sediment and ρ_{sed} is the mass density of the sediment. Also in eq. 7.80 are $K_{sed}(\simeq 0.39)$, a non-dimensional empirical coefficient, the significant wave height at the line of breakers, H_{sb}, and the group velocity at the line of breakers, $c_{gb} \simeq c_b \simeq \sqrt{(g\underline{h}_b)}$, assuming shallow-water conditions. As discussed in Chapter 5, the significant wave height is the average of the one-third highest waves. The reader should note that the beach slope (m) is not explicit in eq. 7.80. The slope is included in the expressions for the breaking wave height, as in eq. 7.1.

Another widely used formula for the mass transport of sediment is that of Kamphuis (1991). His expression is based on both dimensional analysis and experimental data. Kamphuis (2002) modifies his earlier expression, and shows that the results obtained from that expression agree well with measured data. His expression for the longshore transport of sediment mass is

$$\frac{dm_{sed}}{dt} = \rho_{sed} Q_{sed} = 7.9 \times 10^{-4}(\rho_{sed} - \rho)\left(\frac{g}{2\pi}\right)^{1.25} H_{sb}^2 T_p^{1.5} m^{0.75} D_{50}^{-0.25} \sin^{0.6}(2\beta_b) \tag{7.81}$$

where T_p is the peak period (the period corresponding to the peak of the energy period spectrum, as discussed in Chapter 5) and D_{50} is the mean diameter of the sediment. Kamphius (2002) finds that his expression can be used for both regular and irregular waves if the significant wave height, modal period (the peak spectral frequency period), and mean wave angle for irregular waves are replaced by the respective regular wave height, period, and wave angle values. See eq. 5.51 for the various period relationships.

EXAMPLE 7.10: LONGSHORE SEDIMENT TRANSPORT RATE In Example 7.9, the volume rate of water flow is found to be approximately 0.0070 m^3/s, from the formula in eq. 7.79. Let us now attempt to determine the volume rate of sand transport – where $\rho_{sed} = 2.65\rho \simeq 2.65(1{,}030\ kg/m^3) \simeq 2{,}730\ kg/m^3$ for this example – from the formula in 7.81 for the same (regular) wave conditions. Using the relationship in eq. 7.44, assume that the breaking wave height equals $\gamma \underline{h} = 0.82(1.55\ m) \simeq 1.27\ m$. On the uniform 1/50 beach, $h_b = 1.55$ m, $\underline{x}_b = 77.5$ m, $\lambda_b = 30.7$ m, and $\beta_b = 8.83°$. In addition, a mean sand diameter value must be assumed. For the 1/50 beach slope, Figure 4–36 in the *Shore Protection Manual* of the U.S. Army (1984) shows that the approximate range of D_{50} is from 0.2 mm to 0.3 mm for the New Jersey coast. Assuming the 0.2 mm value, the volume rate of sediment transport from eq. 7.81 is $Q_{sed} \simeq 0.0046\ m^3/s$. Comparing this result with that in Example 7.9, we see that the sediment volume flow rate is about 65% of the water flow rate. Concerning the effect of the mean particle diameter, D_{50}: If the value is doubled in this example to 0.4 mm, the sediment mass flow rate is decreased by approximately 16%.

7.5 Closing Remarks

The analyses and discussions presented in this chapter are designed to give the reader an understanding of both the nature and behavior of the hydromechanics of water waves near and in the surf zone. For more recent advances in surf zone phenomena, the reader is referred to the collection of papers edited by Fredsoe (2002).

In Chapter 8, several aspects of the planning phase of shore protection projects are presented. As the reader will see, to make decisions concerning shore protection, a good knowledge of coastal wave mechanics is vital. Chapter 8 covers rather specific coastal engineering topics. The reader should consult Horikawa (1978), U.S. Army (1984), Goda (1985), Sorensen (1997), Dean and Dalrymple (2002), and Herbich (1999) for more extensive and detailed coverage of ocean engineering topics.

8 Coastal Engineering Considerations

The purpose of this chapter is to give a cursory introduction to shore protection and to discuss some topics that are normally neglected in the coastal engineering literature. By *shore protection*, what is meant is the methodologies used in preventing either a net erosion or a net accretion of beach sand due to wave action. Specifically, some of the considerations that are part of the planning phase of shore protection projects are discussed. Shore protection is one of the topics under the broad heading of *coastal engineering*. The reader is referred to the books by Horikawa (1978), U.S. Army (1994), Goda (1975), Sorensen (1997), Dean and Dalrymple (2002), and that edited by Herbich (1999) for a more thorough discussion of the area of ocean engineering. In addition, the proceedings of the International Coastal Engineering Conference contain papers describing advances in both the science and technology applied to the coastal zone. These conferences occur about every two years, and are sponsored in part by the American Society of Civil Engineers (ASCE).

8.1 Shore Protection Methods

When a shoreline is identified as being *unstable*, the term usually means that there is a net loss of sand (erosion) or net gain (accretion). The engineer must determine whether or not the instability is short-term or long-term. Short-term instabilities are common, and are usually seasonal in nature. Winter waves tend to erode beaches, whereas summer waves tend to restore sand to the beaches. So, many beaches experience a winter erosion and a summer accretion, which are considered to be short-term instabilities and are cyclic in nature. Some long-term instabilities are also cyclic. For example, there are regions where sand spits form and disappear over periods ranging up to centuries.

When the volumes of sediment for winter erosion and summer accretion are equal, then the beach is stable. When this is not the case, then the beaches are unstable, and some sort of remedial action might be required to restore the stability. There are two remedial options that are available to the coastal engineer. These options are soft (natural) stabilization and hard (human-made) stabilization.

Soft stabilization methods include the use of sand dunes in combination with beach grasses, and beach nourishment, where sand from offshore deposits called borrows is slurried onto the beach. The dunes can be artificially constructed and

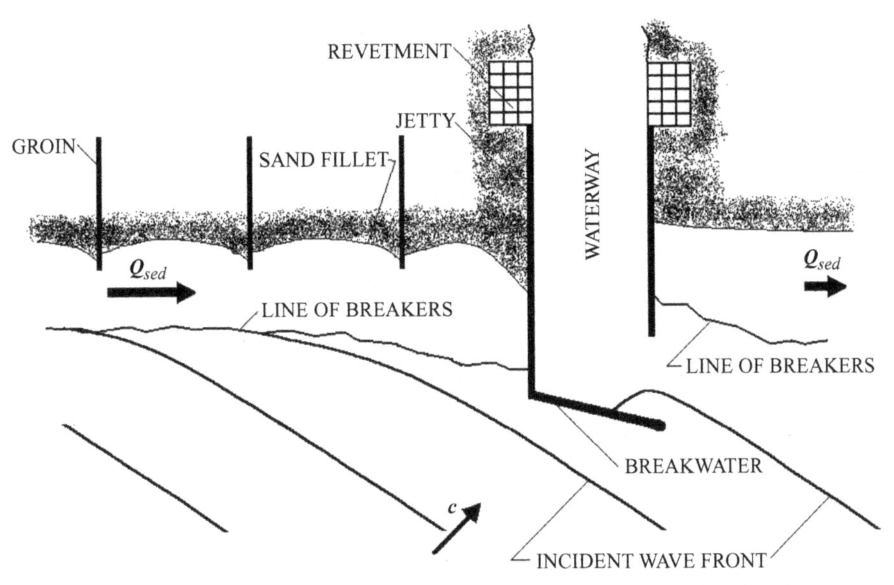

Figure 8.1. *Area Sketch of Four Hard-Stabilization Structures.* In the sketch, three groins, comprising a groin field, are shown up-drift from the up-coast jetty. The groins extend from the back-shore into the surf zone, whereas both the up-coast and down-coast jetties that form the waterway are seen to extend through the surf zone. The breakwater that is sketched is attached to the up-coast jetty. In many cases, breakwaters are unattached, particularly if they are the only hard-stabilization technique used in a shore protection project. The up-coast and down-coast revetments line the respective shores of the inland portion of the waterway.

then covered by deeply rooted beach grass that, in turn, stabilizes the dune. The beach grass grows and subsequently traps wind-blown sand. As the trapped sand piles up on the dune, the beach grasses grow higher, forming a stabilization cycle. The dune then act as a sand supply for both the beach and for transport in the surf zone.

Hard stabilization of a beach is normally accomplished by using one of three types of structures: revetments, groins, and breakwaters. Referring to the area sketch in Figure 8.1, a *revetment* is a structure composed of layered stone or concrete blocks that simply rest on the beach face. One normally finds these on beaches in somewhat protected waters where the wave energy is relatively low, such as in the northern Chesapeake Bay. A *groin* (*groyne* in British literature) is a structure that extends from the shore into but not beyond the design surf zone. Groins are usually oriented at right angles to the design shoreline, and are designed to partially trap part of the sand transported alongshore in the surf zone. When two or more groins are used, the system of groins is called a *groin field*. More will be said of groins later in this section. A *breakwater* is a structure that is seaward of the surf zone, or in other words, seaward of the line of breakers. Breakwaters have the most control of the longshore transport because they shield the waters leeward of the structure (in the shadow zone) from the direct wave action. Wave energy is transmitted into the shadow zone of the structure by the process of diffraction, discussed in Chapter 6. In Figure 8.1, an attached breakwater is sketched. That is, the breakwater shown is attached to the up-coast jetty, protecting the mouth of the waterway from direct wave attack. Also sketched in Figure 8.1 are two revetments that line the inland boundaries of the waterway to stabilize the up-coast and down-coast banks.

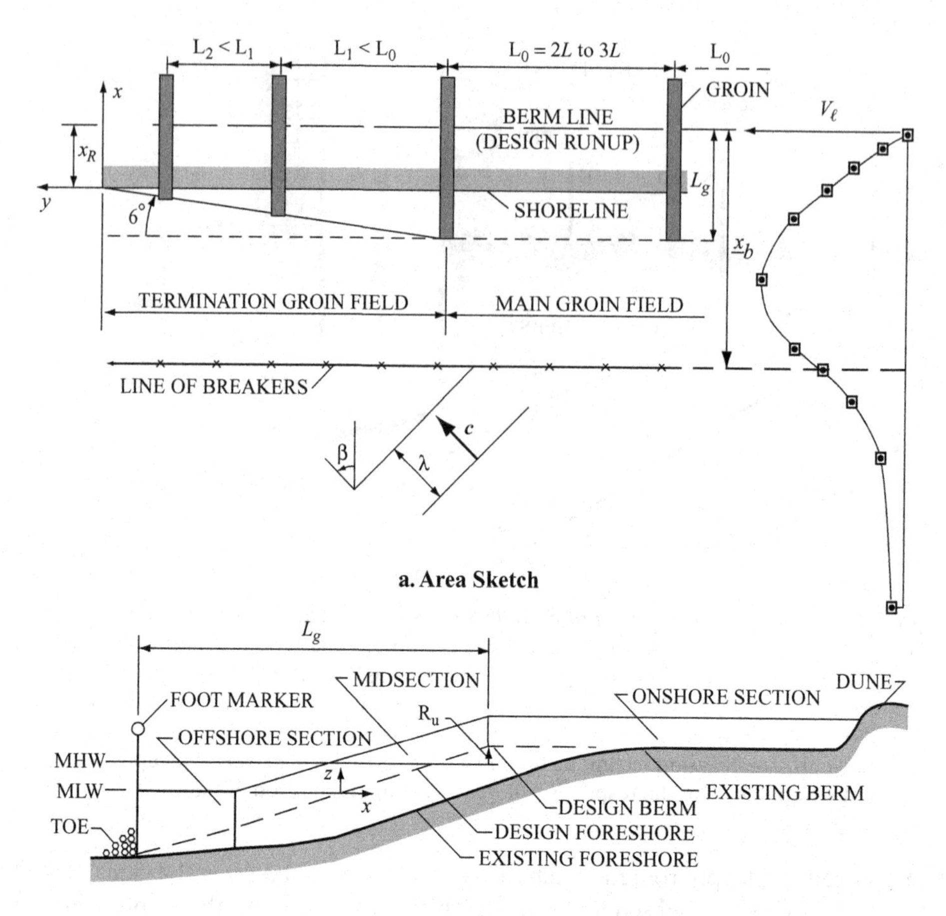

Figure 8.2. *Groin Field Orientation*. The area sketch shows two units of the main groin field and, in addition, a two-unit terminal field. The design position of a foot of a groin in the main field is 40% of the design surf zone width ($\underline{x}_b$) measured from the berm line. The length (L_g) is the distance from the design berm to the toe. The ideal separation distance (L_0) between groins in the main field is two to three times the design groin length. The elevation sketch shows the onshore section terminating at a dune to prevent flanking. The slope of the mid-section cap is parallel to the design foreshore. The design berm is at the end of the design foreshore and at a height equal to the runup measured from the mean high-water (MHW) level, whereas the design shoreline is at the intersection of the mean low-water (MLW) level and the design foreshore. A marker is needed to prevent the foot of the groin from becoming a navigation hazard when submerged. Finally, an example of the averaged longshore velocity distribution is shown at the right of the area sketch.

Referring to the sketches in Figure 8.2, when a groin field has been determined to be the most cost-effective hard-stabilization option for a specific site, the landward extent of each groin will be such that flanking is prevented. *Flanking* is the bypassing of water around the landward end of the groin. At the landward end, a dune or a bulkhead might be required to prevent flanking. However, the design length of the groin is measured from the position of the design berm (at the position of the design run-up) to about 40% of the surf zone width. For a groin at a right angle to the shoreline, $L_g = 0.4\underline{x}_b$, where $\underline{x}_b$ is defined in Figure 7.9. Although the groin's angular orientation with respect to the shoreline can be optimized to produce stable up-drift and down-drift sand fillets, groins are normally at right angles, as sketched in Figure 8.2. Sand fillets are trapped sand deposits adjacent to the groins. In Chapter 7, we find that the wave angle between a breaking wave front and the groin is relatively small. Hence, oblique (non-normal) orientations of groins are not cost-effective for many sites because the oblique groins are both more material-intensive and are more difficult to construct. To minimize down-coast erosion, a terminal groin field is needed, as sketched in Figure 8.2a. As can be seen in that sketch, the foot of a unit in the terminal field is on a line drawn at 6° to the design shoreline. This optimal termination angle has been determined from experimental studies. For thorough discussions of the design and use of groins, the reader is advised to consult the publications by the U.S. Army (1984, 1994), CIRIA (1990), and Fleming (1990).

As mentioned previously, breakwaters are positioned seaward of the line of breakers. These structures can be either attached to jetties (as sketched in Figure 8.1) or other structures, or can be detached. When a detached breakwater is constructed parallel to the shoreline, the response of the shoreline is almost immediate. The initial sand formation is called a *salient*, which is a bulge in the shoreline having a seaward apex that is in the breakwater's shadow zone. Depending on both the length between the breakwater's up-drift and down-drift ends (heads) and the unit's distance form the shoreline, a tombolo might or might not form. A *tombolo* resembles a sand causeway between the detached breakwater and the shoreline, and is formed when the salient grows and attaches to the leeward face of the breakwater. This sand formation can be the cause of erosion of the down-coast beach. The reader is encouraged to consult the publications of CIRIA (1991) and U.S. Army (1984, 1994) for the details of breakwater planning and design.

EXAMPLE 8.1: PLANNING A GROIN FIELD The conditions in Examples 7.2, 7.8, and 7.9 are the following: In deep water, $H_0 = 1\,\text{m}$, $T = 8\,\text{sec}$, $\beta_0 = 30°$, $\lambda_0 \simeq 100\,\text{m}$, and $k_0 \simeq 0.0628\,\text{m}^{-1}$. On the uniform 1/50 beach, we found $h_b = 1.55\,\text{m}$, $\underline{x}_b = 77.5\,\text{m}$, $\lambda_b = 30.7\,\text{m}$, $\beta_b = 8.83°$, and $h_b/\lambda_b \simeq 1/20$, so that the shallow-water approximations are permitted. Also, the breaking wave height is about 1.27 m, from eq. 7.44.

Assume that the turbulent-mixing-to-bed-shear parameter is $C_\varepsilon/C_y = 0.1$; the non-dimensional distribution of the longshore velocity is shown in Figure 7.14. From Example 7.8, the maximum longshore velocity is 0.638 m/s, which is at a distance of $0.65\underline{x}_b \simeq 50.4\,\text{m}$ from the position of the design berm line. The berm line is determined by the runup, the value of which is $R_u \simeq 0.186\,\text{m}$ from Hunt's formula in eq. 7.15, assuming a smooth bed. Because the bed slope is 1/50, the berm line is 9.3 m landward of the shoreline. Hunt's formula is a function of the surf similarity parameter (ξ) defined by eq. 7.9. This parameter is, in turn, a function of the equivalent deep-water wave height, $H_0' = H_0\sqrt{[\cos(\beta_0)]} \simeq 0.931\,\text{m}$. The design length of the groin (L_g in Figure 8.2) is equal to $0.4\underline{x}_b = 31$ m. This is measured seaward from the berm line. The averaged longshore velocity at the foot of a unit in the main field is about 0.531 m/s prior to the construction of the groin field. Of course, the longshore velocity distribution is after the construction of the groins.

In the next section, we apply decision theory to the selection of the most cost-effective hard-stabilization method of shore protection.

8.2 Decision Process in Coastal Protection

Because there are two available remedial options (soft stabilization and hard stabilization) in the stabilization of a beach, the coastal engineer must decide which of the two is most apropos. A course of remedial action will have a probability of success (reliability) that is normally less than unity. The difference (one minus the probability of success) is the probability of failure. Associated with this probability are both material and financial losses. Soft or hard solutions are neither perfect nor have infinite lives, so we can expect losses associated with our remedial decision.

There are a number of considerations that are involved in making such a decision. These involve the states of nature, their probabilities of occurrence, and the expected losses associated with the states of nature.

States of nature are conditions that are beyond the control of the design engineer. For example, the engineer has no control over either the timing or the intensity of storms. Severe storms produce destructive winds, waves, and floods, all of which are probabilistic in nature. Winds, waves, and floods are of primary interest to the coastal engineer in most of the ocean-boarding communities. In some areas, such as the west coast of the contiguous United States, earthquakes must be added to the list. Probabilities of occurrence can be attached to each of the states of nature and, therefore, to their associated expected losses. For example, a rubble-mound breakwater is damaged if any of the armor stones (the protective layer of the structure) are permanently displaced by storm waves. As is discussed in Chapter 5, by examining the history of storm waves at a site, we can assign a probability of occurrence of the height of the destructive wave. To illustrate, consider a storm that statistically occurs once every fifty years and (statistically) produces a maximum wave height of 3 m just seaward of the surf zone. If we use this wave height as a design wave height for a breakwater, then we can expect a loss of, say, from 0% to 5% of the total number of armor stones over fifty years. Hence, as far as the waves are concerned, the expected life of the breakwater is fifty years.

One of the available tools to help the coastal designer logically evaluate the cost-effectiveness of the different methods of beach protection is the *decision tree*. It is one of several tools of *decision theory*, which is widely used in the design process. The use of a decision tree for a project to abate shoreline erosion is demonstrated in the following example. As is seen in the example, the first decision to be made is to either take some remedial action or not.

EXAMPLE 8.2: DECISION TREE FOR SHORELINE EROSION ABATEMENT A small coastal community having a public beach experiences an erosion problem in front of a boardwalk having a length of 1 km. The cause of the erosion has been identified as an up-coast jetty, a structure that extends through the surf zone and is designed to protect a waterway, so the problem is not seasonal or cyclic in nature. The jetty was constructed several years ago approximately 1 km from the beach in question. The shoreline is retreating at a rate of 1 m per year. This rate of retreat has been determined from a photographic survey. At the beginning of the survey, the mean position of the shoreline was about 50 m from the boardwalk. The consequences of the beach erosion are both property damage and a decrease in the number of summer tourists. The monetary loss is estimated to be $2M per year.

The first decision to be made is to determine if spending tax dollars on beach protection is justified. If no action is taken, the losses will then escalate over the years. With no inflation in the monetary losses, which is unrealistic, the total losses will amount to $20M at the end of ten years, In addition, the mean distance between the boardwalk and shoreline at the end of ten years will be 40 m or less, so the decision is made to take action.

The next decision to be made concerns the choice of remedial options (soft or hard). To help in this task, a decision tree is constructed, as in Figure 8.3. The goal of using the decision tree is to determine the most cost-effective option. In the figure, the "bottom line" is the cost per life-year associated with each option. From Figure 8.3, we see that for this erosion problem, the most cost-effective solution is the construction of dunes in front of the boardwalk, where the dunes are stabilized by planted beach grass. One problem is identified with this soft solution, that is, the boardwalk-to-shoreline distance is only 50 m, which is rather small. From the Figure 8.3, we see that an estimated 100,000 m^3 of sand are required. If the dune could be constructed with a rectangular cross-section having a 5-m height, then the dune would extend 20 m from the boardwalk, which is unrealistic. So even though this soft solution has the lowest cost per life-year, it is not feasible.

The most cost-effective hard stabilization option in Figure 8.3 is the groin field having stone and steel construction. This combination groin field is seen to have a bottom-line cost of slightly less than $0.2M per life-year, which is somewhat larger than the dune-beach grass option. The expected life of the groin field is 50 years, with a yearly maintenance cost of about 1% of the construction costs, or $90K per year.

The cost of a hard-stabilization option depends on a number of factors. These include the location of the material supplier, the route of delivery, and permit costs. In addition to these, political and social consideration can also affect the costs. To illustrate, consider using wood as the primary construction material for a groin field on a Southern California beach. This would not be a good choice for several reasons. First, forests are distant. Second, environmentalists would object to cutting down the trees. Finally, because most of the beaches in Southern California are well populated, the route from the supplier to the site would probably travel through very up-scale neighborhoods, making the decision somewhat politically unacceptable. For many coastal structures, quarry stone is the material of choice because of its availability and ease of construction. In the next section, the relationship between the size of the stone and the local wave climate is discussed.

8.3 Rubble-Mound Structures

Because the use of quarry stone in the construction of groins, jetties, breakwaters, revetments, and seawalls is quite common, we shall demonstrate how the selection of the stone weight for such structures is reliant on the local wave conditions. In addition, the reliability of these *rubble-mound* structures is discussed. The popularity of rubble-mound structures is due to both the relative ease of construction and the relatively low cost of materials. The breakwater is chosen as the illustrative structure because this type of structure is under the direct wave attack. In addition

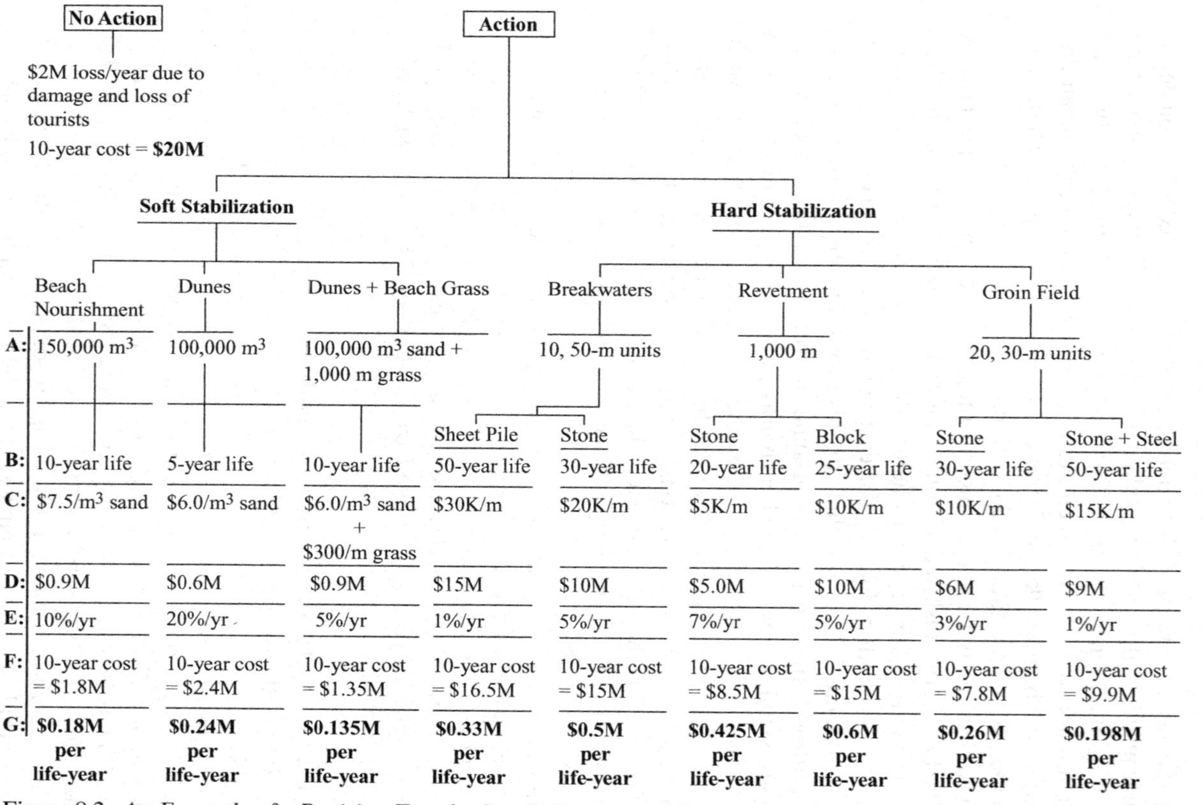

Figure 8.3. *An Example of a Decision Tree for Beach Protection*. The rows are A: units, B: expected life, C: cost per unit, D: total delivered cost, E: maintenance cost per year, F: ten-year costs, and G: cost per life-year, based on the expected life. The "bottom line" is the cost per life-year associated with each decision. The monetary values shown are fictitious and have no relationship to actual values. The reason for using fictitious values is that the actual values depend on location, availability of construction materials, permit costs, and so on, none of which are considered. For the purposes of illustration, the fictitious values are considered to be satisfactory.

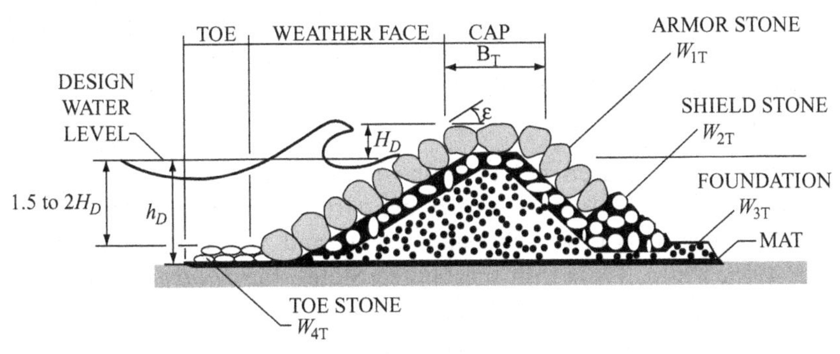

Figure 8.4. *Sketch of a Three-Layer Breakwater Cross-Section.* The depth of the primary armor stone on the weather face and the height of the crest are dependent on the design wave height, H_D. The width of the crest or cap is B_T, whereas the weather face angle is ε. A single layer of armor is shown for the purpose of illustration. Depending on the severity of the wave climate, the number of armor layers can be up to three. For many ocean sites, two layers are considered to be satisfactory. When artificial armor units are used, one layer can be used because the units are designed to be interlocking.

to rubble-mound structures, other types of breakwaters also are used. These are discussed in the *Shore Protection Manual*; see U.S. Army (1984).

A. Stone Selection for Rubble-Mound Breakwaters

Depending on the wave climate at the site, there can be several types of stone that are used in rubble-mound construction. Each stone type has a particular function. For breakwaters subject to direct ocean waves, there are four types, which are armor stone, shield stone, foundation stone, and toe stone. Each of these has a mass range and a particular function. For breakwaters in protected waters, there might only be one type of stone used in the construction. In any case, a mat is normally used to help evenly distribute the weight of the structure over the bed. A three-layer cross-section is sketched in Figure 8.4. In that sketch are the following stone types:

(1) ***Armor Stone:*** This type of stone composing the weather face of the breakwater is called the *primary armor stone*. This is the design stone, because the weights of the other stone types depend on the primary armor weight, W_{1T}. The primary armor is the most massive because it withstands the direct forces of the incident waves. *Secondary armor stone* is that which composes the upper part of the leeward face of the structure. The height of the primary armor layer above the design water level is equal to the design wave height (H_D), whereas the depth of the layer is from $1.5H_D$ to $2H_D$.

(2) ***Shield Stone:*** This type of stone has a weight $W_{2T} < W_{1T}$. It is designed to both distribute the armor weight and to control energy transmission through the structure. For the three-layer cross-section in Figure 8.4, the SPM (U.S. Army, 1984) recommends a shield stone weight range of $W_{1T}/15 \leq W_{2T} \leq W_{1T}/10$.

(3) ***Foundation Stone:*** As the name implies, this small stone of weight $W_{3T} < W_{2T}$ is designed to evenly distribute the structural weight on the bed. It is self-adjusting when settling occurs. A mat is normally used to both ensure the evenness of the weight distribution on the bed and to prevent toe scour. The foundation stone weight range recommended by the SPM is $W_{1T}/6000 \leq W_{3T} \leq W_{1T}/200$.

(4) ***Toe-Berm Stone:*** Toe scour can be a problem in long waves. To reduce the scour, a toe berm is used, as sketched in Figure 8.4. The recommended toe-stone weight is $W_{4T} \simeq W_{1T}/10$; however, in long shallow-water waves, the toe stone is relatively large.

There are two formulas used to determine the average weight of the primary armor. Those are the Irebarren formula (Irrebarren Cavanilles, 1938; Irrebarren Cavanilles and Nogales *y* Olano, 1950) and the Hudson formula (Hudson, 1953, 1959, 1961a, 1961b). In the United States, the *Hudson formula* is used, which is

$$W_{1T} = \frac{\rho_{stone} g H_D^3 \tan(\varepsilon)}{K_{DT}\left(\dfrac{\rho_{stone}}{\rho} - 1\right)^3} \tag{8.1}$$

where ρ_{stone} is the mass density of the stone, ρ is the mass density of salt water (assume 1,030 kg/m^3), ε is the angle of the structure's weather face (measured from the horizontal), and K_{DT} is called the *stability coefficient.* Note that the subscript "T" is used to identify the stability coefficient of the trunk of the breakwater. The *trunk* is the main section of the structure, and the *head* of the breakwater is the structure's free end (or ends). The breakwater sketched in Figure 8.1 is attached to a jetty, so that attached breakwater has one head. When a breakwater is not connected to any other structure, then it is called an unattached breakwater and the trunk is between the two heads. The wave forces on a head are larger than those on the trunk. The reason for this is that the bottom contours near the heads are modified over time because of sand accumulation. Hence, there is a refractive focusing on the heads. For this reason, the stability coefficient for the primary armor on a head is smaller than that for the trunk armor. Values of the stability coefficient for various armor stones and units are presented in Table 7–8 in the SPM. The wave height, H_D, in eq. 8.1 is the design wave height. For random waves, discussed in Chapter 5, this height could be the significant wave height or the extreme wave height occurring over the design life of the structure. This is demonstrated in Example 8.3.

The weights of the other stones depend on the value obtained from eq. 8.1. The weight ranges of these stones are the following:

Shield stone weight:

$$W_{2T} \simeq \frac{W_{1T}}{10} \tag{8.2}$$

Foundation stone weight:

$$W_{3T} \simeq \frac{W_{1T}}{100} \rightarrow \frac{W_{1T}}{10} \tag{8.3}$$

Toe-berm stone weight:

$$W_{4T} \simeq \frac{W_{1T}}{4000} \rightarrow \frac{W_{1T}}{200} \tag{8.4}$$

In some regions, adequate quarries are not available to supply the required armor stone. When this is the situation, artificial armor units (tetrapods, tribars, etc.) are constructed of steel-reinforced concrete. These artificial armor units have stability coefficient values that are greater than those for quarry stone because the artificial armor units are designed to interlock. This feature also allows the artificial armor units to be somewhat lighter than the quarry stone.

Once the armor stone weight is determined, the layer thickness (r_T) of the primary stone must be determined. From the SPM, the formula for the layer thickness is

$$r_T = nk_\Delta \left(\frac{W_{1T}}{\rho_{stone} g} \right)^{1/3} \tag{8.5}$$

where n is the number of layers of armor, k_Δ is called the layer coefficient, and $\rho_{stone} g$ is the specific weight of the stone based on the density of salt water. An accepted (cost-effective) value for n is 2. For rough quarry stone in two layers, $k_\Delta = 1.0$, according to the SPM.

The cap width (B_T in Figure 8.4) is determined from a formula similar to that in eq. 8.5, that is

$$B_T = Nk_\Delta \left(\frac{W_{1T}}{\rho_{stone} g} \right)^{1/3} \tag{8.6}$$

where N is the number of cap stones. The SPM recommends $N = 3$ for stability and to minimize the effects of overtopping.

The breakwater height (h_T, not sketched in Figure 8.4) depends on the amount of overtopping considered to be acceptable. In turn, the overtopping depends on both the runup, R_u, and the cap width, B_T. One formula for the structural height is

$$h_T = h_D + R_u \tag{8.7}$$

where h_D is the design water depth at the toe, equal to the sum of the mean low-water depth and one-half of the tidal range. The runup value depends on both the roughness and porosity of the weather face of the structure, as discussed in the SPM. Referring to the sketch in Figure 8.4, an alternative to the expression in eq. 8.7 is

$$h_T = h_D + H_D \tag{8.8}$$

where H_D is the design wave height.

We now have enough information to illustrate the preliminary design of a rubble-mound breakwater or a rubble-mound seawall.

EXAMPLE 8.3: PRELIMINARY DESIGN OF A RUBBLE-MOUND BREAKWATER The goal is to construct a rubble-mound breakwater on a bed having a slope (m) of 1/20 using the 100-year wave height as our design wave height in eq. 8.1. The worst-case condition of $\beta_0 = 0°$ is assumed because the wave heights will be maximum for this angle. The measured average deep-water wave property values are $H'_{avg0} = 1$ m and $T_{avg} = 8$ sec, where the prime (′) identifies shoaling without refraction. The deep-water wave heights and wavelengths are both found to have Rayleigh probability distributions, as discussed in Chapter 5. With this assumption, the relationship between the extreme wave height (H'_{max0}) and the average wave height (H'_{avg}) in eq. 5.25 is

$$H'_{max0}(100\,\text{years}) \equiv H'_{100} = \frac{2}{\sqrt{\pi}} H'_{avg} \sqrt{\ln(N_{100})} \tag{8.9}$$

Statistically, there are about 6×10^6 waves of engineering importance passing an ocean site each year. So over 100 years, the statistical sample (N_{100}) of waves is 6×10^8. The deep-water 100-year wave height in deep water (H'_{100}) is then approximately 5.07 m. Furthermore, the assumption of a Rayleigh distribution of the deep-water wave heights allows us to combine the expressions in eqs. 5.51 and 5.54 to obtain the following expression for the average wavelength:

$$\lambda_{avg0} = \frac{g}{2\pi}T_{rms}^2 \simeq \left(\frac{1.104}{1.064}\right)^2 \frac{g}{2\pi}T_{avg}^2 \simeq 1.077\frac{g}{2\pi}T_{avg}^2 \tag{8.10}$$

For the 8-sec average period, the average deep-water wavelength is approximately 108 m.

The decision is made to position the toe of the structure at the position of the breaking 100-year wave. To determine the breaking wave height and depth for this wave, we use the results in Figure 7.3 or eq. 7.5. In Figure 7.3, for $m = 1/20$ and $H'_{100}/\lambda_{avg0} = 0.0469$, we find $H'_{100b}/H'_{100} \simeq 1.130$. The breaking 100-year wave height is then $H'_{100b} \simeq 5.73\,\text{m}$. The breaking depth of $h_{100b} \simeq 6.33\,\text{m}$ is determined from either Figure 7.4 or eq. 7.5, for $H'_{100b}/\lambda_{avg0} \simeq 0.0531$. The breaking depth is used here as the design depth ($h_{D.}$) at the toe in Figure 8.4. The design height of the structure (from eq. 8.8) is 12.1 m above the bed at the toe, and the toe of the breakwater is approximately 127 m from the shoreline.

The design face of the structure has a slope if $\tan(\varepsilon) = 1/2$. This is a good value for both stability and cost-effectiveness. The breakwater is to be made of rough quarry stone (assume $\rho_{stone} = 2{,}650\ \text{kg/m}^3$ and $\rho_{stone}g \simeq 26{,}000\ \text{N/m}^3$), which is randomly placed. The value of the stability coefficient for the trunk of the structure for the face angle and stone is 2.0 for a breaking wave, according to the SPM. For breaking and nonbreaking waves on the face of the trunk's structure, the values of K_{DT} are 2.0 and 4.0, respectively. The breaking and nonbreaking values of K_{DT} for the heads of the breakwater are 1.6 and 2.8, respectively, for this face slope and stone, again according to the SPM. For breaking 100-year waves on the trunk, the armor stone weight is approximately 316,000 N or about 32.2 metric tonnes, from eq. 8.1. The shield, foundation, and toe-berm stone weights are obtained from eqs. 8.2 to 8.4.

Assuming a layer coefficient (k_Δ) value of 1.0 and a specific weight of $\rho_{stone}g = 26{,}000\ \text{N/m}^3$, the layer thickness is approximately $r_T \simeq 4.60\,\text{m}$ from eq. 8.5. For a three-stone cap, the cap width is $B_T \simeq 6.90\,\text{m}$ from eq. 8.6. If a leeward slope of 1 is used, then the structure will cover a bed length of 43.2 m if the bed is horizontal beneath the structure. See the sketch of the profile of the breakwater in Figure 8.5.

8.4 Reliability of a Rubble-Mound Structure

The term reliability is synonymous with the probability of success. To be more specific, the *reliability* of an engineering system stands for the probability that the system or system component will perform its specified function for a specified time (life of the system) under specified conditions. It is important to remember that the satisfactory performance, time period, and operating conditions must be clearly defined if reliability is to have a useful, quantitative meaning. For a thorough coverage of engineering reliability, the reader is referred to the book by Ramakumar (1993).

The concept of reliability is easily illustrated when applied to rubble-mound structures. Consider again the breakwater profile sketched in Figure 8.4. As we know, the structure is composed of stones of various classes based on the weight of the primary armor stone. Statistically, the design wave height is that which will ideally move no stones, if the armor is of uniform weight. In actuality, uniform stone

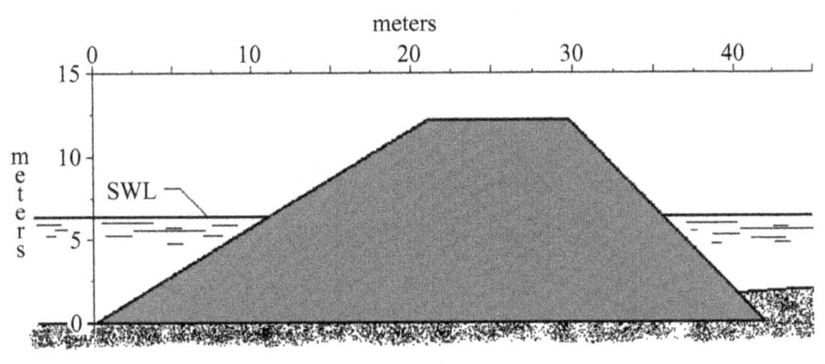

Figure 8.5. *Sketch of the Breakwater Profile of Example 8.3.*

size is not a reality as the armor is usually taken from quarries where the stone is obtained by blasting. Hence, there is a weight distribution of the armor stone, which approximately varies from 75% to 1.25% of the design weight (that in Hudson's formula, eq. 8.1). The designer can then expect that some of the lower-weight stone will be displaced when the design wave height passes. In fact, when Hudson did his original work, he stated that when the design wave height passed, about 5% of the stones of the cover layer would be displaced. Note that the failure mode for a rubble-mound structure is the displacement (not movement) of the primary armor stones on the cover layer.

The following example demonstrates how the reliability of a rubble-mound structure can be determined from scaled experiments in a wave tank.

EXAMPLE 8.4: RELIABILITY OF A RUBBLE-MOUND BREAKWATER A wave-tank study is to be performed on a 100th-scale model of the ocean type of a rubble-mound breakwater in Example 8.3. In that example, the mean weight of the stones on the face cover of the prototype is 164,000 N. The 5.54-m-wide cap of the breakwater is 5.80 m (the design wave height value in Figure 8.4) above the design water level. For this detached breakwater, the designer requires a total of N = 6,000 primary armor stones on the cover layer of the structure.

For the model, the length scale is

$$n_L \equiv \frac{L_m}{L_p} = \frac{1}{100} \tag{8.11}$$

where the subscripts m and p refer to the model dimension and the prototype characteristic lengths, respectively. Because water waves dominate the fluid environment, Froude scaling is used to determine the relationships between the other model and prototype properties. From eqs. 2.109 and 2.116, the Froude number equality is

$$\boldsymbol{F_r} = \frac{V_m}{\sqrt{gL_m}} = \frac{V_p}{\sqrt{gL_p}} \tag{8.12}$$

Dimensionally, $V = L/t$; so eq 8.12 can be rearranged to obtain

$$\frac{V_m/\sqrt{L_m}}{V_p/\sqrt{L_p}} = \frac{n_V}{\sqrt{n_L}} = \frac{(L_m/t_m)/\sqrt{L_m}}{(L_p/t_p)/\sqrt{L_p}} = \frac{\sqrt{n_L}}{n_t} = 1 \tag{8.13}$$

Here, n_V and n_t are the respective velocity-scale factor and time-scale factor. The time-scale factor is of most interest here as it determines the duration of

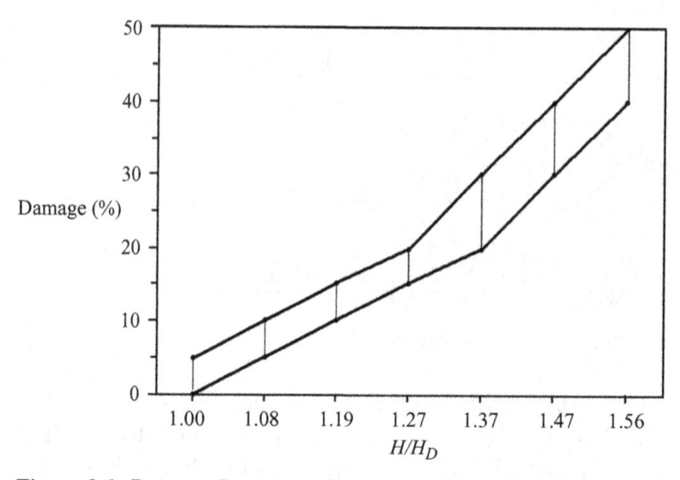

Figure 8.6. *Damage Range as a Function of Wave-Height Ratio.* The curves define the band of uncertainty for the exceedance of the design wave. The curves result from the data presented in the *Shore Protection Manual* (U.S. Army, 1984). *Damage* is defined as the percentage of primary armor stones that are permanently displaced from their as-constructed positions.

the experiment. For the 100th-scale model, $n_t = 1/10$. From this result, we see that each model time duration is a tenth of the prototype duration.

Results of one year of continual testing of the breakwater model in random waves are presented in Table 8.1. The equivalent prototype duration is 10 years from our time scale. The results presented in Table 8.1 are for each month of testing. In that table, the number of failures (f) is the number of model cover-layer armor stones displaced during one month of testing. These are added each month to obtain the cumulative number of failures (Σf). The number of survivors (s = N − Σf) is the next column. The next column is the reliability (R), defined as the number of *survivors* divided by the number of stones at the beginning of the study, that is, s/N. The final column in Table 8.1 is the probability of failure, which is P = 1 − R.

Concerning the armor stability, the data in Table 8.1 show that the worst month of the model study was month 12, corresponding to month 120 on the prototype scale, when 225 primary armor stones were displaced. The cumulative number of failures (885) over the study duration represents about 14.8% of the armor stone sample (6,000). For this breakwater, we could then specify that the normal repair time cycle is 10 years, where (statistically) less than 15% of the armor stone would be restored to the design position.

For the prototype, unpredictable severe storm seas might occur having a maximum wave height that exceeds the design wave height. Following these storms, repairs might be required.

Let us assume that we have been assembling reliability data for breakwaters from many experimental and prototype studies for many years. The data will have scatter, and therefore a resulting band of uncertainty, as in Figure 8.6. Statistical formulas are available to represent both the reliability and probability of failure. One of the most versatile probability formulas available to us is the Weibull probability formula (see Weibull, 1951) discussed in Section 5.5 of this book. Applied to

Table 8.1. *Reliability modal study for a rubble-mound breakwater*

t (month)	f	Σf	s	R(%)	P(%)
0	–	0	6,000	100	0
1	20	20	5,980	99.7	0.3
2	15	35	5,965	99.4	0.6
3	25	60	5,940	99.0	0.1
4	30	90	5,910	98.5	1.5
5	25	115	5,885	98.1	1.9
6	100	215	5,785	96.4	3.6
7	35	250	5,750	95.8	4.2
8	130	380	5,620	93.7	6.3
9	20	400	5,600	93.3	6.7
10	55	455	5,545	92.4	7.6
11	205	660	5,340	89.0	11.0
12	225	885	5,115	85.2	14.8

our problem, the two-parameter Weibull probability formula used to represent the cumulative probability of failure is

$$\mathbf{P}(t) = 1 - e^{-\left(\frac{t}{t_{ref}}\right)^m} \tag{8.14}$$

where m is called the shape parameter, and t_{ref} is some reference time referred to as the scale parameter. Note that if $m = 2$ and $t_{ref} = t_{rms}$, then the expression in eq. 8.14 is a Rayleigh probability, discussed in Section 5.4. The reliability at time t is then

$$\mathbf{R}(t) = 1 - \mathrm{P}(t) = e^{-\left(\frac{t}{t_{ref}}\right)^m} \tag{8.15}$$

Both m and t_{ref} are experimental, and are determined from a process similar to that in establishing Table 8.1.

EXAMPLE 8.5: WEIBULL RELIABILITY OF A RUBBLE-MOUND BREAKWATER The reliability expression in eq. 8.14 is applied to the experimental data in Table 8.1. To determine the two parameters (m and t_{ref}) in the expression, the 6- and 12-month values in the table are used. Our interest is in the prototype and not the model, so we must use the equivalent prototype times for the data, those being 60 and 120 months, respectively. For the prototype data, the scale parameter is $m \simeq 2.13$, and the shape parameter is $t_{ref} \simeq 284$ months. The data corresponding to those in Table 8.1 for the prototype and the results obtained from eqs. 8.14 and 8.15 are presented in Figure 8.7. As expected, the agreement between the results is good because the parametric values are based on the data in the table.

When two or more component reliabilities are considered in the design process, then the system reliability is the product of the reliabilities. In this section, only wave effects on rubble-mound breakwaters are considered. However, we could include tidal currents, storm surges, and a number of other natural events that could affect

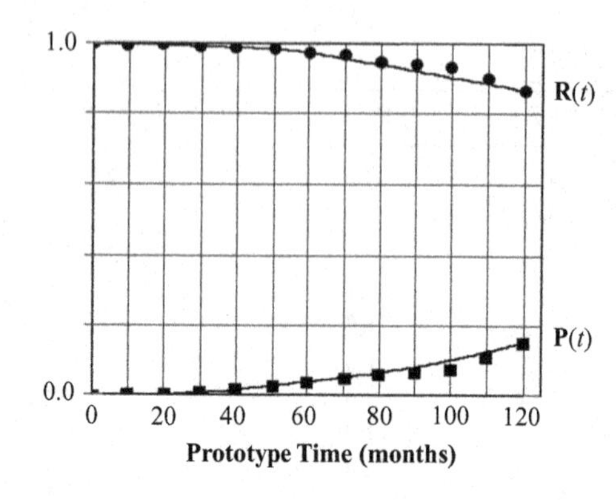

Figure 8.7. *Reliability and Probability of Failure for the Prototype in Example 8.4.* The solid lines represent the Weibull results from eqs. 8.13 and 8.14, whereas the reliability and probability values in Table 8.1 are represented by • and ■, respectively.

the stability of the structure. In that event, for I component reliabilities, the reliability of the system is

$$\mathbf{R} = \boldsymbol{R}_I \boldsymbol{R}_{I-1} \cdots \boldsymbol{R}_1 = \prod_{i=1}^{I} \boldsymbol{R}_i \tag{8.16}$$

A fact of design is that the cost of a system increases with the increase in reliability. Hence, in design we must have trade-offs, that is, we can sacrifice some reliability to keep the cost of the system down. For example, we could use a design wave height with a shorter return period and reduce the armor stone size. This would save money up front, but would add to the service costs later because the reliability of the structure would be reduced, that is, the stones would be displaced sooner. The construction costs are amortized over the life of the structure. Over the life, we can then have an average construction cost. As we increase the reliability of the structure, this cost will increase. The point to be made is that there is some value of reliability that is the most cost-effective.

8.5 Closing Remarks

This chapter is designed to give the reader a brief introduction to shore protection. The topics of decision making and reliability are included because these topics are normally excluded in coastal engineering publications. The primary reference used for this chapter is the *Shore Protection Manual* of the U.S. Army (1984). In addition to this basic design reference, the U.S. Army Corps of Engineers provides a series of Coastal Engineering Technical Notes that are available online. These technical notes are designed to provide information on recent advances in engineering and science related to the coastal zone.

An up-to-date discussion of the damage of stone breakwaters and revetments is presented by Melby (2002). In that reference, empirical formulas are presented to help designers of stone coastal structures. More recently, a two-volume handbook edited by Kim (2009) has become available, in which many of the aspects of coastal engineering are discussed in detail.

9 Wave-Induced Forces and Moments on Fixed Bodies

The majority of engineering problems encountered by ocean engineers involve wave-structure interactions. The waves in these interactions are altered and modified because of both the presence and motions of fixed and floating structures. The analyses of the wave-induced forces and motions of fixed structures require somewhat different mathematical tools than the analyses of wave-structure interactions of floating bodies. Wave-induced forces and motions of floating bodies are introduced in Chapter 10 and discussed in depth in Chapter 11.

The mathematical foundations on which contemporary analyses of wave-structure interactions are based date back to the nineteenth century when Stokes (1851) demonstrated that the total force on a body in an unsteady flow consisted of two components, those being a drag force and an inertial reaction of the fluid. The objects of the Stokes study were pendula (hanging circular cylinders and one with a spherical weight) moving in a viscous fluid where free-surface effects were not considered. Our interest begins with works done in the early part of the twentieth century, when most of the wave-structure interactions involved submerged horizontal circular cylinders. The orientations of the cylinders of interest were horizontal and fully submerged with their axes parallel to the wave fronts, as discussed by Havelock (1917), Lamb (1932), and others. In the Havelock and Lamb references, it is demonstrated how integral transforms and integral equations can be used to represent the free-surface response to a fully submerged, horizontal cylinder. Havelock (1917, 1926) also demonstrates the utility of the method of images in free-surface problems. Most of the significant works of Sir Thomas H. Havelock are reprinted in a volume edited by Wigley (1960).

In the mid-twentieth century, vertical cylinders, both fully submerged and surface-piercing, received much attention from the wave-structure analysts. For this problem, Havelock (1940) presents results of an analytical study of diffraction effects on surface-piercing structures of various cross-sections. A somewhat different approach to the vertical cylinder in waves is taken by MacCamy and Fuchs (1954). The results of that study were slightly modified by Mogridge and Jamieson (1976), and the resulting analytical data were shown to compare well with experimental data. Newman (1962) demonstrates how the forces on a compliant fixed body can be related to the potentials of the exciting incident waves and the radiated waves resulting from the wave-induced motions of the structure. The analysis resulting from the Newman (1962) study contributed to a number of later studies of fixed and floating bodies. Garrett (1971) presents an analysis of the forces on a truncated vertical cylinder in waves. Later, Yeung (1981) used the Garrett force approach to analyze the surging, heaving, and pitching motions of a truncated vertical cylinder in waves.

The study of fixed, submerged and floating bodies of arbitrary shape must be done by using numerical techniques. A series of papers dedicated to the numerical analysis of wave-structure interactions resulted from studies by Garrison (1974, 1975, 1978). Garrison's method is sometimes referred to in the literature as the "fat-body theory." The numerical approach increased in popularity as the memory capacities and speeds of both mainframe and desktop computers increased. Zienkiewicz, Lewis, and Stage (1978) and Schrefler and Zienkiewicz (1988) edited proceedings resulting from conferences devoted to the application of numerical techniques to most aspects of ocean engineering, including fixed and floating structures.

In this chapter, the analytical methods discussed are applied to bodies having rather simple geometries. The reason for this approach is to minimize the need for numerical methods. That is, the body geometries chosen to analyze herein will, for the most part, lead to either closed-form or *quasi*-closed-form analyses. The latter refers to analyses resulting in infinite series functions, such as Bessel functions and numerical integrations. Numerical methods are of great value in modeling problems in ocean engineering. However, they suffer as educational tools.

9.1 Wave-Induced Forces and Moments on a Seawall

In this section, two irrotational waves are used to determine the time-dependent pressures on vertical walls. These are the linear wave and the solitary wave. The former is chosen primarily for the purpose of demonstration, whereas the latter is chosen because of practicality. Also available to analysts is the nonlinear Stokes wave.

A. Pressure, Force, and Moment Resulting from Direct Reflection of Linear Waves

The topic of standing waves resulting from perfect wave reflection from vertical walls is discussed in Sections 3.4 and 6.1. In those discussions, we find that the wave height of the standing linear wave is twice that of the linear incident wave, but the wavelength is unchanged. In this section, the expressions for the wave pressure and the resulting force and moment on a seawall in waters of finite depth are presented.

Assuming that the flow beneath the free surface is irrotational, the pressure at any point in the flow can be obtained from Bernoulli's equation, eq. 2.70. In Section 3.1, we find that the time function, $f(t)$, can be equated to zero when the pressure on the free surface is atmospheric. Furthermore, in Example 3.1 it is shown that the V-squared kinetic energy term in Bernoulli's equation is of second order for waves of small steepness (H/λ), the assumption made here. With these assumptions, the pressure at any point in the flow is

$$p \simeq -\rho\frac{\partial\varphi}{\partial t} - \rho g z = \rho g\left[\frac{\cosh[k(z+h)]}{\cosh(kh)}\eta - z\right] \tag{9.1}$$

where the velocity potential is that in eq. 3.39,

$$\varphi = -\frac{H}{2}\frac{g}{\omega}\frac{\cosh[k(z+h)]}{\cosh(kh)}\cos(kx)\sin(\omega t) \tag{9.2}$$

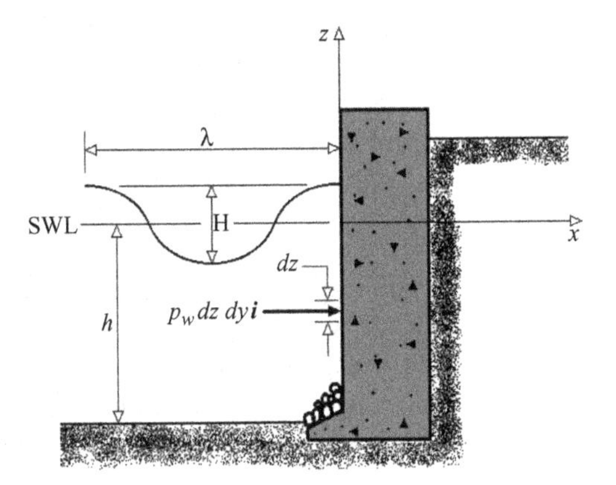

Figure 9.1. *Notation for a Standing Wave at a Seawall.*

and the free-surface expression for the standing wave is that in eq. 3.40,

$$\eta = \frac{\mathrm{H}}{2}\cos(kx)\cos(\omega t) \tag{9.3}$$

In these equations, H is the height of the standing wave, ω is the circular wave frequency, k is the wave number, g is the gravitational constant (9.81 m/s^2), and h is the water depth. When the pressure equation is applied at the free surface ($z = \eta$), we see that the resulting pressure is not equal to zero as it should be from the dynamic free-surface condition. The reason for this discrepancy is that the boundary conditions in eqs. 3.2 and 3.3 are mixed. That is, eq. 3.2 applies to the free surface, whereas eq. 3.3 applies to the SWL. When eq. 9.1 is applied to a standing wave in shallow water, the free-surface condition is satisfied.

Consider a standing linear wave at the seawall, as sketched in Figure 9.1. In that figure, the origin of the coordinate system is at the point of intersection of the SWL and the wall. The pressure at any point on the seawall is

$$p_w(z,t) = \rho g\left\{\frac{\cosh[k(z+h)]}{\cosh(kh)}\eta_w - z\right\} \tag{9.4}$$

from eq. 9.1. The subscript w identifies conditions at the wall. The resulting force on a width B of the wall is found by integrating the pressure expression of eq. 9.4 over the time-dependent wetted surface of the wall. The resulting force expression is

$$F_w(t)\boldsymbol{i} = B\int_{-h}^{\eta_w(t)} p_w(z,t)dz\boldsymbol{i} = \rho g B\left\{\frac{\sinh[k(h+\eta_w)]}{\cosh(kh)}\frac{\eta_w}{k} - \frac{\eta_w^2}{2} + \frac{h^2}{2}\right\}\boldsymbol{i} \tag{9.5}$$

See the sketch in Figure 9.1 for the notation. The corresponding moment about the mud line is obtained from

$$\begin{aligned} M_w(t)\boldsymbol{j} &= -B\int_{-h}^{\eta_w} p(z,t)_w(z+h)dz\boldsymbol{j} \\ &\simeq -\rho g B\left\{\frac{(h+\eta_w)\sinh[k(h+\eta_w)] + 1 - \cosh[k(h+\eta_w)]}{k\cosh(kh)}\eta_w \right. \\ &\quad \left. + \left[\frac{\eta_w^3}{3} + h\frac{\eta_w^2}{2} - \frac{h^3}{6}\right]\right\}\boldsymbol{j} \end{aligned} \tag{9.6}$$

where the positive moment direction is counterclockwise by convention. The unit vectors in the respective x-, y-, and z-coordinate directions are $\boldsymbol{i}, \boldsymbol{j}$, and $\boldsymbol{k}$.

The *center of pressure* is the point on the wall where the force obtained from eq. 9.5 can be applied to produce a moment equal to that obtained from eq. 9.6. From the moment-balance equation, the expression for the position of the center of pressure is

$$z|_{cp} \equiv -Z_w = -\frac{M_w(t)}{F_w} - h \tag{9.7}$$

The respective force and moment expressions in eqs. 9.5 and 9.6 can be simplified if the wave amplitude (H/2) is an order of magnitude less than the water depth at the wall. For this condition, the terms involving the higher powers of the free-surface displacement (η_w^2 and η_w^3) are of second order, and $h + \eta_w \simeq h$. To illustrate the applicability of the approximations, consider the following example.

EXAMPLE 9.1: WAVE FORCE AND MOMENT ON A SEAWALL DUE TO STANDING LINEAR WAVES A wave tank is 100 m in length, 3 m wide, and has a 2-m water depth for a particular study. The tank is equipped with an absorbing wave maker, that is, the wave maker not only creates the waves but also absorbs the waves reflected from opposite end. A wave maker of this type is described by Milgram (1970), among others. The depth of the tank is that in Example 3.5, where traveling waves having a height (H) of 0.5 m and a period (T) of 5 sec perfectly reflect from a vertical seawall. Those wave conditions are produced by the wave maker in this study. The wavelength from eq. 3.31 is approximately 20.9 m. Because the water depth-to-wavelength ratio $h/\lambda = 0.096 \simeq 1/10$, the conditions described are those of intermediate water. Thus, the hyperbolic functions in eqs. 9.5 and 9.6 must be retained. Assuming perfect reflection from a vertical seawall at $x = 0$, the standing wave has a height (H) of 1.0 m. The steepness of the incident traveling wave (H/λ) is 0.0239, and that of the standing wave is 0.0478.

We are to determine the force on the reflecting tank wall at the waterline ($z = 0$) obtained from eqs. 9.5 and 9.6, respectively, when a crest is at the wall, that is, when $\eta_w = \mathrm{H}/2 = 0.5$ m. Our interest is in the force values when the higher-order terms of η_w are both included and excluded. When the η_w terms are included, the maximum force is approximately 92,000 N, and when the terms are neglected, the maximum force is about 88,000 N, or a difference of approximately 4.5%. The percentage difference variation over one period is shown in Figure 9.2. The difference is defined as

$$\frac{\Delta F_w}{F_w} \equiv \frac{F_w - F_W}{F_w} \tag{9.8}$$

where the subscript W refers to the force excluding the higher-order η_w-terms. In Figure 9.2, the maximum difference of about 9.5% occurs at $t = T/2$, when the force values are both minimum.

B. Pressure and Force Resulting from Direct Reflection of a Solitary Wave

A thorough discussion of solitary waves is contained in the book by Wiegel (1964). In that book, Wiegel discusses both direct and oblique reflection of solitary waves, and the appearance of Mach-stem waves that occur when the incident wave angle is less than 45° (see Figure 6.3). In this section, our interest is in the forces produced by directly reflected solitary waves. One consequence of this wave reflection is the

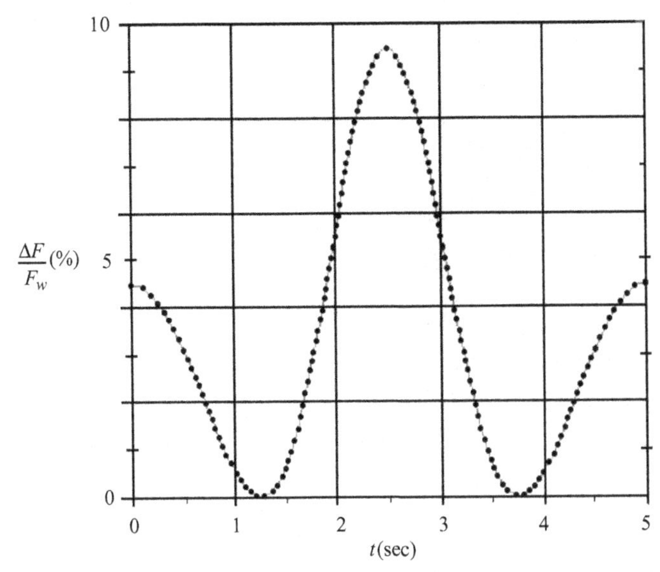

Figure 9.2. *Percentage Difference in the Wave-Induced Wall-Force Values over One Period in Example 9.1.* The forces in question are those obtained from eq. 9.5 with and without the higher-order η_w-terms.

migration of grounded ships, as discussed by Hudson (2001) and McCormick and Hudson (2001). Those authors find that the horizontal force on the mid-section of a broached ship produces migration if the waves are solitary but only rocking of the section occurs if the waves are either linear or Stokian. *Broaching* occurs when the ship's vertical center-plane (from stem to stern) is parallel to the wave front. This condition occurs when the ship loses power. The model used in the Hudson-McCormick studies was slightly embedded in sand in a two-dimensional wave tank.

In Section 4.5, we find that the solitary wave is a limiting case of the cnoidal wave. The former is discussed in Section 4.5B, whereas the latter is discussed in Section 4.5A. Both of these waves are based on the assumption that the flow is irrotational, and the velocity components (u, w) can be represented by a velocity potential, as in eq. 4.84. This assumption allows us to determine the pressure beneath the wave by using Bernoulli's equation, eq. 2.70. Because the solitary wave is infinitely long, we can assume that the term involving the time-derivative of the potential function is of second order compared to the velocity-squared and hydrostatic terms in Bernoulli's equation. We apply Bernoulli's equation to both the free surface $(z = \eta)$, where the pressure is zero-gauge, and at some submerged point $(z < \eta)$ to obtain

$$p + \frac{1}{2}\rho(u^2 + w^2) + \rho g z = \frac{1}{2}\rho g \left[u^2|_{z=\eta} + \left(\frac{d\eta}{dt}\right)^2 \right] + \rho g \eta \tag{9.9}$$

where z is positive above the SWL, as sketched in Figure 9.1.

For a solitary wave, the velocity components (u, w) are obtained from eqs. 4.126 and 4.127, respectively, the free-surface displacement (η) is given by eq. 4.124, and

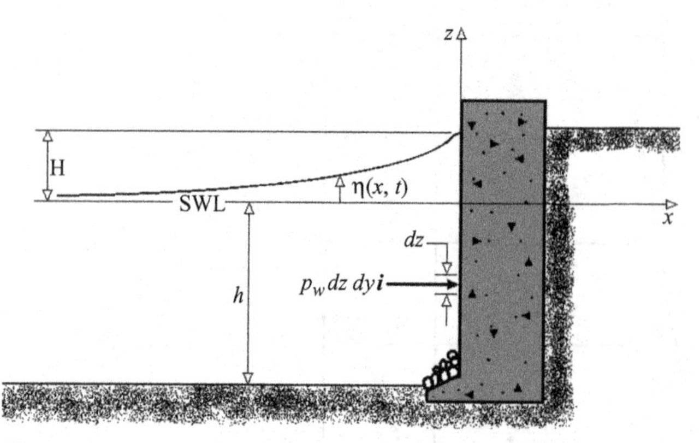

Figure 9.3. *Perfect Reflection of a Solitary Wave from a Vertical Seawall.* For the condition shown, a crest is at the wall. The wave height (H) there is equal to twice the height of the incident wave, that is, H $= 2H$.

the celerity by eq. 4.125. The horizontal velocity component is

$$u = \frac{\partial \varphi}{\partial x} = H\left(1 + \frac{H}{h}\right)\sqrt{\frac{g}{h}}\frac{1}{\cosh^2\left[\sqrt{\frac{3H}{4h^3}}(x - ct)\right]} \simeq H\sqrt{\frac{g}{h}}\frac{1}{\cosh^2\left[\sqrt{\frac{3H}{4h^3}}(x - ct)\right]} \tag{9.10}$$

which is independent of z, so in eq. 9.9, $u = u_\eta$ if the value of x is the same for both sides of the equation. The vertical velocity component is

$$w = \frac{\partial \varphi}{\partial z} = (z + h)\sqrt{\frac{3gH^3}{h^4}}\frac{\sinh\left[\sqrt{\frac{3H}{4h^3}}(x - ct)\right]}{\cosh^3\left[\sqrt{\frac{3H}{4h^3}}(x - ct)\right]} \tag{9.11}$$

The approximation in eq. 9.10 is based on the assumption that the height of the wave is much less than the water depth, that is, $H \ll h$. Again, we note that the horizontal component, u, is independent of the vertical coordinate, z, whereas the vertical component is a linear function of the coordinate. The approximate expression for the horizontal velocity component in eq. 9.10 and the vertical velocity component expression in eq. 9.11 satisfy the equation of continuity, as expressed by Laplace's equation, eq. 2.41.

When the solitary wave reflects from a vertical sea wall, the mirror-image method can be used to obtain the time-dependent wave profile at the wall. The expression in eq. 4.124 can be used to obtain the following free-surface deflection relationship at the wall:

$$\eta_w(t) = 2H\frac{1}{\cosh^2\left[\sqrt{\frac{3H}{4h^3}}ct\right]} = \mathrm{H}\frac{1}{\cosh^2\left[\sqrt{\frac{3H}{4h^3}}ct\right]} \tag{9.12}$$

where H is the height of the wave at the wall, as illustrated in Figure 9.3. At $t = 0$, a crest is at the wall, so $\eta_w = 2H = \mathrm{H}$ in eq. 9.12. As $t \to \infty$ in that equation, $\eta_w \to 0$.

In Bernoulli's equation for a solitary wave, eq. 9.9, the horizontal velocity components at the wall must be zero because, vectorially, we add the horizontal incident and image velocity components, which are equal but opposite in direction. In any event, because $u(x,t) = u_\eta(x,t)$ for any value of x, the terms will cancel as written in eq. 9.9. However, the vertical components are additive because they are both equal and in the same direction. The resulting pressure on the wall at $x = 0$ is obtained from

$$\begin{aligned} p_w &\simeq \frac{3}{2}\rho g \frac{H}{h^2}\left\{1 - \frac{(z+h)^2}{4h^2}\right\}\tanh^2\left[\sqrt{\frac{3H}{4h^3}}(ct)\right]\eta_w^2 + \rho g(\eta_w - z) \\ &\simeq \rho g\,(\eta_w - z) \end{aligned} \tag{9.13}$$

The last approximation is due to the assumption that the wave height is much less than the water depth, that is, $H \ll h$. With this assumption, the celerity in eq. 4.125 is approximately

$$c = \sqrt{gh}\left(1 + \frac{H}{2h}\right) \simeq \sqrt{g\,(h+H)} \simeq \sqrt{gh} \tag{9.14}$$

The celerity of a solitary wave is a linear function of the wave height if the magnitude of the wave height is within an order of magnitude of the wave depth. Before integrating the pressure expression in eq. 9.13 over the wetted face of the wall, the accuracy of the final approximation of that equation is demonstrated in the following example.

EXAMPLE 9.2: PRESSURE DISTRIBUTION ON A SEAWALL BENEATH A SOLITARY WAVE
Consider a 0.5-m, 15-sec wave at a seawall in 2 m of water. Our interest is in the accuracy of the final approximation in eq. 9.13 for the pressure on the wall being subjected to a solitary wave. From the linear theory approximation for the wavelength in eq. 3.38, the wavelength corresponding to the given wave period ($T = 15$ sec) and water depth ($h = 2$ m) is $\lambda \simeq 66.4$ m. Again, from eq. 3.38 and from the last approximation in eq. 9.14, the celerity of this wave is $c \simeq 4.43$ m/s. If the first approximation in eq. 9.14 is used, $c \simeq 4.95$ m/s for this wave (about a 10% difference).

Before applying the solitary theory or any of the wave theories discussed in Chapters 3 or 4, we should test the validity of the theories by using Figure 4.1. The non-dimensional parametric values in that figure are $H/\lambda_0 \simeq 0.00142$ and $h/\lambda_0 \simeq 0.00569$, where from eq. 3.36 the deep-water wavelength is $\lambda_0 \simeq 351$ m. In Figure 4.1, these values correspond to the long-wave region of the cnoidal theory, or approximately to the solitary theory. Hence, the use of the solitary theory is analytically valid. Note that the validity range of the linear and Stokian theories are well away from this point in Figure 4.1.

We shall determine the pressure as a function of time at the intersection of the SWL and the seawall, where $(x, z) = (0, 0)$. The results are presented in Figure 9.4 in normalized form, where $p_w(0,t)/p_{wmax}$ is presented as a function of t/T. The inclusion of the dynamic term in the first approximation in eq. 9.13 results in a peak pressure of 10,890 N/m^2 beneath the crest at the SWL, and the exclusion of the dynamic term in the second approximation yields 10,104 N/m^2. The difference in the two predicted pressures is less than 7.8% over the wave period. Because of this, we can use the second approximation in eq. 9.13 with some confidence.

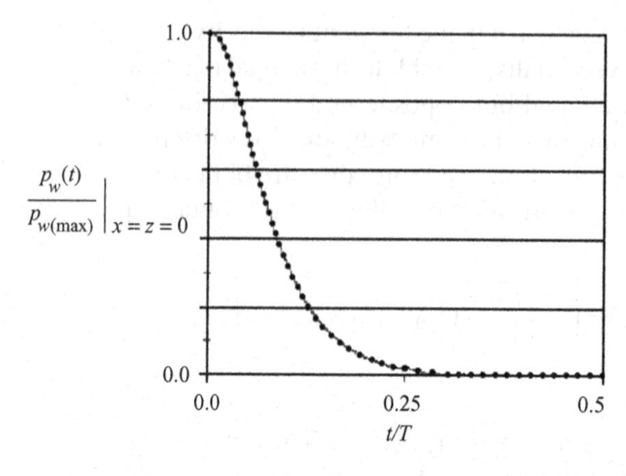

Figure 9.4. *Time-Variation of the Normalized Pressure at the SWL on a Seawall from Example 9.2.* The pressure predicted by the solitary theory is approximate for the wave conditions of the example. A finite period is a given in the example; however, the period for a solitary wave is infinite.

To obtain the force on a wall of width B located at $x = 0$, simply integrate the second approximation of the wall-pressure expression in eq. 9.13 over the wetted height of the wall, that is, from $z = -h$ to $z = \eta_w(t)$. The resulting time-dependent horizontal force expression is

$$F_w(t) = B\int_{-h}^{\eta_w} p_w(z,t)dz = \frac{1}{2}\rho g B(\eta_w + h)^2 \tag{9.15}$$

where the free-surface displacement is in eq. 9.12. Referring to Figure 9.1, the moment about the foot of the seawall (positive in the counterclockwise direction) is

$$M_w(t) = -B\int_{-h}^{\eta_w} p_w(z,t)\,[z+h]\,dz = -\frac{1}{6}\rho g B\,(\eta_w + h)^3 \tag{9.16}$$

The center of pressure for the solitary wave force is

$$z_{cp}(t) = \frac{M_w(t)}{F_w(t)} = -\frac{1}{3}(\eta_w + h) \tag{9.17}$$

This result is as expected from a *quasi*-hydrostatic pressure distribution over the wall.

As previously mentioned, the respective solitary-wave force and moment results in eqs. 9.15 and 9.16 are important in predicting the migration toward shore of a broached, grounded ship.

9.2 Wave-Induced Forces on Submerged and Surface-Piercing Bodies

In this section, we discuss the three types of wave-induced forces on rigid cylinders. Those are the inertial force, the pressure-drag force, and the diffraction force. The inertial force represents the inertial reaction of the added mass. Hence, the concept of the added mass is discussed first.

A. The Concept of Added Mass

Before discussing wave-induced forces *per se*, it is necessary to introduce the concept of *added mass*. This mass is that of the ambient water affected by the presence of a

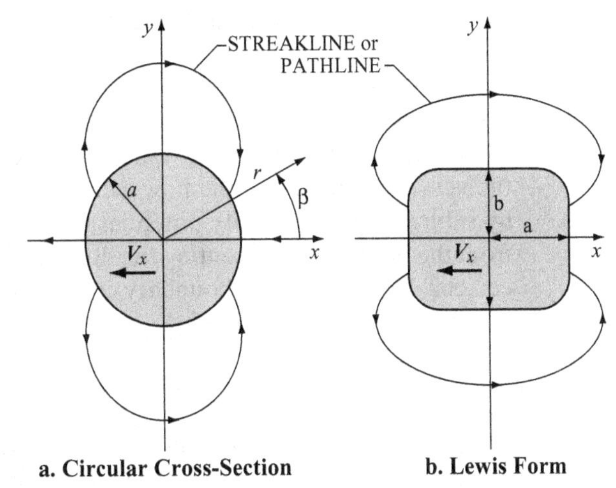

Figure 9.5. *Two-Dimensional Cylindrical Cross-Sections Accelerating in Stationary Fluids.* The sketch in Figure 9.5b is an example of a Lewis form, discussed in Section 9.2A(2).

fixed structure in a moving fluid, or that affected by the motions of a structure in a stationary fluid. For dynamics problems in air, this mass can normally be neglected because the mass density of air (about 1.225 kg/m^2 under standard conditions) is relatively small compared to that of the structural materials. However, in water this is not the case. The sum of the added mass and the moving body mass is sometimes referred to as the *virtual mass*.

Two-dimensional added-mass expressions for two geometries are discussed. The first is the circular-shell geometry sketched in Figures 9.5a and 9.6. The choice of this geometry for the cross-section of a structural element is based on two facts. First, circular cylinders are relatively easy to fabricate. Second, the strength-to-material-volume ratio is relatively large. Both of these make the circular, cylindrical-shell structure cost-effective in applications to marine structures.

The second geometry is referred to as a *Lewis form*. This form is actually a class of geometries that are symmetric in both the flow (or motion) direction and normal to that direction, as described by Lewis (1929). An example of a Lewis form is sketched in Figure 9.5b.

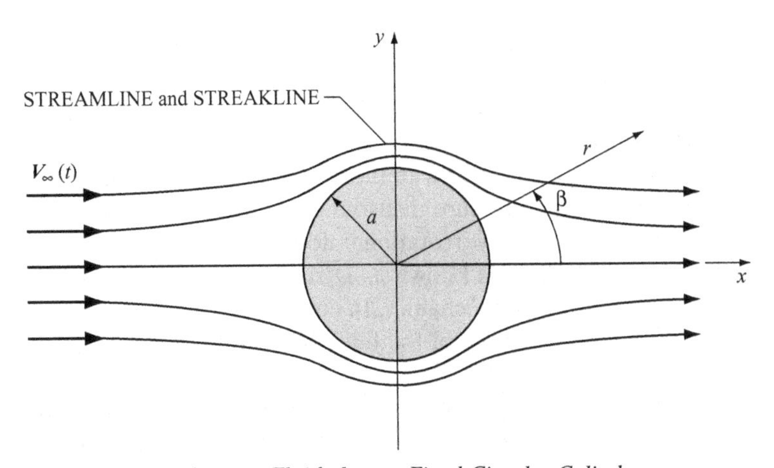

Figure 9.6. *Accelerating Fluid about a Fixed Circular Cylinder.*

(1) Cylinders with Circular Cross-Sections

Consider the sketch in Figure 9.5a. In that figure, a two-dimensional circular cylinder (infinite in length) of radius a is shown accelerating in a standing fluid. The velocity of the cylinder at any time is $V_0(t)$ in the negative x-direction. The expression for the velocity potential of the flow generated by the motion of the cylinder is found by subtracting the velocity potential for a uniform horizontal flow (in Figure 2.9) from the potential representing the flow past a circular cylinder in a uniform flow in eq. 2.62. By applying the boundary condition $\partial\varphi/\partial r|_{r=a,\beta=0} = -V_0(t)$, the following potential expression is found:

$$\varphi = +V_0(t)\frac{a^2}{r}\cos(\beta) \tag{9.18}$$

where r is the radial coordinate having its origin at the center of the cylinder, and β is the angular coordinate measured positively in the counterclockwise direction from the x-axis. Note that in Figure 2.9, the angle θ is used in place of β.

We can use the Cauchy-Riemann relationships of eqs. 2.58 and 2.59 to obtain the following expression for the stream function for this flow:

$$\psi = -V_0(t)\frac{a^2}{r}\sin(\beta) \tag{9.19}$$

Because our analysis is in the Lagrangian frame of reference (as opposed to the Eulerian frame), a zero-value of the stream function does not define the surface of the structure. The positive sign (+) in eq. 9.18 and the negative sign (−) in eq. 9.19 are associated with the cylinder's motion in the negative x-direction. For travel in the positive x-direction, the signs are reversed.

The kinetic energy (per unit depth into the page) of the excited ambient fluid is obtained from

$$E_k' = \frac{1}{2}\rho\int_a^\infty\int_0^{2\pi}(\nabla\varphi)^2 r\,d\theta\,dr = \frac{1}{2}\rho\pi a^2 V_0^2(t) = \frac{1}{2}\,\mathrm{a}_{w1}' V_0^2(t) \tag{9.20}$$

where a_{w1}' is the added mass per unit length of the accelerating cylinder, that is,

$$\mathrm{a}_{w1}' = \rho\pi a^2 \tag{9.21}$$

As is throughout the book, the prime (′) is used to indicate "per unit length" of the cylinder. Physically, the added mass in this case is equal to the mass of displaced fluid per unit length of cylinder.

The derivation of the expression for the added mass of a cylinder in an accelerating fluid is quite different than the derivation for the accelerating cylinder in a stationary fluid. Rather than being concerned with the kinetic energy of the ambient fluid, our attention is focused on the pressure distribution on the cylinder and the resulting force. For a structure in a steady irrotational flow, the in-line (in the flow direction) force on a body is zero. This is known as *D'Alembert's paradox*, and is discussed in detail in most texts on fluid mechanics. In an accelerating flow, there is a net force because the time rate of change of the fluid momentum is not equal to zero.

Referring to the sketch in Figure 9.6, the velocity potential for the unsteady flow about a fixed circular cylinder of radius a in an accelerating fluid is found by combining the potentials for a doublet and a uniform (accelerating) flow. The component potential functions are presented in Figure 2.9. The resulting potential function for accelerating flow about the fixed cylinder is

$$\varphi = V_\infty(t)\left[r + \frac{a^2}{r}\right]\cos(\beta) \tag{9.22}$$

Here, $V_\infty(t)$ is the velocity at $x = \pm\infty$. Combine this relationship with the unsteady pressure expression of the linearized Bernoulli's equation applied on the cylinder. The result is

$$p|_{r=R} = -\rho\frac{\partial\varphi}{\partial t}|_{r=A} = 2\rho a\cos(\beta)\frac{dV_\infty}{dt} \tag{9.23}$$

The positive sign (+) in eq. 9.22 and the negative sign (−) in eq. 9.23 are associated with the fluid traveling in the positive x-direction. When the direction of the flow is reversed, the signs are reversed. Integrate the pressure around the cylinder to obtain the force on the cylinder. The resulting expression is

$$F_x' = -\int_0^{2\pi} p(r,\beta,t)\cos(\beta)a\,d\theta = 2\rho\pi a^2\frac{dV_\infty}{dt} = \text{a}_{w2}'\frac{dV_\infty}{dt} \tag{9.24}$$

The added mass per length of cylinder is then

$$\text{a}_{w2}' = 2\rho\pi a^2 = 2\text{a}_{w1}' \tag{9.25}$$

The added mass in this case is twice that for the situation sketched in Figure 9.5a.

As in any analysis of fluid flows, we try to find non-dimensional numbers that allow us to compare experimental and prototype results. For the added mass, let a_{w_1}' be the reference added mass, and define the added-mass coefficient or *inertial coefficient* as

$$C_i \equiv \frac{\text{a}_w'}{\text{a}_{w1}'} = \frac{\text{a}_w'}{\rho\pi a^2} \tag{9.26}$$

The respective combinations of eqs. 9.21 and 9.25 with eq. 9.26 yield $C_{i1} = 1$ and $C_{i2} = 2$. From the results presented by both Wiegel (1964) and Ippen (1966), experimental values of the inertial coefficient vary from 0.4 to 4.0. This uncertainty is due to the quality of the experimental data.

(2) Cylinders with Noncircular Cross-Sections – Lewis Forms

In this subsection, the analytical method of Lewis (1929) is used to determine the two-dimensional added-mass expressions for cylinders having noncircular cross-sections that are symmetric with respect to the coordinate axes, such as that in Figure 9.5b. In the Lewis paper, the analyses of the inertia of the ambient fluid surrounding two- and three-dimensional bodies are presented. The focus of the paper is on vibrating ship hulls; however, the analytical method of the referenced paper has been found to have a wider applicability. The Lewis analysis incorporates the conformal mapping technique in the broader area of complex variables. For example, see the books by Milne-Thomson (1960), Robertson (1965), Vallentine (1967), Schinzinger and Laura (1991), Mei (1995), and Ablowitz and Fokas (1997), among others, for discussions of conformal mapping and applications to fluid flows.

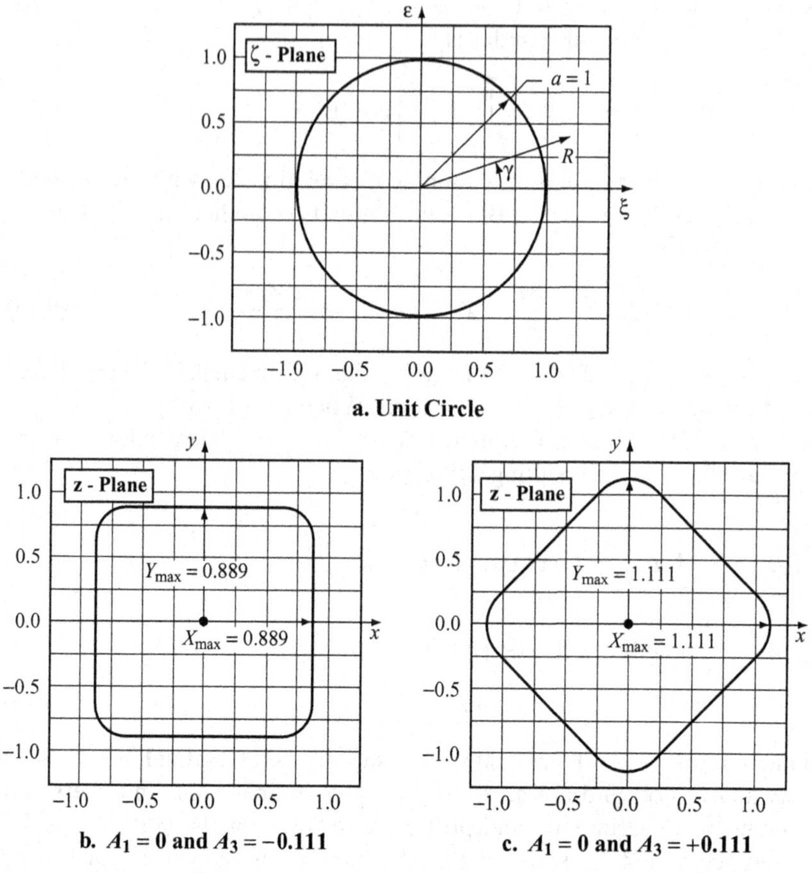

Figure 9.7. *Transformed Geometries in Example 9.3.* These figures in the z-plane are obtained by transforming a unit circle in the ζ-plane, according to eqs. 9.32 through 9.34.

Following the analysis of Lewis (1929), expressions of the velocity potential (φ) and the stream function (ψ) are derived. These expressions can then be used to determine the kinetic energy of a body moving in a standing fluid. The resulting energy expressions are used to determine the expressions for the added mass.

To determine the velocity potential and the stream function, we use the method of complex variables, where the complex variables are in two planes. We refer to these planes as the z-plane (the physical plane) and the ζ-plane (the transformed plane). Note that the non-italicized z is used to represent the complex variable, and the italicized z is a coordinate. The z-plane contains the body shape of interest, whereas the ζ-plane contains the flow about a circle. The complex spatial variable in the z-plane is

$$\mathrm{z} = x + iy = re^{i\beta} \tag{9.27}$$

That is, the x-axis is the real axis and the y-axis is the imaginary axis in the z-plane, as in Figures 9.7a and 9.7b. Similarly, the complex spatial variable in the ζ-plane is

$$\zeta = \xi + i\varepsilon = Re^{i\gamma} \tag{9.28}$$

Here, the ξ-axis is the real axis and the ε-axis is the imaginary axis in the ζ-plane, as in Figure 9.7a.

Our interest is in relating the velocity potential (φ_z) and the stream function (ψ_z) in the z-plane to those respective functions in the ζ-plane, i.e. φ_ζ and ψ_ζ. See Section 2.3 for a discussion of the natures of these functions. The complex potential in the ζ-plane is defined as

$$w_\zeta = \varphi_\zeta + i\psi_\zeta = f_\zeta(\zeta) \tag{9.29}$$

and that in the z-plane is

$$w_z = \varphi_z + i\psi_z = f_z(z) \tag{9.30}$$

In reality, we have created two additional planes called *potential planes*, where the velocity potential in each of these planes is the real axis and the stream function is the imaginary axis. The advantage of using a conformal mapping technique in any fluid analysis is due to the fact that the potentials in the two planes are equal in form.

The process of Lewis (1929) is to first map a *unit circle* in the ζ-plane (Figure 9.7a) onto the z-plane by specifying the functional relationship between z and ζ. Then, the same is done with the w-planes. Finally, the transformed geometry is mapped onto the potential plane (w_z-plane) to obtain the velocity potential and stream function for the flow in the z-plane. For a unit circle, the geometry of the surface in the ζ-plane (Figure 9.7a) is obtained from

$$\sqrt{\xi^2 + \varepsilon^2} = a = 1 \tag{9.31a}$$

On the cylinder, the complex coordinate is then

$$\zeta|_{R=a=1} = R|_{a=1}\, e^{i\gamma} = e^{i\gamma} = \cos(\gamma) + i\sin(\gamma) \tag{9.31b}$$

where γ is the angle measured positively from the ξ-axis and R is the radial coordinate measured from the origin. Note that ζ has units of length, although no length appears in the last two terms. In the following, we shall delay assuming $a = 1$ to avoid confusion later in the added-mass derivation.

The *Lewis transformation* from the ζ-plane to the z-plane is

$$z = x + iy = \zeta + \frac{A_1}{\zeta} + \frac{A_3}{\zeta^3} \tag{9.32}$$

where A_1 and A_3 are real numbers. The expression in eq. 9.32 is a special form of the Laurent series. For example, see Ablowitz and Fokas (1997) for a discussion of this series. The subscripts 1 and 3 in the constants correspond to the powers of ζ in the denominators. The transformation of the circle is then

$$\begin{aligned} z|_a = (x+iy)|_a &= \left[\zeta + \frac{A_1}{\zeta} + \frac{A_3}{\zeta^3}\right]\Bigg|_a = \left[ae^{i\gamma} + \frac{A_1}{a}e^{-i\gamma} + \frac{A_3}{a^3}e^{-3i\gamma}\right] \\ &= \left[\left(a + \frac{A_1}{a}\right)\cos(\gamma) + \frac{A_3}{a^3}\cos(3\gamma)\right] + i\left[\left(a - \frac{A_1}{a}\right)\sin(\gamma) - \frac{A_3}{a^3}\sin(3\gamma)\right] \end{aligned} \tag{9.33}$$

In eqs. 9.32 and 9.33, A_1 and A_3 are parametric constants that determine the body shape. From eq. 9.33, the body shape in the physical plane has the following coordinates:

$$\begin{aligned} x|_a \equiv X &= \left(a + \frac{A_1}{a}\right)\cos(\gamma) + \frac{A_3}{a^3}\cos(3\gamma) \\ y|_a \equiv Y &= \left(a - \frac{A_1}{a}\right)\sin(\gamma) - \frac{A_3}{a^3}\sin(3\gamma) \end{aligned} \tag{9.34}$$

Geometries represented by eq. 9.34 are called *Lewis forms*. Note that the angle γ in the ζ-plane is simply a parameter in the z-plane relating the body coordinates, X and Y. Letting $\gamma = 0$ and $\gamma = \pi/2$, respectively, the longitudinal and transverse maxima are found to be

$$X_{max} = a\left(1 + \frac{A_1}{a^2} + \frac{A_3}{a^4}\right)$$
$$Y_{max} = a\left(1 - \frac{A_1}{a^2} + \frac{A_3}{a^4}\right) \tag{9.35}$$

When both A_1 and A_3 equal zero in eqs. 9.32 through 9.35, then the transformed geometry is a circle, as in the ζ-plane. We can specify the values of a, A_1, and A_3 to obtain the desired body shape. There are practical ranges of the parametric constants, A_1 and A_3, which are discussed in Chapter 11. The transformation from the ζ-plane to the z-plane is illustrated in the following example.

EXAMPLE 9.3: TWO LEWIS FORMS Consider the case where $a = 1$ and the constants A_1 and A_3 in eqs. 9.32 through 9.35 have respective values of 0 and -0.111. The body in the z-plane is presented in Figure 9.7b. When the respective values of A_1 and A_3 are 0 and $+0.111$, the body resulting from those values is presented in Figure 9.7c. In both cases, we have a squarish cylinder with rounded edges. The body in Figure 9.7c has lines of symmetry at 45° to the x-axis.

We note that below the x-axis, the bodies in Figure 9.7b and 9.7c resemble ship hull sections. More is written of this in Chapter 11.

For a two-dimensional cylinder of arbitrary cross-section in a current of velocity $V_\infty(t)$ in the positive x-direction, there are two relationships that can be chosen to relate the complex potentials in eqs. 9.29 and 9.30 with the respective complex variables, ζ and z. The first of these relationships is in the ζ-plane, where the circular cylinder is fixed in a fluid moving in the ξ-direction (similar to that in Figure 9.6). The complex potential describing the flow about the cylinder in the ζ-plane is

$$w_\zeta = \varphi_\zeta + i\psi_\zeta = V_\infty\left(\zeta + \frac{a^2}{\zeta}\right) = V_\infty\left[\left(R = \frac{a^2}{R}\right)\cos(\gamma) + i\left(R - \frac{a^2}{R}\right)\sin(\gamma)\right] \tag{9.36}$$

The body is fixed in a flow that is parallel to the respective ξ-axis at $\zeta = \pm\infty$. Our interest is in the case where the body in the z-plane is moving in a still fluid. The last equality in eq. 9.36 results from the combination of eqs. 2.57 and 2.64 for the respective velocity potential and stream function. Note that the value of the stream function, ψ_ζ, is zero because $R = a$. The stream function is also equal to zero when $\gamma = 0$.

When the body in the z-plane moves at a speed, $V_0(t)$, in the negative x-direction (similar to those in Figure 9.5b), the complex potential is the difference in the complex potential in eq. 9.36 and that representing the horizontal flow in the z-plane, which is $V_0(t)z$. See Figure 2.9a for the velocity potential and stream function describing a parallel flow. In the z-plane, the complex potential in eq. 9.30 is

$$\begin{aligned} w_z = \varphi_z + i\psi_z = w_\zeta - V_0 z = V_0\left(\zeta + \frac{a^2}{\zeta}\right) - V_0\left(\zeta + \frac{A_1}{\zeta} + \frac{A_3}{\zeta^3}\right) \\ = V_0\left[(a^2 - A_1)\frac{1}{\zeta} - \frac{A_3}{\zeta^3}\right] = V_0\left[\frac{1}{R}(a^2 - A_1)\cos(\gamma) - \frac{A_3}{R^3}\cos(3\gamma)\right] \\ + iV_0\left[\frac{1}{R}(a^2 - A_1)\sin(\gamma) - \frac{A_3}{R^3}\sin(3\gamma)\right] \end{aligned} \tag{9.37a}$$

The flow adjacent to the body is found by applying this equation to $r = a$. The result is

$$\begin{aligned} w_z|_{R=a} = (\varphi_z + i\psi_z)|_{R=a} = V_0\left[(a^2 - A_1)\frac{1}{\zeta} - \frac{A_3}{\zeta^3}\right]\Bigg|_{R=a} \\ = V_0\left[\left(a - \frac{A_1}{a}\right)\cos(\gamma) - \frac{A_3}{a^3}\cos(3\gamma)\right] \\ + iV_0\left[\left(a - \frac{A_1}{a}\right)\sin(\gamma) - \frac{A_3}{a^3}\sin(3\gamma)\right] \end{aligned} \tag{9.37b}$$

From eq. 9.37b, the respective velocity potential and stream function expressions for the flow adjacent to the body in the z-plane are

$$\varphi_z|_{R=a} = V_0(t)\,a\left[\left(a - \frac{A_1}{a^2}\right)\cos(\gamma) - \frac{A_3}{a^4}\cos(3\gamma)\right] \tag{9.38}$$

and

$$\psi_z|_{R=a} = -V_0(t)\,a\left[\left(1 - \frac{A_1}{a^2}\right)\sin(\gamma) - \frac{A_3}{a^4}\sin(3\gamma)\right] \tag{9.39}$$

In eqs. 9.38 and 9.39, the $+$ and $-$ signs are associated with the motion of the Lewis form in the negative x-direction. For motion in the positive x-direction, the signs are reversed.

To determine the added mass of the cylinder, the expressions in eqs. 9.38 and 9.39 are combined with that for the kinetic energy of the ambient fluid. In eq. 9.20, the kinetic energy expression for a moving circular cylinder is given. Following Lamb (1932), Milne-Thomson (1960), and others, the area integral for the kinetic energy can be equated to a line integral expression by using Green's theorem of Appendix C, the Cauchy-Riemann relationships of eqs. 2.51 and 2.52, and the integral relationships in eq. 2.43. The resulting integral relationships are

$$\begin{aligned} \iint_S (\nabla\varphi_z)^2\, dS \equiv \iint_S \nabla\varphi_z \cdot \nabla\varphi_z dS \\ = -\iint_S \varphi_z \nabla^2 \varphi_z dS - \int_C \varphi_z \frac{\partial \varphi}{\partial n} dC = 0 - \int_C \varphi_z d\psi_z \end{aligned} \tag{9.40}$$

where S is the area of the fluid and C is the contour of the boundary of the fluid. The cylinder in question is assumed to be in an infinite fluid, so C is simply the line defining the cylinder. The last term in eq. 9.40 results from the Cauchy-Riemann equations of eqs. 2.51 and 2.52, where the orthogonal coordinates on the body are C and n. Because of eq. 9.40, we can express the kinetic energy transferred from the moving body to the ambient fluid (per unit length of cylinder) as

$$E'_k|_{a=1} = \frac{1}{2}\rho \iint_S (\nabla\varphi_z)^2 \, dS|_{R=a} = -\frac{1}{2}\rho \int \varphi_z d\psi_z|_{R=a} = -\frac{1}{2}\rho \int_0^{2\pi} \varphi_z \frac{d\psi_z}{d\gamma} d\gamma|_{R=a}$$

$$= \frac{1}{2}\rho V_0^2(t)\pi a^2 \left[\left(1 - \frac{A_1}{a^2}\right)^2 + 3\frac{A^2}{a^4}\right] = \frac{1}{2}a'_w V_0^2(t) \qquad (9.41)$$

where a'_w is the added mass per unit length of the cylinder. From the last equality in eq. 9.41, the expression for the added mass per unit length for the Lewis form is

$$a'_w = \rho\pi a^2 \left[\left(1 - \frac{A_1}{a^2}\right)^2 + 3\frac{A_3^2}{a^4}\right] \qquad (9.42)$$

The reader can see that when $A_1 = A_3 = 0$, the transformed circle is also a circle of radius a, and the added-mass expression is the same as that in eq. 9.21.

The application of the added-mass expression to a cross-section having a transverse (normal to V_0) dimension, $B_1 = 2Y_{max}|_{a=1}$, is as follows: Following Lewis (1929), first let $a = 1$ in eq. 9.42. Then, multiply both the velocity potential and stream function expressions of eqs. 9.38 and 9.39 by $B_1/2Y_{max}|_{a=1}$ to change the length scale. The added mass per unit length for a Lewis form is obtained by multiplying the expression in eq. 9.42 by $(B_1/2Y_{max}|_{a=1})^2$. The result is

$$a'_w|_{B_1} = \rho\pi\left[(1 - A_1)^2 + 3A_3^2\right]\left(\frac{B_1}{2Y_{max}|_{a=1}}\right)^2 = \rho\pi B_1^2 \frac{\left[(1 - A_1)^2 + 3A_3^2\right]}{4(1 - A_1 + A_3)^2} \qquad (9.43)$$

Here, $Y_{max}|_{a=1}$ in the first equality is replaced by the expression in eq. 9.35, where $a = 1$. The inertial coefficient for a Lewis form from eq. 9.26 is then

$$C_i \equiv \frac{a'_w|_{B_1}}{a'_{w1}} = \frac{a'_w|_{B_1}}{\rho\pi\left(\frac{B_1}{2}\right)^2} = \frac{(1 - A_1)^2 + 3A_3^2}{(1 - A_1 + A_3)^2} \qquad (9.44)$$

The application of Lewis forms to floating bodies is discussed in Chapter 11. In that chapter, we find that the added mass for such oscillating bodies depends on the frequency of oscillation.

In the following example, an application of the method of Lewis (1929) for determining the added mass is presented.

EXAMPLE 9.4: ADDED MASS OF A NONCIRCULAR CYLINDER The cylinders in Figures 9.7b and 9.7c are one of four legs supporting a semi-submergible offshore platform in the open ocean where the mass density (ρ) of the salt water is 1,030 kg/m^3. Our interest is in the added mass per unit length of each leg when the legs move in the x-direction. The transverse dimension (B_{1a}) of the leg is 4 m in Figure 9.7b, and the section of the leg that is of interest is well away from the free surface.

(a) For the leg in Figure 9.7b moving in the negative x-direction, the constants are $A_1 = 0$ and $A_3 = -0.111$. The added mass per unit length of the leg from eq. 9.43 is approximately 16,980 kg/m.

(b) When the leg orientation is at 45° to the X-direction, then $A_1 = 0$ and $A_3 = 0.111$. The transverse dimension (B_{1b}) is not the same as in case (a).

Rather, it is $B_{1a}/\cos(45°) \simeq 5.66\,\text{m}$. The approximate added-mass value for this condition is 21,770 kg/m.

The respective added mass or inertia coefficient values for cases (a) and (b) are approximately 1.312 and 0.840 from eq. 9.44.

In this discussion, we have illustrated an application of the Lewis (1929) method of determining both the geometry and added mass of noncircular bodies. Our application of the Lewis method is to bodies in an infinite still fluid. Lewis applied his method to ship shapes piercing the free surface by determining the added mass of the lower half of the bodies defined by the conformal transformations. The motions of the body were assumed to have no effect on the free surface. This corresponds to the low-wave-frequency condition, $\omega \to 0$, discussed by Frank (1967) and others, where the free surface corresponds to one of the coordinate axes. Brennen (1982) presents a rather thorough review of the transformation techniques used in the determination of the added-mass and the inertia coefficients, such as that of Lewis (1929). There are also limitations on the Lewis transformation, as shown by von Kerczek and Tuck (1969). The limiting conditions are presented in Chapter 11. Finally, the added-mass expressions derived in this section are independent of the frequency of the body motion. The frequency-dependent added mass is discussed in this chapter and in Chapters 10 through 12.

B. Natures of Wave-Induced Forces on Circular Cylinders

When water waves pass a circular cylinder of diameter D $(= 2a)$, the magnitude of the wave-induced force depends primarily on two length ratios, D/λ and D/H. The components of the wave force are the *viscous pressure force* (due to a combination of the boundary layer and the wake), *inertial force* (due to the acceleration of the water, the cylinder, or both), and *diffraction force* (due to scattering – a combination of wave reflection and diffraction of the waves). The realms of these forces are shown in Figure 9.8, which is due to Chakrabarti (1975). In that figure, the horizontal coordinate is the product of the wave number (k) and the radius (a), that is, $ka = \pi D/\lambda$, whereas the vertical coordinate is called the *Keulegan-Carpenter number*, defined as

$$KC \equiv \frac{u_{\max} T}{D} \tag{9.45}$$

where u_{max} is the amplitude of the normal (horizontal) velocity – normal to the centerline of the cylinder. As discussed by Woodward-Clyde (1980), the percentages of the total force that is drag shown in Figure 9.8 are those near the free surface. These percentages change with depth in deep-water waves.

The parameter in eq. 9.45 resulted from experimental studies by Keulegan and Carpenter (1958) at the National Bureau of Standards. To apply the Keulegan-Carpenter number to a vertical or horizontal cylinder positioned normal to a wave front, the velocity u_{max} is the amplitude of the particle velocity at a point under investigation. For a vertical pile in linear waves, the velocity is the maximum horizontal particle velocity at the crest of the wave, obtained from eq. 3.49. For this

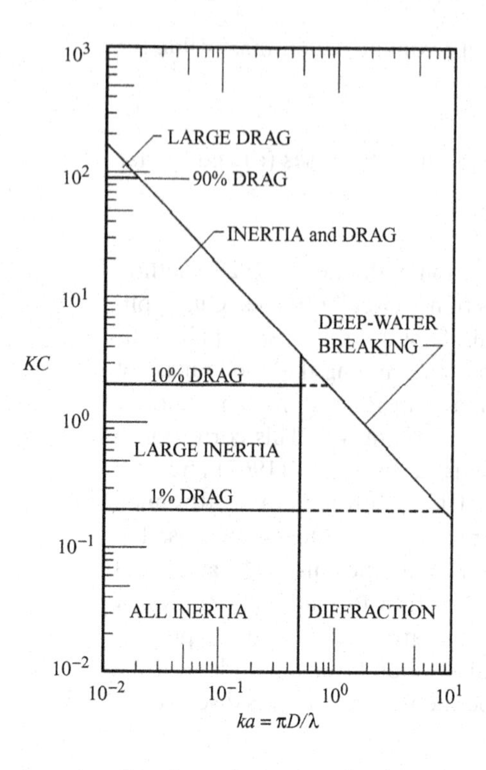

Figure 9.8. *Wave-Induced Force Component Realms.* This chart is of value in the conceptual design phase for offshore structures. The Keulegan-Carpenter number (KC) is defined in eq. 9.45. The figure is due to Chakrabarti (1975). The figure shown is valid in the neighborhood of the free surface. The drag percentage for a given KC value will change with depth, as discussed in the Woodward-Clyde (1980) report.

situation, the approximate expressions for KC in deep water and shallow water, respectively, are

$$KC|_{deep} \simeq \frac{\pi H_0}{D} = \frac{2\pi H_0}{a} \tag{9.46}$$

and

$$KC\,|_{shallow} \simeq H\frac{T}{2D}\sqrt{\frac{g}{h}} \tag{9.47}$$

In the following sections, the wave-induced drag, inertia, and diffraction forces are discussed in terms of the Keulegan-Carpenter number.

C. Wave-Induced Drag Forces

Consider the case of a vertical circular pile piercing the free surface where ka is small and the Keulegan-Carpenter number (KC) is large. Physically, the former indicates that the wavelength (λ) is much greater than the radius (a) whereas the latter can be interpreted as the wave height (H) being much greater than the radius. A vertical, surface-piercing mooring line under tension is an example of a structure satisfying these conditions. From the results in Figure 9.8, the wave-induced force is primarily a drag force, consisting of frictional drag and wake pressure drag.

The empirical analysis of the drag force on a cylinder in waves is more complicated than that for a cylinder in a steady flow – the latter discussed in Section 2.5. The complications arise because the velocity amplitude varies over the length of the cylinder if the cylinder is not both horizontal and normal to the wave. In addition, the flows in both the boundary layer and the wake are continually changing as the wave passes. In fact, the longitudinal position of the wake actually changes with

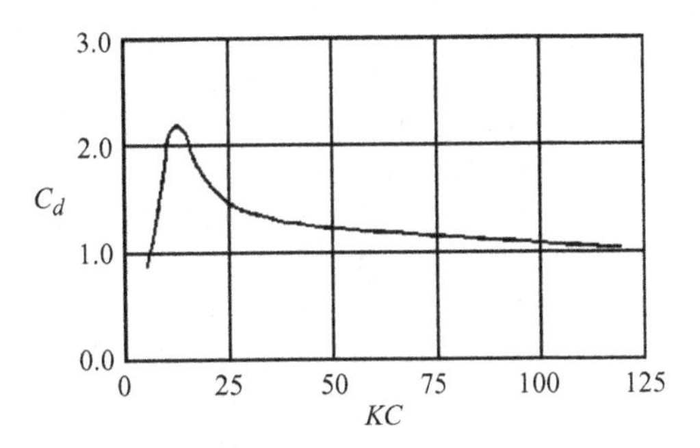

Figure 9.9. *Behavior of the Drag Coefficient for a Smooth Circular Cylinder in an Oscillatory Flow as a Function of the Keulegan-Carpenter Number.* The curve represents an average of the data measured by Keulegan and Carpenter (1958). The results presented in the book by Sarpkaya and Isaacson (1981) show that this behavior is modified by the inclusion of surface roughness.

the wave passage, being leeward over half of the wave period and forward over the other half. In the wakes, vortices form and are either attached (when KC is low) or shed (when KC is high). Vortex shedding can produce both longitudinal, or in-line, and transverse motions of a cylinder. For a mooring line, these motions are referred to as *strumming*. Vortices produced over half a wave period still exist over the other half-period, and can be convected from the side of their origin to the other side. In turn, this cumulatively affects the flow about the cylinder, as demonstrated by Isaacson and Maull (1976) and others.

Returning to the discussion of the small cylinder, where the drag is dominant in the direction of wave travel, the drag force equation for a cylinder that is horizontal and parallel to the wave crest is

$$F_d = \frac{1}{2}\rho u\, |u|\, A_d C_d \tag{9.48}$$

where the absolute value is required to preserve the force direction as the wave passes. As is the case for a steady viscous flow past a cylinder in eq. 2.79, the area (A_d) is that projected onto a plane normal to the celerity, $\mathbf{c}$. The behavior of the *drag coefficient* (C_d) is more complicated for the wave-induced flow, as would be expected. This fact was demonstrated by Keulegan and Carpenter (1958). From that study, the behavior of C_d and as function of KC in Figure 9.9 results. In that figure, we see a peaking of the data in the lower KC range ($KC < 25$). Also determined in the Keulegan-Carpenter study was the behavior of the *inertia coefficient* (C_i) with KC shown in Figure 9.10. In that figure, we see a dip in the data with a minimum value occurring approximately at the same KC value as the peak in the C_d curve. When viewing the data curves in Figures 9.9 and 9.10, one must keep in mind that, first, the curves represent scattered data and, second, another phenomenon is affecting the force readings. That phenomenon is *vortex shedding*.

Approximately two decades after the Keulegan-Carpenter study, Sarpkaya (1976, 1986) performed some of the most definitive experimental studies of the drag

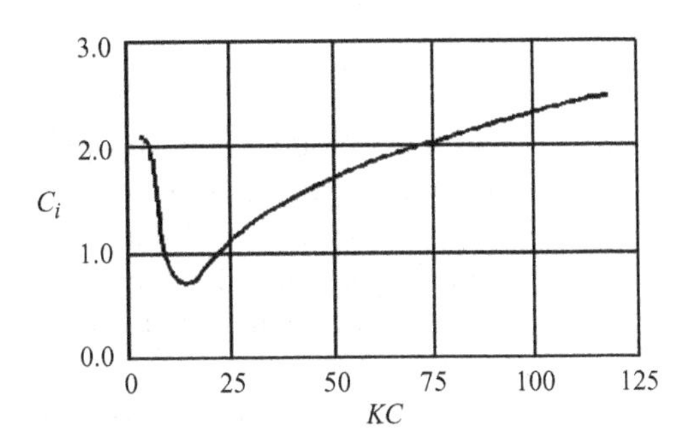

Figure 9.10. *Behavior of the Inertia Coefficient for a Smooth Circular Cylinder in an Oscillatory Flow as a Function of the Keulegan-Carpenter Number.* As for the data in Figure 9.9, the curve represents an average of the data measured by Keulegan and Carpenter (1958). Again, the results presented in the book by Sarpkaya and Isaacson (1981) show that this behavior is modified by the inclusion of surface roughness.

and inertia forces on cylinders in oscillatory flows. He found that both the drag and inertial coefficients are functions of the Reynolds number (R_e in eq. 2.77), the Keulegan-Carpenter number (KC), surface roughness, and a frequency parameter which, in turn, is a function of both R_e and KC. These dependencies cause a change in both the position and magnitude of the peak in the C_d curve and the dip in the C_i curve shown in Figures 9.9 and 9.10, respectively. For those readers interested in the inter-relationships affecting the drag and inertia coefficients, the book by Sarpkaya and Isaacson (1981) is recommended. That book contains an extensive discussion of Sarpkaya's earlier works.

The small-diameter cylinder is the most commonly used structural element in offshore engineering. It is the structural geometry of the vertical pile, where the cylinder is solid, and the basic geometric shape of the cross-brace in offshore structures, where the cylinder is hollow. The legs of most of the earliest offshore towers had dimensions that qualified the structural elements as small-diameter cylinders. The analysis of the forces on cylinders of small diameters involves an equation that incorporates both the drag and inertial forces. The diffraction forces on small cylinders in long waves are of second order, and can be neglected. The resulting equation is called the Morison equation, and is discussed in the next section.

D. The Morison Equation

As previously stated, the total force on a cylinder in waves has three components, those due to the drag, inertia, and scattering (reflection and the subsequent diffraction). The latter is referred to as the diffraction force. Referring to the force domains in Figure 9.8, many of the circular cylinders used in ocean engineering structures experience dominant drag and inertia forces – the diffraction forces being of second order. A number of experimental studies were performed on cylinders in this force domain by engineers at the University of California, Berkeley, after World War II. The results of these studies are discussed by Wiegel (1964). From the Berkeley studies, an expression for the forces on circular piles in the low ka range of Figure 9.8 was formulated by Morison et al. (1950). This expression is called the *Morison equation*, and is still the subject of many studies. An excellent review of the pre-1980 studies of the Morison equation and its applications is found in the Woodward-Clyde (1980) study sponsored by the U.S. Navy's Civil Engineering Laboratory. A comprehensive study of the equation that includes later works is found in the dissertation of Cook (1987), who worked with Emil Simiu at the U.S. National Bureau of Standards.

As is normally used in present-day engineering studies, the Morison equation is an expression for the wave-induced force per unit length in the drag-inertia domains in Figure 9.8. That equation is

$$F' = C_d \frac{1}{2} \rho u \left| u \right| A'_{proj} + C_i \rho \frac{Du}{Dt} \mathcal{V}'_{disp} \tag{9.49}$$

where the prime ($'$) indicates that the term is per unit length. The first term on the right side of the equation is the time-dependent drag force, and the second term is the inertial reaction force of the fluid. This equation is a modified form of that originally presented by Morison et al. (1950). First, the original equation did not account for the change in direction of the horizontal particle velocity so the absolute value of u was added. Second, the original study was directed at a circular cylinder, so the projected area per unit length (A'_{proj}) was simply D. Third, the displaced volume per unit length ($\forall'_{disp}$) was originally the displaced volume per unit length of a circular cylinder, or $\pi D^2/4$. Finally, the total acceleration in eq. 9.49,

$$\frac{Du}{Dt} = \frac{\partial u}{\partial t} + u\frac{\partial u}{\partial x} + w\frac{\partial u}{\partial z} \tag{9.50}$$

was originally the local acceleration, $\partial u/\partial t$. Isaacson (1979) finds that the use of the total derivative in eq. 9.49 yields better results than those obtained by using the local acceleration alone, although he notes that the last convective terms in eq. 9.50 do not included added-mass effects. Sarpkaya and Isaacson (1981) indicate that the convective acceleration (the last two terms in eq. 9.50) is negligible for most practical applications.

There are several other corrections applied to the Morison equation. These include those of Sarpkaya (1981) and Lighthill (1979). The Sarpkaya correction accounts for the effects of vorticity. Lighhill shows that there is an additional non-linear inertial force that is added to the expression in eq. 9.49. Call this force the "Lighthill force." This potential-flow term arises because of the in-line (in the flow direction) gradient of the velocity. As discussed by Cook (1987), the form of the Morison equation in eq. 9.49 must include the Lighthill force. This means that the nonlinear drag has a potential component, and "the Morison equation leads to an erroneous estimation of the force due to viscosity." However, Cook concludes that "the addition of the Lighthill correction term did not improve the Morison equation significantly; in most cases the Morison equation without the Lighthill correction provided a better fit to the measured forces."

By performing a harmonic analysis of experimental data, Sarpkaya (1981) finds that there are two corrective terms that should be added to the expression in eq. 9.49. Those terms correspond to the third and fifth harmonics in his analysis. As presented by Cook (1987), the four-term Morison equation applied to a circular cylinder due to Sarpkaya (1981) is

$$\begin{aligned} F' \equiv \frac{dF}{dz} &= C_d \frac{1}{2}\rho u\,|u|\,D + C_i \rho \frac{\partial u}{\partial t}\pi\frac{D^2}{4} \\ &+ \sqrt{\frac{C_d KC}{2-C_i}}\frac{1}{2}\rho u_{max}^2 D\left[0.01 + 0.10e^{0.08(KC-12.5)^2}\right] \\ &\cdot \cos\left(3\omega t - \sqrt{\frac{C_d KC}{(2-C_i)}}\left[-0.05 - 0.35e^{-0.04(KC-12.5)^2}\right]\right) \\ &+ \sqrt{\frac{C_d KC}{2-C_i}}\frac{1}{2}\rho u_{max}^2 D\left[0.0025 + 0.053e^{-0.06(KC-12.5)^2}\right] \\ &\cdot \cos\left(3\omega t - \sqrt{\frac{C_d KC}{(2-C_i)}}\left[0.25 + 0.60e^{-0.02(KC-12.5)^2}\right]\right) \end{aligned} \tag{9.51}$$

The numerical terms and coefficients in eq. 9.51 are universal constants. A study by Hudspeth and Nath (1985) yielded universal constants that differ from those in the equation. Note that in the denominators of the radical signs in the equation, the constant 2 is the value of the inertia coefficient from eq. 9.26 for the flow sketched in Figure 9.6 – oscillatory potential flow past a fixed, two-dimensional cylinder. The last two terms in eq. 9.51, called the *residue terms*, are of significance in the range

$7 \leq KC \leq 20$, and are approximately zero outside of this region. In Figure 9.8, we see that this range corresponds to that of inertia-dominated flows for $ka \leq 0.5$. As Cook (1987) writes, the residue terms "reflect the role played by the growth and convection of vortices on the in-line force." Finally, Cook writes that the practical value of the expression in eq. 9.51 is in curve-fitting when applying the Morison equation to measured data.

The determination of the drag coefficient, C_d, and the inertial coefficient, C_i, is rather complex as these coefficients have been found to be functions of the Reynolds number based on the diameter,

$$R_{e_D} = \frac{u_{\max} D}{\nu} \tag{9.52}$$

and the Keulegan-Carpenter number, KC, in eq. 9.45. For salt water, the kinematic viscosity (ν) in eq. 9.52 has a value of 1.2×10^{-6} m^2/s when the water temperature is 14°C. The reader is referred to the papers by Sarpkaya (1976, 1986) and the book by Sarpkaya and Isaacson (1981) for experimental and empirical studies of C_d and C_i. For demonstration purposes, the curves of Keulegan and Carpenter (1958) presented in Figures 9.9 and 9.10 are used herein. Furthermore, if we confine our attention to linear waves, then the amplitude of the horizontal particle velocity in eq. 9.52 is obtained from eq. 3.49, that is,

$$u_{\max} = \omega \frac{H}{2} \frac{\cos h[k(z+h)]}{\sinh(kh)} \tag{9.53}$$

Because of the nature of free-surface flows, KC (in eq. 9.45) varies with depth. For a vertical circular cylinder in linear waves in intermediate water, this fact is demonstrated from the expression resulting from the combination of eqs. 9.45 and 9.53. The value of KC might be large at the free surface, and the eddy structure in the neighborhood of the cylinder would be well developed. However, at the base of the cylinder the KC value might be relatively small. For this situation, the horizontal force per unit length would then be drag-dominated near the free surface and inertia-dominated near the base of the structure. The application of the two-dimensional Morison equation must be done in a piecewise fashion to account for the vertical diffusion of the eddies. In the large-KC flow where the vortex production is high, there are "wake-return effects," as coined by Cook (1987). That is, the wake formed by the flow in the wave direction over half of the wave period partially persists over the second half-period, when the flow is in the opposite direction. The remnants of the vortices produced in the first half of the wave period then affect the flow conditions in the second half, and so forth. This historical effect cannot be adequately accounted for by the Morison equation.

In the following example, the expression in eq. 9.51 is applied to a circular pile in shallow-water, long, linear waves.

EXAMPLE 9.5: FORCE AND MOMENT ON A VERTICAL CIRCULAR PILE IN SHALLOW WATER Referring to the sketch in Figure 9.11 for notation, consider a circular pile of 0.3 m diameter ($D = 2a$) in 5 m (h) of water subject to linear swell having a 0.5-m height (H) and a 15-sec period (T). Our goal is to determine the wave-induced force and moment on the pile using the four-term Morison equation in eq. 9.51. For this wave, the deep-water wavelength is $\lambda_0 \simeq gT^2/2\pi \simeq 351$ m, and the wavelength at the site is $\lambda \simeq 104$ m from both the linear theory (eq. 3.31) and Stokes' second-order theory. In the diagram of Figure 4.1, $h/\lambda_0 \simeq 0.0142$

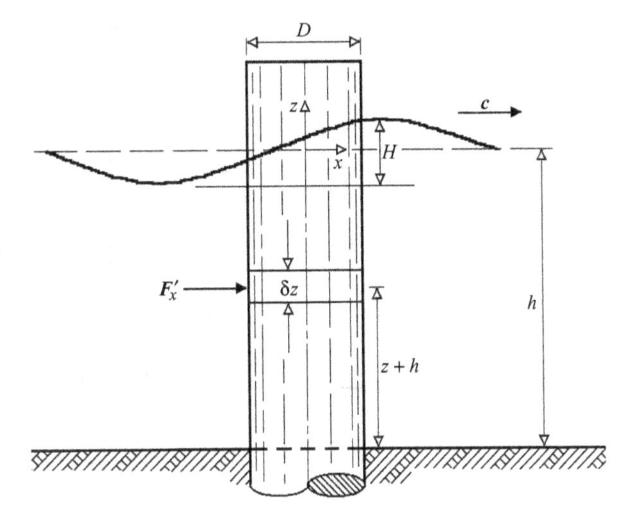

Figure 9.11. *Notation for the Wave-Force Analysis on a Vertical, Circular Pile.*

and $H/\lambda_0 \simeq 0.00142$. For these values, the linear wave theory is marginally valid and is used. The Ursell parameter,

$$U_R \equiv \frac{H\lambda^2}{2h^3} \tag{9.54}$$

in Figure 4.1 has a value of about 21.6. This value corresponds to a long-wave condition because $U_R \gg 1$. Also, the ratio of the wavelength to water depth is approximately 20.8, so the shallow-water approximation is apropos. The shallow-water Keulegan-Carpenter number value from eq. 9.47 is $KC = 17.5$. Also, for the stated conditions, $ka = 0.009$. These values in Figure 9.8 correspond to the drag and inertia force region. Using the KC value in Figures 9.9 and 9.10, respectively, yields $C_d \simeq 1.95$ and $C_i \simeq 0.75$.

The application of the four-term Morison equation (eq. 9.51) to the small-diameter pile in shallow water, where KC is invariant with depth, yields the horizontal wave force in Newtons,

$$\begin{aligned} F_x(t) &= \int_{-h}^{\eta_{CL}(t)} F'(t)dz = [\eta_{CL}(t) + h]\, F'(t) \\ &\simeq [0.25\cos(0.419t) + 5]\,[211\cos(0.419t)\,|\cos(0.419t)| - 8.00\sin(0.419t) \\ &\quad + 0.00226\cos(1.26t + 0.934) + 0.00138\cos(1.26t - 3.21)] \end{aligned} \tag{9.55}$$

and the corresponding wave-induced moment about the base of the pile at the mud line (positive in the counterclockwise direction) in Newton-meters is

$$\begin{aligned} M_w(t) &= -\int_{-h}^{\eta_{CL}(t)} (z+h)\, F'tdz = -\tfrac{1}{2}[\eta_{CL}(t) + h]^2 F'(t) \\ &= -\tfrac{1}{2}[0.25\cos(0.419t) + 5]^2\,[211\cos(0.419t)\,|\cos(0.419t)| \\ &\quad - 8.00\sin(0.419t) + 0.00226\cos(1.26t + 0.934) \\ &\quad + 0.00138\cos(1.26t - 3.21)] \end{aligned} \tag{9.56}$$

where the subscript x indicates the force direction, w refers to the wave-induced moment, and the subscript CL is used to represent the free-surface displacement above the origin of the centerline of the pile. To obtain the values in

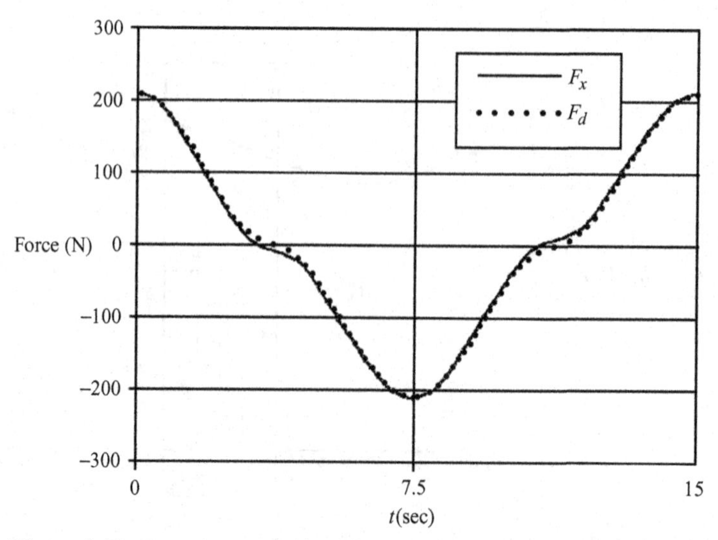

Figure 9.12. *Time History of the Wave-Induced Total Force and Drag Force on the Pile in Example 9.5.* The two drag forces are nearly identical, indicating that the conditions in Example 9.5 are drag-dominated. The inertia force causes the total force (F_x) curve to be asymmetric with respect to the mid-period ($t = T/2 = 7.5$ sec).

eqs. 9.55 and 9.56, the shallow-water maximum horizontal velocity, obtained from eq. 9.53, is used:

$$u_{\max} \simeq \frac{H}{2}\sqrt{\frac{g}{h}} \simeq 0.35 \text{ m/s} \tag{9.57}$$

The center of pressure of the force is obtained from

$$z_{cp}(t) = \frac{M_w(t)}{F_x(t)} - h = -\frac{1}{2}\left[\eta_{CL}(t) - h\right] \tag{9.58}$$

for this shallow-water problem.

Assuming that the pile is in salt water where the ambient water temperature is 14°C and the kinematic viscosity (ν) is 1.2×10^{-6} m^2/s, the value of the Reynolds number based on the pile diameter of eq. 9.52 is 8.75×10^4. For the Keulegan-Carpenter (1958) experiments, the Reynolds-number values were less than or equal to 5×10^4.

Results obtained from eq. 9.55 are presented in Figure 9.12, where the total force on the pile is presented as a function of time. Also in that figure is the component drag force. These results show that the drag force is the dominant component of the total force for the given conditions. The effect of the relatively small inertial component causes the total force curve to be unsymmetric with respect to $t = T/2 = 7.5$ sec. Also, we can conclude that the additional terms in the four-term Morison equation have a small effect on the total force for the stated conditions.

The phase relationships among the drag force, inertial force, and free-surface displacement can be seen in Figure 9.13, where the normalized force components of Morison's equation (eq. 9.49) and the normalized free-surface displacement are presented as functions of the time ratio, t/T.

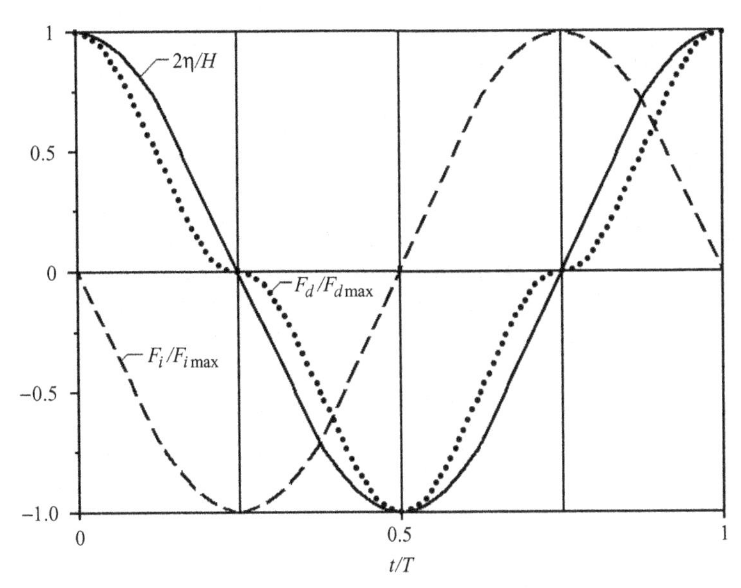

Figure 9.13. *Temporal Behaviors of the Normalized Wave-Induced Drag and Inertia Force Components of Morison's Equation and the Normalized Free-Surface Displacement.*

Finally, for a totally submerged horizontal, circular cylinder (referred to a *brace*) of length ℓ, the Morison equation is

$$F_{x-brace} = \frac{1}{2}\rho u\,|u|\,(D\ell)C_d + \rho\frac{\partial u}{\partial t}\left(\pi\frac{D^2}{4}\right)C_i\ell \tag{9.59}$$

Braces connect legs of an offshore structure, and are used to improve the structural stability. Many cross-braces are not in the plane of the wave front. The application of Morison's equation to these braces is done by breaking the wave-induced horizontal velocity into components that are normal and parallel to the centerline of the brace.

E. Circular Cylinders of Large Diameter – The MacCamy-Fuchs Analysis

Up to this point, our interest has been in small-diameter cylinders subject to either wave-induced drag forces or wave-induced inertia forces. For cylinders of circular cross-section, these forces are dominant for relatively small values of ka ($ka < 0.1$) in the Chakrabarti chart in Figure 9.8. We now direct our attention to the forces for relatively large values of ka (> 0.5) where the dominant wave-induced force is due to diffraction. In terms of the radius-to-wavelength ratio, our interest is in circular cylinders for which $a/\lambda > 0.08$, or a diameter-to-wavelength ratio range of $D/\lambda > 0.16$. Akyildiz (2002) presents experimental diffraction-force data for this range resulting from wave-tank studies of large cylinders in linear waves.

The introduction of structures with large-diameter legs came with the advent of deep-water offshore drilling for oil. These legs are normally found on semi-submersibles and tension-moored and slack-moored spar platforms, the latter discussed by Agarwal and Jain (2003) and others. When the cylinders encounter waves having lengths of the order of magnitude of the diameters, then scattering occurs. *Scattering* is normally referred to as the combination of wave reflection and diffraction. Diffraction, discussed in Section 6.4, can be considered to be the redistribution of wave energy along the wave crest. A scattered wave pattern in the neighborhood

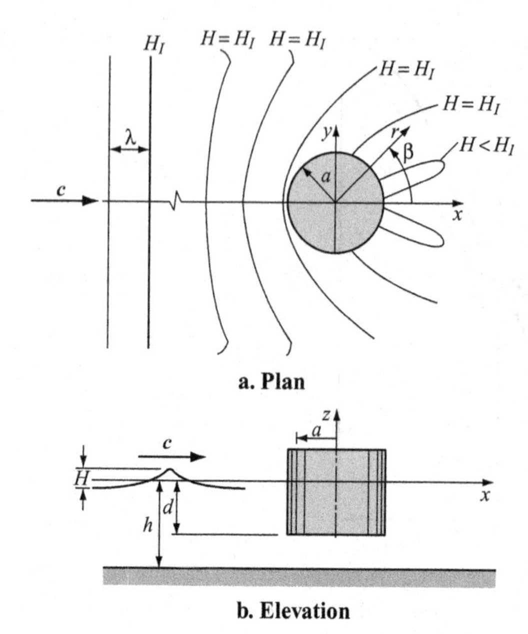

Figure 9.14. *Incident and Scattered Waves for a Fixed, Vertical, Truncated Circular Cylinder of Finite Draft.* The wave-crest pattern, such as that in Figure a, is predicted by eq. 9.69, and depends on the product of the wave number (k) and the radius of the cylinder (a).

of a vertical circular cylinder is sketched in Figure 9.14. In that figure, the incident waves are reflected over the angular range of $\pi/2 < \beta < 3\pi/2$. According to Huygens' principle (Section 6.4A), the reflected waves radiate outward radially from their point of origin and coalesce, so the energy per crest length of the reflected waves decreases as the waves travel away from the body. The height of the reflected wave is maximum at the point on the cylinder where $\beta = \pi$, and decreases to zero at $\beta = \pm\pi/2$. In the range of $-\pi/2 < \beta < \pi/2$, wave energy is transferred into the shadow zone leeward of the cylinder because the shadow zone does not receive direct incident wave energy.

The analysis of wave-structure interactions when a/λ is in the diffraction region of Figure 9.8 is approached using the potential theory. That is, the effects of viscosity are assumed to be insignificant, and other losses are neglected. By assuming that the entire flow field is potential in nature, Havelock (1940) analyzes the wave-induced force on a vertical, circular cylinder of infinite length in waters of infinite depth. For embedded, vertical, circular cylinders in waters of finite depth, MacCamy and Fuchs (1954) present an analysis along the lines of the Havelock study. An excellent engineering discussion of the MacCamy-Fuchs analysis is presented by Mogridge and Jamieson (1976), and they show that the analysis agrees with experimental results. Garrett (1971) applies the potential theory to vertical, cylindrical cylinders of finite draft, and his analytical data are shown to agree well with experimental data of van Oortmerssen (1971).

In this section, we present the MacCamy-Fuchs analysis of the diffraction forces on an embedded vertical, circular cylinder, as modified by Mogridge and Jamieson (1976). The notation for the analysis is presented in Figure 9.15. Following MacCamy and Fuchs (1954), begin by assuming that two irrotational wave patterns exist, those being the incident wave pattern (represented by the velocity potential, φ_I) and the diffraction pattern (represented by φ_D). The total potential is then

$$\varphi = \varphi_I + \varphi_D \tag{9.60}$$

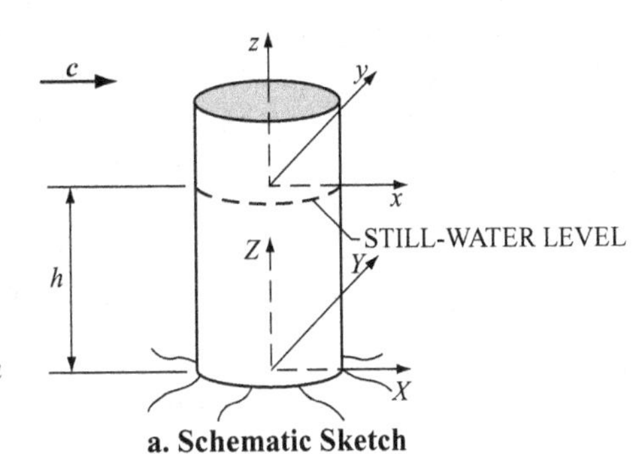

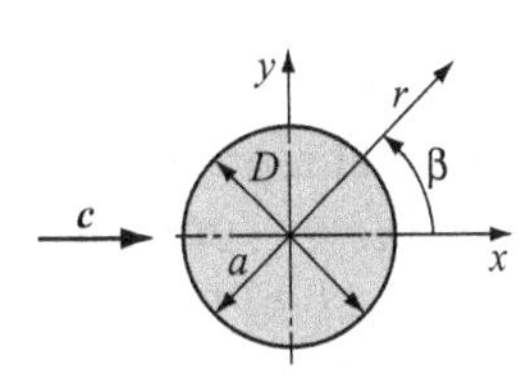

Figure 9.15. *Notation for the Analysis of Diffraction Forces on an Embedded, Vertical Circular Cylinder.*

Because of geometric considerations, the use of polar coordinates is found to be advantageous. In terms of the polar coordinates, the velocity potential representing the incident, right-running linear wave is

$$\varphi_I = \frac{H_I}{2}\frac{g}{\omega}\frac{\cosh[k(z+h)]}{\cosh(kh)}\sin(kx-\omega t) = \frac{H_I}{2}\frac{g}{\omega}\frac{\cosh[k(z+h)]}{\cosh(kh)}\sin[kr\cos(\beta)-\omega t]$$
$$= -\frac{H_I}{2}\frac{g}{\omega}\frac{\cosh[k(z+h)]}{\cosh(kh)}\Re\{ie^{i[kr\cos(\beta)-\omega t]}\} \quad (9.61a)$$

where $\Re$ indicates the real part of the exponential function. The last equality is used so that the temporal exponential function can be separated from the spacial exponential function, and the latter can then be written in terms of Bessel functions. The resulting expression is

$$\varphi_I = -\frac{H_I}{2}\frac{g}{\omega}\frac{\cosh[k(z+h)]}{\cosh(kh)}\Re\left\{i\left[J_0(kr)+2\sum_{m=1}^{\infty} i^m J_m(kr)\cos(m\beta)\right]e^{-i\omega t}\right\}$$
$$= -\frac{H_I}{2}\frac{g}{\omega}\frac{\cosh[k(z+h)]}{\cosh(kh)}\Re\left\{\sum_{m=0}^{\infty} i^{m+1}\varepsilon_m J_m(kr)\cos(m\beta)e^{-i\omega t}\right\} \quad (9.61b)$$

where the relationship between the exponential function in eq. 9.61a and nth-order Bessel functions of the first kind, $J_m(kr)$, in eq. 9.61b can be found in the books of Abramowitz and Stegun (1965) and Gradshteyn and Ryzhik (1965). Also, see Appendix A for a brief discussion of Bessel functions. The reason for representing the incident wave potential in terms of Bessel functions will become apparent later in this derivation. From the last equality in eq. 9.61b, the constants associated with the indices are seen to be $\varepsilon_0 = 1$ and $\varepsilon_m = 2$ for $m > 0$. The notation ε_m is called *Neumann's symbol*, according to Miles and Gilbert (1968).

The potential representing the diffracted wave (scattered wave) must satisfy Laplace's equation in cylindrical coordinates, that is,

$$\nabla^2(\varphi_D) = \frac{1}{r}\frac{\partial}{\partial r}\left(r\frac{\partial\varphi_D}{\partial r}\right) + \frac{1}{r^2}\frac{\partial^2\varphi_D}{\partial\beta^2} + \frac{\partial^2\varphi_D}{\partial z^2} = 0 \quad (9.62)$$

where (r, z, β) are the cylindrical coordinates, as shown in Figure 9.15. The solution of this equation can be obtained by using separation of variables, the method described in Section 3.2. Applying the seafloor boundary condition to the solution results in the following expression for the diffraction potential:

$$\begin{aligned}\varphi_D &= \frac{H_I}{2}\frac{g}{\omega}\frac{\cosh[k(z+h)]}{\cosh(kh)}\Re\left\{\sum_{m=0}^{\infty}\mathrm{E}_m[J_m(kr)+iY_m(kr)]\cos(m\beta)e^{-i\omega t}\right\} \\ &= \frac{H_I}{2}\frac{g}{\omega}\frac{\cosh[k(z+h)]}{\cosh(kh)}\Re\left\{\sum_{m=0}^{\infty}\mathrm{E}_m H_m^{(1)}(kr)\cos(m\beta)e^{-i\omega t}\right\}\end{aligned} \tag{9.63}$$

This potential represents a cylindrical wave radiating away from the origin. In this equation, E_m is a constant associated with a value of the index, m. Also in the equation are the mth-order Bessel function of the second kind, $Y_m(kr)$, and the mth-order Hankel function of the first kind, $H_m^{(1)}(kr)$. From eq. 9.63, we see that the relationship between the Hankel and Bessel functions is

$$H_m^{(1)}(kr) = J_m(kr) + iY_m(kr) \tag{9.64}$$

The Hankel function is also referred to as the Bessel function of the third kind. Again, see Appendix A for the properties of Bessel functions.

The expressions for the velocity potentials representing the incident waves (eq. 9.61b) and the diffracted waves (eq. 9.63) can now be combined with the expression in eq. 9.60 to obtain the expression for the potential of the entire wave field. The result is

$$\varphi = \frac{H_I}{2}\frac{g}{\omega}\frac{\cosh[k(z+h)]}{\cosh(kh)}\Re\left\{\sum_{m=0}^{\infty}\left[-i^{m+1}\varepsilon_m J_m(kr)+\mathrm{E}_m H_m^{(1)}(kr)\right]\cos(m\beta)e^{-i\omega t}\right\} \tag{9.65}$$

The values of E_m are found by applying the boundary condition on the wetted surface of the fixed cylinder. Physically, there is no flow across that surface; hence, the radial component of the fluid velocity on the surface is

$$\left.\frac{\partial\varphi}{\partial r}\right|_{r=a} = 0 \tag{9.66}$$

Applying this condition to the second equality in eq. 9.65 yields the following expression for the constant E_m:

$$\mathrm{E}_m = \varepsilon_m i^{m+1}\frac{dJ_m/dr}{dH_m^{(1)}/dr\,|_{r=a}} \equiv \varepsilon_m i^{m+1}\frac{J_m'(ka)}{H_m^{(1)\prime}(ka)}, \quad m = 0, 1, 2, \ldots \tag{9.67}$$

where the prime ($'$) indicates differentiation with respect to r and, again, ε_m is Neumann's symbol, where $\varepsilon_0 = 1$ and $\varepsilon_m = 2$ for $m \geq 1$. The combination of

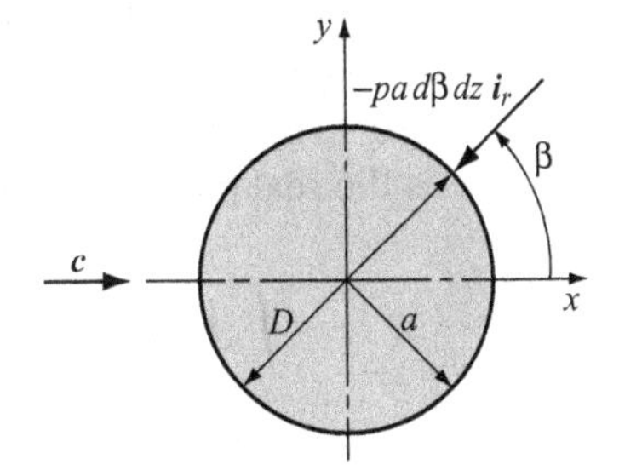

Figure 9.16. *Notation for Wave-Induced Pressure Force on a Surface Element of Cylinder.* In this figure, the unit vector in the radial direction is i_r.

eqs. 9.65 and 9.67 results in the following expression for the potential representing the entire wave field:

$$\varphi = -\frac{H_I}{2}\frac{g}{\omega}\frac{\cosh[k(z+h)]}{\cosh(kh)} \times$$

$$\Re\left\{\sum_{m=0}^{\infty} i^{m+1}\varepsilon_m\left[J_m(kr) - \frac{J'_m(ka)}{H_m^{(1)'}(ka)}H_m^{(1)}(kr)\right]\cos(m\beta)e^{-i\omega t}\right\} \tag{9.68}$$

Because both the incident and scattered waves are linear, the free-surface expression corresponding to the velocity potential in eq. 9.68 is found from application of the linearized dynamic free-surface condition in eq. 3.6. From that combination, the free-surface displacement expression is found to be

$$\eta(r,\beta,t) = \eta_I + \eta_S = -\frac{1}{g}\frac{\partial\varphi}{\partial t}\bigg|_{z=\eta\simeq 0} = i\frac{\omega}{g}\varphi\bigg|_{z=\eta\simeq 0}$$

$$= \frac{H_I}{2}\Re\left\{\sum_{m=0}^{\infty} i^{m}\varepsilon_m\left[J_m(kr) - \frac{J'_m(ka)}{H_m^{(1)'}(ka)}H_m^{(1)}(kr)\right]\cos(m\beta)e^{-i\omega t}\right\} \tag{9.69}$$

The subscripts I and S identify the respective incident and scattered waves. The velocity potential in eq. 9.68 is used to determine the time-dependent pressure on the vertical cylinder sketched in Figure 9.16. From the linearized Bernoulli's equation (eq. 3.70), the dynamic pressure is

$$p|_{r=a} = -\rho\frac{\partial\varphi}{\partial t}\bigg|_{r=a}$$

$$= \rho g\frac{H_I}{2}\frac{\cosh[k(z+h)]}{\cosh(kh)}\Re\left\{\sum_{m=0}^{\infty} i^{m}\frac{\varepsilon_m}{H_m^{(1)'}(ka)}\right.$$

$$\left.\times\left[J_m(ka)\,H_m^{(1)'}(ka) - J'_m(ka)H_m^{(1)}(ka)\right]\cos(m\beta)e^{-i\omega t}\right\}$$

$$= \rho g\frac{H_I}{2}\frac{\cosh[k(z+h)]}{\cosh(kh)}\Re\left\{\sum_{m=0}^{\infty} i^{m+1}\frac{\varepsilon_m}{H_m^{(1)'}(ka)}\left[\frac{2}{\pi ka}\right]\cos(m\beta)\,e^{-i\omega t}\right\} \tag{9.70}$$

The relationship between the bracketed Bessel-function expression in the second equality and the bracketed term in the last equality is obtained by, first, replacing the Hankel functions by the Bessel functions of the first and second kind, as in eq. 9.64, and then using recurrence relationships for the cross products of Bessel functions found in Abramowitz and Stegun (1965). The reader should note that the derivatives of the Bessel and Hankel functions with respect to r in eq. 9.70 have dimensions of m^{-1}. For example, if $f(r) = kr$, then $dJ_n(f)/dr = k\,dJ_n(f)/df$.

To obtain the expression for the wave-induced horizontal force on the vertical cylinder, integrate the pressure expression in eq. 9.70 over the surface. In Figure 9.16, the elemental pressure force on the cylinder is shown. The horizontal force is the real part of the resulting expression, that is,

$$F_x = -\int_{-h}^{0}\int_{0}^{2\pi} p|_{r=a}\cos(\beta)a d\beta dz = \rho\int_{-h}^{0}\int_{0}^{2\pi}\Re\left\{\left.\frac{\partial\varphi}{\partial t}\right|_{r=a}\right\}\cos(\beta)a d\beta dz$$

$$= 2\frac{\rho g H_I}{k}\tanh(kh)\left\{\frac{[J_1'(ka)\cos(\omega t) - Y_1'(ka)\sin(\omega t)]}{J_1'^2(ka) + Y_1'^2(ka)}\right\}$$

$$= 2\frac{\rho g H_I}{k}\tanh(kh)\frac{\sin[\omega t - \sigma(ka)]}{\sqrt{J_1'^2(ka) + Y_1'^2(ka)}} = 2\frac{\rho g H_I}{k^2}\tanh(ka)\Lambda(ka)\sin[\omega t - \sigma(ka)] \tag{9.71}$$

where the amplitude function is

$$\Lambda(ka) = \frac{k}{\sqrt{J_1'^2(ka) + Y_1'^2(ka)}} = \frac{1}{\sqrt{\left[J_0(ka) - \frac{1}{ka}J_1(ka)\right]^2 + \left[Y_0(ka) - \frac{1}{ka}Y_1(ka)\right]^2}} \tag{9.72}$$

Results from this expression are shown in Figure 9.17. The phase angle in eq. 9.71 is

$$\sigma(ka) = \tan^{-1}\left[\frac{J_1'(ka)}{Y_1'(ka)}\right] = \tan^{-1}\left[\frac{J_0(ka) - \frac{1}{ka}J_1(ka)}{Y_0(ka) - \frac{1}{ka}Y_1(ka)}\right] \tag{9.73}$$

Results from this expression are presented in Figure 9.18. The integration of the pressure over the wetted surface in eq. 9.71 contains only one term of the infinite series. This result is due to the β-integration from which the only nonzero result occurs for $m = 1$. The behaviors of the amplitude function, $\Lambda(ka)$, and the mass coefficient, C_M, are shown in Figure 9.17, whereas that of the phase angle, $\sigma(ka)$, is seen in Figure 9.18. The mass coefficient is defined by eq. 9.78.

The wave-induced overturning moment (counterclockwise moment about the base-axis Y of the structure in Figure 9.15), corresponding to the force of eq. 9.71, is obtained from

$$M_Y = \int_{-h}^{0}\int_{0}^{2\pi}(z+h)p|_{r=a}\cos(\beta)a d\beta dz = -\rho\int_{-h}^{0}\int_{0}^{2\pi}(z+h)\Re\left\{\left.\frac{\partial\varphi}{\partial t}\right|_{r=a}\right\}\cos(\beta)a d\beta dz$$

$$= -2\frac{\rho g H_I}{k^2}\frac{[J_1'(ka)\cos(\omega t) - Y_1'(ka)\sin(\omega t)]}{J_1'^2(ka) + Y_1'^2(ka)}\left\{kh\tanh(kh) + \frac{1}{\cosh(kh)} - 1\right\}$$

$$= -2\frac{\rho g H_I}{k^3}\left\{kh\tanh(kh) + \frac{1}{\cosh(kh)} - 1\right\}\Lambda(ka)\sin[\omega t - \sigma(ka)] \tag{9.74}$$

The center of pressure of the diffraction force is located at

$$z|_{cp} = -\frac{M_Y(t)}{F_w(t)} - h \tag{9.75}$$

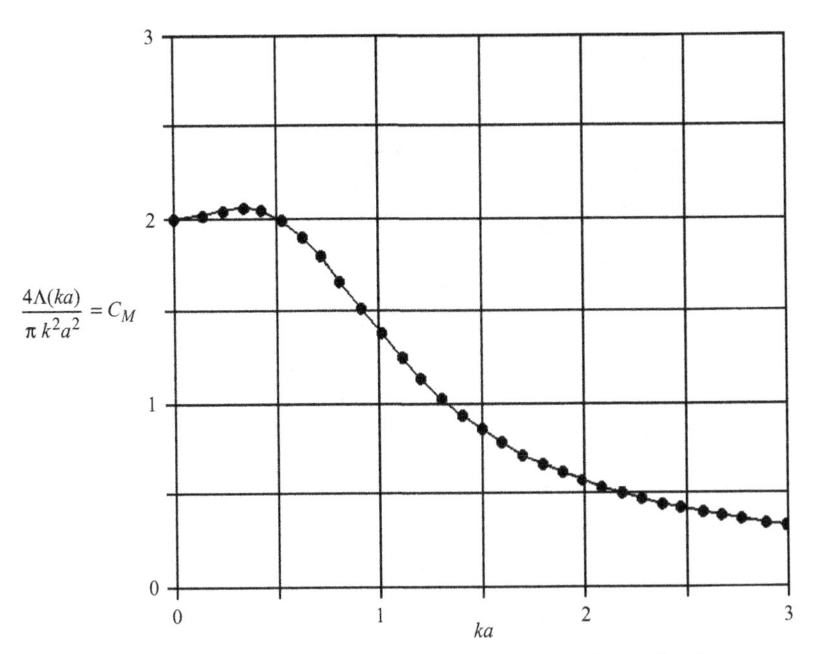

Figure 9.17. *MacCamy-Fuchs Amplitude Function and Mass Coefficient versus Dimensionless Cylinder Radius.* The relationship between the amplitude function $[\Lambda(ka)]$ and the mass coefficient (C_M) results from eq. 9.79. Note that the value of the mass coefficient is 2 for $ka = 0$. Physically, this condition corresponds to $k \to 0$ as $\lambda \to \infty$. This value is that of the inertia coefficient for an infinitely long cylinder with its axis normal to an accelerating fluid, obtained by combining eqs. 9.25 and 9.26. The reader should also note that the region of dominant wave diffraction for a vertical, surface-piercing circular cylinder in Figure 9.8 is $ka \geq 0.5$, according to Chakrabarti (1975).

EXAMPLE 9.6: FORCE AND MOMENT ON A COFFERDAM IN SHALLOW-WATER LINEAR WAVES A second Chesapeake Bay Bridge Tunnel is to be constructed between the Delmarva Peninsula and Virginia Beach parallel and seaward of the first bridge tunnel. East of the planned bridge tunnel is the Atlantic Ocean. Several of the piers of the bridge portion of the thoroughfare are in 3 m of water where the average wave height is 1 m. A relatively long-period wave having a 1-m height and a 12-sec period is common at this site. For this wave, the deep-water wavelength is approximately 225 m. From eq. 3.31, the wavelength at the site is approximately 64.5 m, which results in a depth-to-wavelength value in the shallow-water region of $h/\lambda < 1/20$. To facilitate the construction of these particular piers, single cofferdams of 40-m diameter are used. Our goal is to determine the amplitudes of the horizontal wave force on a cofferdam and the resulting overturning moment.

First, we should consult the Chakrabarti diagram in Figure 9.8 to determine the nature of the wave force. For the conditions at the site, $ka = \pi D/\lambda \simeq 0.56$, which is in the diffraction region of the figure. Because of this, the MacCamy-Fuchs analysis of the wave-induced force is appropriate. The resulting shallow-water wave force and overturning moment expressions are, respectively,

$$F_x = 2\frac{\rho g H_I}{k} h\Lambda(ka) \sin[\omega t - \sigma(ka)] \tag{9.76}$$

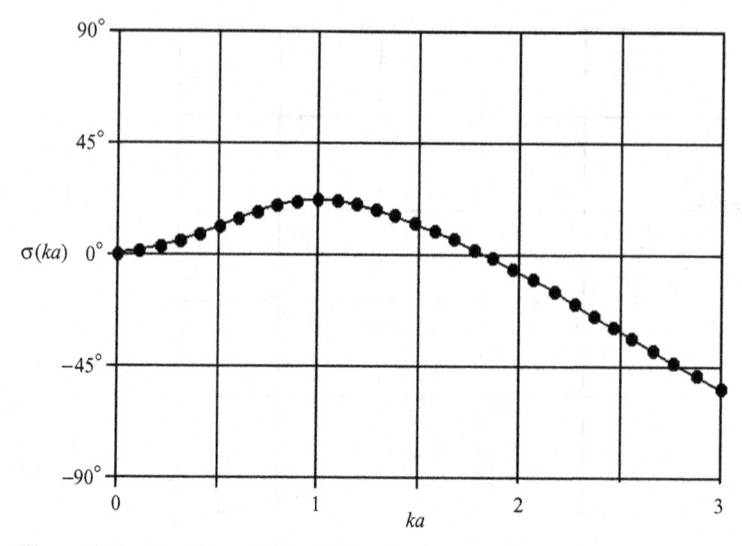

Figure 9.18. *MacCamy-Fuchs Phase Angle versus Dimensionless Cylinder Radius.*

from eq. 9.71, and

$$M_Y = -2\frac{\rho g H_I}{k} h^2 \Lambda(ka) \sin[\omega t - \sigma(ka)] \tag{9.77}$$

from eq. 9.74, where the respective amplitude function and the phase-angle values are approximately 0.482 and 12.3° from eqs. 9.72 and 9.73. The maximum force is 1.04×10^6 N, and maximum moment is -3.12×10^6 Nm. From eq. 9.75, the center of pressure for this shallow-water problem is $z|_{cp} = 0$, that is, at the waterline, and the resulting moment arm about the base is simply h, the water depth.

The diffraction theory of MacCamy and Fuchs (1954) was experimentally tested by Sundaravadivelu, Sundar, and Rao (1999), and the theoretical and experimental results are shown to agree rather well. Hence, the MacCamy-Fuchs analysis can be used with confidence.

F. Mass Coefficient for a Circular Cylinder

The expression in eq. 9.71 for the diffraction force on a vertical, circular cross-section caisson can also be considered to be due to the time rate of change of linear momentum. To obtain a corresponding expression, we must use a depth-averaged horizontal acceleration of the particles within the incident wave, evaluated at the position of the center of the cylinder when the cylinder is not present. Furthermore, we assume that this acceleration is in phase with the horizontal wave force of eq. 9.71. With these assumptions, the following expression for the horizontal force on a vertical circular caisson is obtained:

$$\begin{aligned} F_x &= 2\frac{\rho g H_I}{k^2}\tanh(kh)\,\Lambda(ka)\sin[\omega t - \sigma(ka)] \\ &= \rho\pi a^2 h C_M \frac{dU}{dt} = \rho\pi a^2 h C_M \frac{d}{dt}\left\{\frac{1}{h}\int_{-h}^{0} \left.\frac{\partial \varphi_I}{\partial x}\right|_{x=0} dz\right\} \\ &= \rho\pi a^2 C_M \frac{H_I g}{2}\tanh(kh)\sin[\omega t - \sigma(ka)] \end{aligned} \tag{9.78}$$

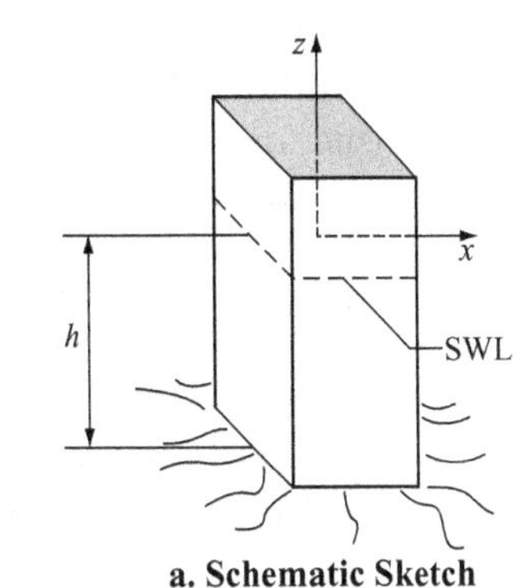

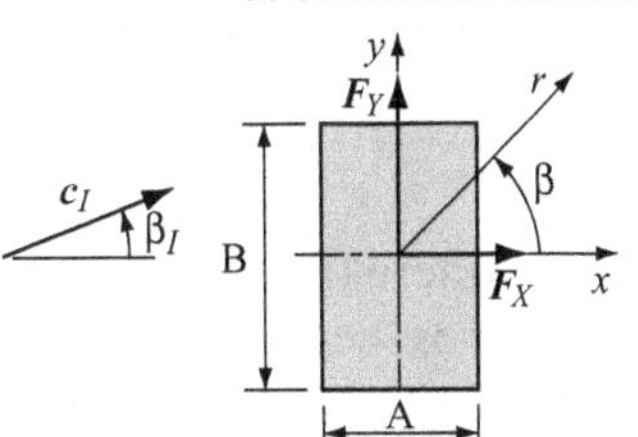

Figure 9.19. *Notation for a Rectangular Cross-Section Caisson in Oblique Waves.*

where C_M is called the *mass coefficient*, as opposed to the inertial coefficient in eq. 9.26. In this book, the difference between the two coefficients is that the mass coefficient is based on the total volume of the submerged portion of the structure, whereas the inertial coefficient is based on the mass per unit depth of submergence. The velocity, U, is a depth-averaged horizontal velocity component of the water particles in the wave direction at the site when the structure is not present. From the comparison of the second and last expressions in eq. 9.78, the following expression for the mass coefficient is obtained:

$$C_M = 4\frac{\Lambda\,(ka)}{\pi k^2 a^2} \tag{9.79}$$

The behavior of the mass coefficient is seen in Figure 9.17. In that figure, we see that $C_M = 2 = C_2$, where in eq. 9.26 $i = 2$. The value of 2 for C_M then corresponds to the case of a wave of infinite length.

In the next section, the Havelock/MacCamy-Fuchs theory is again modified to obtain the diffraction force and overturning moment on a surface-piercing rectangular caisson resting on the bed.

G. Diffraction Force and Moment on a Rectangular Cylinder

Vertical caissons with rectangular cross-sectional areas have received relatively little attention compared to that directed at caissons having circular cross-sections. In Chapter 8 of the book by Dean and Dalrymple (1984), the forces and moments on a fixed, surface-piercing, rectangular solid of finite draft are analyzed. The body, having vertical sides, is subject to linear waves of arbitrary incident angle (β_I in Figure 9.19). In their analysis, Dean and Dalrymple assume that the presence of the body has no effect on the wave field. This assumption is called the *Froude-Krylov hypothesis*, and is discussed later in Chapter 10 of this book. Rahman (1987a, 1987b) presents a method of analysis of diffraction forces on bed-resting, rectangular caissons based on the MacCamy-Fuchs (1954) analysis. The first of the Rahman papers contains a second-order approximation of the MacCamy-Fuchs analysis with applications to both circular and square caissons. The second paper is an extension of the first, where the theory is applied to rectangular caissons. Essentially, Rahman modifies the MacCamy-Fuchs force equation by introducing an "equivalent" circular cylinder. The Rahman method gives rather good results and, because of its relative simplicity, is outlined here.

Following Rahman (1987a, 1987b), the force expression in eq. 9.78 is used as an equivalent force expression for the rectangular caisson. Referring to the sketch in Figure 9.19 for notation, consider waves obliquely approaching the weather face of the rectangular caisson of length A and width B. The angle of approach is β_I, measured from the x-direction. The diffraction force on the rectangular caisson due to these waves can be represented by an expression similar to that in eq. 9.78, that is,

$$F_{x\Box} = \rho \mathrm{AB} h C_{\mathrm{M}} \frac{dU}{dt} \cos(\beta_I) \tag{9.80}$$

where the subscript notation $\Box$ is used to indicate that the forces are on rectangular cross-sectioned structures. In this expression, the subscript M is used to distinguish the force and mass coefficients from those for a circular caisson, for which the subscript is italicized. Let the forces represented by eqs. 9.78 and 9.80 be equal when the approach angle, β_I, is zero. The circular caisson is then equivalent to the rectangular caisson under direct wave attack. By equating the two force equations, an expression for the equivalent radius of the vertical cylinder is obtained. That expression is

$$a_e = \sqrt{\frac{\mathrm{AB}}{\pi} \frac{C_{\mathrm{M}}}{C_M}} \tag{9.81}$$

For an infinitely long caisson in an unbounded fluid, the relationship between the mass coefficients can be approximated by

$$\frac{C_{\mathrm{M}}}{C_M} = \frac{C_{\mathrm{M}}}{4\frac{\Lambda(ka_e)}{\pi k^2 a_e^2}} \simeq 0.478\left(\frac{\mathrm{A}}{\mathrm{B}}\right)^{0.410} + 1 \tag{9.82}$$

where the expression for the circular cylinder mass coefficient (C_M) is found in eq. 9.79. In the range of $0.1 \leq \mathrm{B/A} \leq 10$, the maximum difference in the values obtained from eq. 9.82 and those presented by Saunders (1957) is less than 4%. The Saunders values are based on the Lewis (1929) analysis presented in Section 9.2A(2). Rahman (1987a) assumes $C_{\mathrm{M}} = C_M$ in his second-order diffraction equations, and obtains force and moment results that compare well with experimental results for square caissons for $k\mathrm{B}$ less than approximately 0.19. In his second paper, Rahman (1987b) again assumes $C_{\mathrm{M}} = C_M$, and applies the modified first-order (MacCamy-Fuchs) equation to rectangular caissons. A variation of this study is presented here.

The modified MacCamy-Fuchs expression of eq. 9.80 for a rectangular caisson yields the following diffraction force expression:

$$F_{x\Box} = 2\frac{\rho g H_I}{k^2} \tanh(kh)\, \Lambda(ka_e) \cos(\beta_I) \sin[\omega t - \sigma(ka_e)] \tag{9.83}$$

where $\Lambda(ka_e)$ is defined in eq. 9.72, and $\sigma(ka_e)$ in eq. 9.73. Similarly, the diffraction force in the y-direction is

$$F_{y\Box} = 2\frac{\rho g H_I}{k^2} \tanh(kh)\, \Lambda(ka_e) \cos(\beta_I) \sin[\omega t - \sigma(ka_e)] \tag{9.84}$$

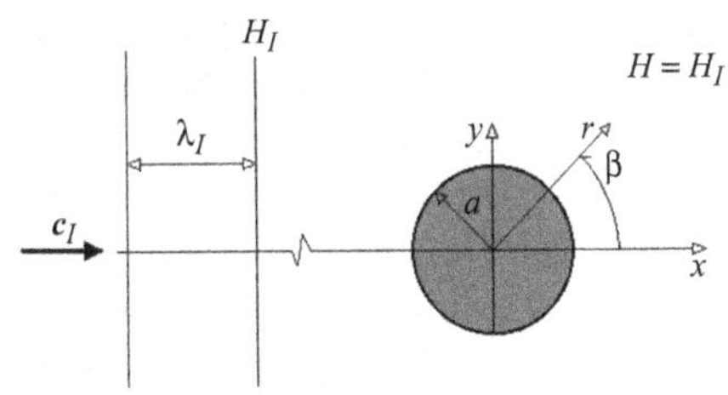

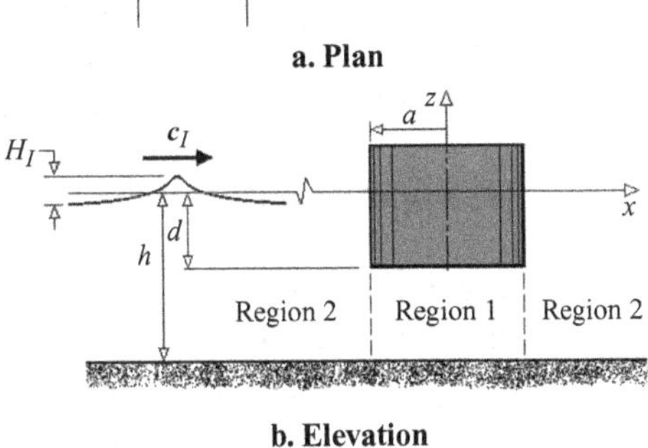

Figure 9.20. *Notation for a Vertical, Truncated, Circular Cylinder.* Also, see Figure 9.16.

The application of these equations is illustrated in the following example.

EXAMPLE 9.7: FORCE AND MOMENT ON A SQUARE CYLINDER IN SHALLOW-WATER LINEAR WAVES For the Chesapeake Bay Bridge Tunnel situation in Example 9.6, we are to replace the circular caissons with square caissons, where the face of the structure subject to direct wave attack ($\beta_I = 0°$) has a width (B) equal to the diameter of the circular caisson (40 m). The water depth is 3 m, and the design wave has a height of 1 m and a period of 12 sec. In Example 9.6, we find that the deep-water wavelength is approximately 225 m, and the wavelength at the site is approximately 64.5 m which qualifies as a shallow-water wave. We shall determine the maximum horizontal force on the caisson, assuming that the diffraction force is dominant.

For the conditions presented, the mass-coefficient ratio for a square caisson is approximately 1.48 from eq. 9.82. The combination of this value with the expression in eq. 9.81 yields an equivalent radius (a_e) value of about 27.4 m. Hence, $ka_e \simeq 2.67$ which, from the Chakrabarti diagram in Figure 9.8, is well in the diffraction region. When the waves approach the structure directly, the maximum wave force is approximately 1.14×10^6 N from eq. 9.83. For the circular caisson, the wave force for this condition is about 0.93×10^6 N. If we assume that the mass coefficient-ratio value equals one, as Rahman (1987a, 1987b) does, then the equivalent radius is about 22.6 m, and $ka_e \simeq 2.20$. For this value, the wave force is about 1.02×10^6 N. That assumption then under-predicts the wave-force amplitude by about 9%.

H. Truncated Circular Cylinder of Large Diameter

The truncated, vertical, circular cylinder, such as that sketched in Figures 9.14 and 9.20, might be found as a leg of a *tension-leg platform* (TLP), as discussed in Chapters 10 and 12. The analyses of the ambient hydrodynamics and the corresponding wave forces on either rigid or moving truncated cylinders are presented by Miles and Gilbert (1968), Garrett (1971), Yeung (1981), Yilmaz and Incecik (1998), and others. The Yilmaz-Incecik study and many of the other analyses are either applications or extensions of the Garrett study. Van Oortmerssen (1971) presents an "engineering-friendly" formula for the horizontal force on a truncated cylinder in waters of finite depth. This formula is based on the works of Havelock (1940) and MacCamy and Fuchs (1954). The analysis of MacCamy and Fuchs (1954), presented in the last section, is also based on the Havelock analysis.

In this section, two approximate analyses are first presented. The first is that of van Oortmerssen (1971) for the wave-induced horizontal force on a truncated circular cylinder. The second is a complimentary analysis presented by McCormick and Cerquetti (2004) approximating the vertical force on that structure. These analyses are followed by the more exact analysis of both the horizontal and vertical forces by Garrett (1971). Results obtained from these three studies are then compared.

(1) Approximations of the Horizontal Force and Resulting Moment

Van Oortmerssen (1971) applies a parametric coefficient to the force expression derived by MacCamy and Fuchs (1954) to obtain an expression for the horizontal diffraction force on a vertical circular cylinder of finite draft in water of finite depth. As discussed in Section 9.2E, the MacCamy-Fuchs force and moment formulas for a vertical circular cylinder resting on a bed in waters of finite depth are based on the Havelock (1940) analysis. The Havelock force formula applies to a cylinder of infinite draft in infinitely deep water.

Referring to the sketches in Figure 9.20 for notation, the MacCamy-Fuchs horizontal diffraction force expression in eq. 9.71 is

$$F_{xMF} = 2\frac{\rho g H_I}{k^2}\tanh(kh)\,\Lambda(ka)\sin[\omega t - \sigma(ka)] \tag{9.85}$$

where the subscript x refers to the direction, and the subscript MF identifies the expression as that of MacCamy and Fuchs. To obtain the horizontal diffraction force on a truncated circular cylinder of draft (d) in waters of finite depth (h), van Oortmerssen (1971) multiplies the force expression in eq. 9.85 by a ratio of draft and depth integrals as follows:

$$F_{(xd)} \simeq \frac{\int_{-d}^{0}\cosh[k(z+h)]dz}{\int_{-h}^{0}\cosh[k(z+h)]dz}F_{xMF} = \frac{\sinh(kh) - \sinh[k(h-d)]}{\sinh(kh)}F_{xMF} \tag{9.86}$$

The moment about the free-surface axis, the y-axis, due to the force in eq. 9.86 is obtained from

$$M_{Y(xd)-} = F_{(xd)}z|_{cp} = F_{(xd)}\frac{\int_{-d}^{0}z\cosh[k(z+h)]dz}{\int_{-d}^{0}\cosh[k(z+h)]dz}$$

$$= F_{(xd)}\frac{1}{k}\frac{\{kd\sinh[k(h-d)] + \cosh[k(h-d)] - \cosh(kh)\}}{\{\sinh(kh) - \sinh[k(h-d)]\}} \tag{9.87}$$

following van Oortmerssen. The moment is positive in the counterclockwise direction. From this expression, we find that the center of pressure is at

$$z|_{cp} = \frac{kd\sinh[k(h-d)] - \cosh(kh) + \cosh[k(h-d)]}{\{k\sinh(kh) - k\sinh[k(h-d)]\}} \tag{9.88}$$

The overturning moment about the bottom of the cylinder is simply the product of the force in eq. 9.86 and $(d + z|_{cp})$. The reader might note that the center of pressure for the cylinder resting on the bed in shallow water is at the waterline, as is the case for the MacCamy-Fuchs formula in Example 9.6.

The expressions in eqs. 9.85 and 9.88 can be applied with relative ease. Van Oortmerssen (1971) presents no expression for the vertical force on the truncated cylinder. Depending on the magnitudes of the wave height, period, water depth, and the cylinder's draft, the variation of the wave-induced pressure over the bottom of the cylinder could be appreciable. The bottom-pressure force could then result in a large component of the overturning moment. To compliment van Oortmerssen's analysis, McCormick and Cerquetti (2004) present an approximate expression for the vertical force. That expression is presented and discussed in the next paragraphs.

(2) Approximations of the Vertical Force and Resulting Moment

Referring to Figure 9.20, the goal of the McCormick and Cerquetti (2004) analysis is to find a vertical force expression resulting from a velocity potential in Region 1 (the gap) that will be finite at the origin of r, satisfy the equation of continuity, and satisfy the boundary condition of no flow across both the bottom of the cylinder and the sea bed. Before introducing such a potential, the potential in Region 2 is assumed to be of the form

$$\varphi_2 = -\frac{H_I}{2}\frac{g}{\omega}\frac{\cosh[k(z+h)]}{\cosh(kh)}\Re\left\{\sum_{m=0}^{\infty} i^{m+1}\varepsilon_m f(kr)\cos(m\beta)e^{-i\omega t}\right\} \tag{9.89}$$

where, again, $\Re$ indicates the real part of the expression and ε_m is Neumann's symbol ($\varepsilon_0 = 1$ and $\varepsilon_m = 2$ for $m \geq 1$). The function $f(kr)$ is to be determined. Comparing the expressions in eqs. 9.68 (the incident and scattered wave potential for a bed-resting cylinder) and 9.89, we see that the form of the latter equation is modeled on the former.

The potential function applicable to Region 1 of Figure 9.20 that is proposed is

$$\varphi_1 = \frac{H_I}{2}\frac{g}{\omega}q(z)\Re\left\{\sum_{m=0}^{\infty} i^{m+1}\varepsilon_m \frac{I_m(kr)}{I_0(ka)}\cos(m\beta)e^{-i\omega t}\right\} \tag{9.90}$$

where $I_m(kr)$ is a modified Bessel function of the first kind, order m. The spacial function $q(z)$ is to be specified. That function must cause the velocity potential to satisfy $\nabla^2\varphi_B = 0$, the continuity equation, and the conditions on the body $(\partial\varphi_B/\partial z|_{z=-d} = 0)$ and sea bed $(\partial\varphi_B/\partial z|_{z=-h} = 0)$.

The radial pressure force over the 1–2 vertical boundary is assumed to be the same for the velocity potentials in eqs. 9.89 and 9.90. From Bernoulli's equation, the radial force is

$$\begin{aligned}
F_r\Big|_{\substack{r=a\\ h<z<-d}} &= -\rho\int_0^{2\pi}\int_{-h}^{-d}\Re\left\{\frac{\partial\varphi_1}{\partial t}\Big|_{r=a}\right\}a\,dz\,d\beta = -\rho\int_0^{2\pi}\int_{-h}^{-d}\Re\left\{\frac{\partial\varphi_2}{\partial t}\Big|_{r=a}\right\}a\,dz\,d\beta \\
&= -\rho g\frac{H_I}{2}2\pi a^2\frac{(h-d)}{a}q_{avg}\cos(\omega t) \\
&= \rho g\frac{H_I}{2}\frac{2\pi a^2\sinh[k(h-d)]}{ka\cosh(kh)}f(ka)\cos(\omega t)
\end{aligned} \tag{9.91}$$

where the only nonzero terms in each of the β-integrals occur when $m = 0$. Because $q(z)$ is not known, we have chosen to spatially average the function over the gap, h-d. From the last equality in eq. 9.91, we find the following expression for that averaged function:

$$q_{avg} = -\frac{1}{k(h-d)}\frac{\sinh[k(h-d)]}{\cosh(kh)}f(ka) \tag{9.92}$$

Although $f(ka)$ is not known, the expression in eq. 9.92 gives us some idea of the form of the spacial function. Assume then that the spacial function can be represented by

$$q(z) = -\frac{1}{k(h-d)}\frac{\sinh[k(h-d)]}{\cosh(kh)}Q(z) \tag{9.93}$$

The potential function in eq. 9.90 is then

$$\varphi_1 = -\frac{H_I}{2}\frac{g}{\omega}\frac{1}{k(h-d)}\frac{\sinh[k(h-d)]}{\cosh(kh)}Q(z)\sum_{m=0}^{\infty} i^{m+1}\varepsilon_m \frac{I_m(ka)}{I_0(kr)}\cos(m\beta)e^{-i\omega t} \tag{9.94}$$

where the real parametric function $Q(z)$ is to be determined. This function can be considered to be empirical in nature.

The vertical force on the bottom of the truncated cylinder that is due to the velocity potential in eq. 9.94 is obtained from

$$\begin{aligned} F_z &= \int_0^{2\pi}\int_0^{a} p|_{z=-d}\, r\,dr\,d\beta = -\rho\int_0^{2\pi}\int_0^{a} \Re\left\{\left.\frac{\partial\varphi_1}{\partial t}\right|_{z=-d}\right\} r\,dr\,d\beta \\ &= \rho g H_I \pi a^2 \frac{a\sinh[k(h-d)]1 I_1(ka)}{(h-d)\cosh(kh)(ka)^2 I_0(ka)}Q(d)\cos(\omega t) \end{aligned} \tag{9.95}$$

The moment about the y-axis in Figure 9.20 corresponding to the vertical force in eq. 9.95 is obtained from

$$\begin{aligned} M_y &= \int_0^{2\pi}\int_0^{a} p|_{z=-d}\, r^2 dr\cos(\beta)d\beta = -\rho\int_0^{2\pi}\int_0^{a} \Re\left\{\left.\frac{\partial\varphi_1}{\partial t}\right|_{z=-d}\right\} r^2 dr\cos(\beta)d\beta \\ &= \rho g H_I \pi a^3 \frac{a}{(h-d)}\frac{\sinh[k(h-d)]}{\cosh(kh)}\frac{1}{(ka)^2}\frac{I_2(ka)}{I_0(ka)}Q(d)\sin(\omega t) \end{aligned} \tag{9.96}$$

The parametric function $Q(d)$ in eqs. 9.95 and 9.96 must be determined. For an infinitely long wave, where $ka \to 0$, we know that the vertical force is simply due to the additional buoyancy caused by the rise and fall in the horizontal free surface. That is, the force must be equal to

$$\lim_{ka\to 0}\{F_z\} = \rho g\eta(t)\pi a^2 = \rho g\frac{H_I}{2}\pi a^2\cos(\omega t) \tag{9.97}$$

where $\eta(t)$ is the free-surface displacement of eq. 3.24, where $x = 0$. For the other extreme, as $ka \to \infty$, the vertical force must vanish. So, $Q(d)$ must either vanish or be finite for this condition to be satisfied. In addition to the ka conditions, the vertical force must vanish if the cylinder rests on the bed, or when $d = h$. Based on

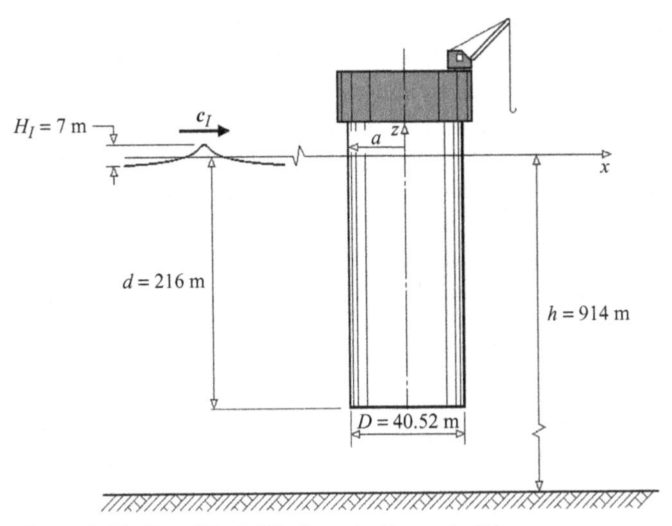

Figure 9.21. *Spar Work Platform in Example 9.8.*

experimental data, the relationship chosen by McCormick and Cerquetti (2004) that satisfies these conditions is

$$Q(d) = 1 - \left(\frac{d}{h}\right)^{\frac{10}{ka}} \tag{9.98}$$

It is demonstrated later in this chapter that the values of the force amplitudes obtained by substituting the expression in eq. 9.98 into eq. 9.95 compare well with those obtained using the more exact force expressions of Garrett (1971) over specific *ka* ranges.

To gain an idea of the applicability of the approximate formulas, we apply the force expressions in eqs. 9.86 and 9.95 (combined with the expression in eq. 9.98) to the spar platform studied experimentally by Agarwal and Jain (2003) in the following example.

EXAMPLE 9.8: WAVE-INDUCED FORCES ON A SPAR WORK PLATFORM Agarwal and Jain (2003) study the effects of linear waves acting on a spar platform in deep water, such as sketched in Figure 9.21. The spar has an approximate 20.3-m radius (a) and a 216-m draft (d). The structure is in regular waves having a 7-m height and a 12.5-sec period. The depth (h) of the water is 914 m; hence, the condition is that of deep water because the wavelength (λ) corresponding to the wave period is approximately 244 m from eq. 3.36. The *ka* value for this wavelength is 0.522 which, from the Chakrabarti chart in Figure 9.8, is approximately at the lower limit of the diffraction force region for the horizontal force. To see if the horizontal viscous and inertia forces might be significant, the deep-water Keulegan-Carpenter number of eq. 9.46 must be evaluated. For the spar in deep water, $KC \simeq \pi H_I / T \simeq 0.543 > 0.5$. For the KC and *ka* values in Figure 9.8, the diffraction force is dominant. The combination of the MacCamy-Fuchs force of eq. 9.71 with the van Oortmerssen force of eq. 9.86 yields a horizontal force amplitude of approximately 4.27×10^8 N. The vertical force from eq. 9.95 is approximately 9.39×10^3 N.

The approximate force and moment formulas of van Oortmerssen (1971) and McCormick and Cerquetti (2004) are analytical. Because of this, the formulas are relatively easy to use in the conceptual design phase of an engineering projects. The mathematically elegant theory of Garrett (1971), presented in the next section, is more exact; however, results from this theory must be obtained numerically. The Garrett theory is presented herein for the sake of completeness.

In the past decade, several studies of wave forces on truncated vertical cylinders have been reported. These studies involve computational fluid dynamics (CFD) techniques and, as a result, are numerical in their natures. Results obtained from these CFD studies are found to compare well with those observed in both the laboratory and the field. The topic of CFD is outside of the scope of this book.

(3) Garrett's Analysis

Garrett's (1971) analysis of the diffraction force on a truncated, vertical, circular cylinder, outlined herein, is a modification of the earlier work by Miles and Gilbert (1968). The major difference in the Garrett and Miles-Gilbert studies concerns the conditions at the boundary separating Regions 1 and 2, sketched in Figure 9.20b. This difference in the boundary conditions is discussed later in this section. Miles and Gilbert (1968) and Garrett (1971) take a somewhat different approach by basing their analyses on the displacement potential (Φ) rather than the velocity potential (φ). The relationship between these two potentials is simply

$$\Phi = \frac{i}{\omega}\varphi \tag{9.99}$$

The gradient of the displacement potential then gives the particle displacement at any point in the fluid. As sketched in Figure 9.20b, the flow field is divided into two regions: Region 1 is the circular cylindrical region beneath the cylinder, where $r \leq a$ and $-h \leq z \leq -d$. Region 2 is external to Region 1 and the cylinder, where $r > a$ and $-h \leq z \leq 0$. In both regions, the equation of continuity for the (assumed) irrotational flow must be satisfied by the velocity and displacement potentials. In cylindrical coordinates, that equation expressed in terms of the displacement potential is

$$\frac{\partial^2 \Phi}{\partial r^2} + \frac{1}{r}\frac{\partial \Phi}{\partial r} + \frac{1}{r^2}\frac{\partial^2 \Phi}{\partial \beta^2} + \frac{\partial^2 \Phi}{\partial z^2} = 0 \tag{9.100}$$

Also satisfied in both regions is the seafloor condition which, in this analysis, is that there is no particle displacement across the flat horizontal bed. From eq. 3.4, that condition is

$$\left.\frac{\partial \Phi}{\partial z}\right|_{z=-h} = 0 \tag{9.101}$$

where the sea bed is assumed to be both flat and horizontal.

The boundary conditions on the cylinder are the following: In Region 1, the vertical particle displacement on the bottom of the fixed cylinder must be zero. Mathematically, this condition is expressed by

$$\left.\frac{\partial \Phi}{\partial z}\right|_{z=-d} = 0, \quad 0 \leq r \leq a \tag{9.102}$$

In Region 2, the radial particle displacement on the cylinder's vertical surface is also zero, that is,

$$\left.\frac{\partial \Phi}{\partial z}\right|_{r=a} = 0, \quad -d \le z \le 0 \tag{9.103}$$

In Region 2, the linearized free-surface condition of eq. 3.7 and the Sommerfeld radiation condition of eq. 6.100 must also apply. The respective expressions for these conditions applied to the displacement potential are

$$\left.\left(\frac{\partial^2 \Phi}{\partial t^2} + g\frac{\partial \Phi}{\partial z}\right)\right|_{z\simeq 0} = 0, \quad r \ge 0 \tag{9.104}$$

and

$$\lim_{r\to\infty}\left[\sqrt{r}\left(\frac{\partial \Phi}{\partial r} - ik\Phi\right)\right] = 0 \tag{9.105}$$

Note that the radiation condition is satisfied by the scattered waves, but not the incident wave.

Region 1: In Region 1, the solution of eq. 9.100, subject to the boundary conditions in eqs. 9.101 and 9.102, consists of two components. Those component equations are

$$\Phi_{1mn(\mathrm{A})} = C_{1mn(\mathrm{A})} I_m(\mathrm{K}_n r)\cos[\mathrm{K}_n(z+h)]\cos(m\beta)e^{-i\omega t}, \quad \begin{cases} m = 0, 1, 2, \ldots \\ n = 1, 2, \ldots \end{cases} \tag{9.106}$$

where $I_m(\kappa_n r)$ is a modified Bessel function of the first kind of order m (see Appendix A), and

$$\Phi_{1m0(\mathrm{B})} = C_{1m0(\mathrm{B})}\left(\frac{r}{a}\right)^m \cos(m\beta)e^{-i\omega t}, \quad n = 0 \tag{9.107}$$

In eqs. 9.106 and 9.107, $C_{1mn(\mathrm{A})}$ and $C_{1mn(\mathrm{B})}$ are coefficients that must be determined from the boundary conditions. The subscript 1 refers to the region of application, whereas the subscripts A and B are used to identify the specific solutions. Also in eq. 9.106 is parameter

$$\mathrm{K}_n = \frac{n\pi}{h-d} \tag{9.108}$$

which results from the application of the no-vertical-displacement boundary conditions on the bed (eq. 9.102) and on the bottom of the cylinder (eq. 9.103). The solution obtained using the separation-of-variables method results in the expression in eq. 9.106 when the conditions on z-behavior are applied. Concerning eq. 9.103: The r-solution in the equation contains $I_m(\mathrm{K}_n r)$; however, $K_m(\mathrm{K}_n r)$, the modified Bessel function of the second kind of order m, is also a solution. At $r = 0$ in Region 1, we require the potential function to be finite. At the origin, $I_m(\mathrm{K}_n r)$ equals either zero (for $m > 0$) or one (for $m = 0$), whereas $K_m(\mathrm{K}_n r)$ is infinite when $r = 0$ for any integer value of m. Hence, $I_m(\mathrm{K}_n r)$ is chosen for the inner region solution. Concerning eq. 9.107: The index, n, in the solution includes the value 0; however, when this value is applied to the separated r-equation, the equation becomes a homogeneous, second-order Cauchy-Euler equation. Again, assuming that the potential is finite at $r = 0$, the solution of Cauchy-Euler equation is eq. 9.106. The full solution of eq. 9.100 must include all values of m and n, and is

$$\begin{aligned}\Phi_1(r,z,t) &= \sum_{m=0}^{\infty}\left\{\Phi_{1m0(B)} + \sum_{n=1}^{\infty}\Phi_{1mn(A)}\right\} \\ &= \sum_{m=0}^{\infty}\left\{\left\langle C_{1m0(B)}\left(\frac{r}{a}\right)^m + \sum_{n=1}^{\infty} C_{1mn(A)} I_m(K_n r)\cos[K_n(z+h)]\right\rangle \cos(m\beta)\right\} e^{-i\omega t} \\ &= \frac{H_I}{2}\left\{P_{10}(r,z) + 2\sum_{m=1}^{\infty} i^m P_{1m}(r,z)\cos(m\beta)\right\} e^{-i\omega t}\end{aligned} \tag{9.109}$$

The last equality is similar to a general expression for the internal potential function used by both Miles and Gilbert (1968) and Garrett (1971). Following Garrett (1971), apply the function $P_{1m}(r,z)$ at $r = a$, where $-h \le z \le -d$, and let this be represented by

$$\begin{aligned}P_{1m}(a,z) \equiv h f_{1m}(z) &= h\left\{\mathscr{F}_{1m0} + 2\sum_{n=1}^{\infty}\mathscr{F}_{1mn}\cos[K_n(z+h)]\right\} \\ &= h\sum_{n=0}^{\infty}\varepsilon_n \mathscr{F}_{1mn}\cos[K_n(z+h)], \quad (-h \le z < -d)\end{aligned} \tag{9.110}$$

We note that the $P_{1m}(a,z)$ function has units of length, whereas $f_{1m}(z)$ is dimensionless. We refer to $f_{1m}(z)$ as the *boundary function.* The last equality in eq. 9.110 can be considered to be a representation of $f_{1m}(z)$ by a Fourier series. As such, the coefficients can be written as

$$\mathscr{F}_{1mn} = \frac{1}{(h-d)}\int_{-h}^{-d} f_{1m}(z)\cos[K_n(z+h)]dz \tag{9.111}$$

The combination of eqs. 9.109 applied at $r = a$ and eq. 9.110 results in the relationships between C_{1mn} and $\mathscr{F}_{1mn}$. By both replacing C_{1mn} by the appropriate $\mathscr{F}_{1mn}$ relationship and requiring $\Phi(a,z,t)$ to be continuous at the boundary separating Regions 1 and 2, the following expression for the *displacement potential* for the fluid particles in *Region 1* results:

$$\Phi_1(r,z,t) = \frac{H_I}{2}h\sum_{m=0}^{\infty}\left\{\mathscr{F}_{1m0}\left(\frac{r}{a}\right)^m + 2\sum_{n=1}^{\infty}\mathscr{F}_{1mn}\frac{I_m(K_n r)}{I_m(K_n a)}\cos[K_n(z+h)]\right\} \times \cos(m\beta)e^{-i\omega t} \tag{9.112}$$

where the expression for K_n is presented in eq. 9.108. The coefficients $\mathscr{F}_{1mn}$ are unknown and must be determined. These coefficients are determined by matching the solutions for Regions 1 and 2 at the interface of the two regions.

Region 2: In Region 2, the displacement potential represents the incident wave, the waves reflected from the cylinder, and the effect of the particle motions represented by Φ_1 at the interface of Regions 1 and 2. The terms representing the incident and reflected waves should resemble the expression in eq. 9.68, and the form representing the displacement across the interface must be determined. To determine the displacement potential (Φ_2) in Region 2, we take the same approach as that leading to eq. 9.109. Following Garrett (1971), eq. 9.100 is solved by the separation-of-variables method and then the respective seafloor, cylinder-boundary, and free-surface conditions of eqs. 9.101, 9.103, and 9.104 are applied to the solution. If the assumed solution is $\Phi_2(r, \beta, z) = \mathrm{R}(r)\mathrm{Q}(\beta)\mathrm{Z}(z)$, then the separated equation for the z-coordinate can be expressed as

$$\frac{d^2\mathrm{Z}}{dz^2} = \alpha^2 \tag{9.113}$$

where the eigenvalue, α, can be real or imaginary. When real, we can write $\alpha = \pm k$, where k is the wave number, and the unique solution of eq. 9.113 is

$$Z_0 \propto \cosh[k(z+h)] \tag{9.114}$$

as in eq. 3.16. The value of k is obtained from the dispersion equation, eq. 3.31, written as

$$\frac{\omega^2}{g} = k_0 = k\tanh(kh) \tag{9.115}$$

where k_0 is the deep-water wave number. Equation 9.115 is a transcendental equation and can be solved by the method of successive approximations, as illustrated in Example 3.3. When α is imaginary, we can write $\alpha = \pm i\kappa$, and eq. 9.115 becomes

$$k_0 = (i\kappa)\tanh(i\kappa h) = -\kappa\tan(\kappa h) \tag{9.116}$$

This equation has an infinite number of κ-values that satisfy this relationship because a trigonometric function is involved. Miles and Garrett (1968) write that a good approximate relationship for the eigenvalue is

$$\kappa_n \simeq \frac{n\pi}{h} - \frac{k_0}{n\pi}, \quad n = 1, 2, \ldots \tag{9.117}$$

According to Miles and Garrett (1968), values obtained from the expression in eq. 9.117 have a maximum error of 1% for $k_0h = 1$, and an error less than 1% for all but the lowest mode ($n = 1$) for $k_0h < 10$.

The β-solution of eq. 9.100 for Region 2 is the same as that for Region 1. However, the r-solution is different in that it involves the modified Bessel function of the second kind, $K_m(\kappa_n r)$. The reason for this choice is that we require the potential function to remain finite as $r \to \infty$. Because $K_m(\kappa_n r) \to 0$ and $I_m(\kappa_n r) \to \infty$ as $r \to \infty$, the choice is obvious.

The complete solution of eq. 9.100 in Region 2 must represent the incident wave and the scattered wave fields. The latter field results from reflection and diffraction and includes a set of waves called either *trapped waves* or *evanescent waves.* The trapped waves are in the near field, and are standing waves having wave heights that decrease with increasing r. The displacement potential representing the incident wave component for Region 2 is

$$\begin{aligned}\Phi_I &= \frac{i}{\omega}\varphi_I \\ &= \frac{H_I}{2}\frac{1}{k_0}\frac{\cosh[k(z+h)]}{\cosh(kh)}\Re\left\{\sum_{m=0}^{\infty}\varepsilon_m i^m J_m(kr)\cos(m\beta)e^{-i\omega t}\right\}\end{aligned} \tag{9.118}$$

where the velocity potential, φ_I, is expressed by eq. 9.61b of the MacCamy-Fuchs derivation in Section 9.2E. A part of the scattered displacement potential should be similar to that in eq. 9.63 of the MacCamy-Fuchs derivation. Hence, we write

$$\Phi_{2S(\mathrm{A})} = -\frac{H_I}{2}\frac{1}{k_0}\frac{\cosh[k(z+h)]}{\cosh(kh)}\Re\left\{\sum_{m=0}^{\infty} E_m H_m^{(1)}(kr)\cos(m\beta)e^{-i\omega t}\right\} \tag{9.119}$$

where the subscript S identifies the scattered potential and the coefficient E_m is determined from a boundary condition. Garrett (1971) assumes that the sum of the potentials in eqs. 9.118 and 9.119 satisfies

$$[\Phi_I + \Phi_{2S(\mathrm{A})}]|_{r=a} = 0, \quad -d \leq z \leq 0 \tag{9.120}$$

$$\frac{\partial}{\partial r}[\Phi_I + \Phi_{2S(\mathrm{A})}]|_{r=a} = 0, \quad -d \leq z \leq 0 \tag{9.121}$$

and

$$\left|\left\{\frac{\partial}{\partial r}[\Phi_I + \Phi_{2S(\mathrm{A})}]\Big|_{r=a}\right\}\right| \geq 0, \quad -h \leq z < -d \tag{9.122}$$

The last of these conditions simply means that the radial velocity is continuous across the boundary separating the regions. The first of the three boundary conditions (eq. 9.120), applied to the side of the cylinder, yields

$$E_m = \varepsilon_m i^m \frac{J_m(ka)}{H_m^{(1)}(ka)} \tag{9.123}$$

The condition in eq. 9.120 is not assumed by Miles and Gilbert (1968). The reader should also note the difference between eqs. 9.67 (of the MacCamy-Fuchs derivation) and 9.123.

Garrett (1971) assumes that the remainder of the scattered field is represented by the first term of the following displacement potential expression:

$$\begin{aligned}\Phi_{2S(\mathrm{B})}(r,z,t) &= \frac{H_I}{2}h\Re\left\{\sum_{m=0}^{\infty}\sum_{\alpha}\mathcal{F}_{2m\alpha}\frac{K_m(\alpha r)}{K_m(\alpha a)}\cos[\alpha(z+h)]\cos(m\beta)e^{-i\omega t}\right\} \\ &= \frac{H_I}{2}h\Re\left\{\sum_{m=0}^{\infty}\left\langle\mathcal{F}_{2mk}\frac{K_m(-ikr)}{K_m(-ika)}\cos[-ik(z+h)]\right.\right. \\ &\quad \left.\left.+\sum_{\kappa}\mathcal{F}_{2m\alpha\kappa}\frac{K_m(\kappa r)}{K_m(\kappa r)}\cos[\kappa(z+h)[\cos(m\beta)\right\rangle e^{-i\omega t}\right\}\end{aligned} \tag{9.124}$$

where the modified Bessel function of the second kind, $K_m(\kappa_n r)$, is chosen because $K_m(\kappa_n r) \to 0$ as $r \to \infty$ and $I_m(\kappa_n r) \to \infty$. The values of $\mathcal{F}_{2m\alpha}$ are to be determined. In eq. 9.124, the first term in the series corresponds to $\alpha = -ik$ and, again, represents the remainder of the scattered field. The rest of the index values are the κ, which are real and greater than zero. Again, a good approximation of κ is in eq. 9.117, where $k_0 h < 10$. The reader should also note that

$$K_m(-ikr) = \frac{1}{2}\pi i^{m+1}H_m^{(1)}(kr) \tag{9.125}$$

and

$$\cos[-ik(z+h)] = \cosh[k(z+h)] \tag{9.126}$$

See the book edited by Abramowitz and Stegun (1965). The application of the displacement potential in eq. 9.124 to the boundary at $r = a$ results in

$$\Phi_{2S(B)}(a, z, t) = \frac{H_I}{2} h\Re\left\{\sum_{m=0}^{\infty}\sum_{\alpha} \mathscr{F}_{2m\alpha} \cos[\alpha(z+h)]\right\} \cos(m\beta)e^{-i\omega t}$$

$$= \frac{H_I}{2} h\Re \sum_{m=0}^{\infty} f_{2m}(z) \cos(m\beta)e^{-i\omega t}, \quad (-h \le z \le 0) \qquad (9.127)$$

where, as in eq. 9.110, $f_{2m}(z)$ is called a *boundary function*. The summation in eq. 9.127 can be considered to be a Fourier expansion similar to that in eq. 9.110. So, the coefficients in eq. 9.127 can be expressed as

$$\mathscr{F}_{2m\alpha} = \frac{1}{h}\int_{-h}^{0} f_{2m}(z) \cos[\alpha(z+h)]dz \qquad (9.128)$$

The use of the Fourier series to represent the boundary functions is rather clever. In both eqs. 9.111 and 9.126, the coefficients are, in effect, the values of the integrands averaged over the respective water depths in Regions 1 and 2, respectively.

The displacement potential in Region 2 can now be expressed as

$$\Phi_2 = \Phi_I + \Phi_{2S(A)} + \Phi_{2S(B)}$$

$$= \frac{H_I}{2} h\Re\left\{\sum_{m=0}^{\infty}\left[\left[\left\langle J_m(kr) - \frac{J_m(ka)}{H_m^{(1)}(ka)} H_m^{(1)}(kr)\right\rangle \frac{1}{kh}\frac{\cosh[k(z+h)]}{\sinh(kh)} \cos(m\beta)\right.\right.\right.$$

$$\left.\left.\left. + \sum_{\alpha} \mathscr{F}_{2m\alpha} \frac{K_m(\alpha r)\cos[\alpha(z+h)]}{K_m(\alpha a)\sqrt{\frac{1}{2}\left[1+\frac{\sin(\alpha h)}{2\alpha h}\right]}} \cos(m\beta)\right]\right] e^{-i\omega t}\right\},$$

$$(r \ge a, -h \le z \le 0) \qquad (9.129)$$

Again, in the α-summation, the first value of α is $-ik$ and the remaining values are equal to κ, representing real numbers. For $\alpha = -ik$, we note the relationships in eqs. 9.125 and 9.126. The potential function in eq. 9.129 satisfies the seafloor condition of eq. 9.101 and the *radiation condition* of eq. 9.105. It must also satisfy the respective cylinder-boundary condition and free-surface condition of eqs. 9.103 and 9.104.

Combined Solutions: Following Garrett (1971), there are three radial boundary conditions that must be satisfied by the displacement potential and its radial derivative (the radial displacement). The first two of these are that the radial displacement potential and its derivative must be continuous across the interface of Regions 1 and 2 of Figure 9.20b. Mathematically, these respective conditions are

$$\Phi_2|_{r=a} = \Phi_1|_{r=a}, \quad (-h \le z < -d) \qquad (9.130)$$

and

$$\left.\frac{\partial \Phi_2}{\partial r}\right|_{r=a} = \left.\frac{\partial \Phi_1}{\partial r}\right|_{r=a}, \quad (-h \le z < -d) \qquad (9.131)$$

As a result, the boundary functions defined in eqs. 9.110 and 9.127 must be equal over the common boundary of Regions 1 and 2, that is,

$$f_{1m}(z) = f_{2m}(z) = f_m(z), \quad (-h \le z < -d) \qquad (9.132)$$

The condition in eq. 9.130 (and in eq. 9.132) is satisfied by taking advantage of the Fourier coefficient expression in eq. 9.111 and the boundary-function relationships in eqs. 9.127 and 9.128. From these relationships, we obtain

$$
\begin{aligned}
\mathscr{F}_{1mn} &= \frac{1}{(h-d)}\int_{-h}^{-d} f_m(z)\cos[\mathrm{K}_n(z+h)]dz \\
&= \frac{1}{(h-d)}\int_{-h}^{-d}\sum_{\alpha}\mathscr{F}_{2m\alpha}\frac{\cos[\mathrm{K}_n(z+h)]}{\sqrt{\frac{1}{2}\left[1+\frac{\sin(2\alpha h)}{2\alpha h}\right]}}\cos[\alpha(z+h)]dz \\
&= \sum_{\alpha}\mathscr{F}_{2m\alpha}\frac{(-1)^n}{[\alpha^2(h-d)^2-n^2\pi^2]}\alpha(h-d)\frac{\sin[\alpha(h-d)]}{\sqrt{\frac{1}{2}\left[1+\frac{\sin(2\alpha h)}{2\alpha h}\right]}}, \quad (-h\le z<-d)
\end{aligned}
\tag{9.133}
$$

where K_n is defined in eq. 9.108. Again, we note that the first α-value is $-ik$, followed by the κ-values, where the latter can be approximated as in eq. 9.117. The continuous displacement boundary condition in eq. 9.131 yields

$$
\begin{aligned}
&\sum_{n=0}^{\infty}\varepsilon_n\mathscr{F}_{1mn}\mathrm{K}a\frac{I'_m(\mathrm{K}_n a)}{I_m(\mathrm{K}_n a)}\cos[\mathrm{K}_n(z+h)] \\
&= \frac{-\frac{2i}{\pi kh}\cosh[k(z+h)]}{H_m^{(1)}(ka)\sinh(kh)} + \sum_{\alpha}\mathscr{F}_{2m\alpha}\frac{\alpha a K'_m(\alpha a)}{K_m(\alpha a)}\frac{\cos[\alpha(z+h)]}{\sqrt{\frac{1}{2}\left[1+\frac{\sin(2\alpha h)}{2\alpha h}\right]}}, \\
&\qquad (-h\le z<-d)
\end{aligned}
\tag{9.134}
$$

The expressions in eqs. 9.133 and 9.134 can be combined by eliminating $\mathscr{F}_{1mn}$. This results is

$$
\begin{aligned}
&\sum_{\alpha}\mathscr{F}_{2m\alpha}\left[\left[\frac{\alpha a K'_m(\alpha a)}{K_m(\alpha a)}\frac{\cos[\alpha(z+h)]}{\sqrt{\frac{1}{2}\left[1+\frac{\sin(2\alpha h)}{2\alpha h}\right]}} - \alpha(h-d)\frac{\sin[\alpha(h-d)]}{\sqrt{\frac{1}{2}\left[1+\frac{\sin(2\alpha h)}{2\alpha h}\right]}}\right.\right. \\
&\left.\left.\times\sum_{n=0}^{\infty}\varepsilon_n\frac{(-1)^n}{[\alpha^2(h-d)^2-n^2\pi^2]}\mathrm{K}a\frac{I'_m(\mathrm{K}_n a)}{I_m(\mathrm{K}_n a)}\cos[\mathrm{K}_n(z+h)\right]\right] \\
&= \frac{\frac{2i}{\pi kh}}{H_m^{(1)}(ka)}\frac{\cosh[k(z+h)]}{\sinh(kh)}, \quad (-h\le z<-d)
\end{aligned}
\tag{9.135}
$$

The last boundary condition is the no-displacement condition that applies to the vertical surface of the truncated circular cylinder. That condition, expressed mathematically by the expression in eq. 9.121, results in

$$\sum_{\alpha} \mathscr{F}_{2m\alpha} \frac{\alpha a K_m'(\alpha a)\cos[\alpha(z+h)]}{K_m(\alpha a)\sqrt{\frac{1}{2}\left[1+\frac{\sin(2\alpha h)}{2\alpha h}\right]}} = \frac{\frac{2i}{\pi kh}}{H_m^{(1)}(ka)} \frac{\cosh[k(z+h)]}{\sinh(kh)}, \quad (-d \le z \le 0) \tag{9.136}$$

The following relationship is used to obtain the results in eqs. 9.134 and 9.136:

$$J_m(ka)H_m^{(1)'}(ka) - J_m'(ka)H_m^{(1)}(ka) = \frac{2i}{\pi ka} \tag{9.137}$$

Equations 9.135 and 9.136 comprise a set of expressions containing the unknown coefficients $\mathscr{F}_{2m\alpha}$. To determine the coefficient values for given wave, water depth, and cylinder-draft conditions, Garrett (1971) multiplies both equations by the function

$$G_\gamma(z) \equiv \frac{1\cos[\gamma(z+h)]}{h\sqrt{\frac{1}{2}\left[1+\frac{\sin(2\gamma h)}{2\gamma h}\right]}} = \frac{\cos[\gamma(z+h)]}{hQ_\gamma} \tag{9.138}$$

integrates each over their respective z-ranges, and adds the respective expressions. The resulting expression is

$$\sum_{\alpha} \mathscr{F}_{2m\alpha}\left\{-\gamma a \frac{K_m'(\gamma a)}{K_m(\gamma a)}\delta_{\alpha\gamma} + \sum_{n=0}^{\infty} \varepsilon_n\left(1-\frac{d}{h}\right)\frac{\alpha\gamma(h-d)^2}{[\alpha^2(h-d)^2 - n^2\pi^2][\gamma^2(h-d)^2 - n^2\pi^2]}\right.$$

$$\left.\cdot\frac{\sinh[\alpha(h-d)]\sin[\gamma(h-d)]}{Q_\alpha Q_\gamma} \mathrm{K}_n a \frac{I_m'(\mathrm{K}_n a)}{I_m(\mathrm{K}_n a)}\right\} = -\frac{2i\sqrt{\frac{1}{2}\left[1+\frac{\sinh(2kh)}{2kh}\right]}}{\pi kh H_m^{(1)}(ka)\sinh(kh)}\delta_{k\gamma} \tag{9.139}$$

In eq. 9.138, we have introduced a coefficient γ. The purpose of this coefficient is to take advantage of the property of orthogonality of the trigonometric functions. Also in eq. 9.138, the parameter Q_γ is defined. In eq. 9.139, the parameter Q_α is that where α is used in place of γ. Due to the orthogonality condition, the Kronecker delta, defined by

$$\delta_{\alpha\gamma} \equiv \begin{cases} 0, & (\alpha \neq \gamma) \\ 1, & (\alpha = \gamma) \end{cases} \tag{9.140}$$

is used in eq. 9.139.

The expression in eq. 9.139 is complex due to the Hankel function that appears when the index $\alpha = -ik$. After separating the real and imaginary terms, Garrett (1971) defines a real matrix resulting from the separation, assumes that the matrix is nonsingular, and introduces a real solution $\Gamma_{m\alpha}$ to the resulting matrix equation. That resulting equation is

$$\sum_{\alpha} \Gamma_{m\alpha}\left\{-\gamma a \frac{K_m'(\gamma a)}{K_m(\gamma a)}\delta_{\alpha\gamma} + \sum_{n=0}^{\infty} \varepsilon_n\left(1-\frac{d}{h}\right)\frac{\alpha\gamma(h-d)^2}{[\alpha^2(h-d)^2 - n^2\pi^2][\gamma^2(h-d)^2 - n^2\pi^2]}\right.$$

$$\left.\cdot\frac{\sin[\alpha(h-d)]\sin[\gamma(h-d)]}{Q_\alpha Q_\gamma} \mathrm{K}_n a \frac{I_m'(\mathrm{K}_n a)}{I_m(\mathrm{K}_n a)} + ka\frac{H_m^{(1)'}(ka)}{H_m^{(1)}(ka)}\delta_{k\alpha}\delta_{k\gamma}\right\} = \delta_{k\gamma} \tag{9.141a}$$

Because this equation is somewhat unwieldy, let us rewrite the equation as

$$\sum_{a} \Gamma_{ma}\left\{-A_m(\gamma a)\delta_{a\gamma} + \sum_{n=0}^{\infty} B_{mn}(a,\gamma) + A_m(ka)\delta_{ka}\delta_{k\gamma}\right\} = \delta_{k\gamma} \tag{9.141b}$$

To help the reader obtain solutions for $\Gamma_{m\alpha}$, the following results from eq. 9.140 are useful:

α	γ	$\delta_{\alpha\gamma}$	$\delta_{k\alpha}$	$\delta_{k\gamma}$
$-ik$	$-ik$	1	1	1
$-ik$	$\neq -ik$	0	1	0
κ	κ	1	0	0
κ	$\neq \kappa$	0	0	0

Restricting the values of γ such that $\gamma = \alpha$, only the first and third lines of the table need be considered. For the first condition in the table, $A_m(-ika) = A_m(ka)$ and eq. 9.141b becomes

$$\Gamma_{mk} \sum_{n=0}^{\infty} B_{mn}(k) = 1 \tag{9.142}$$

For the third line in the table, the following relationship results:

$$\sum_{\kappa} \Gamma_{m\kappa}\left\{-A_m(\kappa a) + \sum_{n=0}^{\infty} B_{mn}(\kappa)\right\} = 0 \tag{9.143}$$

Garrett (1971) finds that the convergence of the n-summation is rather rapid, so he obtains solutions for a finite number of α values. The reader is encouraged to consult that reference for a discussion of the solution method.

The combination of eqs. 9.139 and 9.141a, using the values of Γ_{ma} obtained from eqs. 9.142 and 9.143, results in the following expression for $\mathscr{F}_{2m\alpha}$:

$$\mathscr{F}_{2ma} = -\frac{\dfrac{2i\sqrt{\dfrac{1}{2}\left[1+\dfrac{\sinh(2kh)}{2kh}\right]}}{\pi kh H_m^{(1)}(ka)\sinh(kh)}\Gamma_{ma}}{1 - ka\dfrac{H_m^{(1)'}(ka)}{H_m^{(1)}(ka)}\Gamma_{mk}} \tag{9.144}$$

The reader should note that $\mathscr{F}_{2m\alpha}$ is complex; however, $\Gamma_{m\alpha}$ is real, and is the only term depending on α. Therefore, the *phase* of $\mathscr{F}_{2m\alpha}$ is independent of α. By using the relationship in eq. 9.134, the values of $\mathscr{F}_{1m\alpha}$ are determined from eq. 9.133. The respective displacement potential functions Φ_1 and Φ_2 of eqs. 9.112 and 9.129 are then determined.

The purpose of the Garrett (1971) study was to determine the wave-induced forces and moments on a fixed circular dock (a vertical, truncated circular cylinder). The force and moment expressions are now derived.

Wave-Induced Forces and Moments: The Garrett analysis is linear and based on the assumption of irrotational flow; hence, the wave-induced pressure dynamic on the vertical, truncated, circular cylinder can be obtained from the linearized

Bernoulli's equation (see eq. 3.70). The expression for the dynamic pressure in terms of both the velocity potential (φ) and displacement potential (Φ) is

$$p = -\rho\frac{\partial\varphi}{\partial t} = i\omega\rho\frac{\partial\Phi}{\partial t} \tag{9.145}$$

where the Region 1 displacement potential, Φ_1, is obtained from eq. 9.112, and the Region 2 potential, Φ_2, is expressed by eq. 9.129. Our interest is in the horizontal force (F_x), the vertical force (F_z), and the overturning moment (M_y) caused by the wave-induced dynamic pressure.

Consider first the horizontal force on the cylinder. This is found by integrating the pressure in eq. 9.145 over the side of the cylinder. Referring to Figure 9.16, the resulting force expression is

$$F_{xG} = -\int_0^{2\pi}\int_{-d}^{0} p|_{r=a}\, a\cos(\beta)dzd\beta = -i\omega\rho\int_0^{2\pi}\int_{-d}^{0}\frac{\partial\Phi_2|_{r=0}}{\partial t}a\cos(\beta)dzd\beta \tag{9.146}$$

From the substitution of the displacement potential of eq. 9.129, this following expression for the non-dimensional horizontal force is found:

$$\frac{F_{xG}}{\rho g\frac{H_I}{2}\pi a^2} = -2kh\tanh(kh)\Re\left\{\left\langle i_{\mathscr{F}_{2,1k}}\frac{\sinh(kh)-\sinh[k(h-d)]}{(ka)\sqrt{\frac{1}{2}\left[1+\frac{\sinh(2kh)}{2kh}\right]}}\right.\right.$$
$$\left.\left.+\sum_{\Re\alpha}\mathscr{F}_{2,1\alpha}\frac{\sin(\alpha h)-\sin[\alpha(h-d)]}{(\alpha a)\sqrt{\frac{1}{2}\left[1+\frac{\sin(2\alpha h)}{2\alpha h}\right]}}e^{-\omega t}\right\rangle\right\} \tag{9.147}$$

where the β-integration in eq. 9.146 results in nonzero terms if and only if $m = 1$. Note that the summation is over the real values of α, that is, $\alpha = -ik$ is not included in the summation.

The vertical force on the truncated circular cylinder is obtained from

$$F_{zG} = \int_0^{2\pi}\int_0^{a} p|_{z=-d}\, r\,dr\,d\beta = i\omega\rho\int_0^{2\pi}\int_0^{a}\frac{\partial\Phi_1|_{z=-d}}{\partial t}r\,dr\,d\beta \tag{9.148}$$

The non-dimensional vertical force expression resulting from the combination of eqs. 9.148 and 9.112 is

$$\frac{F_{zG}}{\rho g\frac{H_I}{2}\pi a^2} = 2kh\tanh(kh)\left\langle\frac{1}{2}\mathscr{F}_{1,00} + 2\sum_{n=1}^{\infty}(-1)^n\mathscr{F}_{1,0n}\frac{1}{K_n a}\frac{I_0'(K_n)}{I_0(K_n a)}\right\rangle e^{-i\omega t} \tag{9.149}$$

The β-integration in eq. 9.148 results in nonzero terms only for $m = 0$, and K_n is defined in eq. 9.108.

Finally, the moment about the y-axis (positive in the counterclockwise direction) due to the respective horizontal and vertical forces in eqs. 9.146 and 9.148 is

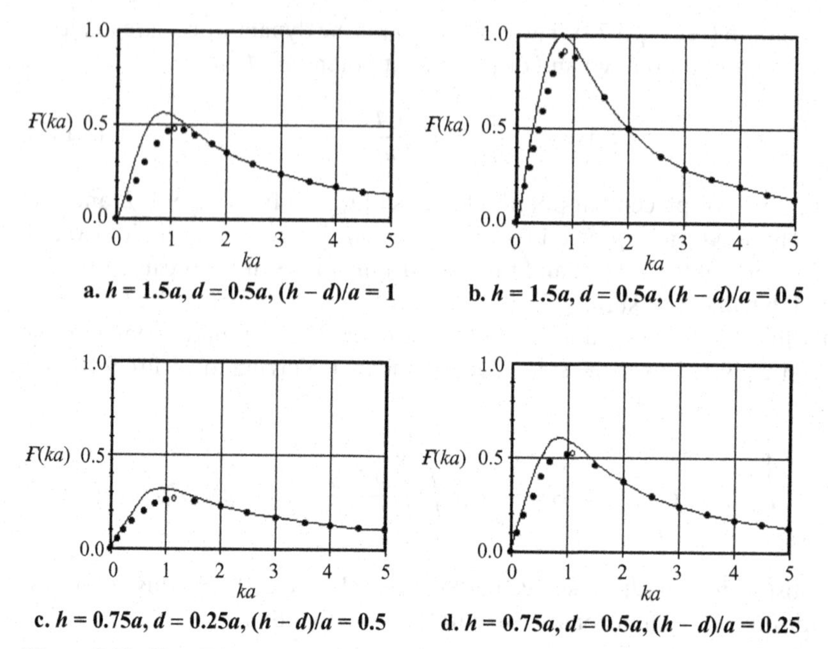

a. $h = 1.5a, d = 0.5a, (h - d)/a = 1$ b. $h = 1.5a, d = 0.5a, (h - d)/a = 0.5$

c. $h = 0.75a, d = 0.25a, (h - d)/a = 0.5$ d. $h = 0.75a, d = 0.5a, (h - d)/a = 0.25$

Figure 9.22. *Non-Dimensional Horizontal Force on a Truncated Vertical Cylinder.* The solid curves are obtained from the van Oortmerssen (1971) expression in eq. 9.85. The data points • result from the Garrett (1971) expression in 9.147, where the peak values of these are denoted by ∘. The non-dimensional force notation, $\mathcal{F}$, is defined in eq. 9.151.

obtained from

$$M_{yG} = \rho\omega\Re\left\{i\left\langle -\int_0^{2\pi}\int_{-d}^{0}\frac{\partial\Phi_2|_{r=a}}{\partial t}a\cos(\beta)z dz d\beta + \int_0^{2\pi}\int_0^{a}\frac{\partial\Phi_1|_{z=-d}}{\partial t}r^2\cos(\beta)dr d\beta\right\rangle\right\} \tag{9.150}$$

The substitution of the displacement potential expressions and the subsequent integrations is left to the reader because, in the following paragraphs, we present comparisons of the forces only.

(4) Results of the Approximate and Garrett Forces

Following McCormick and Cerquetti (2004), results obtained from the approximate force expressions in eqs. 9.86 and 9.95 are compared with the respective Garrett (1971) expressions in eqs. 9.147 and 9.149. The conditions used in the comparisons are those used in the Garrett paper. The force amplitudes are presented in non-dimensional forms, where the forces are divided by the buoyant force of a passing wave of infinite length, that is,

$$\mathcal{F}(ka) \equiv \frac{F}{\rho g\frac{H_I}{2}\pi a^2}. \tag{9.151}$$

The horizontal force results are presented in Figure 9.22, where the following values apply:

Figure 9.22a: $h = 1.5a$, $d = 0.5a$, $(h - d)/a = 1$
Figure 9.22b: $h = 1.5a$, $d = a$, $(h - d)/a = 0.5$

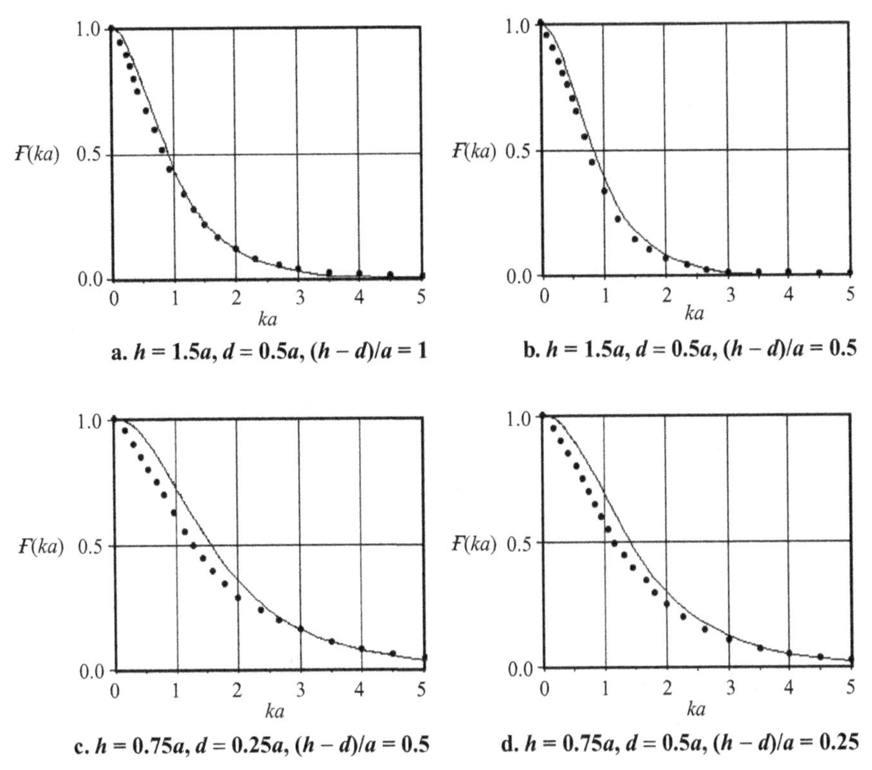

Figure 9.23. *Non-Dimensional Vertical Force Results.* The solid curves are obtained from the McCormick and Cerquetti (2004) expression in eq. 9.95 combined with that in eq. 9.98. The data points • result from the Garrett (1971) expression in eq. 9.147. The non-dimensional force notation, F, is defined in eq. 9.151.

Figure 9.22c: $h = 0.75a$, $d = 0.25a$, $(h - d)/a = 0.5$
Figure 9.22d: $h = 0.75a$, $d = 0.5a$, $(h - d)/a = 0.25$

In Figure 9.23, the dimensionless vertical force results are presented for the following conditions:

Figure 9.23a: $h = 1.5a$, $d = 0.5a$, $(h - d)/a = 1$
Figure 9.23b: $h = 1.5a$, $d = a$, $(h - d)/a = 0.5$
Figure 9.23c: $h = 0.75a$, $d = 0.25a$, $(h - d)/a = 0.5$
Figure 9.23d: $h = 0.75a$, $d = 0.5a$, $(h - d)/a = 0.25$

The ka range for each of these figures is from 0 to 5. From the Chakrabarti chart in Figure 9.8, we see that this range takes us well into the diffraction force region identified as $ka > 0.5$.

In Figure 9.22, the curves of van Oortmerssen (1971) are seen to agree with the Garrett (1971) data for $ka > 1$. For $ka > 1.5$, the curves essentially coalesce. Again, according to the Chakrabarti chart in Figure 9.8, the diffraction force range is $ka > 0.5$. We see that the qualitative agreement over this range is quite good. However, the quantitative agreement somewhat lacks in $0.5 < ka < 1$.

The vertical force results in Figure 9.23, comparing data obtained from Garrett (1971) and from eq. 9.95 (combined with the expression in eq. 9.98), show good qualitative agreement for all depth, draft, and radius relationships considered. The quantitative results are satisfactory over the entire ka range, with the best agreement where $ka > 2$.

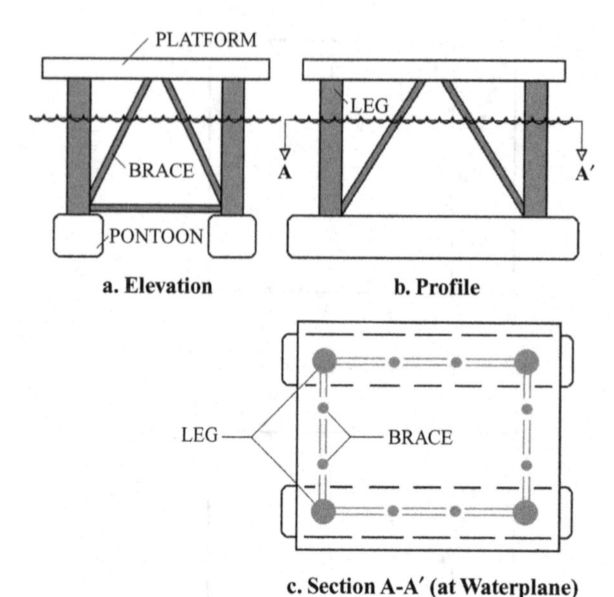

Figure 9.24. *Schematic Sketches of an Offshore Floating Platform.* The legs have diameters that are of the order of magnitude of the wavelength of the incident waves. The wave force on each leg is then primarily a diffraction force. The braces are of relatively small diameter, and are subject to viscous and inertial forces. See the Chakrabarti diagram in Figure 9.8.

The advantage of the using the approximate expressions of van Oortmerssen (1971) and McCormick and Cerquetti (2004) in predicting the respective horizontal and vertical force on a vertical, truncated circular cylinder in waters of finite depth is that the expressions are analytical in nature and relatively easy to apply to engineering projects, particularly in the conceptual design phase. In addition, because of their simplicity the expressions are readily usable in the analysis of wave forces on truncated cylinders in random seas. The subject of random wave forces is discussed in Section 9.3.

In the next section, our attention is focused on the scattering effects of multiple vertical cylinders in waters of finite depth.

I. Scattering Effects of Large-Diameter Leg Arrays

Semi-submerged offshore structures are platforms supported by three or more large legs. The legs usually have circular cross-sections, although other cross-sectional geometries have been used. A sketch of a four-leg semi-submersible is presented in Figure 9.24 for the sake of discussion. The legs of the structure are normally supported by fully submerged pontoons when the structure is onsite in an operational mode. When in transit from the yard to the site, the pontoons are floating and the legs are entirely above the waterplane. There are also floating structures that are totally supported by buoyant vertical legs, that is, the legs "stand alone" without pontoons and the submerged ends of the legs are free. The free-leg configuration might be found on a *tension-leg platform* (TLP). The paper by Söylemez and Yilmaz (2004) contains a discussion of the hydrodynamic design of a TLP, and is recommended for those readers interested in this type of structure. The motions of the TLP are discussed in Chapter 10.

In addition to the legs of the structure, small-diameter cylinders are positioned between both leg pairs and the pontoons to give both longitudinal and transverse support. These support structures are known as braces or cross-braces, and are illustrated in Figure 9.24. In a typical operational seastate, the legs mostly experience

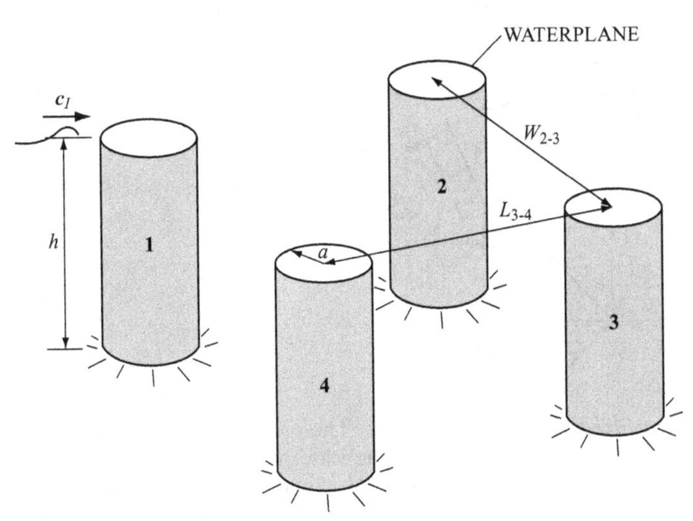

Figure 9.25. *Leg Orientation for a Fixed, Four-Leg Structure.* The orientation sketch is for the submerged portion of the legs, that is, extending from the waterplane to the flat, horizontal bed.

diffraction forces, whereas the braces experience Morison-type forces, that is, viscous and inertial forces discussed in Sections 9.2C and 9.2D.

In this section, we confine our discussion to the sheltering influence on wave-induced diffraction forces on a leg of a bed-mounted, multi-leg structure. That is, our interest is in the diffraction forces on a component leg due to the cumulative wave effects of wave scattering from the neighboring legs of the structure. The analysis present is analytical, and is amenable to calculations using spreadsheets. Wave forces on large-diameter, vertical cylinders in waters of finite depth can also be studied using various numerical techniques such as the finite-element method (FEM), the boundary-element method (BEM), or the volume-of-fluid method (VOF). The reader is referred to the book by Faltinsen (1990) for an excellent introduction to the application of numerical methods to load analyses on ships and offshore structures. We differentiate analytical and numerical techniques as follows: Analytical techniques are those involving either exact or numerical solutions of differential or integral equations. For example, the approximate solution of a differential equation might be obtained by using the Runga-Kutta method, presented in Appendix B. Numerical hydrodynamic techniques involve the FEM, the BEM, or the VOF previously mentioned. Numerical techniques are the tools of the general field known as CFD and, as such, are computationally intensive.

Consider the sketch in Figure 9.25, where four vertical circular legs of a fixed platform are shown without braces. The legs are arranged in a rectangular pattern, where the leg centers are separated length-wise by $L_{j\text{-}k}$ and width-wise by $W_{j\text{-}k}$. The subscripts refer to the numbers of the leg pairs. Initially, an incident wave is scattered by Leg 1 in the sketch. Two wave systems exist just leeward of Leg 1, those being the incident wave system and the scattered wave system from Leg 1. Both systems then encounter the other legs, and subsequent scattering wave systems result at each leg. In turn, these scattered waves are scattered again by the neighboring legs, and so forth. One can see that the wave field within the region defined by the legs rapidly becomes rather unwieldy and difficult to analyze.

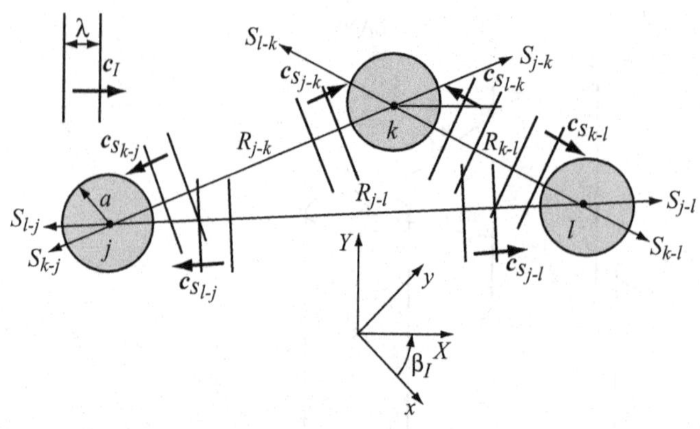

Figure 9.26. *Orientation for a Fixed, Three-Leg Structure.* The orientation sketch is for the submerged portion of the legs, that is, extending from the waterplane to the bed. The scattered waves are shown as plane waves, following the approximation of McIver and Evans (1984). The x-y coordinate system is fixed, where the x is the onshore coordinate and y is the longshore coordinate. The X-Y coordinate system is determined by the wave direction, where the X-coordinate is in the wave direction and Y the crest-wise coordinate. The s-coordinates are in the direction of the arriving scattered waves originating at neighboring legs or caissons.

In this section, the approximate analysis of McIver and Evans (1984) is presented. Our choice of this analysis is based on the simplicity of the analysis and the relatively good agreement with more accurate analyses. The McIver-Evans scattered plane-wave approximation is an extension of that of Simon (1982). For those readers interested in more exact analyses, the software called WAMIT is recommended. This software was developed by Prof. J. N. Newman and his colleagues at the Massachusetts Institute of Technology, and is one of the most accurate of the computational hydrodynamic tools available. For a description of WAMIT, see Lee (1995).

To visualize the primary and secondary scattering, consider three arbitrarily spaced vertical legs each having a radius, a, sketched in Figure 9.26. The legs are exposed to linear, monochromatic waves. Following McIver and Evans (1984), assume that the wave crests of the scattered waves from one leg arriving at a neighboring leg have crest lines that are approximately rectilinear. In other words, the scattered waves are approximately plane waves. The results of the assumption improve as the distance between the legs increases and becomes large compared to the radius, for example, $R_{j-k} \gg a$.

We begin our analysis of the wave field and the resulting forces and moments on the legs by considering a structure supported by two legs, Legs j and k, sketched in Figure 9.27. In that figure, the approximately planar scattered waves from Leg j arrive at Leg k at a celerity $\boldsymbol{c}_{j\text{-}k}$, and those from Leg k arrive at Leg j at a celerity $\boldsymbol{c}_{k\text{-}j}$. These waves are again scattered and as part of the secondary scattering return to legs of origin. The process continues with tertiary scattering, and so on. Note that the sea bed is assumed to be flat and horizontal so that the wavelengths (λ) and celerities of the scattered systems are equal in magnitude to those of the incident waves. The velocity potential for the incident and scattered waves from Leg j can be represented by the MacCamy-Fuchs expression in eq. 9.68. The second term of that expression represents the scattered waves. For an isolated Leg j, the velocity

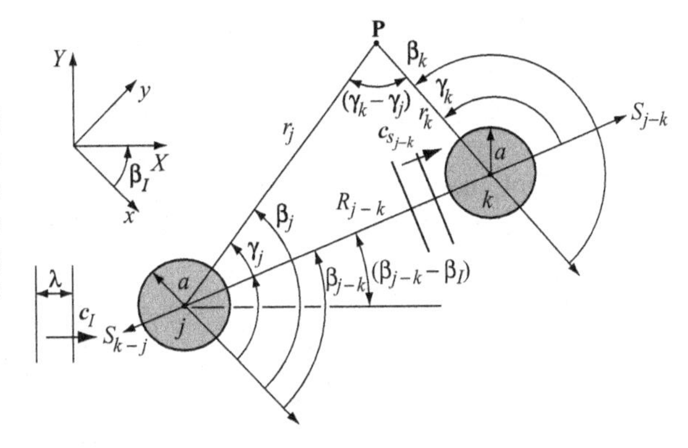

Figure 9.27. *Incident and Approximated Scattered Plane Waves.* The separation length $R_{j\text{-}k}$ of Legs j and k is assumed to be much greater than the radius a to use the plane-wave approximation, following McIver and Evans (1984). The point P is referred to as a *field point.*

MacCamy-Fuchs potential (denoted by the subscript MF) is

$$\varphi_{j,MF} = -\frac{H_I}{2}\frac{g}{\omega}\frac{\cosh[k(z+h)]}{\cosh(kh)} \times \Re\left\{\sum_{m=0}^{\infty} i^{m+1}\varepsilon_m\left[J_m(kr) - \frac{J'_m(ka)}{H_m^{(1)'}(ka)}H_m^{(1)}(kr)\right]\cos(m\beta)e^{-i\omega t}\right\}$$
$$= Z(z)\Re\left\{\sum_{m=0}^{\infty} i^{m+1}\varepsilon_m\left[J_m(kr) - \frac{J'_m(ka)}{H_m^{(1)'}(ka)}H_m^{(1)}(kr)\right]\cos(m\beta)e^{-i\omega t}\right\} \quad (9.152)$$

where the notation Z(z) is introduced for the purpose of brevity. It should be noted that eq. 9.152 can also be written in terms of a summation from $-\infty$ to $+\infty$, as is done by McIver and Evans (1984). Our choice of the semi-infinite summation is consistent with the MacCamy-Fuchs analysis presented in Section 9.2E. In Figure 9.27, the velocity potential of the scattered waves from Leg j is assumed to be

$$\varphi_{j-\mathrm{P}} = -Z(z)\Re\left\{\sum_{m=0}^{\infty} i^{m+1}\varepsilon_m\frac{J'_m(ka)}{H_m^{(1)'}(ka)}H_m^{(1)}(kr_j)\cos(m\gamma_j)e^{-i\omega t}\right\} \quad (9.153)$$

where the radial distance r_j and the angle γ_j are defined in Figure 9.27. The triangle formed by r_j, r_k, and $R_{j\text{-}k}$ has interior angles of γ_j, $\pi - \gamma_k$, and between sides r_j and r_k, $\gamma_k - \gamma_j$. In the sum eq. 9.153, the Addition Theorem for Bessel functions [see eq. 21.8–70 in Korn and Korn (2000)] can be used to replace the product of the Hankel and the cosine functions as

$$H_m^{(1)}(kr_j)\cos(m\gamma_j) = \sum_{n=-\infty}^{\infty}(-1)^n H_{m+n}^{(1)}(kR_{j-k})J_n(kr_k)\cos[n\gamma_k] \quad (9.154)$$

The combination of eqs. 9.153 and 9.154 yields

$$\varphi_{j-\mathrm{P}} = -Z(z)\Re \times \left\{\sum_{m=0}^{\infty} i^{m+1}\varepsilon_m\frac{J'_m(ka)}{H_m^{(1)'}(ka)}\sum_{n=-\infty}^{\infty}(-1)^n H_{m+n}^{(1)}(kR_{j-k})J_n(kr_k)\cos[n\gamma_k]e^{-i\omega t}\right\} \quad (9.155)$$

According to McIver and Evans (1984), for large values of the separation distance R_{j-k}, the Hankel function in eqs. 9.154 and 9.155 can be well represented by the approximation

$$H_{m+n}^{(1)}(kR_{j-k}) \simeq (-i)^n H_m^{(1)}(kR_{j-k})\left[1 + i\frac{(mn+n^2/2)}{kR_{j-k}}\right] + \mathrm{O}\left\{\frac{1}{(kR_{j-k})^{5/2}}\right\} \tag{9.156}$$

Combine the expressions in eqs. 9.155 and 9.156, and assume that $kR_{j-k} \gg 1$ so that the order-of-magnitude term (the last term) in eq. 9.156 is negligible. The result is

$$\begin{aligned}
\varphi_{j-\mathrm{P}} &= \varphi_{j-\mathrm{P1}} + \varphi_{j-\mathrm{P2}} \\
&\simeq -Z(z)\Re\left\{\sum_{m=0}^{\infty} i^{m+1}\varepsilon_m \frac{J_m'(ka)}{H_m^{(1)'}(ka)} H_m^{(1)}(kR_{j-k}) \sum_{n=-\infty}^{\infty} i^n J_n(kr_k)\cos[n\gamma_k]e^{-i\omega t}\right\} \\
&\quad - Z(z)\Re\left\{\sum_{m=0}^{\infty} i^{m+1}\varepsilon_m \frac{J_m'(ka)}{H_m^{(1)'}(ka)} H_m^{(1)}(kR_{j-k})\right. \\
&\quad \left.\times \sum_{n=-\infty}^{\infty} i^{n+1}\frac{(mn+n^2/2)}{kR_{j-k}} J_n(kr_k)\cos[n\gamma_k]e^{-i\omega t}\right\}
\end{aligned} \tag{9.157}$$

The second line of this equation represents scattered plane-wave potential ($\varphi_{j\text{-P1}}$) from Leg j at Leg k. The reader can verify this by noting that the m-summation of the second line of eq. 9.157 is a constant multiplied by the imaginary number, i, and for the negative index values in the n-summation, $J_{-n}(kr_k)$ can be replaced by $(-1)^n J_n(kr_k)$. The potential φ_{j-P1} is then of the form of the incident plane-wave velocity potential in eq. 9.61b. The second line of the equation is called the *plane-wave approximation* of the scattered potential at Leg k, as termed by Simon (1982) and McIver and Evans (1984). McIver and Evans (1984) introduce a first correction to the plane-wave approximation by including the third line of eq. 9.157, represented by $\varphi_{j\text{-P2}}$.

When eq. 9.157 is applied at the center of Leg k (that is, where the field point P is at the center of Leg k), the value of r_k is zero and the only nonzero value of $J_n(0)$ is for $n = 0$. The resulting scattered wave at Leg k originating at Leg j is represented by the following velocity potential:

$$\varphi_{j-k} \simeq -Z(z)\Re\left\{\sum_{m=0}^{\infty} i^{m+1}\varepsilon_m \frac{J_m'(ka)}{H_m^{(1)'}(ka)} H_m^{(1)}(kR_{j-k})e^{-i\omega t}\right\} \tag{9.158}$$

This plane wave is also scattered, and the secondary scattered wave from Leg k arriving at Leg j has a potential φ_{j-k-j}. In summary, at Leg j the velocity potential represents that of the incident wave (φ_I) and the scattered wave (φ_{Sj}) resulting from the incident wave, the primary scattered waves ($\varphi_{N-j,}$) due to the incident wave at the N neighboring legs and the corresponding scattered waves ($\varphi_{SN-j,}$), the secondary scattered waves ($\varphi_{j-N-j,}$) from the neighboring legs and the corresponding scattered waves ($\varphi_{Sj\text{-}N\text{-}j,}$), and so on. The potential at Leg j is then

$$\begin{aligned}
\varphi_j = \varphi_I + \varphi_{Sj} &= \varphi_{1-j} + \cdots + \varphi_{N-1} + \varphi_{S1-j} + \cdots + \varphi_{SN-j} \\
&\quad + \varphi_{j-1-j} + \cdots + \varphi_{j-N-j} + \varphi_{Sj-1-j} + \cdots + \varphi_{Sj-N-j} + \cdots
\end{aligned} \tag{9.159}$$

The scattered waves mentioned here are represented in the respective rows of eq. 9.159. Note, that the first line is the MacCamy-Fuchs velocity potential in eq. 9.152. The potential at each of the neighboring legs has the form of the expression in eq. 9.159.

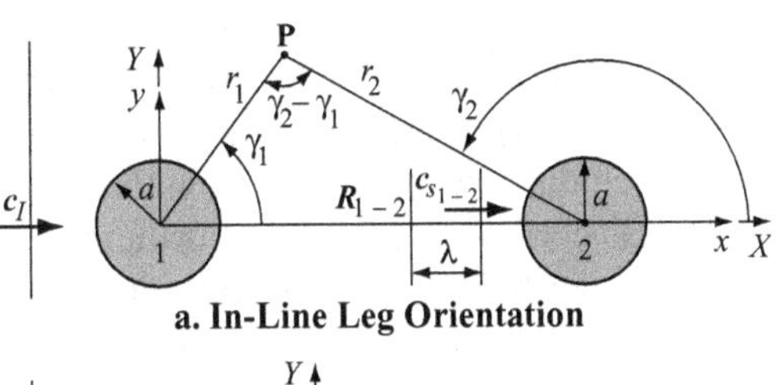

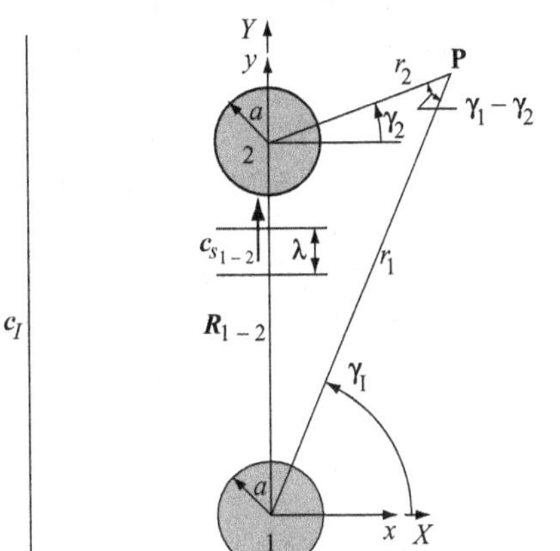

Figure 9.28. *Waterplanes of In-Line and Transverse Leg Structures.* For both orientations, incident wave angle $\beta_I = 0$. In (a), the in-line leg orientation angle is $\beta_{1-2} = 0$. For (b), the transverse orientation angle is $\beta_{1-2} = \pi/2$.

The goal of this section is to discuss the effects of scattering on the wave-induced forces on fixed, multi-leg structures. To this end, we confine our attention to the two specific orientations of a two-leg platform in Figure 9.28. The first of these (the in-line orientation in Figure 9.28a) is that where a vertical plane through the wave crest is normal to the vertical plane containing the centers of the legs. We refer to these respective planes as the *crest plane* and the *centerplane.* This in-line orientation is a case of pure sheltering because the leeward leg (Leg 2 in Figure 9.28a) is not directly exposed to the incident wave, that is, Leg 2 is in the shadow of Leg 1. The second case is that where the crest plane and centerplane are parallel (the transverse orientation in Figure 9.28b). For both cases, assume that the wave approaches the shoreline directly, and that the origins of the inertial (x,y) and wave (X,Y) coordinate systems are both fixed on Leg 1. These two assumptions are made simply to reduce the "bookkeeping" in the analysis. For the in-line orientation in Figure 9.28a, the angles in Figure 9.27 are as follows:

incident wave angle————$\beta_I = 0$
in-line leg orientation————$\beta_{1\text{-}2} = 0$
variable angles——————$\gamma_1 = \beta_1$

For the transverse orientation in Figure 9.28b, the angles are

incident wave angle————$\beta_I = 0$
in-line leg orientation————$\beta_{1\text{-}2} = \pi/2$
variable angles——————$\gamma_1 = \beta_1$

Consider the situation sketched in Figure 9.28a where the legs are in-line. Again, the leeward leg (Leg 2) is sheltered because the leg is in the shadow of Leg 1. If the distance $R_{1\text{-}2}$ is relatively large compared to the leg radius a (a condition required to have scattered plane waves at Leg 2), then the incident wave nearly reforms over R_{1-2} and, as a result, Leg 2 directly encounters the incident wave. Without a loss in generality, we can assume that there is an incident semi-infinite, shallow-water wave group traveling over a bed of uniform depth, h. Because of the shallow-water assumption, the celerity and group velocity are approximately equal, according to eq. 3.64b. Furthermore, because the bed is uniform, the scattered-wave celerity is equal in magnitude to the incident wave celerity, that is, $c_S = c_I = c$. The shallow-water wave group assumption allows us to study the scenario of events from the time that the leading wave of the group arrives at Leg 1. Let that time be $t_a = 0$. The scattered plane wave resulting from this incident wave encounter arrives at Leg 2 at a time $t_b = R_{1-2}/c$. In turn, this wave results in a scattered wave that arrives back at Leg 1 at time $t_c = 2R_{1-2}/c$, and so forth. The heights of the successive scattered plane waves are reduced as time increases due to diffraction. From the MacCamy-Fuchs analysis, the first two of these wave heights are obtained from eq. 9.69. The expression for the initially scattered wave height from that equation applied at Leg 1 is

$$H_{s1}(r_1, \gamma_1) = -\frac{H_I}{2}\Re\left\{\sum_{m=0}^{\infty} i^m \varepsilon_m \frac{J'_m(ka)}{H_m^{(1)'}(ka)} H_m^{(1)}(kr_1)\cos(m\gamma_1)\right\} \tag{9.160}$$

where $H_m^{(1)}$ is the Hankel function of the first kind. Applying this expression at $r_1 = R_{1-2}$ results in the following expression for the height of the primary scattered wave originating at Leg 1 and arriving at Leg 2:

$$H_{s1}(R_{1-2}, 0) \equiv H_{1-2} = -\frac{H_I}{2}\Re\left\{\sum_{m=0}^{\infty} i^m \varepsilon_m \frac{J'_m(ka)}{H_m^{(1)'}(ka)} H_m^{(1)}(kR_{1-2})\right\} \tag{9.161}$$

where, again, $R_{1-2} \gg a$. In terms of the coordinate system at Leg 2, the free-surface displacement representing the incident scattered wave from Leg 1 and the newly scattered wave at Leg 2 is

$$\begin{aligned} &\eta_{1-2}(r_2, \gamma_2, t) + \frac{H_{S1-2}(r_2, \gamma_2)}{2}e^{-i\omega t} \\ &= \frac{H_{1-2}}{2}\Re\left\{\sum_{m=0}^{\infty} i^m \varepsilon_m \left[J_m(kr_2) - \frac{J'_m(ka)}{H_m^{(1)'}(ka)} H_m^{(1)}(kr_2)\right]\cos(m\gamma_2)e^{-i\omega t}\right\} \end{aligned} \tag{9.162}$$

From this equation, the scattered plane wave from Leg 2 arriving at Leg 1 (where $r_2 = R_{1-2}$ and $\gamma_2 = 180°$) has a height of

$$\begin{aligned} H_{1-2-1} &= \frac{H_{1-2}}{2}\Re\left\{\sum_{m=0}^{\infty} i^m \varepsilon_m \frac{J'_m(ka)}{H_m^{(1)'}(ka)} H_m^{(1)}(kR_{1-2})e^{-i\omega t}\right\} \\ &= -\frac{H_I}{2}\Re\left\{\sum_{m=0}^{\infty} i^m \varepsilon_m \frac{J'_m(ka)}{H_m^{(1)'}(ka)} H_m^{(1)}(kR_{1-2})\right\} \\ &\quad \cdot \Re\left\{\sum_{n=0}^{\infty} i^n \varepsilon_m \frac{J'_n(ka)}{H_m^{(1)'}(ka)} H_n^{(1)}(kR_{1-2})e^{-i\omega t}\right\} \end{aligned} \tag{9.163}$$

The reader can see the progression of events as time increases.

Because the energy of the incident wave is distributed among an increasing number of scattered waves as time increases, the wave heights of the successive scattered waves decrease. This reduction in the scattered wave heights results in a corresponding wave force reduction over time. After some time, say t_q, the additional

Table 9.1. *Force scenario for the in-line leg orientation over* $0 \leq t \leq 3R_{1-2}/c$

Time	Force on Leg 1	Force on Leg 2
$t_a = 0$	$\boldsymbol{F}_I + \boldsymbol{F}_{S1} = \boldsymbol{F}_{1,\mathbf{MF}}$ at $t = 0$ (the MacCamy-Fuchs force in eqs. 9.71 and 9.85)	0
$t_b = \boldsymbol{R}_{1-2}/\boldsymbol{c}$	$\boldsymbol{F}_I + \boldsymbol{F}_{S1}$ at t_b	$\boldsymbol{F}_I + \boldsymbol{F}_{S1} + \boldsymbol{F}_{1-2} + \boldsymbol{F}_{S1-2}$ at t_b
$t_c = 2\boldsymbol{R}_{1-2}/\boldsymbol{c}$	$\boldsymbol{F}_I + \boldsymbol{F}_{S1} + \boldsymbol{F}_{1-2-1} + \boldsymbol{F}_{S1-2-1}$ at t_c	$\boldsymbol{F}_I + \boldsymbol{F}_{S1} + \boldsymbol{F}_{1-2} + \boldsymbol{F}_{S1-2}$ at t_c
$t_d = 3\boldsymbol{R}_{1-2}/\boldsymbol{c}$	$\boldsymbol{F}_I + \boldsymbol{F}_{S1} + \boldsymbol{F}_{1-2-1} + \boldsymbol{F}_{S1-2-1}$ at t_d	$\boldsymbol{F}_I + \boldsymbol{F}_{s1} + \boldsymbol{F}_{1-2} + \boldsymbol{F}_{S1-2} + \boldsymbol{F}_{1-2-1-2} + \boldsymbol{F}_{S1-2-1-2}$ at t_d

scattered-wave forces on a leg become negligible for a specific incident wave group. Notations for the forces on Leg 1 and Leg 2 over the time period $0 \leq t \leq 3R_{1-2}/c$ are listed in Table 9.1. In the table, F_I is the force on Leg 1 due to the incident wave, F_{S1} is the force on Leg 1 due to the scattering of the incident wave, and $F_{1\text{-}2}$ is the force on Leg 2 resulting from the scattered plane wave from Leg 1. The latter wave results in both a scattered wave force $F_{S1\text{-}2}$ on Leg 2 and a secondary scattered plane wave traveling to Leg 1 resulting in a force $F_{1\text{-}2\text{-}1}$, and so forth. In general, the notation S in the force subscript identifies the scattered-wave force component on the leg in question. The subscripts not containing S are identified with the scattered plane waves incident on the leg.

Because of the scattered plane-wave approximation, the MacCamy-Fuchs force expression of eqs. 9.71 and 9.85 can be applied each time a scattered wave arrives at a leg. As discussed in the previous paragraph, the wave heights will change, as will the phase relationships. These wave height and phase differences are illustrated in Example 9.9, where a two-leg, in-line orientation is discussed. The mathematical expressions for the wave forces are now formulated by applying the MacCamy-Fuchs relationship to each force pair in Table 9.1.

The expression for the MacCamy-Fuchs force on Leg 1 (at $t_a + t = t$) is

$$(F_I + F_{SI})|_{t_a+t} = (F_I + F_{SI})|_t = 2\frac{\rho g H_I}{k}\tanh(kh)\left\{\frac{[J_1'(ka)\cos(\omega t) - Y_1'(ka)\sin(\omega t)]}{J_1'^2(ka) + Y_1'^2(ka)}\right\}$$
$$= 2\frac{\rho g H_I}{k^2}\tanh(kh)\Lambda(ka)\sin[\omega t - \sigma(ka)] \tag{9.164}$$

from eq. 9.71, where the expressions for $\Lambda(ka)$ and the phase angle $\sigma(ka)$ are found in eqs. 9.72 and 9.73, respectively. In Table 9.1, the force on Leg 2 at time $t_a + t_b + t = 0 + t_b + t = R_{1-2}/c + t$ is

$$\begin{aligned}(F_I + F_{SI} + F_{1-2} + F_{S1-2})|_{t_b+t} &= 2\frac{\rho g H_I}{k^2}\tanh(kh)\Lambda(ka)\sin[\omega(t_b + t) - \sigma(ka)] \\ &\quad + 2\frac{\rho g H_{1-2}}{k^2}\tanh(kh)\Lambda(ka)\sin[\omega(t_b + t) - \sigma(ka)]\end{aligned} \tag{9.165}$$

where the wave height H_{1-2} is obtained from eq. 9.161. The reader can see the progression of forces on the legs as time increases. In eq 9.165, the phase angle is

$$\omega t_b + \sigma(ka) = \omega\frac{R_{1-2}}{c} + \sigma(ka) \tag{9.166}$$

The scattered plane-wave technique is demonstrated in Example 9.9. In that example, a platform is supported by two legs in waters of finite depth. The geometry at the waterplane resembles that in Figure 9.28a. The first *correction* of McIver and Evans (1984) is *not* included in the analysis. That is, in the analysis the scattered waves from one leg to its neighbor are represented by potentials of the form of $\varphi_{j\text{-P1}}$ in eq. 9.157, neglecting the effects of the $\varphi_{j\text{-P2}}$-type waves.

EXAMPLE 9.9: WAVE-INDUCED FORCES ON AN IN-LINE, TWO-LEG PLATFORM A work platform is placed in 10 m of water. The platform is supported by two 5-m-diameter legs, the centers of which are separated by 25 m. The wave-leg orientation is in-line and the waves approach the shoreline directly, as in Figure 9.28a. The incident waves in deep water have a 1-m height and a 5-sec period. To facilitate our analysis, the origins of the inertial and wave coordinate systems (x-y and X-Y) are located at the center of the seaward leg, Leg 1, again, as in Figure 9.28a. Referring to Figure 9.27, the notation and parametric values for this problem are as follows:

leg radius——————————$a = D/2 = 2.5\,\text{m}$
water depth——————————$h = 10\,\text{m}$
leg index numbers——————$j,k = 1,2$
leg separation——————————$R_{1\text{-}2} = 25\,\text{m}$
incident wave angle——————$\beta_I = 0°$
in-line leg orientation—————$\beta_{1\text{-}2} = 0°$
variable angles———————$\beta_1, \beta_2 = \gamma_2 + 1$
deep-water wave height————$H_0 \simeq 1\,\text{m}$
wave period————————$T = 5\,\text{sec}$
deep-water wavelength————λ_0 -$gT^2/2\pi \simeq 39.0\,\text{m}$ (eq. 3.36)

The incident wave height (H_I) at the site must be determined. For this example, the waves are assumed to be both linear and purely shoaled. To determine the wave height at the site, we must first determine the wavelength λ-value. This value is obtained by numerically solving eqs. 3.31 or 3.79, as in Example 3.2. The incident wave height is then determined from the shoaling coefficient in eq. 3.78. The incident wave properties at the site are found to be the following:

wavelength at the site——————$\lambda \simeq 36.6\,\text{m}$ (from 3.31 or 3.79)
wave number at the site————$k = 2\pi/\lambda \simeq 0.1717\,\text{m}^{-1}$
celerity magnitude at the site——$c = \lambda/T \simeq 7.32\,\text{m/s}$
incident wave height at the site—$H_I \simeq 2.88\,\text{m}$ (from eq. 3.78)

For the wave property values and the leg geometry, $ka \simeq 0.429$, $kR_{1\text{-}2} \simeq 4.30$, and $kh \simeq 1.72$. The wave height of the primary scattered wave at Leg 2 is the real part of the expression in eq. 9.161, that is,

$$
\begin{aligned}
H_{1-2} &= -\frac{H_I}{2}\sum_{m=0}^{\infty}\{i^{2m}\{\varepsilon_{2m}E_{2m}J'_{2m}(ka) + \varepsilon_{2m+1}G_{2m+1}J'_{2m+1}(ka)\}\\
&= -\frac{H_I}{2}\{E_0J'_0(ka) + 2G_1J'_1(ka) - 2E_2J'_2(ka) - 2G_3J'_3(ka) + 2E_4J'_4(ka) + \cdots\}\\
&\simeq -\frac{1_I}{2}\{0.014141 + 0.000038 + 0.000000 + \cdots\} \simeq 0.014179\,\text{m}
\end{aligned}
\qquad (9.167)
$$

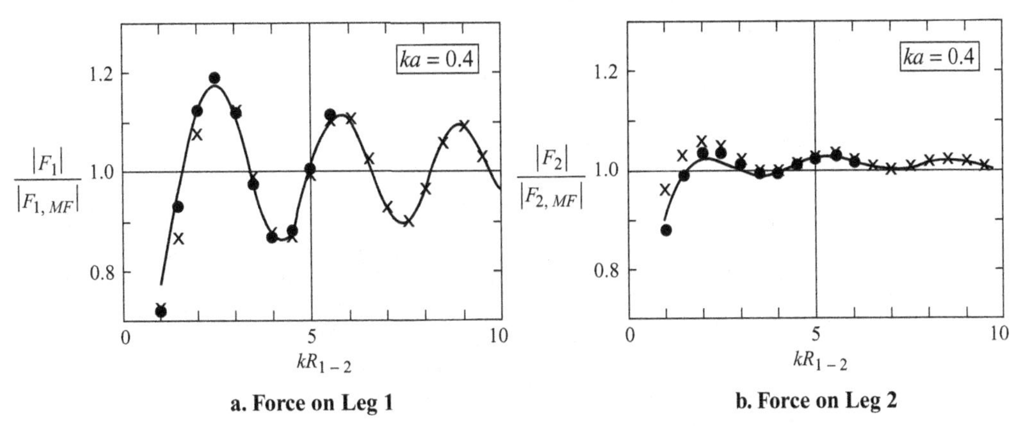

Figure 9.29. *Non-Dimensional Wave Force on Legs versus Leg Separation.* Here, the continuous line represents the Spring-Monkmeyer (1974) exact analysis, × the Simon (1982) plane-wave approximation, and • the McIver-Evans (1984) first correction of the plane-wave approximation.

where Neumann's symbol is $\varepsilon_0 = 1$ and $\varepsilon_m = 2$ for $m > 0$. The numerical values in eq. 9.167 are a bit spurious because we wish to demonstrate the convergence of the series. The derivatives of the Bessel functions of the first kind in the equation are expressed by

$$J'_m(ka) = k\left[J_m(ka) - \frac{1}{ka}J_{m+1}(ka)\right] \tag{9.168}$$

The general forms of the coefficients in eq. 9.167 are

$$E_{2m} = \frac{[J_{2m}(kR_{1-2})J'_{2m}(ka) + Y_{2m}(kR_{1-2})Y'_{2m}(ka)]}{[J'^{2}_{2m}(ka) + Y'^{2}_{2m}(ka)]} \tag{9.169}$$

and

$$G_{2m+1} = \frac{[J_{2m+1}(kR_{1-2})Y'_{2m+1}(ka) - Y_{2m+1}(kR_{1-2})J'_{2m+1}(ka)]}{[J'^{2}_{2m+1}(ka) + Y'^{2}_{2m+1}(ka)]} \tag{9.170}$$

From the numerical values in eq. 9.167, we see that the height of the scattered wave height is approximately two orders of magnitude less than the incident wave height, and the series rapidly converges as m increases. The respective amplitude function (eq. 9.72) and phase angle (eq. 9.73) values are $\Lambda(ka) \simeq 0.066054$ and $\sigma(ka) \simeq 0.01038\,\text{rad}$. Finally, the absolute value of the force amplitude on Leg 2 at time $t_b (= R_{1-2}/c \simeq 3.42s)$ is approximately 2.647 $\times\ 10^5$ N from eq. 9.165. Because the force components in the equation are proportional to the incident and scattered wave heights, we see that the force amplitude due to the incident wave is 2.610×10^5 N and the scattered wave contribution to the total force is 3.7×10^3 N.

McIver and Evans (1984) present non-dimensional force results for three leg orientations. Those are the in-line two-leg structure sketched in Figure 9.28a, a triangular-configured three-leg structure, and a star-configured five-leg structure. The results for the in-line orientation of the two-leg system are presented in Figures 9.29 and 9.30. In the former figure, the absolute value of the force divided by the MacCamy-Fuchs force is presented as a function of kR_{1-2} for the weather and leeward legs, whereas in the latter figure, the force ratio is presented as a function if

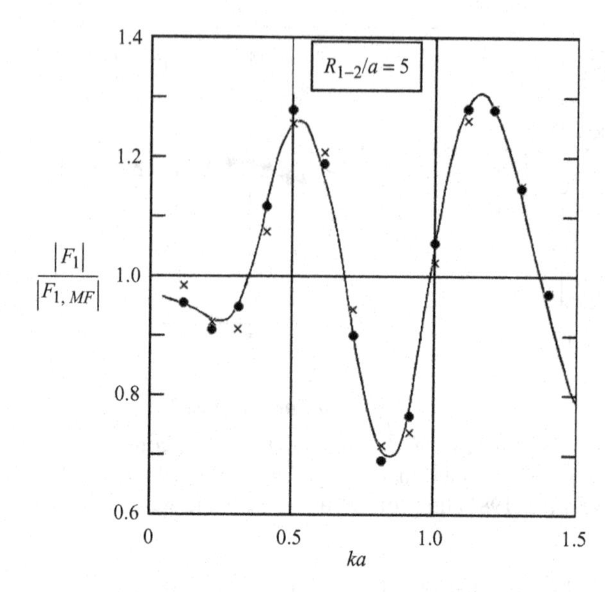

Figure 9.30a. *Non-Dimensional Wave Force on Leg 1 versus Leg Radius.* Here, the continuous line represents the Spring and Monkmeyer (1974) exact analysis, × the Simon (1982) plane-wave approximation, and • the McIver and Evans (1984) first correction of the plane-wave approximation.

ka. For these leg configurations, results obtained from the exact analysis of Spring and Monkmeyer (1974), the *plane-wave analysis* of Simon (1982) demonstrated in Example 9.9, and the McIver-Evans first correction to the plane-wave analysis are compared.

In Figure 9.29, we see that the results obtained from the plane-wave analysis converge on the results obtained from the Spring-Monkmeyer exact method as kR_{1-2} increases. The effect of the McIver-Evans first correction to the plane-wave analysis is evident for $kR_{1-2} < 2$ for Leg 1 and for $kR_{1-2} < 3$ for Leg 2. For the conditions in Example 9.9, $kR_{1-2}4.29 \simeq$ and $|F_2|/2, MF \simeq 0.986$, which agree with the values in Figure 9.29b. The wavy nature of the curves is due to the constructive and destructive interference of the wave components between the legs. That is, a crest from one wave system will be superimposed on a crest from another wave system for the former, and a crest-trough combination produces the latter depending on the separation distance of the legs.

The McIver-Evans results in Figure 9.30 show the effects of the separation distance of the legs on the wave-induced forces. For the forces on Leg 1 in Figure 9.30a, we see that the results from the three analytical methods converge for $ka > 1.2$. No convergence is seen in Figure 9.30b for Leg 2. The forces predicted by the Simon (1982) plane-wave method are the most conservative over the *ka* range in this figure. As in Figure 9.29, the effects of constructive and destructive interference are evident.

For engineering analyses, the results in Figures 9.29 and 9.30 demonstrate that the plane-wave analysis of Simon (1982) is quite good. The results show that the maximum disagreement is within 5% for both the separation-to-wavelength results in the former figure and the radius-to-wavelength results of the latter.

9.3 Wave-Induced Forces and Moments on Bodies in Random Seas

Three types of wave-induced forces are discussed previously in this chapter, those being viscous-pressure forces, inertia forces, and diffraction forces. In the next chapter dealing with wave-induced motions of fixed and floating structures, all three types of wave loading are considered. The regimes of these forces are shown in the

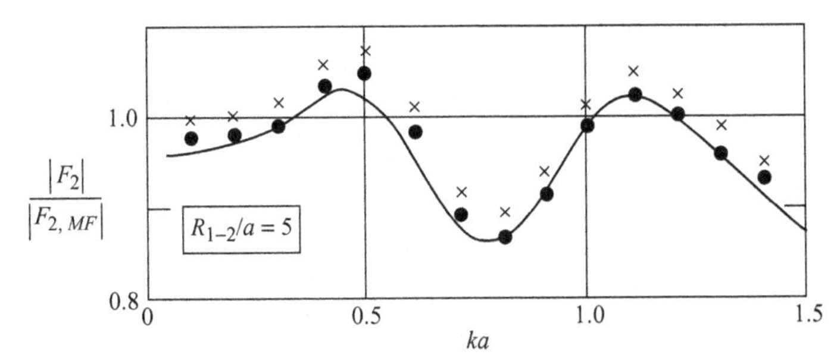

Figure 9.30b. *Non-Dimensional Wave Force on Leg 2 versus Leg Radius.* – represents the Spring and Monkmeyer (1974) exact analysis, × the Simon (1982) plane-wave approximation, and • the McIver and Evans (1984) first correction of the plane-wave approximation.

Chakrabarti force diagram of Figure 9.8 for surface-piercing, circular, cylindrical bodies. In that figure, the three force regimes are defined in terms of the Keulegan-Carpenter number (KC) and the non-dimensional body radius ($ka = 2\pi a/\lambda$). From eqs. 9.45 through 9.47, the relationships for the Keulegan-Carpenter number are found to be

$$KC = \frac{u_{\max} T}{D} \simeq \frac{H}{D} \frac{\pi}{\tanh(kh)} \Rightarrow \begin{array}{l} KC|_{deep} \simeq \dfrac{\pi H}{D} = \dfrac{\pi H}{2a} \\ KC|_{shallow} \simeq \dfrac{HT}{2D}\sqrt{\dfrac{g}{h}} \end{array} \tag{9.171}$$

The results in Figure 9.8 imply that bodies having small radii compared to the wave height are primarily subjected to viscous-pressure forces, large-radius bodies in long-period waves experience dominant inertial forces, and diffraction forces are dominant on large-radius bodies in short-period waves.

In ocean engineering and naval architecture hydrodynamics, there are three flow phenomena that can be random in nature. On the smallest scale, the random nature of turbulence must be considered. Turbulence can cause problems associated with underwater sound generation. For example, the intense turbulent pressure fluctuations on a ship panel can cause high-frequency vibrations of the panel which, in turn, produce near-field sound. Moving up in scale, the vortex shedding from braces of offshore towers can be random if the free-stream flow is random. This is a problem associated with structural-brace vibrations and cable strumming, the former discussed in the next chapter. Finally, on the largest scale of interest to ocean engineers and naval architects, random surface waves must be considered. These are discussed in Chapter 5, and are of interest here. As discussed in Chapter 5, wind waves result from wind turbulence over the free surface of the sea, and because of the nature of their generation, are random. The height and length of the wind waves increase due to the combined actions of wind shear and wind pressure gradients.

In this section, we study the random natures of the viscous pressure, inertia, and diffraction forces. The viscous-pressure and inertia forces are found in the Morison equation, eq. 9.49. The earliest studies of the random force components of the Morison equation include those of Borgman (1965) and Bretschneider (1965, 1967). The Borgman study concentrates on the spectral natures of the forces, whereas the Bretschneider studies focus on the probabilistic natures. Other studies of the Morison forces include those of Borgman (1981), Bostrom and Overvik (1986), Isaacson, Baldwin, and Niminsk (1991), Burrows et al. (1997), Najafian et al. (2000), and O'Kane, Troesch and Thiagarajan (2002). In addition, an excellent summary paper

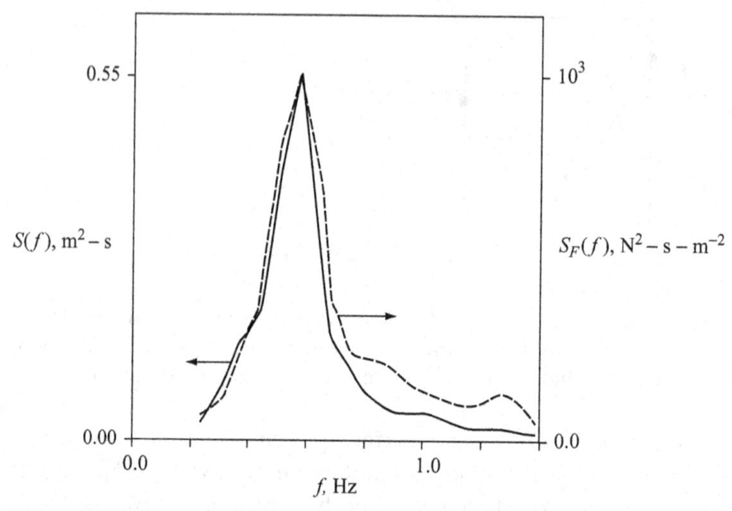

Figure 9.31. *Wave and Force Spectral Data from Field Measurements.* These curves result from measurements reported by Wiegel, Beebe, and Moon (1957), as reported by Borgman (1965). The specifics of the measurements are also found in Chapter 11 of the book by Wiegel (1964). The force data were obtained at $z = -1.92$ m on a 0.305-m-diameter vertical pile in 14.9 m of water. The forces were measured over a 0.305-m length of the pile. The units of force spectral density, $S_F(f)$, indicate that the spectrum is that of the force per unit length.

devoted to wave-induced forces on jack-up structures is that of Cassidy, Eatock Taylor, and Houlsby (2001). The literature on random diffraction forces is not as extensive as that on the Morison-type forces. Most of the recent works are numerical in nature, and will not be discussed herein. An analytic expression for random diffraction forces based on the MacCamy-Fuchs equation is developed later in the chapter.

The random nature of the sea is discussed and analyzed in Chapter 5. Much of the statistical background needed for the discussions in the present section can be found in that chapter. As discussed in Chapter 5, the randomness of a sea is characterized by the wave height (H) and the wave period (T). Although much attention is paid to the probabilistic distribution of H, the distributions of the wavelength (λ), the corresponding wave number ($k = 2\pi/\lambda$), and the circular wave frequency ($\omega = 2\pi/T$) are as important in determining the forces and moments induced by random waves.

A. Spectral Nature of Wave-Induced Viscous-Pressure and Inertia Forces

The Borgman (1965) random-wave force analysis was inspired, in part, by a conclusion of Professor Robert L. Wiegel of the University of California, Berkeley. Wiegel and his colleagues made a number of wave force measurements at a near-shore site off the coast of Davenport, California. From the data, Wiegel concluded that the spectral density of wave forces was similar to the spectral density of the waves producing those forces. The wave spectral density is introduced and discussed in Chapter 5 of this book. Some, but not all, of the results of the Davenport study are reported by Wiegel, Beebe, and Moon (1957). The force and wave spectra from one data set (November 5, 1953) of the Davenport study are presented in Figure 9.31. This data set was given to Borgman by Wiegel, and is not presented in numerical form in the referenced paper. One can see that the spectral shapes are similar, and almost identical for wave frequencies (f) less than 0.7 Hz. Other data presented by Wiegel, Beebe, and Moon (1957) also show a strong similarity of the spectra for the lower frequencies. We note that the wave period, T, is the inverse of the wave frequency, f. The strong spectral similarity occurs for $T > 1.4$ sec in the data in Figure 9.31. The spectral similarity can then be assumed to exist in the range of wave periods that are of prime interest to ocean engineers.

Based on the results of the study described by Wiegel, Beebe, and Moon (1957), Borgman (1965) writes that the force spectral density for the drag and inertial forces (per unit vertical length of a pile) in Morison's equation, eq. 9.49, can essentially be represented by products of transfer functions and the wave spectral density at a site. Following Borgman (1965), the spectral form of Morison's equation can be written as

$$S_F(f)|_z \simeq \left\langle (2\pi f)^2 \mathrm{K}_d + (2\pi f)^4 \mathrm{K}_i \right\rangle \frac{\cosh^2[k(z+h)]}{\sinh^2(kh)} S(f) \equiv T(f)|_z S(f) \qquad (9.172)$$

where $\mathrm{T}(f)|_z$ is the transfer function and $S(f)$ is the wave spectral density. The drag parameter is defined as

$$\mathrm{K}_d = \frac{8}{\pi}\left(C_d \frac{1}{2}\rho A'_{proj} u_{rms}|_z\right)^2 \qquad (9.173)$$

In eq. 9.173, $u_{rms}|_z$ is the root-mean-square value of the horizontal velocity component at a depth z. According to Borgman (1965), the value of the mean-square of the horizontal velocity is obtained from

$$\overline{u2}|_z = u^2_{rms}|_z = 2\int_0^\infty (2\pi f)^2 \frac{\cosh^2[k(z+h)]}{\sinh^2(kh)} S(f)df \qquad (9.174)$$

Also in eq. 9.172 is the *inertial parameter*, defined as

$$\mathrm{K}_i = (C_i \rho \mathrm{V}'_{disp})^2 \qquad (9.175)$$

In eq. 9.173, A'_{proj} is the projected area of the pile per unit length, and in eq. 9.175, V'_{disp} is the displaced volume of the pile per unit length. For a circular cylinder, the former is simply D, and the latter is $\pi D^2/4$, where D is the pile diameter. The wave number (k) and wave frequency (f) are related by the dispersion relationship in eq. 3.30. In the form needed here, that relationship is

$$k \tanh(kh) = \frac{4\pi^2 f^2}{g} = \frac{4\pi^2}{gT^2} \qquad (9.176)$$

The values of k are obtained numerically. One such numerical solution is presented in Example 3.3. In the Borgman (1965) analysis, the drag term is represented by a series expansion, and only the first term of the expansion appears in eq. 9.172. So, the expression of the spectral density of the drag force in eq. 9.172 is a first approximation to the drag-force spectral density. The accuracy of this approximation is further discussed in Example 9.10.

In eq. 9.172, the wave spectral density in the frequency domain is used. The relationship of the frequency spectral density and that in the period domain is

$$S(f) = -S(T)\frac{dT}{df} = \frac{1}{f^2}S(T) = T^2 S(T) \qquad (9.177)$$

This expression results from eqs. 5.44a and 5.44b. The values of $S(f)$ can be obtained directly from onsite measurements of the wave heights and periods, as discussed in Section 5.7, or by using an empirical formula, as discussed in Section 5.8. The Pierson-Moskowitz spectrum (eq. 5.63) can be used directly in the Borgman (1965) analysis if the average wind speed at the site is known. If the Brestschneider spectral formula in eq. 5.59 is used, the relationship in eq. 9.177 is slightly modified according to the relationships in eq. 5.68. The relationship in the frequency domain for these two empirical spectral formulas is

$$S(f) = T^2 S_{PM}(T) = \frac{1}{8} T^2 S_B(T) \tag{9.178}$$

where the subscript PM refers to the Pierson-Moskowitz formula in eq. 5.63 and the subscript B refers to the Bretschneider formula.

EXAMPLE 9.10: PREDICTED AND MEASURED WAVE AND FORCE SPECTRA ON A CIRCULAR PILE In this example, a data set of the Wiegel, Beebe, and Moon (1957) study of a vertical circular pile having a diameter of 0.324 m and located in 14.9 m of water is used to determine the wave and force spectra. This data set is analyzed by Borgman (1965). The force data were measured at $z = -1.92$ m, and the measured wave and force spectra are presented in Figure 9.31. Borgman finds that the root-mean-square value of the horizontal velocity, the least-square fitted values of the drag, and inertial coefficients at $z = -1.92$ m are $u_{rms}|_{z=-1.92m} \simeq 0.334$ m/s, $C_d \simeq 1.88$, and $C_i \simeq 1.73$, respectively. For these values, the drag parameter in eq. 9.173 is $K_d \simeq 2.47 \times 10^4\ N^2s^2/m^4$, and the inertial parameter of eq. 9.175 is $K_i \simeq 1.69 \times 10^4\ N^2s^4/m^4$. The force spectral density expression in eq. 9.172 is then

$$S_F(f)|_{z=-1.92m} \simeq \langle 9.75 \times 10^5 f^2 + 2.63 \times 10^7 f^4 \rangle \frac{\cosh^2(13.0\,k)}{\sinh^2(14.9k)} S(f)$$

where the wave spectral density values, $S(f)$, at the site are measured. The units of the force-per-unit-depth spectral density on the vertical element of the pile are N^2-s/m^2. Note that the units of the total-force spectral density over the entire pile would be N^2-s. The relationship between k and f is found in the dispersion relationship in eq. 9.176. Values obtained from the force-spectrum expression and the measured values are presented in Figure 9.32. One can see from the results in that figure that the agreement is rather good over the lower frequency range, $0 < f < 0.7$. Again, this would be the frequency range of primary interest in engineering problems.

The results presented in Example 9.10 give us confidence in the ability of the expression in eq. 9.172 to predict the force spectral density using measured wave spectra. We now turn our attention to the probabilistic aspects of wave-induced forces. Of particular interest in engineering applications is the maximum statistical wave force that might be encountered at a site. The information used in this discussion is from the papers of Bretschneider (1965, 1967).

B. Probabilistic Nature of the Viscous-Pressure and Inertia Wave Forces

The studies of Borgman (1965) and Bretschneider (1965) were both presented at the Coastal Engineering Conference at Santa Barbara, California. The at-sea data used

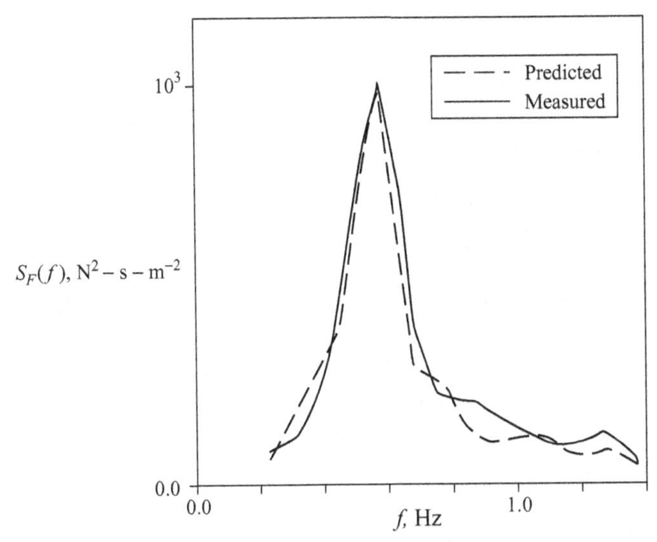

Figure 9.32. *Measured and Predicted Force Spectra.* These curves result from measurements reported by Wiegel, Beebe, and Moon (1957), as reported by Borgman (1965). The measured wave spectrum corresponding to this data set is shown in Figure 9.31 as is the force spectrum. The measured force data were obtained at $z = -1.92$ m on a 0.305-m length of a 0.305-m-diameter vertical pile in 14.9 m of water. The force spectra shown represent those of the force per unit length of pile.

by both investigators were those of Professor Robert Wiegel and his colleagues obtained near the coast of Davenport, California. The focus of the Bretschneider (1965, 1967) study was on the probability of the wave-induced viscous-pressure and inertial forces, referred to as the Morison forces because these are the forces in the Morison equation, eq. 9.49. Results of this study are used herein. In more recent times, the large majority of studies has been devoted to the Morison forces on compliant structures. For example, see the works of Issacson, Baldwin, and Niwinski (1991) and Burrows et al. (1997). Taylor and Rajagopalan (1983) present the results of a definitive study of the load spectra on compliant slender structures. The wave-structure interaction of compliant structures in random seas is discussed in Chapter 10.

As shown in Figure 5.7, the Rayleigh probability distribution is found to agree rather well with the probability distribution of the wave height data presented in Figure 5.2. The distribution of the free surface corresponding to the data in Figure 5.2 would be expected to be Gaussian, as discussed in Section 5.6. This Gaussian-Rayleigh relationship was first shown by Longuet-Higgins (1952) to apply to random seas having wave spectra of narrow band. We refer to such a sea as being *Gaussian.* From the classic book by Crandall and Mark (1963), a *narrow-band process* is "a stationary random process whose mean square spectral density $S(\omega)$ has significant values only in a band or range of frequencies whose width is small compared with the magnitude of the center frequency of the band." In other words, the energy of the sea is confined to a relatively small band of wave frequencies.

When a sea is not Gaussian, other probability distributions must be found. As discussed in Section 5.5, a general formula for the cumulative probability of occurrence is that of Weibull (1951). Following Bretschneider (1965), we apply the two-parameter Weibull formula to the wave heights when considering non-Gaussian sea. That formula is

$$P(0 < H \leq H_J) = 1 - e^{-C\left(\frac{H_J}{H_{rms}}\right)^m} \tag{9.179}$$

In this expression, the parameters are C and m, and are determined empirically. The corresponding probability density of the wave heights is

$$p(H_J) = Cm\left(\frac{H_J^{m-1}}{H_{rms}^m}\right)e^{-C\left(\frac{H_J}{H_{rms}}\right)^m} \tag{9.180}$$

We note that when $C = 1$ and $m = 2$, the expression in eq. 9.180 is the Rayleigh probability density in eq. 5.19.

In Section 5.4D, the expression for the most-probable maximum wave height corresponding to a Rayleigh distribution is presented. Following the logic of that section, the most-probable maximum wave height corresponding to a Weibull distribution of wave heights is found from

$$H_{\max} = H_{rms}\left[\frac{1}{C}\ln(N_M)\right]^{1/m} \tag{9.181}$$

where N_M is the (statistical) number of waves occurring over M years. According to Bretschneider (1965), the statistical number of waves passing a site can be obtained from

$$N_M = N_1 M \simeq \frac{t_M}{T_s} = M\frac{t_1}{T_s} \simeq M\frac{t_1}{T_{avg}} \tag{9.182}$$

In this expression, t_M is the length of time of interest and T_s is the significant wave period (approximately equal to the average period, T_{avg}, as in eq. 5.51), both measured in seconds. If we continually measure waves over one year, then $M = 1$ and $t_1 \simeq 3.15 \times 10^7$ sec. The statistical wave periods are discussed in Section 5.8A of this book. In the remainder of this section, we assume that the wave climate is Gaussian. The corresponding probability density is that of Rayleigh in eq. 5.19.

As discussed previously, our interest is in the relationships of the probabilities of the sea and the probabilities of the associated wave-induced viscous-pressure and inertia forces and moments. For a deterministic sea, these forces are in the Morison equation, eq. 9.49. When applied to a cylinder having a circular cross-section, the Morison equation is

$$F' = F'_d + F'_i = C_d\frac{1}{2}\rho u|u|D + C_i\rho\frac{Du}{Dt}\pi\frac{D^2}{4} \tag{9.183}$$

In this expression, the subscript d identifies the drag or viscous-pressure force, the subscript i identifies the inertia force, and the prime (′) indicates that the forces are per unit length of cylinder. Assuming that the waves are linear, as described in Section 3.2, the maximum values of the component forces are the following: The maximum drag force is

$$F'_{d\max} = C_d\frac{1}{2}\rho D\left\{\frac{\pi}{T}\frac{\cosh[k(z+h)]}{\sinh(kh)}\right\}^2 H^2 \equiv K_1(z)H^2 \tag{9.184}$$

which is in phase with the wave, and the maximum inertia force is

$$F'_{i\max} = C_i\rho\pi\frac{D^2}{4}\left(2\frac{\pi^2}{T^2}\right)\frac{\cosh[k(z+h)]}{\sinh(kh)} \equiv K_2(z)H \tag{9.185}$$

This force is 90° out of phase with the passing wave.

Assume that C_d, C_i, T, and the wave number k in eqs. 9.184 and 9.185 are averaged. By making this assumption, we can relate the probability densities of the force components and the wave heights. Statistically, the assumption concerning the period implies that there is no correlation between H and T. In reality, the force coefficients do change in time and there is a correlation between H and T. For the purpose of this discussion, the assumption is satisfactory. The assumption allows us to obtain the average maximum drag force and the root-mean-square inertia force. These respective forces are

$$F'_{d\max-a} = \mathrm{K}_1(z)H^2_{rms} \tag{9.186}$$

where the subscript a identifies the average value, and

$$F'_{i\max-r} = \mathrm{K}_2(z)H_{rms} \tag{9.187}$$

where r identifies the root-mean-square.

Because the probability densities of H and H^2 are the same, the probability densities of the maximum force components are obtained from

$$p(F'_{d\max})dF'_{d\max} = p(F'_{i\max})dF'_{i\max} = p(H)dH \tag{9.188}$$

where, from eq. 9.184,

$$\frac{dF'_{d\max}}{dH} = 2\mathrm{K}_1(z)H \tag{9.189}$$

and, from eq. 9.185,

$$\frac{dF'_{i\max}}{dH} = \mathrm{K}_2(z) \tag{9.190}$$

In eq. 9.188, assume a Rayleigh probability density for the wave heights, as expressed in eq. 5.19, that is,

$$p(H) = 2\left(\frac{H}{H^2_{rms}}\right)e^{-\left(\frac{H}{H_{rms}}\right)^2} \tag{9.191}$$

Again, this expression is identical with the Weibull expression in eq. 9.180 when $C = 1$ and $m = 2$. By combining eqs. 9.184, 9.189, and 9.191 in the drag relationship of eq. 9.188, one obtains the following expression for the drag probability density function:

$$p(F'_{d\max}) = \frac{1}{F'_{d\max-a}}e^{-\frac{F'_{d\max}}{F'_{d\max-a}}} \tag{9.192}$$

Similarly, the expression for the inertia probability density function is found to be

$$p(F'_{i\max}) = \frac{2F'_{i\max}}{F'^2_{i\max-r}}e^{-\left(\frac{F'_{i\max}}{F'_{i\max-r}}\right)^2} \tag{9.193}$$

The respective expressions in eqs. 9.192 and 9.193 yield the following cumulative probabilities of occurrence:

$$P(0 \le F'_{d\max}) = 1 - e^{-\frac{F'_{d\max}}{F'_{d\max-a}}} \tag{9.194}$$

and

$$P(0 \le F'_{i\max}) = 1 - e^{-\left(\frac{F'_{i\max}}{F'_{i\max-r}}\right)^2} \tag{9.195}$$

Just as the most-probable maximum wave height for a sample of N_M waves is found from eq. 5.25 for a Gaussian-Rayleigh sea to be

$$H_{\max}|_M = H_{rms}\sqrt{\ln(N_M)} \tag{9.196}$$

we find that the most-probable maximum drag force is

$$F'_{d(\max)}|_M = F'_{d\max-a}\ln(N_M) \tag{9.197}$$

and that the most-probable maximum inertia force is

$$F'_{i(\max)}|_M = F'_{i\max-r}\sqrt{\ln(N_M)} \tag{9.198}$$

In the following example, the predicted and measured probabilities of occurrence for a data set of Wiegel, Beebe, and Moon (1957) is presented.

EXAMPLE 9.11: PROBABILITIES OF OCCURRENCE FOR THE MAXIMUM DRAG FORCE ON A CIRCULAR PILE The force time trace for a sample of a data set presented by Wiegel, Beebe, and Moon (1957) and analyzed by Bretschneider (1965, 1967) shows that the in-line force (in the direction of wave travel) is approximately in phase with the free-surface displacement [$\eta(t)$] of the wave. In Figure 9.13, we see that the *drag force* is in phase with $\eta(t)$ and the *inertial force* is 90° out of phase. The time trace is then indicative of a dominant drag force on the pile element. Because of this, Wiegel, Beebe, and Moon (1957) report only the drag coefficient values for the data set in question.

For the data set, Bretschneider (1965, 1967) finds the following: $h \simeq 14.6$ m, $z \simeq -1.68$ m, $D = 0.305$ m, $H_{rms} \simeq 2.90$ m, $F'd_{\text{max-a}} \simeq 80.0$ N, and $T_{\text{avg}} \simeq 14.5$ sec. We choose here to use the root-mean-square period in our analysis rather than the average value. From eq. 5.51, we find that the approximate relationship between the root-mean-square period and the average period is $T_{rms} \simeq 1.039T_{\text{avg}} \simeq 15.2$ sec. The value of the wave number based on this period is $k = 0.0361\ \text{m}^{-1}$ from the dispersion equation, eq. 3.31. Due to scatter in the data, Bretschneider introduces the *correlation drag coefficient*, defined as

$$C_{d-rms} \equiv \left.\frac{F'_{d\max-a}}{\frac{1}{2}\rho\pi Du_{rms}^2}\right|_z \tag{9.199}$$

For the data under consideration, Bretschneider uses $C_{d-rms} \simeq 0.75$. Also in eq. 9.199 is u_{rms}, which is the root-mean-square of the horizontal particle velocity and is considered to be a measured quantity here. Bretschneider also defines the *correlation inertial coefficient* as

$$C_{i-rms} \equiv \left.\frac{F'_{i-rms}}{\rho\pi\dfrac{D^2}{4}\left(\dfrac{du}{dt}\right)_{rms}}\right|_z \tag{9.200}$$

This is not needed in this example because the dominant force is the drag.

The measured probability of occurrence values presented by Bretschneider (1965, 1967) and those obtained from eq. 9.194 are presented in Figure 9.33. In that figure, we see that the agreement is excellent for maximum drag force value ≥ 67 N.

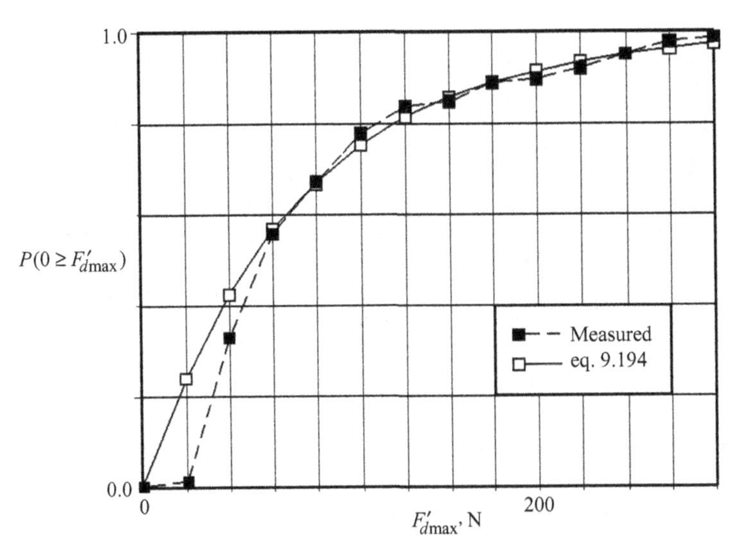

Figure 9.33. *Measured and Predicted Probabilities of Wave-Force Occurrence.* The conditions given in Example 9.11 are $h \simeq 14.6$ m, $z \simeq -1.68$ m, $D = 0.305$ m, $H_{rms} \simeq 2.90$ m, $F \simeq_{d\max-a} \simeq 80.0$ N, and $T_{rms} \simeq 15.2$ sec. The force is drag-dominant because the force and free-surface displacement are in phase.

In the next example, the data used in Example 9.11 are used to determine the 100-year design force on both the cylindrical element and the entire wetted cylinder.

EXAMPLE 9.12: MOST-PROBABLE MAXIMUM FORCE FOR THE 100-YEAR STORM In Example 9.11, we compare the measured and empirically determined probabilities of occurrence for the wave-induced drag force on a segment of a vertical circular cylinder. The empirical formula for the probability is that of eq. 9.194. The assumptions leading to the formula are, first, that the sea is Gaussian having Rayleigh-distributed wave heights, and second, that the wave heights and periods are not correlated. In the analysis of the Wiegel, Beebe, and Moon (1957) data analyzed by Bretschneider (1965, 1967), the following values are used: $h \simeq 14.6$ m, $z \simeq -1.68$ m, $D = 0.305$ m, $H_{rms} \simeq 2.90$ m, $F'_{d\max-a} \simeq 80.0$ N, $C_{d-rms} \simeq 0.75$, and $T_{avg} \simeq 14.5$ sec. In this problem, we shall determine the 100-year drag force on the cylinder.

First, the statistical number of waves encountered by the cylinder must be determined. From eq. 9.186, that number is

$$N_{100} \simeq 100\frac{t_1}{T_{avg}} = 100\frac{3.1536 \times 10^7}{14.5} \simeq 2.17 \times 10^8$$

For this value and the measured $F'_{d\max-a}$, the most probable maximum wave force on the segment is 1.351×10^3 N from eq. 9.197. This force value is about 19.2 times greater than that of the average maximum wave force.

Assume that the 100-year wave is sinusoidal and has an amplitude of $0.5H_{\max}|_{100} \equiv a_{100}$. Integrate eq. 9.197 over the wetted height of the cylinder, that is, the integration is from the mud line ($z = -h$) to the wave amplitude ($z = a_{100}$). The value of the 100-year wave height, $H_{\max}|_{100}$, is obtained

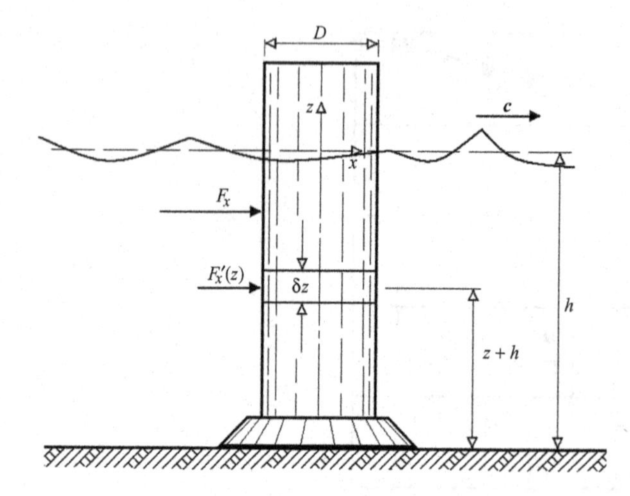

Figure 9.34. *Sketch of a Large-Diameter, Gravity-Type, Cylindrical Structure Resting on a Spread Footing in a Random Sea.* The forces shown are the total diffraction force, F_x, and the diffraction force per unit height, $F \simeq_x$.

from eq. 9.196. The resulting most-probable maximum force is then

$$\begin{aligned}
F_{d(\max)}|_{100} &= \ln(N_{100}) \int_{-h}^{a_{100}} FA'_{d\max-a}(z)dz \\
&= \frac{\ln(N_{100})}{2} C_{d-rms}\rho D \frac{\pi^2}{T_{rms}^2} \frac{H_{rms}^2}{\sinh^2(kh)} \int_{-h}^{a_{100}} \cosh^2[k(z+h)]dz \\
&= \frac{\ln(N_{100})}{8} \frac{C_{d-rms}\rho D\pi^2 H_{rms}^2}{kT_{rms}^2 \sinh^2(kh)} \{2k(a_{100}+h) + \sinh[2k(a_{100}+h)]\} \\
&\simeq 4.28 \times 10^4 \text{ N}
\end{aligned} \tag{9.201}$$

The corresponding overturning moment about the mud line, positive in the counterclockwise direction, is

$$\begin{aligned}
M_{d(\max)}|_{100} &= -\ln(N_{100}) \int_{-h}^{a_{100}} (z+h)F'_{d\max-a}(z)dz \\
&= -\frac{\ln(N_{100})}{16} \frac{C_{d-rms}\rho D\pi^2 H_{rms}^2}{k^2 T_{rms}^2 \sinh^2(kh)} \{[2k(a_{100}+h)\sinh[2k(a_{100}+h)] \\
&\quad - \cosh[2k(a_{100}+h)] + 2k^2(a_{100}+h)^2\} \\
&\simeq -5.20 \times 10^5 \text{ N-m}
\end{aligned} \tag{9.202}$$

Finally, the probabilistic center of force beneath the SWL is determined from

$$z_{100} = -\frac{M_{d(\max)}|_{100}}{F_{d(\max)}|_{100}} - h \simeq -2.45\,\text{m} \tag{9.203}$$

From this, the center of force is also about 12.15 m above the mud line.

In the next section, we focus our attention on the diffraction forces in random seas. This discussion would be applicable to some large tension-leg platforms, semi-submersible structures, and monolithic structures resting on the sea bed in wind-generated seas. An example of the latter structure is sketched in Figure 9.34. As discussed previously in this chapter, diffraction forces on large vertical, circular, cylindrical structures would be dominant where, as shown in Figure 9.8, the

Keulegan-Carpenter number (*KC*) in eq. 9.171 is small and the wave number radius (ka) is large.

C. Random Nature of Diffraction Forces on a Fixed, Vertical Circular Cylinder

Consider the large-diameter, vertical, circular cylinder sketched in Figure 9.34. The structure in that figure is a gravity-type structure resting on a spread footing in an irregular sea. The stability of the structure depends on both the weight and the position of the center of gravity. In the design of such structures, the pressure on the soil-structure interface is uniform. In regular seas, where the waves are linear, the horizontal diffraction wave force on the cylinder is well predicted by the MacCamy-Fuchs equation, eq. 9.71. In this discussion, the following form of that equation is used:

$$F_x = 2\frac{\rho g H}{k^2}\Lambda\tanh(kh)\sin[\omega t - \sigma(ka)] = Q_x H\sin[\omega t - \sigma] \tag{9.204}$$

where H is the incident wave height, and $\Lambda(ka)$ is defined in eq. 9.72 and $\sigma(ka)$ in eq. 9.73. Using the expression in the last equality of this equation, we first determine the spectral density of the diffraction force. The probabilistic nature of the force is then analyzed.

The time-averaged mean-square of the force expression in eq. 9.204 is

$$< F_x^2 > = Q_x^2(ka)\frac{H^2}{2} \tag{9.205}$$

In this expression, we use the notation $< >$ to indicate time-averaging over one wave period. Following the method presented in Section 5.7A devoted to wave spectra, the spectral density of the diffraction force is

$$S_{Fx}(f) = 4Q_x^2(ka)S(f) \tag{9.206}$$

where $S(f)$ is the wave spectral density.

Small-scale experimental diffraction force data are normally obtained using methods similar to those reported by Wiegel, Beebe, and Moon (1957), Burrows et al. (1997), and Najafian et al. (2000). That is, a horizontal element of the structure is isolated and fitted with some type of transducer. Surface-pressure measurements are more common in experimental diffraction force studies, such as that described by Akyildiz (2002). For full-scale studies on large-diameter cylinders, the data are normally obtained from surface-pressure measurements, and the force values are obtained by integrating the pressures around the circumference line upon which the pressure transducers are positioned. To compare the measured diffraction forces on a large-diameter cylindrical tower with those obtained from the MacCamy-Fuchs equation, the following modified form of that equation is required:

$$\begin{aligned} F_x|_z &= -\int_{z_b}^{z_a}\int_0^{2\pi} p|_{r=a}\cos(\beta)a\,d\beta\,dz \\ &= 2\frac{\rho g H_I}{k^2}\left\{\frac{\sinh(kz_a)-\sinh(kz_b)}{\cosh(kh)}\right\}\Lambda(ka)\sin[\omega t-\sigma(ka)] \\ &= q_x(ka)H_I\sin[\omega t-\sigma(ka)] \end{aligned} \tag{9.207}$$

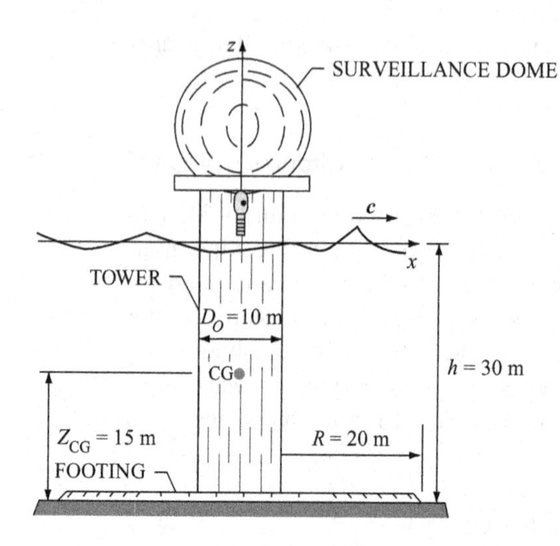

Figure 9.35. *Sketch of a Surveillance Platform Resting on a Spread Footing Foundation in Examples 9.13 and 9.14.*

where z_a and z_b are the respective upper and lower extent of the force gauge. Also in eq. 9.207 is the amplitude function,

$$q_x(ka) = 2\frac{\rho g}{k^2}\left\{\frac{\sinh(kz_a) - \sinh(kz_b)}{\cosh(kh)}\right\}\Lambda(ka)) \tag{9.208}$$

The expression in eq. 9.207 is that of the diffraction force on a gauge of length $z_a - z_b$ whose center is located at z. Again, $\Lambda(ka)$ in eqs. 9.207 and 9.208 is defined in eq. 9.72, and $\sigma(ka)$ in eq. 9.207 is defined in eq. 9.73. The expression for the spectral density of the diffraction force on a segment of a vertical cylinder of radius a is

$$S_{Fx}(f)|_z = 4q_x^2(ka)S(f) \tag{9.209}$$

This expression is similar in form to that in eq. 9.206.

In Example 9.13, the expression in eq. 9.206 is applied to a large vertical tower resting on a flat bed in waters of finite depth. In the example, the Bretschneider (1963) spectral formula in eq. 5.59 is used to represent the sea.

EXAMPLE 9.13: DIFFRACTION FORCES ON MONOLITHIC GRAVITY STRUCTURE To curtail smuggling activities, a series of monolithic observation towers are designed to be deployed approximately 1 km from the coast where the design water depth (h) is 30 m. The circular, cylindrical towers support platforms equipped with a variety of electronic surveillance devices and rest on circular spread footings having radii (R) of 20 m. One such tower is sketched in Figure 9.35. Referring to that sketch, the 40-m-high tower is hollow, having an outside diameter ($D_O = 2a_O$) of 10 m and an inside diameter (D_I) of 8 m. During deployment of the tower, the ends of the structure are capped to allow it to be towed to the site while floating with the axis of symmetry horizontal. During deployment, holes in the end plates are opened and the chamber is flooded. Additional ballast is also added at the base of the vertical tower. The structure is made of steel-reinforced marine concrete. The total dry weight of the platform, tower, and footing is $W = 2.5 \times 10^8$ N, and the center of gravity is located $Z_{CG} = 15$ m above the mud line, as sketched in Figure 9.35. The deep-water wave field seaward of a site has a 2.0-m average wave height (H_{0avg}) and a 5.84-sec average wave period (T_{avg}). Our interest is in the wave-induced force over

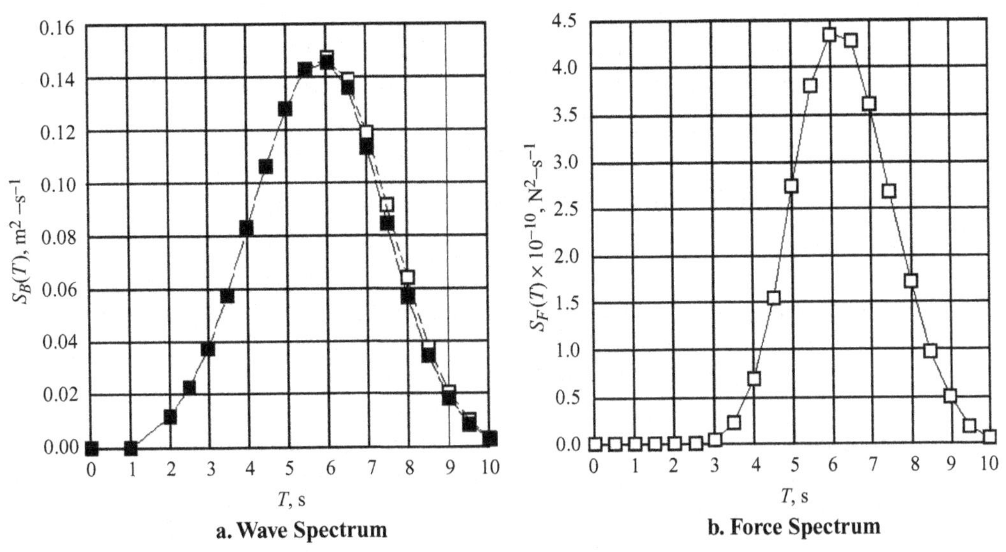

Figure 9.36. *Deep-Water Spectrum, Site Wave Spectrum, and Corresponding Force Spectrum from Example 9.13.* In Figure 9.36a, the deep-water spectrum is denoted by ■ and the spectrum at the site by □. The deep-water spectrum is obtained from the Bretschneider (1963) formula, eq. 5.59, where $S_0(T) = S_{B0}(T)/8$. The spectrum at the site is obtained from $S(T) = K_S^2 S_0(T)$, where K_S is the shoaling coefficient in eq. 3.78. The force spectrum in Figure 9.36b is obtained from eq. 9.210.

a period range of $2\,\text{sec} \leq T \leq 10\,\text{sec}$. Over the 30-m bottom contour, the wavelength (λ) must be less than or equal to about 126 m for the diffraction-dominant range of $ka_o \geq 0.5$ in Figure 9.8. This wavelength value corresponds to a period value of approximately 9.45 sec. For $T \leq 9.45$ sec, then, the diffraction forces are dominant.

We choose to predict the spectrum of the deep-water wave field by using the Bretschneider formula in eq. 5.59. The resulting deep-water spectral density values are shown in Figure 9.36. Because the waves of period greater than 6 sec are in intermediate water at a site, we modify the deep-water Bretschneider formula by multiplying it by the square of the shoaling coefficient in eq. 3.78. Furthermore, the relationship between the wave spectral density in eq. 9.206 and the Bretschneider formula is $S(T) = S_B(T)/8$. The force spectral density at the site is then

$$
\begin{aligned}
S_{Fx}(T) &= \frac{1}{2} K_S^2 Q_x^2(ka) S_{B0}(T) \\
&= 2\frac{\rho^2 g^2}{k^4}\Lambda^2 \frac{\sinh^2(kh)}{[kh + \sinh(kh)\cosh(kh)]} S_{B0}(T) \qquad (9.210)
\end{aligned}
$$

The deep-water wave spectrum, $S_0(T) = S_{B0}(T)/8$, and the wave spectrum at the site, $S(T) = K_S^2 S_0(T)$, are presented in Figure 9.36a, and the corresponding force spectrum obtained from eq. 9.210 is presented in Figure 9.36b. Because of the moderate water-depth value ($h = 30$ m), the effects of shoaling are small for the given wave conditions. The maximum change in the spectrum from deep water to the site is about 5%.

We now focus our attention on the probabilistic aspects of random diffraction forces. To do this, we use the method of Bretschneider (1965, 1967) outlined in

Section 9.3B as applied to the wave-induced inertial forces. Begin by considering the expression for the maximum MacCamy-Fuchs diffraction force. As is done for the inertial force expression in eqs. 9.185, write the force amplitude of eq. 9.200 as

$$F_{x\max} = 2\frac{\rho g}{k^2}\Lambda \tanh(kh)H \equiv \mathrm{K}_3 H \tag{9.211}$$

The root-mean-square of the maximum diffraction force is then

$$F_{x\max-r} = \mathrm{K}_3 H_{rms} \tag{9.212}$$

Assuming a Rayleigh probability distribution of the wave heights, we can write

$$p(F_{x\max})dF_{x\max} = p(H)dH = 2\frac{H}{H_{rms}^2}e^{-\left(\frac{H}{H_{rms}}\right)^2}dH \tag{9.213}$$

By replacing H and H_{rms} in the last term of this equation according to eqs. 9.211 and 9.212, respectively, we obtain the following expression for the probability density of the diffraction forces:

$$p(F_{x\max}) = \frac{dP(F_{x\max})}{dF_{x\max}} = 2\left(\frac{F_{x\max}}{F_{x\max-r}^2}\right)e^{-\left(\frac{F_{x\max}}{F_{x\max-r}}\right)^2} \tag{9.214}$$

where $P(F_{x\max})$ is the probability function of the maximum diffraction force. Because K_3 is a function of the wave number and, therefore, the period, we assume that the value of this parameter is that at the average wave period, that is, $\mathrm{K}_3 = \mathrm{K}_3(T_{\mathrm{avg}})$.

From the last equality in eq. 9.214, the cumulative probability of occurrence is

$$P(0 \le F_{x\max} < F_{x\max-J}) = \int_0^{F_{x\max-J}} p(F_{x\max})dF_{x\max} = 1 - e^{-\left(\frac{F_{x\max-J}}{F_{x\max-r}}\right)^2} \tag{9.215}$$

where $F_{x\max-J}$ is a force value of interest. Similarly, the cumulative probability of exceedance is

$$P(F_{x\max-J} \le F_{x\max} < \infty) = \int_{F_{x\max-J}}^{\infty} p(F_{x\max})dF_{x\max} = e^{-\left(\frac{F_{x\max-J}}{F_{x\max-r}}\right)^2} \tag{9.216}$$

From this expression, the extreme wave force over M years can be determined, as is done for the extreme wave heights in Section 5.4. For the Rayleigh-distributed diffraction forces, the most-probable maximum diffraction force in M years is obtained from

$$F_{x\max}|_M = F_{x\max-r}\sqrt{\ln(N_1 M)} \tag{9.217}$$

In this expression, N_1 is the (statistical) number of waves occurring over one year. From the relationship in eq. 9.182, $N_1\ M\ t_1 \simeq T_{avg}$. The application of the expression in eq. 9.217 is demonstrated in Example 9.14.

EXAMPLE 9.14: EXTREME DIFFRACTION FORCES ON A MONOLITHIC GRAVITY STRUCTURE Our interest here is in the extreme diffraction force and the corresponding overturning moment on the structure sketched in Figure 9.35. The design conditions at the site are those in Example 9.13. Those are $h = 30$ m, $R = 20$ m, $D_O = 2a_O = 10$ m, $W = 2.5 \times 10^8$ N, and $\mathrm{Z} = 15$ m. Referring to the discussion in Section 5.7B, the mean-square of the maximum diffraction force is obtained from the following integral of the expression in eq. 9.210:

$$F_{x\max -r} = \sqrt{\int_0^\infty S_{Fx}(T)dt} = \sqrt{\frac{1}{2}\int_0^\infty [K_S^2 Q_x^2(ka)S_{B0}(T)]dT}$$
$$= \sqrt{\int_0^\infty 2\frac{\rho^2 g^2}{k^4}\Lambda^2 \frac{\sinh^2(kh)}{[kh+\sinh(kh)\cosh(kh)]}\left[3.437\frac{H_{0avg}^2}{T_{avg}^4}T^3 e^{-0.675\left(\frac{T}{T_{avg}}\right)^4}\right]dT} \quad (9.218)$$

For the 2.0-m, 5.84-sec average deep-water waves, the numerical integration of the term under the last radical gives an approximate root-mean-square wave force of 7.46×10^5 N.

From eq. 9.182, the statistical number of waves for the first year is $N_1 \simeq 5.4 \times 10^6$. For year M, the probable maximum diffraction force in Newtons is

$$F_{x\max}|_M = 7.46 \times 10^5 \sqrt{\ln(5.4 \times 10^5 M)}$$

For the average wave period of 5.84 sec, the center of pressure for the diffraction force is obtained from eq. 9.75. From that equation, the expression for the height of the center of pressure above the bed is

$$Z|_{cp} = \frac{1}{k}\left[kh + \frac{1}{\sinh(kh)} - \frac{1}{\tanh(kh)}\right] \quad (9.219)$$

For the site conditions, this expression yields approximately 22.0 m. The force causing the wave-induced moment about the right edge of the footing in Figure 9.35 is

$$F_{x\max}|_M = 7.46 \times 10^5 \sqrt{\ln(5.4 \times 10^5 M)} > \frac{WR}{Z|_{cp}} \simeq 2.27 \times 10^8 \text{ N}$$

For the 100-year wave ($M = 100$), the maximum force is approximately 3.00×10^6 N, and for the 1,000-year wave, the force is approximately 3.26×10^6 N. We conclude then that the structural reliability of the tower as designed is satisfactory.

9.4 Closing Remarks

In this chapter, the natures of wave-induced forces in both deterministic and random seas are discussed. In addition, the concept of added mass has been introduced. It is shown that the dominance of the viscous-pressure, inertial, or diffraction forces depends on the values of both the ratio of the characteristic dimension of a structure and the wavelength and, in addition, on the Keulegan-Carpenter number. Motions excited by the wave forces have not been discussed. In the next chapter, the wave-induced motions of fixed and floating bodies are discussed. The reader is encouraged to consult the references to obtain more detailed discussions of each of the topics covered in this chapter.

10 Introduction to Wave-Structure Interaction

In this chapter, basic analyses of the interactions of waves and compliant fixed and floating structures are presented. The reason for discussing both fixed and floating structures in a single chapter is that several of the analytical techniques are common to both. As in the previous chapters, the structural geometries studied here are those that can be dealt with on an analytical basis. The analyses of wave interactions with complicated structural geometries require the use of numerical techniques, such as finite-element analysis. Such situations are not addressed herein. The initial discussions of each type of structure are based on the assumption that the incident wave field is composed of regular, linear waves. These discussions are followed by considerations of structural motions in irregular (random), linear seas. Depending on water depth at the site, the wind fetch, and the wind duration, the irregular-wave analyses presented in Chapter 5 can be used for seas having a Beaufort Wind Force Scale up to 5. Sea states are used to quantify the severity of a sea, and are discussed in Chapter 1. The various sea-state scales are presented in Table 1.2.

10.1 Basic Concepts

As introduced in Section 9.2, the ambient water mass excited by an unsteadily moving body is called the *added mass*. The magnitude of the added mass is proportional to the inertial reaction force on the body resulting from the body motions. When the body moves close to the free surface, waves are also created. These waves carry energy away from the body, which leads to the damping of the body motions. This damping component is called *radiation damping*. For surface-piercing bodies that are under way, the motion-produced waves are of two types. These are called *divergent waves* and *transverse waves*. Because ships are the largest class of such bodies, the waves are collectively known as *ship waves*. Divergent waves are attached to the ship, whereas transverse waves follow the ship, advancing at the speed of the ship. The creation of these ship waves results in a resistance force on the ship called *wave drag*. This topic is not discussed herein. The reader is referred to the text by Newman (1977) for the specifics of wave drag.

The concepts of added mass and radiation damping are associated with oscillatory motions of both fixed and floating structures. Both of these are related to the reaction forces, or hydrodynamic forces, that act on the moving structures. A knowledge of the nature of wave-structure interactions is needed to determine the

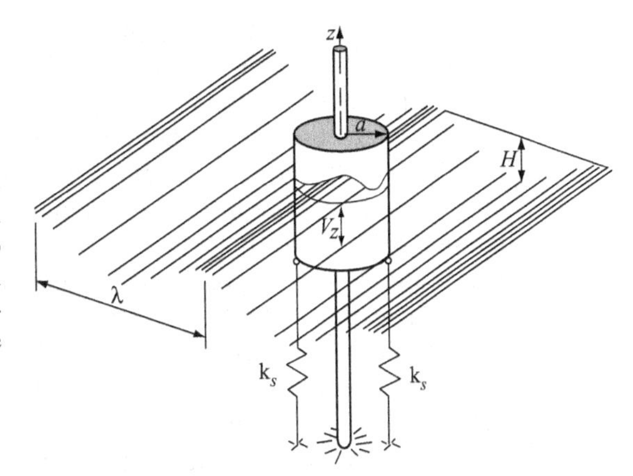

Figure 10.1. *Floating Vertical, Circular Cylinder Constrained to Heave.* The staff through the center of the cylinder is assumed to be rigid and of relatively small diameter when compared to a. The frictional resistance to the heaving motion between the cylinder and the staff is negligible. Each of the two linear-elastic mooring lines is represented by the spring constant k_S.

magnitudes of hydrodynamic forces. To gain this knowledge, a simple (but practical) structural model is used throughout this chapter. That model is a vertical, circular cylinder of finite draft constrained to heave in waters of finite depth, as sketched in Figure 10.1. In our discussions, we assume that the incident waves are linear, as discussed in Chapter 3, and are relatively long compared to the radius of the body. Mathematically, the latter assumption is expressed as $ka = (2\pi/\lambda)a \ll 1$, where k is the wave number, λ is the wavelength, and a is the radius of the cylinder.

A. Equations of Motion

When one refers to an equation of motion, what is normally referred to is an equation expressing Newton's second law of motion. That is, the time rate of change of linear momentum of a body must equal the sum of the external forces on the body. The equation of motion for a purely heaving floating body (such as the cylinder sketched in Figure 10.1) is

$$m\frac{d^2z}{dt^2} = -\,\mathrm{a}_{wz}\frac{d^2z}{dt^2} - \mathrm{b}_{\mathrm{rz}}\frac{dz}{dt} - b_{\mathrm{vz}}\left(\frac{dz}{dt}\right)\left|\left(\frac{dz}{dt}\right)\right|^{\mathrm{N}}$$
$$-\,\mathrm{b}_{\mathrm{pz}}\frac{dz}{dt} - \rho g A_{wp}z - N\mathrm{k}_s z + F_{zo}\cos(\omega t + a_z) \qquad (10.1)$$

In this equation, the inertial force (time rate of change of linear momentum) of the heaving body is on the left side. On the right side, the respective terms are (1) the inertial reaction force of the water, where a_{wz} is the added mass, (2) the radiation damping force, where b_{rz} is the radiation damping coefficient, (3) the viscous damping force, where b_{vz} is the viscous damping coefficient and N, for our purposes, is either 0 or 1, (4) the damping due to power take-off, where b_{pz} is the power take-off coefficient, (5) the hydrostatic restoring force, where A_{wp} is the waterplane area when the body is at rest, (6) the mooring restoring force, where k_s is the effective mooring spring constant of each line, and N is the number of lines, and (7) the wave-induced vertical force, where F_{zo} is the force amplitude. Also in the force term are $\omega = 2\pi/T$, the circular wave frequency, where T is the wave period, and α_z, the phase angle between the wave and the wave-induced heaving force. The first six terms on the right side of eq. 10.1 are negative because they represent some sort of opposition to the heaving motions of the body, whereas the last term is positive because the waves cause the body motions. As discussed in Chapter 2, the hydrostatic restoring force is equal to the weight of the displaced water, as in eq. 2.1. For the circular cylinder sketched in Figure 10.1, the waterplane area is $A_{wp} = \pi a^2$.

The power (N) of the velocity in the viscous damping term depends on the type of flow causing the viscous damping. For our cylinder, if the relative flow about the body is laminar, then N = 0. If the relative flow is turbulent, then N = 1. For the latter, the equation of motion is nonlinear. For the turbulent condition, the viscous damping term in eq. 10.1 is $b_{vz}(dz/dt)|dz/dt|$, where the absolute value of the velocity is required to preserve the direction of the force, as in the Morison equation discussed in Section 9.2D. For the purposes of this book, a linearized form of the equation of motion is preferred. So, the turbulent viscous-damping term is replaced by an equivalent linear damping term. The value of b_{vz} in eq. 10.1 is obtained by equating the viscous drag force in eq. 10.1 to that in eq. 9.48. The result is $b_{vz} = 1/2\rho C_d A_d$, where ρ is the mass density of the fluid, C_d is the drag coefficient, and A_d is the projected area. The notation is that used in Chapter 2.

The power take-off damping represented by the coefficient b_{pz} in eq. 10.1 represents a somewhat special case. In wave-energy conversion, this term might represent a linear-inductance electrical generator wired to the power grid onshore, as discussed by Omholt (1978), McCormick (1983) McCormick et al. (1981), Mueller, Baker, and Spooner (2000), Ivanova et al. (2003), and Baker, Mueller, and Brooking (2003). This is not the most efficient method of wave energy conversion, but it does present a rather interesting way of attenuating wave heights for shoreline protection, as discussed by McCormick, Lazarus, and Speight (2004).

Before applying eq. 10.1 to the vertical circular cylinder in Figure 10.1, let us rearrange the equation to be in the standard form of the differential equation of motion. We shall use the linearized form of the equation, where the viscous damping in the equation is represented by a generic linearized damping coefficient, $b_{\nu N}$. When the flow is turbulent (where N = 1 in eq. 10.1), this coefficient is an equivalent linear damping coefficient, b_v, discussed later in this chapter. The equivalent linear equation of heaving motion is

$$
\begin{aligned}
(m + a_{wz})\frac{d^2z}{dt^2} + (b_{rz} + b_{vz} + b_{pz})\frac{dz}{dt} + (\rho g A_{wp} + Nk_s)z &= F_z(t) \\
&= F_{zo}\cos(\omega t + \alpha_z)
\end{aligned}
\tag{10.2}
$$

The form of the equation in eq. 10.2 has a well-known steady-state solution, which is discussed later in this section. The steady-state solution of eq. 10.2 can be written as

$$
z = Z\cos(\omega t + \alpha_z - \varepsilon_z) \tag{10.3}
$$

where Z is the heaving amplitude and ε_z is the phase angle between the heaving response of the cylinder and the wave-induced force.

B. Added Mass and Radiation Damping

In Section 9.2, the concept of the added mass is introduced. The discussion in that section is focused on the water mass affected by the presence of a body in an accelerating fluid, such as a body in waves. Here, our interest is in the mass of water affected by the body motions, also called the added mass. As stated previously in the chapter, the body motions also result in a transfer energy to the sea, and the transferred

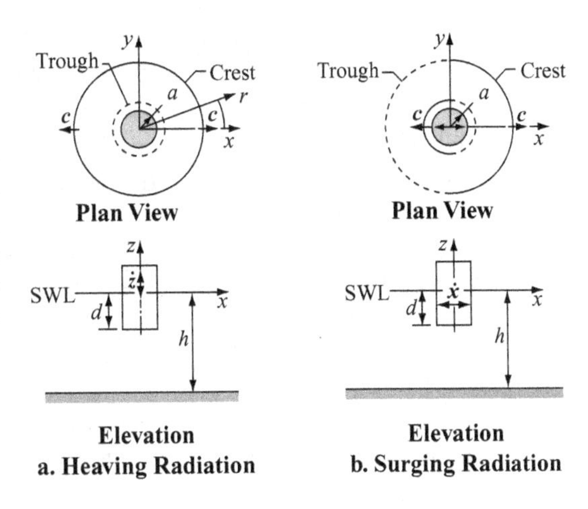

Figure 10.2. *Monopole-Type and Dipole-Type Radiation Due to Heaving and Surging Motions, Respectively*. Both types of radiation are outward; however, the heaving monopole radiation is symmetric with respect to the y-axis, and the surging dipole radiation is asymmetric with respect to the y-axis.

energy flux is away from the body. This results in an energy loss to the body motion. The resulting damping effect on the body motion is called the *radiation damping*.

The best way to introduce the added mass and radiation damping is to consider the cylinder sketched in Figure 10.1 as a radial wave maker. That is, assume that the cylinder is in calm water and undergoing forced vertical motions having a circular frequency of $\omega = 2\pi/T$, where T is the oscillation period and, hence, the period of the generated waves. The general theory of wave makers is well discussed in the book of Dean and Dalrymple (1984). The waves created by the heaving motions of the cylinder are sketched in Figure 10.2a. Waves created by an oscillating surge motion of the cylinder would resemble those sketched in Figure 10.2b. In acoustics and electromagnetics, these respective wave patterns are referred to as being created by a monopole source and a dipole source. See the books by Joos (1986) and Lighthill (1996).

To obtain mathematical expressions for the added mass and radiation damping, we must obtain a potential function representing the water particle motions excited by the body motions. For the vertical circular cylinder of finite draft in water of finite depth, Yeung (1981) presents a comprehensive analysis of the hydrodynamic reaction forces and moments for the planar motions of surge, heave, and pitch. Here, we use approximations of the long-wave equations of Yeung and neglect the high-frequency evanescent waves created by the heaving motions of the cylinder. The evanescent waves created by floating-body motions are discussed in Chapter 11. The long-wave assumption makes use of the vertically integrated form of the potential, that is, the potential function is essentially depth-averaged. Referring to Figure 10.2 for notation, for the stated conditions, Yeung's vertically integrated velocity potential for particle motions under the heaving cylinder is

$$
\begin{aligned}
\varphi_z(r,t) &\simeq -i\omega\frac{Z}{2}e^{-i\omega t}\left\{\frac{a}{2k(h-d)}\frac{H_0^{(1)}(ka)}{H_1^{(1)}(ka)} + \frac{1}{4}\frac{(a^2-r^2)}{(h-d)}\right\} \\
&\xrightarrow[ka\ll 1]{} -i\omega\frac{Z}{2}e^{-i\omega t}\left\{\frac{a}{2k(h-d)}\frac{1+i\dfrac{2}{\pi}\left[\ln\left(\dfrac{ka}{2}\right)+\gamma\right]}{\dfrac{ka}{2}\left\langle 1+i\dfrac{2}{\pi}\left[\ln\left(\dfrac{ka}{2}\right)+\gamma\right]-\dfrac{i}{\pi}\right\rangle} + \frac{1}{4}\frac{(a^2-r^2)}{(h-d)}\right\}
\end{aligned}
\tag{10.4}
$$

where $\gamma = 0.5772157$ is Euler's constant. The last approximation (the second line in eq. 10.4) is based on Yeung's assumptions that $\lambda \gg a$, and that a is the same order of magnitude as h. In eq. 10.4, Z is the heaving amplitude. The functions $H_0^{(1)}(ka)$ and $H_1^{(1)}(ka)$ are Hankel functions of the first kind of zero order and first order, respectively. See Abramowitz and Stegun (1965) for both the details of these functions and the approximation of these functions leading to the second line of the equation. From the linearized Bernoulli's equation (eq. 3.70), the hydrodynamic pressure on the bottom of the cylinder due to the vertically integrated velocity potential is

$$p_{dyn}|_{z=-d} = -\rho \frac{\partial \varphi_z}{\partial t} = -i\omega\rho\varphi_z \tag{10.5}$$

This pressure is integrated over the bottom of the cylinder to obtain the hydrodynamic reaction force. The result is

$$\begin{aligned} F_z(t) &= \int_0^{2\pi}\int_0^a p_{dyn}\, r\,dr\,d\beta = -\mathrm{a}_{wz}\left(-\omega^2 \frac{Z}{2} e^{-i\omega t}\right) - \mathrm{b}_{rz}\omega\left(-i\omega \frac{Z}{2} e^{-i\omega t}\right) \\ &= -\mathrm{a}_{wz}\frac{d^2 z}{dz^2} - \mathrm{b}_{rz}\frac{dz}{dt} \end{aligned} \tag{10.6}$$

By combining the results in eqs. 10.4 through 10.6, we find the expressions for the added mass (hydrodynamic inertial coefficient), a_{wz}, and the radiation damping coefficient, b_{rz}. Those are

$$\mathrm{a}_{wz} = \rho\pi a^3 \left\{ \frac{a}{8(h-d)} - \frac{a}{2h}\left[\gamma + \ln\left(\frac{ka}{2}\right)\right]\right\} \tag{10.7}$$

and

$$\mathrm{b}_{rz} = \rho\pi a^3 \omega \frac{\pi a}{4h} \frac{1}{\left\{1 + (ka)^2\left[\frac{\pi}{2} - \gamma - \ln\left(\frac{ka}{2}\right)\right]\right\}} \tag{10.8}$$

respectively. These equations show that both the added mass and radiation damping coefficient are dependent on the wave number, $k = 2\pi/\lambda$, and because of the dispersion relationship in eq. 3.30, they are functions of the circular frequency of motion, ω. However, only the added mass depends on the buoy draft, d, and water depth, h. Results obtained from the approximate expressions for the added mass in eq. 10.7 and the radiation damping coefficient in eq. 10.8 are presented in Figures 10.3 and 10.4, respectively, for the conditions given in Example 10.1. McCormick and Kraemer (2006) show that values obtained by the application of the long-wave approximations of Yeung (1981) compare well with experimental data over an appropriate wave-number range.

C. Equivalent Viscous Damping Coefficient

Comparing the forms of the equation of motion in eqs. 10.1 and 10.2, we see that the viscous damping coefficient, b_{vz}, in eq. 10.1 has been replaced by an equivalent linear damping coefficient, $\mathrm{b}_{\nu N}$, in eq. 10.2. When the exponent $\mathrm{N} = 0$ in eq. 10.1, the damping is due to a fully laminar flow. When the flow is turbulent, $\mathrm{N} = 1$. The relationship between the two coefficients is established by assuming that the nonlinear

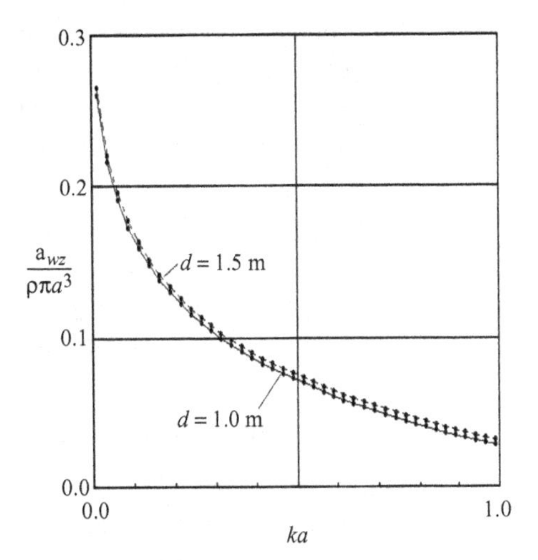

Figure 10.3. *Non-Dimensional Added Mass as a Function of ka in Example 10.1.*

motions are approximately sinusoidal. Neglect the phase angles in eq. 10.3 for now, and assume that the linear and nonlinear heaving displacements are both $Z\cos(\omega t)$. With this assumption, we also assume that the energies lost by both the linear and nonlinear motions over one period are approximately equal. The resulting energy relationship, assuming N = 1, is

$$\int_{z(0)}^{z(T)} \mathrm{b}_{vz}\frac{dz}{dt}dz = \int_{0}^{T} \mathrm{b}_{vz}\left(\frac{dz}{dt}\right)^2 dt = \pi\omega Z^2\mathrm{b}_{vz}$$

$$= \int_{z(0)}^{z(T)} b_{vz}\frac{dz}{dt}\left|\frac{dz}{dt}\right| dz = 2\int_{0}^{T/2} b_{vz}\left(\frac{dz}{dt}\right)^3 dt = \frac{8}{3}\omega^2 Z^3 b_{vz} \tag{10.9}$$

From this equation, we find the relationship between the two damping coefficients. The resulting equivalent linear viscous damping coefficient expression is

$$\mathrm{b}_{vz} = \frac{8}{3}\frac{\omega}{\pi} Z b_{vz} \tag{10.10}$$

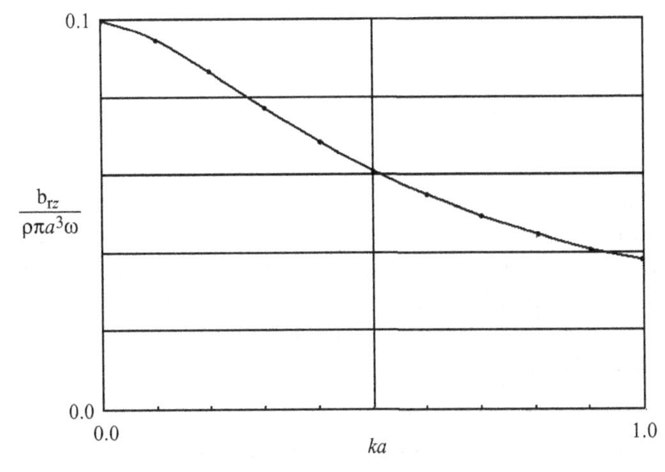

Figure 10.4. *Non-Dimensional Radiation Damping as a Function of ka in Example 10.1.*

For a body in long waves, approximate values of the velocity-squared, viscous damping coefficient can be obtained from experimental values of the steady-state experimental drag coefficient, C_d, such as those reported by Hoerner (1965). As stated previously, the relationship between the nonlinear coefficient and the drag coefficient is

$$b_{vz} \simeq \frac{1}{2}\rho C_d A_d \tag{10.11}$$

where A_d is, again, the projected area. For our heaving circular cylinder, that area is $A_d = \pi a^2$, where a is the radius.

D. Steady-State Solution of the Heaving Equation

The equation of motion in eq. 10.2 has the same form as that describing the forced vibrations of a linear spring-mass system (for example, see Zill, 1986). If the viscous damping is due to turbulence, then N = 1 in eq. 10.1, and the equivalent damping coefficient in eq. 10.10 can be used. The *steady-state solution* of eq. 10.2 for this damping condition is

$$z = Z\cos(\omega t + \alpha_z - \varepsilon_z) = \frac{\dfrac{F_{zo}}{(\rho g A_{wp} + N k_s)}}{\sqrt{\left(1 - \dfrac{\omega^2}{\omega_n^2}\right)^2 + \left[2\dfrac{\omega}{\omega_n}\dfrac{(b_{rz} + b_{vz} + b_{pz})}{b_{cz}}\right]^2}}\cos(\omega t + \alpha_z - \varepsilon_z) \tag{10.12}$$

Three new terms appear in this equation, which are the following: The *natural heaving frequency* is mathematically represented by

$$\omega_n = \frac{2\pi}{T_{nz}} = \sqrt{\frac{\rho g A_{wp} + N k_s}{m + a_{wz}}} \tag{10.13}$$

Here, T_{nz} is the *natural heaving period.* The *critical damping coefficient* is

$$b_{cz} = 2\sqrt{(m + a_{wz})(\rho g A_{wp} + N k_s)} \tag{10.14}$$

Last, the *phase angle* between the force and motion is expressed mathematically as

$$\varepsilon_z = \tan^{-1}\left[\frac{2\dfrac{\omega}{\omega_n}\dfrac{(b_{rz} + b_{vz} + b_{pz})}{b_{cz}}}{\left(1 - \dfrac{\omega^2}{\omega_n^2}\right)}\right] \tag{10.15}$$

Note that the values of the added mass and linearized damping coefficients all depend on the wave excitation frequency. Furthermore, the natural circular frequency, critical damping, and phase angle depend on these coefficients and, subsequently, depend on the wave frequency.

In the following example, the effect of the equivalent damping coefficient for turbulent flow losses is illustrated.

EXAMPLE 10.1: HEAVING MOTION OF A CAN BUOY A 1-m-diameter ($D = 2a = 1$ m) can buoy having a 2.5-m height is constrained to heave in 4 m (h = 4 m) of salt water (see the sketch in Figure 10.1). The mass density of the salt water is $\rho = 1.03 \times 10^3$ kg/m^3. Ballast is added to the buoy to cause it to float with a

1-m draft ($d = 1$ m). The weight of the buoy is then $\rho g \pi a^2 d \simeq 7.94 \times 10^3$ N and its mass (m) is approximately 809 kg. When a single mooring line is attached, the draft of the buoy is $d + \delta d = 1.5$ m. The unstretched mooring line length is $\ell_S = 2.4$ m, so the stretched line length is $\delta\ell_S = (h - d - \delta d - \ell_S) = 0.1$ m, and the effective spring constant of the line is $\mathrm{k}_S = \mathrm{T}_S/\delta\ell_S = 3.97 \times 10^4$ N/m, where the tension in the line is then $\mathrm{T}_S = \rho g \pi a^2 \Delta d \simeq 3.97 \times 10^3$ N.

The buoy is subjected to linear waves having a height $H = 1$ m and a period $T = 7$ sec. The circular wave frequency is then $\omega = 2\pi/T \simeq 0.898$ rad/s. We must determine the length of this wave from eq. 3.31 using the method of successive approximations, as illustrated in Example 3.3. The result is $\lambda \simeq 41.4$ m. For this wavelength, $ka = 2\pi a/\lambda \simeq 0.0759$, which well satisfies our long-wave assumption. From the curves in Figure 9.23, we see that the ka value is relatively small. We can use the approximate vertical formula in eq. 9.97 to represent the wave-induced heaving force, that is,

$$F_z(t) = \rho g \pi a^2 \eta(t) = \rho g \pi a^2 \frac{H}{2} \cos(\omega t) = F_{zo} \cos(\omega t) \tag{10.16}$$

The force amplitude is then $F_{zo} \simeq 3.97 \times 10^3$ N. Under the conditions stated, we see that the force is in-phase with the incident wave, that is, $\alpha_z = 0$ in eqs. 10.1 and 10.2.

For $ka = 0.0759$, we find in Figure 10.3 that the non-dimensional added mass is $\mathrm{a}_{wz}/\rho\pi a^3 \simeq 0.195$, and in Figure 10.4, the non-dimensional radiation damping value is $\mathrm{b}_{rz}/\rho\pi a^3\omega \simeq 0.0958$. The added mass is then $\mathrm{a}_{wz} \simeq 78.7$ kg, and the radiation damping coefficient is $\mathrm{b}_{rz} \simeq 34.8$ N-s-m^{-1}. Assuming a drag coefficient (C_d in eq. 10.11) value of approximately 1 for the heaving cylinder, the time-averaged (nonlinear) turbulent damping coefficient is $b_{vz} = 1/2\rho C_d \pi a^2 \simeq$ 405 N-s^2-m^{-2}. The equivalent turbulent drag coefficient from eq. 10.10 is $\mathrm{b}_{vz} \simeq 308(Z/2)$, having units of N-s-m^{-1}.

With the numerical values of the coefficients in eq. 10.2, the natural circular frequency of the heaving motions is $\omega_n \simeq 7.32$ rad/s from eq. 10.13. The critical damping coefficient from eq. 10.14 is $\mathrm{b}_{cz} \simeq 6.50 \times 10^3$ N-s-m^{-1}.

Because the equivalent viscous damping is linearly proportional to the heaving amplitude, the amplitude of the heaving motion is obtained from eq. 10.13, where the equivalent viscous damping is replaced by the expression in eq. 10.10. The result is

$$Z = \frac{\dfrac{F_{zo}}{(\rho g A_{wp} + N\mathrm{k}_s)}}{\sqrt{\left(1 - \dfrac{\omega^2}{\omega_n^2}\right)^2 + \left[2\dfrac{\omega}{\omega_n}\dfrac{(\mathrm{b}_{rz} + \mathrm{b}_{vz})}{\mathrm{b}_{cz}}\right]^2}} \tag{10.17}$$

where the power take-off damping coefficient is not yet considered here, that is, $\mathrm{b}_{pz} = 0$. If the two damping components in eq. 10.17 are of the same order of magnitude, then the equation must be solved numerically. If either of the damping terms can be neglected, then eq. 10.17 can be solved exactly. For the conditions given, the numerical solution of eq. 10.17 yields a motion amplitude of $Z \simeq$ 0.083 m. The equivalent viscous damping coefficient is then $\mathrm{b}_{vz} \simeq 25.5$ N-s-m^{-1}. We see that this value is about the same as the radiation damping coefficient, $\mathrm{b}_{rz} \simeq 34.8$ N-s-m^{-1}. Finally, the frequency ratio is $\omega/\omega_n \simeq 0.123$, which is well below the resonance condition of $\omega/\omega_{nz} = 1$.

In the following section, it is shown how values of the damping coefficient and, in addition, the added mass and radiation damping coefficient are obtained experimentally.

E. Determination of Added Mass and Resonant Damping Coefficients in Calm Water

The natures of the added mass and radiation damping are discussed in Section 10.1B, and that of the velocity-square viscous damping is discussed in Section 10.1C. The problem addressed here is how to experimentally determine the magnitudes of the added mass and damping coefficients. To illustrate the techniques, we again use the vertical, circular cylinder constrained to heave in waters of finite depth, such as that sketched in Figure 10.1. The radiation pattern resulting from the heaving motions is sketched in Figure 10.2a.

There are two types of experiments that are used to determine the added mass and damping coefficients. The first is a calm-water test, where the body is initially displaced and released. From the resulting motions, the logarithmic decrement (a measure of the decay rate) is determined. The second experiment is in waves, where the absorbed wave power over a range of frequencies is studied. The damping in a resonant mechanical system is inversely proportional to the half-power frequency bandwidth, as discussed later in this chapter.

It must be noted that the experimental methods mask the frequency dependence of both the added mass and the damping coefficients. Hence, the values for the coefficients are approximate.

In this subsection, we shall discuss the logarithmic decrement method. Consider the freely floating, vertical, circular cylinder in Figure 10.2a. In calm water, the body is raised to a height Z_O and released. The resulting linear motions are predicted from the solution of the following homogeneous equation:

$$(m + \mathrm{a}_{wz})\frac{d^2z}{dz^2}(\mathrm{b}_{rz} + \mathrm{b}_{vz})\frac{dz}{dt} + \rho g A_{wp} z = 0 \tag{10.18}$$

Comparing eqs. 10.2 and 10.18, we see in eq. 10.18 that the power take-off damping coefficient, spring constant, and forcing function are absent because the body is freely floating after its release. The viscous damping is represented by the equivalent viscous damping coefficient in eq. 10.10. The solution of the homogeneous equation in eq. 10.18, subject to the initial condition $z(0) = Z_o$, is

$$z(t) = Z_o e^{-\frac{1}{2}\left(\frac{\mathrm{b}_{rz}+\mathrm{b}_{vz}}{m+\mathrm{a}_{wz}}\right)t} \cos(\omega_{dz} t) \tag{10.19}$$

Here, $\omega_{dz} = 2\pi/T_{dz}$ is the damped, natural circular frequency of the decayed heaving motion, and T_{dz} is the damped natural period. The time response of the heaving body would resemble that sketched in Figure 10.5. The time-dependent amplitude of the motion is then

$$|z_{\max}(t_j)| \equiv Z_j = Z_o e^{-\frac{\mathrm{b}_{rz}+\mathrm{b}_{vz}}{2(m+\mathrm{a}_{wz})}t_j} \equiv Z_o e^{-\Delta_z \omega_{nz} t_j} \tag{10.20}$$

where, referring to Figure 10.5, $t_j = jT_d/2\,(j = 1, 2, \ldots)$ is the time corresponding to a maximum displacement. The natural circular frequency is found in eq. 10.13,

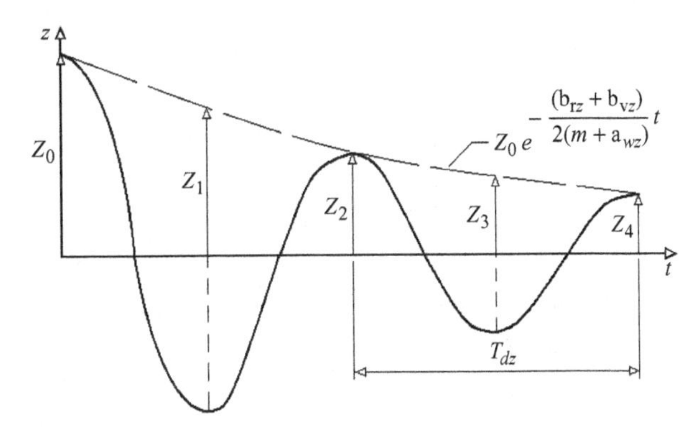

Figure 10.5. *Time-Response of a Freely Floating, Damped, Heaving Circular Cylinder.*

where the spring constant (k_S) is zero for the unmoored body. From the relationship between the exponents in eq. 10.20, we find the expression for the *resonant damping ratio* to be

$$\Delta_z \equiv \frac{b_{rz} + b_{vz}}{b_{cz}} = \frac{b_{rz} + b_{vz}}{2\sqrt{\rho g A_{wp}(m + a_{wz})}} \tag{10.21}$$

that is, Δ_z is the ratio of the damping to the critical damping. The expression for the damped natural frequency is found from the combinations of eqs. 10.20 and 10.21. The result is

$$\omega_{dz} = \frac{2\pi}{T_{dz}} = \sqrt{2\frac{\rho g A_{wp}}{m + a_{wz}} - \left[\frac{b_{rz} + b_{vz}}{2(m + a_{wz})}\right]^2} = \omega_{nz}\sqrt{1 - \Delta_z^2} \tag{10.22}$$

When the expression under the radical sign is positive ($\Delta_z < 1$), the motions are under-damped and will be oscillatory, as sketched in Figure 10.5. When the radicand equals zero ($\Delta_z = 1$), then the motions are critically damped, and the cylinder simply returns to its resting position after being released. Finally, when $\Delta_z > 1$ in eq. 10.22, the expression in the equation is imaginary and the motions are over-damped. That is, the cylinder returns to its resting position less rapidly than for the critically damped case.

Returning to eq. 10.20, the ratio of two successive amplitudes is

$$\frac{Z_j}{Z_{j+1}} = e^{-\Delta_z \omega_{nz}(t_j - t_{j+1})} = e^{\Delta_z \omega_{nz}\left(\frac{T_{dz}}{2}\right)} = e^{\Delta_z \pi \frac{\omega_{nz}}{\omega_{dz}}} = e^{\pi \frac{\Delta_z}{\sqrt{(1-\Delta_z^2)}}} \tag{10.23}$$

The exponent of the last term is called the *logarithmic decrement*, that is,

$$\ln\left(\frac{Z_j}{Z_{j+1}}\right) = \pi \frac{\Delta_z}{\sqrt{(1 - \Delta_z^2)}} \tag{10.24}$$

From this equation, the damping ratio in eq. 10.21 is

$$\Delta_z \equiv \frac{b_{rz} + b_{vz}}{b_{cz}} \equiv \frac{b_z}{b_{cr}} = \frac{\ln(Z_j/Z_{j+1})}{\sqrt{\pi^2 + [\ln(Z_j/Z_{j+1})]^2}} \tag{10.25}$$

By measuring two consecutive amplitudes of the damped heaving motion, we can then determine the value of the combined damping coefficient, b_z.

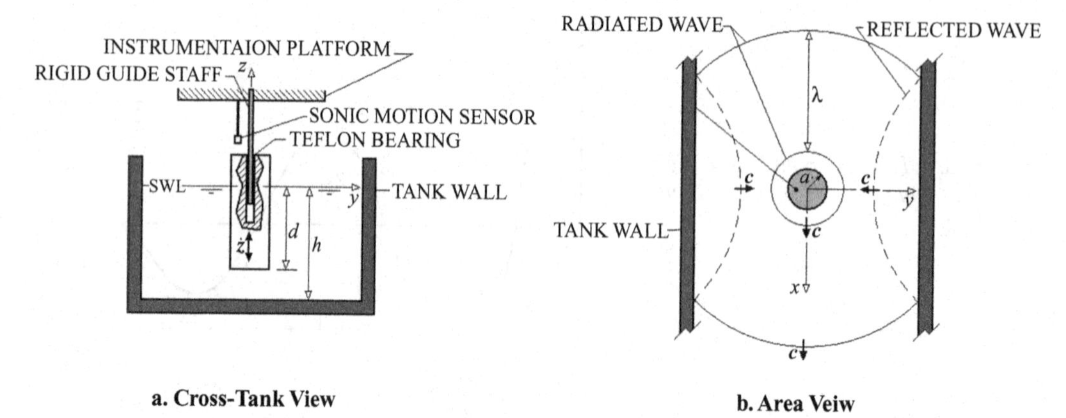

Figure 10.6. *Sketch of Experimental Set-up to Determine the Added Mass and Damping.*

The damped natural period (T_{dz}) is experimentally determined, as can be seen in Figure 10.5. In eq. 10.22, we combine this measured period value and that of the damping ratio (Δ_z in eq. 10.25) to determine the value of the undamped natural frequency, ω_{nz}. Using this frequency value in eq. 10.13, we find that the added mass is

$$a_{wz} = \frac{\rho g A_{wp}}{\omega_{nz}^2} - m \tag{10.26}$$

One final note: As seen in Figures 10.3 and 10.4, respectively, the added mass and the linear damping coefficients are functions of the wave number (k) and, therefore, frequency. If ballast is added to the cylinder, the draft will increase and the natural frequency will change. This change will subsequently alter the values of the added mass and damping coefficients.

EXAMPLE 10.2: EXPERIMENTAL DETERMINATION OF HEAVING ADDED MASS AND DAMPING LOGARITHMIC DECREMENT METHOD In a study similar to that conducted by McCormick, Coffey, and Richardson (1982), motions of a heaving vertical cylinder are studied in a 2.5-m-wide wave tank having a 1.5 m depth. The experimental set-up is sketched in Figure 10.6. A cylinder having a 0.1-m radius ($a = 0.1$ m) and a 1-m draft ($d = 1$ m) is located in the center of the tank. In still water, the cylinder is given a 0.3-m displacement (Z_O) and released. The subsequent time response is similar to that sketched in Figure 10.5. The damped natural period (T_d) is found to be 2.12 sec. The first two amplitude values (Z_1 and Z_2) are 0.220 m and 0.170 m, respectively. From these values, the damped natural circular frequency (ω_{dz}) is 2.96 rad/s and the approximate values of the logarithmic decrement from eq. 10.24 and the corresponding damping ratio (Δz) values from eq. 10.25 are, respectively, 0.258 and 0.0824. Note that the value of the logarithmic decrement based on the initial displacement and the first amplitude (Z_O and Z_1) will be somewhat greater because the initial motions are nonlinear. So, when using the linear equations derived in this section, it is best to use the larger time-value measured double amplitudes, that is, Z_j and Z_{j+1}, where $j \geq 1$. The undamped natural frequency (ω_{nz}) is 2.95 rad/s from eq. 10.22. Because the total damping is small, we see that there is little difference between ω_{nz} and ω_{dz}. The added mass can now be determined from eq. 10.26, where the mass of the structure (m) must equal the mass of the displaced water in that equation. The water in the tank is fresh water having a mass-density (ρ) of 10^3 kg/m^3. Therefore, the displaced mass is approximately 31.42 kg, and the added mass is $a_{wz} = 3.99$ kg. For this experiment, the ratio of the added mass to the body mass is $a_{wz}/m \simeq 0.127$.

The critical damping coefficient (b_{cz}) for the unmoored cylinder is 209 N-s/m from eq. 10.14. The total linear damping coefficient is then $b_z = b_{cz}\Delta_z \simeq$ 17.2 N-s/m. The damping ratio is then $\Delta_z \simeq 0.082$.

F. Bandwidth Determination of Damping in Wave-Induced Heaving Motions

When our vertical, circular cylinder is exposed to monochromatic waves, it is beneficial to consider the rate at which wave energy is both absorbed and subsequently dissipated due to both radiation and viscosity. The time rate of change of the energy is the power, so our attention here is focused on the power gained and lost by the body.

The analysis of the power of the floating cylinder is straightforward. Because power equals the product of force and velocity, multiply the linear equation of motion (eq. 10.2) by the heaving velocity to obtain a time-dependent power equation. The heaving velocity is the time-derivative of the expression in eq. 10.3. The time-averaged power over one wave period is found to be

$$\begin{aligned}\frac{1}{T}\int_0^T b_z\left(\frac{dz}{dt}\right)^2 dt &= \frac{1}{T}\int_0^T \left(\frac{dz}{dt}\right)F_{zo}\cos(\omega t + \alpha_z)\,dt \\ &= \frac{1}{2}\omega^2 Z^2 b_z = \frac{1}{2}F_{zo}\omega Z\sin(\varepsilon_z - \alpha_z)\left[\cos(\alpha_z) - \sin(\alpha_z)\right] \qquad (10.27) \\ &= P_{bz} = P_{wz}\end{aligned}$$

Here, the damping coefficient b_z represents all of the linear damping coefficients in eq. 10.2, where P_{bz} is the power dissipated by the damping. The available wave power is represented by P_{wz}. We note that in the time-averaging process, the inertial and restoring terms of eq. 10.3 vanish because the incident linear waves and the induced body motions are sinusoidal. The phase angle ε_z, introduced in eq. 10.3 and expressed in eq. 10.15, is the phase angle between the motions of the cylinder and the wave force. Physically, eq. 10.27 expresses the fact that the power lost to radiation and viscous damping (P_{bz}) over one wave period must equal the power supplied by the wave (P_{wz}).

Assume that the incident waves are relatively long compared to the radius of the cylinder. That is, we assume that the values of the wavelength λ and the radius a are such that $ka \ll 1$. By making this assumption, the dominant wave force is hydrostatic due to the addition and loss of the hydrostatic force caused by the respective rise and fall of the free surface, $\eta(t)$. The wave-induced force is then

$$\begin{aligned}F_{zo}\cos(\omega t + \alpha_z) &= F_{zo}\left[\cos(\omega t)\cos(\alpha_z) - \sin(\omega t)\sin(\alpha_z)\right] \simeq \rho g A_{wp}\eta(t) \\ &= \rho g A_{wp}\frac{H}{2}\cos(\omega t) \qquad (10.28)\end{aligned}$$

From this equation, we see that the exciting force is approximately in phase with the wave, that is, the phase angle is $\alpha_z \simeq 0$. The resulting power dissipated by the damping in eq. 10.27 is

$$P_{bz} = \frac{1}{2}\omega^2 b_z Z^2 \simeq \frac{1}{2}F_{zo}\omega Z \sin(\varepsilon_z) = \frac{1}{2}F_{zo}\omega Z \frac{2\frac{\omega}{\omega_{nz}}\Delta_z}{\sqrt{\left(1-\frac{\omega}{\omega_{nz}^2}\right)^2 + \left(2\frac{\omega}{\omega_{nz}}\Delta_z\right)^2}} \tag{10.29}$$

The expression for ε_z in this equation is obtained from eq. 10.15, and the expression for the damping ratio, Δ_z, is found in eq. 10.25. At resonance, $\omega = \omega_{nz}$ and $\varepsilon_z = 90°$. The resonant absorbed-power expression is then

$$P_{bzn} = \frac{1}{2}\omega_{nz}^2 b_{zn} Z_n^2 \simeq \frac{1}{2}F_{zo}\omega_{nz} Z_n \tag{10.30}$$

The term "absorbed power" refers to that power lost to damping.

Consider the case where the power absorbed by the heaving circular cylinder is one half of the absorbed-power value of the resonant value of eq. 10.30. The combination of eqs. 10.17 and 10.29 with 0.5-times eq. 10.30 yields the half-power relationship. From this combination, one obtains

$$8\left(\frac{\omega^2}{\omega_{nz}^2}\right)\Delta_z\Delta_{zn} - 4\left(\frac{\omega^2}{\omega_{nz}^2}\right)\Delta_z^2 - \left[1-\left(\frac{\omega^2}{\omega_{nz}^2}\right)\right]^2 \simeq 4\left(\frac{\omega^2}{\omega_{nz}^2}\right)\Delta_{zn}^2 - \left[1-\left(\frac{\omega^2}{\omega_{nz}^2}\right)\right]^2 = 0 \tag{10.31}$$

Here the frequency, ω, is that for which $P_{bz} = P_{bzn}/2$. The approximation results from the assumption that the damping does not vary significantly over the frequency band $|\omega_{nz} - \omega|$. Hence, $\Delta \simeq \Delta_{zn}$. The solution of eq. 10.31 for the frequency ratio is

$$\frac{\omega^2}{\omega_{nz}^2} = 1 + 2\Delta_{zn}^2 \pm 2\Delta_{zn}\sqrt{1+\Delta_{zn}^2} \simeq 1 \pm 2\Delta_{zn} \tag{10.32}$$

Here, the approximation is made on the assumption that the motions are lightly damped, that is, $(\Delta_{zn})^2 \ll 1$. The two half-power frequencies resulting from the approximation are then the lower band frequency,

$$\omega_1 = \omega_{nz}\sqrt{1-2\Delta_{zn}} \tag{10.33}$$

and the upper band frequency,

$$\omega_2 = \omega_{nz}\sqrt{1+2\Delta_{zn}} \tag{10.34}$$

From these relationships, we obtain

$$\frac{\omega_2^2}{\omega_{nz}^2} - \frac{\omega_1^2}{\omega_{nz}^2} = \frac{(\omega_2-\omega_1)}{\omega_{nz}}\frac{(\omega_2+\omega_1)}{\omega_{nz}} = 4\Delta_{zn} \tag{10.35}$$

With the assumption that the lower and upper band frequencies are equidistant from the resonant frequency, we find that the *half-power bandwidth* is

$$\omega_2 - \omega_1 = 2\Delta_{zn}\omega_{nz} \tag{10.36}$$

By measuring the half-power bandwidth, we have a method of approximately determining the resonant damping (Δ_{zn}) of the wave-induced motions of our floating body. Again, we note that the damping ratio value obtained from eq. 10.36 is approximate, as the ratio is actually frequency-dependent, that is, the damping value will vary as the excitation frequency varies. The half-power bandwidth method is illustrated in the following example.

EXAMPLE 10.3: EXPERIMENTAL DETERMINATION OF DAMPING IN WAVE-INDUCED HEAVING-HALF-POWER BANDWITH METHOD In the study of the wave-induced motions of a vertical, circular cylinder described in Example 10.2, a series of tests is conducted where the cylinder is subjected to incident waves of 0.03-m height and various frequencies. The measured double amplitude at resonance (where $\omega = \omega_{nz} = 2.95\,\text{rad/s}$) is 0.109 m. The approximate damping coefficient can be determined from the measured half-power bandwidth. That is, by determining the upper and lower frequencies corresponding the cases where the power lost due to both radiation and viscosity is equal to one half of the power lost at the resonant frequency, the damping ratio can be determined from eq. 10.36. It has been noted that the damping coefficients due to both radiation and viscosity are frequency-dependent. Hence, by using the expression in eq. 10.36 to determine the damping, we are neglecting the frequency behavior. Results of the experiment are presented in Figure 10.7, where the ratio of the absorbed power and the resonant absorbed power are presented as a function of the frequency ratio. The half-power bandwidth ratio is shown to be about 0.082.

Although the frequency dependence of the total damping is masked in the experimental study, we can include the frequency dependence in the theoretical equations. Consider the time-averaged power absorbed by the body by the damping. Combining the results of eqs. 10.15 and 10.17 with eq. 10.29, the expression for the absorbed power of the heaving body is

$$P_{bz} \simeq \frac{1}{2} F_{zo} \omega Z \sin(\varepsilon_z) = F_{zo}^2 \frac{\omega^2}{\omega_{nz}} \frac{\Delta_z}{\rho g A_{wp}} \frac{1}{\left[\left(1 - \frac{\omega^2}{\omega_{nz}^2}\right)^2 + \left(2 \frac{\omega}{\omega_{nz}} \Delta_z\right)^2\right]} \tag{10.37}$$

If our interest in the oscillating buoy was in wave power conversion, we would include the damping coefficient representing the energy conversion system in our damping ratio, Δ_z. Returning to the experimental study, we plot the normalized absorbed power as a function of the frequency ratio, that is

$$\begin{aligned} \frac{P_{bz}}{P_{bzn}} &= 2\left(\frac{Z}{Z_n}\right)\left(\frac{\omega}{\omega_{nz}}\right)^2 \frac{\Delta_z}{\sqrt{\left(1 - \frac{\omega^2}{\omega_{nz}^2}\right)^2 + 4\left(\frac{\omega}{\omega_{nz}}\right)^2 \Delta_z^2}} \\ &\simeq 2\left(\frac{\omega}{\omega_{nz}}\right)^2 \Delta_{zn} \frac{1}{\left[\left(1 - \frac{\omega^2}{\omega_{nz}^2}\right)^2 + 4\left(\frac{\omega}{\omega_{nz}}\right)^2 \Delta_{zn}^2\right]} \end{aligned} \tag{10.38}$$

The results of the approximation are presented in Figure 10.7, where data obtained at the U.S. Naval Academy are presented. In that figure, we see that the half-power frequency bandwidth $(\omega_2 - \omega_1)$ is approximately $0.082\,\omega_{nz}$. From eq. 10.36, the resonant damping ratio is

$$\Delta_{zn} \equiv \frac{b_z}{b_{cz}} = \frac{1}{2}\left(\frac{\omega_2 - \omega_1}{\omega_{nz}}\right) \simeq 0.041 \tag{10.39}$$

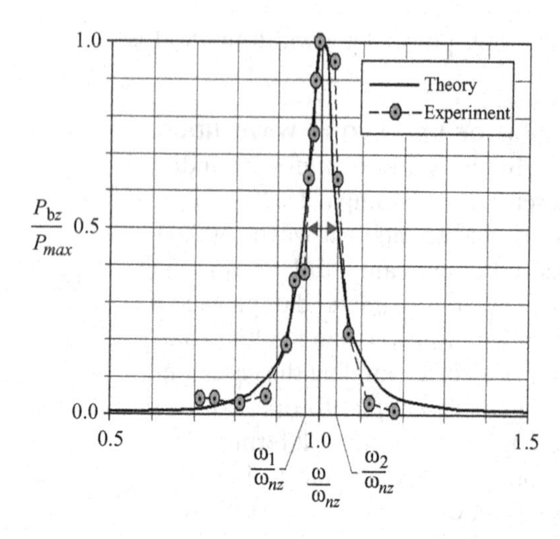

Figure 10.7. *Normalized Absorbed Power Response as a Function of Frequency Ratio.* The half-power bandwidth ratio is approximately 0.082. From eq. 10.36, the damping ratio is approximately 0.041. This value is used in eq. 10.38 to obtain the theoretical curve (see Example 10.3). The experimental data were obtained at the Hydrodynamics Laboratory of the U.S. Naval Academy.

In Example 10.2, our calm-water damping ratio values are 0.258 and 0.0824. The initial displacement in the calm-water tests is 0.3 m. The motions in the first oscillations for this relatively large initial displacement are nonlinear. As time increases in Figure 10.6, the motions become linear, and the damping in the system approaches the value in eq. 10.39. Because the critical damping coefficient (b_{cz}) for the unmoored cylinder is 209 N-s/m from Example 10.2, the total damping value is $b_z \simeq 8.36$ N-s/m.

The damping value obtained from the data in Example 10.3 is the total damping, that is, the damping term, b_z, represents the sum of the radiation and viscous damping as in eq. 10.25. We now address the problem of determining the radiation damping. This is done in Example 10.4.

EXAMPLE 10.4: EXPERIMENTAL DETERMINATION OF COMPONENT DAMPING COEFFICIENTS The circular cylinder in Examples 10.2 and 10.3 is now forced to heave in calm water by attaching an electrical oscillator to the top of the cylinder. Again, the radius (a) of the cylinder is 0.1 m, and the draft is 1 m. The water depth is 1.5 m. The frequency of the oscillator equals the undamped natural frequency, which from Example 10.2 is $\omega_{nz} = 2.95$ rad/s. The magnitude of the force is adjusted so that the amplitude of the heaving motions is that measured in the wave study in Example 10.3, that is $Z_n = 0.035$ m. Referring to the sketch in Figure 10.6b, a wave gauge is placed in the center of the tank at a distance of 1 m from the centerline of the cylinder. This distance corresponds to a value of x, equal to ten times the radius of the cylinder. At this position, a radiated wave height (H_r) of 0.002 m is measured. From eq. 3.72, the corresponding energy flux of the wave is

$$\begin{aligned} P_{rzn} &= \rho g \frac{H_r^2}{8} c_g (2\pi x) \simeq 1000\,(9.81)\,\frac{0.002^2}{8}\,(1.67)\,2\pi 1.0 \simeq 0.0527 \text{ Watts} \\ &= \frac{1}{2}\omega^2 b_{rzn} Z_n^2 = \frac{1}{8}\,(2.95)^2\, b_{rzn}\,(0.109)^2 \end{aligned} \tag{10.40}$$

where the group velocity (c_g) is obtained from eq. 3.63, and the crest width is the circumference of the circular wave front of radius $x = 1.0$ m. The second line in eq. 10.40 comes from eq. 10.27. Solving for the radiation damping coefficient, we obtain $b_{rzn} \simeq 4.01$ N-s/m.

The equivalent viscous damping coefficient is the difference between the total damping coefficient (from Example 10.3) and the radiation damping coefficient, that is, $b_{zn} - b_{rzn} = (8.36 - 4.01)\,\text{N-s/m} = 4.35\,\text{N-s/m}$. Assume that the total damping does not vary significantly over the narrow bandwidth. Finally, the nonlinear viscous damping coefficient is found by solving the expression in eq. 10.10 at the resonant frequency. The resulting coefficient value is $b_v zn \simeq 31.88\,\text{N-s}^2/\text{m}^2$.

Several notes on the experiment described in Example 10.4 are the following: First, the position of the gauge should be at least one body diameter ($D = 2a$) from the center of the cylinder to avoid distorting the reaction forces caused by the small waves reflected from the gauge. Second, the gauge should not be placed close to a tank wall because reflected waves will also be detected by the gauge. Therefore, the best location for the gauge is in the center of the tank. Third, side beaches can be positioned adjacent to the tank walls to absorb most of the radiated wave energy, thereby minimizing side-wall reflection. Most side beaches have an effective frequency band over which they are most efficient. Thus, care must be taken to ensure that the frequencies of motion of the body are within this band.

10.2 Power Take-Off

The damping coefficient representing the power take-off is b_{pz} in eqs. 10.1 and 10.2. The application of the power take-off concept to the float sketched in Figure 10.1 would depend on the purpose of the system. For example, the float might be a self-contained wave measurement system attached to a pier. By self-contained, we mean that the electronics and power take-off system are in the float. The power take-off is then designed to convert the kinetic energy of the wave-induced heaving motion of the float into electrical energy, which in turn powers the electronic measurement system.

To determine the effects of the power take-off, let the damping coefficient b_z in the power expression of eq. 10.27 be the sum of the linear damping coefficients representing the radiation damping (b_{rz}), the equivalent velocity-squared viscous damping (b_{vz}), and the power take-off (b_{pz}). The time-averaged damping power is found to be

$$\begin{aligned} P_p &= P_{wz} - P_{rz} - P_{vz} = \frac{1}{2}\omega^2 b_{pz} Z^2 \\ &= \frac{1}{2} F_{zo}\omega Z \sin(\varepsilon_z - \alpha_z)[\cos(\alpha_z) - \sin(\alpha_z)] - \frac{1}{2}\omega^2 (b_{rz} + b_{vz}) Z^2 \end{aligned} \tag{10.41}$$

The average is over one wave period. In eq. 10.41, the P-terms are the component powers. Again, the available wave power is represented by P_{wz}. For a given wave frequency, the maximum power extracted from the system from this equation occurs when $dP_p/dZ = 0$. Before solving this relationship for Z, it should be noted that the long-wave radiation damping coefficient (b_{rz}) of Yeung (1981) in eq. 10.8 is independent of the amplitude Z. The equivalent viscous damping coefficient of eq. 10.10 is a linear function of Z. Assume that the system is in long waves ($ka \ll 1$) so that $\alpha_z = 0$ as in eq. 10.28. Again, this assumption physically means that the wave and wave-induced force are in phase. By replacing b_{vz} in eq. 10.41 by the expression in eq. 10.10, the maximum power take-off is obtained when the amplitude is

$$Z_{p-\max} = -\frac{\pi}{8\omega}\frac{b_{rz}}{b_{vz}} + \frac{1}{2}\sqrt{\frac{\pi^2}{16\omega^2}\frac{b_{rz}^2}{b_{vz}^2} + \frac{1}{2}\frac{\pi}{\omega^2 b_{vz}} F_{zo}\sin(\varepsilon_z)} \tag{10.42}$$

If the cylinder is of deep draft (d), then the viscous losses are dominant, and the maximum power occurs when

$$Z_{p-\max} = \sqrt{\frac{\pi}{8}\frac{1}{\omega^2 b_{vz}} F_{zo}\sin(\varepsilon_z)} \tag{10.43}$$

If the cylinder is of shallow draft, then the radiation damping is significant, and the maximum power occurs when

$$Z_{p-\max} = \frac{1}{\omega b_z} F_{zo}\sin(\varepsilon_z) \tag{10.44}$$

The maximum power that can be extracted from this wave-induced heaving motion is obtained from eq. 10.41, where the amplitude is represented by eqs. 10.42, 10.43, or 10.44, whichever is appropriate.

For a wave-energy conversion system, one additional requirement is placed on the heaving motions, that is, the motions should be tuned to the frequency of a design incident wave. In other words, the natural heaving frequency in eq. 10.13 should be equal to the incident wave frequency. In that equation, we see properties of the float that can be changed or adjusted in the tuning process. These changes would be in the design phase of the project and would be focused on the waterplane area (A_{wp}), the number (N) of tension lines, and the spring constant (k_S) of the lines and the floating mass (m). As can be seen in the long-wave expression of Yeung (1981) in eq. 10.7, the added mass changes with changes in both geometry and frequency. So, if the system is designed for wave-energy conversion, eq. 10.13 can be used in the conceptual design of the system.

In the following example, the time-averaged heaving power of the heaving cylinder of Example 10.1 is converted into electricity using a linear inductance system.

EXAMPLE 10.5: POWER TAKE-OFF OF A HEAVING CIRCULAR CYLINDER AT RESONANCE The cylinder in Example 10.1 has an internal linear inductance power take-off system designed to charge a battery system that, in turn, powers several navigation aids including a light on the top of the fixed staff and a fog horn. For this application, the system is as sketched in Figure 10.8, where no mooring lines are used ($k_S = 0$). The draft of the unmoored body is 1 m, as stated in Example 10.1. For the resonant condition where $\omega = \omega_n$, we shall determine the optimal power available to the linear inductance system. See Ivanova et al. (2003) for the description of a linear inductance system applied to wave energy extraction.

The frequency of interest in this application is the natural circular frequency of the heaving motions, ω_n. The incident wave height for this frequency is measured to be 0.5 m. Because the body is unmoored, the value of the added mass differs from that in Example 10.1. For the cylinder having a 1-m draft in 4 m of water, the added mass is $a_{wz} \simeq 77.0$ kg from eq. 10.7, and the radiation damping coefficient is again $b_{rz} \simeq 38.4$ N-s/m from eq. 10.8. The system is designed to be in resonance with the incident wave, so the phase angle between the wave-induced force and the heaving motions is $\varepsilon_z = 90°$ from eq. 10.15. From

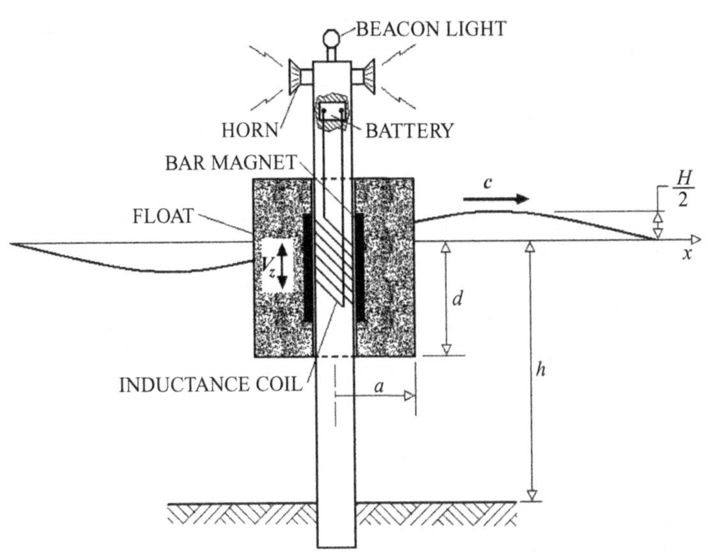

Figure 10.8. *Wave-Powered Navigation Aid.* The bar magnets are placed vertically in the float about the inductance coil, which is attached to the fixed, vertical staff. See McCormick et al. (1981) for details of the inductance power take-off system.

Example 10.1, the nonlinear viscous damping coefficient is $b_{rz} \simeq 405\,\text{N-s}^2/\text{m}^2$. The wave-force amplitude is $F_{zo} \simeq 1.98 \times 10^3\,\text{N}$ for the 0.5-m wave from eq. 10.16. The natural circular frequency is $\omega_{nz} \simeq 2.99\,\text{rad/s}$, and the critical damping is $b_{cz} \simeq 2.65 \times 10^3\,\text{N-s/m}$. With the parametric values, the optimal heaving amplitude is

$$Z_{p-max}|_{opt} = -\frac{\pi}{8\omega_n}\frac{b_{rz}}{b_{vz}} + \frac{1}{2}\sqrt{\frac{\pi^2}{16\omega_{nz}^2}\frac{b_{rz}^2}{b_{vz}^2} + \frac{1}{2}\frac{\pi}{\omega_{nz}^2 b_{vz}}F_{zo}} \simeq 0.453\,\text{m} \qquad (10.45)$$

The optimal power available to the linear inductance system is $P_p = 886\,\text{W}$ from eq. 10.41. Note that for the 0.5-m, 2.10-sec incident wave, the power per crest width is $P' = 522\,\text{W/m}$ from eq. 3.72. Due to diffraction focusing, the effective capture crest width (ℓ) of the wave is $P_p/P' \simeq 1.70\,\text{m}$. Physically, the 1-m-diameter buoy having resonant-heaving motions captures the power of the wave from a crest width that, in this case, is 1.7 times the diameter.

In Example 10.5, the concept of the *capture width* (w_{cw}) is introduced. This is an equivalent crest width from which the power of the incident wave that is captured by the floating body. The wave power is transferred along the crest by the phenomenon of diffraction, which is discussed in Section 6.4. Some wave-energy analysts refer to this phenomenon as diffraction focusing. In terms of linear wave properties, the capture width is mathematically defined by

$$w_{cw} \equiv \frac{P_p}{P'} = \frac{P_p}{\frac{1}{8}\rho g H^2 c_g} \qquad (10.46)$$

Here, c_g is the group velocity, defined in eq. 3.62 and presented in eq. 3.63. The wave power per crest width is P'. The diffraction-focusing phenomenon is a major consideration in the design of wave-energy conversion systems, as discussed by

Falnes (2002). For an overview of wave-energy conversion techniques, the reader is referred to the book by McCormick (2007) and the book edited by John Brooke (2003).

10.3 Random Motions

In Chapter 5, the random nature of the sea is discussed. In that chapter, we assume that seas are composed of superimposed sine waves having various heights and periods that pass a site randomly in time and are distributed randomly in space. When either a floating structure or a fixed compliant structure is located at an ocean site, each component of the wave field will produce a force on the structure that can result in a motion. The design goals are to minimize these wave-induced motions if the structure is a ship or some type of production facility, and to maximize the motions if the structure is a component of a wave-energy conversion system. To achieve either of these design goals, we must have a good understanding of the wave-structure interaction in random seas. In this section, we introduce the analytical method used to describe this interaction, and apply the analysis to a heaving buoy in random waves for the purpose of illustration. The statistical description of the waves is taken from Chapter 5.

Consider the motions of a deep-water spar buoy in a random sea. The notation for this buoy is the same as that for the vertical, circular cylinder sketched in Figure 10.1, where $h = \infty$. Assume that the forcing function is due to random waves. Again, for the purpose of illustration, also assume that the component wavelengths are an order of magnitude greater than the radius (a) of the buoy. Because of the latter assumption, $ka < 0.3$ and the vertical wave-induced force on the buoy is *quasi*-hydrostatic.

For this discussion, assume that the added mass for the heaving motions of the deep-draft spar buoy is independent of frequency and is represented by

$$\mathrm{a}_{wz} \simeq \frac{2}{3}\rho\pi a^3 \tag{10.47}$$

which equals the mass displaced by a hemisphere having a diameter equal to that of the spar. This formula has been used in the conceptual design phase of cylindrical spar buoys (for example, see the paper by Bhattacharyya, Sreekumar, and Idichandy, 2002). The reader is also encouraged to consult the book by Patel (1989) for an additional discussion on the concept of the added mass. For the long-wave approximation of Yeung (1981) plotted in Figure 10.3, it can be seen that the expression in eq. 10.47 is approximately valid where $0 < ka < 0.01$. The damping coefficient used here is the Yeung (1981) long-wave expression in eq. 10.8 and plotted in Figure 10.4. For $0 < ka < 0.01$, the radiation damping coefficient is well approximated by

$$\mathrm{b}_{rz} \simeq 0.1\rho\pi a^3\omega \tag{10.48}$$

The use of the approximate expressions for the added mass in eq. 10.47 and the radiation damping in eq. 10.48 are for demonstration purposes only.

The linear heaving motions of the body excited by monochromatic waves are represented by eq. 10.2, where in this case the radiation damping coefficient is $\mathrm{b}_{rz} \gg \mathrm{b}_{vz} + \mathrm{b}_{pz}$ and the mooring spring constant is $\mathrm{k}_s = 0$. With these assumptions, eq. 10.2 can be written as

$$\left(m+\frac{2}{3}\rho\pi a^3\right)\frac{d^2z}{dz^2}+0.1\rho\pi a^3\omega\frac{dz}{dt}+\rho g A_{wp}z=\Re[F_{zo}e^{-iwt}]\simeq\Re\left[\rho g A_{wp}\frac{H}{2}e^{-i\omega t}\right] \tag{10.49}$$

In this equation, the notation $\Re$ is used to identify the real part of the term. The steady-state solution of eq. 10.49 can be written as

$$z=\Re[Z(\omega)e^{-i\omega t}]=\frac{1}{2}[Z(\omega)e^{-i\omega t}+Z^*(\omega)e^{i\omega t}] \tag{10.50}$$

In eq. 10.50, $Z(\omega)$ is the frequency-dependent complex heaving amplitude and $Z^*(\omega)$ represents its complex conjugate. That is, if $Z(\omega)=Z_{\Re}(\omega)+iZ_{\Im}(\omega)$, then $Z^*(\omega)=Z_{\Re}(\omega)-iZ_{\Im}(\omega)$. The combination of eqs. 10.49 and 10.50 yields the following expression for the complex heaving amplitude:

$$Z(\omega)=\frac{\dfrac{F_{zo}}{\rho g A_{wp}}}{\left(1-\dfrac{\omega^2}{\omega_{nz}^2}\right)-i0.2\dfrac{\omega^2}{\omega_{nz}}\dfrac{\rho\pi a^3}{b_{cz}}}=\mathcal{H}(\omega)\frac{F_{zo}}{\rho g A_{wp}}=\mathcal{H}(\omega)\frac{H}{2} \tag{10.51}$$

where $\mathcal{H}(\omega)$ is called the amplitude response function. The function is also known as the admittance, the frequency response function, and the harmonic response function. The amplitude response function is complex and the wave height is real. The last equality in eq. 10.51 is a special case, where the excitation is due only to the hydrostatic force of the passing long wave.

Our interest is in both the statistical averages and the extreme values of the body displacements. The statistical methodology used to obtain these is presented in Chapter 5, where the statistics of random wave fields are discussed. Consider first the mean-square response of the heaving motions of the spar buoy, defined by

$$\begin{aligned}\overline{z^2}=z_{rms}^2&=\lim_{T\to\infty}\frac{1}{T}\int_0^T z^2dt\simeq\frac{1}{T}\int_0^T z^2dt\\&=\frac{Z(\omega)Z^*(\omega)}{2}=\frac{1}{2}|Z(\omega)|^2\\&=\mathcal{H}(\omega)\mathcal{H}^*(\omega)\frac{1}{T}\int_0^T\eta^2dt=|\mathcal{H}(\omega)|^2\eta_{rms}^2=|\mathcal{H}(\omega)|^2\frac{\overline{H^2}}{8}\end{aligned} \tag{10.52}$$

The last equality comes from eq. 5.35a. In eq. 10.52, T is the time interval. Also in eq. 10.52 are the root-mean-square response (z_{rms}) and the root-mean-square displacement of the free surface (η_{rms}). The integral term containing the limit notation is the true mean-square, and those without the limit are approximations. Confidence in the approximation increases as the time interval (T) of the data measurement increases. Although eqs. 10.49 through 10.52 apply to heaving motions in monochromatic waves, we shall use the expressions in eqs. 10.51 and 10.52 in the following paragraphs devoted to heaving motions in random seas.

In Section 5.7, we find that the energy content of a sea can be represented by the wave spectral density, $S(T)$. This property of the sea is introduced in eq. 5.37 in terms of a discrete wave period, $T_i=2\pi/\omega_i$, and the wave heights associated with that period, $H_{j(i)}$. If we further assume that there are a number of such wave periods, then the expression in eq. 5.37 leads to the expression for the root-mean-square wave height expression in eq. 5.40. Because there are an infinite number of wave periods, the integral expression of eq. 5.41 is obtained in the limiting process.

Our goal is to obtain an expression for the mean-square heaving response of the vertical circular cylinder in terms of the wave spectral density. The expression for the mean-square heaving response of our vertical cylinder can be written in terms of the probability density of the motion amplitude, $p(Z)$, and the amplitude spectral density, $S_z(Z)$. To determine these functions, we use a common property of the cumulative probability of occurrence of eq. 5.3. That is, the cumulative probabilities of occurrence of the excitation (the waves) and the heaving response must satisfy the following relationship:

$$\begin{aligned} 1 &= P(0 < H \leq \infty) = \int_0^\infty p(H)dH \\ &= P(0 < Z \leq \infty) = \int_0^\infty p(Z)dZ = \int_0^\infty p(Z)\frac{dZ}{dH}dH = \int_0^\infty p(Z)\frac{\mathcal{H}(\omega)}{2}dH \end{aligned} \tag{10.53}$$

where the derivative, dZ/dH, is obtained from eq. 10.51. In eq. 10.53, the notation $p(Z)$ is that for the probability density. From eq. 10.53, we find that the relationship between the probability density of the heaving response and the waves is

$$p(Z) = \frac{2}{\mathcal{H}(\omega)}p(H) \tag{10.54}$$

The mean-square amplitude response of the body can now be expressed as

$$\begin{aligned} \overline{Z^2} &= \int_0^\infty p(Z)Z^2dZ = 8\int_0^\infty S_z(T)dT = 8\int_\infty^0 S_z(T)\frac{dT}{d\omega}d\omega = -8\int_\infty^0 S_z(T)\frac{2\pi}{\omega^2}d\omega \\ &= \int_0^\infty p(H)H^2(\omega)\frac{\mathcal{H}(\omega)\mathcal{H}^*(\omega)}{4}dH = \int_0^\infty p(H)H^2(\omega)\frac{|\mathcal{H}(\omega)|^2}{4}\frac{dH}{d\omega}d\omega \\ &= C\int_0^\infty S_J(\omega)\frac{|\mathcal{H}(\omega)|^2}{4}d\omega = C\int_0^\infty S_J(T)\frac{|\mathcal{H}(T)^2|}{4}dT \end{aligned} \tag{10.55}$$

In this equation, $S_J(\omega)$ is the frequency spectral density of the wave field, the subscript J is generic (used to identify the wave spectral formula), and the coefficient C is a constant that depends on the relationship between the spectral density and the mean-square of the wave height. Note that T in the last line is the wave period and not the time interval, T, in eq. 10.52. If the Bretschneider (1963) spectral formula of eq. 5.59 is selected, then $J = B$ and $C = 1$. When the Pierson and Moskowitz (1964) formula of eq. 5.63 is used, then $J = PM$ and $C = 8$. The last relationship in eq. 10.55 results from the relationship of the circular wave frequency and the wave period, $\omega = 2\pi/T$. In Chapter 5, the wave spectra are in the period domain and not the frequency domain. For this reason, it is useful to express the amplitude response function as a function of the wave period. From eq. 10.51, we find

$$\mathcal{H}(T) = \frac{1}{\left(1 - \frac{T_{nz}^2}{T^2}\right) - i0.4\frac{T_{nz}}{T^2}\frac{\rho\pi^2 a^3}{b_{cz}}} \tag{10.56}$$

where the critical damping coefficient of eq. 10.14 for this discussion is

$$b_{cz} - 2\sqrt{\left(m + 2\frac{\rho\pi a^3}{3}\right)(\rho g\pi a^2)} \tag{10.57}$$

In summary, the equations presented are based on the assumptions that the heaving motions of the freely floating, vertical, circular cylinder are excited by randomly occurring long waves ($\lambda \gg a$), where the radiation damping is much greater than the sum of the viscous damping and the damping due to any power take-off. The analysis of the resulting heaving motions is illustrated in Examples 10.6 and 10.7.

EXAMPLE 10.6: ROOT-MEAN-SQUARE HEAVING RESPONSE OF UNDERDAMPED MOTIONS OF A CAN BUOY IN A RANDOM SEA The 0.5-m-radius can buoy in Example 10.1 is now located in deep water ($h = \infty$), where the mass density of the salt water is $\rho = 1.03 \times 10^3$ kg/m^3. Ballast is added to the buoy to cause it to float with a 1-m draft ($d = 1$ m). The weight of the buoy equals that of the displaced water, $\rho g\pi a^2 d \simeq 7.94 \times 10^3$ N, and the body mass (m) is approximately 809 kg. There are no tension lines attached to the buoy. The critical damping coefficient value is $b_{cz} \simeq 5.48 \times 10^3$ N-s/m from eq. 10.57, and the natural heaving period is $T_n \simeq 2.17$ from eq. 10.13. The buoy is subjected to a long-wave random sea consisting of linear waves, where the average height is $H_{avg} = 1.5$ m and the average period is $T_{avg} = 7$ sec. These are the same sea conditions as those in Example 5.14 and in Figure 5.23. In that figure are the results of the application of the Bretschneider spectral formula (eq. 5.59) applied to deep water and to the site conditions. The Bretschneider formula for the stated wave conditions is

$$S_B(T) = H_{rms}^2 p(T) = 3.437\frac{H_{avg}^2}{T_{avg}^4}T^3 e^{-0.675\left(\frac{T}{T_{avg}}\right)^4} \simeq 3.22 \times 10^{-3} T^3 e^{-2.81\times10^{-4}T^4} \tag{10.58}$$

The absolute value of the amplitude response function for the can buoy is found to be

$$\mathcal{H}_{ABS}(T) \equiv |\mathcal{H}(T)| = \sqrt{\mathcal{H}(T)\mathcal{H}^*(T)} \simeq \frac{1}{\sqrt{\left(1 - \frac{4.71}{T^2}\right)^2 + \frac{0.0405}{T^4}}} \tag{10.59}$$

from eq. 10.56. Results obtained from eqs. 10.58 and 10.59 are presented in Figure 10.9 as functions of T/T_{nz}. In that figure, one sees that the can buoy is de-tuned from the waves, in that the peak amplitude response is well away from the modal period, which is the period of the peak spectral value (see eq. 5.45). The peak-value amplitude response function (approximately 23.4 occurring at $T = T_{nz}$) is relatively large because the damping (assumed to be only due to radiation) is small. When $T = T_{nz}$, we find that the ratio of the radiation damping coefficient and the critical damping coefficient is approximately 0.0214, that is, the radiation damping loss is about 2.14% of the critical value. The amplitude response function for this condition is said to be resonance-dominated.

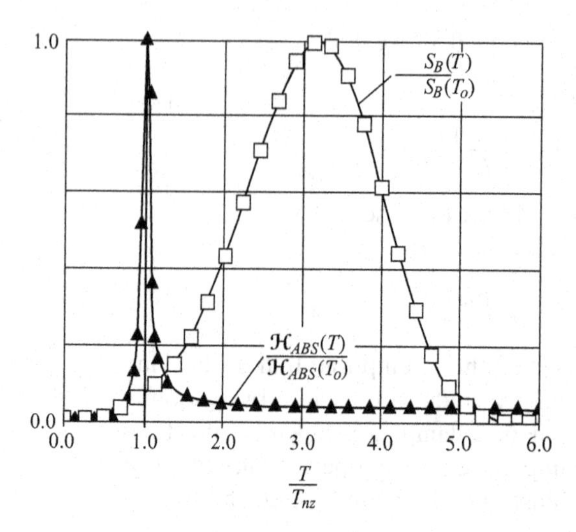

Figure 10.9. *Normalized Bretschneider Wave Spectrum and Amplitude Response Function versus Period Ratio for Example 10.6.* The Bretschneider spectrum is normalized by using the modal period of eq. 5.45, the value of which is approximately 7.19 sec. The absolute value of the amplitude response function is normalized by using its value at the natural heaving period, which is about 2.17 sec. The peak values for the wave spectrum and response function are approximately 0.565 m^2/s and 23.4, respectively.

The combination of the last expressions in eqs. 10.58 and 10.59 with the mean-square expression in eq. 10.55 (where $C = 1$ because we are using the Bretschneider spectral formula) results in the following expression for the mean-square response of the buoy:

$$\overline{Z^2} = Z_{rms}^2 = \int_0^\infty S_B(T)\frac{\mathcal{H}(T)\mathcal{H}^*(T)}{4}dT = \frac{1}{4}\int_0^\infty \frac{3.22\times10^{-3}T^3e^{-2.81\times10^{-4}T^4}}{\left[\left(1-\frac{4.71}{T^2}\right)^2+\frac{0.0405}{T^4}\right]}dT$$

$$\simeq S_B(T_0)\frac{\mathcal{H}(T_0)\mathcal{H}^*(T_0)}{4}\frac{T\,\delta T}{2} + S_B(T_1)\frac{\mathcal{H}(T_1)\mathcal{H}^*(T_1)}{4}\delta T + \cdots$$

$$S_B(T_{N-1})\frac{\mathcal{H}(T_{N-1})\mathcal{H}^*(T_{N-1})}{4}\delta T + S_B(T_N)\frac{\mathcal{H}(T_N)\mathcal{H}^*(T_N)}{4}\frac{\delta T}{2} \tag{10.60}$$

For our parametric values, the mean-square response value is approximately 2.05 m^2, and the root-mean-square of the heaving amplitude in this sea is $Z_{rms} \simeq$ 1.43 m. Because of the complexity of the integrand expression, we have chosen to use a simple numerical integration method to obtain the result. In this, we have used the trapezoidal rule. The value of the period increment (δT) used here is 0.1 sec. Note that there are a number of numerical integration methods available. Because of its simplicity, the trapezoidal formula is widely used by practicing naval architects and ocean engineers. For example, see Chapter 25 of Abramowitz and Stegun (1965).

In the next example, the highly damped heaving motions of a can buoy in a long-wave random sea are analyzed. The damping is due to a horizontal circular plate suspended below the buoy.

EXAMPLE 10.7: ROOT-MEAN-SQUARE AND EXTREME HEAVING AMPLITUDES OF HIGHLY DAMPED MOTIONS OF A CAN BUOY IN A RANDOM SEA An inertial damping plate is attached to a 1-m-diameter ($2a_P = 1$ m) can buoy in deep water, as sketched in Figure 10.10. The sea at this site is that described in Example 10.6, that is, the wave period spectrum is predicted by the Bretschneider formula in eq. 10.58 where $H_{avg} = 1.5$ m and $T_{avg} = 7$ sec. The inertial damping plate is horizontal and is well below the free surface. The net weight (dry weight minus

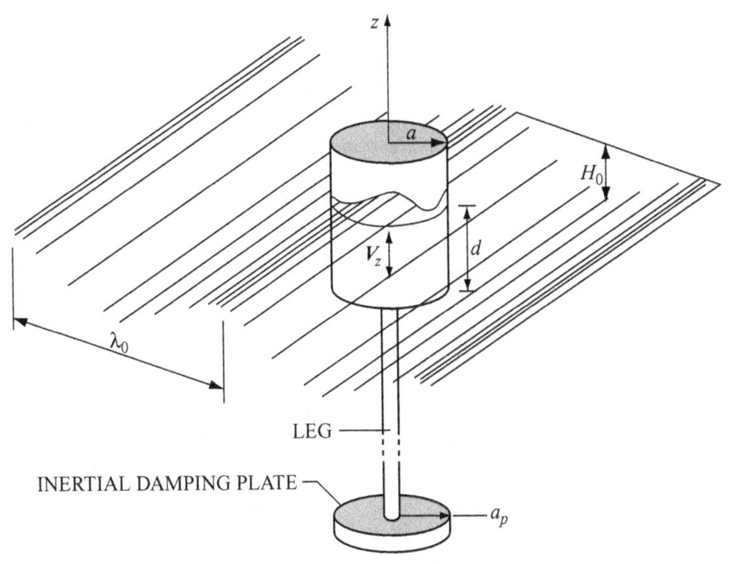

Figure 10.10. *Can Buoy with an Attached Inertial Damping Plate in Deep Water.*

buoyancy) of the buoy-leg-plate structure causes the buoy to have a draft (d) of 3 m. The mass of the buoy system, equal to the displaced mass of water, is $m = \rho\pi a^2 d \simeq 2.43 \times 10^3$ kg, where $\rho = 1.03 \times 10^3$ kg/m^3 is the mass density of salt water. From tests of the system in a large tank prior to deployment, the heaving motions are found to be damped such that $b_z \simeq 0.7b_{cz}$, that is, the mean value of the damping is 70% of the critical damping. Also from these tests, the total added mass (a_{wz}) of the buoy plus plate is found to be 500 kg. From eq. 10.13, we find that the natural heaving period is $T_n \simeq 3.82$ sec. For the highly damped heaving motions, the amplitude response function is

$$\mathcal{H}(T) = \frac{1}{\left(1 - \frac{T_{nz}^2}{T^2}\right) + i2\frac{b_z}{b_{cz}}\frac{T_{nz}}{T}} \simeq \frac{1}{\left(1 - \frac{14.6}{T^2}\right) + i2(0.7)\frac{3.82}{T}}$$

From eq. 10.55, we obtain

$$\mathcal{H}(T)\mathcal{H}^*(T) = \mathcal{H}_{ABS}^2(T) - \frac{1}{\left(1 - \frac{14.6}{T^2}\right)^2 + \frac{28.6}{T^2}}$$

The wave spectrum and the amplitude response function, $\mathcal{H}_{ABS}(T)$, are plotted in Figure 10.11. The peak value of $\mathcal{H}_{ABS}(T)$ occurs at a wave period of approximately 33 sec because of the high damping. That is, the amplitude response function is damping-dominated, as opposed to that of Example 10.6, which is resonance-dominated. Because the peak value of $\mathcal{H}_{ABS}(T)$ occurs at a high wave period, the two normalized functions are presented as functions of T/T_o in Figure 10.11, where T_o is the modal period of the wave spectrum. From an expression similar to that in eq. 10.60, the mean-square heaving amplitude of the system is approximately 0.453 m^2, and the corresponding root-mean-square value is $Z_{rms} \simeq 0.673$ m.

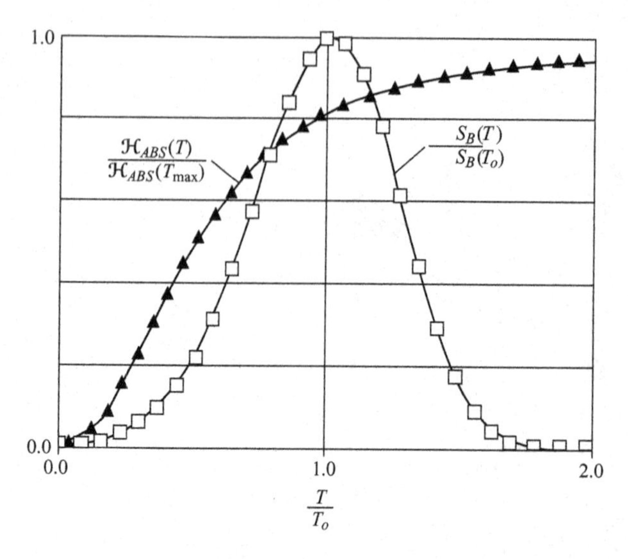

Figure 10.11. *Normalized Bretschneider Wave Spectrum and Amplitude Response Function for Example 10.7.* The Bretschneider spectrum is normalized by using the modal period (T_o) of eq. 5.45, which is approximately 7.19 sec, whereas the absolute value of the amplitude response function is normalized by using its maximum value, which occurs at a period of about 33 sec. The peak values for the wave spectrum and response function are approximately 0.565 m^2/s and 1.00, respectively.

To determine the extreme heaving amplitude of the buoy, we first associate a probability function to the amplitude. Noting that the Bretschneider spectral formula is based on the assumption of Rayleigh-distributed wave heights, the Rayleigh probability density function of eq. 5.19 is used as a model for the probability distribution of the amplitude. The result is

$$p(Z) = \frac{2Z}{Z_{rms}{}^2} e^{-\left(\frac{z}{z_{rms}}\right)^2} \simeq \frac{2Z}{0.673^2} e^{-\left(\frac{z}{0.673}\right)^2} \simeq 4.416 Z e^{-2.208 Z^2} \tag{10.61}$$

The corresponding probability of occurrence needed to determine the extreme or maximum heaving amplitude is

$$P(Z_{\max} < Z \le \infty) = \int_{Z_{\max}}^{\infty} p(Z) dZ = e^{-\left(\frac{Z_{\max}}{Z_{rms}}\right)^2} \simeq e^{-2.208 Z_{\max}^2} \simeq \frac{1}{N_M} \tag{10.62}$$

The last term in this equation is the ratio of the observed maximum amplitude (there can only be one maximum) over a measured sample, N_M, at the site, where M is the number of years during which the measurements are made. Because each passing wave is assumed to cause a heaving motion, we can determine the number of observations over one year ($N_M = N_1$) by noting that the average wave period is 7 sec. The one-year sample based on the average wave period is $N_1 = (60/7) \times 60 \times 24 \times 365 \simeq 4.505 \times 10^6$. By rearranging the expression in eq. 10.62, we find that the expression for the expected maximum heaving amplitude over one year is

$$\begin{aligned} Z_{\max}|_{M=1} &= Z_{rms}\sqrt{\ln(N_M)}|_{M=1} \\ &\simeq 0.673\sqrt{\ln(4.50 \times 106)} \simeq 2.63\,\text{m} \end{aligned} \tag{10.63}$$

For the Rayleigh-distributed wave heights at the site, we find that the expected extreme wave height over one year is $H_{max} \simeq 6.98$ m from eq. 5.25. So, the freeboard of the buoy greater than $H_{\max}/2 - Z_{\max} = 0.86$ m to prevent the deck becoming awash. In reality, the deck of this can buoy would be several meters above the SWL to prevent the buoy from being a navigation hazard.

Several notes concerning this section: The random sea is assumed to be composed of long waves, that is, the wavelengths (λ) of the component waves are assumed to be much greater than the buoy radius, a. The excitation of the heaving motions is then hydrostatic in nature, and the short-wave effects, such as diffraction, are neglected. Furthermore, both the sea and the resulting motions are assumed to be linear. These somewhat idealistic assumptions are made simply to facilitate the discussion of motions excited by random waves. The equations presented can be used by engineers to obtain good first approximations of the forces and motions. For additional reading on the subject of random phenomena, the books by Bendat and Piersol (1971) and Lutes and Sarkami (1997) are recommended.

10.4 Closing Remarks

A segment of the readers might consider some of the topics in this chapter to be a little too "basic." For most readers, the material of the chapter is considered to be more a review rather than an introduction. Many of the mathematical expressions are the foundations for those found in Chapter 11, where the respective wave-structure interactions of floating bodies are discussed.

11 Wave-Induced Motions of Floating Bodies

Equipped with the basic analytical methods presented in Chapters 9 and 10, the wave-induced motions of floating bodies are discussed in this chapter. In Chapter 12, the final chapter of this book, those methods are applied to the wave-structure interactions of fixed structures. In this book, *fixed structures* are those that are either resting on the sea bed or directly supported by foundations in the bed. Floating bodies include ships, floating platforms, buoys, and other specialized bodies that are either under way or maintained in position by moorings. The motion of ships in waves is a topic in the field of naval architecture referred to as *seakeeping*. Thorough coverages of seakeeping are found in the writings of Korvin-Kroukovsky (1961), Newman (1977), Bhattacharyya (1978), Lloyd (1989), and Faltinsen (1990, 2005) among others. Floating bodies discussed in this chapter that are not normally under way are referred to herein as *ocean engineering* bodies, as opposed to ships. The geometry of an ocean engineering body normally has two vertical planes of symmetry, whereas ships have one, called the *centerplane*.

In this chapter, the degrees of freedom (surge, sway, heave, roll, pitch, and yaw) of a floating body are introduced and the coupled heaving and pitching motions are analyzed. The stability of a body in calm water is first discussed. Methods of motion analysis are then introduced that lend themselves to both analytical and simple numerical solutions. Body motions in waves are analyzed using the linear strip theory. This method has its foundation in fluid mechanics, and its application to floating bodies was first done by A. Krylov (1896), according to Pedersen (2000). The strip theory was revised over a half-century later by Korvin-Kroukovsky (1955). As presented by that author, the method was later corrected and refined by Korvin-Kroukovsky and Jacob, (1957). The work of these authors was, in turn, modified by Motora (1964). Those papers describe the application of the strip theory to the planar motions of ships in regular seas. The theory was expanded by Salvesen, Tuck, and Faltinsen (1970). That work is considered to be the cornerstone of contemporary ship-motion analysis using the strip theory. An excellent review of the linear strip theory is presented by Bishop and Price (1979). There have been several formulations of nonlinear strip theories. One of these is the "quadratic theory" of Jensen and Pedersen (1978).

The flows involved in the linear strip theory are two-dimensional and, hence, the hydrodynamic analyses presented are also two-dimensional. Following an introduction to body motions, several methods for obtaining the hydrodynamic coefficients

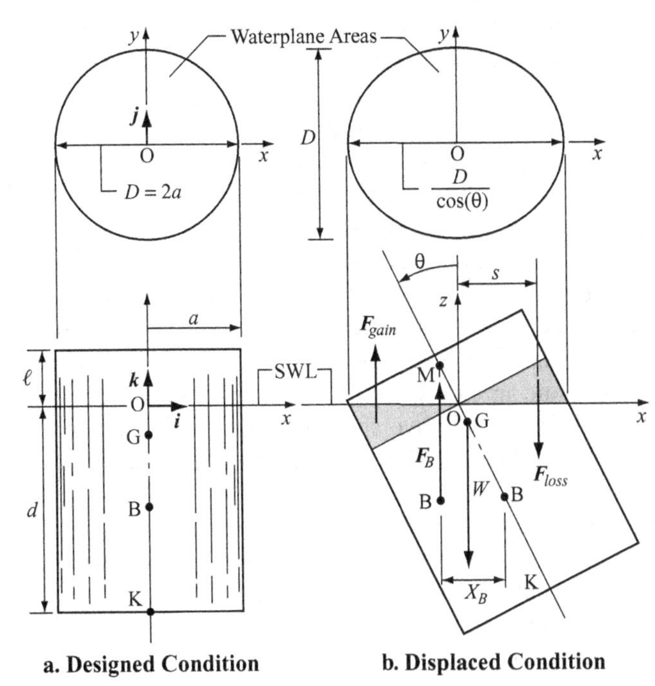

Figure 11.1. *Floating, Vertical, Circular Cylinder in Calm Water.* Note that the waterplane area for the rotationally displaced cylinder is elliptical. In (a) *i,j,k* are unit vectors, and ℓ is the freeboard.

are presented. The first two of these methods are analytical, and are based on the assumption of irrotational flow excited by the body motion. The last method is the Green's function method, which is numerical in nature. Because of the intent of the book, the latter method is simply outlined.

Moorings and their effect on the body motions are then considered. Two types of moorings are discussed. The first is *slack mooring*, where the orientation of the mooring line is dependent on the cable weight and any current that might exist. The second is *tension mooring*, where the geometry of the deployed mooring line is essentially rectilinear. The tension in the line depends on both the buoyancy of the moored body and the mass density of the cable.

A note to the reader: The analytical techniques in this section were developed some time ago. These techniques are used today in analyses and are presented herein for their educational value.

11.1 Hydrostatic Considerations – Initial Stability

The hydrostatic theorem due to the observations of Archimedes is presented in Section 2.1. In Chapter 10, the hydrostatic restoring force inclusion in the equation of heaving motions of a floating, vertical, circular cylinder is discussed. We now direct our attention to the stability of a floating body. The vertical circular cylinder sketched in Figure 11.1 is used in the following development to demonstrate the applicability of the derived equations.

A surface ship will capsize because of a rolling instability. This occurrence of the instability depends on the relative moments of the hydrostatic restoring force and the displaced center of gravity of the ship. Although the analysis of the pitching motion of a ship is similar to that for rolling, pitching instabilities do not normally occur because of the large ship length-to-wavelength ratio. An excellent discussion of the hydrostatic considerations of ships is found in the book by Jensen (2001).

Consider the circular, cylindrical buoy in calm water sketched in Figure 11.1, where the cylinder is shown in the designed (initial) condition in Figure 11.1a, and in the displaced condition in Figure 11.1b. In the design condition, the axis of symmetry is co-linear with the vertical axis, z. In Figure 11.1a, we see that the diameter of the cylinder is $D = 2a$, where a is the radius, and the draft is d. The *waterplane* (the displaced water area at the SWL) is a circle for this body in the design condition. The waterplane area is then $A_c = \pi a^2 = \pi D^2/4$. When the body is given an angular displacement, θ, the waterplane becomes an ellipse, with a major axis of $D/[\cos(\theta)]$ and a minor axis of D. The waterplane area of the displaced cylinder becomes $A_e = \pi D^2/[4\cos(\theta)]$. For the floating cylinder, we have chosen to use the subscript c to identify the circular design waterplane area, and e to identify the displaced elliptical waterplane area. When the body is displaced, there are both a lost buoyancy, due to the part of the body that leaves the water (on the right side of the y-axis), and a gained buoyancy, due to the part that enters the water (on the left side). The net gain in buoyancy ($\boldsymbol{F}_{gain}$) must be equal in magnitude to the net loss ($\boldsymbol{F}_{loss}$) because the displaced volumes are equal for this body. The line of action of a buoyant-force component is at a distance s from the vertical axis. Because the component forces are equal, there is no net change in the buoyant force ($\boldsymbol{F}_{\mathrm{B}}$) acting on the body. We know that the buoyant force must equal the weight of the displaced water (referred to as the *displacement*). For the body to be in static equilibrium, as sketched in Figure 11.1a, the buoyant force and the weight of the body are both equal and co-linear. The expression for the buoyant force is then given by

$$\boldsymbol{F}_{\mathrm{B}} = -\boldsymbol{W} = (\rho g \vee)\boldsymbol{k} = \left(\rho g \pi \frac{D^2}{4} d\right)\boldsymbol{k} = (\rho g \pi a^2 d)\boldsymbol{k} \tag{11.1}$$

Here, $\boldsymbol{k}$ is the unit vector in the vertical direction, and $\vee$ is the volume of the displaced water.

When the body is displaced, the center of buoyancy, B, moves to the point B', as shown in Figure 11.1b. This new position is the centroid of the displaced water volume. The horizontal distance between B and B′ is X_{B}. The line of action of the buoyant force intersects with the centerline of the body at M, which is called the *metacenter*. The *metacentric height* (GM) is defined at the distance between G and M. For the displaced body, we can relate this distance to the displacement angle as

$$\mathrm{GM} = \frac{X_{\mathrm{B}}}{\sin(\theta)} \mp \mathrm{GB} \tag{11.2}$$

The negative sign is used when the center of gravity (G) is above the center of buoyancy (B), and the positive sign when G is below B. The distance $\mathrm{GB} = |\mathrm{OG} - \mathrm{OB}|$ is known because it is a design input.

To determine the horizontal shift (X_{B}) of the center of buoyancy, consider the sketch in Figure 11.2. In the figure, the displaced water volumes are shown. To facilitate the analysis, the x- and z-axes have been rotated through the angle θ in this sketch so that the circular waterplane is horizontal. Referring to Figure 11.2 for

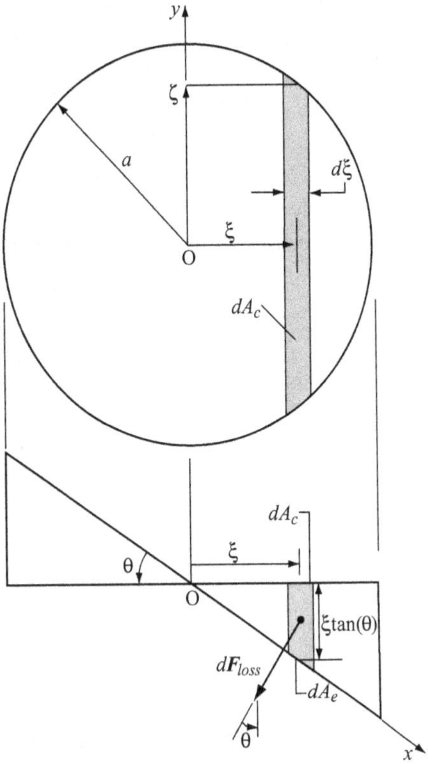

Figure 11.2. *Geometry at the Waterplane for the Displaced Circular Cylinder in Figure 11.1b.* The circular waterplane is horizontal in this figure, and the elliptic waterplane is inclined at the angle θ.

notation, the magnitudes of the gained and lost buoyancies in the displaced condition are

$$|F_{gain}| = |F_{loss}| = \rho g \iint\limits_{A_c/2} \xi \tan(\theta) dA_c$$

$$= \rho g \int_0^a \xi \tan(\theta) 2\zeta d\xi = 2\rho g \int_0^a \xi \tan(\theta)\sqrt{a^2 - \xi^2} d\xi = \frac{2}{3}\rho g a^3 \tan(\theta) \quad (11.3)$$

In this equation, the first line is generic, having applicability to any surface-piercing body, whereas the second line applies to the cylinder sketched in Figure 11.1. The local coordinates (ξ, ζ) are in the undisturbed waterplane, which is a circle of radius a. In Figure 11.2, the elemental area of the circle is dA_c, and that of the displaced elliptical water plane is dA_e. Note that in the design condition, there is no curvature of the body surface in the axial direction at the waterplane. The body is said to be *wall-sided* under this condition. Although the gained and lost buoyancies cancel each other, they each produce buoyant moments, the sum of which equals the moment of the displaced buoyant force in eq. 11.2. The buoyant moment about the point B in Figure 11.1b is then obtained from

$$\iint\limits_{A_c} (\rho g \xi)\xi \tan(\theta) dA_c \cos(\theta) = \rho g \int_{-a}^{a} \xi^2 (2\sqrt{a^2 - \xi^2}) d\xi \sin(\theta) = \rho g \pi \frac{a^4}{4} \sin(\theta)$$

$$= \rho g I_y \sin(\theta) = F_B X_B = (\rho g \vee) X_B \quad (11.4)$$

In this equation, I_y is the second moment of the waterplane area with respect to the y-axis. For the circular cylinder in Figures 11.1 and 11.2, $I_y = \pi a^4/4$. The expression for X_B is obtained from eq. 11.4. The resulting expression is

$$X_B = \frac{I_y}{V}\sin(\theta) \tag{11.5}$$

The combination of this expression with that in eq. 11.2, eliminating X_B, results in the following expression for the metacentric height:

$$GM = \frac{I_y}{V} \mp GB \tag{11.6}$$

This equation applies to any floating body, where again I_y is the second moment of the waterplane area with respect to the y-axis. The negative sign (−) in eq. 11.6 is used when G is above B in Figure 11.1, and the positive sign (+) when B is above G. Referring to Figure 11.1b, we see the following are the conditions of stability for the body:

(a) When $GM > 0$, the body is *stable*, and will return to the design position when released.
(b) When $GM = 0$, the body is *neutrally stable*, and will not move when released.
(c) When $GM < 0$, the body is *unstable*, and will capsize when released.

EXAMPLE 11.1: ROLL STABILITY OF A CAN BUOY A 0.305-m-diameter can buoy is 1.83 m in height. The buoy is designed to be a channel marker over the New York Bight. The salt water at the site has a mass density (ρ) of 1030 kg/m^3. The unballasted weight of the uniform, unballasted buoy is 445 N. A ballast weight of 445 N is added with its center of gravity 0.0305 m above the base (K) of the buoy. For the ballasted buoy, the draft is $d = W/(\rho g \pi a^2) \simeq 1.21$ m. Hence, the *freeboard* (ℓ) is 0.62 m; this is the vertical distance from the SWL to the deck. The position of the center of gravity (G) is 0.737 m below the SWL. This is found by taking moments of the ballast and buoy weights about the origin (O) of the coordinate system in Figure 11.1. The center of buoyancy (B) is at $z = -d/2 = -0.605$ m, and above G. In eq. 11.6, the positive sign then applies. For the circular waterplane, the second moment of area is $I_y \simeq 4.25 \times 10^{-4}$ m^4. From eq. 11.6, the metacentric height is $GM = 0.137$ m. Because $GM > 0$, the buoy is stable.

The transverse stability of a barge is discussed and illustrated in Example 11.11. The hydrostatic restoring force and moment terms in the coupled equations of the heaving and pitching motions of a floating body in waves are derived later in this chapter.

11.2 Floating Body Motions

Consider the floating deep-submergence vehicle (DSV) sketched in Figure 11.3. As illustrated, there are six degrees of freedom. Those are the three rectilinear motions called *surging* (motion in the x-direction), *swaying* (motion in the y-direction), and *heaving* (motion in the z-direction), and the three angular motions called *rolling* (motion about the x-axis), *pitching* (motion about the y-axis), and *yawing* (motion about the z-axis). Three of the degrees of freedom (heaving, rolling, and pitching)

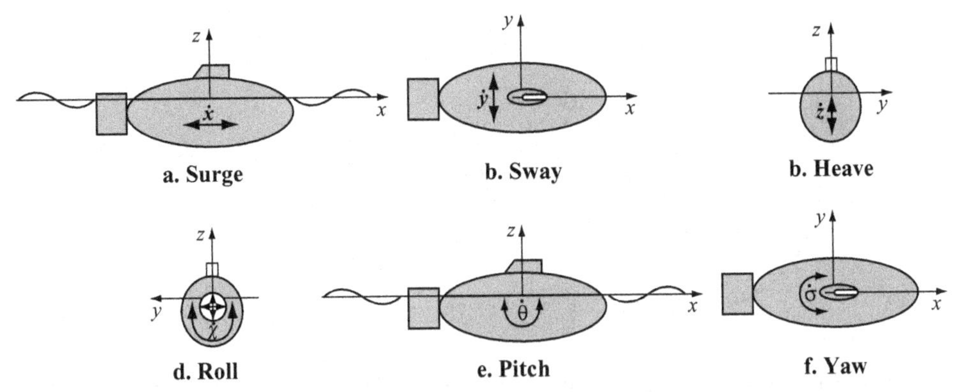

Figure 11.3. *Degrees of Freedom of a Floating Body*. The velocity potentials associated with the waves produced by the body motions are as follows: surge (ϕ_1), sway (ϕ_2), heave (ϕ_3), roll (ϕ_4), pitch (ϕ_5), and yaw (ϕ_6). Here, the body is a deep-submergence vehicle (DSV). Because the design of the DSV hull is based on full-submergence operation, the motions while on the free surface can be quite large, making life for the crew quite uncomfortable.

are oscillatory in nature because they are subject to either a gravitational restoring force (heaving) or a gravitational restoring moment (rolling and pitching). As a result, there are natural periods of motion associated with these oscillatory motions. We could consider these motions analogous to a spring-mass-damper system, as discussed in Chapter 10. In that context, one would consider the floating structure to be to a system having a soft spring because the restoring force due to gravity is relatively small compared to the inertial force of the floating body. The relatively small force ratio results in a rather large natural period and, as a result, the motions of floating bodies are excited by passing waves. See eq. 10.13, which relates both the natural frequency and period of the heaving cylinder (sketched in Figure 10.1) to both the body geometry and the tension-mooring lines.

A. Boundary Condition on the Body

The boundary condition on the floating body, referred to herein as the *body condition*, is the requirement that the normal velocity component of water particles adjacent to a body must be equal to the normal velocity component of the body. The analytical technique used here to obtain the body condition is rather standard in the field of fluid dynamics. According to Lamb (1932), the origin of the technique appears to be in a work by Thomson (1848). In the following paragraphs, we discuss the technique and apply it to several body shapes of interest.

The geometry of any three-dimensional moving body, floating or fixed, can be represented in functional form as

$$S(x, y, z, t) = 0 \tag{11.7}$$

We refer to this in this book as the *body function*. The time, t, appears in the function because the body is in motion. The total time-derivative of the body function is

$$\frac{DS}{Dt} = \frac{\partial S}{\partial t} + \boldsymbol{V} \cdot \nabla S = \frac{\partial S}{\partial t} + u\frac{\partial S}{\partial x} + v\frac{\partial S}{\partial y} + w\frac{\partial S}{\partial z} = \frac{\partial S}{\partial t} + \boldsymbol{V} \cdot \boldsymbol{n}|\nabla S| = 0 \tag{11.8}$$

Here, $\boldsymbol{V} = u\boldsymbol{i} + v\boldsymbol{j} + w\boldsymbol{k}$ is the velocity of the adjacent water particles, and $\boldsymbol{n} = \nabla S/|\nabla S|$ is the normal unit vector on the wetted body surface, outward to the fluid

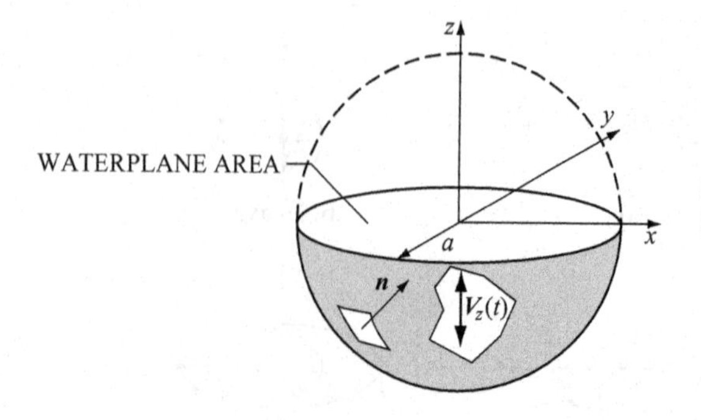

Figure 11.4. *Notation for a Semi-Submerged, Heaving Sphere.*

and inward to the body. One can see that the relationship for the unit vector is obtained directly from the equation. As in eq. 2.65, the notation $D(\ )/Dt$ is used to represent the total time-derivative. The expression for the normal velocity component of the fluid on the surface is obtained from the last equality in eq. 11.8. The result is

$$\boldsymbol{V}\cdot\boldsymbol{n} = (u\boldsymbol{i} + v\boldsymbol{j} + w\boldsymbol{k})\cdot\frac{\nabla S}{|\nabla S|} = \left(u\frac{\partial S}{\partial x} + v\frac{\partial S}{\partial y} + w\frac{\partial S}{\partial z}\right)\left(\frac{1}{|\nabla S|}\right) = -\frac{\dfrac{\partial S}{\partial t}}{|\nabla S|} \tag{11.9}$$

Assuming that the flow about the body is irrotational, the fluid velocity components can be written in terms of the velocity potential. For irrotational flow, the body condition then results in

$$\left.\frac{\partial\phi}{\partial x}\right|_s\frac{\partial S}{\partial x} + \left.\frac{\partial\phi}{\partial y}\right|_s\frac{\partial S}{\partial y} + \left.\frac{\partial\phi}{\partial z}\frac{\partial S}{\partial z}\right|_s = -\frac{\partial S}{\partial t} \tag{11.10}$$

To illustrate the application of eqs. 11.9 and 11.10, the body condition is applied to a semi-submerged, heaving sphere in the following example.

EXAMPLE 11.2: BODY CONDITION FOR A SEMI-SUBMERGED, HEAVING SPHERE The equation for a semi-submerged, floating sphere (wetted hemisphere) sketched in Figure 11.4 is

$$x^2 + y^2 + z^2 = a^2, \quad z < 0 \tag{11.11}$$

The sphere is forced to heave with a circular frequency ω. For this reason, let us isolate the z-coordinate and write eq. 11.11 as

$$z = \sqrt{a^2 - x^2 - y^2} + Z_o e^{-i\omega t}, z < 0 \tag{11.12}$$

The body function of eq. 11.7 is then

$$S(x, y, z, t) = z - \sqrt{a^2 - x^2 - y^2} - Z_o e^{-i\omega t} = 0 \tag{11.13}$$

Here, the amplitude of the body motion is Z_o. The body condition for the semi-submerged, floating sphere in Figure 11.4 is found by combining the expression in eq. 11.13 with that in eq. 11.10. The result is

$$\frac{\partial\phi}{\partial x}\frac{x}{z} + \frac{\partial\phi}{\partial y}\frac{y}{z} + \frac{\partial\phi}{\partial z} = -\frac{z}{a}V_z \tag{11.14}$$

Note that the only imposed velocity is the heaving velocity, $V_z = dz/dt = -i\omega Z_o e^{-i\omega t}$, where Z_o is the motion amplitude.

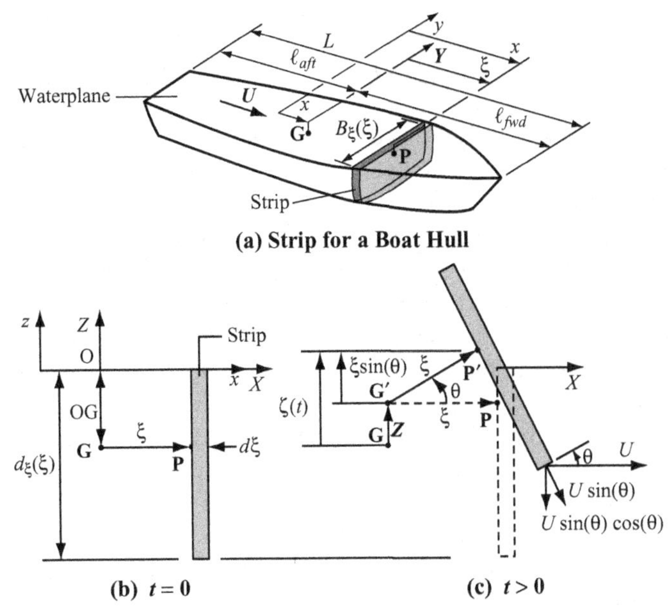

Figure 11.5. *Planar Displacements of a Strip on a Wetted Boat Hull over Time t.* The center of gravity of the body is denoted as G. The axis of rotation (not shown) is through G and parallel to the y-axis. The hull velocity, U, is constant. In the figure for $t > 0$, we see that the vertical velocity component with respect to the water, due to the hull velocity, is in the negative Z direction.

In the books of Stoker (1957) and Dean and Dalrymple (1984), the methodology of this subsection is used to determine both the free-surface and seafloor conditions. In Chapter 3, these boundary conditions are derived by using more heuristic approaches.

B. Heaving and Pitching Equations of Motion

Consider the x-z planar motions of the surface craft sketched in Figure 11.5. These motions are surging, heaving, and pitching of the vessel. In the derivations that follow, we consider the surging motions to be of second order compared to heaving and pitching, and our attention is focused on the heaving and pitching motions. However, the craft is traveling in the x-direction at a velocity $\boldsymbol{U}$, as shown. As sketched in Figure 11.5b, a body coordinate system (X, Z) connected to the body is introduced on the waterline above the center of gravity of the body. The equations of motion for the heaving and pitching are written in terms of the body coordinate system.

If we have a model of the vessel sketched in Figure 11.5, we would find that a forced heaving displacement of the model would also result in an angular (pitching) displacement. Similarly, if we pushed down on the shroud of the propeller to force a pitching displacement, a heaving displacement would also occur. In other words, for that particular hull shape, the heaving and pitching motions are said to be *coupled*. The coupled equations of motion are obtained by applying Newton's second law of motion, which describes the heaving motions of a moored can buoy (see eq. 10.1). The respective equations of motion for a heaving and pitching floating body are

$$m\frac{d^2z}{dt^2} = F_{\text{h}} + F_{\text{v}} + F_{\omega} = F_{\text{h}} + F_{\text{v}} + (F_{\text{w}} + F_{\text{wZ}} + F_{\text{a}} + F_r) \qquad (11.15\text{a})$$

and

$$I_Y \frac{d^2\theta}{dt^2} = M_h + M_v + M_\omega = M_h + M_v + (M_w + M_{wZ} + M_a + M_r) \qquad (11.16a)$$

In eq. 11.15, m is the mass of the body, F_h is the hydrostatic restoring force, F_v is the viscous damping force, and F_ω is the sum of the hydrodynamic forces. The hydrodynamic forces include the wave-induced force (F_w), the wave-body interaction force (F_{wZ}), the inertial reaction force, (F_a), and the radiation damping force (F_r). The inertial reaction and radiation damping forces are reactions to the body motions. The hydrodynamic force components are discussed in Section 11.3. In eq. 11.16, I_Y is the body mass moment of inertia with respect to the axis (Y) through the center of gravity and parallel to the y-axis, and the moments corresponding to the forces in eq. 11.15 are identified by M.

We distinguish between body properties and area properties by using italicized letters to identify the former and nonitalicized letters for the latter. For example, the mass-moment of inertia of a body is represented by I_Y, and the second moment of area by $\mathrm{I_Y}$. The y-axis is fixed on the free surface. The terms on the right side of eq. 11.16a are the counterclockwise moments about the Y-axis corresponding to the forces in eq. 11.15a. The Y-axis is not shown in Figure 11.5.

Equations 11.15a and 11.16a can be rearranged and written respectively as the following second-order, coupled equations of motion for heaving and pitching:

$$(m + \mathrm{a}_w)\frac{d^2Z}{dt} + \mathrm{b}\frac{dZ}{dt} + \mathrm{c}Z + \mathrm{d}\frac{d^2\theta}{dt^2} + \mathrm{e}\frac{d\theta}{dt} + \mathrm{f}\theta = F_w(t) \qquad (11.15b)$$

where a_w is the added mass (the mass of water excited by the body motions), and

$$(I_Y + \mathrm{A}_w)\frac{d^2\theta}{dt^2} + \mathrm{B}\frac{d\theta}{dt} + \mathrm{C}\theta + \mathrm{D}\frac{d^2Z}{dt^2} + \mathrm{E}\frac{dZ}{dt} + \mathrm{F}Z = M_w(t) \qquad (11.16b)$$

where A_w is the added-mass moment of inertia. Note that the script "w" is associated with the added mass, whereas the subscript "w" identifies wave properties. The coefficients in these two equations are defined in the next section. To do so, the strip theory is used, as presented by Korvin-Kroukovsky and Jacobs (1957).

C. Introduction to Strip Theory

Consider the boat traveling at a velocity $\boldsymbol{U}$ in the x-direction, as sketched in Figure 11.5a. Referring to that sketch, the *strip* is a section of the wetted hull having a horizontal elemental thickness equal to the differential $d\xi$. The strip is located at a horizontal distance ξ from the center of gravity (G). The vertical coordinate attached to the waterline above G is Z, and the transverse coordinate is Y. We can imagine that the wetted part of the body is composed of an infinite number of these strips all glued together. The skin of the strip under consideration has curvature in a plane parallel to the y-z plane, as sketched in Figures 11.5a. The keel depth of the strip is d_ξ at a distance ξ from the center of gravity. The maximum value of this depth is the draft, d. The width of the strip at the waterplane (where $Z = 0$) is $B_\xi(\xi)$, and the maximum width is the beam, B. *Strip theory* is a method for determining the hydrodynamic forces on a floating body by, first, determining those forces on a strip, and then summing (integrating) the forces over the length of the body.

Following Korvin-Kroukovsky and Jacobs (1957), the hydrodynamic force, F_ω, is composed of three separate forces. These are the wave-induced force, the force due to the body motions, and the force due to the wave-body interaction. The hydrodynamic forces can be obtained by using one of several methods. In this book, we present three such methods. The first is the analytical method of Lewis (1929), which is based on the motions of a body in a potential flow, as discussed in Section 9.2. This method leads to an added-mass expression that is independent of the excitation frequency. Because Lewis was interested in the vibrations of ships, the method is most applicable to motions at high frequencies. Actually, the frequency can be considered to be infinite and, therefore, no radiation damping coefficient results from this analysis. The second method is, again, based on potential theory, and results in both frequency-dependent added mass and radiation damping coefficients. The last method is called Green's function method, and is a numerical technique for most practical body shapes. The method applies to strips of any shape. For in-depth discussions of the latter method, see Anderson and Wuzhou (1984) and Wang and Miner (1989), among others. As stated previously, when the forces on the strip have been defined, they are integrated over the hull length to obtain the total force on the body and corresponding moment.

The goal of this section is to mathematically describe the coupled heaving and pitching motions of a floating body by focusing on a strip of the hull. If the body in question is a typical buoy, then the form of the hull is symmetric with respect to the X-Y and Y-Z planes. For a boat hull, the symmetry is with respect to only the X-Z plane (the centerplane), such as the boat hull sketched in Figure 11.5a. We shall assume Y-Z body symmetry in the derivation of the equations of motions. For any floating body shape, the waterline length is L in the X-direction. The notation for the draft of a ship used by naval architects is normally T. Because this represents the wave period in this book, d is used to represent the draft. Again, the strip is located at a fixed distance ξ from the center of gravity, G. The center of gravity for this body is at a distance, OG, below the waterplane. For the displaced body, the rotational axis is Y, originating at G and parallel to the y-axis and the Y-axis.

Begin the analysis by considering the forced, X-Z planar motions of the boat hull in calm water, sketched in Figure 11.5. We concentrate on the motions of the center of gravity (G) of the hull and the strip. In Figure 11.5b, the strip is shown at rest at time $t = 0$, and displaced at $t > 0$ in Figure 11.5c. The motions experienced by the strip are assumed to be heaving in the Z-direction and pitching about an axis through the center of gravity (G).

The vertical displacement of the point P on the back face of the strip is

$$\zeta(\xi, t) = Z(t) + \xi \sin[\theta(t)] \simeq Z(t) + \xi\theta(t) \tag{11.17}$$

Here, $\theta = \theta(t)$ is the counterclockwise pitching angle about Y (through G). The approximation in eq. 11.17 arises from the assumption that $\theta(t)$ is small. In the approximation, the angle is in radians. We note that there is a small horizontal displacement, which is $\xi[1 - \cos(\theta]$. Because of the small value of the pitching angle, this displacement approximately equals zero. Our interest is in the relative motions of the body with respect to the still water. Referring to Figure 11.5c, the vertical velocity of the point P on the strip, with respect to the still water, is

$$V_Z(\xi, t) = \frac{\partial \zeta}{\partial t} = \frac{dZ}{dt} + \xi \cos(\theta)\frac{d\theta}{dt} + \frac{d\xi}{dt}\sin(\theta) \simeq \frac{dZ}{dt} + \xi\frac{d\theta}{dt} - U\theta \tag{11.18}$$

The term containing the speed of the craft, U, is in the words of Korvin-Kroukovsky and Jacobs (1957), the "vertical velocity due to the instantaneous angle of trim θ"

(see Figure 11.5c). The approximate vertical acceleration of P with respect to the adjacent water is

$$\frac{\partial V_Z}{\partial t} = \frac{\partial^2 \zeta}{\partial t^2} \simeq \frac{d^2 Z}{dt^2} + \frac{d\xi}{dt}\frac{d\theta}{dt} + \xi\frac{d^2\theta}{dt^2} - U\frac{d\theta}{dt} = \frac{d^2 Z}{dt^2} + \xi\frac{d^2\theta}{dt^2} - 2U\frac{d\theta}{dt} \tag{11.19}$$

Here, the relative velocity due to the time-derivative of ξ is $d\xi/dt = -U$.

The forces on the body and the corresponding moments about the center of gravity are derived in generic forms in the following paragraphs. An introduction to these motion-induced forces is found in Chapter 10.

(1) Hydrostatic Restoring Force and Moment

Consider, again, the sketches in Figure 11.5. Assuming calm water, the hydrostatic restoring force on the displaced strip in Figure 11.5c is

$$dF_{\text{h}} = \rho g \zeta(t)\, B_\xi(\xi)\, d\xi \simeq \rho g\,(Z + \xi\theta)\, B_\xi(\xi)\, d\xi \tag{11.20}$$

Here, the displacement expression is in eq. 11.17 and $B_\xi(\xi)$ is the breadth of the strip in the waterplane. If the strip is wall-sided (having vertical sides at the waterplane), then the strip breadth is not a function of time. The wall-sided condition is assumed herein. The hydrostatic force on the body is obtained by integrating the expression in eq. 11.20 over the body length, L. For the boat hull in Figure 11.5a, the resulting hydrostatic force expression is

$$F_{\text{h}} = -\rho g \int_{-\ell_{aft}}^{\ell_{fwd}} B_\xi(\xi)\, d\xi\, Z - \rho g \int_{-\ell_{aft}}^{\ell_{fwd}} B_\xi(\xi)\, \xi\, d\xi\, \theta = -\text{c}_{\text{h}} Z - \text{f}_{\text{h}}\theta \tag{11.21}$$

In this equation, the body length at the SWL is $L = \ell_{aft} + \ell_{fwd}$, where ℓ_{aft} and ℓ_{fwd} are the respective distances between the center of gravity, G, and the stern and bow at the waterline. The corresponding hydrostatic moment about G is

$$M_{\text{h}} = -\rho g \int_{-\ell aft}^{\ell_{fwd}} B_\xi(\xi)\, \xi\, d\xi\, Z - \rho g \int_{-\ell aft}^{\ell_{fwd}} B_\xi(\xi)\, \xi^2 d\xi\, \theta = -\text{F}_{\text{h}} Z - C_{\text{h}}\theta \tag{11.22}$$

Again, the positive moment direction is in the counterclockwise direction, as is the case for all moments in this book. The negative signs in eqs. 11.21, and 11.22 follow from eqs. 11.15 and 11.16. The last equalities in the force and moment equations are introduced to show the association with the coefficients in eqs. 11.15b and 11.16b. Because the breadth $B_\xi(\xi)$ is a design input, the coefficients of Z and θ are known.

(2) Viscous Damping Force and Moment

The second force type is the damping due to the combination of viscosity and radiation. The natures of the viscous damping force and the radiation damping force are quite different, the former being nonlinear whereas the latter is linear and discussed later in this section. To facilitate the derivation of the strip theory, the viscous force is represented by the equivalent linear viscous damping coefficient, derived in Section 10.1C (see eq. 10.10). By doing so, we avoid the problems associated with the nonlinear (velocity-squared) viscous damping. The elemental, equivalent linear viscous damping force on the strip is then represented by

$$dF_{\mathrm{v}} = \mathrm{b}'_{\mathrm{v}}\frac{\partial\zeta}{\partial t}d\xi = \mathrm{b}'_{\mathrm{v}}\left(\frac{dZ}{dt} + \xi\frac{d\theta}{dt} - U\theta\right)d\xi \tag{11.23}$$

In this expression, $\mathrm{b}'_{\mathrm{v}}(\xi, \omega_e)$ is the equivalent, linear viscous damping coefficient per unit length. The prime ($'$) indicates that the parameter in question is per unit length along the hull (see Figures 11.5 and 11.6). The subscript v refers to the viscous damping. As demonstrated in Section 10.1C, the equivalent linear damping coefficient is a function of the excitation frequency, ω_e, which is the frequency of encounter if the body is under way. Concerning this circular frequency: If the body is advancing with the wave at a velocity of $\boldsymbol{U}$, as in Figures 11.5a and 11.6a, the relative velocity of the body with respect to the wave is the difference between the wave celerity, $\boldsymbol{c}$, and $\boldsymbol{U}$. The *frequency of encounter* is then $\omega_e = k(c - U)$, where $k = 2\pi/\lambda$ is the wave number. See Bhattacharyya (1978) for an excellent discussion of this frequency. More is written of the frequency of encounter in Section 11.3B(1). Returning to the viscous damping, the total viscous-damping force on the body in eq. 11.15a is expressed by

$$\begin{aligned} F_{\mathrm{v}}(\omega_e, t) &= -\int_{-\ell_{aft}}^{\ell_{fwd}} \mathrm{b}'_{\mathrm{v}}(\xi, \omega_e)d\xi\frac{dZ}{dt} - \int_{-\ell_{aft}}^{\ell_{fwd}} \mathrm{b}'_{\mathrm{v}}(\xi, \omega_e)\,\xi d\xi\frac{d\theta}{dt} + U\int_{-\ell_{aft}}^{\ell_{fwd}} \mathrm{b}'_{\mathrm{v}}(\xi, \omega_e)\,d\xi\theta \\ &= -\mathrm{b}_{\mathrm{v}}\frac{dZ}{dt} - \mathrm{e}_{\mathrm{v}}\frac{d\theta}{dt} - \mathrm{f}_{\mathrm{v}}\theta \end{aligned} \tag{11.24}$$

The corresponding radiation damping moment in eq. 11.24 is

$$\begin{aligned} M_{\mathrm{v}}(\omega_e, t) &= -\int_{-\ell_{aft}}^{\ell_{fwd}} \mathrm{b}'_{\mathrm{v}}(\xi, \omega_e)\,\xi d\xi\frac{dZ}{dt} - \int_{-\ell_{aft}}^{\ell_{fwd}} \mathrm{b}'_{\mathrm{v}}(\xi, \omega_e)\,\xi^2 d\xi\frac{d\theta}{dt} + U\int_{-\ell_{aft}}^{\ell_{fwd}} \mathrm{b}'_{\mathrm{v}}(\xi, \omega_e)\,\xi d\xi\theta \\ &= -\mathrm{E}_{\mathrm{v}}\frac{dZ}{dt} - \mathrm{B}_{\mathrm{v}}\frac{d\Theta}{dt} - \mathrm{C}_{\mathrm{v}}\theta \end{aligned} \tag{11.25}$$

Note that viscous damping in the paper of Korvin-Kroukovsky and Jacobs (1957) is neglected. The reason for this is that viscous damping is normally considered to be much less than the radiation damping for small heaving and pitching motions. There are a number of situations in ocean engineering and naval architecture where viscosity cannot be neglected. One example is when a ship is fitted with bilge keels to introduce roll damping. Depending on the orientation of these hull appendages, bilge keels can be the cause of significant damping in both heaving and pitching.

The radiation damping coefficient, b'_{r}, discussed later in the chapter, is due to the creation of traveling waves by the body motions. This damping component is included in the respective hydrodynamic force and moment in eqs. 11.15 and 11.16. Later in the chapter, we shall assume that the radiation damping is much greater than the viscous damping, as is normally done.

(3) Hydrodynamic Forces and Moments

As stated previously, there are three hydrodynamic forces considered. The first is the wave-induced force. The second force is the reaction force of the ambient water

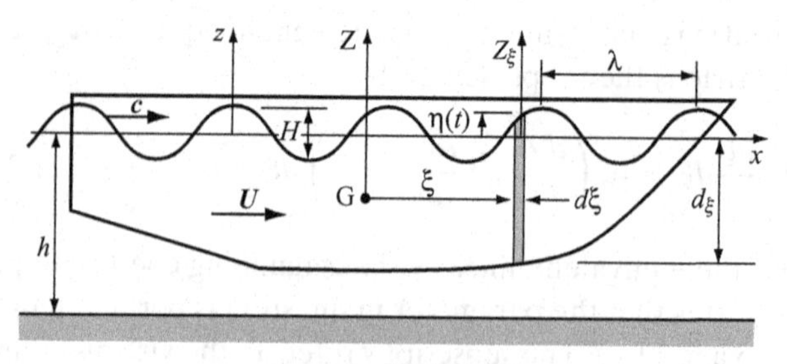

a. Longitudinal Profile

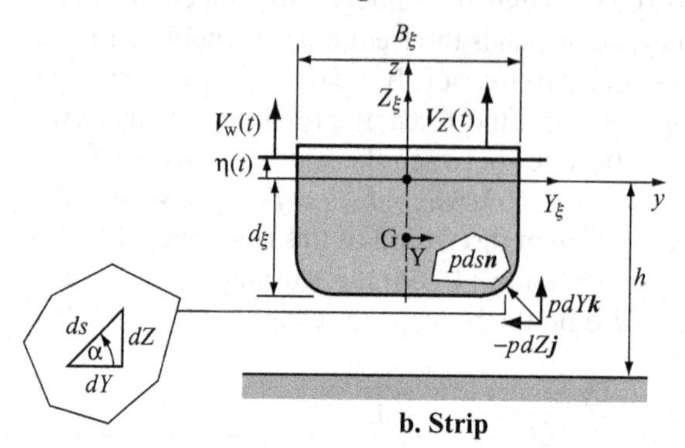

b. Strip

Figure 11.6. *Notation for a Strip of an Advancing Ship in Linear Waves.* The wave celerity, c, is in the direction of the ship velocity, U. This condition is known as a *following sea*, as opposed to a *head sea*, where the directions of the vectors are opposite. The vertical velocity, $V_{\mathrm{w}}(t)$, is that of the free surface. Also in (b) are the unit vectors ***j***, ***k***, and ***n***. The origin of x,y,z is fixed on the waterplane, and the origins of ξ and Y are at the center of gravity (G) of the ship. The planar coordinate system fixed to the strip is Y_ξ, Z_ξ.

due to the motions of the body. The third force is that resulting from the wave-body interaction. In the derivations of these forces herein, we shall rely on the analyses of Korvin-Kroukovsky and Jacobs (1957).

Exciting Force: Consider a craft in a following sea, where the waves and the boat travel in the same direction, as in Figure 11.6. The exciting force is the sum of the wave force (F_{w}) and the wave-body interaction force ($F_{\mathrm{w}Z}$). These exciting-force components are derived in Sections 11.3C(1) and 11.3C(3), respectively. The exciting force is represented by

$$F_{\mathrm{W}}(\omega_e,t) = \int_{-\ell_{aft}}^{\ell_{fwd}} F'_{\mathrm{W}}(\xi,\omega_e,t)\,d\xi = F_{\mathrm{w}}(\omega_e,t) + F_{\mathrm{w}Z}(\omega_e,t)$$

$$= \int_{-\ell_{aft}}^{\ell_{fwd}} \{F'_{\mathrm{w}}(\xi,\omega_e,t) + F'_{\mathrm{w}Z}(\xi,\omega_e,t)\}d\xi \qquad (11.26)$$

The mathematical description of a traveling wave in Chapter 3 is in terms of the inertial coordinates, x, y, and z. However, the force is determined by integrating over the body coordinate ξ, where $x = Ut \pm \xi$. This relationship assumes that G was at $x = 0$ at $t = 0$. The velocity potential describing the flow induced by the passing wave is discussed in Section 11.3.

Returning to eq. 11.26, $F'_W(\xi, \omega_e, t)$ is the exciting force per unit body length, the components of which are derived later in this chapter. The corresponding moment about the center of gravity (G) is

$$M_W(\omega_e, t) = M_w(\omega_e, t) + M_{wZ}(\omega_e, t) = \int_{-\ell_{aft}}^{\ell_{fwd}} F'_W(\xi, \omega_e, t)\,\xi d\xi \tag{11.27}$$

In eqs. 11.27 and 11.28, the subscript W is introduced to represent the sum of the forces associated with the wave encounter.

Inertial Reaction Force: The inertial reaction force on the strip is due to the time rate of change of linear momentum of the ambient water caused by the strip motions. This force is expressed by

$$\begin{aligned}\frac{\partial\left(a'_w \frac{\partial\zeta}{\partial t}\right)}{\partial t} d\xi &= \left[\frac{\partial a'_w}{\partial t}\frac{\partial\zeta}{\partial t} + a'_w \frac{\partial^2\zeta}{\partial t^2}\right] d\xi = \left[\frac{\partial a'_w}{\partial\xi}\frac{d\xi}{dt}\frac{\partial\zeta}{\partial t} + a'_w \frac{\partial^2\zeta}{\partial t^2}\right] d\xi \\ &= \left[-U\frac{\partial a'_w}{\partial\xi}\left(\frac{dZ}{dt} + \xi\frac{d\theta}{dt} - U\theta\right) + a'_w\left(\frac{d^2Z}{dt^2} + \xi\frac{d^2\theta}{dt^2} - 2U\frac{d\theta}{dt}\right)\right] d\xi\end{aligned} \tag{11.28}$$

The vertical velocity and acceleration relationships are from eqs. 11.18 and 11.19, and $d\xi/dt = -U$, as before. The added mass per unit length of the body is represented by $a_w{}'(\xi, \omega_e)$. The respective inertial-reaction force in eq. 11.15 and the total inertial moment in eq. 11.16 are

$$\begin{aligned}F_a &= \int_{-\ell_{aft}}^{\ell_{fwd}} F'_{a\xi}(\xi, \omega_e, t) d\xi = -\int_{-\ell_{aft}}^{\ell_{fwd}} a'_w(\xi, \omega_e) d\xi \frac{d^2Z}{dt^2} + U\int_{-\ell_{aft}}^{\ell_{fwd}} \frac{\partial a'_w}{\partial\xi} d\xi \frac{dZ}{dt} \\ &\quad - \int_{-\ell_{aft}}^{\ell_{fwd}} a'_w(\xi, \omega_e)\xi d\xi \frac{d^2\theta}{dt^2} + U\left[2\int_{-\ell_{aft}}^{\ell_{fwd}} a'_w(\xi, \omega_e) d\xi + \int_{-\ell_{aft}}^{\ell_{fwd}} \frac{\partial a'_w}{\partial\xi}\xi d\xi\right]\frac{d\theta}{dt} \\ &\quad - U^2\int_{-\ell_{aft}}^{\ell_{fwd}} \frac{\partial a'_w}{\partial\xi} d\xi\theta \\ &\equiv -a_w\frac{d^2Z}{dt^2} - b_a\frac{dZ}{dt} - d_a\frac{d^2\theta}{dt^2} - e_a\frac{d\theta}{dt} - f_a\theta = -a_w\frac{d^2Z}{dt^2} - 0 - d_a\frac{d^2\theta}{dt^2} - e_a\frac{d\theta}{dt} + 0\end{aligned} \tag{11.29}$$

and

$$\begin{aligned}M_a &= \int_{-\ell_{aft}}^{\ell_{fwd}} F'_{a\xi}\xi d\xi = -\int_{-\ell_{aft}}^{\ell_{fwd}} a'_w\xi d\xi\frac{d^2Z}{dt^2} + U\int_{-\ell_{aft}}^{\ell_{fwd}} \frac{\partial a'_w}{\partial\xi}\xi d\xi\frac{dZ}{dt} \\ &\quad - \int_{-\ell_{aft}}^{\ell_{fwd}} a'_w\xi^2 d\xi\frac{d^2\theta}{dt^2} + 2U\int_{-\ell_{aft}}^{\ell_{fwd}} a'_w\xi d\xi\frac{d\theta}{dt} + U\int_{-\ell_{aft}}^{\ell_{fwd}} \frac{\partial a'_w}{\partial\xi}\xi^2 d\xi\frac{d\theta}{dt} - U^2\int_{-\ell_{aft}}^{\ell_{fwd}} \frac{\partial a'_w}{\partial\xi}\xi d\xi\theta \\ &\equiv -D_a\frac{d^2Z}{dt^2} - E_a\frac{dZ}{dt} - A_w\frac{d^2\theta}{dt^2} - B_a\frac{d\theta}{dt} - C_a\theta\end{aligned} \tag{11.30}$$

The subscript a identifies a quantity that is dependent on the added mass. Note that in these equations, the integrals of spatial derivatives of the added mass over the hull length vanish because the added mass equals zero at the bow and the stern. The integrals of the moments of these derivatives do not vanish. Hence, $b_a = f_a = 0$ in eq. 11.29, as shown. The determination of the two-dimensional added-mass expressions is discussed in the Section 11.3.

Radiation Damping Force: The last of the hydrodynamic forces is the radiation damping force. This force results from the loss of energy by the floating body to two systems of radiated waves created by the body motions. For the two-dimensional case, these waves must satisfy the radiation condition at $y = \pm\infty$. Simply put, this condition is that the radiated waves must be outward-bound at an infinite distance away from the body. The radiation damping force is assumed to be linear in that it is proportional to the vertical velocity of the strip raised to the first power. This assumption is valid in most practical cases. We can represent the radiation damping force by

$$\begin{aligned} F_r(\omega_e, t) &= -\int_{-\ell_{aft}}^{\ell_{fwd}} b'_r \frac{\partial \zeta}{\partial t} d\xi = \int_{-\ell_{aft}}^{\ell_{fwd}} b'_r(\xi, \omega_e)\, d\xi \frac{dZ}{dt} - \int_{-\ell_{aft}}^{\ell_{fwd}} b'_r(\xi, \omega_e)\, \xi d\xi \frac{d\theta}{dt} \\ &\quad + U \int_{-\ell_{aft}}^{\ell_{fwd}} b'_r(\xi, \omega_e)\, d\xi\, \theta \\ &= -b_r \frac{dZ}{dt} - e_r \frac{d\theta}{dt} - f_r \theta \end{aligned} \tag{11.31}$$

In this expression, $b'_r(\xi, \omega_e)$ is the radiation damping coefficient per unit length. The corresponding radiation damping moment in eq. 11.31 is

$$\begin{aligned} M_r(\omega_e, t) &= -\int_{-\ell_{aft}}^{\ell_{fwd}} b'_r(\xi, \omega_e)\, \xi d\xi \frac{dZ}{dt} - \int_{-\ell_{aft}}^{\ell_{fwd}} b'_r(\xi, \omega_e)\, \xi^2 d\xi \frac{d\theta}{dt} + U \int_{-\ell_{aft}}^{\ell_{fwd}} b'_r(\xi, \omega_e)\, \xi d\xi\, \theta \\ &= -E_r \frac{dZ}{dt} - B_r \frac{d\theta}{dt} - C_r \theta \end{aligned} \tag{11.32}$$

We see that both the radiation damping force and moment are frequency-dependent. As is the case for the added mass, the radiation damping is discussed in Section 11.3.

D. Coupled Heaving and Pitching Equations of Motion

The response forces in eq. 11.15a, the heaving equation of motions, and the corresponding response moments in eq. 11.16a, the pitching equation of motion, have now been defined. These linearized equations describe the coupled heaving and pitching motions of a floating body traveling at a constant velocity (U) in the

x-direction. Those respective equations are written as

$$(m+\mathrm{a}_w)\frac{d^2Z}{dt}+(\mathrm{b_v}+\mathrm{b_r})\frac{dZ}{dt}+\mathrm{c_h}Z+\mathrm{d_a}\frac{d^2\theta}{dt^2}+(\mathrm{e_v}+\mathrm{e_r}+\mathrm{e_a})\frac{d\theta}{dt}+(\mathrm{f_h}+\mathrm{f_v}+\mathrm{f_r})\theta = F_\mathrm{w}(\omega_e,t) \tag{11.33}$$

where m is the mass of the body, and

$$(I_\mathrm{Y}+\mathrm{A}_w)\frac{d^2\theta}{dt^2}+(\mathrm{B_v}+\mathrm{B_r}+\mathrm{B_a})\frac{d\theta}{dt}+(\mathrm{C_h}+\mathrm{C_v}+\mathrm{C_r}+\mathrm{C_a})\theta+\mathrm{D_a}\frac{d^2Z}{dt^2} + (\mathrm{E_v}+\mathrm{E_r}+\mathrm{E_a})\frac{dZ}{dt}+\mathrm{F_h}Z = M_\mathrm{w}(\omega_e,t) \tag{11.34}$$

Again, I_Y is the mass moment of inertia with respect to the athwart-ships axis (Y) through the center of gravity, and ω_e is the frequency of encounter discussed in Section 11.3C(2). The generic expressions for the coefficients in these equations are found in eqs. 11.21 through 11.32. The coupling terms are those having the d, e, and f coefficients in eq. 11.33, and the D, E, and F coefficients in eq. 11.34. More is written of the coupling terms in Section 11.3 and in the examples presented later in this chapter. The coefficients in eqs. 11.33 and 11.34 are tabulated in Table 11.1.

Concerning the coefficients: When a body is symmetric with respect to the center of gravity (G in Figure 11.6), then the added mass (a'_w), body breadth (B_ξ), and viscous and radiation damping terms (b'_v and b'_r) are all even functions of the variable ξ. The products of these terms and ξ are odd functions, as are the spatial derivatives of the functions. Hence, the integrations of the odd functions over the body length equal zero for a symmetric body. Furthermore, when the body is both symmetric and not under way ($U=0$), then all of the coupling terms vanish. That is, the heaving and pitching motions of a symmetric body at rest are uncoupled.

A note concerning the notation in eqs. 11.33 and 11.34: When more than two degrees of freedom are considered, it is common practice to use index notation. If we designate the coordinates x, y, z as x_1, x_2, x_3 and the rotational angles χ, θ, σ as x_4, x_5, x_6 in Figure 11.3, then the coefficients in the six equations of motion are $\mathrm{a_{i,j}}$, where the first subscript indicates direction of motion of the equation and the second subscript indicates the coupled direction. For example, the first coefficient in eq. 11.34 would be $\mathrm{a}_{5,5}$, and the fourth coefficient would by $\mathrm{a}_{5,3}$. The first coefficient then is for the motions in the θ-direction (about the y-axis) due to the motions in the θ-direction. The fourth coefficient is for the motions in the θ-direction due to the motions in the z-direction. For planar motions, such as surging, heaving, and pitching, the notation used in eqs. 11.33 and 11.34 is preferable.

Faltinsen (1974) presents the variations in the coefficients in Table 11.1 due to both Ogilvie and Tuck (1969) and Salvesen, Tuck, and Faltinsen (1970). As previously written, the latter contains the version of the strip theory that is considered to be the preferred version today. As noted in these papers and in that of Loukakis and Sclavounos (1978), the strip theory is not valid near the ends of a floating body.

In the next section, expressions for the respective hydrodynamic forces F_w, F_a, and F_r in eqs. 11.26 and 11.33, and their associated moments, are considered in more detail.

Table 11.1. *Coefficients in the coupled heaving and pitching equations of motion*

$a_w = \int_{-\ell_{aft}}^{\ell_{fwd}} a'_w(\xi, \omega_e) d\xi$	$A_w = \int_{-\ell_{aft}}^{\ell_{fwd}} a'_w(\xi, \omega_e)\, \xi^2 d\xi$
$b_v = \int_{-\ell_{aft}}^{\ell_{fwd}} b'_v(\xi, \omega_e) d\xi$	$B_a = U \left[2 \int_{-\ell_{aft}}^{\ell_{fwd}} a'_w \xi d\xi + \int_{-\ell_{aft}}^{\ell_{fwd}} \frac{\partial a'_w}{\partial \xi} \xi^2 d\xi \right]$
$b_r = \int_{-\ell_{aft}}^{\ell_{fwd}} b'_r(\xi, \omega_e) d\xi$	$B_v = \int_{-\ell_{aft}}^{\ell_{fwd}} b'_v(\xi, \omega_e)\, \xi^2 d\xi$
$c_h = \rho g \int_{-\ell_{aft}}^{\ell_{fwd}} B_\xi(\xi) d\xi$	$B_r = \int_{-\ell_{aft}}^{\ell_{fwd}} b'_r(\xi, \omega_e)\, \xi^2 d\xi$
$d_a = \int_{-\ell_{aft}}^{\ell_{fwd}} a'_w(\xi, \omega_e) \xi d\xi$	$C_a = U^2 \int_{-\ell_{aft}}^{\ell_{fwd}} \frac{\partial a'_w}{\partial \xi} \xi d\xi$
$e_a = U \left[2 \int_{-\ell_{aft}}^{\ell_{fwd}} a'_w(\xi, \omega_e)\, d\xi + \int_{-\ell_{aft}}^{\ell_{fwd}} \frac{\partial a'_w}{\partial \xi} \xi d\xi \right]$	$C_h = \rho g \int_{-\ell_{aft}}^{\ell_{fwd}} B_\xi(\xi)\, \xi^2 d\xi$
$e_v = \int_{-\ell_{aft}}^{\ell_{fwd}} b'_v(\xi, \omega_e)\, \xi d\xi$	$C_v = U \int_{-\ell_{aft}}^{\ell_{fwd}} b'_v(\xi, \omega_e)\, \xi d\xi$
$e_r = \int_{-\ell_{aft}}^{\ell_{fwd}} b'_r(\xi, \omega_e)\, \xi d\xi$	$C_r = U \int_{-\ell_{aft}}^{\ell_{fwd}} b'_r(\xi, \omega_e)\, \xi d\xi$
$f_h = \rho g \int_{-\ell_{aft}}^{\ell_{fwd}} B_\xi(\xi)\, \xi d\xi$	$D_a = \int_{-\ell_{aft}}^{\ell_{fwd}} a'_w \xi d\xi$
$f_v = U \int_{-\ell_{aft}}^{\ell_{fwd}} b'_v(\xi, \omega_e)\, d\xi$	$E_a = U \int_{-\ell_{aft}}^{\ell_{fwd}} \frac{\partial a_w{}'}{\partial \xi} \xi d\xi$
$f_r = U \int_{-\ell_{aft}}^{\ell_{fwd}} b'_r(\xi, \omega_e)\, d\xi$	$E_v = \int_{-\ell_{aft}}^{\ell_{fwd}} b'_v(\xi, \omega_e)\, \xi d\xi$
$F_w(\omega_e, t)$ (See eq. 11.26.)	$E_r = \int_{-\ell_{aft}}^{\ell_{fwd}} b'_r(\xi, \omega_e)\, \xi d\xi$
	$F_h = \rho g \int_{-\ell_{aft}}^{\ell_{fwd}} B_\xi(\xi)\, \xi d\xi$
	$M_w(\omega_e, t)$ (See eq. 11.27.)

11.3 Two-Dimensional Hydrodynamics – Vertical Body Motions

In this section, two-dimensional analyses of the hydrodynamic forces on a vertically moving strip are presented. The analyses result in hydrodynamic coefficients for the strip. The hydrodynamic coefficients are approximated by closed mathematical forms, as are the body shapes under consideration. Such body shapes include the wedge, the conic sections, and the Lewis forms. The latter is introduced in Section 9.2A(2). The method is based on the potential theory of fluid mechanics, where the flow is considered to be irrotational. See Sections 2.3 and 9.2 for the derivations of some of the equations used in this section. In deriving the hydrodynamic coefficients, we follow the methodology of Korvin-Kroukovsky and Jacobs (1957). Essentially, the method involves the potential describing the flow about a two-dimensional body in an infinite fluid. To account for the free surface, half of the flow field is used, that half being beneath the free surface. This method was modified by Motora (1964), who presents correction factors to improve the analysis of the free-surface effects. In addition to the heaving and pitching motions, Motora (1964) analyzes the hydrodynamic coefficients for the swaying of the strip.

After defining the form of the strip using the Lewis conformal mapping method presented in Chapter 9, the wave-induced force on the strip is derived assuming that the body is fixed in a moving sea. Then, the hydrodynamic forces due to the body motions in a calm sea are derived.

In the determination of the added mass of the vertically moving Lewis form, the method of Landweber and Macagno (1957) is used. Finally, the wave-body interaction forces are determined. These correspond to the coefficients presented by Korvin-Kroukovsky and Jacobs (1957).

A. Strip Geometries – Lewis Forms

The Lewis two-dimensional body shapes, or *Lewis forms*, are introduced in Section 9.2A(2) of this book. In that section, it is shown how bodies of certain forms moving in infinite ideal fluids can be represented using the conformal mapping technique employed by Lewis (1929). In this section, we demonstrate the use of the technique in representing floating two-dimensional bodies with application to the strip theory. For discussions of the conformal mapping applications to floating bodies, the reader is referred to the papers of Landweber and Macagno (1957, 1959, 1967), Porter (1960), Macagno (1968), von Kerczek and Tuck (1969), and Faltinsen (1974). In addition, the books of Bishop and Price (1979) and Lloyd (1989) contain informative discussions of the Lewis forms.

In the following derivations, there are two complex planes to be considered. The first of these we refer to as the *transform plane* or *ζ-plane*. The second plane is called the *physical plane* or *z-plane*. The flow in the ζ-plane is conformally transformed to a flow in the z-plane. An excellent discussion of the basics of conformal mapping is found in the book by Schinzinger and Laura (2003).

Consider the horizontal motions of the strip in Figure 11.7b. The horizontal velocity of the strip is $V_Y(t)$ in Figure 11.7b. As stated previously, the strip geometry of interest is a Lewis form. From Section 9.2A, a section of the submerged portion of a craft, such as that sketched in Figure 11.7b, can be defined by coordinates similar to those in eq. 9.34, where the radius of the transformed circle is a, as in Figure 9.7. Referring to Figures 11.7b and 11.7c, the Y- and Z-coordinates are known as the physical coordinates. From the transformation resulting from eq. 9.27, the physical coordinates can be written in terms of $\zeta = Re^{i\gamma} = \varepsilon + i\epsilon$ in Figure 11.7a. The transformed complex variable, ζ, can also be written in terms of z. We are mapping the exterior points on the bodies in Figure 11.7 using a special case of a Laurent series from the theory of complex variables. As a result, we can write the following expressions for the transform pair:

$$
\begin{aligned}
z = Y + iZ &= \zeta + \frac{A_1}{\zeta} + \frac{A_3}{\zeta^3} \\
&= \left(R + \frac{A_1}{R}\right)\cos(\gamma) + \frac{A_3}{R^3}\cos(3\gamma) + i\left[\left(R - \frac{A_1}{R}\right)\sin(\gamma) - \frac{A_3}{R^3}\sin(3\gamma)\right] \quad (11.35a)
\end{aligned}
$$

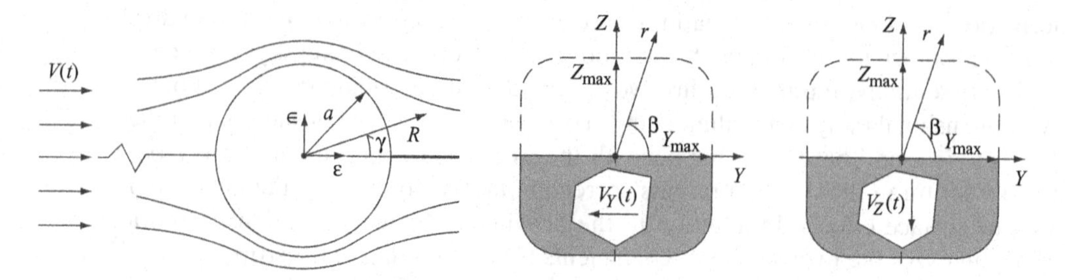

a. Flow about the *a*-circle (ζ–Plane) **b. Horizontal Motion (z–Plane)** **c. Vertical Motion (z–Plane)**

Figure 11.7. *Circle Transformed into a Lewis-Form Strip.* The Lewis parametric values are $A_1 = 0$ and $A_3 = -0.111$ in eqs. 11.35 and 11.36, as in the report of Wendel (1956). The Lewis (1929) method can be used to determine the flows about both horizontally moving strips, as in (b) and vertically moving strips, as in (c).

and

$$\zeta = z - \frac{A_1}{z} - \frac{\left(A_1^2 + A_3\right)}{z^3} \tag{11.35b}$$

In this equation, A_1 and A_3 are called the *Lewis parametric constants* or, simply, *Lewis parameters.* Both constants are real. The angular values of $\gamma = 0$, $\pi/2$, π, $3\pi/2$, and 2π in Figure 11.7a correspond to the same angular values of β in Figures 11.7b and 11.7c. Later, our focus is on the range of γ that corresponds to the submerged strip. That is, if $Z = 0$ is designated as the free surface, then the range of interest for the angular variable is $\pi \leq \beta \leq 2\pi$ in Figures 11.7b and 11.7c. From eq. 11.35, the maximum values of the body coordinates respectively correspond to $\gamma = 0$ and $\gamma = \pi/2$, where $R = a$. In the paper of Lewis (1929), the unit circle ($a = 1$) is transformed. Here, the radius, a, is retained in the derivations of the various equations to remind the reader of the dimensional units of the variables. Following the method of Lewis (1929), the resulting expressions for the maxima are

$$Y_{max} = \left(a + \frac{A_1}{a} + \frac{A_3}{a^3}\right)\bigg|_{a=1} = 1 + A_1 + A_3 \tag{11.36a}$$

and

$$Z_{max} = \left(a - \frac{A_1}{a} + \frac{A_3}{a^3}\right)\bigg|_{a=1} = 1 - A_1 + A_3 \tag{11.36b}$$

We note that these maxima depend on our choice of a and the design choices of the Lewis parameters, A_1 and A_3. We see that A_1 has units of $(\text{length})^2$, whereas A_3 has units of $(\text{length})^4$. The expressions in eqs. 11.35 and 11.36 are used to determine the shape of the strip. Equations 11.36a and 11.36b can be simultaneously solved to obtain expressions for the Lewis parameters in terms of the maximum body coordinates in the physical plane. However, this is somewhat misleading because the selection of A_1 and A_3 determine the body shape. To illustrate: If $A_1 = 0$ and $A_3 \neq 0$, then the maxima in eqs. 11.36a and 11.36b are equal for any finite value of A_3. For example, the strip geometry in Figure 11.8b corresponds to $A_1 = 0$ and $A_3 = -0.111$. For these Lewis parametric values, one finds that $Y_{max} = Z_{max}$, as expected. The strip geometry in Figure 11.8b is also used by Wendel (1954) as an application of the Lewis (1929) conformal mapping method.

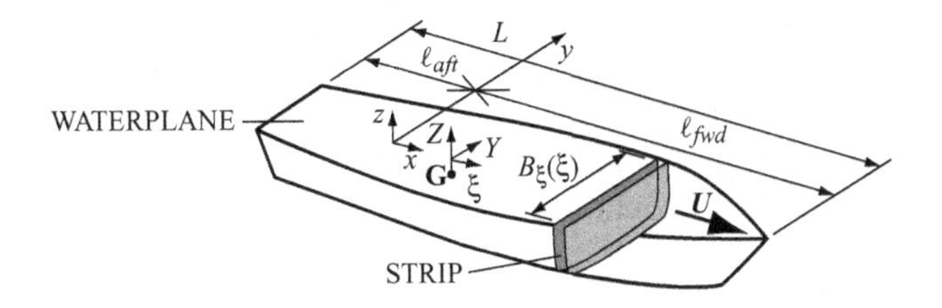

a. Strip in a Boat Hull

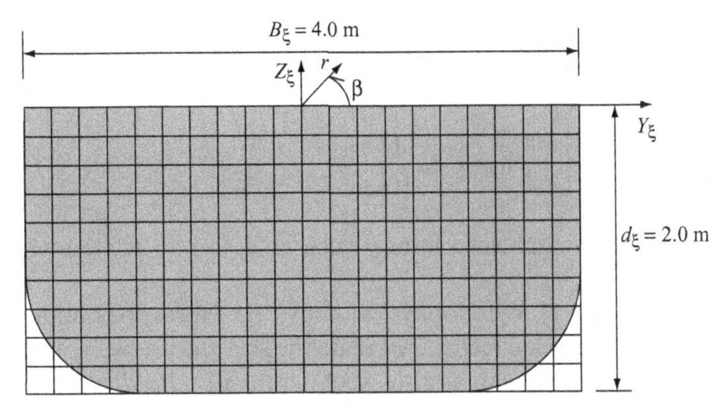

b. Displaced Area of the Strip

Figure 11.8. *Lewis Form Strip for which* $A_1 = 0$ and $A_3 = -0.111$. The coordinate system attached to the strip is Y_ξ, Z_ξ, and the coordinates of the strip are obtained from eqs. 11.38a and 11.38b, multiplied by the scale factor in eq. 11.40b. The design breadth of the strip is $B_\xi(\xi) = 4\,\text{m}$, and the design keel depth is $d_\xi(\xi) = 2\,\text{m}$. From Example 11.3, the area (S_ξ) of the strip is approximately 7.66 m^2.

The expressions for the maxima in eq. 11.36 can be incorporated in eq. 11.35 to obtain the coordinates of any point in the flow field of the z-plane. The results are

$$Y = R\left[\left(1 + \frac{A_1}{R^2}\right)\cos(\gamma) + \frac{A_3}{R^4}\cos(3\gamma)\right]$$

$$= \frac{Y_{\max}}{(1 + A_1 + A_3)} R\left[\left(1 + \frac{A_1}{R^2}\right)\cos(\gamma) + \frac{A_3}{R^4}\cos(3\gamma)\right] \qquad (11.37a)$$

Here, the coefficient in the second equality must be equal to 1 from eq. 11.36a. Similarly, the vertical coordinate is

$$Z = R\left[\left(1 - \frac{A_1}{R^2}\right)\sin(\gamma) - \frac{A_3}{R^4}\cos(3\gamma)\right]$$

$$= \frac{Z_{\max}}{(1 - A_1 + A_3)} R\left[\left(1 - \frac{A_1}{R^2}\right)\sin(\gamma) - \frac{A_3}{R^4}\sin(3\gamma)\right] \qquad (11.37b)$$

The coordinates of the strip profile at a longitudinal position of ξ are then

$$Z_\xi \equiv Z(\xi, \gamma)\,|_{R=a=1} = \frac{Z_{\max}}{(1 - A_1 + A_3)}[(1 - A_1)\sin(\gamma) - A_3\sin(3\gamma)] \qquad (11.38a)$$

and

$$Y_\xi \equiv Y(\xi, \gamma)|_{R=a=1} = \frac{Y_{\max}}{(1 + A_1 + A_3)}[(1 + A_1)\cos(\gamma) + A_3\cos(3\gamma)] \qquad (11.38b)$$

Referring to Figure 11.8a, the subscript ξ identifies the longitudinal position of the strip with respect to the center of gravity, G. The coefficients introduced in eq. 11.38 are obtained from eq. 11.36.

The area of the strip (or *sectional area*) can be obtained using the expressions in eq. 11.38. The displaced *Lewis-form area* is obtained from

$$\mathrm{S}_\xi = \left| \int_{-Y_{\max}}^{Y_{\max}} Z_\xi \, dY_\xi \right| = \left| \int_{\pi}^{2\pi} Z_\xi \frac{dY_\xi}{d\gamma} d\gamma \right| \tag{11.39a}$$

The substitution of the coordinates in eq. 11.38 into the last integrand of this equation results in

$$\mathrm{S}_\xi = \left| -\frac{\pi}{2} Y_{\max} Z_{\max} \frac{(1 - A_1^2 - 3A_3^2)}{[(1 + A_3)^2 - A_1^2]} \right| \tag{11.39b}$$

We note, again, that the nonitalicized letters, such as S, are associated with area properties.

To apply eqs. 11.38 and 11.39 to a physical body, the expressions in the equations must be multiplied by a scale factor. To create the scale factor, we note that the ratio $2Y_{\max}/Z_{\max}$ must equal to the breadth-to-keel-depth ratio, B_ξ/d_ξ, of the desired strip. That is, using the results in eq. 11.36, we find $B_\xi/d_\xi = 2Y_{\max}/Z_{\max} = 2(1 + A_1 + A_3)/(1 - A_1 + A_3)$. By rearranging these equalities, the following expression for the scale factor is obtained:

$$C_{SF-a} = \frac{B_\xi}{2Y_{\max-a}} = \frac{B_\xi}{2a\left(1 + \dfrac{A_1}{a^2} + \dfrac{A_3}{a^4}\right)} = \frac{d_\xi}{Z_{\max-a}} = \frac{d_\xi}{a\left(1 - \dfrac{A_1}{a^2} + \dfrac{A_3}{a^4}\right)} \tag{11.40a}$$

Note that the value of the scale factor depends on our choice of the value of the radius a. Following Lewis (1929), we use the unit circle, $a = 1$. Although the resulting expression appears to be dimensional, it is not. The resulting scale factor, as used in the remainder of the chapter, is

$$C_{SF} \equiv C_{SF-1} = \frac{B_\xi}{2Y_{\max}} = \frac{B_\xi}{2(1 + A_1 + A_3)} = \frac{d_\xi}{Z_{\max}} = \frac{d_\xi}{(1 - A_1 + A_3)} \tag{11.40b}$$

EXAMPLE 11.3: LEWIS SHIP-SHAPE SECTION (STRIP) The shape in Figure 11.8b corresponds to Lewis-parameter values of $A_1 = 0$ and $A_3 = -0.111$; hence, the breadth-to-keel-depth ratio is $B_\xi/d_\xi = 2Y_{\max}/Z_{\max} = 2$, using the expressions in eq. 11.36. Our goal is to design a floating body having breadth of $B_\xi = 4$ m that is uniform from the bow to the stern. Because of the chosen values of A_1 and A_3, the keel-depth value is $d_\xi = 2$ m. The scale factor in eq. 11.40b is $C_{SF} = B_\xi/(1.778) = d_\xi/(0.889) \simeq 2.25$. Multiplying the expressions in eq. 11.38 by the scale factor, the following coordinate expressions of the Lewis form are obtained:

$$Y_\xi = \frac{B_\xi}{2Y_{\max}} \frac{Y_{\max}}{(1 - 0.111)} [\cos(\gamma) - 0.111\cos(3\gamma)]$$
$$= \frac{B_\xi}{1.778} [\cos(\gamma) - 0.111\cos(3\gamma)], \quad \pi \le \gamma \le 2\pi \tag{11.41a}$$

and

$$Z_\xi = \frac{d_\xi}{Z_{max}} \frac{Z_{max}}{(1-0.111)} [\sin(\gamma) + 0.111\sin(3\gamma)]$$
$$= \frac{d_\xi}{0.889} [\sin(\gamma) + 0.111\sin(3\gamma)], \quad \pi \le \gamma \le 2\pi \tag{11.41b}$$

The area of the strip (S_ξ), obtained by multiplying eq. 11.39 by the scale factor squared ($C^2{}_{SF}$), is approximately 7.66 m^2. This strip is sketched in Figure 11.8b.

We must note that the results of the analysis of von Kerczek and Tuck (1969) give us some caution as to the practicality of some of the geometries resulting from the method of Lewis (1929). In Appendix 2 of the von Kerczek-Tuck paper, the authors classify the Lewis forms as re-entrant, bulbous, tunneled-bulbous, and conventional. The last of these types include ship-shapes, such as those in Figures 11.6 through 11.8. However, the first three types are somewhat unconventional, at least in their application to the strip theory. The classification of Lewis forms by von Kerczek and Tuck (1969) is based on two non-dimensional geometrical groups. The first of these is the ratio of the half-breadth to keel depth, called here the *maxima coefficient*. This coefficient is

$$C_{max}(\xi) = \frac{B_\xi}{2d_\xi} = \frac{Y_{max}}{Z_{max}} = \frac{1+A_1+A_3}{1-A_1+A_3} \tag{11.42}$$

where the expressions for the maxima in eq. 11.36 are those resulting from the transformation of the unit circle (where $a = 1$). Note that von Kerczek and Tuck (1969) refer to the ratio in eq. 11.42 as the half-beam-to-draft ratio.

The second is the *sectional area coefficient*,

$$C_{area}(\xi) = \frac{S_\xi}{B_\xi d_\xi} = \frac{\pi}{4}\left[\frac{1-A_1^2-3A_3^2}{(1+A_3)^2-A_1^2}\right] \tag{11.43}$$

obtained from eq. 11.39. The boundaries of the conventional Lewis forms, in terms of the non-dimensional parameters in eqs. 11.42 and 11.43, are presented in Figure 11.9.

In Appendix III of the paper by Lewis (1929), the author derives the expression for the added mass of a section with "sharp bilges." This analysis involves a Schwarz (or Schwarz-Christoffel) transformation leading to body coordinates in terms of elliptic integrals. This type of transformation is used to analyze sectional shapes having sharp edges. For a rather thorough discussion of the Schwarz-Christoffel transformation, see the book by Ablowitz and Fokas (1997). The sharp-bilge analysis of Lewis (1929) is presented in Appendix F of this book.

Once the geometry of the strip has been defined, our attention is directed toward the complex velocity potential describing the flow adjacent to the strip. Following Korvin-Kroukovsky and Jacobs (1957), there are three velocity potentials of interest. Those represent the flow due to a passing wave, the flow induced by the body (strip) motions and, lastly, the wave-body interaction. The potentials are needed to determine the corresponding dynamic pressure distributions over the wetted surface of the strip. The potentials for these are presented in the next section.

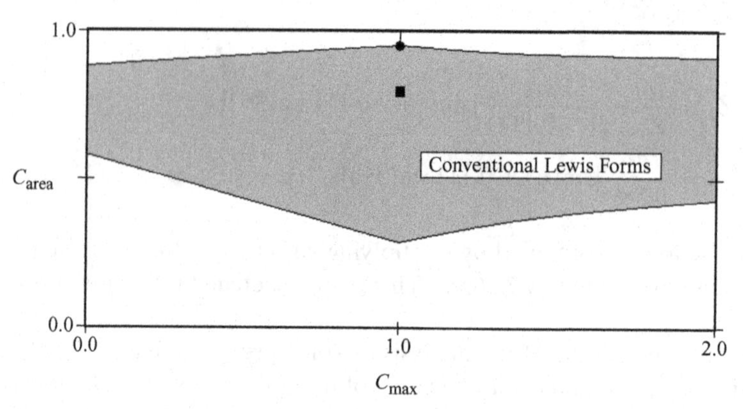

Figure 11.9. *Validity Range for the Maxima and Sectional-Area Coefficients for Lewis Forms.* The respective expressions for C_{max} and C_{area} are in eqs. 11.42 and 11.43. The two data points (• and ■) respectively correspond to the Lewis form in Example 11.3, where $C_{max} = 1$ and $C_{area} \simeq 0.957$, and to a semicircular strip, where $C_{max} = 1$ and $C_{area} \simeq \pi/4$. For the latter, $A_1 = A_3 = 0$. The validity range is due to von Kerczek and Tuck (1969).

B. Velocity Potentials

In hydrodynamic analyses, when the flow can be considered to be irrotational, then the defining variable is the velocity potential. As previously written, the velocity potential representing the flow induced by the combination of the wave field and the wave-induced body motions, according to Korvin-Kroukovsky and Jacobs (1957), consists of three components. These are the incident-wave potential, the body-motion induced potential, and the wave-body interaction potential. In the following subsections, these component potentials are discussed. By establishing the mathematical forms of these potentials, the pressure distribution on the strip and the corresponding vertical force can be determined.

(1) Incident Wave Potential

Referring to the sketch in Figure 11.6b, we consider the velocity potential describing the flow induced by a linear traveling wave at the strip located at ξ. From eq. 3.24, the free-surface displacement can be written as

$$\begin{aligned}\eta &= \frac{H}{2}\cos[k(x-ct)] = \frac{H}{2}\cos[k(\pm Ut + \xi - ct)] \\ &= \frac{H}{2}\cos[k\xi - (c \mp U)t] = \frac{H}{2}\cos(k\xi - \omega_e t)\end{aligned} \tag{11.44}$$

The velocity potential due to the passing wave is given in eq. 3.23. That potential at the strip can be written as

$$\Phi_w(\xi, Z, t) = \frac{g}{\omega_e}\frac{H}{2}\frac{\cosh[k(Z_\xi + h)]}{\cosh(kh)}\sin(k\xi - \omega_e t) \tag{11.45}$$

The origin of the x,y,z coordinate system is fixed in space on the calm-water surface. The origin of the ξ,Y,Z coordinate system is on the waterplane above the center of gravity, G, of the vessel and moves with either a forward speed (U) or backward speed ($-U$) in the x-direction. Also in eqs. 11.44 and 11.45 are the incident wave height (H), the wave number ($k = 2\pi/\lambda$), and the water depth (h).

In eqs. 11.44 and 11.45, the apparent wave frequency, called the circular *frequency of encounter*, is

$$\omega_e = \omega \mp kU = k(c \mp U) = \frac{2\pi}{\lambda}\left(\frac{\lambda}{T} \mp U\right) = \frac{2\pi}{T_e} \tag{11.46}$$

We see that this frequency depends on the velocity of the wave (the celerity, c) and the velocity (U) of the body. If the body is at rest, then the frequency is simply that of the wave, $\omega = kc$. When the body is traveling in the same direction as the wave, as sketched in Figure 11.6a, the condition is called a *following sea*. The frequency of encounter in this case is $\omega_e = 2\pi/T_e = \omega - kU = k(c - U)$, where T_e is the *period of encounter*. When the body and the wave travel in opposite directions, the frequency of encounter is $\omega_e = k(c + U)$, and the condition is called a *head sea*. By rearranging the terms in the frequency-of-encounter expression, we find $T_e = T/(1 \mp U/c)$. Finally, when the body and the wave have the same velocity, the period of encounter becomes infinite because the body simply rides with the wave. Numerical values of the frequency of encounter are found in Example 11.5. The frequency of encounter does not affect the geometric properties of the wave. In the paragraphs that follow, a following-sea condition is assumed.

The vertical component of the wave-induced velocity of the water particle at ξ (where $Z = Z_\xi$) is obtained from

$$V_{wz}(\xi, Z_\xi, t) = \frac{\partial \phi_w}{\partial Z_\xi} = k\frac{g}{\omega_e}\frac{H}{2}\frac{\sinh[k(Z_\xi + h)]}{\cosh(kh)}\sin(k\xi - \omega_e t) \tag{11.47}$$

Referring to Figure 11.6b, the vertical velocity component of the particles on the waterline at either side of the strip is assumed to be equal to the vertical velocity of the free surface, as in eq. 3.3. From this assumption, we can write

$$\begin{aligned} V_{wz}(\xi, 0, t) \equiv V_{w\eta} &= \frac{kg}{\omega_e}\frac{H}{2}\tanh(kh)\sin(k\xi - \omega_e t) = \frac{\omega^2}{\omega_e}\frac{H}{2}\sin(k\xi - \omega_e t) \\ &= \left(\frac{\omega}{\omega_e}\right)^2 \frac{\partial \eta}{\partial t} \end{aligned} \tag{11.48}$$

The third equality results from the application of the dispersion relationship in eq. 3.30. That is, the hyperbolic tangent has been replaced by ω^2/kg.

Following Korvin-Kroukovsky and Jacobs (1957), the expression for the velocity potential in eq. 11.45 is used to determine the wave-induced force on the strip. Furthermore, the velocity expression in eq. 11.47 is needed to determine the wave-structure interaction pressure and force. This force is analogous to the diffraction force discussed in Sections 9.2 and 10.4.

(2) Vertical Motions of Lewis Forms – Velocity Potential and Stream Function

From Chapter 9, the primary advantage of conformal mapping is that the form of the complex potential (w) is the same in both the ς-plane and the z-plane. Hence, when we transform the flow about a fixed circular cylinder of radius a in the ς-plane to the z-plane, the portion of the potential corresponding to the flow about the a-circle has the same form in both planes. Advantage is taken of this property in the following derivation.

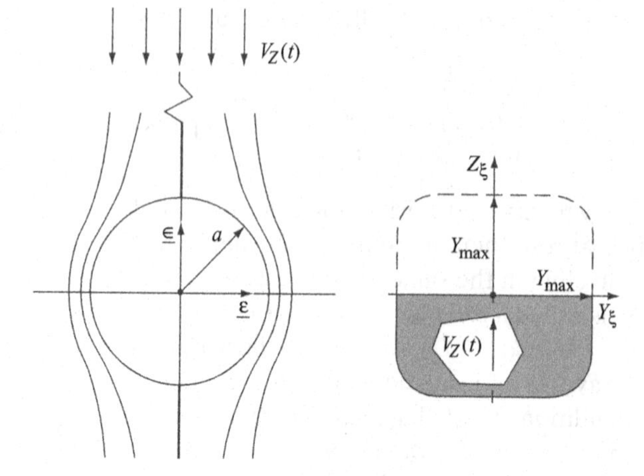

Figure 11.10. *Transformation of a Vertical Flow about the a-Circle in the ζ-plane to Vertical Flow about a Lewis Form in the z-plane.* The flow in the ζ-plane is rotated by eq. 11.49. Note that the subscript ξ signifies that the coordinates are in the z-plane of the strip, as in eq. 11.38.

To transform the flow about the a-circle in the ζ-plane to obtain the flow excited by the vertical motions of a strip in still water in the z-plane, an intermediate conformal relationship is needed. The focus of attention here is on the complex potential at any point in the fluid where $R \geq a$. Then, as sketched in Figure 11.10a, the flow about the a-circle in the ζ-plane and about the body in the z-plane are both rotated through an angle of $-\pi/2$. As in Figure 11.10b, the strip is moving in the Z_ξ (or z) direction with a scaled velocity $V_Z(t)$, that is, the velocity in Figures 11.7a and 11.7c is multiplied by the scale factor in eq. 11.40. Following Section 5.17 of the book by Milne-Thomson (1955), to rotate the flows in the ζ-plane and z-plane so that the body is traveling in the negative ϵ- and z-directions, respectively, we introduce the intermediate complex variables

$$\underline{\zeta} = Re^{i(\gamma + \frac{\pi}{2})} = i\zeta$$
$$\underline{z} = iz \qquad (11.49)$$

The plane in Figure 11.10a is called an *intermediate plane*. The relationship between this plane and the physical plane is similar to that in eq. 11.25, where ζ is replaced by $\underline{\zeta}$. Again, the flow is rotated through an angle of $-\pi/2$ to obtain flow in the $\underline{\zeta}$-plane. The analysis then proceeds as before.

In determining the velocity potential about vertically moving Lewis forms in a free surface, we use some of the aspects of the analysis of Landweber and Macagno (1957). Those investigators present analyses of both horizontally moving strips and vertically moving strips, as respectively sketched in Figures 11.7b and 11.7c. The Landweber-Macagno technique can be used to analyze the potential flow about bodies having other than Lewis forms. Landweber and Macagno (1957) begin by assuming that the relationship between z and ζ is the following infinite series:

$$z = Y + iZ = A_0\zeta + \frac{A_1}{\zeta} + \frac{A_2}{\zeta^2} + \frac{A_3}{\zeta^3} \cdots = A_0\zeta + \sum_{n=1}^{\infty} A_n\zeta^{-n} \qquad (11.50)$$

As in eq. 11.35, the coefficients, A_n, are real numbers. The series in eq. 11.50 represents a body having symmetry about the real axis. Bodies having double symmetry (about both the flow axis and the axis normal to the flow axis) are represented by expansions that include only the odd powers of ζ. The Lewis forms are included in this group. In the paragraphs that follow, we confine our attention to the analysis of the vertically moving strip having Lewis forms. Hence, the transformation described by eq. 11.35 is assumed. Horizontal (swaying) motions of strips having Lewis forms are discussed in Section 11.5.

In Chapter 9, eq. 9.37 is the complex velocity potential describing the flow field about a horizontally moving Lewis form. The form of the potential is similar to that representing the rotated flows in Figure 11.10. Again, the rotation is accomplished by using eq. 11.49. The complex potential for the vertically moving Lewis form in Figure 11.10b is then

$$\begin{aligned}\frac{w_Z}{C_{SF-a}} &= (\phi_Z + i\Psi_Z)\frac{1}{C_{SF-a}} = (w_{\underline{\zeta}} - iV_Z z)\\ &= \left[V_Z\left(i\zeta + \frac{a^2}{i\zeta}\right) - iV_Z\left(\zeta + \frac{A_1}{\zeta} + \frac{A_3}{\zeta^3}\right)\right]\Bigg|_{\zeta=-i\underline{\zeta}}\\ &= -V_Z\left[\frac{1}{R}(a^2 + A_1)\sin(\gamma) + \frac{A_3}{R^3}\sin(3\gamma)\right]\\ &\quad - iV_Z\left[\frac{1}{R}(a^2 + A_1)\cos(\gamma) + \frac{A_3}{R^3}\cos(3\gamma)\right]\end{aligned} \tag{11.51}$$

Concerning the second term in the second equality in this equation: The scaled velocity, $V_Z = V_Z(t)$, originally in the positive direction of the real axis in the z-plane, is rotated according to iV_Z. Also, the scale factor, $C_{SF\text{-}a}$, is that in eq. 11.40a. The results in eq. 11.51 are the same as those of Landweber and Macagno (1957). Those investigators take a somewhat different approach to obtain the complex velocity potential.

From eq. 11.51, the scaled velocity potential is

$$\phi_Z = -C_{SF-a}V_Z\left[\frac{1}{R}(a^2 + A_1)\sin(\gamma) + \frac{A_3}{R^3}\sin(3\gamma)\right] \tag{11.52}$$

We note that $\phi_Z = 0$ on the Y-axis, where $\gamma = 0$. This is a condition that is normally met by velocity potentials representing flows produced by bodies undergoing high-frequency vertical oscillations. The stream function in eq. 11.51 is

$$\psi_Z = -C_{SF-a}V_Z\left[\frac{1}{R}(a^2 + A_1)\cos(\gamma) + \frac{A_3}{R^3}\cos(3\gamma)\right] \tag{11.53}$$

These expressions contain the undesignated radius, a, for a vertically oscillating strip, $a = a(t)$. This time-dependency of the transformed radius is a factor in determining the dynamic pressure on the oscillating strip, discussed later in this chapter. For this reason, we delay the assumption that the transformed circle is a unit circle ($a = 1$) until the final form of the dynamic pressure is determined. In addition, when a is not designated, the scale factor in eq. 11.40a must be used.

To check the validity of the analysis, we apply the respective expressions for the velocity potential and stream function in eqs. 11.52 and 11.53 to a floating, horizontal circular cylinder in the following example.

EXAMPLE 11.4: VELOCITY POTENTIAL AND STREAM FUNCTION FOR A HEAVING CIRCULAR STRIP When the floating body of interest is a circular cylinder, the Lewis constants are $A_1 = A_3 = 0$. We shall transform the a-circle, where $R = a$. If D is the diameter of the cylinder, the scale factor in eq. 11.40a is $C_{SF-a} = D/2a$. The velocity potential in eq. 11.52 for the flow adjacent to the body is then

$$\phi_Z|_{R=a} = -V_Z(t)\frac{D}{2}\sin(\gamma), \quad \pi \le \gamma \le 2\pi \tag{11.54}$$

From eq. 11.52, the corresponding stream function for this flow is

$$\psi_Z|_{R=a=1} = -V_Z(t)\frac{D}{2}\cos(\gamma), \quad \pi \le \gamma \le 2\pi \tag{11.55}$$

Compare these respective expressions with those in eqs. 9.18 and 9.19, where in the Chapter 9 equations, $r = a = D/2$. We find that the respective equations are not identical because the body motion for the cylinder in this example is in the positive imaginary-axis direction, whereas in Figure 9.5 the body motion is in the negative real-axis direction. Note that the results in eqs. 11.54 and 11.55 are also obtained when $a = 1$.

(3) Velocity Potential for the Wave-Body Interaction

The interaction of the motions of the wave and the body motions can be mathematically represented by the wave-body interaction potential, ϕ_{wZ}. In three-dimensional analyses, this is analogous to the diffraction potential. In two dimensions, we have chosen not to use this terminology because diffraction involves wave energy being transferred along the crest, which is a three-dimensional phenomenon. See Section 6.4 for a discussion of wave diffraction.

Following Korvin-Kroukovsky and Jacobs (1957), the velocity potential expression describing the wave-body interaction is based on the form of eq. 11.51. In place of the strip velocity $V_Z(t)$, the vertical component of the wave-induced particle velocity, $-V_w(\xi, Z_\xi, t)$, is used, which is obtained from eq. 11.47. The resulting expression for the wave-body interaction potential at the strip is

$$\begin{aligned}\phi_{wZ} &= V_w(t)C_{SF-a}\left[\frac{1}{R}(a^2 + A_1)\sin(\gamma) + \frac{A_3}{R^3}\sin(3\gamma)\right]\\ &= k\frac{g}{\omega_e}\frac{H}{2}\frac{\sinh[k(Z_\xi + h)]}{\cosh(kh)}\sin(k\xi - \omega_e t)\cdot\\ &\quad C_{SF-a}\left[\frac{1}{R}(a^2 + A_1)\sin(\gamma) + \frac{A_3}{R^3}\sin(3\gamma)\right], \quad \pi \le \gamma \le 2\pi\end{aligned} \tag{11.56}$$

Physically, this potential describes the flow field excited by the strip traveling in the negative Z_ξ direction with a velocity equal to the vertical particle velocity in a wave.

(4) Total Velocity Potential

The total potential describing the flow about the strip at a distance ξ from the center of gravity, including the free surface, is

$$\phi_\xi = \phi_w + \phi_Z + \phi_{wZ} \tag{11.57}$$

The potentials on the right side of the equation are given in eqs. 11.45, 11.52, and 11.56, respectively.

In the following section, the potential in eq. 11.57 is used to determine the dynamic pressure distributions on the strip and the resulting vertical forces.

C. Hydrodynamic Pressures and Forces on the Strip in Deep Water

The dynamic pressure on the strip is found by combining the total velocity potential in eq. 11.57 with the linearized Bernoulli's equation, eq. 3.70. This pressure, called the *hydrodynamic pressure*, is due to the combination of the flow adjacent to the strip, the relative motions of the strip and water particles, and the strip motions. At any point on the strip surface, the resulting hydrodynamic pressure is

$$p_\xi = (p_w + p_Z + p_{wZ})|_{R=a=1} = -\rho \frac{\partial \phi_\xi}{\partial t}\bigg|_{R=a=1} = -\rho \frac{\partial}{\partial t}(\phi_w + \phi_Z + \phi_{wZ})|_{R=a=1} \tag{11.58}$$

Our interest is in both the pressure distribution on the strip and the resulting forces on the strip. To obtain these, we consider separately the contributions of each of the velocity potentials in eq. 11.58. Note that the partial time-derivatives of body-motion potential, ϕ_Z, and wave-body interaction potential, ϕ_{wZ}, are taken prior to the application of the pressure on the surface, as is discussed by Korvin-Kroukovsky and Jacobs (1957). The reason for this is that the radius (a) of the transformed circle in the ζ-plane is a function of time because the strip in the z-plane moves with respect to the physical axes, that is, in the ζ-plane, $a = a(t)$. This is not the case for the potential, ϕ_w, in the wave-induced pressure expression because the strip is assumed to be fixed in the passing wave.

(1) Wave-Induced Pressure and Force

The pressure distribution on the edge of the fixed strip [where $a \neq a(t)$] due to the passing incident wave is obtained from

$$\begin{aligned} p_w(\xi, Z_\xi, t)|_{R=a} &= -\rho \frac{\partial \phi_w(\xi, Z_\xi, t)}{\partial t}\bigg|_{R=a} = \rho g \frac{H}{2} \frac{\cosh[k(Z_\xi + h)]}{\cosh(kh)}\bigg|_{R=a} \cos(k\xi - \omega_e t) \\ &= \rho g \frac{H}{2} \left\{ 1 + \frac{1}{2!}[k(Z_\xi + h)]^2 + \frac{1}{4!}[k(Z_\xi + h)]^4 + \cdots \right\}\bigg|_{R=a} \\ &\times \frac{\cos(k\xi - \omega_e t)}{\cosh(kh)} \end{aligned} \tag{11.59}$$

In the last equality, the frequency of encounter, ω_e, has been replaced by the following-sea equivalent $(\omega - kU)$ found in eq. 11.46. The series representation of the hyperbolic-cosine ratio in the second line of this equation is introduced to facilitate the integration required to determine the vertical force on the strip. From this point on, we shall follow Korvin-Kroukovsky and Jacobs (1957) by assuming that the floating body is in deep water. The ratio of the hyperbolic functions in deep water is approximated by

$$\begin{aligned} \frac{\cosh[k(Z_\xi + h)]}{\cosh(kh)}\bigg|_{h\to\infty} &\simeq \frac{\sinh[k(Z_\xi + h)]}{\cosh(kh)}\bigg|_{h\to\infty} \\ &\simeq e^{k_0 Z_\xi} = 1 + \frac{1}{1!}(k_0 Z_\xi) + \frac{1}{2!}(k_0 Z_\xi)^2 + \cdots \end{aligned} \tag{11.60}$$

The wave-induced pressure distribution over the edge of the strip in deep water is then obtained from

$$\begin{aligned} p_w|_{\substack{h\to\infty \\ R=a}} &\simeq \rho g \frac{H_0}{2} e^{k_0 Z_\xi|_{R=a}} \cos(k_0\xi - \omega_e t) = \rho g \sum_{j=0}^{\infty} \frac{(k_0 Z_\xi)^j}{j!}\bigg|_{R=a} \frac{H}{2}\cos(k_0\xi - \omega_e t) \\ &= \rho g \eta_0(\xi, \omega_e, t) + \sum_{j=1}^{\infty} \frac{(k_0 Z_\xi)^j}{j!}\bigg|_{R=a} \eta_0(\xi, \omega_e, t) \end{aligned} \tag{11.61}$$

where $Z_\xi|_{R=a}$ is the vertical dimensional coordinate of the strip surface in the plane of the strip. The notation for the respective deep-water wave number and deep-water wave height are $k_0(= 2\pi/\lambda_0)$ and H_0, which are used throughout this book. In the second line of eq. 11.61, we see that the first term is simply the hydrostatic pressure due to the deep-water free-surface displacement, $\eta_0(\xi,\omega_e,t)$. As noted by Korvin-Kroukovsky and Jacobs (1957) and Motora (1964), the last term is a modification to the hydrostatic pressure due to the exponential variation of the particle motions with depth. This is known as the *Smith effect*.

The coordinates of the Lewis form are obtained by applying eq. 11.38 at $R = a = 1$ and multiplying the result by the scale factor in eq. 11.40b. The resulting scaled coordinate expressions are

$$Y_\xi|_{a=1} = C_{SF}[(1+A_1)\cos(\gamma) + A_3\cos(3\gamma)] = \frac{B_\xi}{2}\left[\frac{(1+A_1)\cos(\gamma) + A_3\cos(3\gamma)}{1+A_1+A_3}\right], \quad \pi \le \gamma \le 2\pi \tag{11.62a}$$

and

$$Y_\xi|_{a=1} = C_{SF}[(1-A_1)\sin(\gamma) - A_3\sin(3\gamma)] = T_\xi\left[\frac{(1-A_1)\sin(\gamma) - A_3\sin(3\gamma)}{1-A_1+A_3}\right], \quad \pi \le \gamma \le 2\pi \tag{11.62b}$$

Again, the strip is assumed to be rigidly fixed in the passing wave, so these coordinates are not time-dependent as far as the passing wave is concerned. The expression for the scale factor, C_{SF}, is in eq. 11.40b.

Referring to Figure 11.11, where s is the coordinate alongside the strip, the vertical component of the wave-induced force on the strip is

$$F'_{wz}(\xi,\omega_e,t) \equiv \frac{dF_{wz}}{d\xi} = \int p_w|_{R=a=1}\cos(\alpha)ds = \int_{-\frac{B_\xi}{2}}^{\frac{B_\xi}{2}} p_w|_{R=a=1} dY_\xi|_{R=a=1} = \int_{\pi}^{2\pi} p_w|_{R=a=1}\frac{dY_{\xi a}}{d\gamma}d\gamma|_{R=a=1} \tag{11.63a}$$

The combination of this expression with those in eqs. 11.61 and 11.62b results in the following expression for the vertical wave-induced force per unit length in deep water:

$$\begin{aligned} F'_{wz}(\xi,\omega_e,t)|_{h\to\infty} &= \rho g\frac{H_0}{2}C_{SF}\int_{\pi}^{2\pi}\left[1 + \frac{1}{1!}(k_0 Z_{\xi a}) + \frac{1}{2!}(k_0 Z_{\xi a})^2 + \cdots\right] \\ &\quad \cdot [(1+A_1)\sin(\gamma) + 3A_3\sin(3\gamma)]\,d\gamma\cos(k_0\xi - \omega_e t) \\ &= \rho g\frac{H_0}{2}B_\xi\cos(k_0\xi - \omega_e t)\cdot\left\{1 - \frac{\pi k_0 d_\xi}{4}\left[\frac{1 - A_1^2 - 3A_3^2}{1 - A_1^2 + A_3^2}\right]\right. \\ &\quad + \frac{(k_0 d_\xi)^2}{(1 - A_1^2 + A_3^2)(1 - A_1 + A_3)}\cdot\left[\left[\frac{1}{3}[(1-A_1^2)(1-A_1) + A_3^3]\right.\right. \\ &\quad \left.\left.\left. - \frac{1}{15}A_3(1 - 6A_1 + 5A_1^2) - \frac{9}{35}A_3^2(5 - 7A_1)\right]\right]\cdots\right\} \\ &= \rho g B_\xi \eta_0(1 + C_{Smith}) \end{aligned} \tag{11.63b}$$

The *Smith correction factor*, C_{Smith}, results from the grouping of the frequency-dependent (wave number dependents) terms in the expansion. The Smith effect

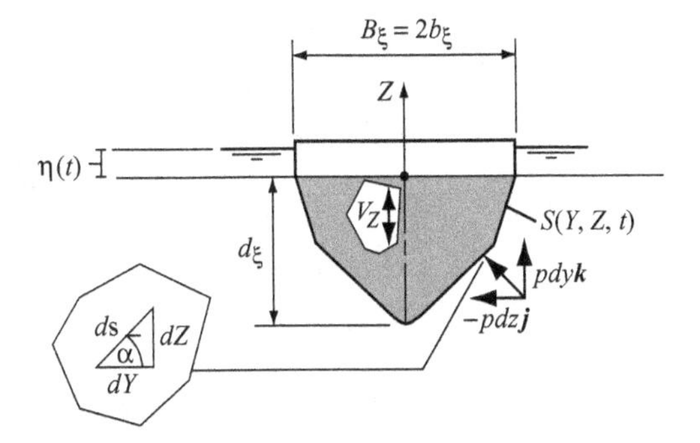

Figure 11.11. *Notation for the Force on a Vertically Oscillating Strip.*

and the Smith correction factor are discussed by Bhattacharyya (1978), Bishop and Price (1979), Jensen (2001), and others. As previously mentioned, the hydrostatic pressure on the strip is modified by the Smith effect to account for the variation of the pressure with depth position on the strip. The wave-induced force in eq. 11.63b is based on the assumption that the wave field is not affected by the presence of the body. This is known as the *Froude-Krilov hypothesis*, and the force is called the *Froude-Krilov force.* The hypothesis is named after two nineteenth-century investigators, W. Froude in England and A. Krylov in Russia. In some Western writings, the last name of the latter investigator is spelled either "Krilov" or "Kryloff." Both Froude and Krylov concluded that if the body dimensions are small when compared to the wavelength, then the pressure field of the wave is not affected by the presence of the body. The works of these pioneer researchers and others are described by S. N. Blagoveshchensky. The book by Blagoveshchensky was first published in Russian in 1954. In 1962, a translation of the book by L. Landweber was published (see Blagoveshchensky, 1962). Finally, we note that because the potential ϕ_{wZ} is included in the analysis, the Froude-Krilov assumption concerning the body-to-wavelength size relationship need not be made.

The expression in eq. 11.63b is now combined with eqs. 11.26 and 11.27 to obtain the respective wave-induced force and moment on the floating body. To perform the integrations in eqs. 11.26 and 11.27, the functional relationship between the keel draft (d_ξ) and the distance (ξ) from the center of gravity of the body must be specified, assuming that the values of the Lewis constants, A_1 and A_3, are uniform over the length of the body, that is, neither A_1 nor A_3 are functions of ξ. In eq. 11.63b, we follow Korvin-Kroukovsky and Jacobs (1957), and retain only those terms containing $(k_0 d_\xi)^n$ where $n \leq 2$ for all of the forces concerned. For application to a semicircular strip, those authors show good agreement with experimental data with this assumption.

In the next example, the wave-induced vertical force in eq. 11.63b is applied to both a semicircular strip (where $A_1 = A_1 = 0$) and a strip having the shape of that in Figure 11.8 (where $A_1 = 0$ and $A_3 < 0$). Our choice of the semicircular shape to illustrate the applicability of the equations is due to the fact that this strip geometry is used by Korvin-Kroukovsky and Jacobs (1957), Motora (1964), and others in their strip-theory derivations.

EXAMPLE 11.5: WAVE-INDUCED VERTICAL FORCES ON TWO STRIP GEOMETRIES An ocean engineer has been tasked to design a long, stiffened oil bladder for deep-water operation. Two sectional areas are to be compared for this task, those being a Lewis form strip in Figure 11.8b and a semicircular strip in Figure 11.12. The design values for the Lewis form are given in Figure 11.8b. The two strips

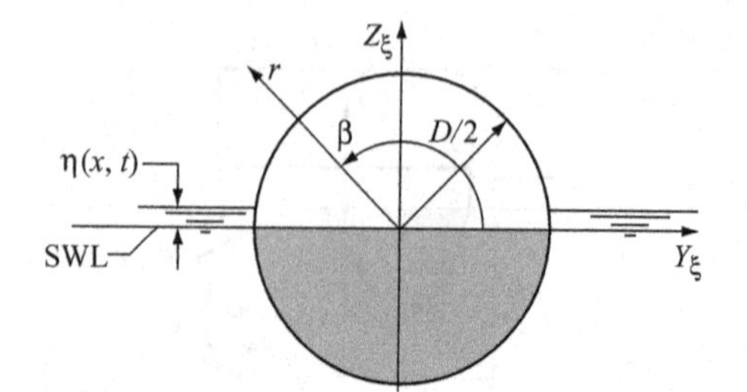

Figure 11.12. *Notation for a Semicircular Strip in a Passing Wave.*

are to displace the same water volume, hence, the strip areas of the two must be approximately equal to 7.66 m^2. This value is obtained from eq. 11.39. Referring to the sketch in Figure 11.12, the diameter (D) of the semicircular strip corresponding to the sectional area is approximately 4.42 m. The operational design sea is one having a wave height (H_0) of 1.5 m and a period (T) of 7 sec in deep water. For this wave-period value, the deep-water wavelength (λ_0) is approximately 76.5 m, and the corresponding wave number (k_0) is approximately 0.0821 m^{-1} (see eq. 3.31). The design towing speed (U) of the bladder is 1 m/s (3.28 ft/s $\simeq$ 1.94 knots). The respective periods of encounter for head-sea and following-sea conditions are approximately 6.41 sec and 7.70 sec from eq. 11.46.

Our interest here is in the maximum wave-induced Froude-Krilov force on the strips when the strips are fixed. For the semicircular strip, the Lewis parameters are $A_1 = A_3 = 0$ in the wave-induced force expression in eq. 11.63b. The expression for the wave-induced vertical force on the semicircular strip from eq. 11.63b is approximated by

$$F'_{wz}(\xi, \omega_e, t)|_{cir} \simeq \rho g D \eta_0(\xi, \omega_e, t)\left[1 - \frac{\pi k_0 D}{8} + \frac{(k_0 D)^2}{12}\right] \tag{11.64}$$

where the salt water mass density (ρ) is approximately 1,030 kg/m^3. For the chosen Lewis form in Figure 11.8b, $A_1 = 0$ and $A_3 = -0.111$ in eq. 11.63b, and the load-waterline breadth and draft are $B_\xi = 4\,\text{m}$ and $d_\xi = 2\,\text{m}$, respectively. That force equation is then approximately

$$F'_{wz}(\xi, \omega_e, t)|_{h\to\infty} = \rho g B_\xi \eta_0\left[1 - \frac{\pi k_0 d_\xi}{4}\left(\frac{(1-3A_3^2)}{1+A_3^2}\right) + \frac{(k_0 d_\xi)^2}{21}\frac{(7-20A_3^3)}{(1+A_3^2)(1+A_3)}\cdots\right] \tag{11.65}$$

In both force equations, the series in the brackets are truncated at the second-degree terms, as is done by Korvin-Kroukovsky and Jacobs (1957). The maximum values for each occur when the cosine term equals 1, or when a crest passes ($\eta_0 = H_0/2$), as expected. The peak forces on the semicircular strip at the passing of a wave crest are approximately 29,090 N/m. For the Lewis form, the peak forces are approximately 26,900 N/m. The respective hydrostatic forces on the semicircle and Lewis form (from the first terms in the brackets of eqs. 11.64 and 11.65) are 33,500 N/m and 30,300 N/m. From these results, we conclude then that the Smith correction factor reduces the predicted wave-induced hydrostatic force on a strip.

One final note: The maximum and hydrostatic force values are based on the assumption that the bodies are approximately *wall-sided* at the waterline, that is, the sides of the strip at the waterline are vertical.

Having derived the wave-induced pressure on a strip, we can now write the general expressions for the total wave-induced force and moment in eqs. 11.26 and 11.27, respectively. These depend on the variations with ξ of the breadth of the body at the waterplane $[B_\xi = B_\xi(\xi)]$ and the keel depth $[d_\xi = d_\xi(\xi)]$. Both of these must be specified. Special cases are dealt with later in the chapter.

(2) Motion-Induced Pressure and Force

The velocity potential representing the flow about the vertically moving, Lewis-form strip in calm water is presented in eq. 11.52. That potential is now combined with the corresponding dynamic pressure component of eq. 11.58 to obtain the dynamic pressure due to the vertical body motion. In obtaining this pressure, we must account for the time dependence of a. That is, because the submerged area of the strip varies in time, the radius of the transformed a-circle must also vary. Noting then that $a = a(t)$, the time-derivative of the motion-induced velocity potential in eq. 11.52 is

$$\frac{\partial \phi_Z}{\partial t} = -C_{SF-a}\left\{\frac{\partial V_Z}{\partial t}\left[\frac{1}{R}(a^2 + A_1)\sin(\gamma) + \frac{A_3}{R^3}\sin(3\gamma)\right] + V_Z\frac{2a}{R}\frac{da}{dt}\sin(\gamma)\right\},\quad \pi \le \gamma \le 2\pi \qquad (11.66)$$

where the scale factor, $C_{SF\text{-}a}$, is defined in eq. 11.40a. The derivation of the time-derivative of a is as follows: Following Korvin-Kroukovsky and Jacobs (1957), we express the derivative in terms of the resting sectional area ($S_{\xi\text{-}a}$) presented in eq. 11.39. The first term of the third line of that equation is used here. In terms of S_ξ, the time-derivative of a is

$$\begin{aligned}\frac{da}{dt} &= \frac{da}{d\xi}\frac{d\xi}{dt} = -\tan(\Theta)U \\ &= \left(\frac{da}{dS_{\xi-a}}\right)\left(\frac{dS_{\xi-a}}{d\xi}\right)\left(\frac{d\xi}{dt}\right) = \frac{1}{\pi C_{SF-a}^2 a\left(1 + \dfrac{A_1^2}{a^4} + 9\dfrac{A_3^2}{a^8}\right)}\left(\frac{dS_{\xi-a}}{d\xi}\right)(-U)\end{aligned} \qquad (11.67)$$

In this expression, $da/dS_{\xi-a}$ is obtained from eq. 11.39, and the angle Θ is that between the bottom (keel) and the longitudinal (x) direction. The sectional area at each ξ-value has the same form but different dimensions. Hence the Lewis parameters (A_1 and A_3) are independent of ξ but a is not. We note that the time-derivative of the radius, a, vanishes when either the sectional area (strip area) is uniform from the bow to the stern or the body is not under way. To obtain the derivative $da/dS_{\xi-a}$, write the sectional area expression in eq. 11.39 incorporating the scaling constant in eq. 11.40a. The resulting area expression is

$$S_{\xi-a} = \left|-\frac{\pi}{2}C_{SF-a}^2 a^2\left(1 - \frac{A_1^2}{a^4} - 3\frac{A_3^2}{a^8}\right)\right| \qquad (11.68)$$

It is important to remember that the scale factor, $C_{SF\text{-}a}$, is constant in both time and space. Hence, the a appearing in eq. 11.40a is the design value that is independent of time. After taking the derivative, $da/dS_{\xi-a}$, the unit value of a can be applied. The derivative $dS_{\xi-a}/d\xi$ in eq. 11.67 is a design parameter, as is illustrated in the following example.

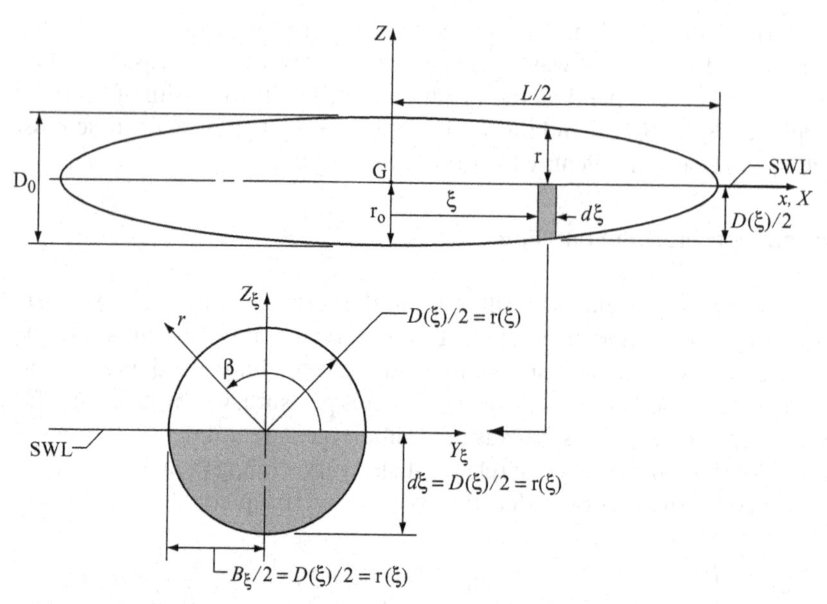

Figure 11.13. *Sketch of the Oil Bladder in Example 11.6.*

EXAMPLE 11.6: SPATIAL VARIATION OF A SEMICIRCULAR SECTIONAL AREA From the results in Example 11.5, the ribbed oil bladder having a circular-cylinder section has been selected. This decision is based on fabrication rather than the magnitude of the wave-induced force. Referring to the sketches in Figure 11.13, the design dimensions of the semi-submerged body at any position ξ are $B_\xi = D(\xi) = 2\mathrm{r}(\xi)$ at the waterplane and $d_\xi = \frac{1}{2}D(\xi) = \mathrm{r}(\xi)$ on the centerplane, where the diameter of any section is $D(\xi)$ and the radius is $\mathrm{r}(\xi)$. The maximum strip radius is $\mathrm{r_o}$ at $\xi = 0$, and the length of the floating bladder is L. The bladder is ballasted such that the center of gravity is at the center of the waterplane. As stated earlier in this section, the circular cylinder is represented by the Lewis form equations when $A_1 = A_3 = 0$. As sketched in Figure 11.13, the design profile in the vertical centerplane is elliptic. The body is symmetric with respect to the x-axis. For this geometry, the strip area (sectional area) at ξ is obtained from

$$S_{\xi-a}(\xi) = \left|-\frac{\pi}{2}C_{SF-a}^2 a^2\right| = \frac{\pi \mathrm{r}^2}{2} = \frac{\pi}{2}\mathrm{r}_0^2\left(1 - 4\frac{\xi^2}{L^2}\right) \tag{11.69}$$

The spatial derivative of the area is then

$$\frac{dS_{\xi-cir}}{d\xi} = \pi \mathrm{r}\frac{dr}{d\xi} = \pi \mathrm{r}\tan(\Theta) = -4\pi\left(\frac{\mathrm{r}_0}{L}\right)^2 \xi \tag{11.70}$$

In this equation, the trim angle, Θ, is that between the longitudinal tangent to the strip side and the x-axis, as in Figure 11.14. In that figure, the trim angle is shown at the centerplane of the body. The results in eq. 11.70 can now be combined with eq. 11.67 to obtain the time-derivative of the radius of the a-circle in the ζ-plane. For the body having semicircular strips, where $A_1 = A_3 = 0$, the result is

$$\left.\frac{da}{dt}\right|_{cir} = \left(\frac{da}{dS_{\xi-cir}}\right)\left(\frac{dS_{\xi-cir}}{d\xi}\right)\left(\frac{d\xi}{dt}\right) = \frac{1}{\pi C_{SF-a}\mathrm{r}}(\pi \mathrm{r}\tan(\Theta))(-U) \tag{11.71}$$

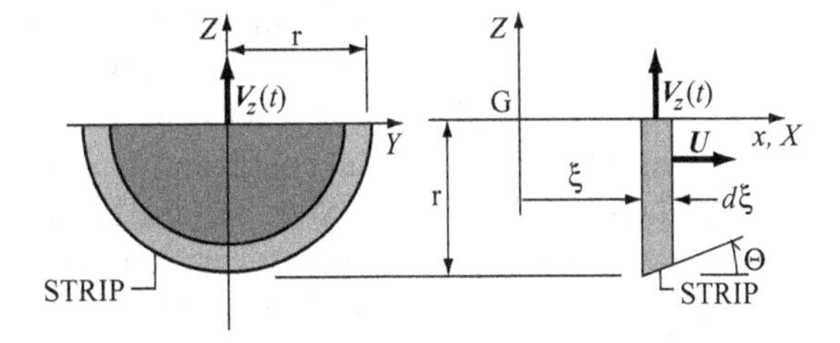

Figure 11.14. *Notation for a Strip Having Longitudinal Curvature.* The trim angle (Θ) in (b) is the angle between the "keel" and the longitudinal direction.

for the semicircular strip, $C_{SF-a}a = \text{r}$. Multiplying eq. 11.71 by $C_{SF\text{-}a}$, and incorporating the results in eq. 11.70, we obtain

$$\frac{d\text{r}}{dt} = \frac{d\text{r}}{d\xi}\frac{d\xi}{dt} = -U\tan(\Theta) = 4U\frac{\text{r}_0}{L^2}\frac{\xi}{\sqrt{1-4\dfrac{\xi^2}{L^2}}} \tag{11.72}$$

The relationship between the first and third terms in this equation are those obtained by Korvin-Kroukovsky and Jacobs (1957). Again, the time-derivative of a vanishes when the vessel either has a horizontal keel from stem to stern ($\Theta = 0$) or when the body is not under way ($U = 0$).

The dynamic pressure resulting from the body motion in deep, still water is

$$\begin{aligned}
p_Z\big|_{\substack{h\to\infty\\R=a=1}}
&= -\rho\frac{\partial\phi_Z(\xi,t)}{\partial t}\Big|_{\substack{h\to\infty\\R=a=1}}\\
&= \rho C_{SF-a}\left\{\frac{\partial V_Z}{\partial t}\left[\frac{1}{R}(a^2+A_1)\sin(\gamma)+\frac{A_3}{R^3}\sin(3\gamma)\right]+V_Z\frac{2a}{R}\frac{da}{dt}\sin(\gamma)\right\}\Bigg|_{R=a=1}\\
&= \rho\left\{\frac{dV_Z}{dt}C_{SF}[(1+A_1)\sin(\gamma)+A_3\sin(3\gamma)]\right.\\
&\quad\left.-2UV_Z\left[\frac{1}{\pi C_{SF-1}(1+A_1^2+9A_3^2)}\left(\frac{dS_{\xi-1}}{d\xi}\right)\right]\sin(\gamma)\right\},\quad \pi\le\gamma\le 2\pi
\end{aligned} \tag{11.73}$$

This expression results from the combination of eqs. 11.66 and 11.67. The spatial derivative of the sectional area ($dS_{\xi-a}/d\xi$) is a design parameter for which $a = 1$, and the scale factor ($C_{SF} \equiv C_{SF-1}$) is from eq. 11.40b. To obtain the vertical force due to the vertical motions of the strip, integrate the expression in eq. 11.73 over the angle γ from π to 2π as

$$\begin{aligned}
F'_Z &\equiv \frac{dF_Z}{d\xi}\\
&= \int_{\pi}^{2\pi} p_Z\frac{dY_{\xi a}}{d\gamma}d\gamma\Big|_{\substack{h\to\infty\\R=a=1}} = -\rho\int_{\pi}^{2\pi}\frac{\partial\phi_Z(\xi,t)}{\partial t}\frac{dY_{\xi a}}{d\gamma}d\gamma\Big|_{\substack{h\to\infty\\R=a=1}}\\
&= \rho\frac{\pi}{2}C_{SF}^2\left\{[(1+A_1)^2+3A_3^2]\frac{\partial V_Z}{\partial t}-2U\tan(\Theta)(1+A_1)V_Z\right\}\\
&= a'_{aZ}\left[\frac{d^2Z}{dt^2}+\xi\frac{d^2\theta}{dt^2}-2U\frac{d\theta}{dt}\right]-\rho\pi C_{SF}^2U\tan(\Theta)(1+A_1)\left[\frac{dZ}{dt}+\xi\frac{d\theta}{dt}-U\theta\right]
\end{aligned} \tag{11.74}$$

In the second and third lines, the vertical velocity (V_Z) and acceleration (dV_Z/dt) are written in terms of the heaving and pitching displacements according to eqs. 11.18 and 11.19, respectively. The coefficient of the first term in the last line of eq. 11.74 is an inertial coefficient, and that of the second term is a velocity-induced damping coefficient. Note that the term $(1 + A_1)$ is $(a + A_1/a)$, where $a = 1$. Hence, the units of the damping term are force per length, as expected.

Because the first term in the last two lines in eq. 11.74 is an inertial-reaction force, a'_{wZ} is the vertical motion-induced added mass per unit body length. Comparing the last two lines of the equation, we see that the added-mass expression for a Lewis form is

$$a'_{wZ}(\xi) = \rho\frac{\pi}{2}C_{SF}^2\left[(1 + A_1)^2 + 3A_3^2\right] = \rho\pi\frac{B_\xi d_\xi}{4}\frac{\left[(1 + A_1)^2 + 3A_3^2\right]}{\left[(1 + A_3)^2 - A_1^2\right]} \tag{11.75}$$

The relationships between the scale factor (C_{SF}), the breadth of the strip (B_ξ), and the keel depth (d_ξ) are found in eq. 11.40b. The added-mass expression is that obtained by Landweber and Macagno (1957). Note that the added-mass expression in eq. 11.75 is not a function of the frequency of encounter (ω_e). As stated by Landweber and Macagno (1957) and others, this added-mass condition corresponds to an infinite motion frequency or a zero motion period. The expression in eq. 11.75 is used as an approximation for high-frequency motions. The frequency-dependent added-mass expressions are discussed later in this chapter.

We are operating under the assumption that the Lewis parameters (A_1 and A_3 in eq. 11.35) do not vary with the hull coordinate, ξ. We make this assumption for the purpose of demonstration. The reader can see that a hull form can be made up of different Lewis sectional forms, where A_1 and A_3 vary with ξ. The calculated component force on each strip is numerically integrated over the hull length to obtain the total force component on the hull.

EXAMPLE 11.7: MOTION-INDUCED VERTICAL FORCES ON TWO STRIP GEOMETRIES
Our goal is to determine the motion-induced forces on the circular cylinder and the Lewis form in Example 11.5. See Figures 11.12 and 11.8b, respectively, for these shapes. The body motions will have the same period as the incident wave in Example 11.5, that is, the period of motion is 7 sec. The diameter of the semicircular strip is $D = 4.42$ m, and the breadth and keel depth of the Lewis form are $B_\xi = 4$ m and $d_\xi = 2$ m, respectively. The relative water speed (U) in the x-direction is 1 m/s which is the design towing speed. The sectional (strip) areas for each geometry are independent of ξ. As a result, $\Theta = 0$ from stem to stern.

For the semicircle, the Lewis parametric values are $A_1 = A_3 = 0$. These values combined with the motion-induced force expression in eq. 11.74 results in

$$\begin{aligned}
\frac{dF_Z}{d\xi}\Big|_{cir} &= \rho\pi\frac{D^2}{8}\frac{\partial V_Z}{\partial t} - \rho\pi\frac{D^2}{4}U\tan(\Theta)V_Z(t)\Big|_{\Theta=0} \\
&= \rho\pi\frac{D^2}{8}\left[\frac{d^2Z}{dt^2} + \xi\frac{d^2\theta}{dt^2} - 2U\frac{d\theta}{dt}\right] - \rho\pi\frac{D^2}{4}U\tan(\Theta)\left[\frac{dZ}{dt} + \xi\frac{d\theta}{dt} - U\theta\right]\Bigg|_{\Theta=0} \\
&\simeq 7.90\times 10^3\left[\frac{d^2Z}{dt^2} + \xi\frac{D^2\theta}{dt^2} - 2U\frac{d\theta}{dt}\right]
\end{aligned} \tag{11.76}$$

in Newtons per meter. The second line is the expression derived by Korvin-Kroukovsky and Jacobs (1957). In this equation, the scale factor is $C_{SF}|_{cir} = D/2 \simeq 2.21$ from eq. 11.40b. Note that the scale factor is non-dimensional because it is the ratio of the radius of the strip to the radius ($a = 1$) of the unit circle. The added mass (per unit length) for the semicircular strip is approximately 7.90×10^3 N-s^2/m/m. This value is obtained from eq. 11.75, where $A_1 = 0$, $A_3 = 0$, $C_{SF}|_{cir} = D/2$, $B_\xi = D$ and $d_\xi = D/2$.

The respective velocity and acceleration terms result from eqs. 11.18 and 11.19. For the Lewis form, $A_1 = 0$ and $A_3 = -0.111$, and the scale factor is $C_{SF}|_{Lew} = B_\xi/[2(1 + A_3)] = d_\xi/(1 + A_3) \simeq 2.25$. The force expression in eq. 11.73 is

$$\begin{aligned}\left.\frac{dF_Z}{d\xi}\right|_{Lew} &\simeq 0.328\rho\pi B_\xi d_\xi \frac{\partial V_Z}{\partial t} - 0.633\rho\pi B_\xi d_\xi U\tan(\Theta)V_Z|_{\Theta=0}\\ &= 0.328\rho\pi B_\xi d_\xi\left[\frac{d^2Z}{dt^2} + \xi\frac{d^2\theta}{dt^2} - 2U\frac{d\theta}{dt}\right]\\ &\quad - 0.633\rho\pi B_\xi d_\xi U\tan(\Theta)\left.\left[\frac{dZ}{dt} + \xi\frac{d\theta}{dt} - U\theta\right]\right|_{\Theta=0}\\ &\simeq 8.49\times 10^3\left[\frac{d^2Z}{dt^2} + \xi\frac{d^2\theta}{dt^2} - 2U\frac{d\theta}{dt}\right]\end{aligned} \tag{11.77}$$

in N/m, assuming $\rho \simeq 1{,}030\,\text{kg/m}^3$ for salt water. The added mass (per unit length) for the Lewis form is seen to approximately be 8.49×10^3 N–s^2/m/m. Comparing the results in eqs. 11.76 and 11.77, we see that the value of the motion-induced added mass per unit body length is greater for the Lewis form.

(3) Wave-Body Interaction Pressure and Force on a Strip

The velocity potential representing the flow resulting from the wave-body interaction at the strip is found in eq. 11.56. The dynamic pressure expression resulting from the flow is

$$\begin{aligned}p_{wz}|_{R=a} &= -\rho\left.\frac{\partial\phi_{wz}}{\partial t}\right|_{R=a}\\ &= -\rho C_{SF-a}\left\{\frac{\partial V_{wz}}{\partial t}a\left.\left[\left(1+\frac{A_1}{a^2}\right)\sin(\gamma) + \frac{A_3}{a^4}\sin(3\gamma)\right]\right|_{R=a}\right.\\ &\quad\left. + 2\frac{a}{a}\frac{da}{dt}V_{wz}|_{R=a}\sin(\gamma)\right\}\\ &= -\rho\frac{d_\xi}{a\left(1-\frac{A_1}{a^2}+\frac{A_3}{a^4}\right)}\left\{\frac{\partial V_{wz}}{\partial t}a\left.\left[\left(1+\frac{A_1}{a^2}\right)\sin(\gamma)+\frac{A_3}{a^4}\sin(3\gamma)\right]\right|_{R=a}\right.\\ &\quad\left. - 2UV_{wz}\tan(\Theta)\sin(\gamma)\right\}, \qquad \pi \le \gamma \le 2\pi\end{aligned} \tag{11.78}$$

The expression for the scale factor, C_{SF-a}, is found in eq. 11.40a. The time-derivative of a is found in eq. 11.67. Again, the subscript wZ is used to identify the wave-body interaction variables, and the subscript wz identifies the vertical velocity of the wave-induced water particle motions. In eq. 11.78 are the wave-induced, vertical, particle velocity (V_{wz}) from eq. 11.47 and the time-derivative of that velocity.

For this deep-water application, that velocity is obtained from

$$\begin{aligned}
&V_{wz}(\xi, Z_\xi, t)\\
&\simeq k_0 \frac{g}{\omega_e}\frac{H_0}{2} e^{k_0 Z_\xi} \sin(k_0\xi - \omega_e t) = \frac{\omega^2}{\omega_e}\frac{H_0}{2}\left[1 + \frac{k_0 Z_\xi}{1!} + \frac{(k_0 Z_\xi)^2}{2!}\cdots\right] \sin(K_0\xi - \omega_e t)\\
&= \left(\frac{\omega}{\omega_e}\right)^2 \left[1 + \frac{k_0 Z_\xi}{1!} + \frac{(k_0 Z_\xi)^2}{2!}\cdots\right]\Bigg|_{R=a} \frac{\partial \eta_0}{\partial t}\\
&= \left(\frac{\omega}{\omega_e}\right)^2 \left[\sum_0^\infty \frac{(k_0 Z_\xi)^i}{j!}\right]\Bigg|_{R=a} \frac{\partial \eta_0}{\partial t}\\
&= \left(\frac{\omega}{\omega_e}\right)^2 \left\{\sum_0^\infty \frac{(k_0 C_{SF-a} a)^i \left[\left(1 - \frac{A_1}{a^2}\right)\sin(\gamma) - \frac{A_3}{a^4}\sin(3\gamma)\right]^j}{j!}\right\}\Bigg|_{R=a=1} \frac{\partial \eta_0}{\partial t}
\end{aligned} \tag{11.79}$$

where the Z_ξ expression is from eq. 11.35a applied at $R = a$. The time-derivative of the velocity is

$$\frac{\partial V_{wz}}{\partial t} \simeq \left(\frac{\omega}{\omega_e}\right)^2 \left[\sum_0^\infty \frac{(k_0 C_{SF} a)^i \left[\left(1 - \frac{A_1}{a^2}\right)\sin(\gamma) - \frac{A_3}{a^4}\sin(3\gamma)\right]^j}{j!}\right]\Bigg|_{R=a} \frac{\partial^2 \eta 0}{\partial t^2} \tag{11.80}$$

where the frequency of encounter, ω_e, for a following sea equals $\omega - k_0 U \simeq 0.815$ rad/s, according to eq. 11.45 for the deep-water conditions. In these equations, the expression for the coordinate Z_ξ is found in eq. 11.38b, and the deep-water dispersion relationship in eq. 3.31 is introduced, where $k_0 g = \omega^2$. As in the wave-induced pressure expression in eq. 11.61, the ratio of the hyperbolic functions is, first, approximated for deep water, and then the resulting expression is replaced by a series expansion. The series expansion facilitates the integration of the pressure over the strip. As is the custom, the subscript 0 in eq. 11.79 identifies deep-water wave properties. Hence, the deep-water free-surface expression in eq. 11.44 is

$$\eta_0 = \frac{H_0}{2}\cos(k_0\xi - \omega_e t) \tag{11.81}$$

The second-order time-derivative of this expression appears in eq. 11.80.

The wave-body interaction force per unit length is obtained by integrating the pressure expression in eq. 11.78 over the wetted surface of the strip, where the respective velocity and acceleration terms in that equation are found in eqs. 11.79 and 11.80. This integration results in the following expression for the vertical wave-body interaction force on the strip:

$$\begin{aligned}
F'_{wZ}|_{h\to\infty} &\equiv \frac{dF_{wz}}{d\xi}|_{h\to\infty} = \int_\pi^{2\pi} p_{wz}\frac{dY_\xi}{d\gamma} d\gamma \Big|_{\substack{h\to\infty\\ R=a}}\\
&= \rho C_{SF-a}^2 \left(\frac{\omega}{\omega_e}\right)^2 \int_\pi^{2\pi} \left\{\frac{\partial^2 \eta_0}{\partial t^2} a \left[\left(1 + \frac{A_1}{a^2}\right)\sin(\gamma) + \frac{A_3}{a^4}\sin(3\gamma)\right]\right.\\
&\quad \left. - 2U\frac{\partial \eta_0}{\partial t}\tan(\Theta)\sin(\gamma)\right\}
\end{aligned}$$

$$\cdot \sum_{j=0}^{\infty} \frac{(k_0 C_{SF-a} a)^j \left[\left(1 - \frac{A_1}{a^2}\right)\sin(\gamma) - \frac{A_3}{a^4}\sin(3\gamma)\right]^j}{j!}$$

$$\cdot a \left[\left(1 + \frac{A_1}{a^2}\right)\sin(\gamma) + 3\frac{A_3}{a^4}\sin(3\gamma)\right]\Bigg|_{R=a} d\gamma$$

$$= \mathrm{a}'_{wZ}\frac{\partial^2 \eta_0}{\partial t^2} + \mathrm{b}'_{wZ}\frac{\partial \eta_0}{\partial t} \tag{11.82}$$

The a's have been included to remind the reader of the dimensions. When $a = 1$, as done by Lewis (1929), the resulting expressions become dimensionally vague. The first term in the last line of eq. 11.82 is an inertia-type term, whereas the second term in that line is a damping-type (*quasi*-damping) term. For this reason, the notations for the respective coefficients are introduced. The second term vanishes under two conditions. The first condition is the body is at rest ($U = 0$), and the second is a uniform keel depth ($\Theta = 0$) from bow to stern. Concerning the expansions: The series representation of the exponential functions in eqs. 11.79 through 11.81 are Maclaurin series. As such, the expansions of the function are about $k_0 Z_\xi = 0$. Physically, this corresponds to a zero frequency condition.

Let us include only two terms in the series in eq. 11.82 ($j = 0, 1$), and determine approximate expressions for the coefficients in the last line of that equation. The results are

$$\mathrm{a}'_{wZ} \simeq \rho C_{SF-a}^2 a^2 \left(\frac{\omega}{\omega_e}\right)^2 \left\{\frac{\pi}{2}\left(1 + \frac{A_1}{a^2}\right)^2 + \frac{3}{2}\pi\left(\frac{A_3}{a^4}\right)^2\right.$$

$$\left. - \frac{4}{15}k_0 C_{SF-a} a \left[\left(1 - \frac{A_1^2}{a^4}\right)\left(5 + 5\frac{A_1}{a^2} - 4\frac{A_3}{a^4}\right) + \left(1 + \frac{A_1}{a^2}\right)^2 \frac{A_3}{a^4} - 5\left(\frac{A_3}{a^4}\right)^2\right]\right\} \tag{11.83}$$

and

$$\mathrm{b}'_{wZ} \simeq 2\rho C_{SF-a}^2 a \left(\frac{\omega}{\omega_e}\right)^2 U\tan(\Theta) \left\{\frac{\pi}{2}\left(1 + \frac{A_1}{a^2}\right) - \frac{4}{15}k_0 C_{SF-a}\right.$$

$$\left. \cdot a \left[5\left(1 - \frac{A_1^2}{a^4}\right) - 2\frac{A_3}{a^4} + 4\left(\frac{A_1}{a^2}\right)\left(\frac{A_3}{a^4}\right)\right]\right\} \tag{11.84}$$

Again, the a's are retained in these equations so that the reader can see that the expressions are dimensionally correct. In an application of the equations, such as in the next example, the analysis is facilitated by letting $a = 1$, that is, the unit circle is transformed. The coefficients in eqs. 11.83 and 11.84 can be a function of ξ if the sectional geometry changes along the body length. That is, the Lewis parameters (A_1 and A_3) can vary over the length, as is the case of a ship.

EXAMPLE 11.8: WAVE-BODY INTERACTION FORCES ON TWO STRIP GEOMETRIES
Consider the problem described in Example 11.5, where an ocean engineer has been tasked to design a long, stiffened oil bladder for deep-water operation, designed to be towed at a speed (U) of 1 m/s in a head sea. We assume here that the body is at rest ($U = 0$ and $\omega_e = \omega = 2\pi/7 \simeq 0.898$ rad/s), and that the sectional (strip) dimensions are uniform from bow to stern; hence, $\Theta = 0$ in eq. 11.84. As a result, the damping-type component of the force equation equals zero. Again, the two sectional areas to be compared are the Lewis form in

Figure 11.8b and the semicircular strip in Figure 11.12. In Example 11.5, we find that the strip areas of the two strip geometries are approximately equal to 7.66 m^2. The diameter (D) of the semicircular strip is approximately 4.42 m, and the breadth (B_ξ) and keel depth (d_ξ) of the Lewis form are 4 m and 2 m, respectively. The forces of interest are those in a deep-water swell, where the wave height (H_0) is 1.5 m and the period (T) is 7 sec. From Example 11.5, the deep-water wavelength (λ_0) is approximately 76.5 m, and the wave number (k_0) is approximately 0.0821 m^{-1}. From eq. 11.40b, the scale factors are $D/2$ for the semicircular strip and $B_\xi/[2(1 + A_3)] = d_\xi/(1 + A_3)$ for the Lewis form, where $A_3 = -0.111$.

The combination of eqs. 11.82 through 11.84 applied to each strip geometry yields the following expressions for the wave-body interaction force: For the semicircular strip at rest, this force is

$$\begin{aligned} F'_{wZ}|_{cir} &\equiv \left.\frac{dF_{wZ}}{d\xi}\right|_{cir} \simeq \rho_0\left(\frac{D}{2}\right)^2\left(\frac{\pi}{2}-\frac{4}{6}k_0 D\right)\frac{\partial^2\eta_0}{\partial t^2} \\ &= \frac{1}{2}\rho\pi\frac{D^2}{4}\left(1-\frac{8}{6\pi}k_0 D\right)\frac{\partial^2\eta_0}{\partial t^2} \\ &= \mathrm{a}'_{wZ}|_{cir}\left(1-\frac{8}{6\pi}k_0 D\right)\frac{\partial^2\eta_0}{\partial t^2} = \mathrm{a}'_{wZ}|_{cir}\frac{\partial^2\eta_0}{\partial t^2} \\ &\simeq -7.90\times 10^3(1-0.154)\frac{\partial^2\eta_0}{\partial t^2} \simeq 6.68\times 10^3\frac{\partial^2\eta_0}{\partial t^2},\quad \frac{\mathrm{N}}{\mathrm{m}} \end{aligned} \tag{11.85}$$

The added mass (a'_{wZ}) in eq. 11.75 is introduced in the second line of the equation. We also note that the vertical acceleration of the free surface with respect to the traveling body is $-\omega^2\eta_0$. For the Lewis form, the wave-body interaction force is

$$\begin{aligned} &F'_{wZ}|_{Lew} \\ &\simeq \rho C^2_{SF-a}a^2\left[\frac{\pi}{2}\left(1+3\frac{A_3^2}{a^8}\right)-\frac{4}{15}k_0 C_{SF-a}\,a\left(5-2\frac{A_3}{a^4}-5\frac{A_3^2}{a^8}\right)\right]\Bigg|_{a=1}\frac{\partial^2\eta_0}{\partial t^2} \\ &= \rho C^2_{SF}\frac{\pi}{2}(1+3A_3^2)\left[1-\frac{8}{15\pi}k_0 C_{SF}\frac{(5-2A_3-5A_3^2)}{(1+3A_3^2)}\right]\frac{\partial^2\eta_0}{\partial t^2} \\ &= \mathrm{a}'_{wZ}|_{Lew}\left[1-\frac{8}{15\pi}k_0 C_{SF}\frac{(5-2A_3-5A_3^2)}{(1+3A_3^2)}\right]\frac{\partial^2\eta_0}{\partial t^2} = \mathrm{a}'_{wZ}|_{Lew}\frac{\partial^2\eta_0}{\partial t^2} \\ &\simeq 8.49\times 10^3(1-0.156)\frac{\partial^2\eta_0}{\partial t^2} \simeq 7.17\times 10^3\frac{\partial^2\eta_0}{\partial t^2},\quad \frac{\mathrm{N}}{\mathrm{m}} \end{aligned} \tag{11.86}$$

In eq. 11.86, the scale factor is $C_{SF-a} = D/2 = 2.21$, and $C_{SF} = d_\xi/0.889 \simeq 2.25$. These values are obtained by applying the Lewis parametric values to eq. 11.40b. By comparing the last terms of eqs. 11.85 and 11.86, we see that the added mass of the force on the Lewis form is greater than that on the semicircle. It is also interesting to note that the coefficients of the free-surface acceleration are equal to the product of the infinite-frequency added mass and a truncated expansion where, in each case, the value of the expansion is about the same for both geometries for the given conditions.

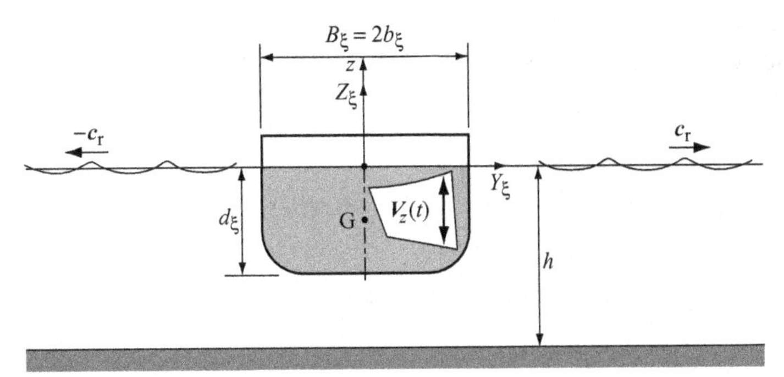

Figure 11.15. *Forced Vertical Motions of a Strip in Calm Water.* The vertical displacement of the strip at any time is $Z(t)$.

As suggested by Jacobs (1958), Motora (1964) essentially assumes that the motions of an infinitely long body subjected to beam waves are proportional to the motions of the free-surface motions at the centerplane. With this assumption, the summation in eq. 11.82 becomes a correction factor to the added mass, which accounts for the wave-induced particle motions. Motora applies the analysis to the uncoupled heaving and swaying motions of the body with some success.

D. Radiation Damping

When a body moves in a still body of water, waves are created. According to the wave-maker theory, as described by Dean and Dalrymple (1984), there are two types of waves resulting from the body motions. For a given body, these are traveling waves, which carry energy away from the body, and a series of standing, evanescent waves, which are attached to the body. The frequencies of the evanescent waves are larger than that of the traveling waves. Hence, if the body motions are excited by low-frequency (high-period) incident waves, then the effects of the evanescent waves can be neglected.

Consider an experimental study where a model of a uniform cross-section is forced to oscillate vertically in a long tank. Although this is a heaving motion, we can apply the results to the strip theory because derivations in the theory are based on the near-vertical motions of the strip. The section sketched in Figure 11.14 is representative of any section of the model. The frequencies of oscillation and the depth of the tank are such that the waves created by the model motions are deep-water waves. In this experiment, the body is "smooth" so that the viscous damping is much less than the radiation damping. Our goal is to determine the expression for the radiation damping coefficient.

The displacement of the model in Figure 11.15 is $Z(t)$, and is assumed to be sinusoidal in time. The period of oscillation is either the wave period (T) or the period of encounter (T_e). The latter is defined in eq. 11.45. If we are modeling a floating body that is under way, such as a ship traveling at a design speed, U, the period of motion in the design condition is the period of encounter. This period is used herein in the derivation of the expression for the radiation-damping coefficient. The force on a strip of the model due to the creation of the waves by the body motions is $b'_{r\xi}\, dZ/dt$, where $b'_{r\xi}$ is the radiation-damping coefficient per unit body length. When this force is multiplied by the vertical velocity, dZ/dt, and the resulting expression is integrated over the period (T_e), the result is the averaged power lost by the body over the period. This power can be equated to the average power radiated from both sides of the body. The wave power is analogous to the energy flux obtained from eq. 3.73. The resulting power expression is then

$$2P_r' = \frac{1}{T_e}\int_0^{T_e} b_r'\left(\frac{dZ}{dt}\right)^2 dt = \frac{1}{2}b_r'\omega^2 Z_o^2 = \frac{1}{2}b_r' V_o^2$$
$$= 2\left[\frac{\rho g H_{r0}^2 c_{r0}}{16}\right] = \frac{\rho g^2 H_{r0}^2 T_e}{16\pi} = \frac{\rho g^2 H_{r0}^2}{8\omega_e} \tag{11.87}$$

The factor of 2 in the wave expressions accounts for the radiation on both sides of the body. In eq. 11.87, the respective deep-water wave height and celerity of these waves are H_{r0} and c_{r0}.

The frequency of the radiated waves is the same as that of the forcing function, that is, the frequency of encounter (ω_e) in eq. 11.45. From eq. 11.87, the heaving radiation damping coefficient (per unit length) is

$$b_r' = \frac{\rho g^2}{\omega_e^3}\left(\frac{H_{r0}}{2Z_o}\right)^2 = \frac{\rho g^2}{\omega_e^3} R_Z^2 \tag{11.88}$$

This equation includes the ratio of the radiated-wave amplitude and the body-motion amplitude, that is, $R_Z \equiv (H_{r0}/2)/Z_o$.

Various mathematical expressions for R_Z have been derived. Some of these include those of Holstein (1937), Havelock (1942), Ursell (1954), Tasai (1959), and Yamamoto, Fujino, and Fukasawa (1980). One of the R_Z-expressions is that of Yamamoto, Fujino, and Fukasawa (1980), referred to herein as the *YFF formula*, that is,

$$R_z \equiv \frac{H_{r0}/2}{Z_o} = 2e^{-\frac{\omega_e^2}{g}Z_{ref}} \sin\left(\frac{\omega_e^2}{g} b_\xi\right) \tag{11.89}$$

according to Petersen and Marnæs (1989). In eq. 11.89, Z_{ref} is a reference draft of the wetted surface of the section, and B_ξ is the breadth. The reference draft, $Z_{ref} = S_\xi / B_\xi$, has been used by a number of investigators over the years. Hence, for a rectangular section having a draft, d_ξ, we see that $Z_{ref} = d_\xi$. The expression in eq. 11.89 has the same form as that of Havelock (1942), where the Havelock expression is obtained by placing a source distribution on the bottom of the section. The Havelock method is introduced in Sections 11.6A and 11.6B.

In the following example, the radiation damping coefficients for two strip shapes are presented. These coefficients result from the combination of the expressions in eqs. 11.88 and 11.89, where $Z_{ref} = S_\xi / B_\xi$, where S_ξ is a sectional area, and where B_ξ is the breadth of the section at ξ.

EXAMPLE 11.9: RADIATION DAMPING COEFFICIENTS USING THE YFF FORMULA Compare the radiation damping values of the semicircular strip and the Lewis form discussed in Example 11.5, where the amplitude ratio is obtained using the YFF formula, eq. 11.89. The Lewis form is sketched in Figure 11.8b. In the example, both shapes are subject to deep-water waves having a 7-sec period, and are towed at a speed $U = 1$ m/s (3.28 ft/s $\simeq$ 1.94 knots) in a following sea. The corresponding wavelength (λ_0) and wave number (k_0) are approximately 76.5 m and 0.0821 m^{-1}, respectively. From Example 11.5, the sectional area is approximately 7.66 m^2, where the diameter of the semicircle (D_ξ) is approximately 4.42 m, the breadth of the Lewis form is 4.00 m, and the draft of the Lewis form is 2.00 m. The respective wavelength-to-breadth ratios for these shapes are then 17.31 and 19.13. For these values, the reference draft for the semicircle is $Z_{ref}|_{cir} = S_\xi / D = 1.73$ m, and that of the Lewis form is $Z_{ref}|_{Lew} = S_\xi / B_\xi = 1.92$ m. The amplitude ratios for the two shapes are

$$\mathrm{R}_Z|_{cir} = \frac{H_0/2}{Z_0}\Big|_{cir} = 2e^{-k_0 \frac{S_\xi}{D_\xi}} \sin(k_0 D_\xi) \simeq 0.616 \tag{11.90}$$

and

$$\mathrm{R}_Z|_{Lew} = \frac{H_0/2}{Z_0}\Big|_{Lew} = 2e^{-k_0 \frac{S_\xi}{B_\xi}} \sin(k_0 B_\xi) \simeq 0.551 \tag{11.91}$$

Because the design towing speed (U) of the bladder is 1 m/s ($\equiv$ 3.28 ft/s $\simeq$ 1.94 knots), the period of encounter for the following-sea condition is approximately 7.70 sec from eq. 11.45. The radiation-damping coefficient values for the respective circular and Lewis form sections, obtained from eq. 11.88, are

$$\mathrm{b}'_r|_{cir} = \frac{\rho g^2}{\omega_e^3}\mathrm{R}_Z^2|_{cir} \simeq \frac{1.03 \times 10^3\,(9.81)^2}{(0.816)^3}(0.616)^2 \simeq 6.92 \times 10^4 \text{ N–s/m} \tag{11.92}$$

and

$$\mathrm{b}'_r|_{Lew} = \frac{\rho g^2}{\omega_e^3}\mathrm{R}_Z^2|_{Lew} \simeq \frac{1.03 \times 10^3\,(9.81)^2}{(0.816)^3}(0.551)^2 \simeq 5.54 \times 10^4 \text{ N–s/m} \tag{11.93}$$

When the bodies are not under way, then the frequency in the equations is that of the incident wave, approximately 0.898 rad/s. For this frequency, the respective damping coefficients are 5.19×10^4 N–s/m and 4.16×10^4 N–s/m. Hence, in a following sea, the radiation damping decreases with U. It is also interesting to note that the vertically oscillating semicircular strip is a stronger radiator than the Lewis form for the same sectional area.

An amplitude ratio formula that is directly applicable to the Lewis forms (discussed in Section 11.3A) is that of Tasai (1959). According to Pedersen (2000) and Jensen (2001), the Tasai formula is

$$\mathrm{R}_Z = \frac{\omega_e^2 \dfrac{B_\xi}{g}}{1 + A_1 + A_3} \int_1^\infty \left[\frac{(1+A_1)}{\varsigma^2} + 3\frac{A_3}{\varsigma^4}\right] \cos\left\{\frac{\omega_e^2 B_\xi}{2g}\left[\frac{(\varsigma^4 + A_1\varsigma^2 + A_3)}{(1 + A_1 + A_3)\varsigma^3} - 1\right]\right\} d\varsigma \tag{11.94}$$

where ς is an integration variable. This equation must be solved numerically for bodies for which A_1 and A_3 are not equal to zero. The combination of eqs. 11.88 and 11.94 is applied to the semi-submerged circular cylinder in the next example.

EXAMPLE 11.10: RADIATION DAMPING COEFFICIENT USING THE TASAI FORMULA Assuming deep water, eqs. 11.88 and 11.94 are now applied to the semi-submerged circular cylinder (where the Lewis parameters are $A_1 = A_3 = 0$) to obtain the radiation damping coefficient. The amplitude ratio obtained from eq. 11.94 is

$$
\begin{aligned}
R_Z|_{cir} &= \frac{\omega_e^2 D_\xi}{g} \int_1^\infty \frac{1}{\varsigma^2} \cos\left[\left(\frac{\omega_e^2 D_\xi}{2g}\right)(\varsigma - 1)\right] d\varsigma \\
&= \frac{\omega_e^2 D_\zeta}{2g} \left\{1 + \frac{\omega_e^2 D_\zeta}{2g} \left[\cos\left(\frac{\omega_e^2 D_\zeta}{2g}\right) \mathrm{si}\left(\frac{\omega_e^2 D_\zeta}{2g}\right)\right.\right. \\
&\quad \left.\left. - \sin\left(\frac{\omega_e^2 D_\zeta}{2g}\right) \mathrm{ci}\left(\frac{\omega_e^2 D_\varsigma}{2g}\right)\right]\right\}
\end{aligned} \tag{11.95a}
$$

In this equation, the diameter of the body at a position ξ is $D_\xi = 2r_\xi = 4.42\,\mathrm{m}$, and si() and ci() are sine and cosine integrals, respectively. See Abramowitz and Stegun (1965) for details of these functions. Consider the case when the body is at rest in the 7-sec waves. For this case, the frequency of encounter is replaced by the circular wave frequency, that is, $\omega_e \to \omega = 2\pi/T \simeq 0.898\,\mathrm{rad/s}$. For this condition, we rewrite eq. 11.95a in terms of the product of the deep-water wave number ($k_0 = \omega^2/g$) and the radius of the body (r_ξ) at the longitudinal position, ξ. The result is

$$
R_Z|_{cir} = 2k_0 r_\xi\{1 + k_0 r[\cos(k_0 r_\xi)\mathrm{si}(k_0 r_\xi) - \sin(k_0 r_\xi)\mathrm{ci}(k_0 r_\xi)]\} \tag{11.95b}
$$

From Abramowitz and Stegun (1965), the sine and cosine integrals can be represented by the following respective expressions:

$$
\mathrm{si}(k_0 r_\xi) = -\mathrm{f}(k_0 r_\xi)\cos(k_0 r_\xi) - g(k_0 r_\xi)\sin(k_0 r_\xi) \tag{11.96}
$$

and

$$
\mathrm{ci}(k_0 r_\xi) = \mathrm{f}(k_0 r_\xi)\sin(k_0 r_\xi) - \mathrm{g}(k_0 r_\xi)\cos(k_0 r_\xi) \tag{11.97}
$$

For the f- and g-functions, the following approximations are valid over $1 \le k_0 r_\xi < \infty$:

$$
\mathrm{f}(k_0 r_\xi) \simeq \frac{1}{(k_0 r_\xi)} \left[\frac{(k_0 r_\xi)^4 + 7.241163\,(k_0 r_\xi)^2 + 2.463936}{(k_0 r_\xi)^4 + 9.068580\,(k_0 r_\xi)^2 + 7.157455}\right] \tag{11.98a}
$$

and

$$
\mathrm{g}(k_0 r_\xi) \simeq \frac{1}{(k_0 r_\xi)^2} \left[\frac{(k_0 r_\xi)^4 + 7.547478\,(k_0 r_\xi)^2 + 1.564072}{(k_0 r_\xi)^4 + 12.723684\,(k_0 r_\xi)^2 + 15.723606}\right] \tag{11.98b}
$$

Unfortunately, for the 7-sec wave, $k_0 r_\xi \simeq 0.182 < 1$, so we cannot use the approximate formulas in eq. 11.98. Instead, the computed values of the sine and cosine integrals found in the book by Abramowitz and Stegun (1965) are used. For the 7-sec wave, the values of the sine and cosine integrals are approximately -1.39 and -1.14, respectively. The amplitude ratio from eq. 11.95b is then approximately 0.285. For this value, the approximate value of the radiation-damping coefficient in eq. 11.94 is 1.11×10^4 N-s/m. This value is approximately one sixth that found in Example 11.9, eq. 11.92, which is 6.92×10^4 N-s/m. Later in this chapter, the theoretical values of the added-mass and radiation-damping coefficients are compared with those found experimentally by Vugts (1968).

In addition to the equations for the amplitude ratios in eqs. 11.89 and 11.94, and the radiation damping coefficient presented in eq. 11.88, Havelock (1927, 1942) presents an integral method of estimating the radiation damping for a ship section (strip). This method is discussed later in this chapter. The Havelock method involves placing a source distribution at some depth on the centerline of the strip which, in turn, creates a wave system similar to that in the wavemaker theory. Korvin-Kroukovsky and Jacobs (1957) use the results of Havelock (1942) to obtain an expression for the radiation-damping coefficient, b'_r. After introducing the source method in two dimensions, Havelock (1942) applies his analysis to a whole ship. He considers two cases, those being pure heaving and pure pitching. In determining the radiation-damping coefficient, Havelock determines the time rate of change of the energy lost to these motions, and equates that energy property to the energy flux of the traveling waves created by the body motions.

Expressions for the vertical-motion hydrodynamic coefficients used in the Korvin-Kroukovsky and Jacobs (1957) strip theory have now been defined. The sectional added mass for the Lewis forms in eq. 11.75 is seen to be independent of frequency. However, the radiation damping coefficient in eq. 11.88 is frequency-dependent. Later in this chapter, the added-mass coefficient (modified to include frequency dependence) and radiation-damping expressions are applied to the experimental data of Vugts (1968).

11.4 Coupled Heaving and Pitching Motions Based on Strip Theory

The equations of motion for coupled heaving and pitching bodies in waves are found in eqs. 11.33 and 11.34. There are a number of methods available to solve the coupled linear system represented by these equations. These methods are discussed in the book by Zill (1986) and others. The coefficients in the equations of motion are represented collectively, as in eqs. 11.15b and 11.16b. This reduces the bookkeeping in the derivations in this section and in Section 11.4. The heaving equation of motion is then

$$
\begin{aligned}
&(m+\mathrm{a}_w)\frac{d^2Z}{dt}+(\mathrm{b_v}+\mathrm{b_r})\frac{dZ}{dt}+\mathrm{c_h}Z+\mathrm{d_a}\frac{d^2\theta}{dt^2}+(\mathrm{e_v}+\mathrm{e_r}+\mathrm{e_a})\frac{d\theta}{dt}+(\mathrm{f_h}+\mathrm{f_v}+\mathrm{f_r})\,\theta \\
&\equiv (m+\mathrm{a}_w)\frac{d^2Z}{dt^2}+\mathrm{b}\frac{dz}{dt}+\mathrm{c_h}Z+\mathrm{d_a}\frac{d^2\theta}{dt^2}+\mathrm{e}\frac{d\theta}{dt}+\mathrm{f}\theta=F_{\mathrm{w}}(\mathrm{w}_e,t) \qquad (11.99)
\end{aligned}
$$

Similarly, the pitching equation of motion is

$$
\begin{aligned}
&(I_{\mathrm{Y}}+\mathrm{A}_w)\frac{d^2\theta}{dt^2}+(\mathrm{B_v}+\mathrm{B_r}+\mathrm{B_a})\frac{d\theta}{dt}+(\mathrm{C_h}+\mathrm{C_v}+\mathrm{C_r}+\mathrm{C_a})\,\theta+\mathrm{D_a}\frac{d^2Z}{dt^2} \\
&+(\mathrm{E_v}+\mathrm{E_r}+\mathrm{E_a})\frac{dZ}{dt}+\mathrm{F_h}Z \\
&\equiv (I_{\mathrm{Y}}+\mathrm{A}_w)\frac{d^2\theta}{dt^2}+\mathrm{B}\frac{d\theta}{dt}+\mathrm{C}\theta+\mathrm{D_a}\frac{d^2Z}{dt^2}+\mathrm{E}\frac{dZ}{dt}+\mathrm{F}Z=M_w(\omega_e,t) \qquad (11.100)
\end{aligned}
$$

The last three terms on the left side of each equality of these equations are the coupling terms. The subscripted coefficients are defined in Table 11.1. In the following analysis, the second lines of eqs. 11.99 and 11.100 are used for the sake of brevity. The body mass in eq. 11.99 is represented by m, and the mass-moment of inertia of the body in eq. 11.100 is I_{Y}. At this point, the reader is reminded that the origin of the coordinate system x, y, z is a fixed point on the calm-water surface, whereas X, Y, Z are on the calm-water surface above the center of gravity (G) of the body (see Figures 11.5 and 11.6). Furthermore, Y is through G and parallel to y and Y.

The exciting force in eq. 11.99 is the sum of the wave-induced force discussed in Section 11.3C(1), eq. 11.63b, and the wave-body interaction force derived in Section 11.3C(3), eq. 11.82. That heaving force, introduced in eq. 11.26, is

$$
\begin{aligned}
F_W(\omega_e, t)\,|_{h\to\infty} &= \int_{-\ell_{aft}}^{\ell_{fwd}} (F'_{wz} + F'_{wz})\,\xi = \int_{-\ell_{aft}}^{\ell_{fwd}} \Bigg\{ a'_{wz}\frac{\partial^2 \eta_0}{\partial t^2} + b'_{wz}\frac{\partial \eta_0}{\partial t} \\
&\quad + \rho g B_\xi\,(1 + C_{Smith})\,\eta_0 \Bigg\} d\xi \\
&= \int_{-\ell_{aft}}^{\ell_{fwd}} \{[-a'_{wz}\omega_e^2 + \rho g B_\xi\,(1 + C_{Smith})]\cos(k_0\xi - \omega_e t) \\
&\quad + b'_{wz}\omega_e \sin(k_0\xi - \omega_e t)\}\frac{H_0}{2} d\xi \\
&= F_{wc}\cos(\omega_e t) + F_{ws}\sin(\omega_e t) = F_{wo}\cos(\omega_e t + \alpha_F)
\end{aligned}
\tag{11.101}
$$

where in deep water, the wave-induced force expression is in eq. 11.63b, and the wave-body interaction force is in eq. 11.82. The parametric constants in the last line are

$$
F_{wc} = \int_{-\ell_{aft}}^{\ell_{fwd}} \{[-a'_{wz}\omega_e^2 + \rho g B_\xi\,(1 + C_{Smith})]\cos(k_0\xi) + b'_{wz}\omega_e \sin(k_0\xi)\,]\}d\xi \tag{11.102}
$$

and

$$
F_{ws} = \int_{-\ell_{aft}}^{\ell_{fwd}} \{[-a'_{wz}\omega_e^2 + \rho g B_\xi\,(1 + C_{Smith})]\sin(k_0\xi) - b'_{wz}\omega_e \cos(k_0\xi)\,]\}d\xi \tag{11.103}
$$

Also in the last line of eq. 11.101 are the force amplitude,

$$
F_{wo} = \sqrt{F_{wc}^2 + F_{ws}^2} \tag{11.104}
$$

and the phase angle between the wave and the force,

$$
\alpha_F = -\tan^{-1}\left(\frac{F_{ws}}{F_{wc}}\right) \tag{11.105}
$$

The exciting moment in eq. 11.100 is the sum of the moments corresponding to the force components in eqs. 11.63b and 11.82. The pitching moment about the center of gravity corresponding to the total heaving force in eq. 11.101 is

$$
\begin{aligned}
M_w(\omega_e, t)\,|^+|_{h\to\infty} &= \int_{-\ell_{aft}}^{\ell_{fwd}} (F'_{wz} + F'_{wz})\,\xi d\xi \\
&= \int_{-\ell_{aft}}^{\ell_{fwd}} \{[-a'_{wz}\omega_e^2 + \rho g B_\xi\,(1 + C_{Smith})]\cos(k_0\xi - \omega_e t) \\
&\quad + b'_{wz}\omega_e \sin(k_0\xi - \omega_e t)\Bigg\}\frac{H_0}{2}\xi d\xi \\
&= M_{wc}\cos(\omega_e t) + M_{ws}\sin(\omega_e t) = M_{wo}\cos(\omega_e t + \alpha_M)
\end{aligned}
\tag{11.106}
$$

As is the case for all moments in this book, the positive moment direction is in the counterclockwise direction. In eq. 11.106, the parametric coefficients are

$$M_{wc} = \int_{-\ell_{aft}}^{\ell_{fwd}} \{[-a'_{wz}\omega_e^2 + \rho g B_\xi(1 + C_{Smith})]\cos(k_0\xi) + b'_{wz}\omega_e \sin(k_0\xi)]\}\xi d\xi \tag{11.107}$$

and

$$M_{ws} = \int_{-\ell_{aft}}^{\ell_{fwd}} \{[-a'_{wz}\omega_e^2 + \rho g B_\xi(1 + C_{Smith})]\sin(k_0\xi) - b'_{wz}\omega_e \cos(k_0\xi)]\}\xi d\xi \tag{11.108}$$

Also in eq. 11.106 are the moment amplitude,

$$M_{wo} = \sqrt{M_{wc}^2 + M_{ws}^2} \tag{11.109}$$

and the phase angle between the wave and the moment,

$$\alpha_M = -\tan^{-1}\left(\frac{M_{ws}}{M_{wc}}\right) \tag{11.110}$$

There are several methods that can be used to solve the equations of motion in eqs. 11.99 and 11.100. Two of these are the operator method, as suggested by McCormick (1973), which is a time-domain solution. The second method uses the complex notation, and is a frequency-domain solution. These methods are described in the book by Zill (1986) and other books on linear differential equations. The time-domain solutions of the homogeneous equations of eqs. 11.112 and 11.113 result in two fourth-degree, linear auxiliary equations. The time-domain solutions of eqs. 11.94 and 11.95 are presented in the following paragraphs.

Write the respective heaving and pitching equations in eqs. 11.99 and 11.100 using operator notation as

$$L_{11}Z + L_{12}\theta = F_w(\omega_e, t) \tag{11.111}$$

and

$$L_{21}Z + L_{22}\theta = M_w(\omega_e, t) \tag{11.112}$$

From the second lines of eqs. 11.99 and 11.100, the operators in these equations are

$$L_{11} \equiv (m + a_w)\frac{d^2}{dt^2} + b\frac{d}{dt} + c_h \tag{11.113}$$

$$L_{12} \equiv d_a\frac{d^2}{dt^2} + e\frac{d}{dt} + f \tag{11.114}$$

$$L_{21} \equiv D_a\frac{d^2}{dt^2} + E\frac{d}{dt} + F \tag{11.115}$$

and

$$L_{22} \equiv (I_Y + A_w)\frac{d^2}{dt^2} + B\frac{d}{dt} + C \tag{11.116}$$

Rewrite eqs. 11.111 and 11.112 in matrix form as

$$\begin{bmatrix} L_{11} & L_{12} \\ L_{21} & L_{22} \end{bmatrix} \begin{Bmatrix} Z \\ \theta \end{Bmatrix} = \begin{Bmatrix} F_w(\omega_e, t) \\ M_w(\omega_e, t) \end{Bmatrix} \tag{11.117}$$

By using Cramer's rule for linear equations, we obtain

$$\begin{vmatrix} L_{11} & L_{12} \\ L_{21} & L_{22} \end{vmatrix} Z = \begin{vmatrix} F_w & L_{12} \\ M_w & L_{22} \end{vmatrix} \tag{11.118}$$

and

$$\begin{vmatrix} L_{11} & L_{12} \\ L_{21} & L_{22} \end{vmatrix} \theta = \begin{vmatrix} L_{11} & F_w \\ L_{22} & M_w \end{vmatrix} \tag{11.119}$$

The expanded forms of eqs. 11.118 and 11.119 are, respectively,

$$(L_{11}\, L_{22} - L_{12}\, L_{21})\, Z = L_{22} F_w - L_{12} M_w \tag{11.120}$$

and

$$(L_{11}\, L_{22} - L_{12}\, L_{21})\, \theta = L_{11} M_w - L_{21} F_w \tag{11.121}$$

In eqs. 11.118 through 11.121, the frequency and time dependencies of the displacements, forces, and moments are assumed. Equations 11.120 and 11.121 are fourth-order, linear differential equations that can be solved by using well-established methods.

The following example is presented to give the reader an idea of the magnitudes of the force, moment, and displacements a floating body would experience in linear waves. The static stability of the body is also considered.

EXAMPLE 11.11: TRANSVERSE STABILITY AND PLANAR MOTIONS OF A BARGE IN LINEAR WAVES Referring to Figure 11.16a, a barge having a rectangular waterplane is designed to carry dredged sand over deep water where the average wave height is 1.5 m and the average period is 7 sec. The deep-water wavelength (λ_0) is approximately 76.5 m, and the approximate wave number (k_0) value is about 8.21×10^{-2} m^{-1}. The length (L) of the barge is 40 m, the beam (B_{max}) is 4 m, and the draft (d_{max}) is 2 m. When fully loaded, the sectional area of the barge is the Lewis form sketched in Figures 11.8b and 11.16b, and is uniform over the barge length. The sectional area (S_ξ) is 7.66 m^2, and is obtained from eq. 11.39, where the Lewis parameters are $A_1 = 0$ and $A_3 = -0.111$. The displacement (the weight of the displaced water) of the loaded barge in salt water is $W \simeq 3.10 \times 10^6$ N, or about 316 metric tons. One metric ton (or tonne) equals 10^3 kg. The center of gravity (G) due to the barge structure, ballast, and sand is located 1.5 m below the waterline (see Figure 11.16b). The mass moment of inertia (I_Y) with respect to the Y-axis through G is approximately 4.22×10^7 N-m-s^2/rad.

Transverse Stability: Although our interest is in *x-y* planar motions of the barge, for the sake of safety, we must first consider the transverse stability of the loaded barge in a calm-water condition. A discussion of the stability of floating bodies is presented in Section 11.1. The goal is to ensure that the barge will not capsize due to the relative positions of the center of gravity (G) and the center of buoyancy (B). For the barge in question, the sectional area is uniform from the bow to the stern. The center of buoyancy will be at the depth of the center of area of any strip. From a course in statics, the centroid of the displaced volume of water

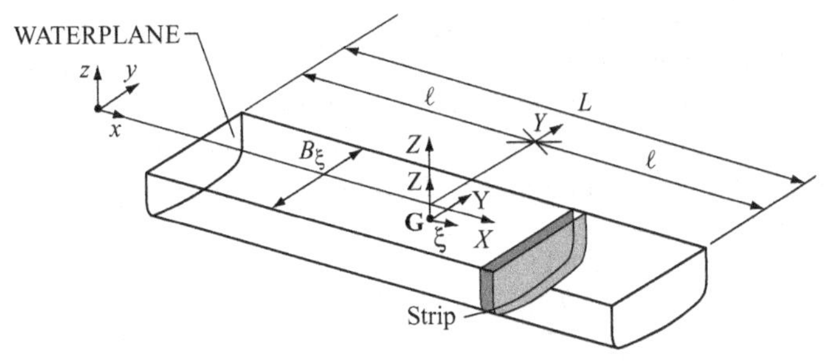

a. Displaced Volume

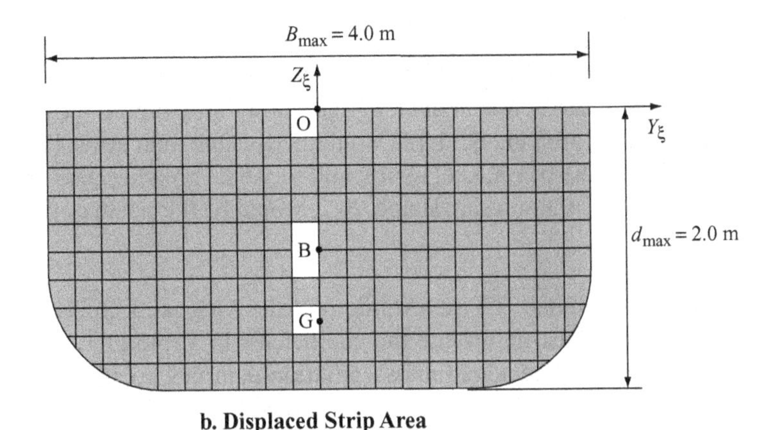

b. Displaced Strip Area

Figure 11.16. *Displaced Barge Having a Uniform Lewis Form Sectional Area over the Length in Example 11.11*. In (b) are the origin of the coordinate system (O), the center of area of buoyancy (B), and the center of gravity (G). The Lewis parameters for the strip are $A_1 = 0$ and $A_3 = -0.111$. As shown, there are four coordinate systems. Those are x,y,z (fixed to the waterplane), X,Y,Z (fixed to the waterplane above G), ξ,Y,Z (fixed to G), and Y_ξ,Z_ξ (fixed to the strip). The spatial variable ξ is then the directed distance from G to the strip.

is found by dividing the first moment of volume (with respect to the X-axis in Figure 11.16a) by the displaced volume. Hence, the position center of buoyancy is at a depth determined from

$$\mathrm{OB} = \left| \frac{\int_{\ell_{aft}}^{\ell_{fwd}} \int_{\pi}^{2\pi} \frac{Z_\xi}{2} (Z_\xi) \frac{dY_\xi}{d\gamma} d\gamma\, d\xi}{\int_{\ell_{aft}}^{\ell_{fwd}} \int_{\pi}^{2\pi} Z_\xi \frac{dY_\xi}{d\gamma} d\gamma\, d\xi} \right| = \left| \frac{\int_{\ell_{aft}}^{\ell_{fwd}} \left[\frac{1}{2} \int_{\pi}^{2\pi} Z_\xi^2 \frac{dY_\xi}{d\gamma} d\gamma \right] d\xi}{\vee} \right|$$

$$\simeq \frac{1}{2} \left(\frac{14,510\,(40)}{7.66\,(40)} \right) \simeq 0.947\,\mathrm{m} \tag{11.122}$$

In this equation, $\vee$ is the displaced water volume, which equals the product of the Lewis form area ($S_\xi = 7.66\,\mathrm{m}^2$) an the body length ($L = 40$ m). Again, for the Lewis form in Figure 11.16b, $A_1 = 0$ and $A_3 = -0.111$, and the position of the center of buoyancy (B) is at $\xi = X = 0$, and at a distance OB $\simeq 0.947$ m from the free surface. As previously stated, the center of gravity due to the weight of the barge and sand load is OG = 1.5 m below the free surface. Hence, G is below B and GB = |OG − OB| = 0.553 m. The metacentric height is obtained from eq. 11.6, where the positive sign (+) in that equation is used because G is below B. For the barge having a rectangular waterplane area, the metacentric height for the barge in question is

$$\mathrm{GM}|_{barge} \simeq \frac{\mathrm{I}_X}{\mathrm{V}} + \mathrm{GB} = \frac{\frac{LB_{max}^3}{12}}{\mathrm{S}_{max} L} + \mathrm{GB} \simeq 1.25 \text{ m} \tag{11.123}$$

Here, I_X is the second moment of the rectangular waterplane area with respect to the x-axis. Because the metacentric height is greater than zero, the body has transverse stability. If our interest was in the longitudinal stability (about the Y-axis), the second moment of area in eq. 11.123 would be I_Y. One final note: The subscript "max" is used for the beam, keel depth, and sectional area because all are uniform from bow to stern. For a ship shape, B_{max} would be the beam of the ship.

Heaving and Pitching Motions: To determine the heaving and pitching responses of the barge, we must determine the coefficients in Table 11.1. The integrands in the table involve the added mass, radiation-damping coefficients, linear-equivalent viscous damping coefficient, body breadth, and the speed of the body. The flow is assumed to be inviscid, the body is not under way, and the added mass, radiation-damping coefficient, and breadth of the body are invariant over the body length. As a result, the spatial derivatives of these quantities vanish in the coefficient expressions and, consequently, all of the coupling terms vanish. Physically, the barge can experience a displacement in one degree of freedom without having a displacement occur in the other degree of freedom.

We now determine the coefficients in the uncoupled equations of motions. From eq. 11.75, the added mass per unit length for the barge is

$$\mathrm{a}'_w(\xi) \equiv \mathrm{a}'_w = \rho\pi \frac{B_{max} d_{max}}{4} \frac{(1 + 3A_3^2)}{(1 + A_3)^2} \simeq 8.49 \times 10^3 \text{ kg/m} \tag{11.124}$$

The remaining coefficients are presented in Table 11.2. Combining this added-mass value with the first relationship in Table 11.1 yields the added mass of the barge. That value is $\mathrm{a_a} = \mathrm{a}'_w L \simeq 3.40 \times 10^5$ kg = 340 tonne, and is the first relationship in Table 11.2. The added mass is then slightly larger than the mass of the ship, which is 316 tonne. Because the added mass per unit length is not a function of the longitudinal position, we remove ξ from the notation, as in eq. 11.124.

The damping of the vertical motions of the strip is due to the radiation on both sides of the strip. The radiation damping coefficient in eq. 11.88, combined with the YFF amplitude-ratio formula in eq. 11.89a, results in

$$\mathrm{b}'_r(\xi) \equiv \mathrm{b}'_r = 4 \frac{\rho g^2}{\omega^3} e^{-2k_0 \frac{S_{max}}{B_{max}}} \sin^2\left(k_0 \frac{B_{max}}{2}\right) \simeq 1.070 \times 10^4 \text{ N–s/m}^2 \tag{11.125}$$

Table 11.2. *Coefficients for the coupled heaving and pitching motions in Example 11.11*

$m \simeq 3.16 \times 10^5$ kg or N-S^2/m	$I_Y \simeq 4.22 \times 10^7$ N-m-s^2/rad
$a_w = a'_w L \simeq 3.40 \times 10^5$ kg or N-s^2/m	$A_w = a'_w \frac{L^3}{12} \simeq 4.53 \times 10^7$ N-m-s^2/rad
$b_v = 0$	$B_a = 0$
$b_r = b'_r L \simeq 4.28 \times 10^5$ N-s/m	$B_v = 0$
$c_h = \rho g B_{max} L \simeq 1.62 \times 10^6$ N/m	$B_r = b'_r \frac{L^3}{12} \simeq 5.71 \times 10^7$ N-m-s/rad
$d_a = 0$	$C_a = 0$
$e_a = 0$	$C_h = \rho g B_{max} \frac{L^3}{12} \simeq 2.16 \times 10^8$ N-m/rad
$e_v = 0$	$C_v = 0$
$e_r = 0$	$C_r = 0$
$f_h = 0$	$D_a = 0$
$f_v = 0$	$E_a = 0$
$f_r = 0$	$E_v = 0$
$F_w(\omega, t) \simeq 6.31 \times 10^5 \cos(0.898t)$ N	$E_r = 0$
	$F_h = 0$
	$M_w(\omega, t) \simeq 8.58 \times 10^6 \sin(0.898t)$ N-m

The total radiation damping coefficient for the barge is $b_r = b'_r L \simeq 4.28 \times 10^5$ N–s/m, and is the third relationship in Table 11.2.

The hydrostatic restoring coefficients, c_h in heave and C_h in pitch, depend on the breadth of the strip. Because the breadth is uniform over the length of the barge, we replace B_ξ by B_{max}, the beam of the barge. For the 4-m beam, the hydrostatic restoring coefficient for the heaving motions is $c_h = \rho g B_{max} L \simeq 1.62 \times 10^6$ N/m, and is the fourth relationship in Table 11.2.

The excitation force in eq. 11.99, consisting of the wave force and the wave-body interaction force, is represented by the fourth equality in eq. 11.101. The expression for the heave-exiting force acting on the stationary (but not fixed) Lewis form barge in deep water is

$$F_w(\omega, t)|_{h\to\infty} = \int_{-\frac{L}{2}}^{\frac{L}{2}} (F'_{wZ} + F'_{wz}) d\xi \simeq \int_{-20\text{m}}^{20\text{m}} (-7.17 \times 10^3 \omega^2 + 4.04 \times 10^4) \left(\frac{H_0}{2}\right) \times \cos(k\xi - \omega t)\, d\xi$$
$$= (-7.17 \times 10^3 \omega^2 + 4.04 \times 10^4) \left(\frac{H_0}{k_0}\right) \sin\left(k_0 \frac{L}{2}\right) \sin(\omega t)$$
$$= F_{wo} \cos(\omega t) \simeq 6.31 \times 10^5 \cos(0.898t)\,, \text{N} \tag{11.126}$$

The values of the coefficients for the vertical wave force component (F'_{wz}) and the wave-body interaction force (F'_{wZ}) are from Examples 11.5 and 11.8, respectively. The body is not under way ($U = 0$), and the strip geometry is uniform from bow to stern ($\theta = 0$). For either of these conditions, $b'_{wZ} = 0$ in eq. 11.84. In addition, because the body is symmetric about the midsection (where $\xi = 0$), the force amplitude F_{ws} in eqs. 11.101 and 11.103 is also equal to zero. Subsequently, in eq. 11.101, the phase angle α_F in eq. 11.105 is equal to zero, meaning that the force is in phase with the wave.

From eq. 11.106, the corresponding pitch-excitation moment is

$$M_w(\omega,t)|_{h\to\infty} = \int_{-\frac{L}{2}}^{\frac{L}{2}} (F'_{wZ} + F'_{w\dot{z}})\xi d\xi$$

$$= [-a'_{wZ}\omega^2 + \rho g B_\xi (1 + C_{Smith})]\frac{H_0}{2}\int_{-20\mathrm{m}}^{20\mathrm{m}} \cos(k_0\xi - \omega t)\,\xi d\xi$$

$$= \left[-a'_{wZ}\omega^2 + \rho g B_\xi (1 + C_{Smith})\right]\frac{H_0}{k_0^2}\left[\sin\left(k_0\frac{L}{2}\right) - k_0\frac{L}{2}\cos\left(k_0\frac{L}{2}\right)\right]\sin(\omega t)$$

$$= M_{ws}\sin(\omega t) = M_{wo}\cos\left(\omega t - \frac{\pi}{2}\right) \simeq 8.58\times10^6\sin(0.898t),\ \text{N–m} \quad (11.127)$$

As is the case of all moments in this book, the pitch-excitation moment is positive in the counterclockwise direction. The phase angle between the wave and exiting moment in eq. 11.110 is equal to $-\pi/2$, that is, the moment lags the wave by 90°.

Applying the results in Table 11.2 to eqs. 11.99 and 11.100 results in the following respective uncoupled equations of motion:

$$\begin{aligned}(m + \mathrm{a}_w)\frac{d^2Z}{dt^2} &+ \mathrm{b_r}\frac{dZ}{dt} + \mathrm{c_h}Z \\ &= F_{wo}\cos(\omega t) \\ &= (6.56\times10^5)\frac{d^2Z}{dt^2} + (4.28\times10^5)\frac{dZ}{dt} + (1.62\times10^6)Z \\ &= 6.31\times10^5\cos(0.898t)\ \text{N}\end{aligned} \quad (11.128)$$

and

$$\begin{aligned}(I_Y + \mathrm{A}_w)\frac{d^2\theta}{dt^2} &+ \mathrm{B_r}\frac{d\theta}{dt} + \mathrm{C_h}\theta \\ &= M_{wo}\cos(\omega t) \\ &= (8.75\times10^7)\frac{d^2\theta}{dt^2} + (5.71\times10^7)\frac{d\theta}{dt} + (2.16\times10^8)\theta \\ &= 8.58\times10^6\sin(0.898t)\ \text{N–m}\end{aligned} \quad (11.129)$$

The steady-state solution of eq. 11.128 yields the expression for the uncoupled heaving response, that is,

$$Z(\omega,t) = \frac{\dfrac{F_{wo}}{\mathrm{c_h}}\cos(\omega t - \epsilon_Z)}{\sqrt{\left(1 - \dfrac{\omega^2}{\omega_{Zn}^2}\right)^2 + \left(2\dfrac{\omega}{\omega_{Zn}}\dfrac{\mathrm{b_r}}{\mathrm{b}_{cr}}\right)^2}} = Z_o\cos(\omega t - \epsilon_Z)$$

$$\simeq 0.390\frac{\cos(\omega t - 0.340)}{\sqrt{(1 - 0.406\omega^2)^2 + 0.0700\omega^2}}\Bigg|_{\omega\simeq0.898\mathrm{rad}},\ \text{meters} \quad (11.130)$$

where Z_o is the heaving amplitude. Also in this equation are the following:

Static Heaving Displacement: This is the displacement that would occur if a static vertical force equal to the amplitude of the wave-induced force was applied to the center of gravity. This is obtained from eq. 11.130 when $\omega = 0$. The result is

$$Z_{stat} = \frac{|F_{wo}|}{c_h} \simeq 0.390 \text{ m} \tag{11.131}$$

Undamped Natural Heaving Frequency: This frequency is a function of the hydrostatic restoring coefficient and the inertial coefficient. Physically, it is the natural heaving frequency that the body would experience if released from a static displacement if no damping is present. Mathematically, this frequency is

$$\omega_{Zn} = \frac{2\pi}{T_{Zn}} = \sqrt{\frac{c_h}{m + a_w}} \simeq 1.57 \text{ rad/s} \tag{11.132}$$

Critical Damping in Heave: This is the limit damping coefficient value for which heaving oscillations exist. It is obtained from

$$b_{cr} = 2\sqrt{c_h(m + a_w)} \simeq 2.06 \times 10^6 \text{ N–s/m} \tag{11.133}$$

That is, when $b_r < b_{cr}$, the body will experience at least one heaving cycle when released from a static position. When $b_r = b_{cr}$, the system is said to be *critically damped*. More is written of this condition later in this section.

Phase Angle in Heave: The phase angle is that between the body motion and the wave-induced force, that is

$$\epsilon_Z |_{\omega \simeq 0.898 \text{ rad/s}} = \tan^{-1}\left(\frac{2\dfrac{\omega}{\omega_{Zn}}\dfrac{b_r}{b_{cr}}}{1 - \dfrac{\omega^2}{\omega_{Zn}^2}} \right)\Bigg|_{\omega \simeq 0.898 \text{ rad/s}}$$

$$\simeq \tan^{-1}\left(\frac{0.265\omega}{1 - 0.406\omega^2} \right)\Bigg|_{\omega \simeq 0.898 \text{ rad/s}} \simeq 0.340 \text{ rad} \tag{11.134}$$

From eq. 11.132, the approximate natural heaving period value is 4.00 sec. We note that the pitching period is somewhat below the average wave period of the sea, which is 7 sec. Comparing the value of the critical damping in eq. 11.133 with the radiation damping value of 4.28×10^5 N-s/m in Table 11.2, we see that the heaving motions of the barge are underdamped. Concerning the phase angle expression eq. 11.134: When a resonance condition occurs, where $\omega = \omega_{Zn}$, the phase angle is $\epsilon_Z = 90°$. That is, the motions lag the wave-induced force by 90°. For the given conditions, the resulting phase angle between the wave-induced force and the heaving motions obtained from eq. 11.134 is about 0.338 rad, or 18.9°. The heaving amplitude (Z_o) is found to be approximately 0.546 m.

The solution of the pitching equation of motion in eq. 11.129 results in the following expression for the uncoupled pitching response of the barge:

$$\theta(\omega,t) = \frac{\dfrac{M_{wo}}{C_h}\cos(\omega t - \epsilon_\theta)}{\sqrt{\left(1 - \dfrac{\omega^2}{\omega_{\theta n}^2}\right)^2 + \left(2\dfrac{\omega}{\omega_{\theta n}}\dfrac{B_r}{B_{cr}}\right)^2}} = \theta_o \cos(\omega t - \epsilon_\theta)$$

$$\simeq 0.0397\frac{\cos(\omega t - 0.339)}{\sqrt{(1 - 0.405\omega^2)^2 + 0.0699\omega^2}}, \quad \text{radians} \tag{11.135}$$

where θ_o is the pitching amplitude. Also, in eq. 11.135 are the following:

Static Pitching-Angle Displacement:

$$\theta_h = \frac{|M_{wo}|}{C_h} \simeq 0.0397\,\text{rad} \tag{11.136}$$

Natural Pitching Frequency:

$$\omega_{\theta n} = \frac{2\pi}{T_{\theta n}} = \sqrt{\frac{C_h}{I_Y + A_w}} \simeq 1.57\,\text{rad/s} \tag{11.137}$$

Critical Damping in Pitch:

$$B_{cr} = 2\sqrt{C_h(I_Y + A_w)} \simeq 2.75 \times 10^8\,\text{N–m–s/rad} \tag{11.138}$$

Phase Angle in Pitch: This is the phase angle between the pitching displacement and the wave-induced pitching moment. It is obtained from

$$\epsilon_\theta = \tan^{-1}\left(\frac{2\dfrac{\omega}{\omega_{\theta n}}\dfrac{B_r}{B_{cr}}}{1 - \dfrac{\omega^2}{\omega_{\theta n}^2}}\right)\Bigg|_{\omega \simeq 0.898\,\text{rad/s}} \simeq \tan^{-1}\left(\frac{0.264\omega}{1 - 0.406\omega^2}\right)\Bigg|_{\omega \simeq 0.898\,\text{rad/s}}$$

$$\simeq 0.339\,\text{rad} \tag{11.139}$$

The physical meanings of the terms in eqs. 11.136 through 11.139 are analogous to those in eqs. 11.131 through 11.134. The approximate natural pitching period value is 4.00 sec, which is the same as that for the heaving motions. Also, because the damping coefficient value in Table 11.2 (5.71×10^7 N-m-s/rad) is much less than B_{cr}, the pitching motions of the barge are underdamped. Finally, the phase angle obtained from eq. 11.139 is approximately 0.339 rad, or 19.4°, which is about the same as that for the heaving motions. The pitching amplitude (θ_n) is approximately 3.19°.

The results for the barge having symmetry with respect to the *x-z* and *y-z* planes described in Example 11.11 are for uncoupled body motions. That is, a displacement can be imposed in either heave or pitch when the barge is in calm water without causing a displacement in the other mode. For coupled motions, a displacement in one mode will produce a displacement in the other mode. A ship shape having symmetry only with respect to the *x-z* plane experiences coupled motions. When the motions are uncoupled, the heaving and pitching motions are analogous to a forced spring-mass-damper system, as we can see from the nonhomogeneous, second-order, linear differential equations in eqs. 11.128 and 11.129. From the solutions of these equations, we define the respective *magnification factors* for the heaving and pitching motions as

$$Z_Z \equiv \frac{Z_o}{F_{wo}/C_h} = \frac{1}{\sqrt{\left(1 - \dfrac{\omega^2}{\omega_{Zn}^2}\right)^2 + \left(2\dfrac{\omega}{\omega_{Zn}}\dfrac{b_Z}{b_{cr}}\right)^2}} = \frac{1}{\sqrt{\left(1 - \dfrac{\omega^2}{\omega_{Zn}^2}\right)^2 + \left(2\dfrac{\omega}{\omega_{Zn}}\Delta_Z\right)^2}} \tag{11.140}$$

and

$$Z_\theta \equiv \frac{\theta_0}{M_{wo}/C_h} = \frac{1}{\sqrt{\left(1-\frac{\omega^2}{\omega_{\theta n}^2}\right)^2 + \left(2\frac{\omega}{\omega_{\theta n}}\frac{B_\theta}{B_{cr}}\right)^2}} = \frac{1}{\sqrt{\left(1-\frac{\omega^2}{\omega_{\theta n}^2}\right)^2 + \left(2\frac{\omega}{\omega_{\theta n}}\Delta_\theta\right)^2}} \tag{11.141}$$

In these equations, the Δ represents the *damping ratio* for each mode, and is defined as the ratio of the damping to critical damping, that is, $\Delta_Z = b_Z/b_{cr}$ in eq. 11.140, and $\Delta_\theta = B_\theta/B_{cr}$ in eq. 11.141. The magnification factor equations can be plotted as a function of the frequency ratio, ω/ω_n, where ω_n is the undamped natural frequency of the motion of interest. A generic magnification factor is plotted in Figure 11.17a. In eqs. 11.140 and 11.141, we see that the value of the magnification factor is infinite at resonance (when $\omega = \omega_n$) when the system is undamped ($\Delta = 0$). The physical interpretation of this is that the system at resonance continues to store the supplied energy without losing any of the energy to damping. For the curves in Figure 11.17a corresponding to finite damping, the peak values in the curves occur at frequencies less than the resonant frequency. The peak value occurs at the damped natural frequency, which is obtained from

$$\omega_d = \omega_n\sqrt{1-\Delta^2} \tag{11.142}$$

This expression is obtained by letting the derivative of the magnification factor with respect to the frequency ratio be equal to zero, and solving for the value of the frequency ratio.

In addition to the magnification factors, the respective phase angles for the heaving and pitching motions in eqs. 11.134 and 11.139 are seen to be functions of the frequency ratio, and also to depend on the damping ratio. The generic phase angle (ϵ) is plotted in Figure 11.17b as a function of the generic frequency ratio (ω/ω_n). For low values of the frequency ratio (ω/ω_n), we see that the phase angle between the body motions and the wave-induced force approaches zero. The force and the body are then nearly in phase. The condition is referred to as *wave-riding*. For the heaving motions, the body and the free surface of the wave rise and fall with the same amplitudes and at the same velocities under a wave-riding condition. For the pitching motions, the average slope of the free surface at any time and the pitching angle are equal. At the other end of the frequency range, we see that the high-frequency waves do not excite body motions. Note that curves similar to those in Figure 11.17 can be found in books on mechanical vibrations.

Some design ramifications of the equations describing the undamped natural frequencies and the critical damping coefficients are evident. Both the undamped natural frequencies in eqs. 11.132 and 11.137 and the critical damping coefficients in eqs. 11.133 and 11.139 are seen to depend on the restoring coefficient and the total inertial coefficients. The restoring coefficient, in turn, depends on the waterplane area. By decreasing the waterplane area, the value of the undamped natural frequency is reduced. Hence, altering the waterplane area can be used to de-tune the heaving and pitching motions from the sea. This fact led to the creation of a class of both floating structures and marine vehicles called *small-waterplane-area* platforms and ships. The motions of the body can also be de-tuned by increasing the inertial coefficient in the heaving and pitching equations of motion. Usually, the mass of the structure is specified. This is an input value that normally relates to the purpose of

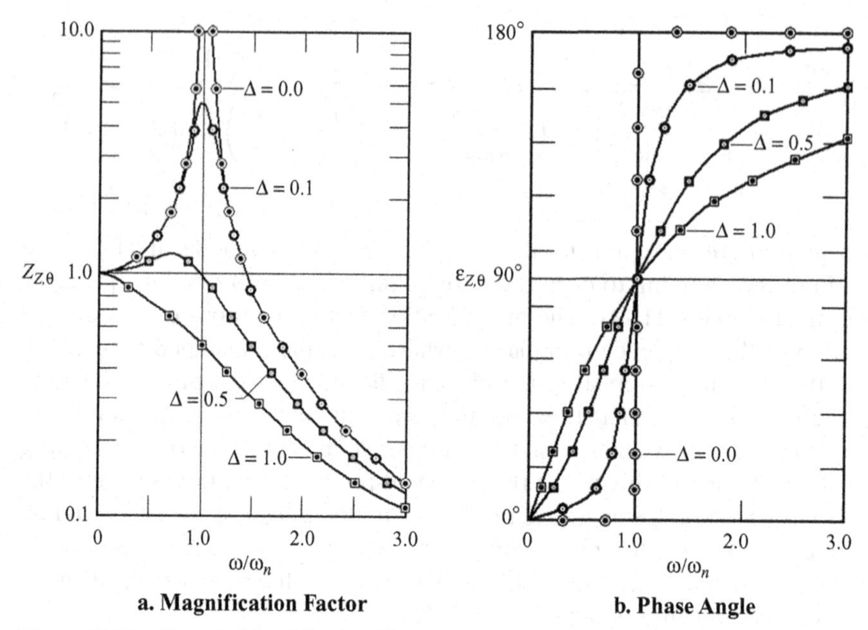

a. Magnification Factor **b. Phase Angle**

Figure 11.17. *Generic Magnification Factor and Phase Angle as Functions of Frequency Ratio.* The parameter Δ is the ratio of the linear damping and critical damping coefficients for the motion of interest (see eqs. 11.140 and 11.141). The curves apply to the uncoupled heaving and pitching of any floating body having x-z and y-z symmetries that are not under way. Resonance ($\omega/\omega_n = 1$) is a condition to be avoided in the design of barges, but is a desirable condition in wave-energy conversion. See McCormick (2007) for a discussion of resonant wave-energy conversion techniques.

the body. Although the body mass is usually fixed in the conceptual design phase, the added-mass component of the inertial coefficient is available as a design tool. For a Lewis form, the frequency-independent added mass per unit body length is presented in eq. 11.83. Because this is a shape-dependent parameter, the hull geometry can be modified to change the added-mass value and, subsequently, that of the natural frequency. In reality, the dynamically based design of floating structures is more complex. The natural frequency expressions in eqs. 11.132 and 11.139 are of value in the conceptual design phase of the design process.

The forces on bodies in random seas are discussed in Section 9.3, and the nature of the motions induced by random-wave forces is introduced in Section 10.3. The discussion in Section 10.3 concerns single-degree-of-freedom systems. The barge in Example 11.11, experiencing uncoupled heaving and pitching motions, is such a system. The analysis of random waves is introduced in Chapter 5. To apply the techniques in Section 10.3 to the uncoupled motions of a floating body, one must define the amplitude response function, as in eq. 10.51. Then, simply follow the analytical procedure following that equation. For ship motions in random or "confused" seas, St. Denis and Pierson presented a classical analytical approach to the problem in 1953 (see St. Denis and Pierson, 1953). Twenty years later, to commemorate the publication of the St. Denis-Pierson paper, the H-7 Panel of the Society of Naval Architects and Marine Engineers (SNAME) sponsored a symposium to update the science and technology involved in the prediction of ship motions in random seas. The results are presented in the publication by SNAME (1974). The book by Price and Bishop (1974) is also recommended for those readers interested in the motions of ships in random seas.

In the next section, the added-mass and radiation-damping coefficient expressions are compared with the experimental results of Vugts (1968).

11.5 Experimental and Theoretical Hydrodynamic Coefficient Data

To establish a range of applicability of the Lewis form added-mass coefficients and determine the accuracy of the radiation-damping coefficient expressions, we apply the expressions to the experimental conditions of Vugts (1968). The two sectional geometries chosen for this comparison are the rectangular section and the semi-circular section. The added-mass expression for the former is found in Appendix H. For both geometries, a modifier is applied to account for the frequency dependence of the added-mass coefficients.

A. Modification of the Lewis Added-Mass Coefficients to Include Frequency Dependence

We find in Section 11.3D that the radiation-damping coefficient for a strip depends on the frequency of excitation of the vertical motions of the strip. However, the added-mass coefficients in eq. 11.75 and in Appendix H determined by Lewis (1929) are independent of the motion frequency for any strip geometry. To modify the Lewis coefficients to account for the frequency, the coefficients can be multiplied by a *frequency coefficient* (k_4). The non-dimensional frequency coefficient is thoroughly discussed by Petersen (2000). Also, see Ursell (1949), Tasai (1959), Korvin-Kroukovsky (1955), Korvin-Kroukovsky and Jacobs (1957), and others for discussions of k_4. Following Petersen (2000), the added-mass coefficient that accounts for the frequency effects is

$$a'_w(\xi, \omega) = \rho S_\xi k_2 k_4 \tag{11.143}$$

where ω is the excitation frequency. Here, S_ξ is the sectional (strip) area. Also in eq. 11.143 is the *shape parameter* (k_2). This parameter is defined as the frequency-independent added mass divided by the displaced mass for a given strip. For the Lewis form described in Section 11.3A, that parameter is

$$k_2 = \frac{\left[(1 + A_1)^2 + 3A_3^2\right]}{\left[(1 + A_3)^2 - A_1^2\right]} \tag{11.144}$$

where A_1 and A_3 are the Lewis parametric constants introduced in eq. 11.35. For the rectangular strip of breadth $B_\xi = 2b_\xi$ and draft d_ξ, as presented in Appendix H, the shape parameter is

$$k_2 = C_2 \frac{\pi}{2}\left(\frac{b_\xi}{d_\xi}\right) \tag{11.145}$$

Here, the value of coefficient C_2 depends on the whether or not the body is fully submerged and the water depth. From Petersen (2000), the frequency coefficient (k_4) in eq. 11.143 is found to have various expressions, depending on the range of values of

$$q(\omega) \equiv \frac{\omega^2 B_\xi}{2g} \tag{11.146}$$

Table 11.3. *Expression for the frequency coefficient, k_4, after Petersen (2000)*

For $d_\xi/B_\xi \le 6.1$, the following apply:

(a) $0 < q \le q_a$,

$$k_{4a} = -\frac{8}{\pi}\ln\left[0.795\left(1+\frac{2d_\xi}{B_\xi}\right)q\right] \quad (11.147)$$

where

$$q_a = -\frac{1.3503}{\left(\frac{d_\xi}{B_\xi}\right)^{-0.9846}+2.3567} + 0.5497 \quad (11.148)$$

(b) $q_a < q \le 1.388$,

$$k_{4b} = 0.2367q^2 - 0.4944q + 0.8547 + \frac{0.01}{q+0.0001} \quad (11.149)$$

(c) $1.388 < q \le 7.31$,

$$k_{4c} = 0.4835 + \sqrt{-0.0484 + 0.0504q - 0.001q^2} \quad (11.150)$$

(d) $7.31 < q$,

$$k_{4d} = 1 \quad (11.151)$$

For $d_\xi/B_\xi > 6.1$: $q_a = 0$, and

$$\begin{aligned} k_{4b} &= k_{4b}, \quad \text{where } 0 < q \le 1.388 \\ k_{4c} &= k_{4c}, \quad \text{where } 1.388 < q \le 7.31 \\ k_{4d} &= k_{4d}, \quad \text{where } 7.31 < q \end{aligned} \quad (11.152)$$

From Petersen (2000), we present validity ranges for k_4 in terms of $q(\omega)$ in Table 11.3. In that table, we see that the applicability ranges of k_4 depend, first, on the draft-to-breadth ratio (d_ξ/B_ξ) of the strip. For the two Lewis forms in Example 11.8 (the semicircle, where $A_1 = A_3 = 0$, and that for which $A_1 = 0$ and $A_3 = -0.111$), $d_\xi/B_\xi = 0.5 < 6.1$ for both, and k_{4a} through k_{4b} in Table 11.3 apply.

B. Vertical Motions of a Rectangular Section

In this section, the added-mass and radiation-damping coefficients for a strip having a rectangular geometry are presented. The experimental data are from Vugts (1968), and the theoretical data result from the applications of eq. 11.143 for the added-mass coefficient and eq. 11.88 for the radiation-damping coefficient. In eq. 11.143, the results in eqs. 11.147 and 11.149 through 11.152 in Table 11.3 are can be substituted. In eq. 11.88, the amplitude ratio in eq. 11.89 is included, where $Z_{ref} = S_\xi/B_\xi = d_\xi$ for the rectangular strip. The rectangular model studied by Vugts (1968) had a breadth (B_ξ) of 0.4 m and a draft (d_ξ) of 0.2 m. Hence, $d_\xi/B_\xi = 0.5 < 6.1$ in Table 11.3. For the experiment, the upper bound of $q(\omega)$ for k_{4a} is $q_a \simeq 0.238$. The coefficient in the shape parameter expression in eq. 11.145 is $C_2 = 0.75$ from Appendix H, assuming that the rectangular body is in deep water.

The non-dimensional added-mass results for the rectangular strip are presented in Figure 11.18, and the radiation damping results are found in Figure 11.19. In both figures, the abscissa is $q^{1/2} = \omega\sqrt{(B_\xi/2g)}$. In Figure 11.18, we see that the results of combining eqs. 11.143, 11.145, and the frequency coefficient (k_4) expressions in Table 11.3, although somewhat predicting the behavior of the non-dimensional added mass, improve the agreement over only a small range of $q^{1/2}$. For the rectangular strip, one can conclude that the application of the frequency coefficient lacks in

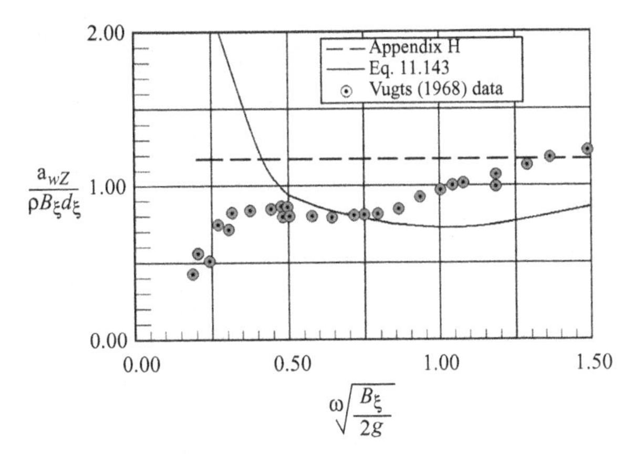

Figure 11.18. *Added-Mass Coefficients for a Heaving Rectangular Section.* For the rectangular section, the beam (B_ξ) is 0.4 m, and the draft (d_ξ) is 0.2 m. Also, see the results in Figure 11.25.

achieving the goal of improving the agreement between the predicted and observed added mass. For the higher frequency range, $\omega\sqrt{(B_\xi/2g)} > 1.50$, the added mass for the rectangular section is well predicted by the infinite frequency expression in Appendix H. The behavior of the non-dimensional radiation-damping coefficient is well predicted by the combination of eqs. 11.88 and 11.89, although the predicted and observed peak values vary by about 33%. It should be noted that the viscous effects and flow separation at the corners of the rectangular geometry are included in the experimental data, but not in the analytical curves. According to Vugts (1968), "The influence of viscosity is negligible, perhaps with the exception of large bulb-shaped sections where separation may occur at the upper side of the bulb."

C. Vertical Motions of a Semicircular Section

The semicircular section studied by Vugts (1968) had a radius (r_ξ) of 0.15 m. Hence, for this model, $B_\xi = 0.30\,\text{m}$ and $d_\xi = 0.15\,\text{m}$. This combination results in $d_\xi/B_\xi = 0.5 < 6.1$, and the frequency coefficient, k_{4a}, is used for $q_a \leq 0.238$ (obtained from eq. 11.148). It is interesting to note that the k_4 segments in Table 11.3 are the same for both the rectangular strip, discussed previously, and the semicircular strip because d_ξ/B_ξ and q have the same numerical values for each of the bodies studied by Vugts (1968).

The predicted and observed non-dimensional added-mass values for semicircular geometry are presented in Figure 11.20, and the non-dimensional radiation-damping coefficient results are found in Figure 11.21. In Figure 11.20, the agreement between the theoretical and experimental values is quite good, except for one experimental data point. It should be noted that Vugts (1968) states that "in the low frequency range ($\omega\sqrt{(B/2g)} < 0.5$) deviations appear, especially in the added mass.

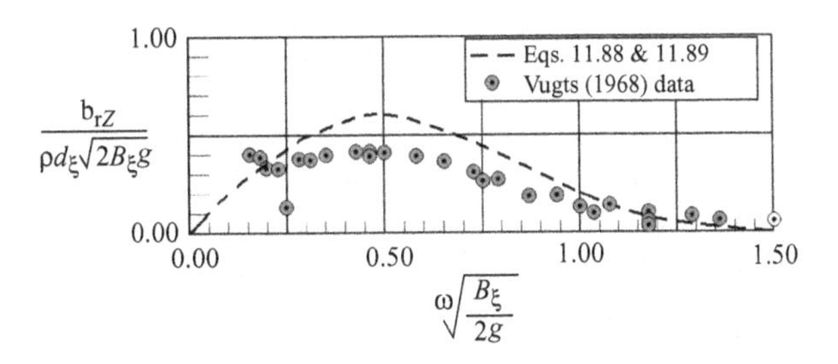

Figure 11.19. *Radiation Damping Coeffients for a Heaving Rectangular Section.* For the rectangular section, the beam (B_ξ) is 0.4 m, and the draft (d_ξ) is 0.2 m. The theoretical curve is also predicted by the Havelock (1942) amplitude-ratio expression in eq. 11.166.

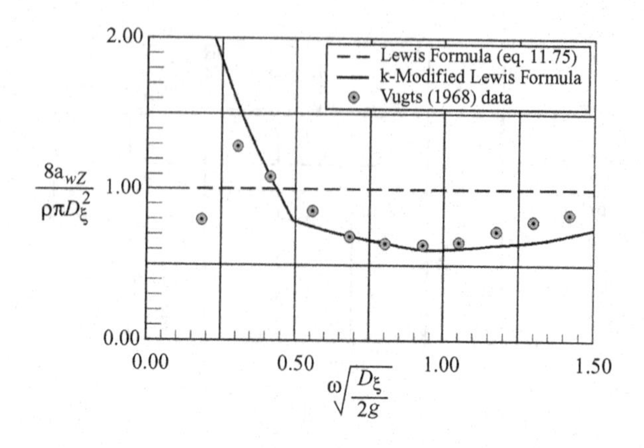

Figure 11.20. *Added-Mass Coeffients for a Heaving Semicircular Section.* For the semicircular section, the diameter (D_ξ) is 0.3 m and, hence, the radius (R_ξ) is 0.15 m.

This is due to experimental inaccuracies." The agreement between the analytical and experimental radiation coefficient results in Figure 11.21 is excellent.

From the results presented in this section, one can conclude that the Lewis formula is satisfactory for predicting the added mass for sections undergoing high-frequency vertical oscillations. When modified for low to moderate frequencies, where the Lewis formulas for the Lewis forms and the rectangular sections in Appendix H are multiplied by the frequency coefficient (k_4) expressions in Table 11.3, the added mass results are mixed, depending on the geometry of the body.

In the next section, the singularity method of determining the hydrodynamic coefficients is introduced. The method includes the frequency dependence of the added-mass and radiation-damping coefficients.

11.6 Singularity Method of Determining Hydrodynamic Coefficients

As discussed in the previous section, the added-mass expressions obtained by using the analysis of Lewis (1929) are not frequency-dependent. This is also discussed in Sections 9.2A(2) and 11.3. Physically, this frequency independence is commonly interpreted as the strip oscillating at an infinite frequency. In Section 11.5, a frequency coefficient (k_4), well described by Petersen (2000), is introduced to account for the frequency dependence, with a modicum of success. We now direct our attention to another analytical method for the determination of the added-mass and radiation-damping coefficients for arbitrary frequencies.

There are a number of methods designed to predict the frequency dependence of the hydrodynamic coefficients (added mass and radiation damping). These include both analytical and numerical techniques. The analytical techniques are of primary interest here, and include those of Havelock (1942), Ursell (1954), Korvin-Kroukovsty and Jacobs (1957), Grim (1960), Motora (1964), Frank (1967),

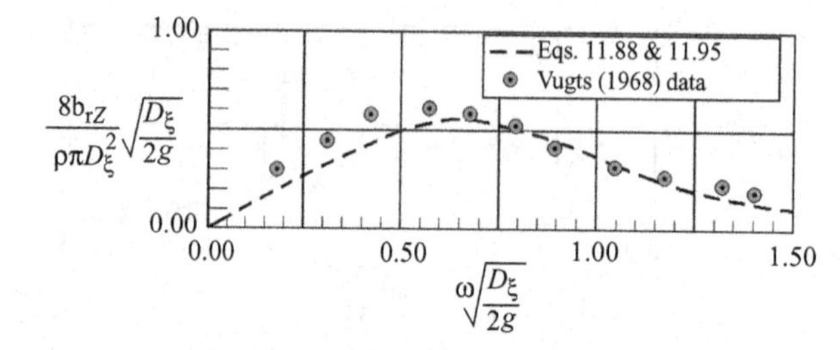

Figure 11.21. *Radiation-Damping Coeffients for a Heaving Semicircular Section.* For the semicircular section, the diameter (D_ξ) is 0.3 m and, hence, the radius (R_ξ) is 0.15 m.

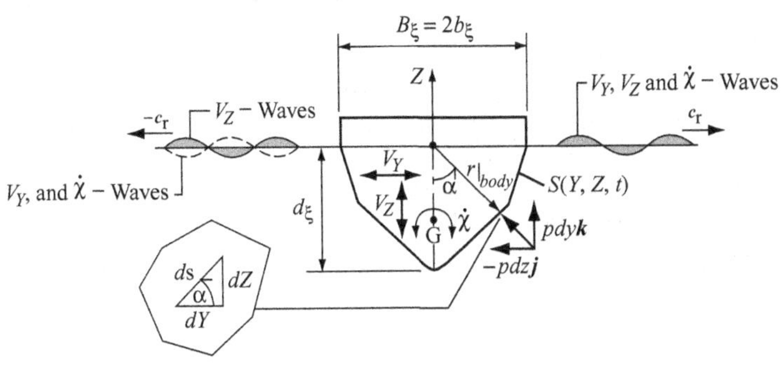

Figure 11.22. *Y-Z Planar Motions of a Strip in Deep Water.* The body motions create waves that are radiated away from the body. Because the body is of infinite length and moves uniformly over its length, the subscript ξ is not needed to identify the coordinates (Y,Z) attached to the strip, as is done in the previous sections. The geometrical shape of the wetted surface of the strip is denoted by the function $S(Y,Z,t)$. The waves on the left side produced by the vertical (heaving) motions are in-phase with those on the right side. For the horizontal (swaying) and angular (rolling) motions, the waves on the sides are 180° out of phase.

and others. Havelock (1942) represents the strip by a distribution of sources with a goal of determining the energy lost to the strip motions due to radiation damping. In addition, Havelock (1927, 1942), Ursell (1954), and Frank (1967) introduce integral forms of the velocity potentials representing the distribution of these singularities in their derivations. For three-dimensional motions, Yeung (1981) presents an expansion method incorporating the Fourier series. His application is to swaying, heaving, and rolling truncated circular cylinders.

As in Section 11.3, the problem is to determine the velocity potential describing the flow excited by the body motions. In this section, the integral method of Havelock (1942) is presented. The Havelock papers are also found in the collection of that author's papers edited by Wigley (1963). For simple sectional (strip) geometries, such as the rectangle and the semicircle, the method leads to *quasi*-analytical expressions for the added-mass and radiation-damping coefficients. Frank (1967) applies the source-distribution technique to several of these strip geometries. For more complicated sectional geometries, the method must be combined with numerical techniques.

Excellent summaries of the various analytical methods applied to the strip theory are found in the reports of McTaggart (1996), Phelps (1997), and Petersen (2000), and in the papers by Fossen and Smogeli (2004) and Arribas and Fernádez (2006).

In the following derivations, the body is assumed to be excited artificially without regard to incident waves. This assumption is in line with the analyses of Havelock (1942), Ursell (1954), and Frank (1967), and the experiments of Vugts (1968). The excitations produce inertial reactions of the ambient water and losses in energy due to radiation. Referring to the strip in Figure 11.22, when the body motions are vertical, then the radiated wave pattern is symmetric with the centerplane, or X-Z plane. When the motions are horizontal, then the created wave patterns are asymmetric with respect to the centerplane. The vertical motions of the section (or strip) could be considered to be due to either the heaving or small-angle pitching of the body or a combination thereof, as in Section 11.3. Similarly, horizontal motions of the strip could be due to either swaying, small-amplitude yawing, or a combination of the two. Only the vertical motions are discussed in detail herein.

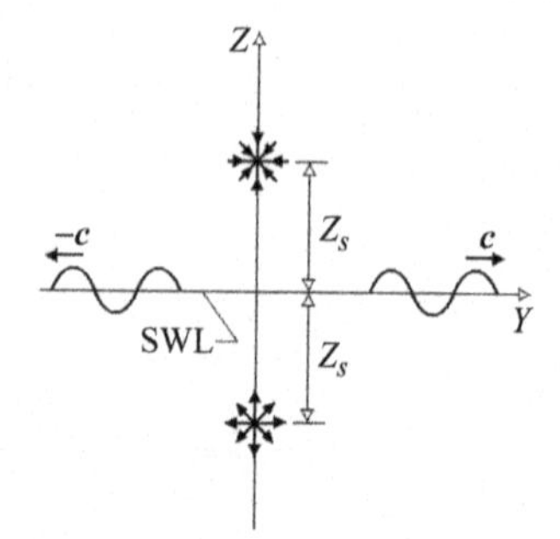

Figure 11.23. *Submerged, Pulsating Source.* The source at $(0, -Z_s)$ has a time-dependent strength, $M(t) = M_o e^{-i\omega t}$, where ω is the circular frequency of the pulsation. As is shown, the negative source (sink) at the image point above the horizontal axis is 180° out of phase in time with the submerged source. Over one half of a period ($T = 2\pi/\omega$), the source becomes a sink, and vise versa.

Referring to the arbitrary strip geometry sketched in Figure 11.22, we begin by assuming that the strip is forced to oscillate vertically, the motions producing monochromatic, linear waves that radiate away from the body. The waves, sketched in Figure 11.22, are right-running in the positive Y-direction and left-running in the negative direction. The strip can be considered to be that of an infinitely long cylinder and of uniform cross-section; hence, the analysis presented is two-dimensional. In Figure 11.22, the d_ξ is the draft, or keel depth, of the body.

To derive the expressions for the frequency-dependent added-mass and radiation-damping coefficients for an oscillating body in the free surface, the velocity potential describing the motion-excited flow must be determined. In the following subsections, the singularity method of Havelock (1942), which results in an integral form of the potential, is described.

A. The Source Pair

To introduce the reader to the singularity method, we present a portion of one of the earlier analyses employing singularities and their images in hydrodynamic analyses, that of Havelock (1927, 1942). The Havelock analyses demonstrate how the *method of images* (the coupling of a submerged singularity and its out-of-phase image) can be used to determine the properties of waves created by submerged bodies. Our goal is to demonstrate how to derive an integral expression for a velocity potential at a point in the flow field at a distance from a singularity located at a depth beneath a free surface. The singularity, used here to demonstrate the method, is a pulsating source of strength $M(t) = M_o e^{-i\omega t}$ at a point $(0, -Z_s)$ in Figure 11.23, where ω is the circular frequency of the pulsation. In addition, a source that is out of phase with the submerged pulsating source is located at $(0, Z_s)$. The subscript s identifies properties and dimensions related to the two sources. See the book by Katz and Plotkin (2001), among others, for a discussion of other applications of singularities in flow analyses.

The velocity potential representing the two sources mentioned in the last paragraph and sketched in Figure 11.23 is

$$\phi_s = M_o e^{-i\omega t} \ln\left[\frac{\sqrt{Y^2 + (Z + Z_s)^2}}{\sqrt{Y^2 + (Z - Z_s)^2}}\right] = \frac{M_o}{2} e^{-i\omega t} \ln\left[\frac{Y^2 + (Z + Z_s)^2}{Y^2 + (Z - Z_s)^2}\right] \qquad (11.153)$$

where, again, the subscript s identifies the source properties. We recall from Chapter 2 that a source of negative strength is a sink. Hence, each source becomes a sink over half of a pulsation period. From the expression in eq. 11.153, the following expression for the vertical velocity is obtained:

$$\frac{\partial\phi_s}{\partial Z} = M_o e^{-i\omega t}\left\{\left[\frac{Z+Z_s}{Y^2+(Z+Z_s)^2}\right]-\left[\frac{Z-Z_s}{Y^2+(Z+Z_s)^2}\right]\right\} \tag{11.154}$$

Applying eqs. 11.153 and 11.154 at $Z = 0$, the waterline conditions for velocity potential and the vertical velocity are obtained. For the out-of-phase source pair, these conditions are

$$\phi_s|_{Z=0} = 0 \quad \text{and} \quad \left.\frac{\partial\phi_s}{\partial Z}\right|_{Z=0} = 2M_o e^{-i\omega t}\left(\frac{Z_s}{Y^2+Z_s^2}\right) \tag{11.155}$$

It is interesting to note that the velocity potential for the Lewis forms in eq. 11.52 vanishes along $Z = 0$, where $\gamma = 0$ in that equation. This is a condition associated with infinite-frequency body oscillations. Because a free surface is assumed to exist, we introduce an additional velocity potential to represent the generated waves. This potential associated with the free surface can be written in a generalized integral form as

$$\phi_{fs} = e^{-i\omega t}\int_0^\infty F(k)e^{kZ}\cos(kY)dk \tag{11.156}$$

Here, the subscript *fs* identifies the free-surface potential. In eq. 11.156, $F(k)$ represents the to-be-determined free-surface function associated with the waves produced by the source pair. The other terms in the integrand of eq. 11.156 result from our knowledge of linear surface waves obtained from Chapter 3. We can also represent the velocity potential in eq. 11.153 in an integral form similar to that in eq. 11.156. Using the integral relationship found on p. 493, number 3.951(3), of the book by Gradshteyn and Ryzhik (1980), we can express the potential in eq. 11.153 as

$$\phi_s = M_o e^{-i\omega t}\int_0^\infty \frac{1}{k}\left[e^{-k(Z_s-Z)} - e^{-k(Z_s+Z)}\right]\cos(kY)dk \tag{11.157}$$

This equivalent integral expression is valid provided that both $(Z_s + Z)$ and $(Z_s - Z)$ are greater than zero. Physically, this condition is satisfied between the sources in Figure 11.23. The purpose of representing the source pairs by integral expressions is to determine the free-surface function, $F(k)$; hence, this condition is not considered to be a limiting restriction. The complete velocity potential representing the source pairs and the free surface is the sum of those in eqs. 11.156 and 11.157, that is,

$$\phi = M_o e^{-i\omega t}\int_0^\infty \frac{1}{k}\{(e^{-k(Z_s-Z)} - e^{-k(Z_s+Z)}\}\cos(kY)dk + e^{-i\omega t}\int_0^\infty F(k)e^{kZ}\cos(kY)dk \tag{11.158}$$

Combine the velocity potential in eq. 11.158 with eq. 3.7, the linearized free-surface condition. In the resulting expression, assume that the sum of the integrands equals zero. From this assumption, the following is obtained:

$$F(k) = -2M_o\frac{e^{-kZ_s}}{(k-k_0)} \tag{11.159}$$

where $k_0 = \omega^2/g$ is the deep-water wave number. For the out-of-phase source pair and resulting free-surface flows, the velocity potential expression in eq. 11.158

becomes

$$\phi = \phi_s + \phi_{fs} = \frac{M_o}{2} e^{-i\omega t} \ln\left[\frac{Y^2 + (Z + Z_s)^2}{Y^2 + (Z - Z_s)^2}\right] - 2M_o e^{i\omega t} \int_0^\infty \frac{e^{-k(Z_s - Z)}}{k - k_0} \cos(kY) dk \tag{11.160}$$

This is the first equation in the paper by Havelock (1942). The free-surface integral in eq. 11.160 appears in one form or another in a number of seakeeping analyses based on singularity distributions. For example, see the paper by Grim (1960). The inclusion of the alternative form of Havelock's free-surface integral is beyond the scope of this book.

The free-surface displacement to the right of the source pair in Figure 11.23 is found by combining eqs. 3.6 and 11.160, noting that the source potential (φ_s) vanishes on the SWL. The resulting free-surface displacement expression is

$$\begin{aligned}
\eta(Y, t)|_{Y \geq 0} &= -\Re\left[\frac{1}{g}\frac{\partial \phi}{\partial t}\bigg|_{z \simeq 0}\right] = -\Re\left\{2i\frac{\omega}{g} M_o e^{-\iota\omega t} \int_0^\infty \frac{e^{-kZ_s}}{k - k_0} \cos(kY) dk\right\} \\
&= \lim_{\epsilon \to 0}\left\langle -\Re\left\{2i\frac{\omega}{g} M_o e^{-i\omega t} \int_0^\infty \frac{e^{-kZ_s}}{k - k_0 + i\sigma} \cos(kY) dk\right\}\right\rangle \\
&= 2\pi\frac{\omega}{g} M_o e^{-kZ_s} \cos(k_0 Y - \omega t) \\
&\quad - 2\frac{\omega}{g} M_o \int_0^\infty \frac{e^{-k_0 Y}}{k^2 + k_0^2} [k\cos(kZ_s) - k_0 \sin(kZ_s)] \sin(\omega t) dk
\end{aligned} \tag{11.161}$$

The last equality of the equation is the result of a contour integration. That is, in the integral of the second line of eq. 11.161, k is treated as a complex variable. The artifice, $\sigma > 0$, is introduced in the second line of eq. 11.161 by Havelock (1942) to provide a singular point off of the imaginary axis. The importance of the equation for our purposes is not in the result per se, but in the physical meaning of the results of the last equality. As indicated by the time functions, the first term of that equality represents a traveling right-running wave, whereas the last integral represents standing evanescent waves. The expression for the traveling-wave component is of particular interest in determining the radiation damping of a heaving body, as demonstrated later in this section.

In the next sections, the point velocity potential in eq. 11.160 is used in determining the reaction forces on a rectangular strip undergoing oscillatory vertical motions.

B. Distributed Sources – Rectangular Strip

We now apply the singularity method to a rectangular strip by distributing the sources over the bottom and top of the strip, as sketched in Figure 11.24. The formulation of the equations is best done using this simplest of strip geometries. The evaluation of the resulting motion-induced forces on other strip geometries must be done numerically. The velocity potential representing the source distribution over the bottom of the rectangular section in the figure is

$$\Phi = \Phi_S + \Phi_{FS} = \int_{-b_\xi}^{b_\xi} \phi_s|_{Z=-d_\xi} dy + \int_{-b_\xi}^{b_\xi} \phi_{fs}|_{Z=-d_\xi} dy \tag{11.162}$$

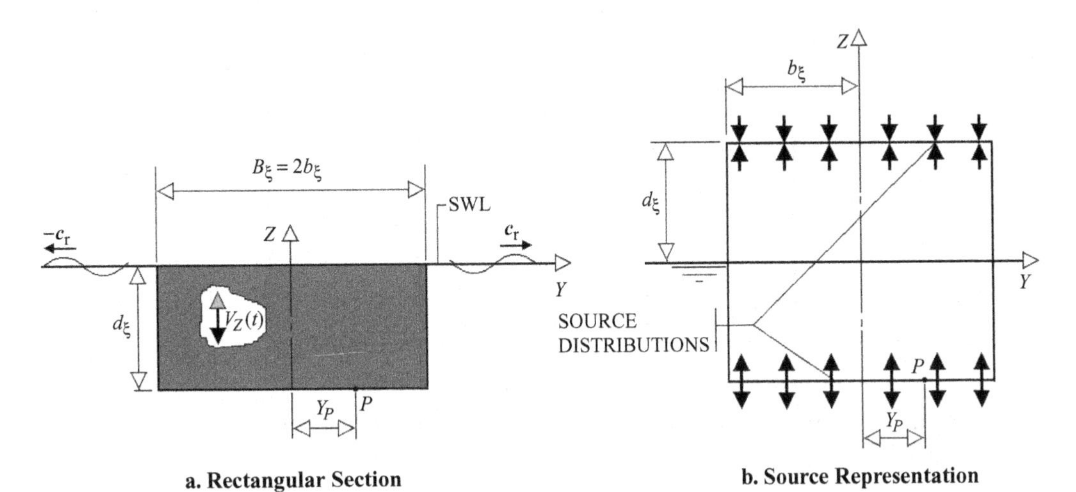

Figure 11.24. *Source Distribution Representation for a Vertically Oscillating Rectangular Strip*. This model is used by Havelock (1942) to obtain an approximate relationship for the radiation-damping coefficient for a ship having a rectangular section. According to Havelock, the model is valid when the radiated wavelength is greater than the beam, B_ξ.

where the source-distribution potential component, identified by the subscript S, is

$$\begin{aligned}\Phi_S &= \frac{M_o}{2} e^{-i\omega t} \int_{-b_\xi}^{b_\xi} \ln\left[\frac{(Y-Y_P)^2+(Z+d_\xi)^2}{(Y-Y_P)^2+(Z-d_\xi)^2}\right] dY_P \\ &= \frac{M_o}{2} e^{-i\omega t} \left\{(Y+b_\xi)\ln\left[\frac{(Y+b_\xi)^2+(Z+b_\xi)^2}{(Y+b_\xi)^2+(Z-b_\xi)^2}\right]\right. \\ &\quad -(Y-b_\xi)\ln\left[\frac{(Y-b_\xi)^2+(Z+b_\xi)^2}{(Y-b_\xi)^2+(Z-b_\xi)^2}\right] \\ &\quad +2(Z+d_\xi)\left[\tan^{-1}\left(\frac{Y+b_\xi}{Z+d_\xi}\right)-\tan^{-1}\left(\frac{Y-b_\xi}{Z+d_\xi}\right)\right] \\ &\quad \left.-2(Z-d_\xi)\left[\tan^{-1}\left(\frac{Y+b_\xi}{Z-d_\xi}\right)-\tan^{-1}\left(\frac{Y-b_\xi}{Z-d_\xi}\right)\right]\right\} \end{aligned} \tag{11.163}$$

and the free-surface potential component, identified by the subscript FS, is

$$\begin{aligned}\Phi_{FS} &= -2M_o e^{-i\omega t} \int_{-b_\xi}^{b_\xi}\int_0^\infty \frac{e^{-k(d_\xi - Z)}}{k-k_0}\cos[k(Y-Y_p)]dkdY_P \\ &= 2M_o e^{-i\omega t}\int_0^\infty \frac{e^{-k(d_\xi - Z)}}{k(k-k_0)}\{\sin[k(Y-b_\xi)]-\sin[k(Y+b_\xi)]\}dk \\ &= -2\frac{M_o}{k_0} e^{-i\omega t}\int_0^\infty \left[\frac{1}{k}-\frac{1}{k-k_0}\right] e^{-k(Z-d_\xi)}\{\sin[k(Y-b_\xi)]-\sin[k(Y+b_\xi)]\}dk \end{aligned} \tag{11.164}$$

Here, we note that the sequence of the integrations can be interchanged without affecting the outcome. The last line of eq. 11.164 results from the application of the partial fractions technique for simplifying the integrand. The integrals in eq. 11.164

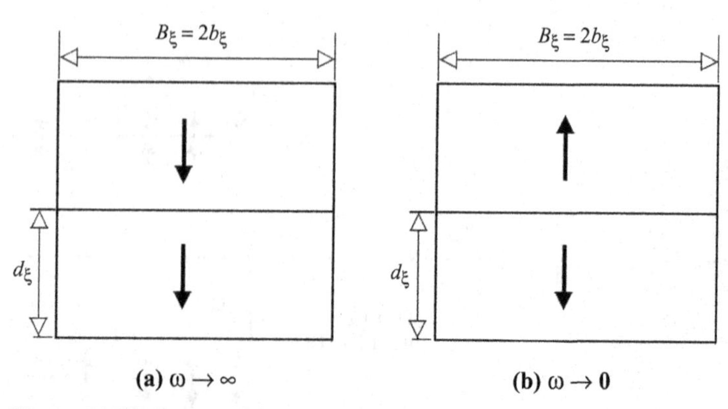

Figure 11.25. *Strip and Image Motions Corresponding to Extreme Frequencies.* After Newman (1977).

can be solved numerically. The integral in last line of eq. 11.164 can be expressed in terms of exponential integrals. The use of that alternative form of the wave-number integrals is outside the scope of this book.

Concerning the method of images, Newman (1977) gives physical interpretations for the two limiting frequency conditions for heaving and swaying bodies. Those limiting conditions are $\omega \to \infty$ and $\omega \to 0$. The former is the case for which the Lewis (1929) formulas apply. The condition for which $\omega \to \infty$ corresponds to the out-of-phase source distributions shown in Figure 11.25b. For this condition, the strip and its image are moving in phase, that is, both bodies are either rising or falling, as illustrated in Figure 11.24a. When $\omega \to 0$, the source distributions are in phase, and the resulting strip and image motions are out of phase, as illustrated in Figure 11.25b. The combined strip and its image resemble a pulsating body. The added mass corresponding to this condition must be infinite because the entire unbounded liquid is excited by the motions. An excellent discussion of the limiting frequency conditions is presented by Bishop and Price (1979).

In the following subsection, the velocity potential expression in eq. 11.163 is used to determine the infinite-frequency added mass of the vertically oscillating rectangular strip. Then, the potential expression in eq. 11.164 is used to obtain the properties of the motion-induced waves well away from the body and, thereby, to determine the radiation-damping coefficient, $b_{r\xi}$, in eq. 11.88. Later in the chapter, the velocity potential representing these far-field waves is combined with the Haskind relationships to determine the exciting force on the heaving strip.

(1) Alternative Infinite-Frequency Added Mass of a Heaving Rectangular Strip

The analysis leading to the Lewis (1929) added-mass expression for a "sharp-bilge" body is outlined in Appendix G of this book. The assumption used in that analysis is that the body is oscillating in some mode with an infinite frequency. One application of the analysis is to the heaving strip having a rectangular geometry, which is the geometry of a barge section. Other results obtained from the Lewis (1929) theory are found in Appendix H. The equations resulting from the Lewis sharp-bilge analysis are in terms of elliptic integrals. Because of this, the equations are not conducive to what we might call "field analyses." As an alternative to the Lewis analysis, the source-distribution method, described by Havelock (1942) and Frank (1967), can be used to obtain expressions for the infinite-frequency added mass. In the following paragraphs, we demonstrate this method by applying the source-distribution method to the infinite-frequency heaving motions of a rectangular strip. The alternative analysis results in a closed-form expression for the added mass. The alternative added-mass expression is applied to the experimental conditions of Vugts (1968). The results of the application are then compared to those obtained from the Lewis analysis and the experimental results.

Consider an infinite-frequency heaving of the rectangular strip, sketched in Figures 11.24a and 11.25a. To mathematically represent the body, the source distributions sketched in Figure 11.24b are used. The velocity potential for the source distribution shown in Figure 11.24a is presented in eq. 11.163. Because the condition of an infinite frequency implies a zero wavelength, the velocity potential in eq. 11.64 does not enter into the derivation. Hence, only the source potential in eq. 11.163 is used here. Combine the potential in eq. 11.163 with the dynamic pressure expression in the linearized Bernoulli's equation to obtain the dynamic pressure on the bottom of the rectangular strip in Figure 11.24. The result is

$$\begin{aligned} p_S|_{Z=-d_\xi} &= -\rho \frac{\partial \Phi_S}{\partial t}\bigg|_{Z=-d_\xi} = i\omega\rho\,\Phi_S|_{Z=-d_\xi} \\ &= i\omega\rho \frac{M_o}{2} e^{-i\omega t} \left\{ (Y+b_\xi)\ln\left[\frac{(Y+b_\xi)^2}{(Y+b_\xi)^2+(-2d_\xi)^2}\right] \right. \\ &\quad - (Y-b_\xi)\ln\left[\frac{(Y-b_\xi)^2}{(Y-b_\xi)^2+(-2d_\xi)^2}\right] \\ &\quad \left. + 2d_\xi\left[\tan^{-1}\left(\frac{Y+b_\xi}{-2d_\xi}\right) - \tan^{-1}\left(\frac{Y-b_\xi}{-2d_\xi}\right)\right]\right\} \end{aligned} \tag{11.165}$$

Before integrating this result over the bottom of the strip in Figure 11.24b, we must first specify the strength of the source distribution. To do so, we use the results of Katz and Plotkin (2001), that is, let the strength of the source distribution be $M_{So} = V_o/\pi$. By substituting this expression in eq. 11.165, the complex vertical reaction force (per unit body length) is obtained by integrating the dynamic pressure over the bottom of the strip. The result is

$$\begin{aligned} F_S' &= \int_{-b_\xi}^{b_\xi} p_S|_{Z=-d_\xi}\,dY = -\rho\frac{\partial \Phi_S}{\partial t}\bigg|_{Z=-d_\xi} = i\omega\rho\,\Phi_S|_{Z=-d_\xi} \\ &= i\omega\rho\frac{M_{So}}{2}e^{-i\omega t}\int_{-b_\xi}^{b_\xi}\left\{(Y+b_\xi)\ln\left[\frac{(Y+b_\xi)^2}{(Y+b_\xi)^2+(-2d_\xi)^2}\right]\right. \\ &\quad - (Y-b_\xi)\ln\left[\frac{(Y-b_\xi)^2}{(Y-b_\xi)^2+(-2d_\xi)^2}\right] \\ &\quad \left.+ 2d_\xi\left[\tan^{-1}\left(\frac{Y+b_\xi}{-2d_\xi}\right) - \tan^{-1}\left(\frac{Y-b_\xi}{-2d_\xi}\right)\right]\right\}dY \\ &= 2\rho\frac{b_\xi^2}{\pi}(i\omega V_0 e^{-i\omega t})\left[\left(\frac{d_\xi^2}{b_\xi^2}\right)\ln\left(\frac{d_\xi^2}{b_\xi^2}\right) + \left(1-\frac{d_\xi^2}{b_\xi^2}\right)\ln\left(1+\frac{d_\xi^2}{b_\xi^2}\right)\right. \\ &\quad \left. + 4\frac{d_\xi}{b_\xi}\tan^{-1}\left(\frac{b_\xi}{d_\xi}\right) - i\pi\right] \end{aligned} \tag{11.166}$$

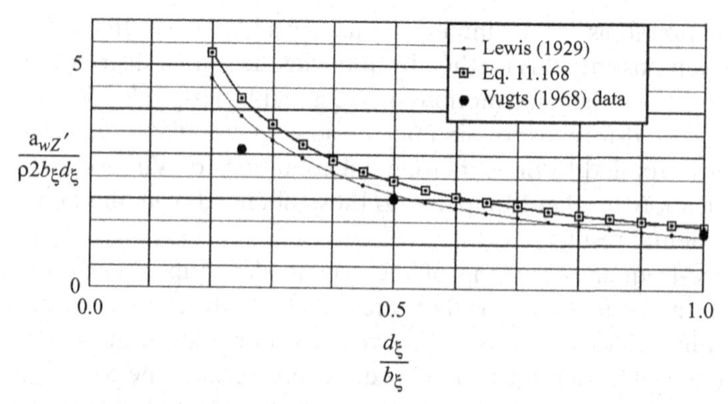

Figure 11.26. *Comparison of Theoretical and Experimental Heaving Added-Mass Coefficients for a Rectangular Strip*. The Vugts (1968) data correspond to the highest test frequency. The models studied had a beam ($B_\xi = 2b_\xi$) of 0.4 m. After McCormick and Hudson (2009).

The term in the first bracket of the last line of the equation is the complex vertical acceleration of the body. The vertical reaction force magnitude obtained from the expression is

$$F'_{so} \equiv |F'_s| = a'_w \omega V_o$$
$$= 2\rho\frac{b_\xi^2}{\pi}\left\{\sqrt{\left[\left(\frac{d_\xi^2}{b_\xi^2}\right)\ln\left(\frac{d_\xi^2}{b_\xi^2}\right)+\left(1-\frac{d_\xi^2}{b_\xi^2}\right)\ln\left(1+\frac{d_\xi^2}{b_\xi^2}\right)+4\frac{d_\xi}{b_\xi}\tan^{-1}\left(\frac{b_\xi}{d_\xi}\right)\right]^2+\pi^2}\right\}\omega V_0 \tag{11.167}$$

Here, a'_w is the added mass (per unit length) for a heaving rectangular strip heaving with an infinite frequency, and ωV_o is the amplitude of the heaving acceleration. The expression for the infinite-frequency added mass for a rectangular strip (section) is then

$$a'_w = 2\rho\frac{b_\xi^2}{\pi}\sqrt{\left\{\left[\left(\frac{d_\xi^2}{b_\xi^2}\right)\ln\left(\frac{d_\xi^2}{b_\xi^2}\right)+\left(1-\frac{d_\xi^2}{b_\xi^2}\right)\ln\left(1+\frac{d_\xi^2}{b_\xi^2}\right)+4\frac{d_\xi}{b_\xi}\tan^{-1}\left(\frac{b_\xi}{d_\xi}\right)\right]^2+\pi^2\right\}} \tag{11.168}$$

This is a closed-form expression that can be used to obtain approximate values for high-frequency vertical motions of a rectangular strip. Results obtained from the application of this added-mass expression to the experimental data of Vugts (1968) are presented in Figure 11.26. The Vugts data correspond to the highest experimental frequency studied. The results are presented in non-dimensional form, where the added mass divided by the mass of the displaced water is presented as a function of the draft-to-semi-beam ratio. Also in Figure 11.26 are the results of the application of the Lewis (1929) analysis, as outlined in Appendix G. The results in Figure 11.26 are after the technical note by McCormick and Hudson (2009).

EXAMPLE 11.12: ALTERNATIVE EXPRESSION FOR THE ADDED MASS FOR A HEAVING RECTANGULAR STRIP Consider the lowest draft-to-beam ratio of the rectangular-section models studied by Vugts (1968). In Figure 11.24a, this value is seen to be $d_\xi/b_\xi = 0.25$. From the Vugts paper, the highest test frequency data for this section corresponds to the frequency parameter value

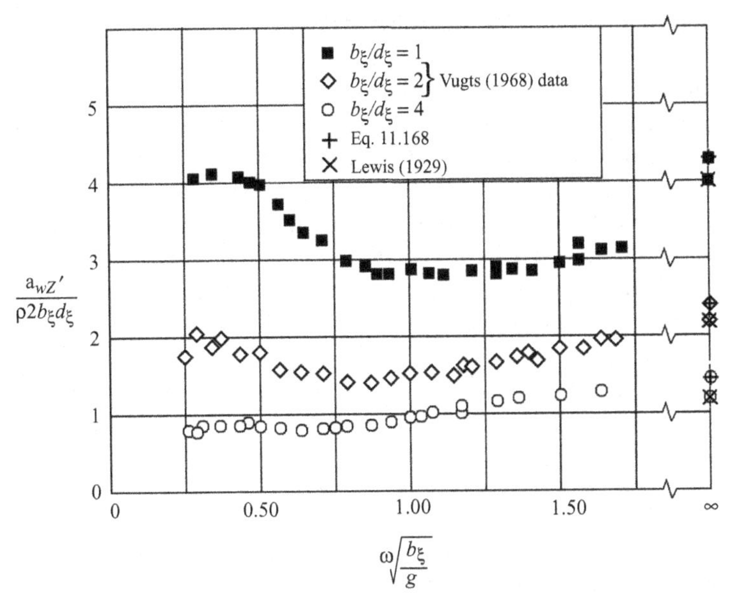

Figure 11.27. *Vugts Added-Mass Coefficients for a Rectangular Section.* The Vugts (1968) data are for three beam-to-draft ratios, $B_\xi/d_\xi = 2b_\xi/d_\xi = 2, 4,$ and 8. The data points below $\omega\sqrt{(b/d_\xi)} = 0.25$ are omitted because of scatter. After McCormick and Hudson (2009).

of $\omega\sqrt{(b_\xi/g)} \simeq 1.7$. Let us apply the Vugts results to a barge deployed in salt water ($\rho = 1.03 \times 10^3$ kg/m^3) having a beam ($B_\xi = 2b_\xi$) of 3 m and a draft (d_ξ) of 0.375 m. From the frequency parameter value, the heaving frequency for this barge is approximately 4.35 rad/s. The heaving period is then about 1.44 sec. The non-dimensional added-mass value (obtained from the application of eq. 11.168 to the rectangular barge section) is approximately 4.29, whereas the experimental value is about 3.1. The predicted added-mass value is 9.94×10^3 kg/m, and that corresponding to the experimental value is 7.18×10^3 kg/m. The percentage difference in the values is about 38.4%. The Lewis result in Figure 11.26 overpredicts the added mass by approximately 26%. In Figure 11.26, it can be seen that the agreement between the theory and experiment increases with d_ξ/b_ξ. The Vugts (1968) data actually have a trend toward larger added-mass values, as can be seen in Figure 11.27. The draft-to-half-beam (B_ξ/d_ξ) value for that figure is 1. From this trend, one can conclude that the agreement between theory and experiment is somewhat better than that in Figure 11.26.

Vugts' (1968) non-dimensional added-mass data over a frequency-parameter range of $0.25 \le \omega\sqrt{(b_\xi/g)} \le 1.71$ are presented in Figure 11.27 with the results from eq. 11.168 and the Lewis (1929) analysis in Appendix G. The experimental data below $\omega\sqrt{(b_\xi/g)} = 0.25$ were omitted from the figure due to scatter. Furthermore, the data presented by Vugts (1968) for each b_ξ/d_ξ value are for three different motion amplitudes. However, the data appear to be insensitive to the amplitudes studied. From the results in Figure 11.27, one can see that the added-mass values corresponding to an infinite-frequency excitation overpredict the values by as much as 40% over the finite frequency range. This observation appears to hold true for the three beam-to-draft ratios studied, $B_\xi/d_\xi = 2b_\xi/d_\xi = 2, 4,$ and 8.

(2) Radiation Damping of a Heaving Rectangular Strip

In the last equality of eq. 11.161, it is seen that there are two wave components resulting from a pulsating source pair situated on the vertical axis. The first term in the equality represents traveling waves, and the second represents evanescent, standing waves. Well away from the vertical axis, the evanescent wave system is of second order, and can be neglected. Consider now the far-field waves excited by the vertical motions of the rectangular strip in Figure 11.24a. Following Havelock (1942), the free-surface displacement well away from the body due to the distributed sources over the bottom of the heaving rectangular strip in the figure is

$$\eta_{r0}(Y,t)|_{Y\gg b_\xi} = \frac{H_{r0}}{2}\cos(k_0 Y - \omega t) = 2\pi\frac{\omega}{g}\mathrm{M}_{\mathrm{FS}_o}e^{-k_0 d_\xi}\int_{-b_\xi}^{b_\xi}\cos[k_0(Y-Y_p)-\omega t]dY_p$$

$$= 4\frac{\pi}{g}\frac{\omega}{k_0}\mathrm{M}_{\mathrm{FS}o}e^{-k_0 d_\xi}\sin(k_0 b_\xi)\cos(k_0 Y - \omega t) \tag{11.169}$$

where H_{r0} is the deep-water height of the radiated wave and k_0 is the deep-water wave number. The waves are generated by the motions of the strip having a displacement described by eq. 11.17 and a vertical velocity obtained from eq. 11.18. In the present situation, the strip is oscillating in pure heave. Hence, the strip displacement at any time is

$$\zeta(t) = Z_0 e^{-i\omega t} \tag{11.170}$$

where Z_o is the amplitude of the heaving motion. The heaving velocity is then

$$V_Z(t) = \frac{d\zeta}{dt} = V_o e^{-i\omega t} = -i\omega Z_o e^{-\omega t} \tag{11.171}$$

Following Havelock (1942), the source strength for the free surface is $\mathrm{M}_{\mathrm{FSo}} = V_o/2\pi$. The reader should note that this strength is half that for the infinite-frequency condition discussed in the previous section. The reason for this difference in the source strength is as follows: Well away from the body on the free surface, the vertical motions of the strip and its image appear to be those of a doublet or dipole; hence, the choice of the strength expression. The substitution of the source strength expression into eq. 11.169 results in an expression from which the ratio of the radiated-wave amplitude and the body-motion amplitude can be obtained. That heaving amplitude ratio is

$$\mathrm{R}_Z \equiv \frac{H_{r0}/2}{Z_0} = 2e^{-k_0 d_\xi}\sin(k_0 b_\xi) \tag{11.172}$$

which is analogous to the expression in eq. 11.89. The *Havelock radiation damping coefficient* is obtained by replacing the heaving amplitude ratio in eq. 11.88 by the expression in eq. 11.172 to obtain

$$\mathrm{b}_r' = \frac{\rho g^2}{\omega^3}\mathrm{R}_Z^2 = 4\frac{\rho g^2}{\omega^3}e^{-2k_0 d_\xi}\sin^2(k_0 b_\xi) \tag{11.173}$$

Compare this equation with the combination of eqs. 11.88 and 11.89. We see that Z_{ref} in eq. 11.89 is d_ξ in eq. 11.173, and the frequency-of-encounter expression ($\omega_e{}^2/g$) in eq. 11.89 has been replaced by the deep-water wave number (k_0) in eq. 11.173. Results obtained from the application of eq. 11.173 to the Vugts (1968) experimental data are the same as those in Figure 11.19. In that figure, we see that the experimental and theoretical results agree rather well. Hence, we can have some confidence in the singularity method in predicting the radiation-damping coefficient.

The energy flux or radiated power due to the vertical oscillations of the rectangular strip is obtained from eq. 3.73 applied to the traveling-wave term in the free-surface expression in eq. 11.161, where $Y \gg b_\xi$ in Figure 11.24a. Using the $\mathrm{M_{FSo}}$ expression in eq. 11.169 and combining the results with the energy flux expression in eq. 11.87 results in the following expression for the radiated wave power per unit length (in the X-direction):

$$2P_r' = 2\frac{\rho g H_{r0}^2 c_0}{16} = \frac{1}{8}\rho g H_{r0}^2 \frac{\omega}{k_0} = \mathrm{b}_r' \frac{V_0^2}{2}$$
$$= 8\rho\pi^2 \frac{\omega}{k_0^2} \mathrm{M}_{\mathrm{FS}0}^2 e^{-2k_0 d_\xi} \sin^2(k_0 b_\xi)$$
$$= 2\rho \frac{\omega}{k_0^2} V_0^2 e^{-2k_0 d_\xi} \sin^2(k_0 b_\xi) = 2\rho \frac{\omega^3}{k_0^2} Z_0^2 e^{-k_o d_\xi} \sin^2(k_0 b_\xi) \qquad (11.174)$$

As in eq. 11.87, the factor of 2 results from the fact that waves are created on both sides of the Z-axis, as illustrated in Figure 11.24a. The last equality in eq. 11.171 is that of Havelock (1942).

EXAMPLE 11.13: RADIATION DAMPING COEFFICIENT FOR A HEAVING RECTANGULAR STRIP As in Example 11.12, we consider the lowest draft-to-breadth ratio of the rectangular-section models studied by Vugts (1968), which is $d_\xi/b_\xi = 0.25$. Let the body in that example be oscillating in salt water at a frequency corresponding to the frequency parameter value of $\omega\sqrt{(b_\xi/g)} \simeq 1.7$. The barge in Example 11.12 has a beam ($B_\xi = 2b_\xi$) of 3 m and a draft (d_ξ) of 0.375 m. The heaving period is approximately 1.44 sec. The value of the damping coefficient from eq. 11.173 from these conditions is approximately 0.738 N-s/m. The approximate heaving amplitude ratio value from eq. 11.172 is 0.0249.

In summary, the importance of the singularity method is that it can be used to numerically determine the hydrodynamic forces on bodies of any geometry by distributing the singularities along the wetted surface of the body. We have chosen the rectangular geometry for the strip to illustrate. Other geometries simply require defining $Z = f(Y)$ for the body in question and then performing the required integrations.

In addition to a source as the singularity, a doublet can also be used. In three-dimensional analyses, the singularities include sources and multipoles. According to Frank (1967), the use of multipole distributions to represent heaving floating bodies in contemporary seakeeping analyses began with Ursell (1949). For a discussion of the coupling of multipoles expansions and conformal mapping techniques applied to strip theory, the paper by Grim (1960) is recommended.

It should also be noted here that for the infinite frequency condition, such as that for the use of the Lewis (1929) forms and the source-distribution method leading to the added-mass expression in eq. 11.168, the deep-water wave number (k_0) also is infinite. Hence, the free-surface potential in eq. 11.164 vanishes, leaving only the distributed-source potential in eq. 11.163.

In the following section, the expression for the potential for the far-field radiated waves is used to determine the forces on the oscillating body. This is possible because of the works of both Haskind (1957) and Newman (1962b). Again, the rectangular strip is used to illustrate the application of the method.

11.7 Two-Dimensional Haskind Force Relationships

In the early 1960s, a simplified method of prediction of wave forces and moments on relatively large, fixed, and movable bodies was introduced to engineers in the Western Hemisphere when a paper by Haskind (1957) was translated by Newman (1962a). In his analysis, Haskind (1957) assumes that the wave-induced fluid motions are irrotational and, by doing so, is able to take advantages of Green's theorem. See Appendix C for the derivation of Green's theorem. In a follow-on study, Newman (1962b) obtained the results of Haskind using a slightly different approach. The Haskind relationships are widely used by naval architects and ocean engineers. In this section, the derivation of the Haskind relationships is both outlined and discussed.

According to Newman (1962b), for bodies of arbitrary geometry, the Haskind (1957) method allows us to derive the expressions for the wave-induced forces and moments on floating and submerged bodies in terms of the far-field velocity potentials representing waves produced by the forced motions. The derived expressions do not depend on the diffraction potential or on the disturbance of the incident waves by the presence of the body. Prior to the introduction of the Haskind method, the Froude-Krylov approach was widely used, where no alteration of the wave field by the presence of the body was assumed. The Haskind force relationships give the analyst a method for determining the exciting forces on bodies that would normally be intractable. In the following paragraphs, the basics of the Haskind analysis, as given by Newman (1962b), are presented. The method is then applied to the oscillatory vertical motions of a rectangular strip, using the motion-induced far-field radiated waves discussed in previous the section.

A. Newman's Formulation

Consider any two-dimensional floating body exposed to deep-water waves. According to Newman (1962b), there are two independent problems to solve. Those are the diffraction problem caused by the (assumed linear) waves moving past the body, and the radiation problem caused by the forced oscillations of the body in calm water. Begin by assuming that the flow in the ambient fluid is irrotational and that the oscillations are small. The velocity potentials representing the incident waves, the wave diffraction, and the radiated wave obey the basic equations of hydrodynamics. Hence, for the body in Figure 11.28, any of these velocity potentials can be represented by

$$\Phi = \phi(Y, Z)\, e^{-i\omega t} \tag{11.175}$$

This potential must satisfy the equation of continuity in the form of Laplace's equation, eq. 3.8, that is,

$$\nabla^2\Phi = \nabla^2\phi = 0 \tag{11.176}$$

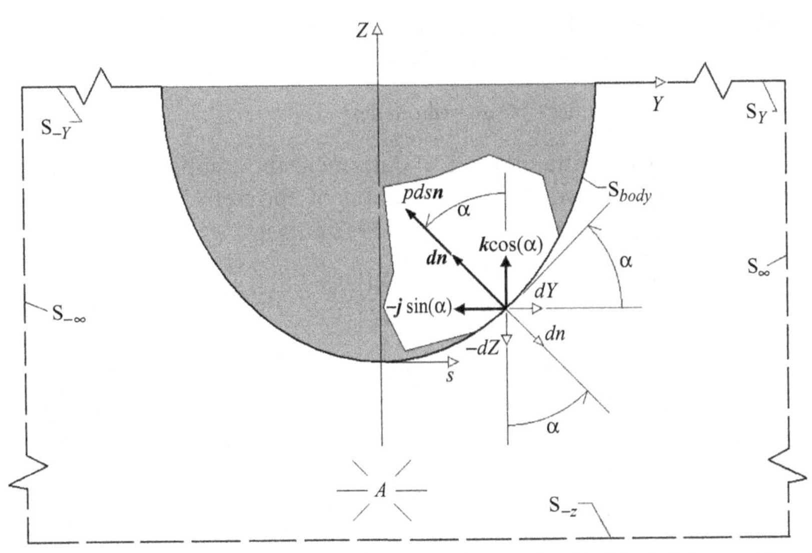

Figure 11.28. *Strip Geometry for the Derivation of the Haskind Relationships.* The pseudo-boundaries $S_{-\infty}$ and S_{∞} are in the far field, whereas for the deep-water application, the pseudo-boundary S_{-Z} is well below the region of influence of the body motions or the waves. The bounded (control) area is A.

In addition, the linear free-surface condition in eq. 3.7 must be satisfied by the potential. That condition is

$$\left(\frac{\partial \Phi}{\partial Z} - k_0 \Phi\right)\bigg|_{Z=\eta\simeq 0} = 0 \tag{11.177}$$

Here, $k_0 = \omega^2/g$ is the deep-water wave number.

Our interest is in a strip having an arbitrary geometry, such as that sketched in Figure 11.28. As is done in eq. 11.7, let the two-dimensional body in that figure be described by the function

$$S_{body}(Y, Z, t) = S_{body}(Y, Z) - s_o e^{-i\omega t} = 0, \quad 0 \leq Y < b_\xi \tag{11.178}$$

In this expression, s_o is the amplitude of body motion when the body is excited in calm water. For example, if the body is excited in heave, then $s_o = Z_o$. The velocity potential representing the flow then must satisfy the following body condition, as in eq. 11.8, that is,

$$\begin{aligned}\frac{DS_{body}}{Dt} &\equiv \frac{\partial S_{body}}{\partial t} + \boldsymbol{V}_{S_{body}} \cdot \nabla S_{body} \\ &= \frac{\partial S_{body}}{\partial t} + \nabla \Phi_{S_{body}} \cdot \boldsymbol{n}|\nabla S_{body}| = \frac{\partial S_{body}}{\partial t} \\ &+ \left(\frac{\partial \Phi}{\partial Y}\boldsymbol{j} + \frac{\partial \Phi}{\partial Z}\boldsymbol{k}\right)|S_{body} \cdot \boldsymbol{n}|\nabla S_{body}| = 0\end{aligned} \tag{11.179}$$

The subscript *body* identifies a function applied to the wetted strip boundary, and $\boldsymbol{j}$ and $\boldsymbol{k}$ are the unit vectors in the respective horizontal and vertical directions.

The water motions beneath the incident and reflected waves are represented by the respective velocity potentials, Φ_I and Φ_R. Note that in the three-dimensional case, we would be dealing with both wave reflection and diffraction. The combination of the two in three-dimensional analyses is normally combined in what is termed the diffraction potential, Φ_D. Returning to the two-dimensional flow, the velocity potential representing the wave excitation is

$$\Phi_e \equiv \Phi_I + \Phi_R = (\phi_I + \phi_R)\, e^{-i\omega t} \tag{11.180}$$

Later in this section, this potential is used to determine the exciting force on the body. Because the body is fixed, the combination of the velocity potentials in eq. 11.180 and the boundary condition in eq. 11.179 results in

$$\left.\frac{\partial \phi_e}{\partial n}\right|_{S_{body}} = \left.\frac{\partial \phi_I}{\partial n}\right|_{S_{body}} + \left.\frac{\partial \phi_R}{\partial n}\right|_{S_{body}} = 0 \tag{11.181}$$

When the body is excited in still water, as in Section 11.6, the radiation velocity potentials due to the swaying (Y), heaving, (Z) and rolling (χ), illustrated in Figure 11.21, are represented by Φ_Y, Φ_Z, and Φ_χ, respectively. The velocity potential at any point in the fluid is the sum of these, that is,

$$\Phi_Y + \Phi_Z + \Phi_\chi = (\phi_Y + \phi_Z + \phi_\chi)\, e^{-i\omega t} \tag{11.182}$$

For three-dimensional problems, we would include the velocity potentials associated with the surging (Φ_X), pitching, (Φ_θ) and yawing (Φ_σ) motions of the floating body, as illustrated in Figure 11.3. In seakeeping analysis in the field of naval architecture, the subscripts of the velocity potentials are usually numerical. That is, Φ_0 represents the incident-wave velocity potential, $\Phi_1, \Phi_2, \ldots,$ and Φ_6 represent the respective potentials due to the surging, swaying,..., and yawing motions and, lastly, Φ_7 represents the potential due to wave diffraction and reflection. Because the motions considered herein are two-dimensional, the alphabetical system is used.

The total flow field due to the incident and reflected waves and the waves excited by the body motions can be represented by the following velocity potential:

$$\Phi = \Phi_I + \Phi_R + \Phi_Y + \Phi_Z + \Phi_\chi = \Phi_e + \Phi_Y + \Phi_Z + \Phi_\chi \tag{11.183}$$

We can express the spatial derivative of each of the motion-induced potentials as

$$\frac{\partial \Phi_j}{\partial n} = \frac{\partial \Phi_j}{\partial q_j}\frac{\partial q_j}{\partial n} = V_j \frac{\partial q_j}{\partial n} \tag{11.184}$$

In this equation, $q_j = Y\,(j = 2), Z\,(j = 3)$, or $\chi\ (j = 4)$ for the two-dimensional (strip) application.

In the two-dimensional situation, the component potentials in eq. 11.183 satisfy the radiation condition, where an outgoing traveling wave exists as $Y \to \pm\infty$. For a three-dimensional situation, the radiation condition is that the outgoing waves vanish as $r = \sqrt{(X^2 + Y^2)} \to \infty$. The incident wave potential, Φ_i, does not satisfy the condition in three dimensions. All of the potentials in eq. 11.183 must individually satisfy the free-surface condition in eq. 11.177. Finally, from eq. C8 in Appendix C, each of the component velocity potentials satisfies Green's theorem, mathematically represented by

$$\int_S \left\{\phi_j \frac{\partial \phi_k}{\partial n} - \phi_k \frac{\partial \phi_j}{\partial n}\right\} ds = \int\!\!\int_A \{\phi_j \nabla^2 \phi_k - \phi_j \nabla^2 \phi_k\} dA = 0 \tag{11.185}$$

where j and k represent any of the potentials in eq. 11.183. In the integral of eq. 11.185, s is the curvilinear coordinate along the boundary of the fluid of area A.

Referring to Figure 11.28, we see that the boundary of A is

$$S = S_{body} + S_{-Y} + S_{-\infty} + S_{-Z} + S_{\infty} + S_Y \tag{11.186}$$

Assume that the waves are in deep water and that the boundary S_{-Z} is well below the influence of the body motion and the free surface.

The dynamic pressure on a strip boundary due to the excitation potential Φ_e (defined in eq. 11.180) is

$$P_e|_{S_{body}} = -\rho \left.\frac{\partial \Phi_e}{\partial t}\right|_{S_{body}} = i\omega\rho\,\phi_e|_{S_{body}} e^{-i\omega t} = i\omega\rho\,\Phi_e|_{S_{body}} \tag{11.187}$$

This expression is from the linearized Bernoulli's equation in eq. 3.70. Referring to the sketch in Figure 11.28, we can write the expressions for the excitation forces and rolling moment (per unit length) as follows: In the horizontal direction, the force is obtained from

$$F'_Y = -\int_{S_{body}} p_e|_{S_{body}} \sin(\alpha) ds = i\omega\rho \int_{S_{body}} \phi_e|_{S_{body}} \sin(\alpha) ds e^{-i\omega t} \tag{11.188}$$

In the vertical direction, the force expression is

$$F'_Z = \int_{S_{body}} p_e|_{S_{body}} \cos(\alpha) ds = -i\omega\rho \int_{S_{body}} \phi_e|_{S_{body}} \cos(\alpha) ds e^{-i\omega t} \tag{11.189}$$

The moment about the X-axis, positive in the counterclockwise direction, is

$$\begin{aligned} M'_X &= \int_{S_{body}} p_e|_S [Y\cos(\alpha) + Z\sin(\alpha)] ds \\ &= -i\omega\rho \int_S \phi_e|_{S_{body}} [Y\cos(\alpha) + Z\sin(\alpha)] ds e^{-i\omega t} \end{aligned} \tag{11.190}$$

We note here that the coordinate Z is negative on the wetted boundary. Hence, on the body in Figure 11.28, a negative (clockwise) moment results from the Z-term in eq. 11.190. The sine and cosine terms in eqs. 11.188 through 11.190 can be written in terms of the radiation potentials of eq. 11.182. This is done by noting

$$\frac{\partial \phi_Y}{\partial n} = \frac{\partial \phi_Y}{\partial Y}\frac{\partial Y}{\partial n} = V_{Yo}\frac{\partial Y}{\partial n} = -V_{Y_O}\sin(\alpha) \Rightarrow \sin(\alpha) = -\frac{1}{V_{Y_O}}\frac{\partial \phi_Y}{\partial n} \tag{11.191}$$

and

$$\frac{\partial \phi_Z}{\partial n} = \frac{\partial \phi_Z}{\partial Z}\frac{\partial Z}{\partial n} = V_{Zo}\frac{\partial Z}{\partial n} = V_{Z_O}\cos(\alpha) \Rightarrow \cos(\alpha) = \frac{1}{V_{Z_O}}\frac{\partial \phi_Z}{\partial n} \tag{11.192}$$

In these last two equations, the amplitudes of the respective horizontal and vertical body motions are V_{Yo} and V_{Zo}. By replacing the sine and cosine expressions in eqs. 11.188 through 11.190 with the expressions in eqs. 11.191 and 11.192, we obtain

$$F'_Y = -i\frac{\omega\rho}{V_{Yo}}\int_{S_{body}} \phi_e|_{S_{body}} \frac{\partial \phi_Y}{\partial n} ds e^{-i\omega t} = -i\frac{\omega\rho}{V_{Yo}}\int_{S_{body}} (\phi_I + \phi_R)|_{S_{body}} \frac{\partial \phi_Y}{\partial n} ds e^{-i\omega t} \tag{11.193}$$

$$F'_Z = -i\frac{\omega\rho}{V_{Zo}}\int_{S_{body}} \phi_e|_{S_{body}} \frac{\partial \phi_Z}{\partial n} ds e^{-i\omega t} = -i\frac{\omega\rho}{V_{Zo}}\int_{S_{body}} (\phi_I + \phi_R)|_{S_{body}} \frac{\partial \phi_Z}{\partial n} ds e^{-i\omega t} \tag{11.194}$$

and

$$M'_X = -i\omega\rho \int_{S_{body}} \phi_e|_{S_{body}} \left(\frac{Y}{V_{Zo}} \frac{\partial \phi_Z}{\partial n} - \frac{Z}{V_{Yo}} \frac{\partial \phi_Y}{\partial n} \right) ds e^{-i\omega t}$$

$$= -i\omega\rho \int_{S_{body}} (\phi_I + \phi_R)|_{S_{body}} \left(\frac{Y}{V_{Zo}} \frac{\partial \phi_Z}{\partial n} - \frac{Z}{V_{Yo}} \frac{\partial \phi_Y}{\partial n} \right) ds e^{-i\omega t} \tag{11.195}$$

The velocity potential representing the reflected wave can be eliminated by using the results of Green's theorem in eq. 11.185. As a result, we can make the following substitution in the integrand of eqs. 11.193 through 11.195:

$$\int_{S_{body}} \phi_R \frac{\partial \phi_j}{\partial n} ds = \int_{S_{body}} \phi_j \frac{\partial \phi_R}{\partial n} ds = -\int_{S_{body}} \phi_j \frac{\partial \phi_I}{\partial n} ds \tag{11.196}$$

Here, the last equality is a result of the boundary condition in eq. 11.181. The substitution changes the respective eqs. 11.193 through 11.195 to

$$F'_Y = -i\frac{\omega\rho}{V_{Yo}} \int_{S_{body}} \left(\phi_I \frac{\partial \phi_Y}{\partial n} - \phi_Y \frac{\partial \phi_I}{\partial n} \right)\Bigg|_{S_{body}} ds e^{-i\omega t} \tag{11.197}$$

$$F'_Z = -i\frac{\omega\rho}{V_{Zo}} \int_{S_{body}} \left(\phi_I \frac{\partial \phi_Z}{\partial n} - \phi_Z \frac{\partial \phi_I}{\partial n} \right)\Bigg|_{S_{body}} ds e^{-i\omega t} \tag{11.198}$$

and

$$M'_X = -i\omega\rho \int_{S_{body}} \left[\left(\phi_I \frac{\partial \phi_Z}{\partial n} - \phi_Z \frac{\partial \phi_I}{\partial n} \right)\Bigg|_{S_{body}} \frac{Y}{V_{Zo}} - \left(\phi_I \frac{\partial \phi_Y}{\partial n} - \phi_Y \frac{\partial \phi_I}{\partial n} \right)\Bigg|_{S_{body}} \frac{Z}{V_{Yo}} \right] ds e^{-i\omega t} \tag{11.199}$$

Following Newman (1962b), we note that Green's theorem applies to the entire control area, A, in Figure 11.28. We have previously noted that in deep water there is no motion at the boundary S_{-Z}. Furthermore, because of the linearized free-surface condition in eq. 11.177, eq. 11.185 is satisfied on the far-field boundaries, S_{-Y} and S_Y. Hence, to fully satisfy Green's theorem, the following must hold true:

$$\int_{S_{body}} \left(\phi_I \frac{\partial \phi_j}{\partial n} - \phi_j \frac{\partial \phi_I}{\partial n} \right)\Bigg|_{S_{body}} ds + \int_{-\infty}^{0} \left(\phi_I \frac{\partial \phi_j^-}{\partial Y} - \phi_j^- \frac{\partial \phi_I}{\partial Y} \right)\Bigg|_{Y\to-\infty} dZ + \int_{-\infty}^{0} \left(\phi_I \frac{\partial \phi_j^+}{\partial Y} - \phi_j^+ \frac{\partial \phi_I}{\partial Y} \right)\Bigg|_{Y\to\infty} dZ = 0 \tag{11.200}$$

Here, the superscripts $-$ and $+$ identify the waves traveling away from the body on the respective left and right sides in Figure 11.28. Replacing the *body* integrals in eqs. 11.197 through 11.199 with the far-field integrals dictated by this equation, we obtain what we term the *Haskind force relationships*. For two-dimensional situations, these are the following:

Horizontal Force:

$$F'_Y = i\frac{\omega\rho}{V_{Yo}} \int_{-\infty}^{0} \left[\left(\phi_I \frac{\partial \phi_Y^-}{\partial Y} - \phi_Y^- \frac{\partial \phi_I}{\partial Y} \right)\Bigg|_{Y\to-\infty} + \left(\phi_I \frac{\partial \phi_Y^+}{\partial Y} - \phi_Y^+ \frac{\partial \phi_I}{\partial Y} \right)\Bigg|_{Y\to\infty} \right] dZ e^{-i\omega t} \tag{11.201}$$

Vertical Force:

$$F'_Z = i\frac{\omega\rho}{V_{Zo}}\int_{-\infty}^{0}\left[\left(\phi_I\frac{\partial\phi_Z^-}{\partial Y} - \phi_Z^-\frac{\partial\phi_I}{\partial Y}\right)\bigg|_{Y\to-\infty} + \left(\phi_I\frac{\partial\phi_Z^+}{\partial Y} - \phi_Z^+\frac{\partial\phi_I}{\partial Y}\right)\bigg|_{Y\to\infty}\right] dZe^{-i\omega t} \tag{11.202}$$

Rolling Moment:

$$M'_X = -i\omega\rho\int_{-\infty}^{0}\left[\left(\phi_I\frac{\partial\phi_Y^-}{\partial Y} - \phi_Y^-\frac{\partial\phi_I}{\partial Y}\right)\frac{Z}{V_{Yo}}\bigg|_{Y\to-\infty} + \left(\phi_I\frac{\partial\phi_Y^+}{\partial Y} - \phi_Y^+\frac{\partial\phi_I}{\partial Y}\right)\frac{Z}{V_{Yo}}\bigg|_{Y\to\infty}\right.$$
$$\left. - \left(\phi_I\frac{\partial\phi_Z^-}{\partial Y} - \phi_Z^-\frac{\partial\phi_I}{\partial Y}\right)\frac{Y}{V_{Zo}}\bigg|_{Y\to-\infty} + \left(\phi_I\frac{\partial\phi_Z^+}{\partial Y} - \phi_Z^+\frac{\partial\phi_I}{\partial Y}\right)\frac{Y}{V_{Zo}}\bigg|_{Y\to\infty}\right] dZe^{-i\omega t} \tag{11.203}$$

B. Wave-Induced Vertical Force on Rectangular Section

To illustrate the use of the Haskind force relationships, consider the vertical motions of the rectangular strip sketched in Figure 11.24. The vertical force acting on the strip is obtained from eq. 11.202. In that equation, two velocity potentials appear. The first is that of the incident wave. For a right-running incident wave in deep water, the velocity potential from eq. 3.29 is found to be

$$\Phi_I = \phi_I e^{-i\omega t} = -i\frac{H_0}{2}\frac{g}{\omega}e^{k_0 Z}e^{i(k_0 Y - \omega t)} \tag{11.204}$$

The second potential in eq. Eq 11.202 is the radiation potential resulting from vertical body motions. Well away from the body (in the far field) the waves are outgoing, and the potentials can be written as

$$\Phi_Z^{\pm} = \phi_Z^{\pm}e^{-i\omega t} = -i\frac{H_{r0}}{2}\frac{g}{\omega}e^{k_0 Z}e^{\pm i(k_0 Y \mp \omega t)}$$
$$= -i[2Z_o e^{-k_0 d_\xi}\sin(k_0 b_\xi)]\frac{g}{\omega}e^{k_0 Z}e^{\pm i(k_0 Y \mp \omega t)} \tag{11.205}$$

The relationship between the wave height of the radiated wave and the amplitude of the body displacement is that of Yamamoto, Fujino, and Fukasawa (1980) found in eq. 11.89. In that equation, the reference draft, $Z_{ref} = S_\xi / B_\xi = d_\xi$ when applied to a rectangular section. The replacements of the spatial velocity potentials in eq. 11.202 by the expressions in eqs. 11.204 and 11.205 and the subsequent integration of the result produces the following expression for the vertical force on the vertically oscillating rectangular strip:

$$F'_Z = -\rho g H_0\frac{e^{-k_0 d_\xi}}{k_0}\sin(k_0 b_\xi)e^{-i\omega t} \tag{11.206}$$

A note on this formula: The formula is derived by considering the vertical strip motions excited by beam waves. Hence, there is reflection, and in the three-dimensional case, there is diffraction. For a body in a head or following sea, the diffraction effects are near the bow and stern of the body.

EXAMPLE 11.14: WAVE-INDUCED FORCES ON A HEAVING RECTANGULAR STRIP Consider again the barge in Examples 11.12 and 11.13. The section of the barge is rectangular, having a beam of ($B_\xi = 2b_\xi$) of 3 m and a draft (d_ξ) of 0.375 m.

The barge is subjected to monochromatic beam waves in deep water having a height of 1.5 m and a period of 7 sec. The deep-water wavelength of the incident wave is approximately $\lambda_0 \simeq gT^2/2\pi \simeq 76.5\,\text{m}$. Hence, the wave number is $k_0 \simeq 2\pi/\lambda_0 \simeq 0.0821\,\text{m}^{-1}$. Our interest is in the wave-induced vertical force on the rectangular strip of the barge. From eq. 11.206, we find that the wave-induced vertical force is

$$F'_Z = -\rho g H_0 \frac{e^{-k_0 d_\xi}}{k_0} \sin(k_0 b_\xi) e^{-i\omega t} \simeq -2.20 \times 10^4 e^{-i0.898t} \left(\frac{\text{N}}{\text{m}}\right)$$

Compare this value to the hydrostatic wave force, which is

$$F'_{Z0} = -\rho g \frac{H_0}{2} (2b_\xi)\, e^{-i\omega t} \simeq -2.27 \times 10^4 e^{-i0.898t} \left(\frac{\text{N}}{\text{m}}\right)$$

One can see that the force amplitude due to the Haskind equation is slightly less than the hydrostatic wave force. The reason for this good agreement of the two is that the deep-water wavelength is over an order of magnitude greater than the beam $(2b_\xi)$ of the strip.

11.8 Closing Remarks

The formulations presented in this chapter are primarily based on the wave-induced floating body motions as determined by strip theory. The focus is on the vertical motions of a strip, which in turn leads to the hydrodynamic coefficients in the coupled equations of motions for a heaving and pitching body. The strip theory does have its failings, particularly when predicting the hydrodynamic coefficients near the bow and stern of a body, where diffraction effects are significant. The reason for the concentration on the strip theory is that a good understanding of the physical phenomena can be obtained by following the derivations involved in the theory.

12 Wave-Induced Motions of Compliant Structures

12.1 Compliant Structures

Fixed ocean structures fall into three major types: rigid structures, compliant structures, and spread-footing structures. There are also two hybrid types of structures, called the *tension-leg platform* (TLP), which is a floating body held in place by high-tension mooring lines, and the *articulated-leg platform* (ALP), which is a *quasi*-rigid cylindrical hull attached to a universal joint on a sea-bed foundation. Rigid structures are usually designed for near-shore operations. Drilling structures used to tap the oil deposits in the shallow near-shore waters are normally of this type. As oil exploration moved further offshore, rigid structures became expensive and their cost-effectiveness decreased. To reduce drilling costs in the deeper water, compliant towers were constructed. One such structure is the Lena guyed tower, which was deployed in the Gulf of Mexico in waters of about 300 m in the 1980s. One advantage of the compliant structure is that its foundation is much less expensive compared to that of the rigid structure. The reason for this economy is that the wave loads are partially absorbed by the elasticity of the structure. In the North Sea, where severe storms occur over the entire year and the waters are relatively shallow, the spread-footing structure has been used to tap oil and gas deposits beneath the sea bed. These structures are monolithic structures, and simply rest on the bed. The structural load is distributed over the soil, reducing the normal load on the bed. As a result, the structures are stable, even on unconsolidated soils.

The three components of wave loads on fixed structures are the viscous-pressure (drag) forces, the inertia forces, and the diffraction forces, all discussed in Chapter 9. The occurrences of these forces depend on two ratios involving the characteristic dimension of a structure and the wave properties. As discussed in Chapter 9, for a specific vertical, surface-piercing, circular cylinder, the wave force regimes are presented in the Chakrabarti (1975) diagram in Figure 9.8. In that figure, the non-dimensional parameters are the Keulegan-Carpenter number of eq. 9.46 ($KC = u_{max}T/D$) and $ka = kD/2$. In these parameters, u_{max} is the maximum horizontal particle velocity (obtained from eq. 3.49), T is the wave period, $k = 2\pi/\lambda$ is the wave number (where λ is the wavelength), and $D = 2a$ is the diameter of the structure (a being the radius of the cylinder). From Figure 9.8, the rigid and compliant structures composed of small-diameter legs and cross-braces would be normally

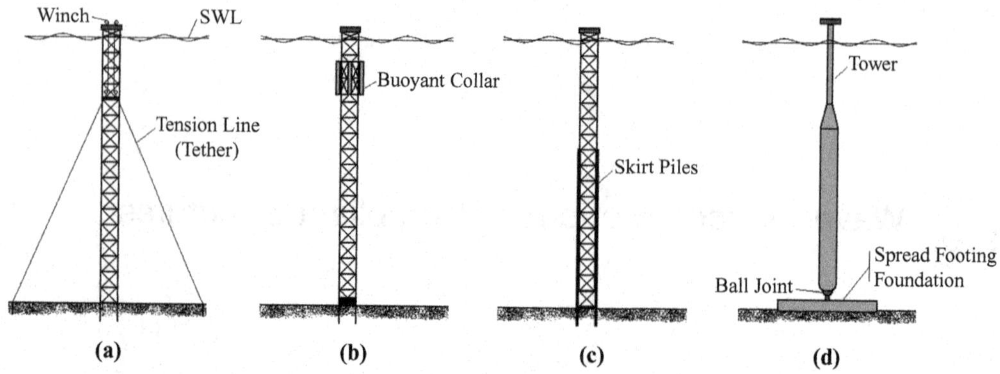

Figure 12.1. *Four Configurations for a Deep-Water Compliant Tower.* In (a), neutrally buoyant tension lines (tethers) are used to limit the movements of the tower. The tension in the lines is controlled by deck winches. This configuration is referred to as a "guyed tower." In (b), we see the "flexible tower," a buoyant collar that is used to increase the vertical tension in the platform legs. In (c), the legs are encased in skirt piles designed to stiffen the lower part of the structure. The upper portion of the tower is compliant, but is rigidly connected to the top of the skirt piles. The articulated-leg platform (ALP) in (d) is attached to a spread-footing foundation by a universal joint, which is the tower's point of articulation.

subject to both drag and inertial forces. The large-diameter monolithic structures are subject to inertial and diffraction forces.

The purpose of this chapter is to present the basics of the wave-structure-bed interactions associated with moored, rigid, and compliant ocean structures. Concerning the first of these: In addition to the tension moorings, the effects of slack moorings are also discussed. An understanding of mooring mechanics is important in the design of offshore structures. Tall compliant offshore towers in moderately deep water might have mooring lines attached below the tower platform in a fail-safe design. For example, the offshore tower sketched in Figure 12.1a is compliant because its height-to-width ratio (assuming a rectangular cross-section) is extremely large. When the tower is subjected to a wave spectrum having a high energy content in the neighborhood of the natural bending frequency of the structure, the platform might experience large deflections. To reduce the deflection, the designer can attach tension moorings, as sketched in Figure 12.1a; attach a buoyant collar, as in Figure 12.1b; or weld "skirt piles" to the legs, as illustrated in Figure 12.1c. The problem with the collars is that more of the wave loads are transferred to the foundation. In turn, this drives up the cost of the system. In Figure 12.1d is an ALP that is connected to a spread-footing foundation by a *universal joint.* The hull of the ALP is relatively rigid, but the structure is compliant in that it is allowed to rotate due to the universal joint. In this case, there is no moment at the foundation due to the loads on the surface-piercing tower. As discussed by Maus, Finn, and Danaczko (1986), Bayazitoglu, Jones, and Hruska (1987), and Will, Morrison, and Calkins (1988), one of the purposes of designing in structural compliance is to relieve the load on the foundation, thereby reducing the overall cost of the structure. The section by Chakrabarti, Capanoglu, and Halyard (2005) in the handbook edited by Chakrabarti (2005) is recommended to the reader. In that chapter is an extensive discussion on the various configurations of offshore structures.

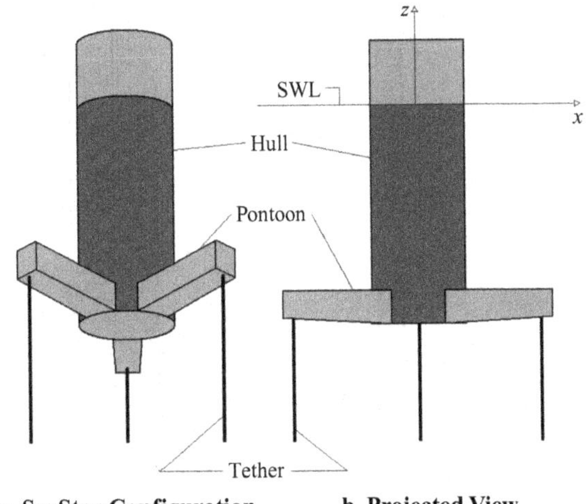

Figure 12.2. *Simplified Schematic of a SeaStar-Type Tension-Leg Platform (TLP).* A tower and working decks (not shown) are located on the top of the hull. This configuration is referred to as either a mono-column TLP or a mini-TLP. A SeaStar TLP deployed in the Gulf of Mexico is described by Kibbee and Snell (2002). Two SeaStar prototypes and two one-fiftieth models of these prototypes are described by Bhattacharyya, Sreekumar, and Idichandy (2003).

12.2 Basic Mooring Configurations

As written in the preamble of this chapter, there are two classifications of mooring systems. The first incorporates *taut (high-tension) moorings*, where the tension in the line is affected by the net weight (weight minus buoyancy) of the mooring line, the buoyancy of the attached body, and a winch designed to control the tension. The second type of mooring system involves a *slack mooring*, where the line tension is due to the net weight of the line. The effective weight of a slack line is often increased by inserting heavy chains at various locations along the line length. Tension lines are designed to minimize the wave-induced motions of the structure, whereas slack moorings are designed to allow some structural motion. The practical aspects of mooring systems are discussed by Brown (2005) in the handbook edited by Chakrabarti (2005). In addition, the paper by Low and Langley (2008) is recommended for a discussion of the dynamics of slack-mooring systems.

A. Taut Moorings

A *tension-leg platform* (*TLP*) is a floating platform held in its design configuration by taut moorings (high-tension moorings) called *tethers*. The draft of a TLP is due to the platform weight, equipment, ballast, and the tension in the mooring lines. The tension in the lines is usually controlled by winches. The TLP is designed to have little or no heaving or pitching motions in the design sea, but is allowed to surge. Depending on the wave direction, the platform might also experience sway and yaw.

Consider the class of TLP called the "SeaStar," which is designed to be deployed in depths of 215 m to 1,000 m. The system is similar to that sketched in Figure 12.2. In that figure are three lines attached to three "pontoons." The pontoon configuration inspired the naming of the system. The pontoons are attached to a hull that is a vertical, circular, cylindrical hull. Atop the cylindrical hull (but not shown) are a tower, working decks, and equipment. The tethers for this TLP are neutrally buoyant. This design feature allows the tension to be uniform along the line length.

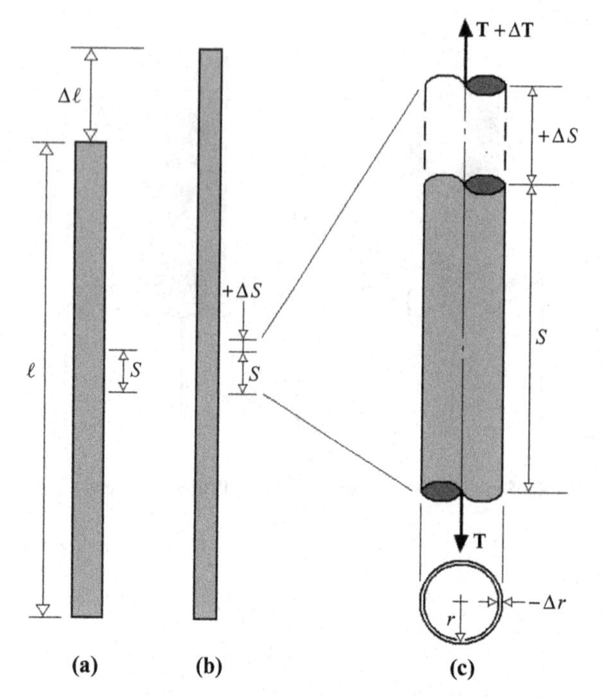

Figure 12.3. *Notation for a Mooring Line under Tension.* Shown are (a) the relaxed line, (b) the stretched line, and (c) an element of the stretched line. In Figure (c) are sketched the longitudinal and radial displacements. One can see that the longitudinal strain ($\Delta s/s$) is positive and the radial strain ($\Delta r/r$) is negative. The latter is considered to be of second order and, therefore, negligible.

For discussions of the SeaStar, see the papers by Kibbee and Snell (2002) and Bhattacharyya, Sreekumar, and Idichandy (2003). Later in the chapter, the SeaStar configuration is used for the purpose of illustration.

The line profile is a function of the net weight (weight minus buoyancy) of the line. For the mooring line configuration of the fixed tower in Figure 12.1a, the lines are neutrally buoyant, that is, the net weight of each line is zero. As a result, the lines are straight. For a line having a positive net weight (negative buoyancy), lines would "sag" and the tension in the line would vary over the length.

Our goal here is to determine the form of the effective spring constant (K_s) of a neutrally buoyant tether, such as those sketched in Figures 12.1a and 12.2. Referring to Figure 12.3, the relaxed tether length is ℓ, as sketched in Figure 12.3a. When under an axial load (T), the tether is elongated by an amount $\Delta\ell$, as in Figure 12.b. For the entire cable, we can then express the modulus of elasticity (Young's modulus) as

$$E_s \equiv \frac{\textit{axial stress}}{\textit{axial strain}} = \frac{\tau_{ss}}{\epsilon_s} = \frac{T/\pi r^2}{\Delta\ell/\ell} = \frac{T/A_s}{\Delta\ell/\ell} \tag{12.1}$$

Here, the subscript s identifies the properties associated with the effective spring effect of the tether, and the effective cross-sectional area is A_s. From eq. 12.1, we obtain the extension or elongation of the tether, which is

$$\Delta\ell = \frac{T\ell}{E_s\pi r^2} = \frac{T\ell}{E_s A_s} \tag{12.2}$$

The spring constant for a single mooring line or tether is then

$$K_s = \frac{\textit{tension}}{\textit{elongation}} = \frac{T}{\Delta\ell} = \frac{E_s\pi r^2}{\ell} \tag{12.3}$$

In many applications, the tether is not homogeneous from the structure to the anchor, either materially or geometrically. For the nonhomogeneous tether, it is beneficial to work with a "local" spring constant by considering the element in Figure 12.3c. In this case, the local length, s, and the elongation, Δs, replace ℓ and $\Delta\ell$, respectively, in eqs. 12.1 and 12.2. We note in Figure 12.3c that the radius of the cable is reduced when the tension (T) is applied. The radial strain, $\Delta r/r$, is small, and can be neglected in our analysis. This negative radial strain is a result of the conservation of mass.

EXAMPLE 12.1: TENSION IN A SEASTAR TETHER In the paper by Bhattacharyya, Sreekumar, and Idichandy (2002), the specifics of two SeaStar prototypes are given (see Figure 12.2 for the SeaStar configuration). A unit is to be deployed in 215 m of water, and the other in 1,000 m. Bhattacharyya, Sreekumar, and Idichandy (2002) also give the particulars of one-fiftieth scale models of the prototypes. From Section 2.7, the length-scale factor is $n_L = 1/50$. Here, we shall determine the effective spring constant of the prototype having a three-tether mooring configuration, where the water depth (h) is 215 m. The tethers are neutrally buoyant.

As the reader can imagine, the tethers and the anchors must be set in place before being connected to the platform. Because the tethers are neutrally buoyant, buoyant collars are attached to the top of the lines to produce a relatively small axial tension that ensures a vertical tether profile. The platform is subsequently positioned over the vertical lines, ballasted such that the pontoons are fully submerged, and the lines are attached. The ballast is then reduced, causing the tethers to be in tension. For the deployment of a three-tethered structure in 215 m of water, the following values are used for each tether:

(a) Relaxed length (ℓ): 175 m
(b) Effective tether area (A_s): 1 m^2 (diameter $\simeq$ 1.128 m)
(c) Modulus of elasticity (E_s): 2×10^{11} N/m^2
(d) Pre-Tension (T): 1,333 metric tons (tonne), the tonne is usually defined in mass units (1,000 kg)

In eqs. 12.1 through 12.3, the units must be in terms of force. Hence, in force units, a tonne is 9,810 N. The tension in a tether is one third of the difference between the displacement (displaced water weight) and the dry weight of the structure.

By substituting these values in eq. 12.2, we find that the elongation ($\Delta\ell$) of the tether is approximately 0.0114 m. From eq. 12.3, the spring constant (K_s) is about 1.15×10^9 N/m (1.17×10^5 tonne/m). Note that the mooring stiffness is large to ensure the previously stated design goals. Those goals are to minimize both the heaving and pitching motions of the TLP in a design sea. From the small elongation value, the reader can see that the radial strain ($\Delta r/r$) is negligible.

Later in this chapter, the TLP is discussed in more detail. For that structure, the combination of waves and wind will cause the structure to surge. Because of the high tension in a tether, the modal frequencies will be large when compared to those of the sea. Although those frequencies should be well away from the high-energy portion of the wave-force spectrum, they can be in the high-energy portion of the vortex-shedding spectrum for the tether. In the design of such moorings, the cable response should be de-tuned from these spectra.

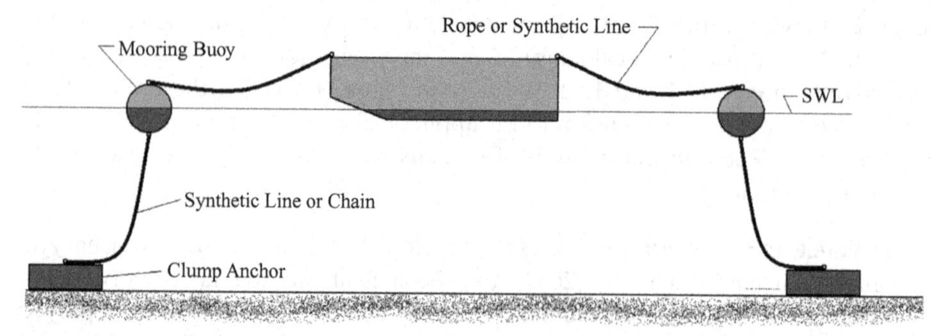

Figure 12.4. *Static Slack Mooring Configurations.* Whether in air or fully submerged, the free portions of the lines have a catenary shape. The actual shape is a function of the net weight (weight minus buoyancy) of the line. The term "line" is generic as used herein.

In the next section, slack moorings are discussed. As noted previously, if a mooring system has a positive net weight (negative buoyancy), then there will be some sag in the cable, although the tension in the line might be large. Slack moorings are those having both a line tension and a profile that are dictated by the weight line intensity (net weight per unit length of line).

B. Slack Moorings

When significant horizontal motions of a moored floating body are allowed, the body will be slack moored. Two such moorings are sketched in Figure 12.4. In that figure, the bow and stern mooring configurations are identical because the condition shown is in static equilibrium. Hence, there are no wind, wave, or current loads with which to contend. In this chapter, our concern is only with wave-induced cable tensions.

The analysis of the effects of steady currents on a slack line, based on works in the 1950s, 1960s, and mid-1970s, is presented by McCormick (1973). The often-referenced paper by Niedzwecki and Casarella (1976) addresses the practical aspects of deep-water moorings. More recent papers covering such topics as linear and nonlinear statics and dynamics of moorings, and the applications thereof, include those of Chiou and Leonard (1991), Brown and Mavrakos (1999), Smith and MacFarlane (2001), and Jordánr and Beltrán-Aguedo (2004).

In the discussions that follow, the term *line* is a generic term representing rope, synthetic lines, or chain. For the lines sketched in Figure 12.4, the shapes are those of catenaries because their elevation profiles are dictated by the net weights (weight minus buoyancy) of the lines. A *catenary* has the shape of a hyperbolic cosine. The derivation of the equations associated with a catenary can be found in books covering either linear differential equations or mechanics (statics).

As in the previous section, our goal is to determine the effective spring constant of the line to determine the effects of the line on the motions of the moored body. To do so, consider the inelastic line in static equilibrium, as sketched in Figure 12.5. The eye points are at (0,0) and (X_2,Z_2), where the origin of the X,Y,Z coordinate system is at the left eye of the mooring. The position of the vertex of the mooring line is X_1,Z_1. As is the case in the prior chapters, the origin of the x,y,z system is fixed at a point on the SWL, as in Figure 12.2 and elsewhere throughout the text.

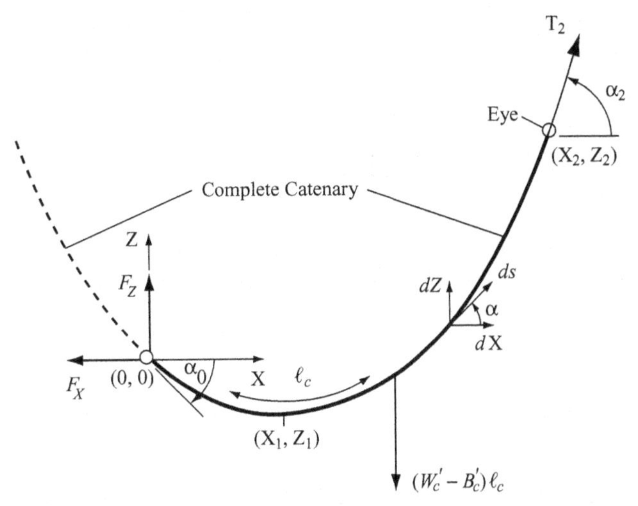

Figure 12.5. *Notation for an Inelastic Slack Line in Equilibrium.* The net weight $(\mathrm{W}_c'\ell_c)$ is the line weight $(W_c'\ell_c)$ minus the line buoyancy $(B_c'\ell_c)$.

The horizontal equilibrium condition for the line is

$$\mathrm{T}_2 \cos(\alpha_2) = \mathrm{T}_0 \cos(\alpha_0) = \mathrm{T}\frac{d\mathrm{X}}{ds} = F_\mathrm{X} \tag{12.4}$$

Here, the tension and angle without subscripts represent any point on the line, and the horizontal force component of the eye located at the origin is F_X. The differential of the line coordinate, s, is shown in Figure 12.5. The vertical equilibrium condition is

$$F_Z + \mathrm{T}_2 \sin(\alpha_2) = (W_c' - B_c')\ell_c \tag{12.5}$$

where F_Z is the vertical force of the eye at the origin. The relationships in eq. 12.4 show that the horizontal component of the tension is uniform over the length of the line. Also, F_X is in the opposite direction of the horizontal tension component at the right eye. In eq. 12.5, the weight density per line length is W_c', and the buoyancy of the line per unit length is B_c'. For brevity, let the net weight per unit length be represented by W_c', that is, $\mathrm{W}_c' = W_c' - B_c'$. The length of the line is ℓ_c. It is well known that the segment of the line sketched in Figure 12.5 is a segment of a catenary.

From the analysis of the catenary in Figure 12.5, the following relationships are found:

$$\frac{dZ}{d\mathrm{X}} = \sinh\left(\frac{\mathrm{W}_c'}{F_\mathrm{X}}\mathrm{X} + C_a\right) = \tan(\alpha) \tag{12.6}$$

where C_a is a to-be-determined constant. The integration of this equation results in the following expression for the vertical coordinate of any point along the centerline of the catenary:

$$\mathrm{Z} = \frac{F_\mathrm{X}}{\mathrm{W}_c'}\cosh\left(\frac{\mathrm{W}_c'}{F_\mathrm{X}}\mathrm{X} + C_a\right) + C_b \tag{12.7}$$

Here, the constant C_b is to be determined. In this equation, there are three unknowns, F_X, C_a, and C_b. In addition, the position $(\mathrm{X}_1, \mathrm{Z}_1)$ of the vertex of the line is not known. We have three conditions that can be applied to eq. 12.7 and its spatial derivative because the eye locations at (0, 0) and (X_2, Z_2) are specified and the spatial derivative at the vertex equals zero. Applying these conditions, we find that eq. 12.7 becomes

$$Z = \frac{F_X}{W'_c}\left\{\cosh\left[\frac{W'_c}{F_X}(X - X_1)\right] - 1\right\} - Z_1 \tag{12.8}$$

In this equation, X_1, Z_1, and F_X are to be determined. To accomplish this, we must use the cable axial coordinate, s. The origin of s is at the left side eye in Figure 12.5. From that figure, we see that the differential of this variable is

$$\frac{ds}{dX} = \sqrt{1 + \left(\frac{dZ}{dX}\right)^2} = \sqrt{1 + \sinh^2\left[\frac{W'_c}{F_X}(X - X_1)\right]} = \cosh\left[\frac{W'_c}{F_X}(X - X_1)\right] \tag{12.9}$$

This equation can be integrated over the known length (ℓ_c) of the cable. The result is

$$\ell_c = \frac{F_X}{W'_c}\left\{\sinh\left[\frac{W'_c}{F_X}(X_2 - X_1)\right] + \sinh\left(\frac{W'_c}{F_X}X_1\right)\right\} \tag{12.10}$$

Now, apply eq. 12.8 to both the left and right eyes in Figure 12.5, and combine the results by eliminating Z_1. The result is

$$Z_2 = \frac{F_X}{W'_c}\left\{\cosh\left[\frac{W'_c}{F_X}(X_2 - X_1)\right] - \cosh\left(\frac{W'_c}{F_X}X_1\right)\right\} \tag{12.11}$$

Equations 12.10 and 12.11 comprise a system of two equations having two unknowns, F_X and X_1, because Z_2 and ℓ_c are input (design) values. The two equations are both transcendental with respect to these unknowns; hence, the equations must be solved numerically. The process can be somewhat simplified by taking advantage of the relationships between the hyperbolic functions. Specifically, we can eliminate one of these using $\cosh^2(\ ) = 1 + \sinh^2(\ )$. Then, after several manipulations, we obtain two transcendental relationships. The first of these is

$$\frac{1}{2}\left(\frac{W'_c}{F_X}\right)^2 (\ell_c^2 - Z_2^2) = \cosh\left(\frac{W'_c}{F_X}X_2\right) - 1 \tag{12.12}$$

which can be numerically solved for the horizontal tension component (F_X). With this force value, the vertex coordinate, X_1, is obtained from either eq. 12.10 or 12.11. We can rewrite the latter equation as

$$\frac{W'_c}{F_X}\ell_c = \sinh\left(\frac{W'_c}{F_X}X_2\right)\sqrt{1 + \sinh^2\left(\frac{W'_c}{F_X}X_1\right)} + \left[1 - \cosh\left(\frac{W'_c}{F_X}X_2\right)\right]\sinh\left(\frac{W'_c}{F_X}X_1\right) \tag{12.13}$$

Substitution of the X_1-value into eq. 12.8 results in the vertical coordinate (Z_1) of the vertex. The solutions of eqs. 12.12 and 12.13 can be obtained by using the method of *successive approximations*. See Example 3.3 of Section 3.3 for an application of this numerical method.

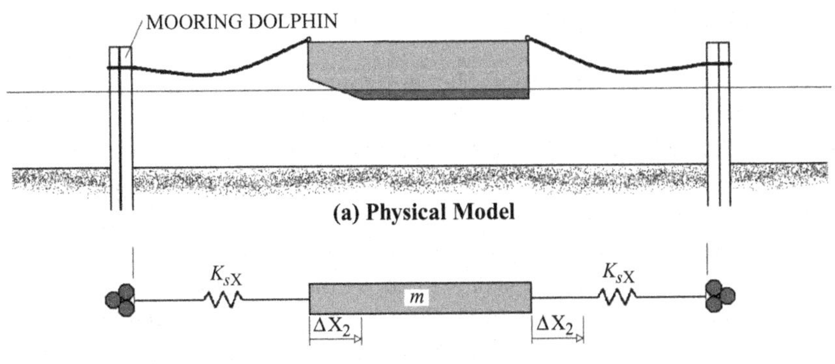

Figure 12.6. *Equivalent Spring-Mass Representation of a Moored Body between Two Dolphins.* For this mooring system, the springs representing the moorings are in parallel. Hence, the equivalent spring constant for the system is the sum of the component spring constants. The mooring dolphins in the figure are vertical.

The effective horizontal spring constant (K_{sX}) of the slack line for the motions in the horizontal direction is obtained by taking the spatial derivative of the expression in eq. 12.12. The result is

$$K_{sX} = \frac{dF_X}{dX_2} = \frac{F_X^2 \sinh\left(\frac{W_c'}{F_X}X_2\right)}{F_X X_2 \sinh\left(\frac{W_c'}{F_X}X_2\right) - W_c'(\ell_c^2 - Z_2^2)} \tag{12.14}$$

Again, F_X is obtained from eq. 12.12, and Z_2 is a design value.

The line tension at any point can now be determined by combining eqs. 12.4 and 12.9. From this combination, the tension is obtained from

$$T = F_X \frac{ds}{dX} = \frac{F_X}{\cos(\alpha)} = F_X \cosh\left[\frac{W_c'}{F_X}(X - X_1)\right] \tag{12.15}$$

From this equation, we see that the maximum tension for the line configuration in Figure 12.5 occurs where $|(X - X_1)|$ is maximum. When the right eye (at X_2) is above the left eye (at $X = 0$), as in the figure, then the maximum tension occurs at the right eye. The maximum tension multiplied by a safety factor is used to determine the design strength of the line.

EXAMPLE 12.2: EFFECTIVE SPRING CONSTANT FOR SLACK LINE Consider a deck barge having a length of 82 m, a beam of 10 m, a draft of 1 m, and a freeboard (waterline-to-deck height) of 3 m. The barge is moored as sketched in Figure 12.6. The mooring attachments to the dolphins are 2 m above the waterline. Hence, the vertical distance (Z_2) between the mooring attachments is 1 m.

The barge is moored using three-strand manila rope having a diameter of 7.64 cm (3-in line). Manila rope has been popular in marine applications because of its cost and availability. The net weight per unit length (W_c') is about 35 N/m. As we shall determine, the line is entirely in air, where the buoyancy is negligible. As a result, the net weight is equal to the material weight ($W_c' = W_c'$). We should also note that manila rope absorbs water; hence, W_c' for a wet rope will increase by about 25%. The mooring dolphins are 25 m forward and aft of the barge. Hence, in eq. 12.9, $X_2 = 25$ m. The free length (ℓ_c) of each line is 25.5 m.

Our first task is to determine the horizontal force component, F_X in eq. 12.12. This is easily done by using the method of *successive approximations*, as in Example 3.3 of Section 3.3. The horizontal force (F_X), determined numerically, is approximately 12.64×10^3 N. The horizontal position (X_1) of the vertex of the line is about 11.05 m, from eq. 12.13. When this value is combined with eq. 12.8, the vertical position of the vertex with respect to the left eye is approximately 1.66 m. Hence, the vertex is 0.34 m above the SWL. Equation 12.15 can now be used to obtain the tension in the line at any point. Applying this equation to the right eye, where $X_2 = 25$ m, one obtains $T \simeq 12.67 \times 10^3$ N. Because of the relatively close values of ℓ_c and Z_2, there is little sag in the line, and the apex of the line is effectively at $X = 0$, where $Z = 0$. Finally, the effective horizontal spring constant is $K_{sX} \simeq 527$ N/m. For the system shown in Figure 12.6, the springs in the effective system are in parallel. The equivalent spring constant for the parallel effective springs is the sum of the component constants. Hence, the parallel spring constant for the two-line mooring system is

$$K_{sp} = 2K_{sX} \simeq 1.05 \times 10^3 \frac{\text{N}}{\text{m}} \tag{12.16}$$

For the stated barge dimensions, the spring system would be considered to be "soft." The significance of this statement will be addressed later in the chapter. The equilibrium angle of the mooring line at the barge ($X = X_2$) can be determined from eq. 12.15. With the values of horizontal force and horizontal position of the vertex of the line, the angle at the barge is $\alpha_2 \simeq 4.0°$.

For the inelastic line, the tension and the effective spring constant will change with a change in the line diameter. For a line composed of inelastic spiral-wound strands, the line will partially unwind as the tension in the line is increased. The resulting restoring torsion in the line can be analyzed by using an analogous solid line that is elastic.

For a fully submerged line, our choice of lines would be either a chain or a synthetic or a combination of both. For example, consider the barge moored as in Figure 12.7. In that figure, we see that the forward and after mooring buoys are coupled to embedded anchors using chains and synthetic lines. Because the profile of the mooring-line component is determined solely by the net weight, each component shape is a portion of a catenary. Equations 12.4 through 12.15 can then be applied to each line component, with the conditions at the shackles acting as boundary conditions.

Assuming that the mooring-buoy mass (m_m) is negligible compared to the barged mass (m), the *series spring constant* for the mooring system on either side of the barge is obtained from

$$\frac{1}{K_{ss}} = \frac{1}{K_{s0a}} + \frac{1}{K_{sab}} + \frac{1}{K_{sbm}} + \frac{1}{K_{sm}} \tag{12.17}$$

from the theory of vibrations. For example, see the book by Thomson and Dahleh (1997). The determination of the component spring constants requires the solution of simultaneous equations. Those equations are eqs. 12.10 and 12.11 applied to each mooring-line component. For the left side of the barge in Figure 12.7, we note that the origin of the mooring system is not at the anchor. Rather, it is the point where the anchor chain rises from the sea bed. As the barge and mooring buoy move to

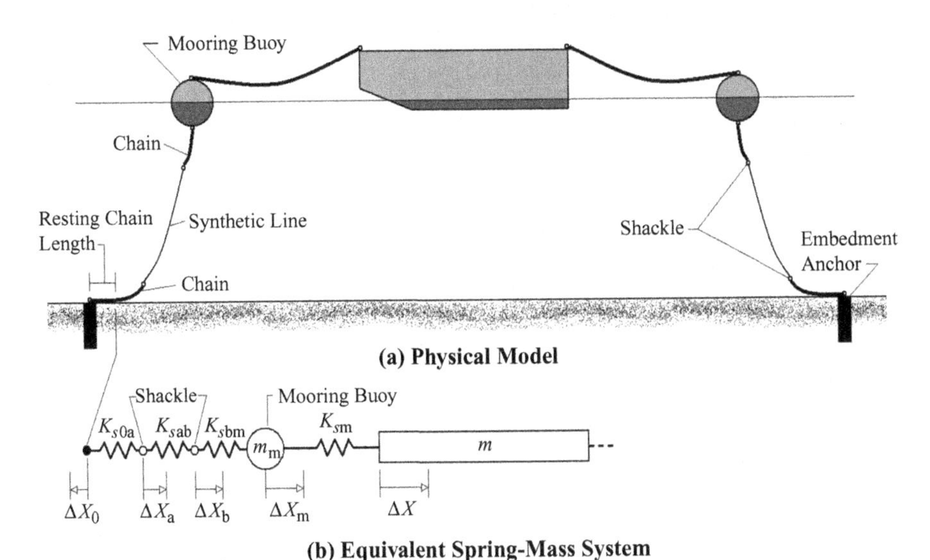

Figure 12.7. *Fully Submerged Mooring Configuration.* The mooring buoys are attached to an embedded anchor with chain-synthetic-chain systems. The effective springs are in series.

the right, the chain lift-off point (also called the touchdown point) moves to the left. This point can be fixed in place by attaching a clump anchor to the chain at the touchdown point. An empirical model of this type of mooring system is presented by Han and Grosenbaugh (2004). To apply eqs. 12.6 through 12.15 to the compound anchor mooring, the vertex points of the complete catenaries (Figure 12.5) must be determined for each component, as must the horizontal force component (F_X). When the system is at equilibrium, F_X is the same at the cable eyes, shackles, and anchor for the mooring-line components in series.

The mooring configurations sketched in Figures 12.4, 12.6, and 12.7 are not fail-safe. That is, if the bow mooring fails, then the resulting barge motions could result in extensive damage. Safety can be increased by designing redundancy into the mooring system. This is done by adding mooring lines to both the bow and stern. To illustrate, four identical moorings are shown in Figure 12.8a. Hence, the swing of the barge following the failure of one of the component lines would be controlled by the remaining lines.

Our interest is in the *equivalent spring constant* of the redundant mooring system. Consider the deformation of the line shown at an angle β to the X-direction in Figure 12.8b. The line is stretched because of the horizontal barge displacement, ΔX, which is assumed to be relatively small. Let the horizontal length of a component line be ξ. The angle of the stretched line is $\beta - \Delta\beta$. So, the horizontal component of the stretch in the line is $\Delta\xi$, which must be determined. From the geometry in Figure 12.8b, we find

$$\frac{d\beta}{d\xi} \simeq \frac{1}{\xi}\tan(\beta) \tag{12.18}$$

This is found be passing to the limit, where $\Delta\xi \to d\xi$ and the product $d\xi(d\beta)$ is negligible. Using this relationship, we find the relationship between the barge displacement and the stretch in the line, that is,

$$\frac{dx}{d\xi} \simeq \frac{1}{\cos(\beta)} \tag{12.19}$$

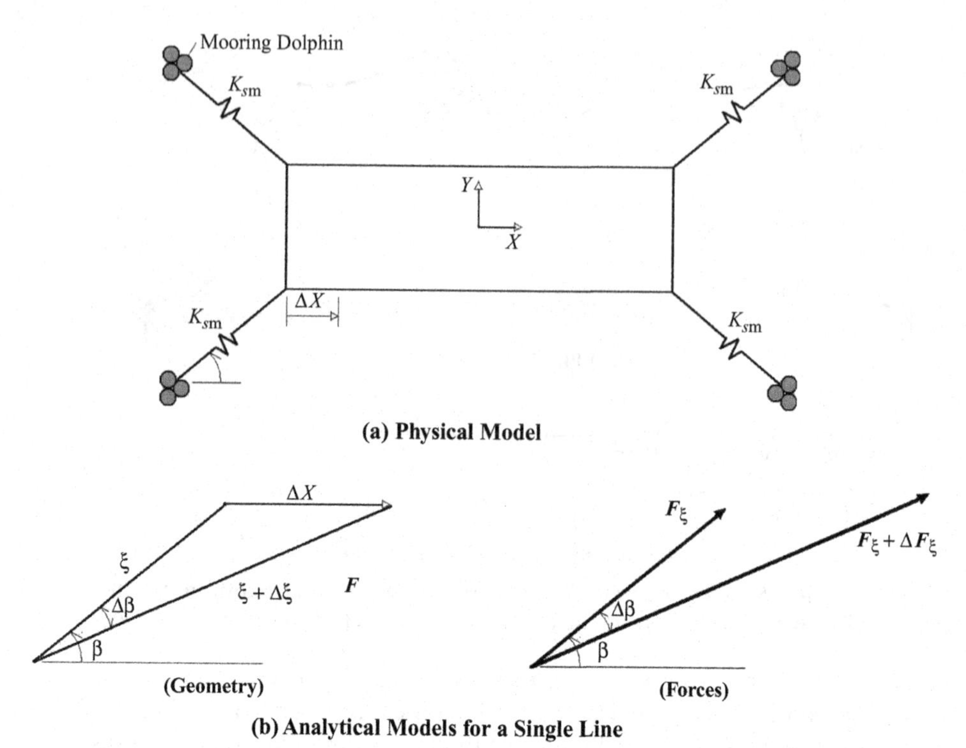

Figure 12.8. *Redundant Mooring Configuration.* The additional bow and stern lines increase the safety of the system. In the theory of structures, a "redundant member" is one that can be removed without causing the structure to fail in a design condition. The use of the term here is not quite the same.

From the force relationships in Figure 12.8b, the following is obtained:

$$K_{sX} \equiv \frac{dF_X}{dX} \simeq F_\xi \sin(\beta)\frac{d\beta}{dX} + \cos(\beta)\frac{dF_\xi}{dX} = \frac{F_\xi}{\xi}\frac{\sin^2(\beta)}{\cos(\beta)}a + K_{s\xi}\cos^2(\beta) \quad (12.20)$$

This is the effective spring constant for motions of the barge in the X-direction. The horizontal force, F_ξ, and the effective spring constant, $K_{s\xi}$, are found from eqs. 12.10, 12.11, and 12.14, where the known horizontal length, ξ, replaces X_2 in those equations. Note that the position of the vertex in the line must also be determined from these equations. For the barge movement in the X-direction, the effective spring constant for the redundantly moored system is

$$K_X = 4K_{sX} \quad (12.21)$$

The same methodology is used to determine the effective parallel spring constant, K_Y, for displacements in the Y-direction. Both springs K_X and K_Y are needed for waves or currents obliquely approaching the barge. For this case, the barge would rotate about its vertical axis. Hence, the barge would experience surging (X), swaying (Y), and yawing motions (about the Z-axis).

EXAMPLE 12.3: EFFECTIVE SPRING CONSTANT FOR A BARGE MOORED WITH FOUR SLACK LINES The barge in Example 12.2 is now moored with four identical manila lines, where each line has the same dimensions as those in the example. Each line is at an angle $\beta = 30°$ to the X-direction. From Example 12.2, the line diameter is 7.64 cm, and the net weight per unit length (W_c') is about 35 N/m.

The mooring dolphins are 25 m from the barge. Hence, in eq. 12.9, $\xi = 25\,\text{m}$. Again, the length (ℓ_c) of each line is 25.5 m. The horizontal force (F_X), from Example 12.2, is 12.64×10^3 N, and the horizontal position (ξ_1) of the vertex of the line is at the mooring dolphin. The tension in each line is $T \simeq 12.67 \times 10^3$ N, and the horizontal spring constant in a line in the static equilibrium condition is $K_{s\xi} \simeq 527$ N/m. With this value, $K_{sX} \simeq 456$ N/m, and the effective parallel spring constant for the mooring system is $K_X \simeq 1.82 \times 10^3$ N/m.

We have only touched on the subject of moorings. The topic has many facets that are not discussed herein. The *quasi*-static analysis of two-dimensional mooring lines in steady currents is presented in the book by McCormick (1973). For two-dimensional moorings lines experiencing unsteady and random loading, the paper by Sannasiraj, Sundar, and Sundaravadivelu (1997) is recommended. For contemporary design considerations of mooring systems, the reader is referred to the paper by Grosenbaugh et al. (2002). Finally, the writings of Brown (2005) are strongly recommended for a comprehensive discussion of the practical aspects of moorings.

12.3 Soil-Structure Interactions

In this section, two types of soil-structure interactions are discussed. First, embedded structures, such as drilled piles and suction anchors, are discussed. These structures are used to resist both the high-tension vertical and lateral loads due to taut moorings. The second type of structure is the spread-footing structure, which is designed to uniformly spread a large vertical load over a large bed area. The subject of marine geotechnology is rather complicated because most sea-bed soils react in a nonlinear manner when excited by structural motions. A good introduction to the subject of soil-structure interactions in and on the sea bed is presented by McClelland (1969) in the handbook edited by Myers, Holm, and McCallister (1969). Comprehensive discussions of both the analysis and design of pile foundations are presented by Poulos and Davis (1980) and Reese (2003). The latter appears in the book edited by Wilson (2003). Offshore structures having large spread footings are normally referred to as *gravity-base structures*, or more simply, *gravity structures*. These and other marine structures are topics discussed by Gerwick (1999). The basics of dynamics of spread-footing-type structures can be found in the writings of Moan, Syvertsen, and Haver (1977), Wolf and Song (1999), McCormick and Hudson (2001), Hudson (2001), and McQuillan (2002). The studies leading to the 2001 publications were directed at the motions of grounded ships. Note that in the following, nonitalicized letters are used to represent soil properties, and italicized letters represent the properties of the structure.

Soils support structures and partially resist structural motions due to cohesion, due to molecular bonding, and due to friction, which depends on the normal load on the soil-structure interface. These are represented in the shear or soil strength, which is mathematically obtained from the *Coulomb equation*,

$$\tau_s = c_s + \sigma \tan(\phi_s) \tag{12.22}$$

In this equation are the apparent cohesion (c_S), the normal stress or pressure (σ) on the shear plane, and the friction angle (ϕ_s). This angle is also referred to as the angle of repose. For a discussion of the significance of this angle in noncohesive soils (where $c_s = 0$), the paper by Modaressi and Evesque (2001) is recommended.

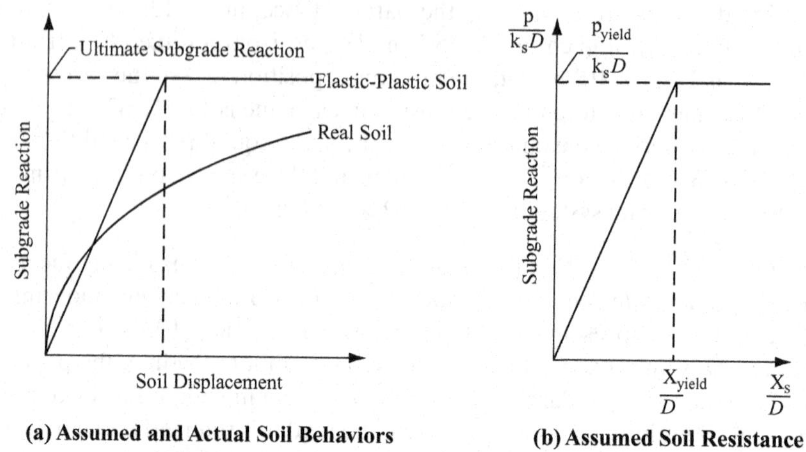

Figure 12.9. *Elastoplastic Soil Behavior Model and Real-Soil Behavior*. The soil resistance (p) is non-dimensionalized by the product of the effective elastic modulus (k_s) of the soil and the pile diameter (D). For the elastoplastic soil model, the non-dimensional resistance is shown as a function of the ratio of the soil displacement (X_s) and D. Note that the nonitalicized letters are used to identify sea-bed properties.

The types of soils composing an offshore bed are clays, sands, and silts. For clays, the first term in eq. 12.22 is dominant, whereas the second term is dominant for sands. According to McClelland (1969), the ranges of these parameters in the equation are $0.03 \leq c_s \leq 0.08$ for clays, and $30° \leq \phi_s \leq 40°$ for sand. Again, note that nonitalicized letters are used to represent the sea-bed properties.

The analytical difficulties in marine geotechnology arise because of the change in the soil properties with depth, and the plasticity of bed materials. For clays, the cohesion increases with depth due to the weight of the material. For sands, the average pores (voids) between the grains decrease in size with depth. These changes are further compounded by the possible layering of materials. For example, one might encounter a thick layer of silt over a clay base. To determine the makeup of the sea bed, core samples must be taken. Soundings should also complement the coring.

In the following subsections, embedded structures and spread-footing structures are discussed. The associated soil mechanics are analyzed using basic methods. That is, the plastic behavior of the soil is approximated by an elastoplastic model. The elastoplastic and plastic behaviors of a soil are illustrated in Figure 12.9a. The assumed soil reaction on a vertical circular cylinder of diameter D is presented in Figure 12.9b. In that figure, the soil resistance (p) is proportional to the soil displacement (X_s) up to the ultimate or yield value. Then, the soil is assumed to deform plastically. The *soil resistance* for the elastoplastic soil can be represented as

$$p = N_s D \tau_s, \quad \text{where } X_s > X_{yield} \text{ (plastic)} \tag{12.23a}$$

and

$$p = k_s X_s, \quad \text{where } X_s \leq X_{yield} \text{ (elastic)} \tag{12.23b}$$

In eq. 12.23a, N_s is a force coefficient, k_s is the effective elastic modulus (obtained from Figure 12.9b), and τ_s is the soil strength defined in eq. 12.22. Equation 12.23

Table 12.1. *Parametric values for soft clay and sand, after Dawson (1980)*

	Soft clay	Sand
a (N/m^2)	9.6×10^{3}*	–
b (N/m^3)	1.6×10^{3}*	–
D(m)	0.323*	0.610**
EI (N–m^2)	3.15×10^{7}*	1.61×10^{8}**
k_s(N/m^2)	2.20×10^{7}	8.30×10^{6}
N_s	3.5	2.5
ϕ_s(degrees)	–	39**
$\gamma_{s\ and}$ (N/m^3)	–	1.04×10^{4}**
R_{pas}	–	4.40

* Matlock (1970).
** Reese, Cox, and Koop (1974).

applies to either soft clays or sand. The soil displacement X_{yield} is that at the transition from the elastic to plastic behaviors. Dawson (1980) applies the elastoplastic model to solid surrounding deeply driven piles. In the following paragraphs, we draw on the Dawson paper for information on both soft clays and sand.

From Dawson (1980), the undrained shear strength in eq. 12.23 for soft clays can be written as

$$\tau_s|_{clay} = a + bZ \tag{12.24}$$

where, using the Dawson notation, a and b are called the shear strength constants. From this relationship, we see that the soil resistance for a soft clay is found to increase with depth from a surface value of a. The transition response of the soil at a soil depth Z_s can be found from the combination of eqs. 12.23 and 12.24. The result is

$$X_{yield}|_{clay} = \frac{N_s D}{k_s}(a + bZ_s) \tag{12.25}$$

Following Dawson (1980), the shear or soil strength for sands can be written as

$$\tau_s|_{sand} = R_{pas}\gamma_{sand}Z_s \tag{12.26}$$

where R_{pas} is the *Rankine passive earth coefficient* from soil mechanics and γ_{sand} is the weight-density of the sand. In terms of the friction angle, ϕ_s in eq. 12.22, the Rankine passive earth coefficient is

$$R_{pas} = \frac{1 + \sin(\phi_s)}{1 - \sin(\phi_s)} \tag{12.27}$$

The combination of eqs. 12.26 and 12.27 with the soil resistance for the elastoplastic sand results in the following transition-response expression:

$$X_{yield}|_{sand} = \frac{N_s D R_{pas}\gamma_{sand}}{k_s} Z_s \tag{12.28}$$

Drawing on information in the papers by Matlock (1970) and Reese, Cox, and Koop (1974), Dawson (1980) presents both constants and parametric values for a soft clay and sand. Those values are presented in Table 12.1. With the information on the

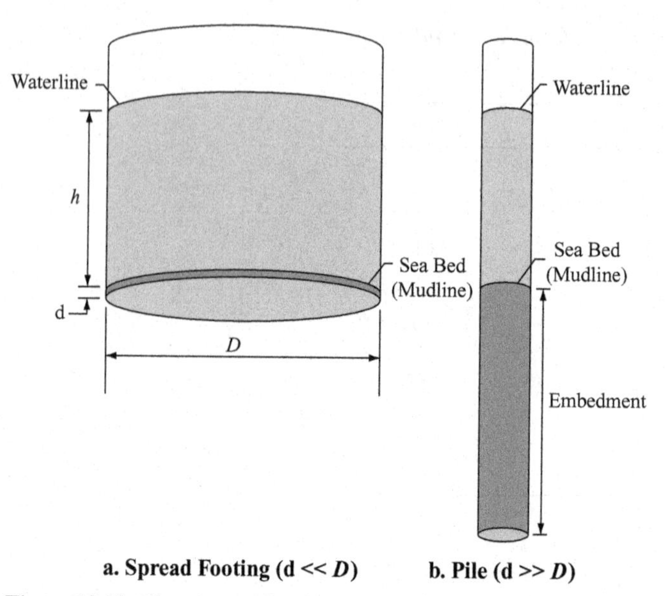

Figure 12.10. *Notation and Conditions for Two Classes of Embedded Foundation Structures.*

soils and soil response, the relationships between the pile deformation and the soil resistance (p) are discussed in the following subsection.

A. Embedded Structures

Marine foundations are composed of piles, spread footings, and combinations of the two. All foundations are embedded to some extent, some more so than others. Referring to the sketches of vertical circular cylindrical structures in Figure 12.10, a structure can be slightly embedded ($d \ll D$), moderately embedded ($d \simeq D$), or deeply embedded ($d \gg D$). The first of these is the spread-footing structure, as sketched in Figure 12.10a. A *spread-footing structure* is designed to uniformly distribute a vertical load over a large bed area, thereby reducing the normal stress on the bed. The soil resistance to the vertical load is dominated by the normal stress over the base, and the frictional resistance on the side of the structure, which is negligible. The resistance to sliding is by the friction at the horizontal soil-structure interface. Most deeply embedded structures are piles, as sketched in Figure 12.10b. These structures are subjected to significant frictional resistance on the embedded sides, and end-bearing resistance. If the pile penetrates to a solid surface, then the end-bearing resistance will be dominant. In this section, our attention is focused on the pile. The discussions of Poulos and Davis (1980) and Reese (2003) are recommended for thorough coverages of the practical aspects of piles.

The forces on offshore structures can be classified as short-term static (*quasi-*steady winds), long-term static (steady water currents), cyclic (waves and vortex shedding), and dynamic (earthquakes). An offshore structure might experience several of these simultaneously. For example, a semi-submersible structure moored in the Florida Current during a hurricane would be exposed to severe wind loads on the platform and superstructure, strong ocean currents on the legs and pontoons, and high surface waves. The resulting forces and moments (long-term static and cyclic) would be resisted by the moorings and anchoring system.

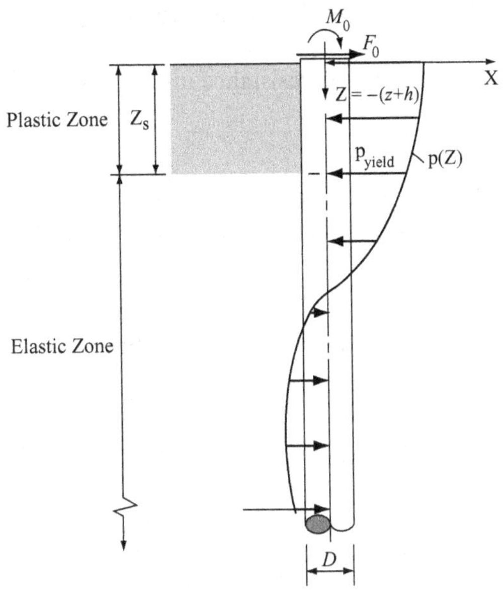

Figure 12.11. *Two-Zone Reaction Model.* Referring to Figure 12.9b, the soil displacement caused by the pile deformation is X_{yield} at $Z = Z_s$, the interface of the plastic and elastic zones.

For piled structures, Reese (1984) reports the following: The static equations from structural mechanics do not work well for the computation of bending deflections in soils due to cyclic lateral loads. However, the static equations can be used to handle the axial loads on piles. The most accurate analytical methods for soil-pile interactions are numerical. For example, see Ellis and Springman (2001). The goal of this section is to introduce the reader to soil-structure interactions. Hence, we shall confine our discussion to analytical techniques. One such technique is that discussed by Dawson (1980), who presents an accurate, simplified analysis of offshore piles exposed to cyclic loads in soft clays and sand. The analysis presented herein is modeled on that presented in the Dawson paper.

Begin by considering a deeply embedded circular cylinder. By deeply embedded, we mean that the cylinder extends far enough into the bed so that the structural displacement below some depth is negligible. Following Dawson (1980), assume that the elastic and plastic ranges in Figure 12.9 correspond to the deep-soil zone and the subsurface-soil zone, respectively, as illustrated in Figure 12.11. The thickness of the subsurface plastic zone is d_I. To facilitate the analysis, the bed coordinate $Z = z + h$ is introduced in Figure 12.11, where z is the coordinate on the SWL and h is the water depth, as in previous chapters. Hence, the plastic zone is $0 \leq Z \leq Z_s$, and the elastic zone is $Z > Z_s$. For a prismatic pile (having a uniform cross-section over the length of the pile), the bending equation is

$$E_{pile} I_Y \frac{d^4X}{dZ^4} = -p(Z) \tag{12.29}$$

This equation applies to both the plastic and elastic zones in Figures 12.9 and 12.11. In the equation, the flexural rigidity is the product of the modulus of elasticity (Young's modulus), E_{pile}, and the second moment of area, I_Y. The mathematical form of the soil resistance, p(Z), depends on both the bed material and the zone of interest in Figure 12.11.

(1) Bending Deflection in the Plastic Zone

The soil resistance in the plastic (yield) zone can be represented by

$$\mathrm{p(Z) = P + QZ} \tag{12.30}$$

This relationship applies to both soft clays and sand. According to Dawson (1980), the parameters in eq. 12.30 for a pile of diameter D in soft clay are

$$\mathrm{P_{clay}} = \mathrm{N_s}D\mathrm{a}, \quad \mathrm{Q_{clay}} = \mathrm{N_s}D\mathrm{b} \tag{12.31}$$

and in sand, the parameters are

$$\mathrm{P_{sand}} = 0, \quad \mathrm{Q_{sand}} = \mathrm{N_s}D\mathrm{R_{pas}}\gamma_{\mathrm{sand}} \tag{12.32}$$

The first equality in eq. 12.32 comes from the assumption that the sand is non-cohesive. The force coefficient ($\mathrm{N_s}$) is introduced in eq. 12.23, and the Rankine passive earth coefficient (R_{pas}) is obtained from eq. 12.27. See Table 12.1 for values of these coefficients.

Combine eqs. 12.29 and 12.30 and integrate the resulting expression with respect to Z. From beam theory, as discussed by Gere (2001) and others, the first integration results in the shear force, and the second integration produces the bending moment. We can apply the results of the integrations to the head of the pile in Figure 12.11, where the applied force and moment are F_0 and M_0, respectively. These are the boundary conditions, represented mathematically by

$$E_{pile}\mathrm{I_Y}\frac{d^3\mathrm{X}_{pile}}{d\mathrm{Z}^3}\bigg|_{\mathrm{Z}=0} = F_0$$

$$E_{pile}\mathrm{I_Y}\frac{d^2\mathrm{X}_{pile}}{d\mathrm{Z}^2}\bigg|_{\mathrm{Z}=0} = M_0 \tag{12.33}$$

Two additional integrations of eq. 12.29, using the soil-resistance expression in eq. 12.30, yield the expression for the pile displacement in the plastic zone. That expression is

$$\mathrm{X}_{pile}|_{\mathrm{Z}<\mathrm{Z_s}} = \frac{1}{E_{pile}\mathrm{I_Y}}\left(-\frac{\mathrm{P}}{24}\mathrm{Z}^4 - \frac{\mathrm{Q}}{120}\mathrm{Z}^5 + \frac{F_0}{6}\mathrm{Z}^3 + \frac{M_0}{2}\mathrm{Z}^2 + C_a\mathrm{Z} + C_b\right) \tag{12.34}$$

In this equation are two additional constants (C_a and C_b), which are to be determined. The pile slope at any point in the plastic zone, from eq. 12.34, is

$$\frac{d\mathrm{X}_{pile}}{d\mathrm{Z}}\bigg|_{\mathrm{Z}<\mathrm{Z_s}} = \frac{1}{E_{pile}\mathrm{I_Y}}\left(-\frac{\mathrm{P}}{6}\mathrm{Z}^3 - \frac{\mathrm{Q}}{24}\mathrm{Z}^4 + \frac{F_0}{2}\mathrm{Z}^2 + M_0\mathrm{Z} + C_a\right) \tag{12.35}$$

Using the displacement expression in eq. 12.34, the respective shear force and bending moment expressions evaluated at $Z = Z_s$ are

$$E_{pile}\mathrm{I_Y}\frac{d^3\mathrm{X_{pile}}}{d\mathrm{Z}^3}\bigg|_{\mathrm{Z}=\mathrm{Z_s}} = -F_{\mathrm{Zs}} = -E_{pile}\mathrm{I_Y}\left(-\mathrm{PZ_s} - \frac{\mathrm{Q}}{2}\mathrm{Z_s^2} + F_0\right)$$

$$E_{pile}\mathrm{I_Y}\frac{d^2\mathrm{X_{pile}}}{d\mathrm{Z}^2}\bigg|_{\mathrm{Z}=\mathrm{Z_s}} = M_{\mathrm{Zs}} = E_{pile}\mathrm{I_Y}\left(-\frac{\mathrm{P}}{2}\mathrm{Z_s^2} - \frac{\mathrm{Q}}{6}\mathrm{Z_s^3} + F_0\mathrm{Z_s} + M_0\right) \tag{12.36}$$

where $\mathrm{Z_s}$, F_{Zs}, and M_{Zs} are unknowns at this point in the analysis.

(2) Bending Deflection in the Elastic Zone

The soil resistance in the elastic zone in Figure 12.9b is mathematically represented by

$$p(Z)|_{Z>Z_s} = k_s X_s, \quad \text{where } X_s \leq X_{yield} \tag{12.37}$$

The combination of this expression with the differential equation in eq. 12.29 results in

$$E_{pile} I_Y \frac{d^4 X_{pile}}{dZ^4}\bigg|_{Z>Z_s} = -k_s X_s, \quad \text{where } X_s \leq X_{yield} \tag{12.38}$$

Note that X_{pile} and X_s are here the same in that a displacement of the pile is assumed to equal a corresponding displacement of the soil at the pile's surface. Equation 12.38 can be solved using operator notation, where $d^n X/dZ^n \equiv D^n$. The resulting algebraic equation is factored, and the components are algebraically solved. This procedure reduces the order of the resulting differential equations to $n = 2$, and the equations can be solved using the standard techniques discussed in books on ordinary differential equations. From Dawson (1980), the solution of eq. 12.38 is

$$X_{pile}|_{Z>Z_s} = \frac{1}{E_{pile} I_Y} e^{-\left(\frac{k_s}{E_{pile} I_Y}\right)^{1/4}(Z-Z_s)} \left\{ C_c \cos\left[\left(\frac{k_s}{E_{pile} I_Y}\right)^{1/4}(Z-Z_s)\right] + C_d \sin\left[\left(\frac{k_s}{E_{pile} I_Y}\right)^{1/4}(Z-Z_s)\right]\right\} \tag{12.39}$$

where Z_s is the depth of the plastic zone below the mud line. We note that both pile displacement in eq. 12.39 and its derivative vanish if $Z \to \infty$, as expected. According to Dawson (1980), for the deeply driven pile, the total depth of the pile (d in Figure 12.10) must satisfy

$$d \geq Z_s + 3\left(\frac{E_{pile} I_Y}{k_s}\right)^{1/4} \tag{12.40}$$

The slope of the pile in the elastic zone is the derivative of the deflection in eq. 12.39, that is,

$$\frac{dX_{pile}}{dZ}\bigg|_{Z>Z_s} = -\frac{1}{E_{pile} I_Y}\left(\frac{k_s}{E_{pile} I_Y}\right)^{1/4} e^{-\left(\frac{k_s}{E_{pile} I_Y}\right)^{1/4}(Z-Z_s)} \cdot \left\{ C_c \cos\left[\left(\frac{k_s}{E_{pile} I_Y}\right)^{1/4}(Z-Z_s)\right] + C_d \sin\left[\left(\frac{k_s}{E_{pile} I_Y}\right)^{1/4}(Z-Z_s)\right]\right\} \tag{12.41}$$

We now use the displacement expression in eq. 12.39 to obtain the shear force and bending moment in the pile at $Z = Z_s$. The process is similar to that in obtaining eq. 12.35. The result is

$$E_{pile} I_Y \frac{d^3 X_{pile}}{dZ^3}\bigg|_{Z=Z_s} = -F_{Z_s} = -2\left(\frac{k_s}{E_{pile} I_Y}\right)^{3/4}(C_c + C_d)$$
$$E_{pile} I_Y \frac{d^2 X_{pile}}{dZ^2}\bigg|_{Z=Z_s} = M_{Z_s} = -2\left(\frac{k_s}{E_{pile} I_Y}\right)^{1/2} C_d \tag{12.42}$$

(3) Complete Bending Solution

There are five parametric constants to be determined. Those are the thickness of the plastic zone, Z_s, the constants C_a and C_b in eq. 12.34, and the constants C_c and C_d

in eq. 12.39. Hence, five equations are required to solve the problem. We find that four of the five required equations apply to the pile at the depth of the interface of the plastic and elastic zones. That is, at $Z = Z_s$, the following equalities apply: the pile displacements obtained from eqs. 12.34 and 12.39, the pile slopes obtained from eqs. 12.35 and 12.41, the shear forces predicted by the first lines in eqs. 12.36 and 12.42, and the bending moments predicted by the second lines in eqs. 12.36 and 12.42. The last equation depends on the sea bed material. For soft clays, that equation is eq. 12.25 applied at $Z = Z_s$, and for sand, eq. 12.28 is applied at $Z = Z_s$.

The following are determined from the solution of the simultaneous equations: In eq. 12.34, the constants are

$$C_c = \frac{1}{2}\left(\frac{E_{pile}I_Y}{k_s}\right)^{1/2}\left[M_{Z_s} - \left(\frac{E_{pile}I_Y}{k_s}\right)^{1/4} F_{Z_s}\right] \tag{12.43}$$

and

$$C_d = -\frac{1}{2}\left(\frac{E_{pile}I_Y}{k_s}\right)^{1/2} M_{Z_s} \tag{12.44}$$

The shear force (F_{Z_s}) and bending moment (M_{Z_s}) at the interface of the elastic and plastic zones are found in

$$2(P + QZ_s) = \left(\frac{k_s}{E_{pile}I_Y}\right)^{1/4}\left[\left(\frac{k_s}{E_{pile}I_Y}\right)^{1/4} M_{Z_s} - F_{Z_s}\right] \tag{12.45}$$

where P and Q depend on the soil type. For soft clays, these parameters are found in eq. 12.31, and in eq. 12.32 for sand. This equation with those in eq. 12.36 comprise three equations with three unknowns, those being Z_s, F_{Z_s}, and M_{Z_s}. By replacing the force and moment in eq. 12.45 by the expressions in eq. 12.36, we obtain the following expression for the interfacial depth:

$$2(P + QZ_s) = (E_{pile}I_Y)^{3/4}k_s^{1/4}\left[\left(\frac{k_s}{E_{pile}I_Y}\right)^{1/4}\left(\frac{P}{2}Z_s^2 - \frac{Q}{6}Z_s^3 + F_0 Z_s + M_0\right) - \left(-PZ_s - \frac{Q}{2}Z_s^2 + F_0\right)\right] \tag{12.46}$$

For a give-soil (soft clay or sand), this cubic equation in Z_s can be solved using the *method of successive approximations* described in Section 3.3, Example 3.3. With the value of Z_s known, the shear force and bending moment at the interface are found from eq. 12.36. In turn, these values are substituted into eqs. 12.43 and 12.44 to obtain the values of C_c and C_d. Finally, the coefficients in the pile-displacement expression in eq. 12.34 are found to be

$$C_a = -\frac{1}{2}\left(\frac{E_{pile}I_Y}{k_s}\right)^{1/2}\left[2\left(\frac{k_s}{E_{pile}I_Y}\right)^{1/4} M_{Z_s} - F_{Z_s}\right] + \frac{P}{6}Z_s^3 + \frac{Q}{24}Z_s^4 - \frac{F_0}{2}Z_s^2 - M_0 Z_s \tag{12.47}$$

and

$$C_b = \frac{1}{2}\left(\frac{E_{pile}I_Y}{k_s}\right)^{1/2}\left[1 + 2\left(\frac{k_s}{E_{pile}I_Y}\right)^{1/4} Z_s\right] M_{Z_s} - \frac{1}{2}\left(\frac{E_{pile}I_Y}{k_s}\right)^{3/4}\left[1 + \left(\frac{k_s}{E_{pile}I_Y}\right)^{1/4} Z_s\right] F_{Z_s} - \frac{P}{8}Z_s^4 - \frac{Q}{30}Z_s^5 + \frac{F_0}{3}Z_s^3 + \frac{M_0}{2}Z_s^2 \tag{12.48}$$

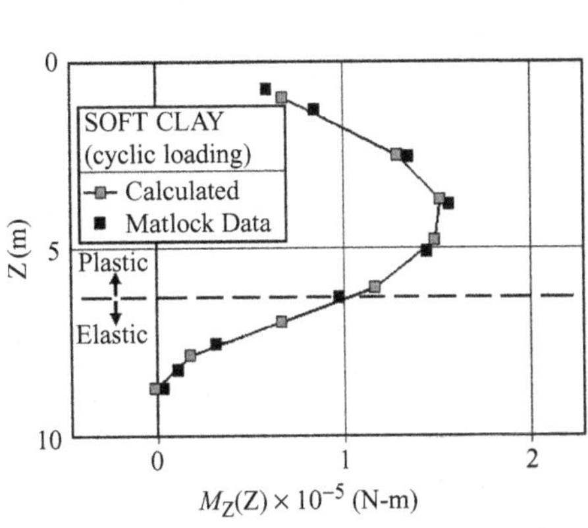

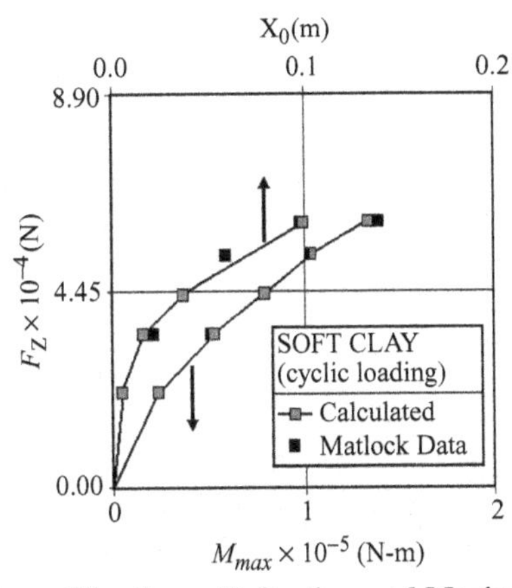

a. Bending Moment-Depth Behavior **b. Groundline Force, Deflection, and Maximum Moment**

Figure 12.12. *Moment, Force, and Displacement Relationships: Deeply Driven Pile in Soft Clay.* The data are those of Matlock (1970) and are presented in Table 12.1 for the 0.323-m pile. The cyclic load, applied at $Z = 0.305$ m, has an amplitude of 6.00×10^4 N. The interface in Figure 12.12a is at $Z = Z_s \simeq 6.40$ m. See Dawson (1980) for a further discussion.

(4) Comparison of Analysis and Data

Dawson (1980) applies the analysis to the field measurements of Matlock (1970) in a soft clay, and to the field measurements of Reese, Cox, and Koop (1974) in sand. The data for the soils for these two studies are presented in Table 12.1. In the studies, hollow piles were used, where the outside diameter of the Matlock pile was 0.323 m, and that of the Reese pile was 0.610 m. For this pile cross-section, the second moment of area is $I_Y = \pi(D_{out}^4 - D_{in}^4)/64$, where the respective diameters are the outer diameter and the inner diameter of the pipe cross-section. The results of the Dawson (1980) analysis and the field data are presented in Figures 12.12 and 12.13. The data in Table 12.1 and in these figures can be considered to be examples, in that the data are not presented in non-dimensional forms. The importance of the results

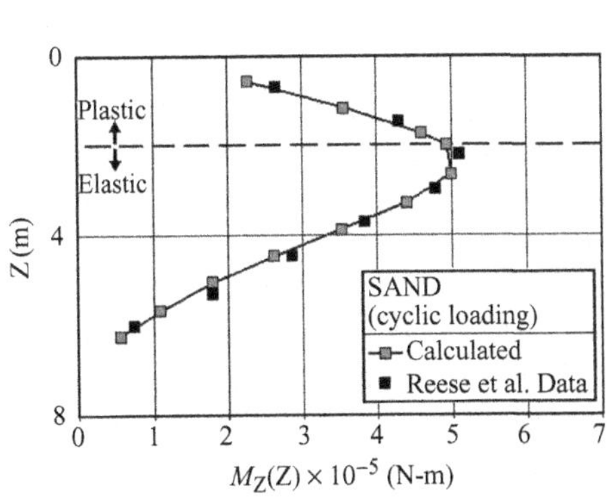

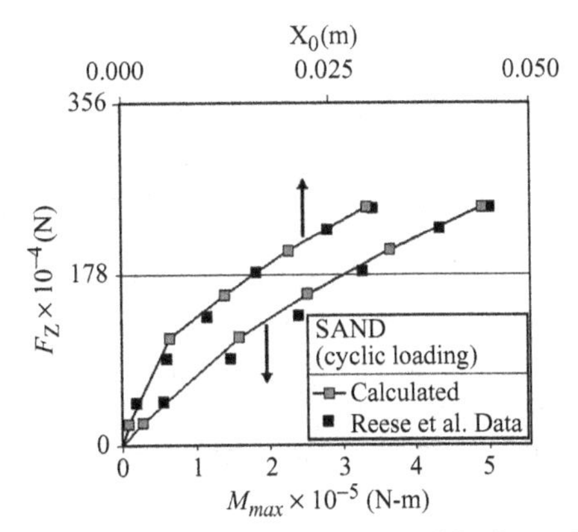

a. Bending Moment-Depth Behavior **b. Groundline Force, Deflection, and Maximum Moment**

Figure 12.13. *Moment, Force, and Displacement Relationships: Deeply Driven Pile in Sand.* The data are those of Reese, Cox, and Koop (1974) and are presented in Table 12.1 for the 0.610-m pile. The cyclic load, applied at $Z = 0.305$ m, has an amplitude of 2.44×10^5 N. The interface in Figure 12.12a is at $Z = Z_s \simeq 2.13$ m. See Dawson (1980) for a further discussion.

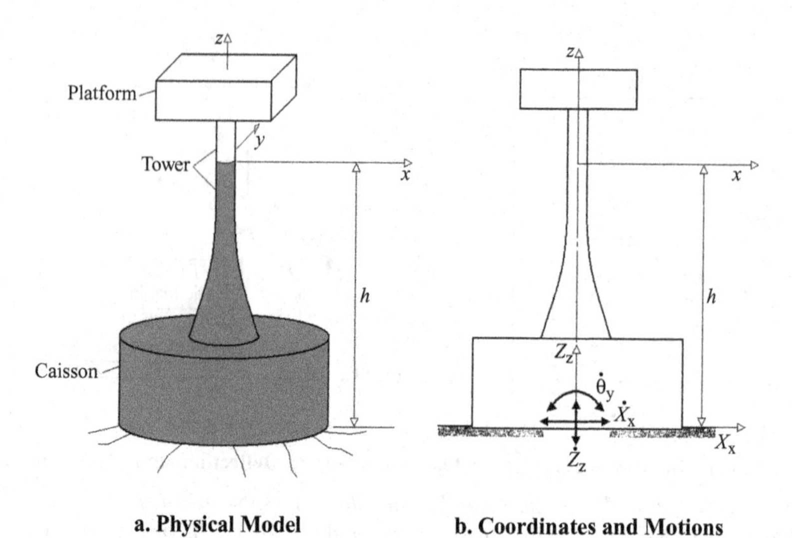

Figure 12.14. *Sketches of a Single-Tower Gravity Structure.* The caisson is normally a concrete structure, whereas the tower is made of concrete and steel. The motions illustrated in (b) are sliding (X), heaving (Z), and rocking (θ).

in the figures is that the analytical points and the field data for both the soft clay and the sand agree rather well. For this reason, we can have confidence in the analysis.

The paper by Dawson (1980) is both concise and well written. The reader is encouraged to consult this paper for further discussions on both the analysis and the field data of Matlock (1970) and Reese, Cox, and Koop (1974).

B. Spread Footings – Gravity Structures

The discovery of oil beneath the Gulf of Mexico, the North Sea, and other moderately deep bodies of water fostered a type of offshore structure called a *gravity structure.* This type of structure is characterized by a large flat interface with the sea bed. Referring to the sketch in Figure 12.14a, the gravity structure is normally characterized by a large *caisson* resting on the bed. The top of the caisson is well beneath the free surface of the water. Because of this design feature, the caisson experiences relatively small wave-induced forces. According to Moan, Syvertsen, and Haver (1977), the height of the caisson is between $h/3$ and $h/2$, where h is the water depth (see Figure 12.14). Extending from the top of the caisson and extending through the free surface is a tower, designed to both support a platform and to protect drilling equipment from the environment. One such platform is the Draugen CONDEEP platform, installed in 1993 in 250 m of water. Other configurations include the two-tower Oseberg A CONDEEP platform, installed in 1988 in about 110 m of water, and the three-tower Sleipner A CONDEEP platform, installed in 1993 in about 80 m of water. The CONDEEP configurations have multicomponent caissons, where each component comprising the caisson can be independently ballasted.

The primary motions of a gravity structure are normally classified as sliding (X_x) and rocking (θ_y) with, possibly, some heaving (Z_z), as illustrated in Figure 12.14b. A combination of motions applied to a grounded ship can result in migration in the direction of the incident waves, as discussed by Hudson (2001). In what follows, the

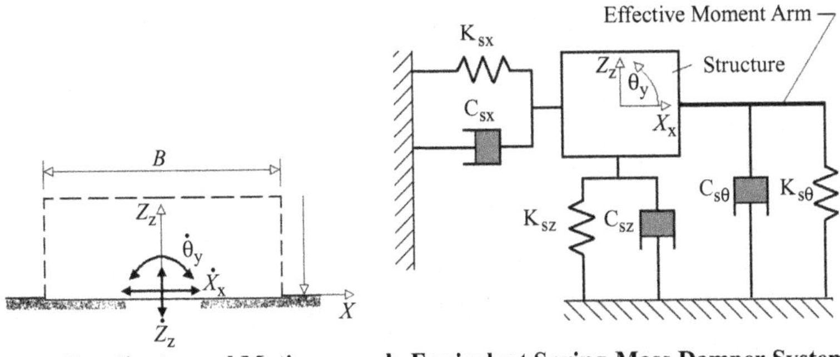

Figure 12.15. *Equivalent Spring-Mass Damper Systems for Soil Reactions to Gravity Structure Motions.* The motions are assumed to be linear, as discussed by Moan, Syvertsen, and Haver (1977) and by Wolf (1988, 1994). The springs and dampers for the respective sliding, heaving, and rocking motions are identified by X_x, Z_z, and θ_y in the subscripts. The model footing in (a) has a uniform width (into the page).

ramifications of the gravity-structure motions on the soil are discussed. In the analysis, a sea-bed coordinate system is used, where the horizontal coordinate is $X_x = x$, and the vertical coordinate is $Z_z = z + h$. Note that in the discussion of embedded structures, the downward vertical coordinate is used. That is, the vertical bed coordinate in Figures 12.12 and 12.13 is $Z = -Z_z$. Our goal here is to determine the forms of the soil-related components in the equations of motion of the structure.

The motions of the gravity structure illustrated in Figure 12.14b excite the bed material, which in turn contributes to both the stiffness and the damping terms in the equations of motion. The exciting forces and moment are assumed to be cyclic in nature. The responses of the bed are assumed to be *quasi*-linear, in that each can be represented by equivalent spring-mass damper systems, as illustrated in Figure 12.15b. In that figure are the effective soil spring constants (K_{sx}, K_{sz}, and $K_{s\theta}$) and soil damping coefficients (C_{sx}, C_{sz}, and $C_{s\theta}$) for the sliding, heaving, and rocking motions. Any constants and coefficients related to these respective motions are identified by X_x, Z_z, and θ_y in the subscripts.

Following Moan, Syvertsen, and Haver (1977), the determination of the spring constant and damping coefficients begins with the stress-strain relationship. For embedded structures, this relationship is the elastoplastic relationship shown in Figure 12.9. For the spread footing, the initial stress-strain relationship is assumed to be linear. That is, the stress on the soil at the soil-structure interface is proportional to the strain in the soil. The soil is then treated as a linear, isotropic elastic half-space (extending to $Z_z = -\infty$ over $-\infty < X_x < \infty$). In the analysis that follows, we shall assume that the heaving motions are of second order when compared to the sliding and rocking motions and, therefore, can be neglected. Furthermore, the spread footing has a rectangular horizontal plane having a breadth B in the x-direction. The width of the footing into the plane is considered to be much greater than the breadth. As a result, the analysis is essentially two-dimensional.

For the sliding and rocking motions of a spread footing, the damping constants are restoring coefficients that can be related to the soil properties following the approximate methods of Wolf (1988, 1994), as done by McCormick and Hudson (2001). For the situation in Figure 12.15, excluding the heaving motions, the following apply:

Sliding Motions: The *sliding spring constant* (soil rectilinear stiffness coefficient) is

$$K_{sx} = G_s(1 + 5\nu_s^2) \tag{12.49}$$

where G_s is the shear modulus of the soil and ν_s is the soil Poisson's ratio. In terms of the effective modulus of elasticity of the soil (E_s), the shear modulus is

$$G_s = \frac{E_s}{2(1+\nu_s)} \tag{12.50}$$

The damping coefficient for sliding (soil rectilinear damping coefficient) is related to the sliding spring constant as

$$C_{sx} \simeq \frac{1}{2} K_{sx} B(2 - 2.2\nu_s)\sqrt{\frac{\rho_s}{G_s}} \tag{12.51}$$

In this equation, B is the breadth of the spread footing, ν_s is Poisson's ratio for the soil, and ρ_s is the mass density of the soil. Concerning the soil mass excited by the sliding motion: Wolf (1988) states that the inertial reaction due to the sliding soil mass is of second order if the embedment depth (d) is much less than $B/2$. This is the case for the offshore spread-footing structure and is assumed herein.

Rocking Motions: The *rocking spring constant* (soil rotational stiffness coefficient) is obtained from

$$K_{s\theta} \simeq \tau_s \frac{B^2}{4}\left[2.38 + \left(2\frac{d}{B}\right) + 1.2\left(2\frac{d}{B}\right)^2\right] \tag{12.52}$$

The expression for the *rocking damping coefficient* (soil rotational damping coefficient) is

$$C_{s\theta} \simeq \frac{1}{2} K_{s\theta} B\sqrt{\frac{\rho_s}{G_s}}\left[0.11 + 0.35\left(2\frac{d}{B}\right) + 0.1\left(2\frac{d}{B}\right)^2\right] \tag{12.53}$$

The inertial reaction to the rocking motions is proportional to the *soil mass-moment of inertia*, which is obtained from

$$I_{s\theta} \simeq 0.15\frac{\rho_s K_{s\theta} B}{G_s} \tag{12.54}$$

The approximate expressions in eqs. 12.51 through 12.54 are used by McCormick and Hudson (2001) and Hudson (2001) with some success in predicting the motions and migration of a grounded ship. In the studies leading to those writings, the geometry of the ship section used in both the experiments and theoretical analyses was rectangular, and was uniform across the wave tank. Hence, the study was two-dimensional in nature, and similar to that sketched in Figure 12.15a.

In the following sections, the information presented in Part A of this section is used in the analysis of the wave-induced motion of a TLP. This is followed by the application of the equations in Part B to spread-footing structures.

12.4 Motions of a Tension-Leg Platform (TLP)

In the preamble of Section 12.2, the mini-TLP SeaStar is described. According to Bhattacharyya, Sreekumar, and Idichandy (2003), the SeaStar configuration "combines the simplicity of a spar and (the) favorable response features of a TLP." The

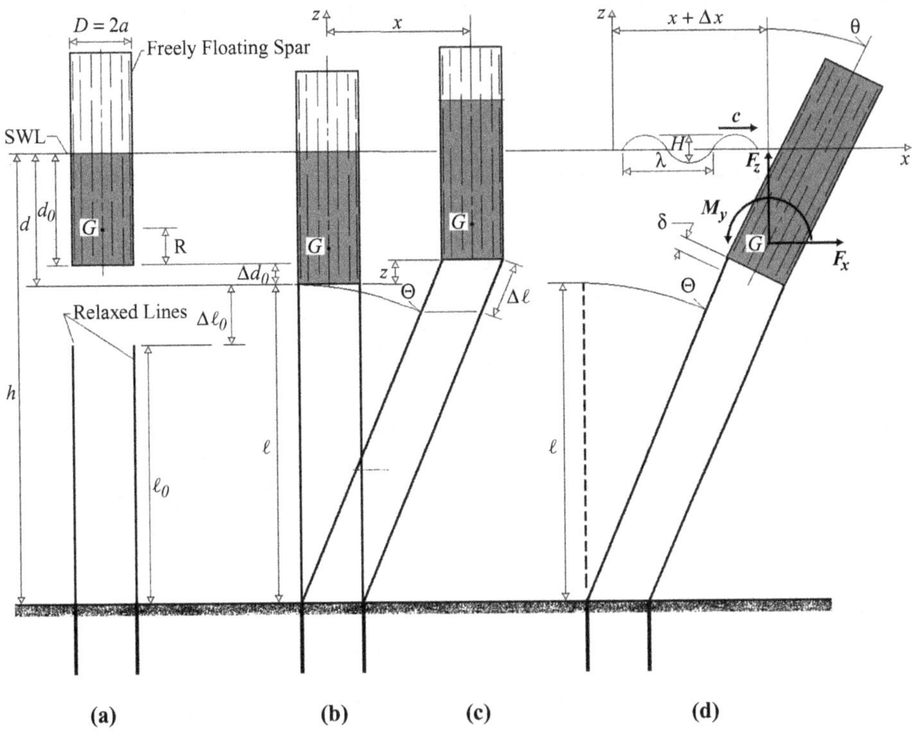

Figure 12.16. *Nomenclature for a Tension-Moored Spar.* In (a), the spar is not coupled to the lines. The flexible lines are slightly buoyant, and are shown in their relaxed position. In (b), the moored spar is shown in static equilibrium, where the draft is $d = d_0 + \Delta d_0$. In (c), the moored spar is displaced both horizontally and vertically. Both lines in this condition are at an angle of Θ to the vertical axis. Finally, in (d) the moored spar is rotated due to an applied moment, M_y. The forces F_x and F_z are due to a passing wave and the change in the displacement.

authors also distinguish between the spar and a TLP as follows: The spar has a deep-draft cylindrical hull, and may have an inner opening exposed to the sea. The opening is referred to as a "moon pool." The TLP is a shallow-draft body equipped with pontoons. The paper by Kibbee and Snell (2002) contains an excellent discussion of the various TLP configurations, including that of the SeaStar. The purpose of this section is to introduce the reader to the coupling of large floating platforms, taut moorings, and embedment foundations. These are essentially the components of a TLP.

Consider a vertical, circular, cylindrical, spar-type hull that is moored by two ($N_s = 2$) flexible lines in tension, as sketched in Figure 12.16. The spar is subject to incident linear waves of height H and period T. Before considering the influence of the wave, we first analyze the effects on the mooring-line displacements by considering the phases shown in Figure 12.16. In Figure 12.16a, the unmoored spar and the relaxed lines are sketched. The lines are considered to be slightly buoyant so that they are vertical in the relaxed condition. With respect to the incident wave direction, the lines are attached at the forward and after positions on the bottom of the spar. In Figure 12.16a, the displacement (equal to the spar weight) is $W = \rho g \pi d_0 D^2/4$, where d_0 is the draft and D is the diameter. The relaxed lines have a relaxed length ℓ_0 and a radius of r. In the moored equilibrium condition in

Figure 12.16b, the line length is ℓ, and the line radius is approximately r. The radial strain in the line is of second order, as discussed in Section 12.2A.

A. Tethers

From the orientation of the tethers (mooring lines) sketched in Figure 12.16c, the following geometric relationships are obtained:

$$\sin(\Theta) = \frac{x}{\ell + \Delta\ell} \tag{12.55}$$

and

$$\cos(\Theta) = \frac{\ell + z}{\ell + \Delta\ell} \tag{12.56}$$

In these equations, x is the surging displacement and z is the set-down. Also, for the equilibrium orientation in Figure 12.16b and displaced orientation of the spar in Figure 12.16c, and from the definition of the modulus of elasticity, we can write

$$E_s = \frac{T_0 \ell_0}{\pi r^2(\Delta\ell_0)} = \frac{T_\ell \ell_0}{\pi r^2(\Delta\ell)} \tag{12.57}$$

Rearrange this equation to obtain the expression for the tension in the line when the body is displaced as sketched in Figure 12.16c, that is,

$$T_\ell = E_s \pi r^2 \left(\frac{\Delta\ell}{\ell_0}\right) \simeq T_0 = E_s \pi r^2 \left(\frac{\Delta\ell_0}{\ell_0}\right) \tag{12.58}$$

The approximation is valid if the angular displacement (Θ) is small. We note here that both the additional draft (Δd_0) and the line stretch ($\Delta\ell_0$) are specified in the design. As a result, all of the terms in eqs. 12.57 and 12.58 are known. In eqs. 12.55 through 12.58, the line stretch in Figure 12.16c is

$$\Delta\ell = \frac{\ell + z}{\cos(\Theta)} - \ell \tag{12.59}$$

where z is the heaving displacement.

In Figure 12c, the spar is displaced horizontally and vertically. However, the centerline of the spar remains vertical. For a TLP, this is the design orientation. As sketched, the displaced mooring lines are at an angle (Θ) to the vertical. This orientation results in both an increased line tension and stretch in each line. The net vertical force component of the line tension results in the vertical displacement, z. As in eq. 12.58, there is an additional tension in the line due to the additional stretch $\Delta\ell$, as sketched in Figure 12.16c. Finally, there will be a rotation of the spar, as sketched in Figure 12.16d. The angle from the vertical direction of the spar rotation is Θ. For a TLP, $\Theta \simeq 0$ by design. This condition is assumed in the analysis presented herein. Our goal is to determine the expressions for the effective spring constants for the surging and heaving motions of the spar.

The horizontal component of the tension in the mooring lines sketched in Figure 12.16c is

$$T_{\ell x} = T_\ell \sin(\Theta) = T_\ell \frac{x}{(\ell + \Delta\ell)} \simeq \frac{T_\ell}{\ell} x = \left(\frac{T_0}{\ell_0 + \Delta\ell_0}\right) x = K_{sx} x \tag{12.60}$$

where $K_{sx} = T_\ell/(\ell_0 + \Delta\ell_0)$ is the horizontal spring constant. The last approximation is based on the assumption that the tension in Figures 12.16b and 12.16c are approximately equal.

The vertical component of the additional tension is

$$\mathrm{T}_z = \mathrm{T}_\ell \cos(\Theta) = E_s \pi r^2 \frac{\Delta \ell}{\ell} \cos(\Theta) = \frac{E_s \pi r^2}{\ell}[\ell - \ell \cos(\Theta) + z]$$
$$\simeq \frac{E_s \pi r^2}{\ell} z = \left(\frac{E_s \pi r^2}{\ell_0 + \Delta \ell_0}\right) z = K_{sz} z \tag{12.61}$$

The last equality contains the vertical spring constant, $K_{sz} = E_s \pi r^2/\ell$, where $\ell = \ell_0 + \Delta\ell_0$. The assumption leading to the approximation in eq. 12.60 is that the angle Θ is small, an assumption used in the remainder of this section.

With the spring constants defined, the linear heaving and swaying motions of the spar-type TLP can now be addressed. For those interested in the derivation of the spring constants for all degrees of freedom of a taut-moored body, the book of Patel (1989) is recommended. Patel includes the nonlinearities resulting from finite values of Θ and the subsequent coupling of the motions.

EXAMPLE 12.4: EFFECTIVE SPRING CONSTANTS FOR A TLP Consider a SeaStar type of TLP, as sketched in Figures 12.2 and 12.17. We model the TLP after one of the prototypes analyzed by Bhattacharyya, Sreekumar, and Idichandy (2003), referred to as Prototype A by those investigators. Prototype A is to be moored in 215 m of water ($h = 215$ m).

The tethers are composed of steel strands; hence, the tethers are not neutrally buoyant. From Bhattacharyya, Sreekumar, and Idichandy (2003), the tethers have the following dimensions and properties:

(a) Relaxed length (ℓ_0): 175 m
(b) Tether area (A_s): 0.0876 m^2 ($r \simeq 0.167$ m)
(c) Modulus of elasticity (E_s): 2×10^{11} N/m^2
(d) Pre-tension ($3\mathrm{T}_0$): 4,000 metric tons (tonne) or 3.924×10^7 N
(e) Total three-tether weight: 360 metric tons or 3.532×10^6 N

The Prototype A SeaStar sketched in Figure 12.17 displaces 16,355 metric tons (1.604×10^8 N) of salt water when deployed, and weighs 12,355 metric tons (1.212×10^8 N). The difference in the displacement and the weight is then the previous pre-tension value. From the first equality in eq. 12.57, the stretch in the tethers when the system is deployed is approximately $\Delta\ell_0 = \mathrm{T}_0\ell_0/E_s A_s \simeq 0.392$ m. Hence, $\ell \simeq \ell_0 = 175$ m.

From eqs. 12.60 and 12.61, the respective horizontal and vertical spring constants are found to be $K_{sx} = \mathrm{T}_0/\ell \simeq 22.4 \times 10^4$ N/m and $K_{sz} = E_s \pi r^2/\ell \simeq 1.001 \times 10^8$ N/m. From these results, we see that the moored TLP is extremely "stiff" in heave.

B. Soil Reactions

The spar-type TLP sketched in Figure 12.16 is moored to two embedment-type anchors. These can be treated simply as deeply driven piles. Hence, the soil-structure interactions are those described in Section 12.3A. See Figure 12.11 for the reaction to the lateral load on such a structure. To include the soil reactions in the time-dependent forces induced by the TLP motions is a bit difficult. For practical purposes, one can assume that the time-dependent vertical reaction of the soil is

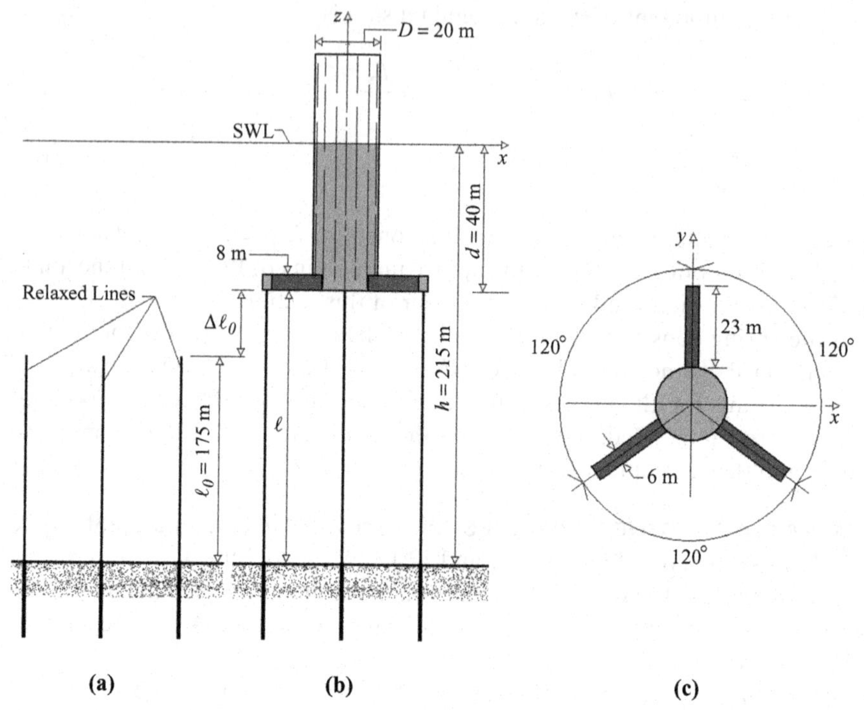

Figure 12.17. *TLP Configuration in Example 12.4.* In (a), the steel tethers are shown in their pre-deployment length. In (b), the SeaStar model is shown in elevation in the deployed configuration. The areal view of the TLP is sketched in (c).

of second order when compared to the equilibrium reaction. That is, in Figures 12.16b and 12.16c, the vertical soil reactions are approximately the same. However, the soil reactions to the lateral loads should be included. Again, for practical purposes, we see in Figure 12.12 (clay) and 12.13 (sand) that the soil pile displacements at the groundline for respective pile diameters of 0.323 m and 0.610 m are relatively small for applied lateral loads of about 5 tonnes for the former and approximately 250 tonnes for the latter. Hence, as a first approximation, the time dependency of the pile at the groundline can be neglected. As a result, the point of attachment of the mooring line and the embedment anchor is considered to be fixed in time.

C. Wave-Induced Forces

Turning our attention to the moored spar in Figure 12.16c, we must determine the type of force that is dominant in a sea. For the circular cylindrical spar, we use the Chakrabarti (1975) results presented in Figure 9.8. That figure shows the dominance ranges of the drag, inertial, and diffraction forces as determined by the Keulegan-Carpenter number ($KC = u_{max}T/D$) and the dimensionless radius ($ka = kD/2 = 2\pi a/\lambda$) of the surface-piercing circular cylinder. In the KC expression are the maximum horizontal particle velocity (u_{max}) in the wave, the wave period (T), and the diameter ($D = 2a$), whereas in the ka expression are the wave number (k) and radius (a). The maximum particle velocity is presented in eq. 9.53.

For linear waves, that velocity expression is

$$u_{\max} = \pi \frac{H}{T} \frac{\cosh[k(z+h)]}{\sinh(kh)} \tag{12.62}$$

The reader should be reminded that when the structure is exposed to extreme waves, the choice of the wave theory to be used in the analysis is influenced by the results presented in Figure 4.1. Assume that the TLP is moored in deep-water waves. As a result of this assumption, the approximate Keulegan-Carpenter number in eq. 9.46 can be used, that is, $KC \simeq \pi H_0/D$, where the subscript "0" identifies deep-water wave properties.

The two extremes in Figure 9.8 are the cases corresponding to $KC \geq 100$, where $ka \simeq 0.01$, and $KC \simeq 0.01$, where $ka \geq 0.5$. In the figure, we see that the former corresponds to a dominant drag force, and the latter corresponds to a dominant diffraction force.

(1) Drag Force

The drag force is dominant when the wavelength is much greater than the body length. For the spar, the length is the diameter. The drag force acting on the sphere is expressed by

$$F_d(t) = \frac{1}{2}\rho D \int_{-d}^{0} (u|u|)_{x=0} C_{dD} dz \simeq \frac{1}{2}\rho D C_{dD} \int_{-d}^{0} (u|u|)_{x=0} dz \tag{12.63}$$

The first approximation is due to the application of the horizontal velocity expression at the centerline of the structure, and the variation of the free-surface displacement is negligible. The second approximation results from Figure 2.15, where the drag coefficient (C_{dD}) is approximately constant on the upper bound of the uncertainty band for $\log_{10}(R_{eD})$ values greater than 4. Here, we assume $C_{dD} \simeq 1$. Using the linear wave formula in eq. 3.49, the horizontal velocity expression in eq. 12.63 is

$$u|_{x=0} = \pi \frac{H}{T} \frac{\cosh[k(z+h)]}{\sinh(kh)} \cos(\omega t) = u_{\max} \cos(\omega t) \tag{12.64}$$

The combination of eqs. 12.63 and 12.64 results in the following expression for the wave-induced drag force:

$$\begin{aligned} F_d(t) &= \frac{1}{32}\rho D H^2 \frac{\omega^2}{k} \frac{1}{\sinh^2(kh)} \{2kd + \sinh(2kh)[1 - \cosh(2kd)] \\ &\quad + \cosh(2kh)\sinh(2kd)\} \cos(\omega t)|\cos(\omega t)| \\ &= F_{do} \cos(\omega t)|\cos(\omega t)| \end{aligned} \tag{12.65}$$

where F_{do} is the force amplitude function. Again, $C_{dD} \simeq 1$ is assumed. The approximate drag-force expression for a taut-moored spar (sketched in Figure 12.16b) in deep-water waves is

$$\begin{aligned} F_{do}(t) &= \frac{1}{16}\rho g D H_0^2 [1 - \cosh(2k_0 d) + \sinh(2k_0 d)] \cos(\omega t)|\cos(\omega t)| \\ &= F_{do0} \cos(\omega t)|\cos(\omega t)| \end{aligned} \tag{12.66}$$

In this equation, we have used the results obtained from applying the dispersion relationship of eq. 3.31 to deep water. From the second equality in that equation, $\omega^2/k_0 = g$.

The drag-force expressions in eqs. 12.65 and 12.66 are nonlinear in time. When the sea is composed of random long waves, it is advantageous to replace the nonlinear drag force with an equivalent linear force. To obtain the equivalent linear force, we use an analytical technique that is similar to that in Section 10.1C. In that section, a method is presented to obtain the equivalent linear damping coefficient for viscous damping. The application of the technique to the wave-induced drag force is done by equating the energies supplied by both the nonlinear force and equivalent linear force over one wave period. Assume that the swaying displacement in Figure 12.16 is $x = x_o \sin(\omega t)$. Following the thought process leading to eq. 10.9, we can write the following energy equation:

$$\int_{x(0)}^{x(T)} F_{\mathrm{d}} dx = \int_0^T F_{\mathrm{d}} \frac{dx}{dt} dt = \int_0^T F_{\mathrm{d}o} \cos(\omega t) \frac{dx}{dt} dt = F_{\mathrm{d}o} \pi x_o = \int_{x(0)}^{x(T)} F_d dx$$

$$= 4 \int_0^{T/4} F_d \frac{dx}{dt} dt = 4 \int_0^{T/4} F_{do} \omega x_o \cos^3(\omega t) dt = \frac{8}{3} F_{do} x_o \tag{12.67}$$

Note that the amplitudes of the linear force and nonlinear force are, respectively, $F_{\mathrm{d}o}$ and F_{do}. In eq. 12.67 is the relationship between the two force amplitudes. The resulting expression for the equivalent linear drag force for a spar moored in deep water is

$$\begin{aligned} F_{\mathrm{d}0} &= F_{\mathrm{d}o0} \cos(\omega t) \\ &\simeq \frac{8}{3\pi} F_{do0} \cos(\omega t) = \frac{1}{6\pi} \rho g D H_0^2 [1 - \cosh(2k_0 d) + \sinh(2k_0 d)] \cos(\omega t) \end{aligned} \tag{12.68}$$

The subscript "0" indicates deep water, as is done throughout the book.

(2) Diffraction Force

From the results of Chakrabarti (1975) in Figure 9.8, the diffraction force is dominant for the spar in Figure 12.16 if the spar diameter satisfies $D \geq \lambda/2\pi$. As is shown in Figure 9.22, the force and moment approximations of van Oortmerssen (1971) for vertical truncated cylinders in the diffraction force range are rather good. The approximation is based on the MacCamy-Fuchs (1954) analysis of the horizontal diffraction force presented in Section 9.2E. Here, we use the deep-water force expression in eq. 9.85 applied to deep water, which is

$$F_{xMF0} = 2 \frac{\rho g H_0}{k_0^2} \Lambda(k_0 a) \sin[\omega t - \sigma(k_0 a)] \tag{12.69}$$

where a is the radius of the spar, the subscript x refers to the direction, and the subscript MF identifies the expression as that of MacCamy and Fuchs. From eq. 9.72, the function $\Lambda(k_0 a)$ is

$$\Lambda(k_0 a) = \frac{1}{\sqrt{\left[J_0(k_0 a) - \frac{1}{k_0 a} J_1(k_0 a)\right]^2 + \left[Y_0(k_0 a) - \frac{1}{k_0 a} Y_1(k_0 a)\right]^2}} \tag{12.70}$$

where the J and Y terms are Bessel functions of the first and second kind, respectively. See Appendix A for a discussion of these functions. The phase angle $\sigma(k_0 a)$ in eq. 12.69 is obtained from eq. 9.73. Applied to deep water, the phase angle is

$$\sigma(k_0 a) = \tan^{-1}\left[\frac{J_0(k_0 a) - \dfrac{1}{k_0 a} J_1(k_0 a)}{Y_0(k_0 a) - \dfrac{1}{k_0 a} Y_1(k_0 a)}\right] \tag{12.71}$$

See Figures 9.17 and 9.18 for the behaviors $\Lambda(k_0 a)$ and $\sigma(k_0 a)$, respectively.

From the van Oortmerssen (1971) expression in eq. 9.86, the horizontal diffraction force on a truncated circular cylinder of draft d (in Figure 12.16) in deep water is obtained by multiplying the force expression in eq. 12.69 by a ratio of draft-integral and depth-integral. The result is the following deep-water diffraction force expression:

$$F_{(x0)} \simeq \frac{\int_{-d}^{0} e^{k_0(z+h)} dz}{\int_{-\infty}^{0} e^{k_0(z+h)} dz} F_{xMF0} = (1 - e^{-k_0 d}) F_{xMF0} \tag{12.72}$$

D. Hydrodynamic Coefficients for a Spar

The large-diameter vertical, circular cylinder is a basic component of many offshore structures. As discussed by Agarwal and Jain (2003), the spar type of structure is essentially a vertical cylinder. The hydrodynamic coefficients (added mass and radiation damping) for surging, heaving, and pitching motions of a vertical cylinder of finite draft are derived in the paper by Sabuncu and Calisal (1981) and Yeung (1981), among others. The Yeung analysis follows that of Garrett (1971) presented in Section 9.2H(3). Garrett predicts the wave-induced forces on a fixed circular, cylindrical dock. Using the Garrett method, Yeung (1981) applies the boundary conditions associated with the planar motions of the vertical, circular cylinder. The Yeung analysis is applied to the motions of cylinders in proximity to reflecting walls by Teng, Ning, and Zhang (2004). Those authors also give a rather complete derivation of the single-cylinder equations. In the subsections that follow, analyses of the reaction forces are presented that are somewhat simplified.

In the heaving analysis that follows, the method used by McCormick (1982) is modified for application in waters of finite depth. That method is based on the Lindsay (1960) Green's function approach to the problem. For the surging motion analyzed herein, the cylindrical wave maker theory of Dalrymple and Dean (1972) is modified using the van Oortmerssen (1971) approximation method, presented in Section 9.2H(1). The van Oortmerssen method is used to determine the wave-induced diffraction forces on truncated cylinders in that section. In Figure 9.22, results obtained using the approximation and the exact analysis of Garrett (1971) are presented. From the good agreement in that figure, some confidence can be gained in the approximation method. The analysis of the coupling of the surging and pitching motions is not presented because, for motions of large spar-type platforms in operational seas, the coupling is of second order.

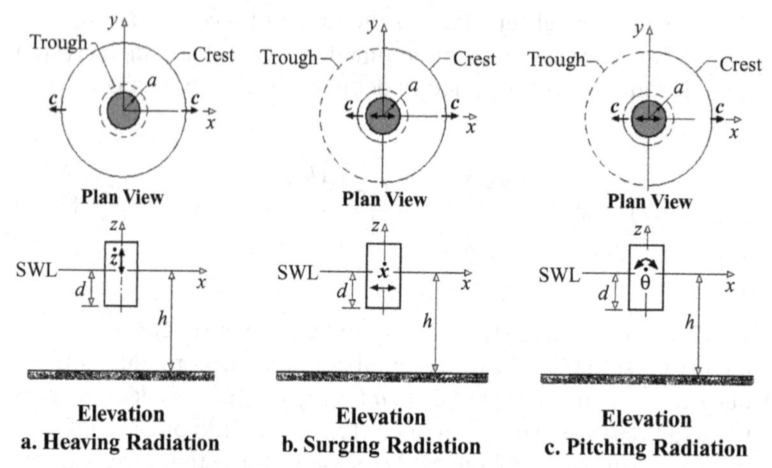

Figure 12.18. *Radiation Due to Heaving, Surging, and Pitching Motions*. The radiation patterns shown are predicted by the theory of Yeung (1981).

(1) Heaving Added Mass and Radiation Damping

As previously mentioned, Yeung (1981) uses a separation-of-variables technique in solving for the velocity potentials representing the flows excited by the surging, heaving, and pitching motions of a truncated, vertical circular cylinder, as sketched in Figure 12.18. The analytical regions used by Yeung (1981) and others are sketched in Figure 12.19. Here, a somewhat simplified approach is taken in the determination of the added-mass and radiation-damping coefficients.

Consider two points on the bottom face of a heaving circular, cylindrical hull in Figure 12.20. The first point, *0*, is the source of the pressure disturbance that passes point *1*, located at a distance s from *0* on the face. For the hull having a heaving velocity of $V_{zo}e^{-i\omega t}$, the velocity potential for the excited fluid at point *0* is

$$d\phi_1 = \frac{V_{zo}}{2\pi}e^{-i\omega t}\frac{e^{iks}}{s}\frac{\cosh[k(h-d)]}{\cosh(kh)}dA_0 \qquad (12.73)$$

The term $1/s$ is the Green's function, described in Appendix D. The ratio of the two hyperbolic signs comes from the expression in eq. 3.29, which is the equation for the velocity potential describing the flow in linear waves. The area element, dA_0, surrounds the source point (*0*), as shown in Figure 12.20. The velocity potential at *1*

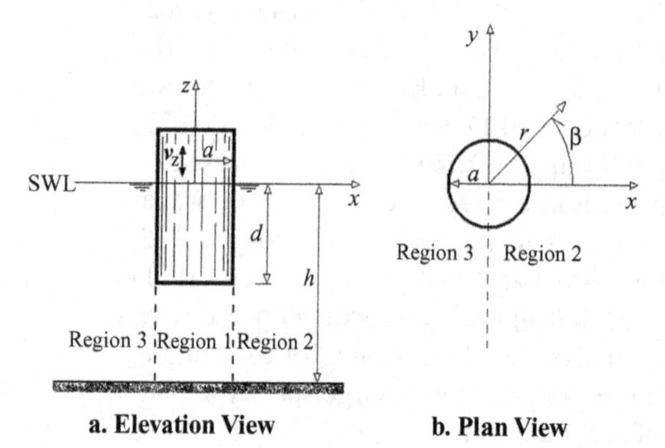

Figure 12.19. *Analytical Regions for the Added Mass and Radiation Damping Analyses*. For the heaving motions in Figure 12.18a, the radiated waves in Regions 2 and 3 are in phase. For the respective surging and pitching motions in Figures 12.18b and 12.18c, the radiated waves in these regions are 180° out of phase.

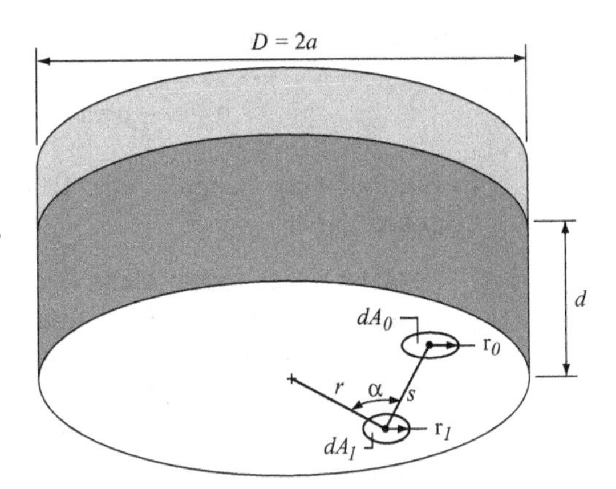

Figure 12.20. *Notation for Heaving Added-Mass and Radiation-Damping Coefficient Derivations.*

resulting from all of the source points on the bottom face is

$$\phi_1 = \frac{V_{zo}}{2\pi} e^{-i\omega t} \frac{\cosh[k(h-d)]}{\cosh(kh)} \int_{A_0} \frac{e^{iks}}{s} dA_0$$
$$= \frac{V_{zo}}{2\pi} e^{-i\omega t} \frac{\cosh[k(h-d)]}{\cosh(kh)} \frac{\pi}{k} \{\mathbf{H_0}(2kr) + i[1 - J_0(2kr)]\} \tag{12.74}$$

The function $\mathbf{H_0}(2kr)$ is called a *Struve function of zero order*, and $J_0(2kr)$ is a Bessel function of the first kind, zero order, as described in Appendix A. From Abramowitz and Stegun (1965), the Struve function is defined as

$$\mathbf{H_0}(2kr) = \frac{2}{\pi}\left[(2kr) - \frac{(2kr)^3}{1^2 \cdot 3^2} + \frac{(2kr)^5}{1^3 \cdot 3^3 \cdot 5^3} - \cdots\right] \tag{12.75}$$

Details of the integration in eq. 12.74 are presented by Lindsay (1960).

The dynamic pressure at point *1*, from Bernoulli's equation, is

$$p_1 = -\rho \frac{\partial \phi_1}{\partial t} = i\omega\rho \frac{V_{zo}}{2\pi} \frac{\cosh[k(h-d)]}{\cosh(kh)} e^{i\omega t} \frac{\pi}{k} \{\mathbf{H_0}(2kr) + i[1 - J_0(2kr)]\} \tag{12.76}$$

Hence, the reaction force due to heaving is

$$F_z = \int_A p_1 dA = i\omega\rho \frac{V_{zo}}{k} \frac{\cosh[k(h-d)]}{\cosh(kh)} e^{-i\omega t} \int_0^a 2\pi r \{\mathbf{H_0}(2kr) + i[1 - J_0(2kr)]\} dr$$
$$= -\rho\pi a \frac{1}{k^2} \frac{\cosh[k(h-d)]}{\cosh(kh)} \mathbf{H_1}(2ka) i\omega V_{zo} e^{-i\omega t} + \frac{\omega}{k} \rho\pi a^2 \frac{\cosh[k(h-d)]}{\cosh(kh)}$$
$$\times \left[1 - \frac{J_1(2kr)}{ka}\right] V_{zo} e^{-i+++}$$
$$= -\mathrm{a}_{wz} i\omega V_{zo} e^{-i\omega t} + \mathrm{b}_{rz} V_{zo} e^{-i\omega t} \tag{12.77}$$

Here, from Abramowitz and Stegun (1965), the Struve function of order one is

$$\mathbf{H_1}(2kr) = \frac{2}{\pi}\left[\frac{(2kr)^2}{1^2 \cdot 3} - \frac{(2kr)^4}{1^2 \cdot 3^2 \cdot 5} + \cdots\right] \tag{12.78}$$

The last line of eq. 12.77 contains the heaving added-mass coefficient, which is

$$\mathrm{a}_{wz} = \rho\pi a \frac{1}{k^2} \frac{\cosh[k(h-d)]}{\cosh(kh)} \mathbf{H_1}(2ka) \tag{12.79}$$

and the heaving radiation-damping coefficient,

$$b_{rz} = \frac{\omega}{k}\rho\pi a^2 \frac{\cosh[k(h-d)]}{\cosh(kh)}\left[1 - \frac{J_1(2ka)}{ka}\right] \tag{12.80}$$

In the following example, eqs. 12.79 and 12.80 are applied to a vertical, circular cylindrical hull.

EXAMPLE 12.5: ADDED-MASS AND RADIATION-DAMPING COEFFICIENTS FOR A HEAVING VERTICAL CYLINDRICAL HULL IN WATER OF FINITE DEPTH A vertical, circular cylindrical hull having a radius (a) of 20 m and a draft (d) of 10 m is forced to heave in 100 m of water. The heaving period (T) is 9 sec. For this period, the deep-water wavelength is approximately 126 m from eq. 3.36. Because this value is less than twice the water depth, the waves created by the motion are in deep water. The product of the wave number (k) and the hull radius is approximately 1.0. For this hull, the added mass obtained from eq. 12.79 is about 0.312×10^7 kg, and the radiation-damping coefficient value is about 4.64×10^6 N-s/m from eq. 12.80. The displaced mass of the resting hull is 1.29×10^7 kg in salt water, where the mass density is assumed to be $\rho = 1.03 \times 10^3$ kg/m^3.

For the circular, cylindrical hull, eqs. 12.79 and 12.80 are applied over a ka range from 0 to 1. This range corresponds to long waves. The resulting hydrodynamic coefficients (added mass and radiation damping) are presented in Figure 12.21 in dimensionless forms.

(2) Surging Added Mass and Radiation Damping

The hydrodynamic coefficients for a surging, circular cylinder have been analyzed by Yeung (1981), Bhatta and Rahman (2003), and others. Here, we shall present an approximate method based on the van Oortmerssen (1971) technique, presented in Section 9.2H. Results from the approximate analysis are compared with those obtained from the Yeung (1981) analysis.

Begin by assuming that the surging cylinder in Figure 12.18b extends from just above the sea bed through the free surface, that is, $d \simeq h$. Physically, this is equivalent to the cylindrical wave maker termed by Dalrymple and Dean (1972) and Dean and Dalrymple (1984) as one experiencing "piston motions." For the cylindrical wave maker, the velocity potential describing the flow excited by a cylindrical wave maker is obtained by using a separation of variables method to solve Laplace's equation, eq. 3.8. The resulting potential expression is

$$\phi(r, \beta, z, t) = \left\{ C_0 H_1^{(1)}(kr)\cosh[k(z+h)] + \sum_{n=1}^{\infty} C_n K_1(\mathrm{K}_n r)\cos[\mathrm{K}_n(z+h)] \right\} \cos(\beta) e^{-i\omega t} \tag{12.81}$$

In this equation, C_0 and C_n ($n = 1, 2, \ldots$) are to-be-determined constants, $H_1^{(1)}(kr)$ is a first-order Hankel function of the first kind, and $K_1(\mathrm{K}_n r)$ is a first-order modified Bessel function of the second kind. These functions are discussed in Appendix A. The first of the two terms on the right represent an outgoing traveling wave, and the second term represents an infinite number of standing evanescent waves that are attached to the body. The wave number (k) for the traveling wave system is

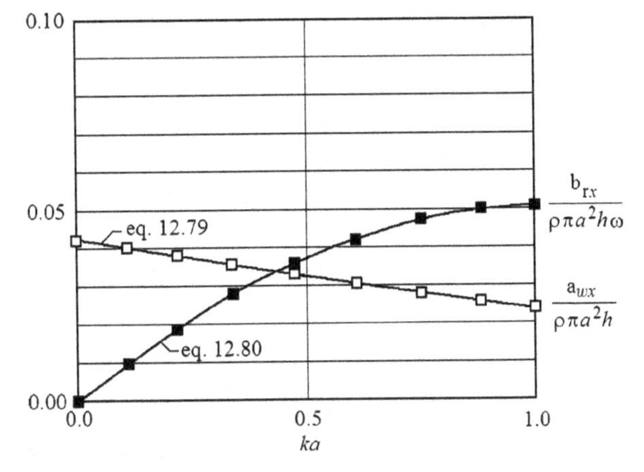

Figure 12.21. *Dimensionless Added-Mass and Radiation-Damping Coefficients for the Heaving, Circular, Cylindrical Hull in Example 12.5.* Referring to Figure 12.18a, the radius (a) of the heaving cylinder is 20 m, the draft (d) is 10 m, and the water depth (h) is 100 m.

determined from the dispersion equation, eq. 3.31, that is,

$$\omega^2 = kg\tanh(kh) \tag{12.82}$$

Similarly, for the standing wave system, the analogous expression is

$$\omega^2 = -K_n g\tan(K_n h) \tag{12.83}$$

where, again, $n = 1, 2, \ldots$. The constants, C_0 and C_n, in eq. 12.81 are determined by first applying the boundary condition at the body surface, where $r = a$. Then, the orthogonality condition due to eqs. 12.82 and 12.83 is applied. The boundary condition is mathematically described by

$$\left.\frac{\partial\phi}{\partial r}\right|_{r=a} = v_r|_{r=a} = V_x(t)\cos(\beta) = V_{xo}e^{-i\omega t}\cos(\beta) \tag{12.84}$$

Here, V_{xo} is the amplitude of the surging velocity (V_x), which is uniform in the vertical direction. Because the body motion is uniform with respect to z, that variable does not appear in the relationship. Applying the orthogonality condition to the expression resulting from the combination of eqs. 12.81 through 12.84 results in the following expressions for the constants:

$$\begin{aligned} C_0 &= \frac{4V_{xo}\sinh(kh)}{\left.\dfrac{dH_1^{(1)}(kr)}{dr}\right|_{r=a}[2kh+\sinh(2kh)]} = \frac{4aV_{xo}\sinh(kh)}{[kaH_0^{(1)}(ka) - H_1^{(1)}(ka)][2kh+\sinh(2kh)]} \\ &= \frac{4aV_{xo}\sinh(kh)}{[2kh+\sinh(2kh)]}\left\{\frac{[kaJ_0(ka)-J_1(ka)] - i[kaY_0(ka)-Y_1(ka)]}{[kaJ_0(ka)-J_1(ka)]^2 + [kaY_0 - Y_1(ka)]^2}\right\} \\ &= (C_{0\Re} + iC_{0\Im})aV_{xo} \end{aligned} \tag{12.85}$$

and

$$\begin{aligned} C_n &= -\frac{4V_{xo}\sin(K_n h)}{\left.\dfrac{dK_1(K_n r)}{dr}\right|_{r=a}[2K_n h + \sin(2K_n h)]} \\ &= -\frac{4aV_{xo}\sin(K_n h)}{[K_n a K_0(K_n a) - K_1(K_n a)][2K_n h + \sin(2K_n h)]} = C_n a V_{xo} \end{aligned} \tag{12.86}$$

Note that the Hankel functions, $H_N^{(1)}(ka)$, in eq. 12.85 are complex functions, related to the Bessel functions of the first and second kinds by $H_N^{(1)}(ka) = J_N(ka) + iY_N(ka)$, where N is the order of the function. The modified Bessel functions, $J_N(K_N a)$, in eq. 12.86 are real functions. We shall see the significance of these observations in the added-mass and radiation-damping expressions. The velocity potential in eq. 12.81 is now completely defined.

The combination of the velocity potential in eq. 12.81 and the dynamic pressure results in the pressure distribution over the cylinder, that is,

$$\begin{aligned} p_{dyn}|_{r=a} &= -\rho\frac{\partial\phi}{\partial t} = i\rho\omega\phi|_{r=a} \\ &= i\rho\omega\left\{C_0 H_1^{(1)}(ka)\cosh[k(z+h)]\right. \\ &\quad \left. + \sum_{n=1}^{\infty} C_n K_1(\mathrm{K}_n a)\cos[\mathrm{K}_n(z+h)]\right\}\cos(\beta)e^{-i\omega t} \end{aligned} \tag{12.87}$$

This expression is now integrated over h to obtain the hydrodynamic force on the surging, cylindrical wave maker, which is

$$\begin{aligned} F_{xh} &= \int_0^{2\pi}\int_{-h}^{0} p_{dyn}|_{r=a}\, a\cos(\beta)dz d\beta \\ &= i\rho\omega a\int_0^{2\pi}\int_{-h}^{0}\left\{C_0 H_1^{(1)}(ka)\cosh[k(z+h)]\right. \\ &\quad \left. + \sum_{n=1}^{\infty} C_n K_1(\mathrm{K}_n a)\cos[\mathrm{K}_n(z+h)]\right\}\cos^2(\beta)dz d\beta e^{-i\omega t} \\ &= i\rho\omega\pi a^2 V_{xo}\left\{\mathrm{C}_0\frac{H_1^{(1)}(ka)}{k}\sinh(kh) + \sum_{n=1}^{\infty}\mathrm{C}_n\frac{K_1(\mathrm{K}_n a)}{\mathrm{K}_n}\sin(\mathrm{K}_n h)\right\}e^{-i\omega t} \end{aligned} \tag{12.88}$$

The reader should note the relationships of C_0 and C_0 and of C_n and C_n in eq. 12.88. The force in eq. 12.88 is the reaction force on a vertical, circular cylinder oscillating in the surging mode. The cylinder extends to the sea bed.

We now apply the van Oortmerssen (1971) approximation to the wave maker. Physically, the approximation is based on the assumption that the ratio of the integrals of the pressures over the truncated cylinder and over the cylinder extending to the sea bed is proportional to the corresponding wave-force ratio (see eq. 9.86). Here, we are dealing with the reaction forces due to the surging motions.

Replacing the MacCamy-Fuchs (1954) force (F_{xMF}) in eq. 9.86 by the expression in eq. 12.88, we obtain the following expression for the approximate hydrodynamic surging force on a circular, cylindrical spar:

$$\begin{aligned} F_{xd} &\simeq \frac{\int_{-d}^{0}\cosh[k(z+h)]dz}{\int_{-h}^{0}\cosh[k(z+h)]dz}F_{xh} = \frac{\sinh(kh) - \sinh[k(h-d)]}{\sinh(kh)}F_{xh} \\ &= -\mathrm{a}_{wx}\frac{dV_x}{dt} - \mathrm{b}_{rx}V_x = -(-i\omega\mathrm{a}_{wx} + \mathrm{b}_{rx})V_x = (i\omega\mathrm{a}_{wx} - \mathrm{b}_{rx})V_{xo}e^{-i\omega t} \end{aligned} \tag{12.89}$$

The second line relates the hydrodynamic force to the added-mass (a_{wx}) and the radiation-damping coefficients (b_{rx}). The negative signs in that line result from the motion-induced force being reactionary.

The expressions for the respective approximate surge added-mass and radiation-damping coefficients are obtained by combining eqs. 12.79, 12.80, 12.82, and 12.83 to obtain

$$a_{ax} = \rho\pi a^2 \left\{ \frac{\sinh(kh) - \sinh[k(h-d)]}{\sinh(kh)} \right\} \cdot \{[C_{o\Re} J_1(ka) - C_{o\Im} Y_1(ka)]\} \frac{\sinh(kh)}{k} + \sum_{n=1}^{\infty} C_n K_1(K_n h) \frac{\sin(K_n h)}{K_n} \tag{12.90}$$

and

$$b_{rx} = \rho\pi a^2 \frac{\omega}{k} \{\sinh(kh) - \sinh[k(h-d)]\}[C_{o\Re} Y_1(ka) + C_{o\Im} J_1(ka)] \tag{12.91}$$

We note that the summation, which is a real term, does not appear in the radiation-damping coefficient, as would be expected, because the energy lost to the motion is to the traveling wave system.

To test the accuracy of the approximations in eqs. 12.90 and 12.91, both equations are applied to a cylinder extending from the free surface to a point just above the sea bed. This is a case studied by Yeung (1981). The cylindrical wave maker in question is in a water depth equal to five times the cylinder radius ($h = 5a$). The non-dimensional added-mass and radiation-damping coefficients obtained from eqs. 12.90 and 12.91, respectively, are presented with those obtained from the Yeung analysis in Figure 12.22. In Figure 12.22a, the "one-term solution" for the added mass refers to the application of eq. 12.90 without any of the terms in the summation. Physically, this solution neglects the traveling waves created by the surging motions. The "three-term solution" in Figure 12.22a shows the effect of including the first two terms in the summation. These terms correspond to two of the evanescent standing waves attached to the body. As more terms in the summation are included, the agreement between the results from the "exact theory" of Yeung (1981) and the approximate expression in eq. 12.90 improves.

In Figure 12.22b, the radiation-damping coefficient values are presented. One can see that the agreement between the eq. 12.91 results and those obtained from the Yeung analysis are rather good. Note that the value of the draft van Oortmerssen (1971) correction factor (on the first line of eq. 12.89) is unity in this application because the wave maker extends down to just above the sea bed. Over the ka-range in Figure 12.22, the results are quite good. Hence, in the conceptual design phase of an engineering design, eqs. 12.90 and 12.91 are considered to be satisfactory.

With the information presented in this section, we are now prepared to analyze the design planar motion of a TLP.

E. Surging Motions in Regular Seas

Referring to the sketches in Figures 12.16b and 12.16c, we see that the primary motion of a TLP is surging. The heaving motion is of second order because the stiffness in the tethers is large, making the natural "springing" frequency due to

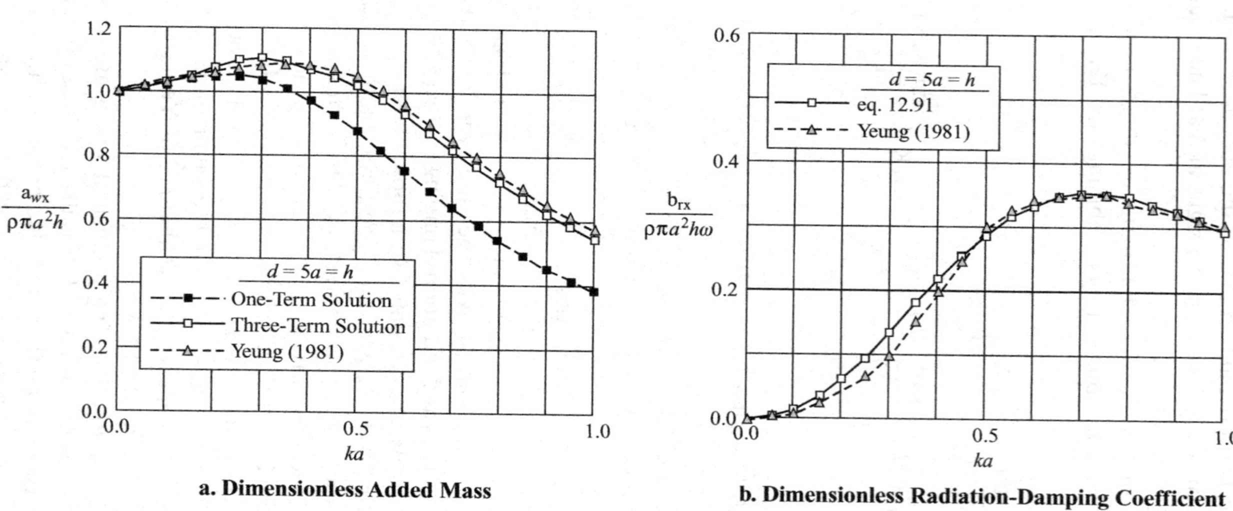

a. Dimensionless Added Mass **b. Dimensionless Radiation-Damping Coefficient**

Figure 12.22. *Dimensionless Added-Mass and Radiation-Damping Coefficients for Surging, Circular, Cylindrical Spars.* The cylinder extends from the free surface to just above the sea bed, that is, $z = -d > -h$. Approximations of the added-mass expression in eq. 12.90 are obtained by including only the traveling-wave term (one-term solution) and two of the evanescent wave terms, referred to as the "three-term solution." The solution of Yeung (1981) for the cylinder extending to just above the sea bed is also presented. The radiation-damping coefficient values are obtained from eq. 12.91.

the lines rather high, and well away from those of the high-energy waves. See the numerical values in Example 12.4. In that example, we see that the spring constant for the heaving motions is of the order of 10^8 N/m, and that of the surging motion is of the order of 10^4 N/m. For heaving, the tethers are extremely stiff.

The surging equation of motion for the TLP is

$$(m+\mathrm{a}_{wx})\frac{d^2x}{dt^2}+\mathrm{b}_{rx}\frac{dx}{dt}+b_{vx}\left(\frac{dx}{dt}\right)\left|\left(\frac{dx}{dt}\right)\right|+NK_{sx}x=F_x(\omega,t)=F_{xo}\sin(\omega t+\alpha_x) \quad (12.92)$$

In this equation are the mass of the structure (m), the added mass (a_{wx}) excited by the surging motions, the radiation-damping coefficient (b_{rx}), the viscous-damping coefficient (b_{vx}), the number (N) of mooring lines, each line having a spring constant (K_{sx}), the wave-force amplitude (F_{xo}), and the phase angle (α_x) between the passing wave and the wave force. Note that in the equation, the wave force has been linearized, as is done in Section 12.4C(1). The viscous-damping term can also be replaced by an equivalent viscous-damping coefficient, as is done for the heaving motion of the cylinder in Figure 10.1. That is, following the energy analysis leading to eq. 10.10, we can write the equivalent linear viscous-damping coefficient as

$$\mathrm{b}_{vx}=\frac{8}{3}\frac{\omega}{\pi}x_o b_{vx} \quad (12.93)$$

where ω is the wave frequency and x_o is the surging amplitude.

EXAMPLE 12.6: SURGING MOTIONS OF A TLP Consider again the SeaStar sketched in Figure 12.17 to be subject to waves having a height (H) of 2 m and a period (T) of 8 sec, as sketched in Figure 12.23. From Chapter 3, the deep-water wavelength (λ_0) corresponding to this period is approximately 100 m, which is less than the water depth (h) of 215 m. Hence, the SeaStar is moored in deep-water waves because $h/\lambda > 1/2$.

From Example 12.4, we find that the relaxed tether length (ℓ_0) is 175 m, the weight (W) of the structure is 1.212×10^8 N, and the surge spring constant is $K_{sx} = \mathrm{T}_0/\ell \simeq 7.474 \times 10^4$ N/m. The mass of the structure is $\mathrm{m} = W/g \simeq 1.24 \times 10^7$ kg.

For the sake of completeness, Figure 4.1 is used to check on the validity of the linear wave theory in this application. According to Le Méhauté (1969), the linear theory presented in Chapter 3 can be used because the Ursell number, U_R from Ursell (1953), is much less than unity, that is, as defined in Figure 4.1, $U_R \equiv H\lambda^2/2h^3 \simeq 0.1 \ll 1$. However, we note that for the coordinate values in that figure, the deep-water Stokes' second-order theory presented in Sections 4.2 through 4.4 would be apropos. We shall follow the Le Méhauté recommendation here, and use Airy's linear wave theory.

We must also determine the types of wave-induced forces that are dominant. To do so, we consult the Chakrabarti diagram in Figure 9.8, where the deep-water Keulegan-Carpenter number value is $KC = 2\pi H_0/a \simeq 1.26$ and $ka \simeq 0.628$. In Figure 9.8, we see that these values correspond to the region of the diffraction force. The MacCamy-Fuchs equation for the horizontal diffraction

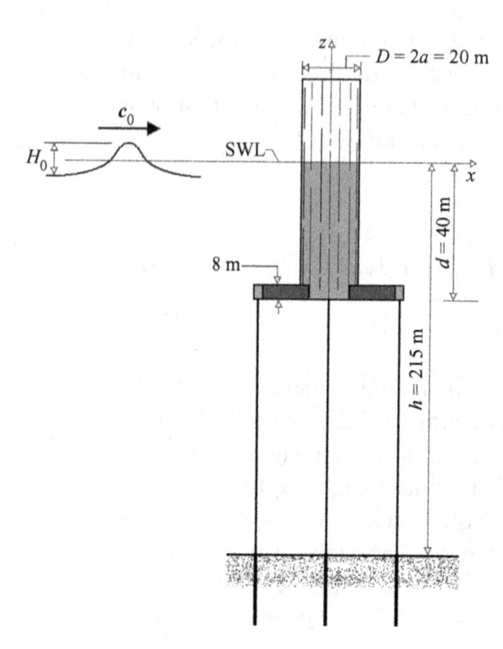

Figure 12.23. *Sketch of the SeaStar-Type TLP in Example 12.6 (Not to Scale).*

force, eq. 9.71, is modified for use here. Applied to deep water, the MacCamy-Fuchs horizontal force is

$$F_{xMF} \simeq 2\frac{\rho g H_0 a^2}{(k_0 a)^2}\left\{\frac{\tanh(k_0 h)}{\sqrt{\left[J_0(k_0 a) - \frac{1}{k_0 a}J_1(k_0 a)\right]^2 + \left[Y_0(k_0 a) - \frac{1}{k_0 a}Y_1(k_0 a)\right]^2}}\right\}$$
$$\times \sin[\omega t - \sigma(k_0 a)]$$
$$= \mathrm{F}_{xMFo}(\omega)\frac{H_0}{2}\sin[\omega t - \sigma(k_0 a)] \qquad (12.94)$$

where, on the second line, the force amplitude function, $\mathrm{F}_{xMFo}(\omega)$, appears. This function will be of use in the analysis of the surging response in a random sea. From Chapter 9, the expression for the phase angle between the wave and the force is

$$\sigma(k_0 a) = \tan^{-1}\left[\frac{J_0(k_0 a) - \frac{1}{k_0 a}J_1(k_0 a)}{Y_0(k_0 a) - \frac{1}{k_0 a}Y_1(k_0 a)}\right] \qquad (12.95)$$

To apply the MacCamy-Fuchs analysis, we simply change the lower limit of the z-integration in eq. 9.71 from $-h$ to $-d$. When this is done, the modified MacCamy-Fuchs equation is

$$F_x \simeq 4\frac{\rho g a^2}{(k_0 a)^2}\left\{\frac{\left\{\tanh(k_0 h) - \frac{\sinh[k_0(h-d)]}{\cosh(k_0 h)}\right\}}{\sqrt{\left[J_0(k_0 a) - \frac{1}{k_0 a}J_1(k_0 a)\right]^2 + \left[Y_0(k_0 a) - \frac{1}{k_0 a}Y_1(k_0 a)\right]^2}}\right\}$$
$$\times \frac{H_0}{2}\sin[\omega t - \sigma(k_0 a)]$$
$$= F_{xo}(\omega)\sin[\omega t - \sigma(k_0 a)] = \mathrm{F}_{xo}(\omega)\frac{H_0}{2}\sin[\omega t - \sigma(k_0 a)] \qquad (12.96)$$

As is similar to eq. 12.94, the frequency-dependent amplitude function, $F_{xo}(\omega)$, is introduced for later use. In Figure 12.24 are the surge-force amplitude for the SeaStar hull, obtained from eq. 12.96, and the MacCamy-Fuchs force amplitude for a bed-resting hull having the same radius, obtained from eq. 12.94. In that figure, the force amplitudes are presented as functions of the wave period, T. One can see that the peak diffraction force on the TLP occurs at about 8.5 sec, where the diffraction force is approximately 2.76×10^6 N. The value of the phase angle, $\sigma(k_0a)$ in eq. 12.96, is approximately 0.253 radians, or about 14.5°. This is the phase angle between the force and wave; hence, $\alpha = -\sigma(k_0a)$ in eq. 12.92, the equation of the surging motion.

We now must determine the values of the added-mass and the radiation-damping coefficients. For the values given, $ka \simeq 0.6288$, $kh \simeq 13.52$, and the non-dimensional added-mass coefficient value is $a_{wx}/(\rho\pi a^2 h) \simeq 0.2297$. The non-dimensional radiation-damping coefficient value is $b_{rx}/(\rho\pi a^2 h\omega) \simeq 0.0751$. Hence, surge added mass is $a_{wx} \simeq 1.60 \times 10^7$ kg, and the surge radiation-damping coefficient is $b_{rx} \simeq 4.10 \times 10^6$ N-s/m. The hydrostatic restoring coefficient in eq. 12.86 is $\rho g\pi a^2 \simeq 3.17 \times 10^6$ N/m.

The horizontal particle velocity on the free surface as a crest passes is obtained from eq. 3.49. In deep water, the maximum particle velocity is $u_{max} \simeq 0.7848$ m/s. The Reynolds number in eq. 2.108 for this velocity is $R_{eD} = u_{max}D/\nu \simeq 1.31 \times 10^7$, where the kinematic viscosity of salt water is about 1.2×10^{-6} m²/s at 14°C. For this Reynolds number value in Figure 2.15, we are in a region where there are no experimental data. Hence, because of the behavior of the drag coefficient up to $R_{eD} = 10^6$ in that figure, we can assume a drag coefficient value $C_{dD} \simeq 1.0$. If the velocity was applied over the entire hull, the drag force would be about 2.54×10^5 N, which is two orders of magnitude less than the maximum diffraction force. For this reason, we neglect the viscous drag effects on the motion, and the coefficient in eq. 12.93 can be neglected.

We can now write eq. 12.92, the equation of the surging motion, in terms of its numerical coefficients as

$$[(1.24 + 1.60) \times 10^7]\frac{d^2x}{dt^2} + [4.10 \times 10^6]\frac{dx}{dt} + [3(7.474 \times 10^4)]x$$
$$= [2.76 \times 10^6]\sin(\omega t - 0.253)$$

The units of this equation are Newtons. The surging equation of motion is similar to the heaving equation in eq. 10.2. The expression for the surging displacement as a function of time is then similar to that for the heaving displacement in eqs. 10.3 and 10.12.

In expressions analogous to those in eqs. 10.13 through 10.15, respectively, we find the natural surging frequency is

$$\omega_{nx} = \frac{2\pi}{T_{nx}} = \sqrt{\frac{NK_{sx}}{m + a_{wx}}}$$
$$\simeq \sqrt{\frac{3(7.474 \times 10^4)}{(1.24 + 1.60) \times 10^7}} \simeq 0.0888\ \frac{\text{rad}}{\text{s}} \tag{12.97}$$

From this expression, the natural surging period is $T_{nx} \simeq 70.7$ sec. It is worth noting here that the similar equation for the heaving natural period, using the

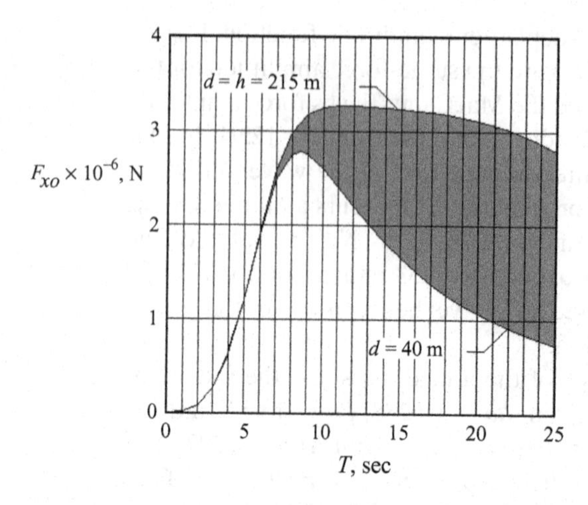

Figure 12.24. *Diffraction Surge-Force Amplitude as a Function of Wave Period.* The conditions associated with the force plot are given in Example 12.6. The structure is the hull of the TLP, SeaStar, where the draft is $d = 40$ m. For reference, the MacCamy-Fuchs diffraction force on a bed-resting cylinder ($d = h$) exposed to the same wave is shown. For wave periods less than 16.6 sec, deep-water wave conditions can be assumed.

added-mass results obtained from eq. 12.79 and the heaving spring constant value in Example 12.4, is approximately 3.0 rad/s. For this value, the heaving natural period value is slightly greater than 2.0 sec.

The critical damping coefficient for the surging motions is

$$\begin{aligned} \mathrm{b}_{cx} &= 2\sqrt{(m + \mathrm{a}_{wx})(NK_s)} \\ &\simeq 2.52 \times 10^6 \, \frac{\text{N–s}}{\text{m}} \end{aligned} \tag{12.98}$$

The critical damping ratio value is then $\mathrm{b}_{rx}/\mathrm{b}_{cx} \simeq 1.62$. Because this value is greater than 1, the surging motions are over-damped. The expression for the force-motion phase angle is

$$\begin{aligned} \epsilon_x &= \tan^{-1}\left[\frac{2\dfrac{\omega}{\omega_{nx}}\dfrac{\mathrm{b}_{rx}}{\mathrm{b}_{cx}}}{\left(1 - \dfrac{\omega^2}{\omega_{nx}^2}\right)}\right] \\ &\simeq -20.4^\circ \simeq -0.356\,\text{rad} \end{aligned} \tag{12.99}$$

The surging motion displacement expression is similar to that for heaving in eq. 10.12. With the values in eqs. 12.97 through 12.99, the surging motion displacement is found to be

$$\begin{aligned} x &= X_o \sin(\omega t + \sigma - \epsilon_x) = \frac{\dfrac{F_{xo}}{(NK_{sx})}}{\sqrt{\left(1 - \dfrac{\omega^2}{\omega_{nx}^2}\right)^2 + \left[2\dfrac{\omega}{\omega_{nx}}\dfrac{(\mathrm{b}_{rx})}{\mathrm{b}_{crz}}\right]^2}} \sin(\omega t + \sigma - \epsilon_x) \\ &\simeq X_o \sin(0.785t - 0.253 + 0.356) \simeq 0.158 \sin(0.785t + 0.103) \end{aligned} \tag{12.100}$$

Hence, the surging motion amplitude is $X_o \simeq 0.158$ m.

For the actual SeaStar mini-TLP, Bhattacharyya, Sreekumar, and Idichandy (2003) theoretically find that the surging amplitude is about 0.9 m in a 2-m, 20-sec sea. We note that the 20-sec period is closer to the natural surging period of the system, which is 70.7 sec. Hence, a larger amplitude is expected.

Using the equations in this example, the amplitude that we find is about 0.415 m. Bhattacharyya, Sreekumar, and Idichandy (2003) use a finite-element computer code and assume a Morison-type of wave loading, discussed in Section 9.2D.

One final note: The surging force on a TLP will be resisted by the piled foundation. The advantage of the TLP is that there is little or no bending moment acting at the connection of the tether and the pile, although there will be bending moments below the soil-water interface due to the soil reactions. The lack of a bending moment at the connection results in a reduced cost of the foundation, which is normally a large component of the total cost of an offshore structure.

The TLP is a structure normally used in extremely deep water. For waters of moderate depth, the articulated-leg platform (ALP) is used. This structure is discussed in the next section.

F. Surging Motions in Random Seas

The basics of wave-induced motions of fixed and floating bodies in random seas is presented in Section 10.3. In this section, those basics are applied to the surging motions of a TLP. We have a choice in the domains of our analysis between the frequency domain, as discussed in Section 10.3, and the time domain. For the latter, numerical techniques are best, whereas for the former, *quasi*-analytical techniques can be effectively used. For a discussion of the two domain analyses applied to the TLP, the paper by Pradnyana and Taylor (1997) is strongly recommended. The paper by Vandiver (1981) is also recommended as an excellent introductory discussion of the analysis of the motions of taut-moored bodies (including the TLP) in random seas. In that paper, the author demonstrates how the Haskind relationships presented in Section 11.7 can be used to analyze the body motions in a directional random sea.

Here, we begin by writing the surging solution in eq. 12.100 in the complex form as

$$x(\omega, t) = \mathrm{X}(\omega)e^{-i\omega t} \tag{12.101}$$

where $\mathrm{X}(\omega)$ is the complex amplitude. Similarly, let the forcing function in the random sea be written as

$$F_x(\omega, t) = F_{xo}(\omega)e^{-i\omega t} = \mathrm{F}_{xo}(\omega)\frac{H_0(\omega)}{2}e^{-i\omega t} \tag{12.102}$$

In eqs. 12.101 and 12.102, the real parts of the complex expressions are assumed. The combination of eqs. 12.92, 12.101, and 12.102 yields the following expression for the complex surging amplitude:

$$\mathrm{X}(\omega) = \frac{\dfrac{F_{xo}(\omega)}{NK_{xs}}}{\left(1 - \dfrac{\omega^2}{\omega_{nx}^2}\right) - i2\dfrac{\omega}{\omega_{nx}}\dfrac{\mathrm{b}_{rx}}{\mathrm{b}_{cx}}} = \mathrm{H}(\omega)\frac{F_{xo}(\omega)}{NK_{sx}} = \mathrm{H}(\omega)\frac{\mathrm{F}_{xo}(\omega)}{NK_{sx}}\frac{H_0(\omega)}{2} \tag{12.103}$$

where $\mathrm{H}(\omega)$ is the amplitude response function, b_{cx} is the critical damping in eq. 12.98, and ω_{nx} is the natural circular frequency for the surging motions in eq. 12.97. Also in eq. 12.103 is the frequency-dependent wave height, $H_0(\omega)$. We can now write the expression for the mean-square surging response as

$$\overline{x^2} = x_{rms}^2 = \lim_{T\to\infty}\frac{1}{T}\int_0^T x^2 dt = \frac{1}{2}|X^2(\omega)| = H(\omega)H^*(\omega)\frac{F_{xo}(\omega)F_{xo}(\omega)^*}{(NK_{sx})^2}\lim_{T\to\infty}\frac{1}{T}\int_0^T \eta^2 dt$$

$$= H(\omega)H^*(\omega)\frac{F_{xo}(\omega)F_{xo}(\omega)^*}{(NK_{sx})^2}\eta_{rms}^2 = |H(\omega)|^2\frac{|F_{xo}(\omega)|^2}{(NK_{sx})^2}\frac{H_0^2}{8} \quad (12.104)$$

As is the case in eq. 10.52, the integration is over a time interval, T. This is not to be confused with the wave period, T. In eq. 12.104 is the root-mean-square surging response, x_{rms}. The relationship between η_{rms} and the mean-square of the wave height comes from eq. 5.35a. Also, from 5.35b, the relationship between the root-mean-square free-surface displacement and the root-mean-square deep-water wave height is found to be

$$\eta_{rms} = \frac{H_{rms}}{2\sqrt{2}} \simeq 0.3536 H_{rms} = 0.3536\sqrt{\overline{H_0^2}} \quad (12.105)$$

In eqs. 12.102 through 12.104, the form of the frequency-dependent force amplitude function, $F_{xo}(\omega)$, depends on the nature of the force. For the conditions in Example 12.6, we find that the diffraction force is dominant. From eq. 12.96, we then find the following expression for the absolute value of the diffraction force amplitude function, which is

$$|F_{xo}(\omega)| = 4\frac{\rho g a^2}{(k_0 a)^2}\frac{\left\{\tanh(k_0 h) - \dfrac{\sinh[k_0(h-d)]}{\cosh(k_0 h)}\right.}{\sqrt{\left[J_0(k_0 a) - \dfrac{1}{k_0 a}J_1(k_0 a)\right]^2 + \left[Y_0(k_0 a) - \dfrac{1}{k_0 a}Y_1(k_0 a)\right]^2}} \quad (12.106)$$

In the following example, the root-mean-square surging response of a TLP is determined.

EXAMPLE 12.7: ROOT-MEAN-SQUARE SURGE RESPONSE OF MOTIONS OF A TLP At the open-ocean site in Example 5.6, we find the following wave-height averages:

$$H_{avg} = 1.5\,\text{m}, \quad H_{rms} = 1.69\ \text{m}$$

The associated significant wave height expression of eq. 5.24 yields $H_s = 2.39$ m, and the corresponding one-year extreme wave height value is $H_{\max} = 6.63$ m from eq. 5.25. Consider the case at this site where the average wave period (T_{avg}) is 8 sec, as in Example 12.6. For the root-mean-square wave height of 1.69 m, the root-mean-square surge displacement of the SeaStar in Example 12.6 is

$$x_{rms} = |H(\omega)|\frac{|F_{xo}(\omega)|}{(NK_{sx})}\frac{H_{rms}}{2\sqrt{2}} \simeq 0.0128\frac{2.75\times 10^6}{3(7.474\times 10^4)}\frac{1.69}{2\sqrt{2}} \simeq 0.0938\ \text{m} \quad (12.107)$$

where the surge amplitude response function, from eq. 12.103, is

$$H(\omega) = \left[\left(1 - \frac{\omega^2}{\omega_{nx}^2}\right) - i2\frac{\omega}{\omega_{nx}}\frac{b_{rx}}{b_{cx}}\right]^{-1} = \left[\left(1 - \frac{T_{nx}^2}{T^2}\right) - i2\frac{T_n}{T}\frac{b_{rx}}{b_{cx}}\right]^{-1}$$

$$= \left[1 - \left(\frac{70.7}{8}\right)^2 - i2\left(\frac{70.7}{8}\right)\frac{4.10\times 10^6}{2.52\times 10^6}\right]^{-1} \quad (12.108)$$

From this expression, $|H(\omega)| \simeq 0.0128$.

Following the methodology leading to eq. 10.55, the spectral density of the surge response can be written in terms of the spectral density of the wind-generated sea as

$$S_x(\omega) = \mathrm{H}(\omega)\mathrm{H}^*(\omega)S(\omega) = |\mathrm{H}^2(\omega)|S(\omega) \tag{12.109}$$

where $S(\omega)$ is the wind-wave spectral density in the circular frequency domain. It is advantageous to perform our surge motion analysis in the period domain. The relationship between the frequency and period spectral densities is

$$S(\omega) = -S(T)\frac{dT}{d\omega} = \frac{T^2}{2\pi}S(T) \tag{12.110}$$

Noting that $\omega = 2\pi f = 2\pi/T$, the first equality comes from eq. 5.44b. The generic expression for the wind-wave spectral density in the period domain is presented in eq. 5.44a, that is,

$$S(T) = A_B T^3 e^{-B_B T^4} \tag{12.111}$$

In this expression, the coefficients A and B depend on the spectral formula used. For example, for the Bretschneider formula, discussed in Section 5.8B and presented in eq. 5.59, the coefficients are, from eq. 5.60,

$$A_B = 3.437\frac{H_{avg}^2}{T_{avg}^4} \tag{12.112}$$

and from eq. 5.61,

$$B_B = 0.675\left(\frac{1}{T_{avg}^4}\right) \tag{12.113}$$

In the period domain, the spectral density of the surge response is then obtained by combining the expressions in eq. 12.109, 12.110, and 12.111 to obtain

$$S_x(T) = |\mathrm{H}^2(\omega)|S(T) = |\mathrm{H}^2(T)|A_B T^3 e^{-B_B T^4} \tag{12.114}$$

This results from the fact that the energies in both the frequency and period domains must be equal. The surge amplitude response function, $\mathrm{H}(\omega) = \mathrm{H}(T)$, is found in eq. 12.108. The spectral density of the response is a measure of the energy absorbed in the surging motion of the TLP.

EXAMPLE 12.8: SPECTRAL DENSITY OF THE SURGE RESPONSE OF MOTIONS OF A TLP
For the wind-generated sea conditions in Example 12.7, the respective average wave height (H_{avg}) and average wave period (T_{avg}) are 1.5 m and 8.0 sec. For these conditions, the coefficients for the Bretschneider spectral density formula, respectively presented in eqs. 12.112 and 12.113, are $A_B = 1.888 \times 10^{-3}$ m^2/s^4 and $B_B \simeq 1.648 \times 10^{-4}$ s^{-4}. The amplitude response function from eq. 12.108 is

$$\mathrm{H}(T) = \left[\left(1 - \frac{T_{nx}^2}{T^2}\right) - i2\frac{T_{nx}}{T}\frac{\mathrm{b}_{rx}}{\mathrm{b}_{cx}}\right]^{-1} = \left[1 - \left(\frac{70.7}{T}\right)^2 - i2\left(\frac{70.7}{T}\right)\frac{4.10 \times 10^6}{2.52 \times 10^6}\right]^{-1} \tag{12.115}$$

The surge response spectral density expression from eq. 12.114 is

$$S_x(T) \simeq \frac{1.888 \times 10^{-3} T^2 e^{-1.648\times10^{-4}T^4}}{\left(1 - \frac{5.00 \times 10^3}{T^2}\right)^2 + \frac{5.293 \times 10^4}{T^2}} \tag{12.116}$$

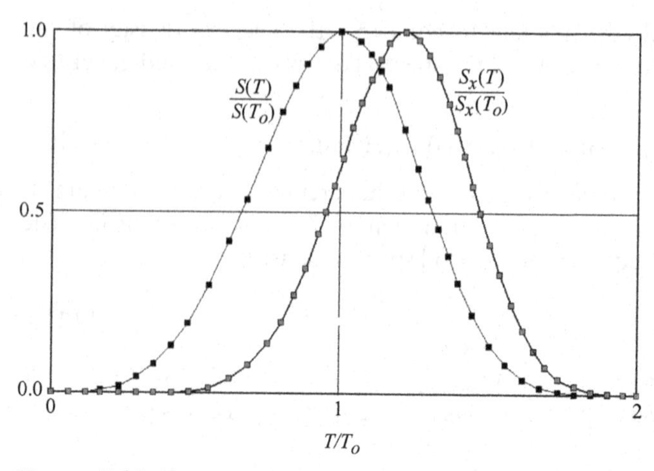

Figure 12.25. *Normalized Wave and Surge Response Spectra for the Conditions in Example 12.8.*

The peak value of the wave spectrum is approximately 0.50 m^2/s, which occurs at the modal period (period of the spectral peak) in eq. 5.45. For the Bretschneider spectral formula, that period is

$$T_o = \left(\frac{3}{4B_B}\right)^{1/4} = \left(\frac{3}{4}\frac{T_{avg}^4}{(0.675)}\right)^{1/4} \simeq 1.03\,T_{avg} \tag{12.117}$$

For the 8-sec average period, the modal period value is approximately 8.24 sec. At the modal period, the peak spectral value is $S(T_o) \simeq 0.494$ m^2/s. The peak value of the surge response spectrum is approximately 1.124×10^{-3} m^2/s. This value occurs at a wave period of $T \simeq 10.0$ sec. The peak value of the surge response shows that relatively little energy is being absorbed by the motions when compared with that of the incident wave. Using the peak spectral values to normalize the wave and surge-response spectra, the results of eqs. 12.111 (combined with the Bretschneider coefficients in eqs. 12.112 and 12.113) and 12.116 are plotted in Figure 12.25 as functions of T/T_o. One can see that the spectral shapes are similar. The response shape is broad because the radiation damping is relatively large, that is, the TLP acts as a good wave maker.

In this section, a number of the concepts that are presented in the previous chapters are applied to the surging motion of a TLP. This compliant type of structure is ideal for demonstrating the application of many of these concepts.

12.5 Motions of an Articulated-Leg Platform (ALP)

As written previously, offshore operations in moderate to extreme depths are expensive. One of the major costs of an offshore structure is that of the foundation. The foundation requirements can be somewhat reduced by designed compliance of some type for the structural design. Discussed in the previous section, The TLP is a compliant structure normally intended for deep-water operations. For moderate water depths, the *articulated-leg platform* (ALP) is sometimes chosen. See the sketch of an ALP in Figure 12.1d.

The use of the ALP in offshore oil production dates back to the 1970s. From that era, the papers by Chakrabarti and Cotter (1979, 1980) and Kokkinowrachos and

Mitzlaff (1981) are noteworthy. The nonlinearities in the motions of the ALP are analyzed by Choi and Lou (1991), Kim and Chen (1994), Nagamanit and Ganapathy (1996), and Islamia and Ahrnad (2003). In addition, a rather interesting discussion on a motion control system of an ALP is found in the paper by Suneja and Datta (1998). As is shown by Liaw, Shankar, and Chua (1989) and others, the motions of an ALP are not planar. Depending on the magnitude and type of exciting force (drag, inertia, or diffraction), the nonplanar motions can be significant. If passing energetic waves produce high-energy vortices, then large out-of-plane motions might occur. The purpose of this section is to introduce the reader to the ALP. For this reason, the motion analysis is confined to the x-z plane.

Referring to Figure 12.26, assume that each of the components of the ALP is positively buoyant. So, for a component j, we assume that $B_j > W_j$, where the notations are for the respective buoyant force and the weight of the component. Although the angular (or "rocking") displacement (θ) from the vertical is exaggerated in Figure 12.26b, it is assumed to be small enough so that the tower is out of the water. The equation of angular motion is that which represents the moments about the ball joint that connects the ALP to the spread-footing foundation. For the system having J components, with the Jth component being the tower, the rocking equation of motion is

$$\begin{aligned}
&\left(\sum_{j=1}^{J} I_{yj} + \sum_{j=1}^{J-1} \mathrm{A}_{wyj}\right)\frac{d^2\theta}{dt^2} + \sum_{j=1}^{J-1}(\mathrm{B}_{ryj} + \mathrm{B}_{vyj})\frac{d\theta}{dt} \\
&\quad + \sum_{j=1}^{J-1}[(B_j - W_j)\ell_j - W_5\ell_5]\sin(\theta) \\
&= \sum_{j=1}^{J-1} M_j \simeq \left(\sum_{j=1}^{J} I_{yj} + \sum_{j=1}^{J-1} \mathrm{A}_{wyj}\right)\frac{d^2\theta}{dt^2} + \sum_{j=1}^{J-1}(\mathrm{B}_{ryj} + \mathrm{B}_{vyj})\frac{d\theta}{dt} \\
&\quad + \sum_{j=1}^{J-1}[(B_j - W_j)\ell_j - W_5\ell_5]\theta = \sum_{j=1}^{J-1} F_j(\omega, t)\ell_j
\end{aligned} \tag{12.118}$$

For the sketch in Figure 12.26, $J = 5$. In the second line of eq. 12.118, the small-angle assumption is made to linearize the equation. Also in the second line is the force term, which is averaged over the component. The terms in eq. 12.118 are described in the following paragraphs.

I_{yj} is the mass moment of inertia of the jth component with respect to the ball joint. Each component is assumed to be a solid vertical circular cylinder. The mass moment of inertia with respect to a line through the center of the body that is parallel to the y-axis is

$$I_{yj} = \frac{1}{12}\frac{W_j}{g}(L_j^2 + 3a_j^2) \tag{12.119}$$

where L_j is the height of the component and a_j is the component radius. The moment of inertia about an axis through the ball joint is obtained using the parallel-axis theorem. For the jth component, the moment of inertia in eq. 12.118 is

$$I_{yj} = \frac{W_j}{g}\ell_j^2 + I_{Yj} = \frac{W_j}{g}\ell_j^2 + \frac{1}{12}\frac{W_j}{g}(L_j^2 + 3a_j^2) \tag{12.120}$$

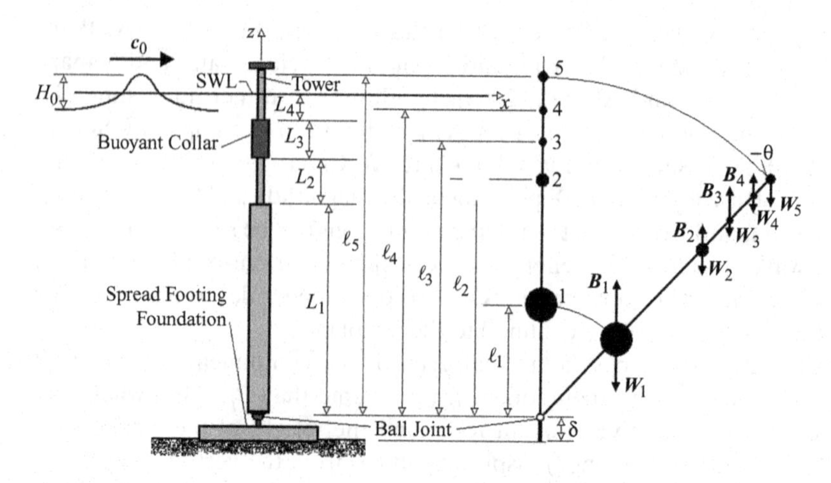

Figure 12.26. *Schematic Drawing of an ALP and Mathematical Model.* The ALP is assumed to be stiff, having only wave-induced rigid-body motions. The water depth is $h = \delta + \ell_4 + d/2$, assuming that the center of mass and buoyancy of the wetted portion of the tower are at $z = -d/2$. It is also noted that the positive angular displacement is in the counterclockwise direction, as is the norm.

The centers of mass and buoyancy of the $j = 4$ component are assumed to be at $d/2$ (not shown) above the SWL. That is, the tower portion of the structure is partially dry and partially wet, where the center of mass of the tower is at d above the SWL.

The second inertial term in eq. 12.118 is the sum of the component added-mass moment of inertia terms. We can modify the analysis presented in Section 12.4D(2) to obtain these terms and the moment due to the radiation damping, which is the third summation from the left side of eq. 12.118. We begin by writing the moment about the ball joint due to the force in eq. 12.88. Refer to the sketch in Figure 12.27 for the notation used in the analysis. In that figure, we see the height (δ) of the ball joint above the sea bed. In the analysis, assume that this height has a small value and can, therefore, be omitted. With this assumption, the moment on the component of the ALP is obtained from

$$
\begin{aligned}
M_{jxh} &= -\int_0^{2\pi}\int_{\ell_j-\frac{L_j}{2}}^{\ell_j+\frac{L_j}{2}} p_{dyn}|_{r=a_j} Z a_j \cos(\beta)\, dZ d\beta \\
&\simeq -i\rho\omega a_j^2 \int_0^{2\pi}\int_{\ell_j-\frac{L_j}{2}}^{\ell_j+\frac{L_j}{2}} \left\{ C_0(i\omega Z\theta_o) H_1^{(1)}(ka_j) Z\cosh(kZ) \right. \\
&\quad \left. + \sum_{n=1}^{\infty} C_n(i\omega Z\theta_o) K_1(K_n a_j) Z\cos(K_n Z) \right\} \cos^2(\beta)\, dZ d\beta e^{-i\omega t}
\end{aligned}
\qquad (12.121a)
$$

We note that the velocity amplitude (V_{xo}) in eq. 12.89 has been replaced by $i\omega Z\theta_o$, where θ_o is the amplitude of the angular displacement. As in Figure 12.26, the angular displacement is positive in the counterclockwise direction. The velocity amplitude occurs in eq. 12.89 due to $aV_{xo}C_0 = C_0$ and $aV_{xo}C_n = C_n$. The substitution is

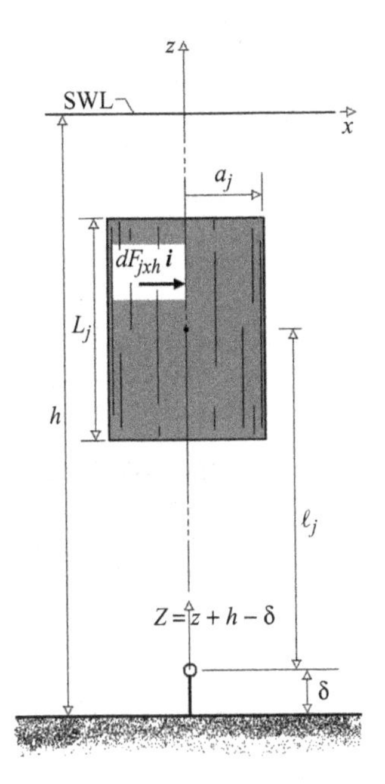

Figure 12.27. *Notation for the Hydrodynamic Force Analysis of an ALP Component.*

required because the linear velocity of the ALP increases with the variable Z above the ball joint. Upon integration, the hydrodynamic rocking moment expression is found to be

$$\begin{aligned} M_{jxh} &= -\rho\omega^2\pi a_j^2\theta_o\left\{C_0 H_1^{(1)}(ka_j)\,f_0(k\ell_j) + \sum_{n=1}^{\infty} C_n K_1(K_n a_j)\,f_n(K_n\ell_j)\right\} e^{-\omega t} \\ &= -\rho\omega^2\pi a_j^2\theta_o\left\{(C_{0\Re} + iC_{0\Im})[J_1(ka_j) + iY_1(ka_j)]\,f_0(k\ell_j)\right. \\ &\quad \left. + \sum_{n=1}^{\infty} C_n K_1(K_n a_j)\,f_n(K_n\ell_j)\right\} e^{-i\omega t} \\ &= A_{wyj}\frac{d^2\theta}{dt^2}B_{ryj}\frac{d\theta}{dt} = -(i\omega^2 A_{wyj} + \omega B_{ryj})\theta \\ &= -(i\omega^2 A_{wyj} + \omega B_{ryj})\theta_o e^{-i\omega t} \end{aligned} \tag{12.121b}$$

where

$$\begin{aligned} f_0(k\ell_j) = \frac{1}{k^3}&\left\{\left[k^2\left(\ell_j + \frac{L_j}{2}\right)^2 + 2\right]\sinh\left[k(\ell_j + \frac{L_j}{2})\right]\right. \\ &- 2k\left(\ell_j + \frac{L_j}{2}\right)\cosh\left[k\left(\ell_j + \frac{L_j}{2}\right)\right] \\ &- \left[k^2\left(\ell_j - \frac{L_j}{2}\right)^2 + 2\right]\sinh\left[k\left(\ell_j - \frac{L_j}{2}\right)\right] \\ &\left. + 2k\left(\ell_j - \frac{L_j}{2}\right)\cosh\left[k\left(\ell_j - \frac{L_j}{2}\right)\right]\right\} \end{aligned} \tag{12.122}$$

and where

$$f_n(\mathrm{K}_n\ell_j) = \frac{1}{\mathrm{K}_n^3}\left\{\left[\mathrm{K}_n^2\left(\ell_j+\frac{L_j}{2}\right)^2-2\right]\sin\left[\mathrm{K}_n\left(\ell_j+\frac{L_j}{2}\right)\right]\right.$$
$$+2\mathrm{K}_n\left(\ell_j+\frac{L_j}{2}\right)\cos\left[\mathrm{K}_n\left(\ell_j+\frac{L_j}{2}\right)\right]$$
$$-\left[\mathrm{K}_n^2\left(\ell_j-\frac{L_j}{2}\right)^2-2\right]\sin\left[\mathrm{K}_n\left(\ell_j-\frac{L_j}{2}\right)\right]$$
$$\left.-2\mathrm{K}_n\left(\ell_j-\frac{\ell_j}{2}\right)\cos\left[\mathrm{K}_n\left(\ell_j-\frac{L_j}{2}\right)\right]\right\} \quad (12.123)$$

In the last line of eq. 12.121b are the added-mass moment of inertia,

$$\mathrm{A}_{wyjj} = \rho\pi a_j^2\left\{[C_{0\Re}J_1(ka_j) - C_{0\Im}Y_1(ka_j)]\,f_0(k\ell_j) + \sum_{n=1}^{\infty}\mathrm{C}_nK_1(\mathrm{K}_na_j)\,f_n(\mathrm{K}_n\ell_j)\right\} \quad (12.124)$$

and the radiation-damping moment coefficient,

$$\mathrm{B}_{ryj} = \rho\pi a_j^2\omega[C_{0\Im}J_1(ka_j) + C_{0\Re}Y_1(ka_j)]\,f_0(k\ell_j) \quad (12.125)$$

The latter is found in the angular-velocity-dependent terms in eq. 12.118.

The viscous damping term in eq. 12.118, the second of the angular-velocity dependent terms, is the result of linearization of the viscous-damping force. Assuming that the wave-induced rocking motions of the ALP are approximately sinusoidal, we can use the equivalent linear damping coefficient for each component. Our interest is in the moment of the drag force per unit height of the component. From eq. 12.93, for the jth component, the elemental equivalent linear viscous-damping coefficient can be obtained. The elemental viscous-force moment acting on the component is

$$dM_{vxj} = -ZdF_{vxj} \simeq -Z\left(b_{vxj}\frac{dV_{xj}}{dt}\right) = -Z\left(\frac{8}{3}\frac{\omega}{\pi}Z\theta_o\mathrm{b}_{vxj}\right)Z\frac{d\theta}{dt}$$
$$= -Z^3\left(\frac{8}{3}\frac{\omega}{\pi}\theta_o\frac{1}{2}\rho C_{dD_j}D_j\right)\frac{d\theta}{dt}dZ = -Z^3\left(\frac{8}{3}\frac{\omega}{\pi}\theta_o\rho C_{dD_j}a_j\right)\frac{d\theta}{dt}dZ \quad (12.126)$$

where the linear displacement amplitude (x_o) in eq. 12.93 has been replaced by $-Z\theta_o$. The units of the nonlinear damping coefficient, b_{vxj}, are N-s^2/m^2. The relationship of the nonlinear damping coefficient and the drag coefficient (C_{dDj}) in eq. 2.78, as used here, is

$$\mathrm{b}_{vxj} = \frac{1}{2}\rho C_{dD_j}D_jdZ = \rho C_{cD_j}a_jdZ \quad (12.127)$$

Here, the projected area element of the component is D_jdZ. The behavior of the drag coefficient with respect to the Reynolds number based on diameter is presented in Figure 2.15. Values obtained from that figure are satisfactory for *quasi*-steady flows, as might be experienced in long-period waves. Many design engineers assume a drag coefficient value of $C_{dDj} \simeq 1.0$. This assumption is due to the relatively constant behavior of the upper bound of drag-coefficient values over the Reynolds number range of approximately 5×10^2 to 2×10^5 in Figure 2.15. As in Figure 9.9, the drag coefficient has been found to depend on the Keulegan-Carpenter number,

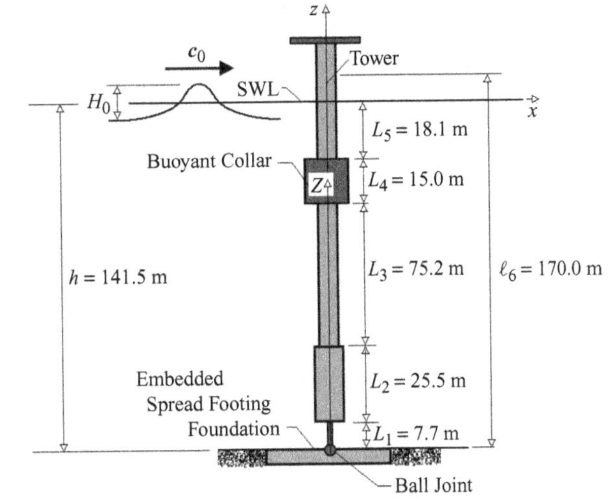

Figure 12.28. *Sketch of the ALP Studied by Suneja and Datta* (1998). The ALP is not proportionally drawn. Furthermore, the moment arm (ℓ_6) of the tower is larger than that shown.

KC in eq. 9.45. For our deep-water assumption, the approximation of KC in eq. 9.46 can be used. An excellent discussion of this dependence is found in Chapter 3 of the book by Sarpkaya and Isaacson (1981). The integration of eq. 12.126 over the height of the jth component results in the following expression for the viscous-damping moment:

$$M_{vyj} = -\left(\frac{2}{3}\frac{\omega}{\pi}\theta_o\rho\, C_{dD_j} a_j\right)\left[\left(\ell_j + \frac{L_j}{2}\right)^4 - \left(\ell_j - \frac{L_j}{2}\right)^4\right]\frac{d\theta}{dt} = -\mathrm{B}_{vyj}\frac{d\theta}{dt} = -B_{vyj}\theta_o\frac{d\theta}{dt} \tag{12.128}$$

Here, B_{vyj} is the viscous-damping coefficient found in eq. 12.118.

The hydrostatic restoring moment terms in eq. 12.118 are simply the net buoyancies of the submerged components. It is assumed that the free-surface effect on the buoyancy of the surface-piercing component, component 4 in Figure 12.26, is of second order for the ALP.

The diameter of an ALP might be found to be somewhat smaller than the leg diameter of a TLP. As a result, the type of the wave-induced force on the ALP might be drag- or inertia-dominant. We shall examine this by considering the ALP studied by Suneja and Datta (1998). The purpose of the study leading to that paper is to model an active control system for an ALP, designed to be deployed in 141.5 m of water. The ball joint (referred to as a "universal joint" by the authors) is located at the mud line. Hence, $\delta = 0$ in Figure 12.26. Referring to the sketch in Figure 12.28, the components in that figure have the following diameters: $D_1 = 2.3$ m, $D_2 = 10.5$ m, $D_3 = 6.3$ m, $D_4 = 15.0$ m, and $D_5 = D_6 = 6.0$ m. Component 5 pierces the free surface and has a height of $L_5 = 18.1$ m, as shown in the figure. The tower is 57 m above the SWL, and is Component 6.

For our illustration, we shall assume that the sea has an average wave height of 1.5 m and an average period of 8 sec. For this wave period in 141.5 m of water, the wave conditions are those in deep water. Hence, the wavelength (λ_0) is approximately 100 m. For these conditions, the motions of the ALP are determined. Using the Chakrabarti (1975) diagram in Figure 9.8, the type of wave-induced forces on the structure are first determined. To use the diagram, we need the Keulegan-Carpenter

Table 12.2. *Weights and buoyant forces of the Suneja and Datta (1998) ALP in Figure 12.28*

Component	Diameter (m)	Length (m)	Weight (N)	Buoyant Force (N)
1	2.3	7.7	840	320
2	10.5	25.5	40,140	24,700
3	6.3	75.2	22,580	23,390
4	15.0	15.0	7,210	27,170
5	6.0	18.1	1,740	5,220
6	6.0	57.0	6,540	0
			79,050	80,800

number value of eq. 9.45, which is approximated for deep-water waves in eq. 9.46. That is, in Figure 9.8,

$$KC|_j \simeq \frac{\pi H_0}{D} = 1.5\frac{\pi}{D_j} \tag{12.129}$$

The average wave excites water-particle motions to a depth of $\lambda_0/2 \simeq 50$ m. Furthermore, the wave-induced force on Components 1, 2, and 3 in Figure 12.28 would be expected to be rather small because the wave-induced particle velocity at the bottom of Component 4 is about 12.5% of that on the free surface, according to linear theory. Hence, for the average wave properties, we can focus on the wave-induced forces on Components 4 and 5.

The component diameters for the ALP in Figure 12.28 are presented in Table 12.2. The Keulegan-Carpenter number values for Components 4 and 5 are $KC_4 \simeq 0.314$ and $KC_5 \simeq 0.785$. The corresponding wave number-diameter products are $k_0a_4 \simeq 0.471$ and $k_0a_5 \simeq 0.188$. In Figure 9.8, these values indicate that forces are primarily a combination of drag and inertia forces. As a result, the wave-induced forces on the ALP can be represented by the Morison equation, presented in Section 9.2D. The Woodmond-Clyde (1980) report contains a thorough discussion of this equation. The force per unit length of the vertical cylindrical components in Figures 12.26 and 12.28 is obtained from eq. 9.49. In terms of the component radius, a_j, the Morison equation is

$$\frac{dF_j}{dz} = \frac{dF_j}{dZ} \simeq \left\{ C_{dD_j}\rho a_j u|u| + C_{ij}\rho\pi a_j^2 \frac{\partial u}{\partial t} \right\}\Bigg|_{x=0} \tag{12.130}$$

The drag coefficient (C_{dDj}) is obtained from Figure 9.9 for the value of the Keulegan-Carpenter number (KC), and the inertial coefficient (C_{ij}) is obtained from Figure 9.10. The horizontal velocity, u, is determined from eq. 3.49. Applied to deep water at the axis of the structure, this equation is

$$u|_{x=0} = \omega\frac{H_0}{2}e^{kz}\cos(\omega t) = \omega\frac{H_0}{2}e^{k(Z-h+\delta)}\cos(\omega t) \tag{12.131}$$

The application of eq. 12.130 at the centerline of the ALP is based on the assumption that the spatial variation of the horizontal velocity around the structure is small. The combination of the velocity expression in eq. 12.131 and eq. 12.130, results in the following form of the Morison equation:

$$\frac{dF_j}{dZ} = \frac{dF_{dj}}{dZ} + \frac{dF_{ij}}{dZ}$$
$$\simeq C_{dD_j}\frac{1}{4}\rho a_j \omega^2 H_0^2 e^{2k(Z-h+\delta)} |\cos(\omega t)| \cos(\omega t) - C_{ij}\frac{1}{2}\rho\pi a_j^2 \omega^2 H_0 e^{k(Z-h+\partial)} \sin(\omega t) \tag{12.132}$$

The effects of the relative motion of the body and fluid are accounted for in eq. 12.118, the equation of motion, because the added mass and viscous damping are included in the θ-terms of the equation. The drag force on the vertical cylinder in Example 9.5 is shown as a function of time in Figure 9.12. In that figure, the drag curve resembles a distorted cosine curve. Because of the similarities between the two curves, we can linearize the drag term in the Morison equation by using the same technique that is used in deriving the equivalent viscous-damping coefficient expression in eq. 10.10. See Section 10.1C for the details. What is assumed is that the time-averaged energies associated with the nonlinear drag and the linearized drag are equal, as in eq. 10.9. Mathematically, we can express time-average energies per unit component length as

$$\left\{\int_{\xi(0)}^{\xi(T)} \frac{dF_{dj}}{dZ} d\xi(\omega,t)\right\}\Bigg|_{x=0} = \left\{\int_0^T \frac{dF_{dj}}{dZ}\frac{d\xi(\omega,t)}{dt}dt\right\}\Bigg|_{x=0} = 4C_{dD_j}\rho a_j \int_0^{\frac{T}{4}} [u^2|u|]|_{x=0} dt$$
$$= \frac{1}{3} C_{dD_j}\rho a_j \omega^2 H_0^3 e^{3k(Z-h+\delta)} = 4\mathbb{C}_{dD_j}\rho a_j \int_0^{\frac{T}{4}} u^2|_{x=0} dt$$
$$= \frac{1}{4}\mathbb{C}_{dDj}\rho a_j \pi\omega H_0^2 e^{2k(Z-h+\delta)} \tag{12.133}$$

The water-particle displacement is ξ, and is represented by eq. 3.53. Hence, the time-derivative of ξ is simply the horizontal particle speed u about the centerline of the structure. The last two terms in eq. 12.133 are the linear equivalent terms, where $\mathbb{C}_{dDj}$ is the equivalent linear drag coefficient. From the second line of eq. 12.133, the expression for the equivalent linear drag coefficient is found to be

$$\mathbb{C}_{dDj} = \frac{4}{3\pi} C_{dD_j}\omega H_0 e^{k(Z-h+\delta)} \tag{12.134}$$

The reader should note that $\mathbb{C}_{dDj}$ has dimensions of m/s. The linearized Morison equation for the jth component of the structure in Figure 12.28 is then

$$\frac{dF_j}{dZ}$$
$$\simeq \left\{\mathbb{C}_{dDj}\rho a_j u + C_{ij}\rho\pi a_j^2 \frac{\partial u}{\partial t}\right\}\Bigg|_{x=0} = \left\{\frac{4}{3\pi} C_{dD_j}\rho a_j \omega H_0 e^{k(Z-h+\delta)} u + C_{if}\rho\pi a_j^2 \frac{\partial u}{\partial t}\right\}\Bigg|_{x=0}$$
$$= \left\{\frac{2}{3\pi} C_{dD_j}\rho a_j \omega^2 H_0^2 e^{2k(Z-h+\delta)} \cos(\omega t) - \frac{1}{2} C_{ij}\rho\pi a_j^2 \omega^2 H_0 e^{k(Z-h+\delta)} \sin(\omega t)\right\} \tag{12.135}$$

It might trouble the reader that the deep-water wave conditions, identified by the subscript 0, and the water depth, h, are in the same equations. Within the region of the free surface where the wave effects are pronounced, $Z-h$ $(=z)$ is finite provided that h is finite. Also note that if the ball joint is displaced above the sea bed by an amount δ, then the ball-joint coordinate is related to the free-surface coordinate by $Z = z + h - \delta$.

For the jth component, we can now write the moment of the linearized wave-induced moment about the ball joint as

$$
\begin{aligned}
M_j &= \int_{\ell_j - \frac{L_j}{2}}^{\ell_j + \frac{L_j}{2}} \frac{dF_j}{dZ} Z dZ = M_{doj}\cos(\omega t) + M_{ioj}\sin(\omega t) = M_{oj}\sin(\omega t + \alpha_\theta) \\
&\simeq \frac{1}{12\pi} C_{dD_j} \rho a_j \frac{\omega^2}{k_0^2} H_0^2 e^{2k(\ell_j - h + \delta)} [(2k_0\ell_j - 1)\sinh(k_0 L_j) \\
&\quad + k_0 L_j \cosh(k_0 L_j)]\cos(\omega t) \\
&\quad - \frac{1}{4} C_{ij} \rho \pi a_j^2 \frac{\omega^2}{k_0^2} H_0 e^{k(\ell_j - j + \delta)} \left[(k_0 \ell_j - 1)\sinh\left(k_0 \frac{L_j}{2}\right) \right. \\
&\quad \left. + k_0 \frac{L_j}{2} \cosh\left(k_0 \frac{L_j}{2}\right)\right] \sin(\omega t)
\end{aligned}
\tag{12.136}
$$

The drag- and inertial-moment amplitudes are simply the coefficients of the time functions. The phase angle in eq. 12.136 is

$$
\alpha_\theta = \tan^{-1}\left(\frac{M_{doj}}{M_{ioj}}\right) \tag{12.137}
$$

and the amplitude of the wave-induced moment is

$$
M_{oj} = \sqrt{M_{doj}^2 + M_{ioj}^2} \tag{12.138}
$$

From the second line of eq. 12.118, the linearized rocking equation of motion for the ALP is

$$
\begin{aligned}
&\left(\sum_{j=1}^{J} I_{yj} + \sum_{j=1}^{J-1} \mathrm{A}_{wyj}\right) \frac{d^2\theta}{dt^2} + \sum_{j=1}^{J-1} (\mathrm{B}_{ryj} + \mathrm{B}_{vyj}) \frac{d\theta}{dt} + \sum_{j=1}^{J-1} [(B_j - W_j)\ell_j - W_5 \ell_5]\theta \\
&= \sum_{j=1}^{J-1} M_{oj} \sin(\omega t + \alpha_\theta)
\end{aligned}
\tag{12.139}
$$

where the wave-induced moment is due to the linearized Morison equation forces. All of the terms in eq. 12.139 are known.

In the following, the analysis presented in this section is applied to the ALP studied by Gernon and Lou (1987).

EXAMPLE 12.9: WAVE-INDUCED MOTIONS OF AN ALP Gernon and Lou (1987) perform a motion analysis on a single-component ALP designed as a single-point mooring and oil terminal for ships. Because there is only one component, the terms in Figure 12.26 having the subscript 1 apply. Because of this, the subscripts can be omitted. The ALP is in a water depth (h) of 91.4 m. The axis of rotation of the ball joint is at $\delta = 3$ m above the sea bed. There is little free board; hence, we can neglect the weight of the structure above the SWL. The respective length and diameter of the hull are $L_1 \equiv L = 88.4$ m and $D_1 \equiv D = 2a = 6.1$ m.

The weight and buoyancy of the structure are $W_1 \equiv W = 4{,}877$ metric tons and $B_1 \equiv B = 5{,}080$ metric tons, respectively. The center of gravity of the hull is $\ell = 44.2$ m above the ball joint. Our interest is in the angular motion in a 1.5-m, 8-sec sea, as in Example 12.8. For the water depth of $h = 91.4$ m, the wavelength (λ) is obtained from the dispersion equation, eq. 3.31, by using the method of successive approximations, as illustrated in Example 3.3. The wavelength value is found to be approximately 99.9 m. One half of this is about 50 m $< h$. Hence, our deep-water assumption is valid.

From eq. 12.120, the mass moment of inertia with respect to the ball-joint axis is $I_y \simeq 1.30 \times 10^9$ N-m-s²/rad. The added-mass moment of inertia and the radiation-damping moment coefficients are now determined from eqs. 12.124 and 12.125, respectively. In those equations are the functions in eqs. 12.122 and 12.123. For this problem, we note that $\ell = L/2$. Hence, the second lines in eqs. 12.122 and 12.123 both vanish. The added-mass moment of inertia is found to be $\mathrm{A}_{wy} \simeq 1.32 \times 10^9$ N-m-s²/rad, and the radiation-damping moment coefficient is $\mathrm{B}_{ry} \simeq 5.89 \times 10^7$ N-m-s/rad.

The linearized viscous-damping moment coefficient in eq. 12.128, which depends on the wave properties, is

$$\begin{aligned}\mathrm{B}_{vy} &= B_{vy}\theta_o = \left(\frac{2}{3}\frac{\omega}{\pi}\rho a L^4\right)\theta_o C_{dD} \\ &\simeq 2.53.197 \times 10^9 \theta_o C_{dD} \simeq 3.54 \times 10^9 \theta_o \end{aligned} \tag{12.140}$$

The units for this coefficient are N-m-s/rad. We note that the amplitude of the rocking angle, θ_o, in eq. 12.140 is to be determined. This is addressed later in the example. The value of the drag coefficient, $C_{dD} \simeq 1.4$, is obtained from Figure 9.9 for the deep-water Keulegan-Carpenter number value of $KC|_{deep} \simeq 0.773$.

The restoring coefficient in eq. 12.139 is simply $(B - W)\ell = (203)44.2$ tonne-meters per radian, or approximately 8.80×10^7 N-m/rad.

The last term to be determined in the equation of motion, eq. 12.139, is the linearized wave-induced moment of eq. 12.136. In the last line of that equation, we again note that $\ell = L/2 = 44.2$ m. The components of the moment amplitude are

$$\begin{aligned}M_{do} &= \frac{1}{12\pi} C_{dD}\rho a \frac{\omega^2}{k_0^2} H_0^2 e^{2k(\ell-h+\delta)}[(2k_0\ell - 1)\sinh(k_0 L) + k_0 L\cosh(k_0 L)] \\ &\simeq 2.05 \times 10^5 \text{ N–m}\end{aligned} \tag{12.141}$$

and

$$\begin{aligned}M_{io} &= -\frac{1}{4} C_i \rho\pi a^2 \frac{\omega^2}{k_0^2} H_0 e^{k(\ell-h+\delta)} \left[(k_0\ell - 1)\sin\left(k_0\frac{L}{2}\right) + k_0\frac{L}{2}\cosh\left(k_0\frac{L}{2}\right)\right] \\ &= 4.70 \times 10^6 \text{ N–m}\end{aligned} \tag{12.142}$$

In eq. 12.141, the drag coefficient is 1.4 for $KC = 0.773$, as previously stated. The inertial coefficient value for this KC-value is $C_j \simeq 1.2$ from Figure 9.10. Comparing the results in the last two equations, we see that the inertia force is more than an order of magnitude greater than the drag force. This is as expected from the Chakrabarti diagram, Figure 9.8. From the drag- and inertia-moment amplitudes in eqs. 12.141 and 12.142, respectively, the phase angle in eq. 12.137

is approximately $-2.50°$, and the wave-induced moment amplitude of eq. 12.138 is approximately 4.70×10^6 N-m.

For the single-component ALP in the Gernon and Lou (1987) paper, we can now write the rocking equation of motion, eq. 12.139, as

$$(I_y + A_{wy})\frac{d^2\theta}{dt^2} + (B_{ry} + B_{vy})\frac{d\theta}{dt} + [(B - W)\ell]\theta = M_o \sin(\omega t + \alpha_\theta)$$
$$(I_y + A_{wy})\frac{d^2\theta}{dt^2} + (B_{ry} + B_{vy}\theta_o)\frac{d\theta}{dt} + [(B - W)\ell]\theta = M_o \sin(\omega t + \alpha_\theta) \quad (12.143)$$

Replacing the coefficients in the second line of this equation by their numerical values results in

$$(1.30 \times 10^9 + 1.32 \times 10^9)\frac{d^2\theta}{dt^2} + (5.89 \times 10^7 + 3.54 \times 10^9\theta_o)\frac{d\theta}{dt} + (8.80 \times 10^7)\theta$$
$$\simeq (2.62 \times 10^9)\frac{d^2\theta}{dt^2} + (5.89 \times 10^7 + 3.54 \times 10^9\theta_o)\frac{d\theta}{dt} + (8.80 \times 10^7)\theta$$
$$\simeq 4.70 \times 10^6 \sin\left(0.785t - \frac{2.50°}{180°}\pi\right) \simeq 4.70 \times 10^6 \sin(0.785t - 0.0139\pi)$$

In the damping term, we note the presence of the rocking angular amplitude, θ_o. With the exception of the presence of θ_o, the equation of motion with the numerical coefficients is similar to that in eq. 10.2. Hence, as in eqs. 10.12 through 10.14, we can determine the rocking angular response, natural rocking frequency, critical damping, and the phase angle between the wave-induced moment and the response of the ALP. The rocking angular response of the ALP is obtained from

$$\theta = \theta_0 \sin(\omega t + \alpha_\theta - \epsilon_\theta)$$
$$= \frac{\dfrac{M_0}{(B - W)\ell}}{\sqrt{\left(1 - \dfrac{\omega^2}{\omega_{n\theta}^2}\right)^2 + \left[2\dfrac{\omega}{\omega_{n\theta}}\dfrac{(B_{ry} + B_{vy}\theta_0)}{B_{cr}}\right]^2}} \sin(\omega t + \alpha_\theta - \epsilon_\theta) \quad (12.144)$$

In this equation are the natural circular frequency,

$$\omega_{n\theta} = \frac{2\pi}{T_{n\theta}} = \sqrt{\frac{(B - W)\ell}{I_y + A_{wy}}} \simeq 0.183\ \frac{\text{rad}}{\text{s}} \quad (12.145)$$

and the critical damping,

$$B_{cy} = 2\sqrt{(I_y + A_{wy})[(B - W)\ell]} \simeq 1.44 \times 10^8\ \frac{\text{N-m-s}}{\text{rad}} \quad (12.146)$$

From eq. 12.144, the rocking angular amplitude is

$$\theta_o = \frac{\dfrac{M_o}{(B - W)\ell}}{\sqrt{\left(1 - \dfrac{\omega^2}{\omega_{n\theta}^2}\right)^2 \left[2\dfrac{\omega}{\omega_{n\theta}}\dfrac{(B_{ry} + B_{vy}\theta_o)}{B_{cr}}\right]}} = \Theta\frac{M_o}{(B - W)\ell}$$
$$\simeq \frac{0.0535}{\sqrt{301 + 3.42{*}10^{-15}(5.89 \times 10^7 + 3.54 \times 10^9\theta_o)^2}} \simeq 0.0038\ \text{rad} \simeq 0.0882° \quad (12.147)$$

This equation is solved by using the method of *successive approximations*, as in Example 3.3 in the determination of the wavelength from the dispersion equation, eq. 3.31. The result in eq. 12.147 shows that response at the 8-sec wave period is extremely small well away from the natural period, $T_{n\theta} \simeq 34.3$ sec, from eq. 12.145.

At resonance, where $T = T_{n\theta}$ or $\omega = \omega_n$, we find that $\theta_o \simeq 0.0258$ rad. For this amplitude, the amplitude of the horizontal displacement at the SWL is $x_o|_{z=0} = L\theta_o \simeq 2.28$ m. The average linear speed at the SWL is $L\theta_o/T_{n\theta} \simeq 0.0665$ m/s. The ratio of the total damping to critical damping $[(B_{ry} + B_{vy})/B_{cr}]$ at resonance is approximately 1.04; hence, the system is slightly over-damped. The magnification factor (Θ in eq. 12.147) at resonance is approximately 0.175.

Finally, the phase angle between the wave-induced moment and the motion for the ALP in the 8-sec waves is

$$\epsilon_\theta = \tan^{-1}\left[\frac{2\dfrac{\omega}{\omega_{n\theta}}\dfrac{(B_{ry} + B_{vy}\theta_o)}{B_{cy}}}{\left(1 - \dfrac{\omega^2}{\omega_{n\theta}^2}\right)}\right] \simeq -0.474 \text{ rad} \simeq -27.2^\circ \tag{12.148}$$

The study of Gernon and Lou (1987) includes the dynamics of a ship moored to the ALP, and the dynamics of the mooring hawser.

It should be emphasized that the advantage of the ALP is that there is no overturning moment transferred to the foundation. This results in a significant reduction in the material and engineering costs of the structure.

12.6 Motions of Flexible Towers

This section is devoted to the prediction of wave-induced motions of the class of compliant towers called a *flexible offshore tower* (FOT). This type of tower has a fixed pile foundation, such as sketched in Figure 12.1b. The FOT is one of the compliant tower concepts discussed by Maus, Finn, and Danaczko (1986). Those authors define a *compliant tower* as "a slender, tubular steel structure, with a relatively constant cross section over its height. In contrast to a rigid jacket or gravity structure, it is designed such that the fundamental natural period (sway period) is greater than the periods of ocean waves, i.e., greater than about 25 sec." From Figure 12.1b, the reader can see that the foundation must be designed to resist an overturning moment. However, this moment is far less than that of an equivalent rigid tower because the inertia of the tower resulting from the compliance resists the wave loads. Like the TLP, this type of tower resulted from our thirst for oil taking us into deeper waters to exploit the resource. The deployment philosophies of this type of structure, the TLP, and others are presented by Regg et al. (2000). The primary purpose of that publication is to address the environmental issues associated with offshore structures. In addition, the papers of Bayazitoglu, Jones, and Hruska (1987), and Will, Morrison, and Calkins (1988) address the design philosophy of the compliant tower. Again, we shall concentrate on a version of the FOT having a fixed pile base. As a result of the fixed base, the dynamics of the tower are similar to those of a cantilevered beam.

There are numerous papers devoted to mathematical modeling and testing of compliant towers. Those include the papers of Foster (1967), Edge and Mayer

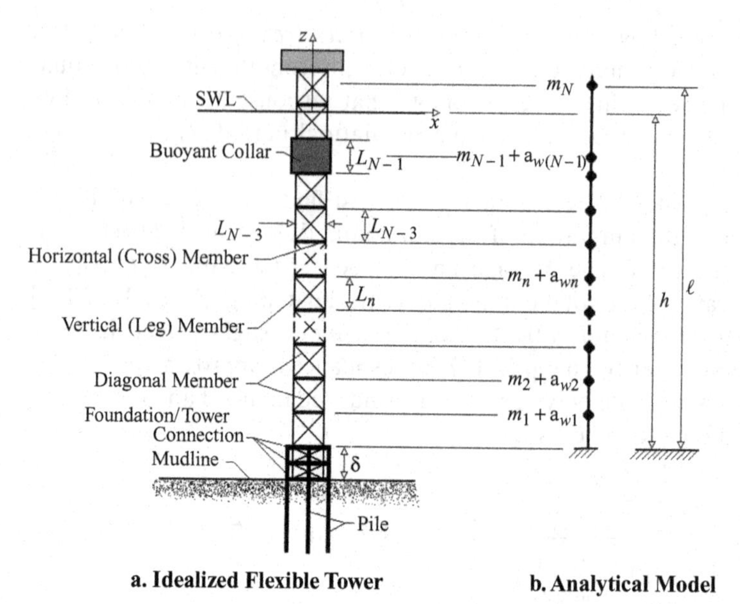

Figure 12.29. *Sketch of an Idealized Flexible Tower with a Fixed Pile Base.* The analytical model in (b) is a lumped-mass model of the tower. This model is used for both the bending motions and the swaying motions of the tower.

(1969), Vugts and Hayes (1979), Arockiasamy et al. (1983), Hanna (1988), Ng and Vickery (1989), Molin and Legras (1990), Sellers and Niedzwecki (1992), and Bishi and Jain (1998).

The motion analysis of a FOT depends, naturally, on the structural design. In Figure 12.29a and in the remainder of this section, assume that all of the elements are tubular and are of small outside diameter. If the horizontal elements are relatively stiff and short when compared to the vertical members ($\mathrm{L}_j \ll L_j$), then the structural motions can be analyzed using the lumped-mass method. In this case, we concentrate the mass of each panel (bay or frame) on the horizontal member above the panel, as in Figure 12.29b. The structure can be treated as a *quasi*-space frame where the bending mode would be dominant. The restoring of the displaced panel to its original shape is due to the axial displacements of the vertical members, where the members on the elongated side of the platform displacement are in tension, and those on the other side would be in compression.

If the horizontal members in Figure 12.29a are approximately of the same stiffness and length of the vertical members, then the motions might be building type, where the vertical member is always at right angles to the horizontal members at the nodes. For this condition, each vertical member can be considered to be two joined cantilevers, where a zero-bending-moment condition is at the mid-height of the vertical member. Thomson and Dahleh (1998) analyze several situations involving this type of motion.

An interesting variation in the geometry of the FOT is the *tension-restrained articulated platform* (TRAP), discussed and analyzed by Sellers and Niedzwecki (1992). This structure is composed of large, vertical, stiff segments (not panels) that are connected by pins at the points of articulation. Referring to the ALP sketched in Figure 12.26, the TRAP can be visualized by imagining that each segment in that

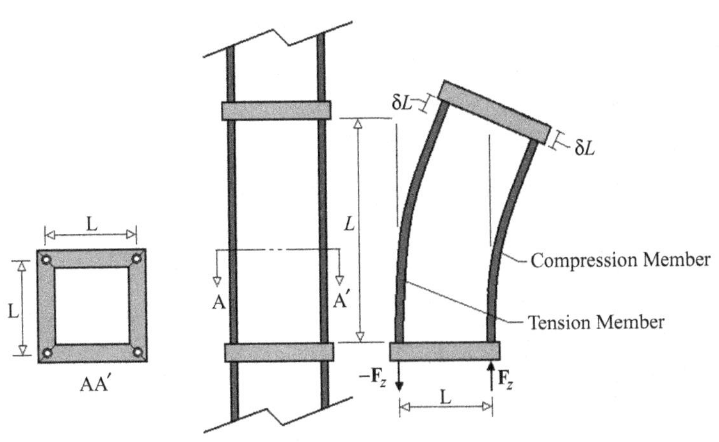

Figure 12.30. *Panel Deformation in Pure Bending.* The vertical members (leg members) in pure bending are deformed equally. That is, the axial strain values for the right and left members are equal. The axial forces acting on the bottom horizontal (cross) member are shown. These forces produce a counterclockwise moment on the bottom cross member.

figure is pinned to the adjacent segment. At each pin is a rotational spring constant. The analysis of the TRAP is presented in some detail later in this section.

A. Effective Spring Constants

In each of the cases described, the major structural task is defining the effective spring constants. These are linear spring constants for the *quasi*-space frame structure, rotational spring constants for the TRAP, and a combination of the two for the building-type motions.

For pure bending of a *quasi*-space-frame type of structure, as in Figure 12.30, where the stiff horizontal (or cross) member experiences a rigid-body rotation, the restoring forces, F_z, are the axial forces along the centerlines of the vertical members. For the axial strain, $\delta L/L$, the axial spring constant in a vertical member is

$$K_{sz} = \frac{\mathrm{F}_z}{\delta L} = \frac{EA_y}{L} \tag{12.149}$$

Here, E is Young's modulus for the member material, and A_y is the cross-sectional area of the vertical (leg) member. Sample values of E are presented in Table 12.3. The restoring moment due to the combined axial loads in the four leg members is

$$M_{yz} = 2\left(2\mathrm{F}_z\frac{\mathrm{L}}{2}\right) = 2\mathrm{F}_z\mathrm{L} = 2EA_y\frac{\mathrm{L}}{L}(\delta L) \simeq EA_y\frac{\mathrm{L}^2}{L}\sin(\Theta) \simeq EA_y\frac{\mathrm{L}^2}{L}\Theta = K_{s\Theta}\Theta \tag{12.150}$$

Table 12.3. *Mid-range specific weights and Young's modulus values for structural materials*

Material	Weight density (N/m^3)	E(N/m^2)
Low Carbon Steel	77,100	2.00×10^{11}
Stainless Steel	78,700	1.93×10^{11}
Aluminum	26,700	0.73×10^{11}
Titanium	44,100	1.10×10^{11}

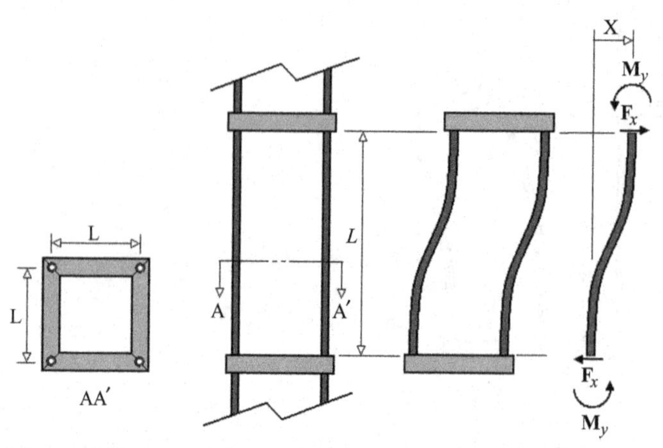

Figure 12.31. *Panel Deflection under Shear.* The horizontal or cross members are stiff, and do not either deform or rotate. The upper and lower half of the legs act as cantilevered members. In the right side sketch are the bending moment, M_y, and the shear force, F_x, acting on the vertical (leg) member.

Here, L represents both the effective width and the effective breadth of the structure having a square cross-section, as sketched in Figure 12.30. The angle Θ is measured from the vertical direction, and is positive in a clockwise direction. The coefficient of the last equality is the effective bending, rotational spring constant for the four-leg unit, that is,

$$K_{s\Theta} = EA_y \frac{L^2}{L} \tag{12.151}$$

From this expression, we see that the structure stiffens as either L increases or L decreases. Note that the two-dimensional, two-leg structural components are referred to herein as *panels*. These are also referred to as bays and frames by some authors. The three-dimensional, four-leg components are referred to herein as *units*. If the panel is subjected to a shear load, then the deformation will resemble that in Figure 12.31. This could be considered to be a panel in a building-type motion. Each vertical member in the unit can be considered to be composed of two cantilevered members, one extending from the upper horizontal member and the other extending from the lower horizontal member. As such, the bending moment is zero at the center of the vertical member. Assuming that the deflection is resisted by the shear, we can write the shear force as

$$F_x = \frac{12EI_y}{L^3} X = \frac{1}{4} K_{sx} \tag{12.152}$$

as derived in books on both strength of materials and vibrations. For example, see p. 173 of Thomson and Dahleh (1998). In eq. 12.152, X_M is the relative displacement related to the bending moment of the upper node. That is, if the node in question is the nth node, then $X = X_{Mn} - X_{Mn-1}$. In eq. 12.154 is the effective shear spring constant for a four-leg unit,

$$K_{sx} = \frac{48EI_y}{L^3} \tag{12.153}$$

For a structure having N vertical members per unit and undergoing a shear deflection such as in Figure 12.31, the spring constant is

$$K_{sx} = N\frac{12EI_y}{L^3} \tag{12.154}$$

Figure 12.32. *Expressions for the Second Moment of Area* (I_{yn}) *for Various Leg Cross-Sections.*

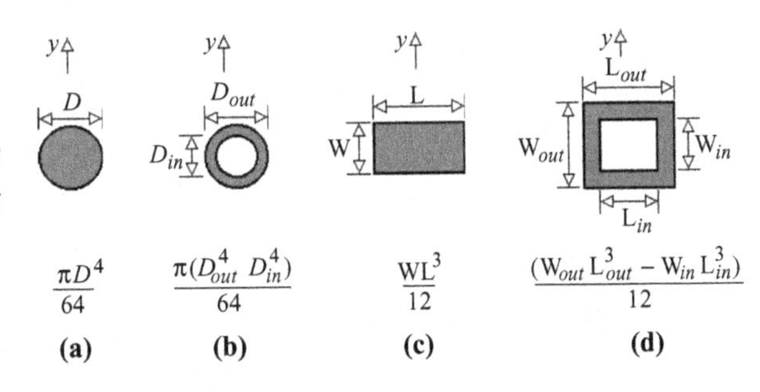

For example, see the book by Crede (1965). In eqs. 12.152 through 12.154 is the second moment of area, I_y, with respect to the horizontal bending axis, y. Expressions for typical member cross-sections are presented in Figure 12.32.

The insertion of diagonal members, as sketched in Figure 12.29a, stiffens the structure because both bending and shear deformations axially and transversely deform these members. To illustrate, consider the diagonal members in the shear situation sketched in Figure 12.33. These members have the same cross-section and material of the vertical members. Assume that the deformations of the transverse members are predominately axial. The axial force expression in the transverse member is similar to that in eq. 12.152. For the horizontal deflection shown in Figure 12.33, the axial force in the diagonal member 1 is

$$F_{\alpha 1} = \frac{12EI_y}{L_{\alpha 3}}\delta_{\alpha 1} = \frac{12EI_y}{L^3}\sin^3(\alpha)\delta_{\alpha 1} \simeq \frac{12EI_y}{L^3}\sin^3(\alpha)X\cos(\alpha) \qquad (12.155)$$

Here, the diagonal axial change is $\delta_{\alpha 1}$. The component of the axial force in the x-direction is

$$F_{\alpha 1x} = F_{\alpha 1}\cos(\alpha) \simeq \frac{12EI_y}{L^3}\sin^3(\alpha)X\cos^2(\alpha) = \frac{3EI_y}{L^3}\sin(\alpha)\sin(2\alpha)X = \frac{K_{s\alpha}}{4}X \qquad (12.156)$$

The last equality defines the diagonal spring constant. For the box-shaped unit, the structural unit is assumed to have two parallel 1-diagonals and two parallel 2-diagonals. Hence, for the unit shown in Figure 12.33, the effective, horizontal, spring constant for the system of diagonals is

$$K_{s\alpha} = \frac{12EI_y}{L^3}\sin(\alpha)\sin(2\alpha) \qquad (12.157)$$

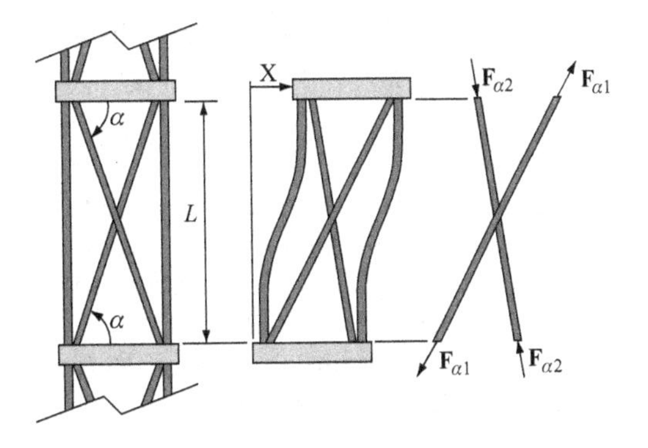

Figure 12.33. *Deformation of Diagonal Members in a Shear Deflection of a Panel.* There will be some bending of the diagonals. For analytical purposes, the deformations of the diagonals are assumed to be axial, and the changes in the angle α during the deformation are considered to be small. For a flexible offshore tower in deep water, these assumptions are valid.

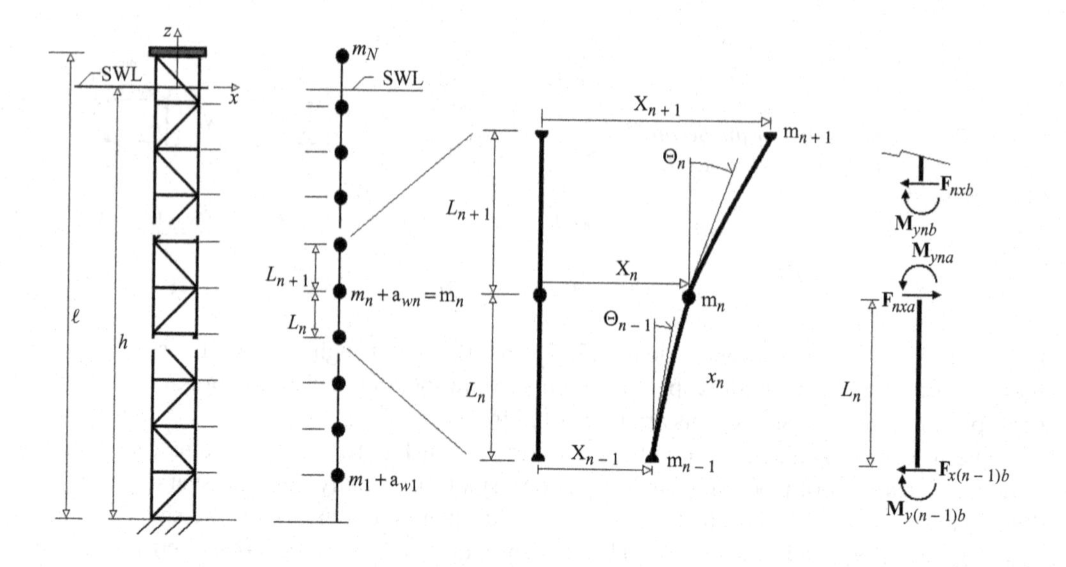

Figure 12.34. *Lumped-Mass Model of a Flexible Offshore Tower.*

With the expressions of the spring constants now defined, we can proceed with the analysis of the FOT.

B. Analysis of the Motions of a Flexible Offshore Tower (FOT)

Consider the FOT in Figure 12.34a. The lumped-mass analysis is based on the model sketched in Figures 12.34b and c, where the shear forces and bending moments on a leg of the n-unit are sketched in Figure 12.34c. As can be seen in Figure 12.34b, there are both horizontal and rotational displacements of the lumped mass, m_n. Those are the horizontal displacement, X_n, and the rotational displacement, Θ_n. The lumped mass includes the structural mass, m_n, and the added mass, a_{wn}, averaged about the node. In Figure 12.34c, the shear force and bending moment on the top of the nth leg are identified by the subscript a, and those resisting the force and moment on the adjacent upper leg are denoted by the subscript b. Our interest is in the first bending mode and the swaying (shear) mode. For a four-leg structure having units that are identical, except for that at the base of the structure and that supporting the platform, we can assume that the spring constants for these respective modes are given by eqs. 12.151 and 12.153, respectively. The analysis of each mode is done independently, and the displacements resulting from each mode can simply be added together to obtain the total displacement. That is, in terms of the moment displacement, X_M, and the shear displacement, X_F, the total displacement of the nth node is

$$X_n = X_{n-1} + L_n\Theta_{n-1} + X_{nM} + X_{nF} \qquad (12.158)$$

Here, the assumption is made that the angular displacement, Θ_{n-1}, is small if the nodes are relatively close together. This is the case when the leg-member length, L_n, is small when compared to the overall height, ℓ, of the tower. The notation is defined in Figure 12.34.

Concerning the modes: Referring to Figure 12.35, the swaying mode is one for which the tower is assumed to respond in a pure shear deflection, as illustrated in

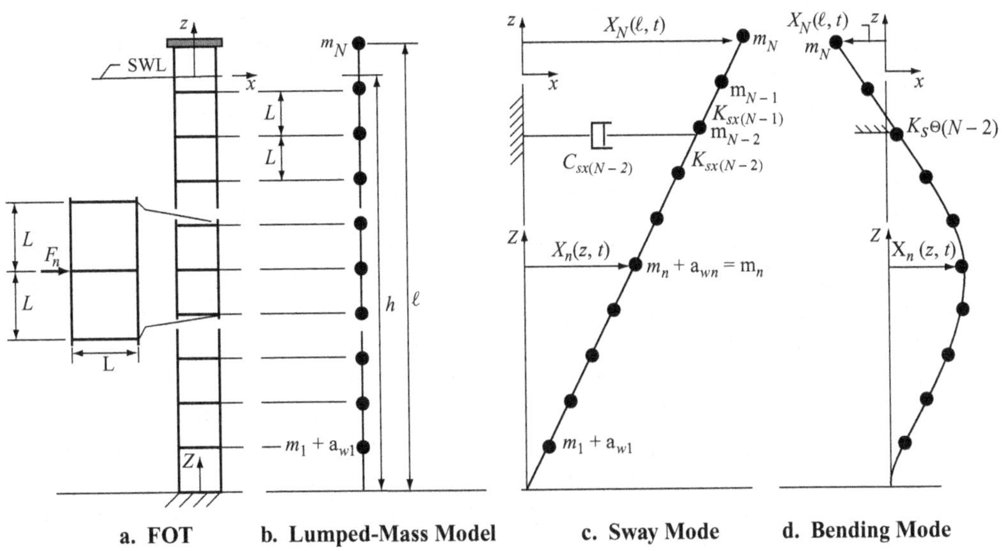

Figure 12.35. *Simplified Models of the Modal Deflections of a Flexible Tower with a Fixed Base.* The models are discussed by Maus, Finn, and Danaczko (1986). The sway mode deflection at any point above mass m_n is $X_n(z,t)$, whereas the bending mode deflection at the same position is $X_n(z,t)$. These deflections are due to the spatially averaged, wave-induced force (F_n) applied to m_n. The truss geometry is one having square panels; hence, the height and width are both L. The cross-section at any Z is square. The linear spring constant and linear damping coefficient for the N-2 node are shown. The hydrodynamic damping is assumed to be much greater than the structural damping.

Figure 12.35c. This mode is considered to be the fundamental mode, or design mode. For the displacement shown in Figure 12.35c, the horizontal cross members shorten under compression but remain horizontal, as sketched in Figure 12.31. The bending mode is illustrated in Figure 12.35d. In this case, the right vertical members are in tension, whereas the left vertical members are in compression for the horizontal forces acting at the cross members. Ideally, for pure bending there are no forces in the cross members. Furthermore, if diagonal members are used, there are no forces on these members in the pure bending mode. The horizontal cross members in pure bending do not remain parallel to each other, as sketched in Figure 12.30.

In the analysis of the tower motion, assume that the vertical, horizontal, and diagonal members are all of the same material, and all are tubular members having an outside diameter, D_{out}, and an inside diameter, D_{in}. See Figure 12.32b for the cross-sectional geometry of a member and the second moment of area (I_{yn}) of this cross-section.

(1) Swaying Motions of a FOT

Consider the compliant tower in Figure 12.35. The masses of the various tower panels are lumped, as sketched in Figure 12.35b. In the analysis, the masses are considered to be concentrated on a vertical line through the centers of the horizontal members. This assumed mass orientation allows us to use the lumped-mass method in determining the motions of the tower. This is a vibration analysis method that is well described by Thomson and Dahleh (1998), and is presented in most books devoted to mechanical and structural vibrations. The tower in Figure 12.35a is on an extended pile foundation. This is referred to as a *fixed base*, and as such the piles are considered to be both rigid and rigidly fixed in the soil. In other words, the tower is "clamped" at the base.

Because the purpose here is to both discuss and demonstrate the analytical method used in determining the tower motions, a simple design is considered where diagonal members are omitted. The horizontal cross-section of the tower is also assumed to be square. Two such panels are sketched in the left side of the tower in Figure 12.35a. In that sketch, the horizontal wave-induced force component on panel n is F_n. This force is the sum of the spatially averaged horizontal wave-induced forces acting on the horizontal and vertical members. Over the latter, the forces are averaged from $L/2$ above node n to $L/2$ below the node for each leg. The equivalent lumped-mass model for the swaying motions is sketched in Figure 12.35b. In that figure, we see that the lumped masses (composed of the structural mass of the panel and the added mass) are positioned along the vertical centerline of the structure at the levels of the cross members or nodes. The legs of each panel are represented in the model by a single flexure having a spring constant, K_{sx}, also referred to as the stiffness of the flexure.

The tower is rigidly attached to the base and free to move where $z > -h + \delta$, where δ is the height of the foundation above the mud line. Assume here that δ is negligible, and let the coordinate (Z) of the tower have its origin at $z = -h$. Hence, $Z = z + h$. For the lumped mass, m_n, the swaying nodal equation of motion is

$$\mathrm{m}_n \frac{d^2\mathrm{X}_n}{dt^2} + C_{sxn}\frac{d\mathrm{X}_n}{dt} + K_{sxn}(\mathrm{X}_n - \mathrm{X}_{n-1}) - K_{sx(n+1)}(X_{n+1} - X_n) = F_n \qquad (12.159)$$

This equation shows the coupling of the modal motions due to the spring constants. The damping coefficient, C_{sxn}, represents only the hydrodynamic damping (radiation and viscosity). In this derivation, the structural damping is assumed to be of second order when compared to the hydrodynamic damping. Like the restoring force, the structural damping is a relative quantity for each node. That is, the structural damping force depends on the difference in the velocities of the adjacent modes.

For the system having uniform panels from the foundation to the tower, eq. 12.159 simplifies to

$$\mathrm{m}_n \frac{d^2\mathrm{X}_n}{dt^2} + C_{sxn}\frac{d\mathrm{X}_n}{dt} + K_{sx}(2\mathrm{X}_n - \mathrm{X}_{n+1} - \mathrm{X}_{n-1}) = F_n \qquad (12.160)$$

In this expression, the spring constant is that in eq. 12.153. For the entire tower consisting of N identical panels, we can write the following swaying matrix equation of motion:

$$[\mathrm{m}]\left\{\frac{d^2\mathrm{X}}{dt^2}\right\} + [C_{sx}]\left\{\frac{d\mathrm{X}}{dt}\right\} + [K_{sx}]\{\mathrm{X}\} = \{F\} \qquad (12.161)$$

The mass matrix, [m], represents the structural mass of the tower and the added mass. The damping matrix, $[C_{sx}]$, represents the hydrodynamic damping, assuming that the structural damping is of second order. Equation 12.161 represents a system of linear, second-order differential equations. This type of equation also results from the bending motions of a FOT, as discussed in the next section. Before presenting the solutions of an equation of the form of eq. 12.161, the determination of the fundamental frequency of the swaying motions are first discussed.

If the nodal masses and the panels in eq. 12.35 are approximately uniform from the base to platform, the undamped modal frequencies can be determined by using the numerical analysis technique known as *difference equations*. Essentially, this is the same method leading to eqs. 12.158 and 12.160 with the assumption that the nodal masses and spring constants are the same for all panels. This assumption for a flexible tower in deep water is valid, as is demonstrated later in an example. To determine the modal frequencies for such a structure, we consider the tower to be in calm water, so that there is no wave-induced force. We also assume that the damping is zero over the structure. Assume that the top of the platform is slightly displaced and then released. We know that the tower will oscillate at specific frequencies. Following Thomson (1965), we can then assume that $X_n = X_{no}e^{-i\omega t}$. Under our assumptions, eq. 12.160 becomes

$$m\frac{d^2X_n}{dt^2} + K_{sx}(2X_n - X_{n+1} - X_{n-1}) = [-\omega^2 mX_{no} + K_{sx}(2X_{no} - X_{(n+1)o} - X_{(n-1)o})]e^{-i\omega t} = 0 \qquad (12.162)$$

The coefficient of the exponential function must be identically zero and, as such, can be rearranged as

$$X_{(n+1)o} - 2\left[1 - \omega^2\left(\frac{m}{2K_{sx}}\right)\right]X_{no} + X_{(n-1)o} = e^{i(n+1)\gamma} - 2\left[1 - \omega^2\left(\frac{m}{2K_{sx}}\right)\right]e^{in\gamma} + e^{i(n-1)\gamma} = 0 \qquad (12.163)$$

where $X_{no} = e^{n\gamma}$, and so on. The angle, γ, is determined from the following relationships obtained from the last equality, where

$$\frac{e^{in\gamma} + e^{-in\gamma}}{2} = \cos(n\gamma) = 1 - \omega^2\frac{m}{2K_{sx}} \qquad (12.164)$$

and, therefore,

$$\omega^2\frac{m}{2K_{sx}} = 1 - \cos^2(n\gamma) = 2\sin^2\left(\frac{n\gamma}{2}\right) \qquad (12.165)$$

In these equations, $n = 0$ at the base where $X_0 = 0$, assuming a "clamped" condition. At the platform, $n = N$. From eq. 12.163, the platform displacement is found to be

$$X_N = \frac{X_{N-1}}{1 - \omega^2\frac{m_N}{2K_{sx}}} \simeq \frac{X_{N-1}}{1 - \omega^2\frac{m}{2K_{sx}}} \qquad (12.166)$$

Here, the platform mass, m_N, is assumed to be approximately equal to the effective mass, m (structural mass plus added mass), at a node. Apply the last equalities in eqs. 12.164 and 12.165 to the tower panel, and solve simultaneously. Then, from eq. 12.165 the circular modal frequency expressions are obtained from

$$\omega_{nx} = 2\sqrt{\frac{K_{sx}}{m}}\sin\left[\frac{(2\mathrm{n} - 1)\pi}{2(2N + 1)}\right] \qquad (12.167)$$

for the swaying motions of the tower. Note that the subscript n identifies the mode, whereas n is used in Figure 12.35 to identify a panel.

EXAMPLE 12.10: FIRST MODAL SWAYING FREQUENCY OF A FOT A FOT is to be placed in 800 m of water. The cross-section of the steel tower is a 75 m × 75 m

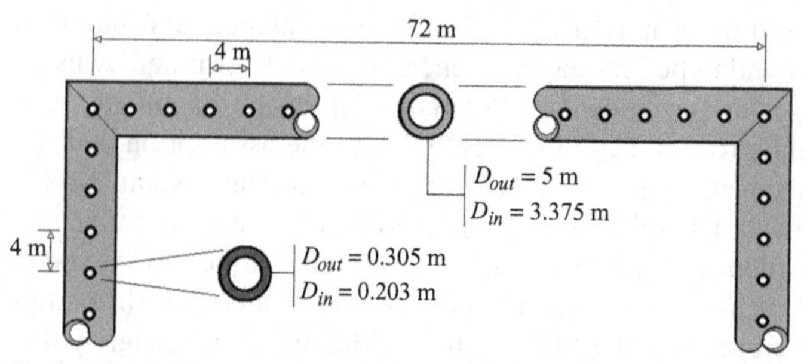

Figure 12.36. *Sketch of the Cross-Section of the FOT in Example 12.10.* There are 72 legs and 10 panels. The tower weight is 50,000 tonnes.

square, and is uniform from the base to the platform. See the cross-sectional layout in Figure 12.36. The platform is supported by 72 legs having an outside diameter of 0.304 m, an inside diameter of 0.203 m, and a length of 75 m. Hence, between the SWL and the base there are ($N=$) 40 panels. Between the four corner legs, there are 17 evenly distributed legs. The effective length of the segment leg is 15 m. The corner legs are horizontally separated along a cross member by 72 m. Hence, the centerlines of the intermediate legs are separated by 4 m. The tower mass is 80,000 metric tons and is evenly distributed over 40 panels, where the top panel is partially dry and supports the platform. The horizontal cross members have an outside diameter of 5 m.

The effective mass of each panel is 2.00×10^6 kg, and the added mass, assuming a Lewis-form inertial coefficient of unity ($C_i = 1$ in eq. 9.44) for both the legs and the cross members, is approximately 3.11×10^6 kg. The added mass includes all of the legs, and two of the four cross members at each node. Hence, the effective mass for a panel is $m \simeq 5.11 \times 10^6$ kg. For the steel legs, $E = 2.00 \times 10^{11}$ N/m^2 from Table 12.3, and the second moment of area for the leg cross-section is $\mathrm{I}_y = 3.31 \times 10^{-4}$ m^4 from Figure 12.32b. With these values, the spring constant for the legs is

$$K_{sxn} = 72\frac{12E\mathrm{I}_y}{L^3} \simeq 1.70 \times 10^7 \ \frac{\mathrm{N}}{\mathrm{m}}$$

from eq. 12.154. The first-modal swaying circular frequency from eq. 12.167 is $\omega_1 \simeq 0.136$ rad/s. The corresponding swaying period is then $T_1 = \omega_1/2\pi \simeq$ 46.2 sec. This period is well away from the high-energy spectra of wind-generated seas.

The displacement of the platform atop the 40-panel FOT in Example 12.10 requires the solution of a 40×40 matrix equation. Let us modify the design of the structure in the example so that only three springs are used for the compliance. This is done in the following example.

EXAMPLE 12.11: FIRST MODAL SWAYING FREQUENCY OF A THREE-PANEL FOT The tower in Example 12.10 is modified such that the base of the tower is cross-braced using relatively stiff, small-diameter diagonals to a height of 35 m above the mudline. Referring to the sketch in Figure 12.37, at $Z = 35$ m the first set of 72 compliant legs are placed, each of which is 15 m between clamped connections to the cross members. The steel legs have an outside diameter of 0.305 m and an inside diameter of 0.203 m, as in Example 12.10. The cross-sectional layout is similar to that in Figure 12.36. Above the braced foundation, the weight

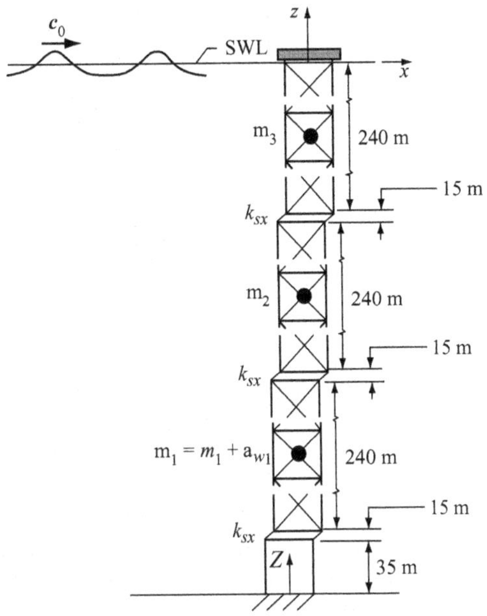

Figure 12.37. *Sketch of the Shear Deflection of the Three-Panel FOT in Example 12.11.* As in Example 12.10, there are 72 compliant legs connecting to each 235-m-high panel.

of the structure is 75,000 metric tons, and is evenly distributed to three sections. The Lewis-form added-mass coefficient for the legs and wave-facing horizontal members is $C_i = 1$. The added mass of each section is primarily due to the two 5-m (outside diameter) horizontal members in the y-direction. There are twelve structural components per panel in Figure 12.37. For a nodal section, the added mass is then

$$\mathrm{a}_{wn} = 12\left[2C_i\rho\frac{\pi}{4}D_{out}^2(\mathrm{L} - 2D_{cross}) + 72C_i\rho\frac{\pi}{4}D_{\mathrm{leg}}^2 L_{\mathrm{leg}}\right] \qquad (12.168)$$

The numerical value obtained from this equation is 4.20×10^7 kg, or 42,000 tonnes. As sketched in Figure 12.30, L is the width of the square cross-section. The mass (m_n) of a panel is 25,000 tonnes. The total effect mass of the panel is then $\mathrm{m_n} = 6.70 \times 10^7$ kg. From Example 12.10, the spring constant for the swaying motions is $K_{sxn} = 1.7 \times 10^7$ N/m. The total panel mass and spring constant values in eq. 12.167 result in a first-modal natural frequency of $\omega_1 \simeq 0.224$ rad/s. The corresponding swaying period is then $T_1 = \omega_1/2\pi \simeq 28.0$ sec. In eq. 12.167, the number of panels is $N = 3$, as sketched in Figure 12.37. A comparison of the first modal period with that in Example 12.10 shows that the period has been significantly reduced by reducing the nodes from forty to three by rigidly bracing the structure.

(2) Bending Motions of a TRAP

In this section, we analyze the motions of the flexible tower called the *tension-restrained articulated platform* (TRAP) that was proposed by the offshore industry for operation in deep water, according to Sellers and Niedzwecki (1992). By "deep," it is meant that the water depth (h) is of the order of 1,000 m. According to Sellers and Niedzwecki (1992), the TRAP "combines the proven technology of steel jacket construction with an interior tensioned cable design which allows articulation at several different elevations." Those authors present a rather complete motion analysis of the structure. They derive the equations of motion of the structure using

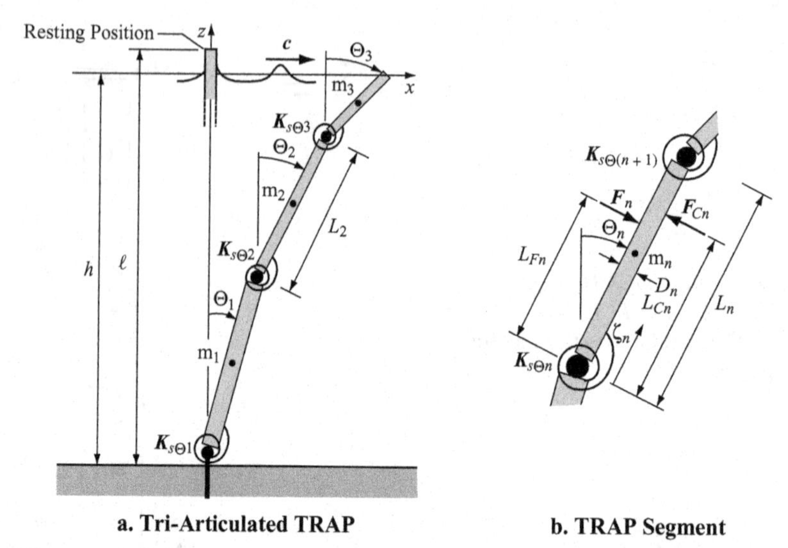

Figure 12.38. *Sketch of an Articulated Tension-Restrained Articulated Platform (TRAP).* Each segment of the tri-articulated TRAP is tubular, having a cross-section such as that sketched in Figure 12.32b. The wave-induced force, F_n, and the hydrodynamic damping force, F_{Cn}, are results of the integrations of pressures over the segment length, L_n.

the Lagrange equations. The TRAP is essentially a tower having multiple segments connected by rotational springs at the points of articulation, where the segments are connected by pins in the two-dimensional analysis. For a three-dimensional analysis, the pins would be replaced by ball joints or swivels. In the Sellers-Niedzwecki analysis, the springs represent any type of restoring mechanism, including buoyancy. The version of the TRAP that is studied here is sketched in Figure 12.38. This is a simplified version of that used in the Sellers-Niedzwecki analysis. Sellers and Niedzwecki (1992) use a "tri-articulated design" for the purpose of demonstration. Their segments are not assumed to be uniform in the axial direction because "the uneven distribution of equipment, buoyancy tanks, deck loads, ballast, *etc.* may be incorporated into the model." The Sellers-Niedzwecki analysis does not include damping. Linear damping is included in the derivation that follows but not in the application, the reason being that the inclusion of damping in the TRAP motions, or for that matter, the motions of any multi-articulated tower, requires extensive matrix manipulations. If damping would be included, the complexity of the matrix algebra would be a distraction.

The model sketched in Figure 12.38 is assumed to consist of uniform segments, where each segment is a capped, tubular structure having a cross-section, such as that sketched in Figure 12.32b. Hence, the centers of gravity and the centers of buoyancy are at the same locations on the centerlines of each segment. The bending matrix equation of motion for the TRAP is

$$[I_y]\left\{\frac{d^2\Theta}{dt^2}\right\} + [C_\Theta]\left\{\frac{d\Theta}{dt}\right\} + [K_{s\Theta}]\{\Theta\} = \{M\} \tag{12.169}$$

In this equation, the $[I_y]$ is the inertial matrix, $[C_\Theta]$ is the damping matrix, $[K_{s\Theta}]$ is the restoring matrix (which includes the buoyancy terms), and {M} is the wave-induced moment matrix. Except for the damping matrix, the derivation of the terms comprising the inertial, restoring, and moment matrices are derived by Sellers and

Niedzwecki (1992) in detail. The respective matrices applied to the TRAP configuration in Figure 12.38 are defined as follows:

Inertial Matrix: From the Sellers-Niedzwecki analysis, the expression for the inertial matrix is

$$[I_y] = \begin{bmatrix} I_1 + \mathrm{m}_1\frac{L_1^2}{4} + \mathrm{m}_2 L_1^2 + \mathrm{m}_3 L_1^2 & \mathrm{m}_2 L_1\frac{L_2}{2} + \mathrm{m}_3 L_1 L_2 & \mathrm{m}_3 L_1\frac{L_3}{2} \\ \mathrm{m}_2 L_1\frac{L_2}{2} + \mathrm{m}_3 L_1 L_2 & I_2 + \mathrm{m}_2 + \frac{L_2^2}{4} + \mathrm{m}_3 L_2^2 & \mathrm{m}_3 L_2\frac{L_3}{2} \\ \mathrm{m}_3 L_1\frac{L_3}{2} & \mathrm{m}_3 L_2\frac{L_3}{2} & I_3 + \mathrm{m}_3\frac{L_3^2}{4} \end{bmatrix} \tag{12.170a}$$

Later in this section, it will be more convenient to write the inertial matrix as

$$[I_y] = \begin{bmatrix} I_{11} & I_{12} & I_{13} \\ I_{12} & I_{22} & I_{23} \\ I_{13} & I_{23} & I_{33} \end{bmatrix} \tag{12.170b}$$

In eq. 12.170a, m_n is the structural mass plus the added mass of the segment. Also in eq. 12.170a, I_n is the sum of the respective structural mass moment of inertia (with respect to the horizontal line parallel to the y-axis through the n-pin) and the added-mass moment of inertia. Following Sellers and Niedzwecki (1992), assume that the inertial coefficient in eq. 9.26 has a value of 1. Hence, the respective effective mass for the circular cylindrical segment is

$$\mathrm{m}_n = m_n + \mathrm{a}_{wn} = \rho_s\pi\frac{D_n^2}{4}L_n + C_{in}\rho\pi\frac{D_n^2}{4}L_n \simeq \pi\frac{D_n^2}{4}L_n(\rho_s+\rho) = \rho\pi\frac{D_n^2}{4}L_n\left(\frac{\rho_s}{\rho}+1\right) \tag{12.171}$$

Here, ρ is the mass density of the ambient water, ρ_S is the mass density of the structural material, and C_{in} is the inertial coefficient. For a circular cylinder, $C_{in} = 1$, as in Section 9.2A(1). The mass moment of inertia with respect to the rotational axis through a pin is

$$I_n = I_{yn} + \mathrm{A}_{yn} \simeq \rho\pi\frac{D_n^2}{48}L_n^3(\rho_s+\rho) = \rho\pi\frac{D_n^2}{48}L_n^3\left(\frac{\rho_s}{\rho}+1\right) \tag{12.172}$$

In this expression, I_{yn} is the mass moment of inertia of the segment, and A_{yn} is the added-mass moment of inertia about the pin n-axis. The expression is based on the assumption that $D_n \ll L_n$. The mass-density ratio (ρ_S/ρ) can be considered to be a measure of the buoyancy, as is done by Sellers and Niedzwecki (1992). It should be noted here that the ratio of the freeboard $(\ell - h)$ to the water depth (h) is relatively small in practical applications; hence, we assume that eqs. 12.171 and 12.172 apply to all of the segments, as written. There will be some error introduced in the overestimation of the added-mass terms for the top segment.

Damping Matrix: Here, we assume that the damping is linear, and the linear damping coefficient for the segment n is $C_{\theta n}$. Because the damping is not included in the paper of Sellers and Niedzwecki (1992), the damping terms are derived here. The nature of the damping is discussed later in this section. Referring to Figure 12.38b for notation, the generalized moments about the pins due to the damping forces in

Lagrange's equation are

$$\begin{aligned} M_{C1} &= \{F_{C1}L_{C1} + F_{C2}L_{C2} + F_{C3}L_{C3} + F_{C2}L_1\cos(\Theta_2 - \Theta_1) \\ &\quad + F_{C3}L_1\cos(\Theta_3 - \Theta_1) + F_{C3}L_2\cos(\Theta_3 - \Theta_2)\} \\ M_{C2} &= \{F_{C2}L_{C2} + F_{C3}L_{C3} + F_{C3}L_2\cos(\Theta_3 - \Theta_2)\} \\ M_{C3} &= F_{C3}L_{C3} \end{aligned} \tag{12.173}$$

We note that the positive-moment direction is clockwise because the positive-angle direction is the same. In this equation, F_{Cn} is the net damping force on a segment acting at a point L_{Cn} above pin n. As is normally done, the angles are assumed to be small enough to allow the approximation $\cos(\Theta_n - \Theta_{n1}) \simeq 1$. The negative signs appear in eq. 12.173 because the positive moment direction is counterclockwise. The linearized damping forces are obtained from

$$F_{Cn} = C_{\Theta n}\int_0^{L_n} v_n(\zeta_n)d\zeta_n = C_{\Theta n}\int_0^{L_n} \zeta_n d\zeta_n \frac{d\Theta_n}{dt} = C_{\Theta n}\frac{L_n^2}{2}\frac{d\Theta_n}{dt} \tag{12.174}$$

where $C_{\Theta n}$ is the linearized damping coefficient. The coordinate, ζ_n, has its origin at the n-pin connecting the segments in Figure 12.38b. The moment arms of the damping forces about the pins connecting the segments are obtained from

$$L_{Cn} = \frac{|M_{Cn}|}{F_{Cn}} = \frac{\int_0^{L_n} \zeta^2 d\zeta}{\int_0^{L_n} \zeta\, d\zeta} = \frac{2}{3}L_n \tag{12.175}$$

The combination of eqs. 12.173 through 12.175 results in the damping matrix in eq. 12.169,

$$[C_\Theta] = \begin{bmatrix} C_{\Theta 1}\frac{L_1^3}{3} & C_{\Theta 2}\frac{L_2^2}{2}\left(L_1 + \frac{2}{3}L_2\right) & C_{\Theta 3}\frac{L_3^2}{2}\left(L_1 + L_2 + \frac{2}{3}L_3\right) \\ 0 & C_{\Theta 2}\frac{L_2^3}{3} & C_{\Theta 3}\frac{L_3^2}{2}\left(L_2 + \frac{2}{3}L_3\right) \\ 0 & 0 & C_{\Theta 3}\frac{L_3^3}{3} \end{bmatrix} \tag{12.176}$$

There are two sources of the linear damping. These are equivalent linear viscous damping, discussed in Section 10.1C, and radiation damping, discussed in Section 12.4D. The coefficients representing the former depend on the effective angular amplitude at a position on the segment. The radiation-damping coefficients depend on the dimensions of the cylinder, the wave number (and circular wave frequency), and the water depth. For both types of damping, finite-element analysis (FEA) is best suited to the analysis. This method is beyond the scope of this book. Hence, as is done by Sellers and Niedzwecki (1992), the damping is neglected.

Restoring Matrix: As previously written, the restoring matrix represents both the rotational springs and the segment buoyancy. The respective notation for these are $K_{s\Theta n}$ and Δ_n. The latter equals $\rho g \vee_n$, where $\vee_n$ is the displaced volume of segment n. For the circular cylindrical segments in Figure 12.38, $\Delta_n = \rho g \pi L_n D_n^2/4$. The force acts through the center of the displaced water mass, $L_n/2$ above pin n, as does the weight of the segment. Hence, the net buoyancy acting on segment n is

$$F_{Bn} = \Delta_n - W_n \tag{12.177}$$

With this notation, the restoring matrix for the TRAP in Figure 12.38 is

$$[K_{s\Theta}] = \begin{bmatrix} \left[F_{B1}\dfrac{L_1}{2} + (F_{B2} + F_{B3})L_2 \right. & -K_{s\Theta2} & 0 \\ \left. +K_{s\Theta1} + K_{s\Theta2}\right] & & \\ -K_{s\Theta2} & \left[F_{B2}\dfrac{L_2}{2} + F_{B3}L_2 \right. & -K_{s\Theta3} \\ & \left. +K_{s\Theta2} + K_{s\Theta3}\right] & \\ 0 & -K_{s\Theta3} & \left[F_{B3}\dfrac{L_3}{2} + K_{s\Theta3}\right] \end{bmatrix} \tag{12.178a}$$

Similar to the inertial matrix, it is advantageous to write the restoring matrix as

$$[K_{s\Theta}] = \begin{bmatrix} K_{11} & -K_{12} & 0 \\ -K_{12} & K_{22} & -K_{23} \\ 0 & -K_{23} & K_{33} \end{bmatrix} \tag{12.178b}$$

Wave-Induced Moment Matrix: For the TRAP sketched in Figure 12.38a, the wave-induced moment matrix is

$$\{M\} = \begin{Bmatrix} F_{W1}L_{W1} + F_{W2}L_1 + F_{W3}L_1 \\ F_{W2}L_{W2} + F_{W3}L_2 \\ F_{W3}L_{W3} \end{Bmatrix} = \begin{Bmatrix} M_{W1} \\ M_{W2} \\ M_{W3} \end{Bmatrix} \tag{12.179}$$

We must now determine the nature of the wave force (F_{Wn}) on segment n. This force acts at a distance L_{Wn} above pin n. To determine the force, we rely on the Chakrabarti diagram in Figure 9.8. As are other compliant towers, the TRAP is designed to resonate with long-period waves. Hence, in Figure 9.8 the assumption that $kD_n/2$ is small is valid. For an operational condition, $kD_n/2$ is of the order of magnitude of 1. The Keulegan-Carpenter number for deep-water waves is $KC \simeq \pi H_0/D_n$. The outer diameter of the structure in operational conditions would normally be much greater than the height of the deep-water wave. For these conditions, we see that the wave-induced force can be either inertial or diffractive in nature, or a combination of the two. In this derivation, we assume that the force is inertial for the purpose of demonstration.

The inertial force per unit length, as in the Morison equation, eq. 9.49, is

$$F'_{Wn} = C_{in}\rho\frac{\partial V_n(\zeta_n, t)}{\partial t}\forall'_n < TB: vspacespace = "2pt"/ > \tag{12.180}$$

Here, $V_n(\zeta_n,t)$ is the water-particle velocity normal to the segment at a distance ζ_n above the pin, C_{in} is the inertial coefficient, and $\forall' = \rho g\pi D_n^2/4$ (the volume per unit length) for the segment volume (see Figure 12.38). For small angular motions, as is assumed, $V_n(\zeta_n, t) \simeq u(\zeta_n, t)$. That is, the normal velocity is approximately equal to the horizontal particle velocity at $z = -z_n + \zeta_n$, where $z_n \geq 0$ is the depth of pin n, and $0 \leq \zeta_n \leq L_n$, We can then approximate the wave-induced force in eq. 12.180 by

$$F'_{Wn} \equiv \frac{dF_{Wn}}{d\zeta_n} \simeq C_{in}\rho\left.\frac{\partial u(\zeta, t)}{\partial t}\right|_{\substack{x=0 \\ z=z_n+\zeta_n}} \pi\frac{D_n^2}{4} \simeq \frac{1}{8}C_{in}\rho\pi D_n^2\omega^2 H_0 e^{-k_0(z_n-\zeta_n)}e^{-i\omega t} \tag{12.181}$$

Here, the complex deep-water approximation for the horizontal particle velocity is used, that is, the cosine expression in eq. 3.49 is the real part of $e^{-i\omega t}$. The total wave force over the entire n-segment is

$$F_{Wn} \simeq C_{in}\rho g \pi D_n^2 H_0 e^{-k_0 z_n}[e^{k_0 L_n} - 1]e^{-i\omega t} = F_{Wno}e^{-i\omega t} \tag{12.182}$$

The center of the force on segment n is obtained from

$$L_{Wn} = \frac{\int_0^{L_n} \zeta_n e^{k_0\zeta_n} d\zeta_n}{\int_0^{L_n} e^{k_0\zeta_n} d\zeta_n} = \frac{1}{k_0}\left[\frac{e^{k_0 L_n}(k_0 L_n - 1) + 1}{e^{k_0 L_n} - 1}\right] \tag{12.183}$$

Angular Response Matrix: The response matrix is the following column matrix:

$$\{\Theta\} = \begin{Bmatrix} \Theta_1 \\ \Theta_2 \\ \Theta_3 \end{Bmatrix} = \begin{Bmatrix} \Theta_{1o} \\ \Theta_{2o} \\ \Theta_{3o} \end{Bmatrix} e^{-i\omega t} = \{\Theta_o\}e^{-i\omega t} \tag{12.184}$$

Hence, the respective angular velocity and acceleration matrices are $-i\omega\{\Theta_o\}$ and $-\omega^2\{\Theta_o\}$.

The undamped natural periods of the TRAP motions are found from the undamped, homogeneous form of eq. 12.169. Assuming an oscillatory motion, the homogeneous, undamped equation of motion for the TRAP is

$$[I_y]\left\{\frac{d^2\Theta}{dt^2}\right\} + [K_{s\Theta}]\{\Theta\} = -\omega_m^2[I_y]\{\Theta_o\} + [K_{s\Theta}]\{\Theta_o\} = \{0\} \tag{12.185}$$

The frequency ω_m is the modal frequency. A nontrivial solution of the equations comprising the matrix equation, eq. 12.185, is obtained only if the determinant of the coefficient matrices vanishes. That is,

$$|[K_{s\Theta}] - \omega_m^2[I_y]| = 0 \tag{12.186}$$

The expansion of this matrix equation results in a third-order algebraic equation in ω_m^2 called the *characteristic equation*. The three roots (ω_1^2, ω_2^2, and ω_3^2) of the equations are the *eigenvalues*.

For the undamped, monochromatic wave-induced motions of the three-segment TRAP, the equation of motion is obtained from eq. 12.169, where $[C_\Theta] = [0]$ and the forces in the moment matrix of eq. 12.179 are represented by those in eq. 12.182. The response matrix is that in eq. 12.184. The resulting equation is

$$\{[K_{s\Theta}] - \omega^2[I_y]\}\{\Theta_o\}e^{-i\omega t} = \{M_o\}e^{-i\omega t} = \begin{Bmatrix} M_{W1o} \\ M_{W2o} \\ M_{W3o} \end{Bmatrix} e^{-i\omega t} \tag{12.187a}$$

Here, $\omega = 2\pi/T$ is the circular wave frequency, where T is the wave period. This matrix equation represents the following system of equations:

$$\begin{aligned} &(K_{11} - \omega^2 I_{11})\Theta_{1o} - (K_{12} + \omega^2 I_{12})\Theta_{2o} - \omega^2 I_{13}\Theta_{3o} = M_{W1o} \\ &-(K_{12} + \omega^2 I_{12})\Theta_{1o} + (K_{22} - \omega^2 I_{22})\Theta_{2o} - (K_{23} + \omega^2 I_{23})\Theta_{3o} = M_{W2o} \\ &-\omega^2 I_{13}\Theta_{1o} - (K_{23} + \omega^2 I_{23})\Theta_{2o} + (K_{33} - \omega^2 I_{33})\Theta_{3o} = M_{W3o} \end{aligned} \tag{12.187b}$$

From this system of equations, we can solve simultaneously for the angular amplitudes (Θ_{1o}, Θ_{2o}, and Θ_{3o}). From these amplitude values, the general modal shapes

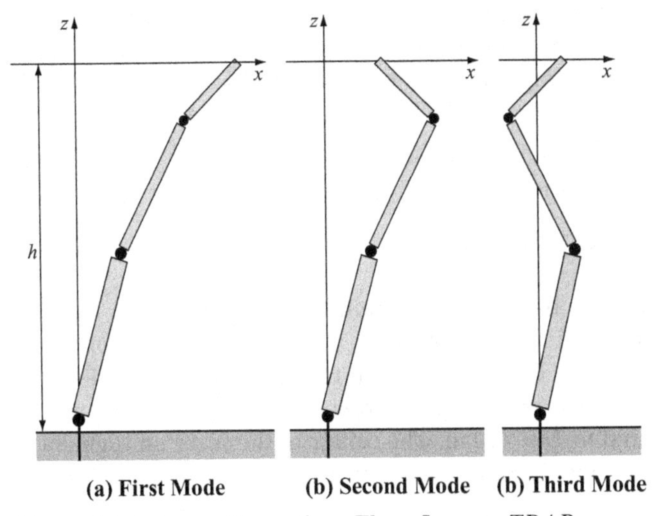

Figure 12.39. *Modal Shapes for a Three-Segment TRAP.*

resemble those sketched in Figure 12.39. Results of a parametric study of the three-segment TRAP by Sellers and Niedzwecki (1992) are presented in Figure 12.40. In the following example, some parametric values studied by those authors are used in eq. 12.187, and the resulting angular deflections of the three-segment TRAP are obtained.

EXAMPLE 12.12: DEFLECTION OF A TRAP IN A DESIGN SEA A three-segment TRAP is to be deployed at a site where the averaged wave height at the site is $H_{avg} = 1.00$ m. The design wave is one having an expected maximum height corresponding to a 100-year storm. Assuming a Gaussian-Rayleigh sea, as discussed in Chapter 5, the highest expected wave height at the site over 100 years is

$$H_{\max} = H_{rms}\sqrt{\ln(N_M)} = \frac{2}{\sqrt{\pi}} H_{avg}\sqrt{\ln(N_M)}$$
$$= \frac{2}{\sqrt{\pi}}(1)\sqrt{\ln(6 \times 10^6 \times 100)} \simeq 5.07 \text{ m}$$

from eq. 5.25. At the site, the average wave period at the site is $T_{avg} \simeq 8.00$ sec. From eq. 5.51, the root-mean-square wave period is approximately 8.31 sec. The average wavelength is obtained from eq. 5.56. For the 8.00-sec average wave period, the average wavelength at the deep-water site is

$$\lambda_{0avg} = \frac{g}{2\pi} T_{avg}^2 \simeq 99.9 \text{ m} \tag{12.188}$$

The wave number corresponding to this value is $k_{0avg} = 2\pi/\lambda_{0avg} \simeq 0.0629\,\text{m}^{-1}$. For the average wave period, only the top segment ($n = 3$) will be subject to the design wave because $L_3 > \lambda_{0avg}/2$. For the design wave, a three-segment TRAP is to be deployed, where the parametric values are those in Figure 12.40a. Referring to Figure 12.38, the height (ℓ) of the TRAP above the sea bed is 330 m, and each segment length (L_n) is 110 m. The three segments are identical. Hence, we can replace L_n by L and D_n by D in the analysis. Each segment has a circular cross-section with an outside diameter of 10 m. For this diameter, $k_{0avg}D/2 \simeq 0.314$. The deep-water Keulegan-Carpenter number from eq. 9.46 is

$KC_n \simeq \pi H_{max}/D \simeq 1.59$. From the Chakrabarti diagram in Figure 9.8, we see that the wave-induced forces are primarily due to both inertia and drag. Because less than 10% of the force is drag according to Figure 9.8, we shall assume that the inertial force is dominant. The inertial force on the top segment is

$$F_{W3} \simeq \frac{1}{8} C_{in} \rho g \pi D^2 H_{max} e^{-k_{0avg} z_3} [e^{k_{0avg} z_3} - 1] e^{-i\omega t}$$
$$= F_{W3o} e^{-i\omega t} \simeq 2.00 \times 10^6_{e^{-i0.758t}}, \text{ Newtons} \tag{12.189}$$

Here, the inertial coefficient is $C_{in} = 1$, assuming a Lewis form [see Section 9.2A(1)]. Pin 3 in Figure 12.38 is at a depth of $z_3 = 85$ m. The force expression is slightly different than that in eq. 12.182, in that the integration of the force is over z_3 rather than L. The reason is that the freeboard ($L - z_3 = 25$ m) is not negligible when compared to $L = 110$ m. The center of the force on segment 3 is obtained from a slight modification of the expression in eq. 12.190. That is, because of the significant freeboard, the location of the center of the force is

$$L_{W3} = \frac{\int_0^{z_3} \zeta_n e^{k_{0avg} \zeta_3} d\zeta_3}{\int_0^{z_3} e^{k_{0avg} \zeta_3} d\zeta_3} = \frac{1}{k_{0avg}} \left[\frac{e^{k_{0avg} z_3}(k_0 z_3 - 1) + 1}{e^{k_{0avg} z_3} - 1} \right] \simeq 69.5 \text{ m} \tag{12.190}$$

above pin 3. The wave-induced moment about pin 3 is then

$$M_{W3} = F_{W3} L_{W3} = M_{W3o} e^{-i0.786t} \simeq 1.39 \times 10^8_{e^{-i0.786t}} \text{ N-m} \tag{12.191}$$

Let the mass-density ratio (ρ_S/ρ) value be 0.25, as in Figure 12.40a. Hence, the TRAP is positively buoyant. Because the three segments of the TRAP are assumed to be identical, a segment's total mass and moment-of-inertia terms in eqs. 12.171 and 12.172 are, respectively,

$$\mathrm{m}_n \equiv \mathrm{m} = \rho \pi \frac{D^2}{4} L \left(\frac{\rho_S}{\rho} + 1 \right) \simeq 8.90 \times 10^6 \left(\frac{\rho_S}{\rho} + 1 \right) = 1.11 \times 10^7 \text{ kg} \tag{12.192}$$

and

$$I_n \equiv I = I_y + A_y = \rho \pi \frac{D^2}{48} L^3 \left(\frac{\rho_S}{\rho} + 1 \right) = \mathrm{m} \frac{L^2}{12} \simeq 1.12 \times 10^{10} \frac{\text{N-m-s}}{\text{rad}} \tag{12.193}$$

where $D = 10$ m, $L = 110$ m, and $\rho_S/\rho = 0.25$. The inertial matrix in eq. 12.170a is then

$$[I_y] = 2\left(\mathrm{m}\frac{L^2}{12}\right) \begin{bmatrix} 14 & 9 & 3 \\ 9 & 8 & 3 \\ 3 & 3 & 2 \end{bmatrix}$$
$$\simeq 2.24 \times 10^{10} \begin{bmatrix} 14 & 9 & 3 \\ 9 & 8 & 3 \\ 3 & 3 & 2 \end{bmatrix} \frac{\text{N-m-s}}{\text{rad}} \tag{12.194}$$

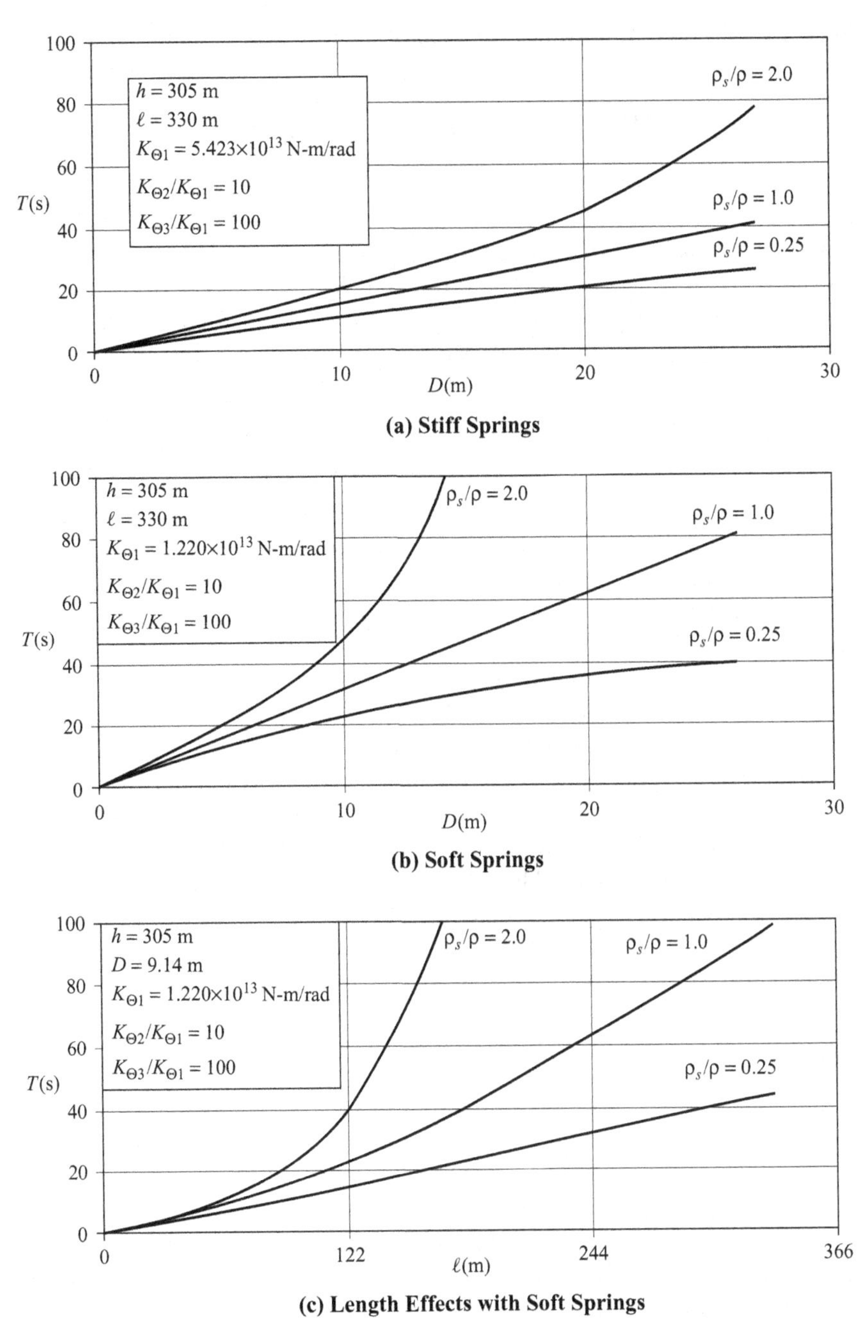

Figure 12.40. *Period Predictions for Parametric Variations of a Three-Segment TRAP.* These results are part of those presented by Sellers and Niedzwecki (1992). The material mass density is denoted by ρ_S, and that of the salt water is denoted by ρ. The value of the latter is approximately 1,030 kg/m^3. The density ratios show the effect of the net buoyancy, where the respective values of 2.0, 1.0, and 0.25 correspond to structures that are negatively buoyant, neutrally buoyant, and positively buoyant.

From eq. 12.177, the net buoyancy for each of the identical segments is

$$F_B = \Delta - W = \rho g \pi \frac{D^2}{4} L - \rho_s g \pi \frac{D^2}{4} L = mg \frac{\left(1 - \frac{\rho_s}{\rho}\right)}{\left(1 + \frac{\rho_s}{\rho}\right)} \simeq 6.55 \times 10^7 \text{ N} \tag{12.195}$$

In Figure 12.40a, we see that the spring constant values are $K_{s\Theta 1} = 5.423 \times 10^{13}$ N-m/rad, $K_{s\Theta 2} = 5.423 \times 10^{12}$ N-m/rad, and $K_{s\Theta 3} = 5.423 \times 10^{11}$ N-m/rad. The substitution of these values and those of the buoyant force of eq. 12.194 results in the following restoring matrix for the three-segment TRAP:

$$[K_{s\Theta}] \begin{bmatrix} 5.97 \times 10^{13} & -5.42 \times 10^{12} & 0.00 \\ -5.42 \times 10^{12} & 5.98 \times 10^{12} & -5.423 \times 10^{11} \\ 0.00 & -5.423 \times 10^{11} & 5.46 \times 10^{11} \end{bmatrix} \frac{\text{N–m}}{\text{rad}}$$

The angular amplitudes can now be determined because all of the coefficients are defined in eq. 12.187b. From the solution of the simultaneous equations, we obtained $\Theta_1 \simeq 2.54 \times 10^{-6}$ rad, $\Theta_2 \simeq 2.79 \times 10^{-5}$ rad, and $\Theta_3 \simeq 2.82 \times 10^{-4}$ rad. If the spring constants are reduced by two orders of magnitude to $K_{s\Theta 1} = 5.423 \times 10^{11}$ N-m/rad, $K_{s\Theta 2} = 5.423 \times 10^{10}$ N-m/rad, and $K_{s\Theta 3} = 5.423 \times 10^{9}$ N-m/rad, then the amplitudes increase to $\Theta_1 \simeq 1.18 \times 10^{-4}$ rad, $\Theta_2 \simeq 1.34 \times 10^{-3}$ rad, and $\Theta_3 \simeq 1.62 \times 10^{-2}$ rad. The latter value is about 1°. It is of interest to see what the linear deflection of the platform is for this soft-spring condition. For pins 2 and 3, the rectilinear amplitudes in the x-direction are $X_2 \simeq L\Theta_1 \simeq 0.0130$ m and $X_3 \simeq L\Theta_2 \simeq 0.147$ m. The platform amplitude is $X_p \simeq L\Theta_3 \simeq 1.780$ m. The total deflection of the platform is then the sum of the three, or about 1.940 m. The total horizontal excursion of the platform over one design period (8 sec) is 3.88 m. Hence, the average horizontal velocity of the platform is 0.485 m/s $\simeq$ 0.5 m/s. This might make some workers on the platform a little uncomfortable.

12.7 Closing Remarks

As in the previous chapters, we have attempted to present the analytical "tools" that are needed to deal with problems involved with both moorings and compliant structures. In some cases, the examples might appear to be a little unrealistic to the reader. This is probably due to the simplicity of the examples. The author believes that simplicity is required for basic learning. The readers that have a firm grasp of the materials presented in this section will be able to apply those materials to far more complex situations.

Appendices

A. Bessel Functions

The Bessel functions are solutions to a second-order differential equation called *Bessel's equation*. That equation is

$$x^2\frac{d^2y}{dx^2} + x\frac{dy}{dx} + (x^2 - v^2)y = 0 \tag{A1}$$

where v is real constant. This equation is often encountered in the analyses of physical phenomena involving waves. The equation is named after Friedrich Bessel, a mathematician whose work was published in the nineteenth century. The general solution of eq. A1 is found by assuming a summation solution of the form

$$y = \sum_{k=0}^{\infty} C_j x^{k+r} \tag{A2}$$

where $r = \pm v$. If v is a positive integer, n, then the solution of eq. A1 is the nth-order *Bessel function of the first kind*, which is

$$J_n(x) = \sum_{k=0}^{\infty} \frac{(-1)^k}{k!(n+k)!}\left(\frac{x}{2}\right)^{n+2k} \tag{A3}$$

For our purposes, the integer order (n) is satisfactory. A second solution of eq. A1 is called the *Bessel function of the second kind*, and is defined as

$$Y_n(x) = \frac{J_n(x)\cos(nx) - J_{-1}(x)}{\sin(nx)} \tag{A4}$$

A third type of solution of eq. A1 results from linear combinations of the first two. This type of solution is called the *Hankel function*, after Hermann Hankel, a nineteenth century German mathematician. There are actually two Hankel functions, those being the Hankel functions of the first and second kinds, which are, respectively,

$$H_n^{(1)}(x) = J_n(x) + iY_n(x) \tag{A5}$$

and

$$H_n^{(2)}(x) = J_n(x) - iY_n(x) \tag{A6}$$

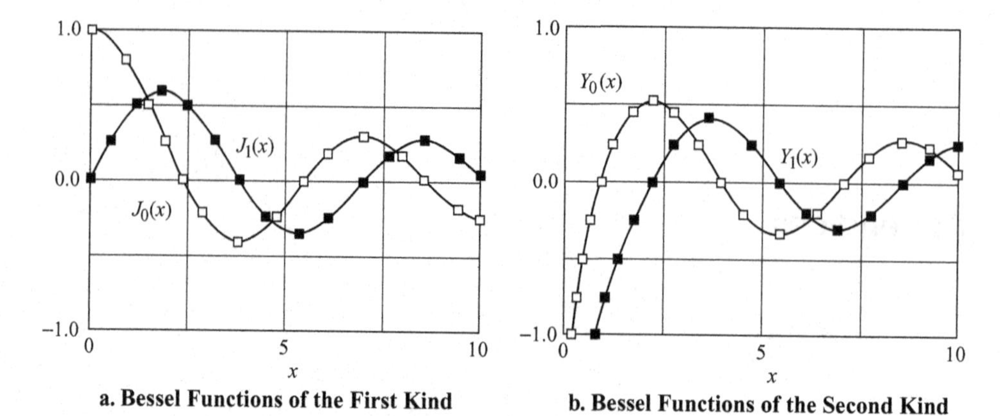

Figure A1. *Bessel Functions of the First and Second Kinds, Zero and First Orders.*

Of particular interest are the zeroth-order ($n = 0$) and first-order ($n = 1$) functions. The behaviors of these functions can be seen in Figure A1.

As is shown by Abramowitz and Stegun (1965), a Bessel function of these orders can be well approximated by polynomials over specific ranges of x. Subroutines for the Bessel function of the first and second kinds of any integer order are included in commercially available spreadsheets. Results obtained by a spreadsheet are presented in Figure A1, where the Bessel functions of the first and second kinds of both zero and first order are presented for argument values up to 10.

B. Runga-Kutta Solution of Differential Equations

Occasionally, we encounter a differential equation that cannot be solved using the standard analytical solution methods. For these equations, there exists an excellent numerical procedure is that called the *Runga-Kutta method.* The name of the method comes from the co-founders of the method, Carl Rung and Wihelm Kutta, who were mathematicians of the nineteenth and twentieth centuries. The method is based on a differential equation of the first order. As is shown here, it can also be applied to differential equations of higher order. The derivation of the Runga-Kutta method can be found in Section 4.11 of the book by Adey and Brebbia (1983).

Consider the first-order differential equation,

$$\frac{dy}{dx} = f(x, y) \tag{B1}$$

This function $y = g(x)$ is associated with the curve in Figure B1. This is an initial-value problem, where the y-value at x_0 in the figure is known, that is, $y(x_0) = y_0$. The discrete y-values are separated by $\Delta x = \mathrm{h}$, which is a constant determined by the analyst. From the figure, we see that $x_N = N\mathrm{h}$. The fourth-order Runga-Kutta formula for the solution of eq. B1 is

$$y_{\mathrm{n}+1} = y_{\mathrm{n}} + \frac{\mathrm{h}}{6}(\mathrm{k}_a + 2\mathrm{k}_b + 2\mathrm{k}_c + \mathrm{k}_d) + \mathrm{O}[\mathrm{h}^5] \tag{B2}$$

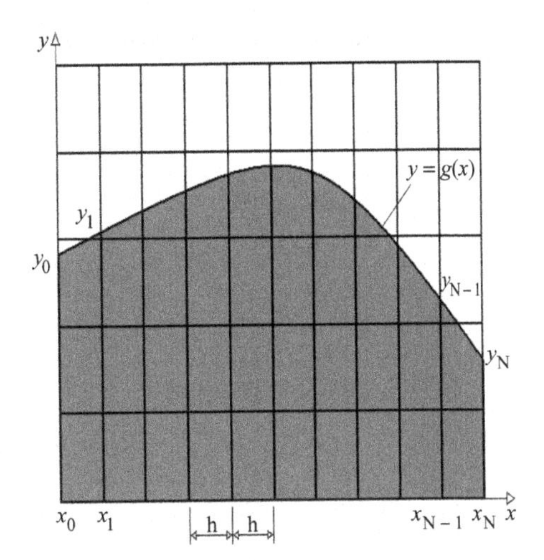

Figure B1. *Discretized Arbitrary Function Illustrating Runga-Kutta Method.*

One can see that the accuracy of the equation increases as h decreases. In this equation are

$$
\begin{aligned}
k_a &= f(x_n, y_n) \\
k_b &= f\left(x_n + \frac{h}{2}, y_n + h\frac{k_a}{2}\right) \\
k_c &= f\left(x_n + \frac{h}{2}, y_n + h\frac{k_b}{2}\right) \\
k_d &= f(x_n + h, y_n + hk_c)
\end{aligned}
\tag{B3}
$$

where the function, $f(x,y)$, is known from eq. B1.

As shown by Adey and Brebbia (1983), the fourth-order Runga-Kutta method can be applied to higher-order functions. Those authors demonstrate the method by applying the method to a second-order differential equation. This is also done here. Consider a second-order equation of the form

$$\frac{d^2y}{dx^2} + A(y)\frac{dy}{dx} + B(y)x = C(x) \tag{B4}$$

Here, the functions $A(y)$, $B(y)$, and $C(x)$ are defined. We can replace the second-order term by a first-order equivalent term as

$$\frac{d^2y}{dx^2} = \frac{d}{dy}\left(\frac{dy}{dx}\right) = \frac{d\xi}{dx} = f(x, y, \xi) \tag{B5}$$

The combination of eqs. B4 and B5 yields

$$\frac{d\xi}{dx} + A(y)\xi + B(y)x = C(x) \tag{B6}$$

This equation can also be written in the form of eq. B1, where

$$\frac{d\xi}{dx} = f(y, \xi, x) = C(x) - A(y)\xi - B(y)x \tag{B7}$$

The fourth-order Runga-Kutta solution of this equation is similar to that in eq. B2, that is,

$$\xi_{n+1} \equiv \frac{dy_{n+1}}{dx} = \xi_n + \frac{h}{6}(\kappa_a + 2\kappa_b + 2\kappa_c + \kappa_d) + O[h^5] \tag{B8}$$

As derived by Adey and Brebbia (1983), the relationships between the k and the κ are

$$\begin{aligned}
\mathrm{k}_a &= f(x_\mathrm{n}, y_\mathrm{n}, \xi_\mathrm{n}) \\
\mathrm{k}_b &= f\left(x_\mathrm{n} + \frac{\mathrm{h}}{2}, y_\mathrm{n} + \mathrm{h}\frac{\mathrm{k}_a}{2}, \xi_\mathrm{n} + \mathrm{h}\frac{\kappa_a}{2}\right) \\
\mathrm{k}_c &= f\left(x_\mathrm{n} + \frac{\mathrm{h}}{2}, y_\mathrm{n} + \mathrm{h}\frac{\mathrm{k}_b}{2}, \xi_\mathrm{n} + \mathrm{h}\frac{\kappa_b}{2}\right) \\
\mathrm{k}_d &= f(x_\mathrm{n} + \mathrm{h}, y_\mathrm{n} + \mathrm{h}\mathrm{k}_c, \xi_\mathrm{n} + \mathrm{h}\kappa_c)
\end{aligned} \tag{B9}$$

where the κ-terms are

$$\begin{aligned}
\kappa_a &= \mathrm{f}(x_\mathrm{n}, y_\mathrm{n}, \xi_\mathrm{n}) \\
\kappa_b &= \mathrm{f}\left(x_\mathrm{n} + \frac{\mathrm{h}}{2}, y_\mathrm{n} + \mathrm{h}\frac{\mathrm{k}_a}{2}, \xi_\mathrm{n} + \mathrm{h}\frac{\kappa_a}{2}\right) \\
\kappa_c &= \mathrm{f}\left(x_\mathrm{n} + \frac{\mathrm{h}}{2}, y_\mathrm{n} + \mathrm{h}\frac{\mathrm{k}_b}{2}, \xi_\mathrm{n} + \mathrm{h}\frac{\kappa_b}{2}\right) \\
\kappa_d &= \mathrm{f}(x_\mathrm{n} + \mathrm{h}, y_\mathrm{n} + \mathrm{h}\mathrm{k}_c, \xi_\mathrm{n} + \mathrm{h}\kappa_c)
\end{aligned} \tag{B10}$$

Here, the functional relationship $\mathrm{f}(x, y, \xi)$ is defined in eq. B7. The approximate solution of the second-order differential equation in eq. B4 is now complete.

C. Green's Theorem

George Green, who lived between 1793 and 1833, gave us several analytical tools that are most useful in water-wave mechanics and, more generally, in the general area of potential theory. Two of these tools of interest to us are *Green's theorem* and *Green's function*. Green's first writings were directed toward electricity and magnetism; however, later in his brief career, he wrote papers on water waves, acoustical waves, optics, and other topics in mechanics. His writings can be found in his collected works edited by Ferrers (1871). In the Preface, Ferrers writes that the most important of Green's papers was that delivered in 1828, "An Essay on the Mathematical Analysis of the Theories of Electricity and Magnetism," in which Green introduced the term *potential* to "denote the result obtained by adding together the masses of all the particles of a system, each divided by its distance from a given point," in the words of Ferrers. In this appendix, Green's theorem is discussed, and in Appendix D, Green's function is introduced.

C1. Three-Dimensional Green's Theorem

Green's theorem, the derivation of which can be found in books covering topics of vector analysis, relates surface integrals and line integrals. To derive this theorem, we begin with the *divergence theorem of Gauss*. The best way to introduce this theorem is to consider a flow issuing from a volume $\vee$ defined by the surface area S with a velocity vector

$$\boldsymbol{V} = u\boldsymbol{i} + v\boldsymbol{j} + w\boldsymbol{k} \tag{C1}$$

where u,v,w are the components in the respective x,y,z coordinate directions, and $\boldsymbol{i,j,k}$ are the associated unit vectors. By paraphrasing Karamcheti (1966), the divergence theorem applied to fluid mechanics can be physically interpreted as follows: The outflow of a vector field ($\boldsymbol{V}$) through a closed surface (S) equals the volume integral of the divergence of the vector field over the region enclosed by S. Mathematically, we can write this as

$$\iiint_{\vee} \nabla \cdot \boldsymbol{V} d\vee = \iint_{S} \boldsymbol{V} \cdot \boldsymbol{n} dS \tag{C2}$$

Now, let us represent the velocity vector by two functions, F and G, that are continuous and can be twice-differentiated in space. We require that

$$\boldsymbol{V} = \nabla\Phi = G\nabla F = F\nabla G \tag{C3}$$

where ∇, the Cartesian operator *del*, is defined as

$$\nabla \equiv \frac{\partial}{\partial x}\boldsymbol{i} + \frac{\partial}{\partial y}\boldsymbol{j} + \frac{\partial}{\partial z}\boldsymbol{k} \tag{C4}$$

The combination of the expression in eq. C2 and first equality in eq. C3 yields

$$\iiint_{\vee} \{\nabla G \cdot \nabla F + G\nabla^2 F\} d\vee = \iint_{S} G\frac{\partial F}{\partial n} dS \tag{C5}$$

Here, $\boldsymbol{n}$ is the outward unit normal vector on the surface (S) of the volume ($\vee$). Similarly, the second equality of eq. C3 combined with eq. C2 yields

$$\iiint_{\vee} \{\nabla F \cdot \nabla G + F\nabla^2 G\} d\vee = \iint_{S} F\frac{\partial G}{\partial n} dS \tag{C6}$$

The right sides of eqs. C5 and C6 result from the following identities: $\nabla() \cdot d\mathbf{S} = \nabla() \cdot \boldsymbol{n} dS = [\partial()/\partial n] dS$. Subtract eq. C6 from eq. C5 to obtain

$$\iiint_{\vee} \{G\nabla^2 F - F\nabla^2 G\} d\vee = \iint_{S} \left\{G\frac{\partial F}{\partial n} - F\frac{\partial G}{\partial n}\right\} dS \tag{C7}$$

This is one form of Green's theorem, called *Green's theorem of the second form.* The theorem is used in diffraction theory and in other wave-structure interaction problems.

C2. Two-Dimensional Green's Theorem

In the analyses of the diffraction of water waves and the forces on two-dimensional bodies, it is advantageous to use the *two-dimensional Green's theorem.* In this case, the left side of eq. C7 becomes an integral over the surface area, S, and the right side becomes an integral over the boundary, s, of S. The resulting equation is

$$\iint_{S} \{G\nabla^2 F - F\nabla^2 G\} dS = \int_{s} \left\{G\frac{\partial F}{\partial n} - F\frac{\partial G}{\partial n}\right\} ds \tag{C8}$$

Here, s is the linear coordinate along the boundary line. See the applied mathematics book by Dettman (1962) or the low-speed aerodynamics book by Katz and Plotkin (2001) for the details leading to eq. C8.

C3. Green's Theorem Applied to an Irrotational Flow

By representing the velocity vector by the potential function, the flow is assumed to be irrotational. As introduced in eq. 2.38, the relationship between the velocity and the velocity potential is

$$V = G\nabla F = F\nabla G = \begin{cases} \nabla\varphi, & \text{two dimensions} \\ \nabla\Phi, & \text{three dimensions} \end{cases} \tag{C9}$$

Assume that the functions F, G, ϕ, and Φ are all continuously differentiable and harmonic everywhere. The latter assumption refers to their satisfying Laplace's equation, eq. 2.41. Because of the harmonic assumption, and because there are no singular points in the region of interest, eq. C7 becomes

$$\iint_s \left\{ G\frac{\partial F}{\partial n} - F\frac{\partial G}{\partial n} \right\} dS = 0 \tag{C10}$$

for a three-dimensional irrotational flow. For a two-dimensional irrotational flow, eq. C8 under the harmonic assumption is

$$\int_s \left\{ G\frac{\partial F}{\partial n} - F\frac{\partial G}{\partial n} \right\} ds = 0 \tag{C11}$$

D. Green's Function

George Green introduced the function bearing his last name in *An Essay on the Application of Mathematical Analysis to the Theories of Electricity and Magnetism*, written in 1828. This paper and others by Green are found in his mathematical papers, edited by Ferrers (1871). Green first applied his function to the conduction of electricity.

Green's function, called an *auxiliary function* or a *resolving kernel*, is a very useful tool used to solve certain types of equations. This method of solution is called the *integral equation method*. The reader is referred to the excellent books by Lindsay (1960) and Dettman (1962) for more thorough discussions of the Green's function and integral equations.

D1. Three-Dimensional Green's Function

(1) Three-Dimensional Flow Source

Consider the sketch in Figure D1, where an incompressible flow is issuing from the three-dimensional point source, P. In that figure, point P is located in the volume bounded by the surface area, S_1. The fluid travels to a point Q located on the interior of S_1. In Figure 2.18b, we find that the velocity potential for such a three-dimensional point source is

$$\Phi_{source} = -\frac{M}{R} = -MG_{source} \tag{D1}$$

where M is the strength of the source, and

$$G_{source} = \frac{1}{R} \tag{D2}$$

is called the *Green's function* for the point source. Note the differences in the radius notation here and in Figure 2.18. By comparing the expressions in eqs. D1 and D2, the reader can see that the Green's function represents a source of unit strength.

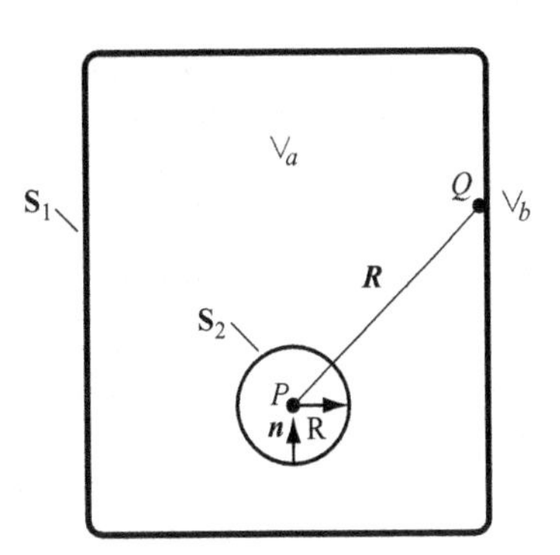

Figure D1. *Three-Dimensional Point Source within an Enclosed Volume.*

The form of the Green's function in eq. D2 is that first introduced by Green. This form satisfies the Laplace equation in the volume $\vee_a$ except at point P because a singularity exists at the point $R = 0$. That is, in $\vee_a$, Laplace's equation in spherical coordinates is

$$\nabla^2 G_{source}\Big|_{R>0} = \frac{1}{R^2}\left(R^2 \frac{dG_{source}}{dR}\right)\Big|_{R>0} = 0 \tag{D3}$$

and is identically satisfied by the expression in eq. D2. The Green's function in eq. D2 also satisfies Laplace's equation everywhere in volume $\vee_b$ because there are no singularities in the external volume.

Let us focus on the Green's function in eq. D2, and let $G \equiv G_{source}$ in eq. C3. Because G_{source} is the spatially variable part of a potential function and satisfies eq. D3, the function F in eq. C3 must also be a potential-type function. This function must also satisfy Laplace's equation, that is,

$$\nabla^2 F = 0 \tag{D4}$$

The reader can prove this by taking the divergence of the vector in eq. C3 and setting the resulting expression equal to zero, as that expression is the equation of continuity for an incompressible flow, eq. 2.21.

We obtain the solution of eq. D4 by using the integral equation approach. Begin by applying Green's formula in eq. D4. Again, because the origin of R is at the source point P (a singularity point), that point is excluded from the integrations of eq. C7 within volume $\vee_a$. To exclude the point P, let the point be encased in a small sphere having a radius R and a surface area S_2, as shown in Figure D1. The surface areas S_1 and S_2 then define the interior volume, $\vee_a$. We now can write eq. C7 in terms of F and G_{source} $(= 1/R)$ as

$$\iiint_{\vee_a}\left[\left(\frac{1}{R}\right)\nabla^2 F - F\nabla^2\left(\frac{1}{R}\right)\right]d\vee = \iint_{S_1}\left[\left(\frac{1}{R}\right)\frac{\partial F}{\partial R} - F\frac{\partial(1/R)}{\partial R}\right]dS$$
$$+\iint_{S_2}\left[\left(\frac{1}{\mathrm{R}}\right)\frac{\partial F}{\partial \mathrm{R}} - F\frac{\partial(1/\mathrm{R})}{\partial \mathrm{R}}\right]dS \tag{D5}$$

Because of the results in eqs. D3 and D7, the left side of eq. D5 must equal zero. Comparing the expressions in eqs. C7 and D5, we see that the derivative in the inward normal direction to the spherical surface, S_2, is $\partial(\,)/\partial n = -\partial(\,)/\partial R|_{R=\mathrm{R}}$. If Ω is a solid angle associated with $dS|_2$ of the surface of the sphere encasing P, then the differential surface element is $dS_2 = \mathrm{R}^2 d\Omega$. The requirements of the function F

are that both F and $\partial F/\partial R$ must be finite as $R \to 0$. Because of these requirements, eq. D5 becomes

$$\int_{S_1} \left[\left(\frac{1}{R}\right)\frac{\partial F}{\partial R} - F\frac{\partial(1/R)}{\partial R}\right] dS - \lim_{R\to 0}\left\{\int_0^{4\pi}\left[\left(\frac{1}{R}\right)\frac{\partial F}{\partial R} - F\frac{\partial(1/R)}{\partial R}\right] R^2 d\Omega\right\}$$
$$= \int_{S_1} \left[\left(\frac{1}{R}\right)\frac{\partial F}{\partial R} - F\frac{\partial(1/R)}{\partial R}\right] dS - 4\pi F_P = 0 \qquad \text{(D6)}$$

where F_P is the value of F at $R = 0$. From the last equality of eq. D6, the value of F at point P is found to be

$$F_P = \frac{1}{4\pi}\int_{S_1} \left[\left(\frac{1}{R}\right)\frac{\partial F}{\partial R} - F\frac{\partial(1/R)}{\partial R}\right] dS \qquad \text{(D7)}$$

In the region that is external to the volume $\vee_a$, we see that eq. C7 is identically zero because there are no singularities in $\vee_b$.

From eq. D7, we can calculate the value of the function F at any point of the internal flow field by knowing the boundary values of F and $\partial F/\partial n$ on S_1. We have arrived at this by using the Green's function in eq. D2. The boundary values depend on the properties of the surface. The surface can be totally reflective or partially or totally absorbent. These are discussed in Chapter 6 in the section devoted to diffraction, and also in Chapter 11 in the discussion of the frequency-dependent hydrodynamic coefficients.

(2) Three-Dimensional Wave Source

The source potential can also be oscillatory, representing a wave being radiated outward from the origin of R located at P. In that case, the potential assumes the form of

$$\Phi_{wave} = -\frac{M}{R}e^{i(kR-\omega t)} = -MG_{wave}e^{-i\omega t} \qquad \text{(D8)}$$

where the Green's function for the wave is

$$G_{wave} = \frac{1}{R}e^{ikR} \qquad \text{(D9)}$$

For this form of the Green's function, Laplace's equation is not satisfied. Instead, the function must satisfy the *Helmholtz equation*,

$$[\nabla^2 G_{wave} + k^2 G_{wave}]|_{R>0} = 0 \qquad \text{(D10)}$$

which must be satisfied everywhere in the internal wave field except at the singular point (P) and at all points in the external region.

D2. Two-Dimensional Green's Function

If our interest is in the two-dimensional form of Green's theorem in eq. C8, then the Green's function needed to solve the integral equation is different than that in eq. D2. To illustrate, consider the two-dimensional flow from the point P in

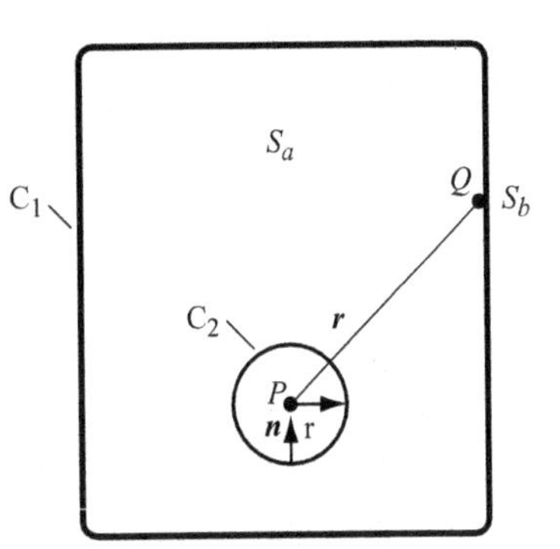

Figure D2. *Flow from a Two-Dimensional Point Source in an Enclosed Surface Area.*

Figure D2. Similar to that in Figure 2.9b, the velocity potential for a two-dimensional source flow is

$$\varphi_{source} = \frac{Q}{2\pi}\ln(r) = \frac{Q}{2\pi}\mathrm{G}_{source} \tag{D11}$$

where the two-dimensional Green's function is

$$\mathrm{G}_{source} = \ln(r) \tag{D12}$$

G_{source} satisfies the Laplace equation in two dimensions. Now, combine the expression in eq. D12 with the integral eq. C8. Following the same solution process leading to eq. C10, the point P is now enclosed in a circle of radius r and circumference C_2.. The interior surface area of the circle is πr^2 and the line element of integration is $rd\theta$, where θ is the planar angle from an arbitrary direction. The solution of the resulting integral equation is

$$\int_{C_1}\left[\ln(kr)\frac{\partial F}{\partial r} - F\frac{\partial \ln(kr)}{\partial r}\right]dC + 2\pi F_P = 0 \tag{D13}$$

So, the expression for the function F at the singular point, P, is

$$F_P = \frac{1}{2\pi}\int_{C_1}\left[\ln(kr)\frac{\partial F}{\partial r} - F\frac{\partial \ln(kr)}{\partial r}\right]dC \tag{D14}$$

By the introduction of the Green's function of eq. D12, the value of F at any point on the surface is known if the boundary values of F and $\partial F/\partial r$ on C_1 are known.

E. Solutions of Laplace's Equation

In Section 2.3B, the equation of continuity for an incompressible, irrotational flow is derived. The result of this derivation is Laplace's equation, eq. 2.41. Of interest here are the equation in the Cartesian-coordinate form,

$$\nabla^2\Phi = \frac{\partial^2\Phi}{\partial x^2} + \frac{\partial\Phi^2}{\partial y^2} + \frac{\partial^2\Phi}{\partial z^2} = 0 \tag{E1}$$

and in the polar-coordinate form,

$$\nabla^2\Phi = \frac{\partial^2\Phi}{\partial r^2} + \frac{1}{r}\frac{\partial\Phi}{\partial r} + \frac{1}{r^2}\frac{\partial^2\Phi}{\partial\beta^2} + \frac{\partial^2\Phi}{\partial z^2} = 0 \tag{E2}$$

Refer to the sketch in Figure E1 for notation. Both of the expressions of Laplace's equation are elliptic partial differential equations and, as such, can be solved using product solutions.

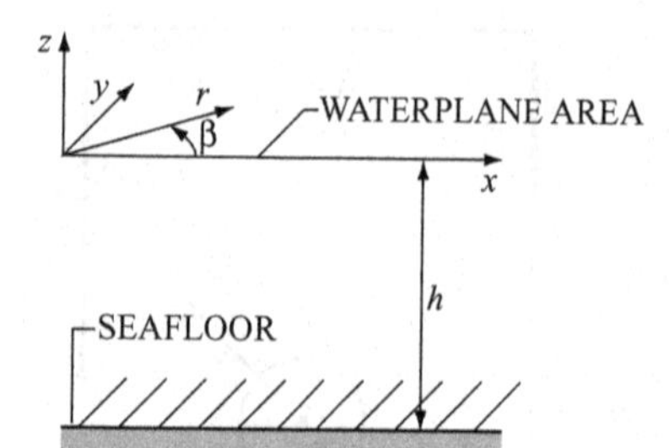

Figure E1. *Coordinate Systems and Notation for Laplace's Equation.*

E1. Cartesian Coordinates

For eq. E1, assume a product solution for the three-dimensional velocity potential of the form

$$\Phi = X(x)Y(y)Z(z)T(t) \tag{E3}$$

Combining this expression with that in eq. E1 results in the following separated equation:

$$\frac{1}{X}\frac{d^2X}{dx^2} + \frac{1}{Y}\frac{d^2Y}{dy^2} + \frac{1}{Z}\frac{d^2Z}{dz^2} = 0 \tag{E4}$$

This equation can be rewritten as

$$\frac{1}{X}\frac{d^2X}{dx^2} + \frac{1}{Y}\frac{d^2Y}{dy^2} = -\frac{1}{Z}\frac{d^2Z}{dz^2} = \mathrm{K}^2 \tag{E5}$$

Here, K is a parameter. Using this parameter, K, the separated equation can also be written as

$$\frac{1}{X}\frac{d^2X}{dx^2} - \mathrm{K}^2 = -\frac{1}{Y}\frac{d^2Y}{dy^2} = \mathrm{N}^2 \tag{E6}$$

where N is a second parameter. The number of solutions of the ordinary differential equations in x, y, and z depends on the both the natures and values of K and N. These parameters can be real, imaginary, or equal to 0. For a two-dimensional problem in the x-z plane, such as the case of the linear wave analysis presented in Section 3.2, $\mathrm{N} = 0$.

For the sake of illustration, let us consider the two-dimensional Laplace's equation, where the two-dimensional velocity potential is represented by ϕ. In the x-z plane, eq. E5 is

$$\frac{1}{X}\frac{d^2X}{dx^2} = -\frac{1}{Z}\frac{d^2Z}{dz^2} = \mathrm{K}^2 \tag{E7}$$

When K is real ($= K$), the general solution for the velocity potential is

$$\varphi_{\Re} = \mathrm{A}\cosh(Kx + C_{\mathrm{A}x})\cos(Kz + C_{\mathrm{A}z})T(t) \tag{E8}$$

In this equation, the subscript $\Re$ signifies that K ($= K$) is real. The constants A, $C_{\mathrm{A}x}$, and $C_{\mathrm{A}z}$ are to be determined. When K is imaginary ($= iK$), the solution is

$$\varphi_{\Im} = \mathrm{B}\cos(Kx + C_{\mathrm{B}x})\cosh(Kz + C_{\mathrm{B}z})T(t) \tag{E9}$$

where the subscript $\Im$ signifies that K is imaginary, and B, $C_{\mathrm{B}x}$, and $C_{\mathrm{B}z}$ are to be determined. Finally, when $\mathrm{K} = 0$, the solution is

$$\varphi_O = (A_{xO}x + B_{xO})(A_{zO}z + B_{zO}) \tag{E10}$$

Because Laplace's equation is linear, the solutions in eqs. E8 through E10 can be superimposed to form a general solution. The result is

$$\begin{aligned}\varphi &= \varphi_{\Re} + \varphi_{\Im} + \varphi_{o} \\ &= [\mathrm{A}\cosh(Kx + C_{\mathrm{A}x})\cos(Kz + C_{\mathrm{A}z}) + \mathrm{B}\cos(Kx + C_{\mathrm{B}x})\cosh(Kz + C_{\mathrm{B}z}) \\ &\quad + (A_{xO}x + B_{xO})(A_{zO}z + B_{zO})]T(t) \qquad \text{(E11)}\end{aligned}$$

The constants and parameters in this equation are determined by applying the boundary conditions. For example, the seafloor is normally assumed to be horizontal beneath the free surface and beneath floating bodies. So, the vertical velocity component of the water particles adjacent to the seafloor must vanish if the seafloor is fixed. Mathematically, this can be written as

$$\begin{aligned}\frac{\partial\varphi}{\partial z}\Big|_{z=-h} &= K[-\mathrm{A}\cosh(Kx + C_{\mathrm{A}x})\sin(-Kh + C_{\mathrm{A}z}) \\ &\quad + \mathrm{B}\cos(Kx + C_{\mathrm{B}x})\sinh(-Kh + C_{\mathrm{B}z})]T(t) \\ &\quad + A_{zO}(A_{xO}x + B_{xO})T(t) = 0 \qquad \text{(E12)}\end{aligned}$$

To avoid a trivial solution, $C_{\mathrm{A}x} = C_{\mathrm{A}z} = Kh$ and $A_{zO} = 0$. The resulting velocity potential expression is

$$\begin{aligned}\varphi &= \{\mathrm{A}\cosh(Kx + C_{\mathrm{A}x})\cos[K(z + h)] \\ &\quad + \mathrm{B}\cos(Kx + C_{\mathrm{B}x})\cos[K(z + h)] \\ &\quad + (\mathrm{A}_{xO}x + \mathrm{B}_{xO})\}T(t) \qquad \text{(E13)}\end{aligned}$$

Note that the constants in the last line of this equation are the results of consolidation of $A_{xO}\ B_{zO}$ and $A_{zO}\ B_{zO}$.

E2. Cylindrical Coordinates

Assume a product solution of Laplace's equation of the form

$$\Phi = \mathrm{P}(r)\mathrm{B}(\beta)\mathrm{Z}(z)\mathrm{T}(t) \qquad \text{(E14)}$$

The substitution of this expression for the velocity potential in eq. E2 leads to the following separated equation:

$$\frac{1}{\mathrm{P}}\frac{d^2\mathrm{P}}{dr^2} + \frac{1}{r\mathrm{P}}\frac{d\mathrm{P}}{dr} + \frac{1}{r^2\mathrm{B}}\frac{d^2\mathrm{B}}{d\beta^2} + \frac{1}{\mathrm{Z}}\frac{d^2\mathrm{Z}}{dz^2} = 0 \qquad \text{(E15)}$$

The z-term can be found by rearranging eq. E15 as

$$\frac{1}{\mathrm{P}}\frac{d^2\mathrm{P}}{dr^2} + \frac{1}{r\mathrm{P}}\frac{d\mathrm{P}}{dr} + \frac{1}{r^2\mathrm{B}}\frac{d^2\mathrm{B}}{d\beta^2} = -\frac{1}{\mathrm{Z}}\frac{d^2\mathrm{Z}}{dz^2} = \mathrm{K}^2 \qquad \text{(E16)}$$

where, again, the parameter, K, can be real, imaginary, or equal to zero. To obtain the r- and β-solutions, rewrite eq. E16 as

$$\frac{r^2}{\mathrm{P}}\frac{d^2\mathrm{P}}{dr^2} + \frac{r}{\mathrm{P}}\frac{d\mathrm{P}}{dr} - \mathrm{K}^2r^2 = -\frac{1}{\mathrm{B}}\frac{d^2\mathrm{B}}{d\beta^2} = M^2 \qquad \text{(E17)}$$

where M is a constant. The physics of the problems of interest to us require that M be either real or equal to zero. Because of this, the β-solution will be either sinusoidal or a first-degree function of x. The r-solution is obtained from

$$r^2\frac{d^2\mathrm{P}}{dr^2}+r\frac{d\mathrm{P}}{dr}-(\mathrm{K}^2r^2+M^2)\mathrm{P}=0 \tag{E18}$$

This is a form of the Bessel differential equation (see Abamowitz and Stegun, 1964). Again, the reader should remember that K can be real, imaginary, or equal to zero. When K is imaginary, the solution of eq. E18 involves the Bessel functions, and when K is real, the solution is composed of the modified Bessel functions, $I_M(\mathrm{K}r)$ and $K_M(\mathrm{K}r)$. The solution for the velocity potential in cylindrical coordinates, subject to the seafloor condition, is

$$\begin{aligned}\Phi = \{&[\mathrm{A}_I I_M(\mathrm{K}r) + A_K K_M(\mathrm{K}r)]\cos[\mathrm{K}(z+h)]\\ &+[\mathrm{B}_J J_M(\mathrm{K}r)+\mathrm{B}_Y Y_M(\mathrm{K}r)]\}\cosh[\mathrm{K}(z+h)]\}\cos(M\beta + C_\beta)T(t)\end{aligned} \tag{E19}$$

The coefficients of the Bessel functions and the phase angle are determined by the boundary conditions.

F. Fourier Transforms

The book by Tolstov (1962), among others, is recommended for a thorough discussion of the Fourier series and the Fourier transform. The derivations of the equations contained here are found in that reference. In addition, Tolstov presents excellent examples and applications of the equations.

Consider the known function $f(z)$. For example, this might be the distribution of the horizontal velocity caused by a passing wave. The *Fourier transform* of the function is defined as

$$\mathscr{F}(\zeta)=\frac{1}{\sqrt{2\pi}}\int_{-\infty}^{\infty} f(z)e^{-i\zeta z}dz \tag{F1}$$

and the *inverse transform* of $\mathscr{F}(\zeta)$ is

$$f(z)=\frac{1}{\sqrt{2\pi}}\int_{-\infty}^{\infty}\mathscr{F}(\zeta)e^{-iz\zeta}d\zeta \tag{F2}$$

The exponential terms in the integrands can also be written in terms of sinusoidal functions as

$$e^{\pm iz\zeta}=\cos(z\zeta)\pm i\sin(z\zeta) \tag{F3}$$

We can combine eqs. F1 and F2 and arrive at the following relationship:

$$f(z)=\frac{1}{2\pi}\int_{-\infty}^{\infty} f(\xi)\left\{\int_{-\infty}^{\infty} e^{i\zeta(z-\xi)}d\zeta\right\}d\xi \tag{F4}$$

This relationship is used in a number of the analyses of the hydrodynamics of floating bodies (for example, see the paper by Havelock, 1929). That paper can also be found in the bound collection of Havelock's papers edited by Wigley (1963).

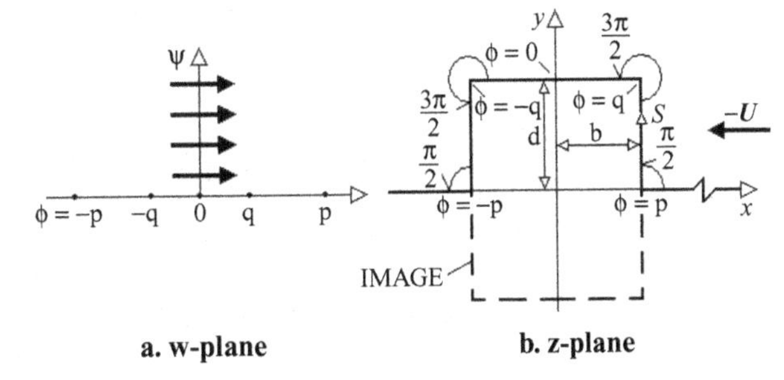

Figure G1. *Potential and Physical Planes Used in the Lewis (1929) Analysis in Transforming a Parallel Flow into Flow about a Fixed Body.*

G. Lewis Sharpe-Bilge Analysis

Here, using the Lewis (1929) conformal mapping technique, we derive the velocity potential for a heaving rectangle. For a thorough discussion of conformal mapping, the text by Schinzinger and Laura (2003) is recommended. Lewis uses the Schwarz (now, the Schwarz-Christoffel) transformation in his analysis. Referring to Figure G1, the complex variables used in the analysis are those in the potential plane ($\mathrm{w} = \varphi + i\psi$), where φ is the velocity potential and ψ is the stream function, and the physical plane ($\mathrm{z} = x + iy$). The rectilinear flow in the potential plane (parallel to the φ-axis) is transformed into a flow about a polygonal boundary (in the x-direction) by

$$\frac{dz}{dw} = (\mathrm{w} - \varphi_1)^{\frac{\alpha_1}{\pi}-1}(\mathrm{w} - \varphi_2)^{\frac{\alpha_2}{\pi}-1}\cdots(\mathrm{w} - \varphi_N)^{\frac{\alpha_N}{\pi}-1} \tag{G1}$$

Applying this equation to the rectangular boundary sketched in Figure G1b, we find $\alpha_1 = \alpha_4 = \pi/2$ and $\alpha_2 = \alpha_3 = 3\pi/2$, where the angles are measured positively in the counterclockwise direction. We specify that the velocity potential values in Figure G1a are the following point values of the velocity potential: $\varphi_1 = -p, \varphi_2 = -q, \varphi_3 = q$, and $\varphi_4 = p$. Where $\varphi_1 = 0$ in Figure G1b, the angle is π, so that point in eq. G1 is represented by the number 1. Note that Lewis (1929) specifies that $p = 1$ in his analysis. The arbitrary value of this potential point used herein is to avoid confusion on the units of the quantities. With the potential point values, the expression in eq. G1 applied to Figure G1 is

$$\frac{dz}{dw} = (\mathrm{w}+\mathrm{p})^{-\frac{1}{2}}(\mathrm{w}+\mathrm{q})^{\frac{1}{2}}(\mathrm{w}-\mathrm{q})^{\frac{1}{2}}(\mathrm{w}-\mathrm{p})^{-\frac{1}{2}} = \frac{\sqrt{\mathrm{w}^2-\mathrm{q}^2}}{\sqrt{\mathrm{w}^2-\mathrm{p}^2}} \tag{G2}$$

Upon integration, we find

$$\mathrm{z} = x + iy = \mathrm{C}\int \frac{\sqrt{\mathrm{w}^2-\mathrm{q}^2}}{\sqrt{\mathrm{w}^2-\mathrm{p}^2}}dw = \mathrm{C}\int \frac{\sqrt{\mathrm{q}^2-\mathrm{w}^2}}{\sqrt{\mathrm{p}^2-\mathrm{w}^2}}dw \tag{G3}$$

Here, the to-be-determined constant, C, has units of time over length.

Following Lewis (1929), our interest is in the rectangular boundary that corresponds to $\psi = 0$ in Figure G1a. On the body, the complex potential is $\mathrm{w}|_{\psi=0} = \varphi$. Combine this with eq. G3 and integrate along the sides of the body in the first quadrant of Figure G1b. Along the top from $y = 0$ to $x = \varphi < q$, where $y = d$, we obtain

$$z = C\int_0^{\varphi} \frac{\sqrt{\varphi^2 - q^2}}{\sqrt{\varphi^2 - p^2}} d\varphi = C\int_0^{\varphi} \sqrt{\frac{q^2 - \varphi^2}{p^2 - \varphi^2}} d\varphi$$

$$= Cp\left\{E\left[\sin^{-1}\left(\frac{\varphi}{q}\right), Q\right] - (1 - Q^2)F\left[\sin^{-1}\left(\frac{\varphi}{q}\right), Q)\right]\right\} \tag{G4}$$

Here, $Q = q/p$. The functions $E(\varepsilon, Q)$ and $F(\varepsilon, Q)$ are called *elliptic integrals* of the first and second kind, respectively. See Gradshteyn and Ryzhik (1980) or Abramowitz and Stegun (1965) for discussions of these functions. In the last integral in eq. G4, we note that $\varphi < q < p$. Hence, the integral is real. The integration method is presented by Lewis (1929), and a similar integral is found in Section 3.169 of the book by Gradshteyn and Ryzhik (1980).

Now, integrate the product of $i = \sqrt{-1}$ and eq. G3 along the right side of the solid in Figure G1b from q to φ, where $x = b$, to obtain

$$z = -C\int_q^{\varphi} \frac{\sqrt{\varphi^2 - q^2}}{\sqrt{p^2 - \varphi^2}} d\varphi = -Cp\left\{E\left[\sin^{-1}\left(\frac{p}{\varphi}\sqrt{\frac{\frac{\varphi^2}{p^2} - Q^2}{1 - Q^2}}\right), \sqrt{1 - Q^2}\right]\right.$$

$$\left. - \frac{q^2}{p^2}F\left[\sin^{-1}\left(\frac{p}{\varphi}\sqrt{\frac{\frac{\varphi^2}{p^2} - Q^2}{1 - Q^2}}\right), \sqrt{1 - Q^2}\right] - \frac{1}{\varphi}\sqrt{\left(1 - \frac{\varphi^2}{p^2}\right)\left(\frac{\varphi^2}{p^2} - Q^2\right)}\right\} \tag{G5}$$

The integral on the top line of this equation also appears in Section 3.169 of Gradshteyn and Ryzhik (1980).

The half-breadth of the body in Figure G1b is b. At $x = b$ and $y = d$, eq. G4 yields

$$b = Cp\left[E\left(\frac{\pi}{2}, Q\right) - (1 - Q^2)F\left(\frac{\pi}{2}, Q\right)\right] = Cp[E(Q) - (1 - Q^2)K(Q)] \tag{G6}$$

In this equation are the complete elliptic integrals of the first and second kind, which are $\boldsymbol{K}(Q) \equiv F(\pi/2, Q)$ and $\boldsymbol{E}(Q) \equiv E(\pi/2, Q)$, respectively.

Apply eq. G5 at $x = b$ and $y = d$ to obtain

$$d = -Cp\left[E\left(\frac{\pi}{2}, \sqrt{1 - Q^2}\right) - Q^2F\left(\frac{\pi}{2}, \sqrt{1 - Q^2}\right)\right] = Cp[\boldsymbol{E}(\sqrt{1 - Q^2}) - Q^2\boldsymbol{K}(\sqrt{1 - Q^2})] \tag{G7}$$

We note that the product Cp can be treated as a single constant, C. Hence, eqs. G6 and G7 are a system of two equations with two unknowns, the latter being C and Q. These parameters can be determined by solving eqs. G6 and G7 simultaneously.

As shown by Abramowitz and Stegun (1965), for a parameter Q, where $0 \leq Q \leq 1$, the complete elliptic integrals in eqs. G6 and G7 are well approximated by

$$\boldsymbol{K}(Q) \simeq [1.38629 + 0.11197(1-Q) + 0.07252(1-Q)^2] + [0.5 + 0.12134(1-Q) + 0.02887(1-Q)^2]\ln\left(\frac{1}{1-Q}\right) \quad \text{(G8)}$$

and

$$\boldsymbol{E}(Q) \simeq [1 + 0.46301(1-Q) + 0.10778(1-Q)^2] + [0.24527(1-Q) + 0.04124(1-Q)^2]\ln\left(\frac{1}{1-Q}\right) \quad \text{(G9)}$$

To determine the low-frequency added mass of the body in motion, we begin by focusing our attention on the portion of the body in the first quadrant of Figure G1b. When the body in Figure G1b moves in the positive y-direction in a still fluid, the flow is represented by the velocity potential Φ , which is $\varphi - Ux$, and the stream function Ψ. As before, the velocity potential component φ represents the flow about the body when the body is fixed in a flow that is parallel to the y-axis at $x = \pm\infty$. The velocity potential representing the parallel flow is $-Ux$. The kinetic energy of the flow adjacent to the moving body is

$$E_k' = \frac{1}{2}\rho \iint_S (\nabla\Phi)^2 dS = -\frac{1}{2}\rho \int \Phi d\Psi|_s = \frac{1}{2}\rho \int \Phi \frac{\partial \Psi}{\partial s}|_s ds = -\frac{1}{2}\rho \int \Phi \frac{\partial \Phi}{\partial n}|_s ds \quad \text{(G10)}$$

The relationship of the integrals in the second row of eq. G10 results from the Cauchy-Riemann relationships, where s is the coordinate tangent to the body surface and n is the outward normal coordinate. On the side at $y = b$ and $0 < z < T$ in the first quadrant of Figure G1b, $ds = dy$ and

$$-\frac{\partial \Phi}{\partial n}|_{y=\mathrm{b}} = \frac{\partial \Phi}{\partial y}|_{y=\mathrm{b}} = U \quad \text{(G11)}$$

Along the first-quadrant side, the application of eq. G10 using the results in eq. G11 yields

$$E_k'|_{\substack{x=b \\ 0<y<d}} = -\frac{1}{2}\rho \int_0^d (\varphi - U\mathrm{b})(-U)dy = \frac{1}{2}\rho U \int_0^d \varphi dy - \frac{1}{2}\rho \mathrm{db} U^2 = \frac{1}{2}\mathrm{a}_{wx}' U^2 \quad \text{(G12)}$$

Here, a_{wx}' is the added mass (per unit length) excited by the motions in the x-direction. Because the top of the body is parallel to the motion direction, the kinetic energy per unit length equals zero over the top. All that remains is to determine the potential φ. To do so, the differential form of the expression leading to eq. G5 is used. That is, on the right side of the body,

$$dz|_s = ds = -\mathrm{C}\frac{\sqrt{\varphi^2 - \mathrm{q}^2}}{\sqrt{\mathrm{p}^2 - \varphi^2}} d\varphi \quad \text{(G13)}$$

The combination of this expression with the last integral of eq. G12 results in the following total kinetic energy per unit length of the entire body:

$$4E'_k|_{\substack{x=b \\ 0<y<d}} = 4\left[\frac{1}{2}\rho UC\int_0^q \varphi\frac{\sqrt{\varphi^2 - q^2}}{\sqrt{p^2 - \varphi^2}}d\varphi - \frac{1}{2}\rho bdU^2\right]$$
$$= 4\left[\frac{1}{2}\rho UC\frac{\pi}{4}(p^2 - q^2) - \frac{1}{2}\rho bdU^2\right]$$
$$= \frac{1}{2}a'_{wx}U^2 \tag{G14}$$

The integration is performed by substituting the variable $t = \sqrt{(p^2 - \varphi^2)}$. The expression for the low-frequency added mass of the total two-dimensional body is

$$a'_{wx} = \rho\left[\frac{\pi}{U}C(p^2 - q^2) - 4bd\right] = \rho\left[\frac{\pi}{U}Cp^2(1 - Q^2) - 4bd\right] \tag{G15}$$

If we assume that the body in the lower half of the plane is both submerged and oscillating such that there is little reaction on the free surface (where $y = 0$), then the added-mass expression for the motions of the half-submerged body is one half that in eq. (G15), or

$$a'_{wx} = \frac{\rho}{2}\left[\frac{\pi}{U}C(p^2 - q^2) - 4bd\right] = \frac{\rho}{2}\left[\frac{\pi}{U}Cp^2(1 - Q^2) - 4bd\right] \tag{G16}$$

We have used the arbitrary p as the upper bound of the potential on the body. Lewis (1929) lets $p = 1$, thereby setting the reference for the potential values on the body. By following Lewis, eqs. G6, G7, and G15 have two unknown parameters, C and Q, that are determined by the simultaneous solutions of eqs. G6 and G7. The reason for maintaining the p-notation throughout the derivations herein is to be conscious of the units in each of the equations.

For this case, we set $p = 1$ in eqs. G6 and G7 and isolate the constant C in both to obtain

$$C = \frac{b}{\boldsymbol{E}(Q) - (1 - Q^2)\boldsymbol{K}(Q)} \tag{G17}$$

H. Infinite-Frequency Added-Mass Expressions

In the following sections, the infinite-frequency added-mass expressions are presented. Most of the expressions presented in this appendix are presented by Brennen (1982).

H1. Two-Dimensional Added Mass

(1) Motions in an Infinite Liquid

In Figure H1 are four basic two-dimensional shapes: The straight line, circle, ellipse, and rectangle. These are assumed to be in a liquid having no bounds. The motions of the shapes are rectilinear with a speed, V, and moving with an angular speed, ω. For these shapes, the added mass and added-mass moment of inertia (both per

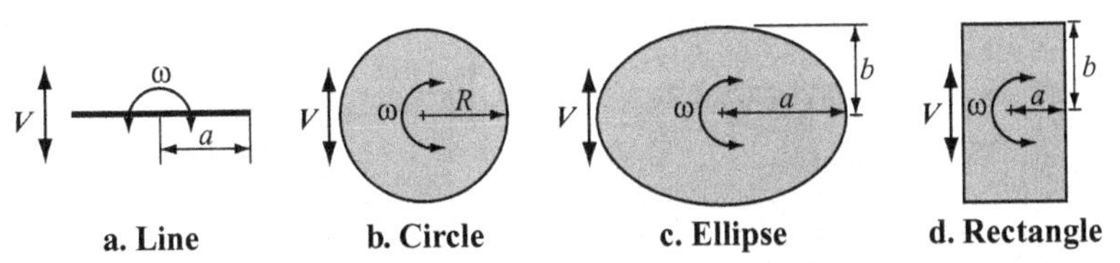

Figure H1. *Four Basic Two-Dimensional Shapes Undergoing Rectilinear and Angular Motions.*

unit length into the page) are give by Wendel (1956). The expressions are the following:

Line or Two-Dimensional Flat Plate in Figure H1a:
Added Mass per Unit Length (Attributed to Horace Lamb by Kurt Wendel):

$$\mathrm{a}'_w = \rho\pi a^2 \tag{H1}$$

Added-Mass Moment of Inertia per Unit Length:

$$\mathrm{A}'_w = \frac{1}{8}\rho\pi a^4 \tag{H2}$$

Circle in Figure H1b:
Added Mass per Unit Length:

$$a'_w = \rho\pi R^2 \tag{H3}$$

Added-Mass Moment of Inertia per Unit Length:

$$\mathrm{A}'_w = 0 \tag{H4}$$

Ellipse in Figure H1c:
Added Mass per Unit Length (Attributed to Horace Lamb by Kurt Wendel):

$$\mathrm{a}'_w = \rho\pi a^2 \tag{H5}$$

Added-Mass Moment of Inertia per Unit Length:

$$\mathrm{A}'_w = \frac{1}{8}\rho\pi(a^2 - b^2)^2 \tag{H6}$$

Rectangle in Figure H1d:
Added Mass per Unit Length:

$$\mathrm{a}'_w = C_\mathrm{a}\rho\pi a^2 \tag{H7}$$

Added-Mass Moment of Inertia per Unit Length:

$$\mathrm{A}'_w = C_\mathrm{A}\rho\pi b^4 \tag{H8}$$

In these expressions, the values of the coefficients C_a and C_A depend on the ratio *a/b*. The values presented by Wendel (1956) are as follows:

$a/b = 0.10$	0.20	0.50	1.00	2.00	5.00	10.00
$C_a = 2.23$	1.98	1.70	1.51	1.36	1.21	1.14
$C_A = 0.147$	0.15	0.15	0.234	0.15	0.15	0.147

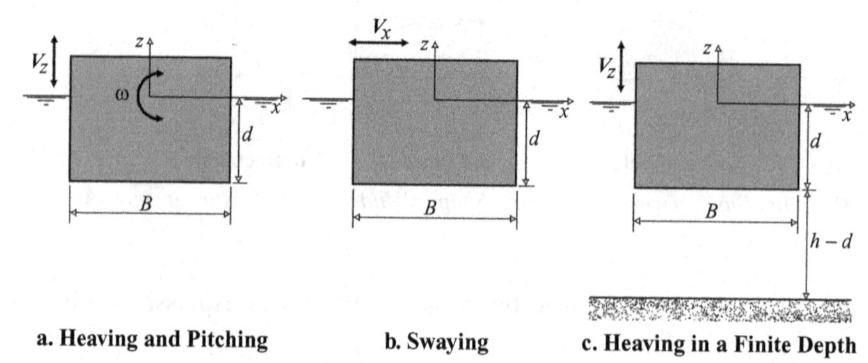

a. Heaving and Pitching **b. Swaying** **c. Heaving in a Finite Depth**

Figure H2. *Two-Dimensional Motions of Rectangular Sections in a Rigid Free Surface.*

(2) Motions of a Rectangular Section in Liquid with a Free Surface

In Figure H2a is a sketch of a rectangular section undergoing both heaving and rolling motions in a liquid having a free surface and an infinite depth. In Figure H2b is the sketch of rectangular section swaying in a liquid of infinite depth. In Figure H2c, the sketched rectangular section is shown heaving in a liquid of finite depth. The added mass and added-mass moment of inertia expressions (both per unit length into the page) for rectangular section motions in Figure H2 are presented.

Heaving and Pitching in a Liquid of Infinite Depth in Figure H2a:
Added Mass per Unit Length [From Lewis (1929) in Appendix G]:

$$\mathrm{a}'_w = \frac{3}{4}\rho\pi\left(\frac{B}{2}\right)^2, \quad \text{where } d = \frac{B}{2} \tag{H9}$$

Added-Mass Moment of Inertia per Unit Length [From Wendel (1956)]:

$$\mathrm{A}'_w = 0.117\rho\pi\left(\frac{B}{2}\right)^4, \quad \text{where } d = \frac{B}{2} \tag{H10}$$

Swaying in a Liquid of Infinite Depth in Figure H2b:
Added Mass per Unit Length [From Wendel (1956)]:

$$\mathrm{a}'_w = \frac{1}{4}\rho\pi\left(\frac{B}{2}\right)^2, \quad \text{where } d = \frac{B}{2} \tag{H11}$$

Heaving in a Liquid of Finite Depth in Figure H2c:
Added Mass per Unit Length:

$$\mathrm{a}'_w = C_a\rho\pi\left(\frac{B}{2}\right)^2, \quad \text{where } d = \frac{B}{2} \tag{H12}$$

The values of C_α are presented here for depth-to-draft ratios:

$\frac{(h-d)}{d} = \infty$	2.60	1.80	1.50	0.50	0.25
$C_a = 0.755$	0.83	0.89	1.00	1.35	2.00

The values associated with $(h-d)/d = \infty$, 2.60, and 1.80 are from Wendel (1956), and the remaining are due to the Lewis (1929) analysis.

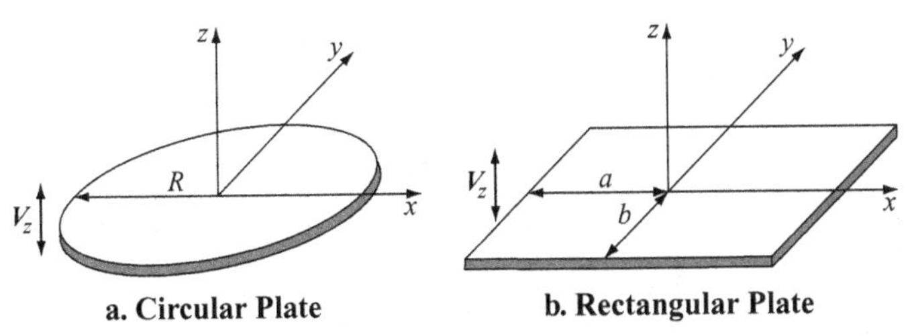

Figure H3. *Circular and Rectangular Plates Moving Rectilinearly in an Infinite Liquid.*

H2. Three-Dimensional Added Mass

(1) Flat Plate Motions in an Infinite Liquid

A circular plate is sketched in Figure H3a, and a rectangular plate is shown in Figure H3b. For both plates, the thicknesses are considered to be much smaller that the planar dimensions. That is, if the plate thickness is t, then $t \ll R$ in Figure H3a and $t \ll a$ and b in Figure H3b. The added-mass expressions for these plates are as follows:

Circular Plate in Figure H3a:
Added Mass [From Lamb (1932)]:

$$a'_w = \frac{8}{3}\rho \pi R^3 \tag{H13}$$

Rectangular Plate in Figure H3b:
Added Mass [From Kennard (1967)]:

$$a'_w = C_A \frac{1}{4}\rho \pi a^2 b \tag{H14}$$

The values of C_A are presented here for aspect ratio, $2a/b$:

$2a/b = \infty$	4.00	3.50	3.00	2.50	2.00	1.50	1.00
$C_A = 1.00$	1.00	1.00	1.00	0.953	0.840	0.860	0.478

These values are also reported by Brennan (1982). It is interesting to note that the added mass for the circular plate is four times that of the sphere having the same radius and moving in an infinite fluid, as presented in eq. H15.

(2) Motions of a Sphere in an Infinite Liquid and beneath a Free Surface

In Figure H4a is sketched a sphere undergoing rectilinear motions in an infinite liquid. In Figure H4b, the liquid has a free surface and is infinitely deep ($h = \infty$). The center of the sphere is at a depth of $d \geq 0$. For both spheres in Figure H4, R is the radius of the sphere and D is the diameter. The added-mass expressions for the two conditions are as follows:

Sphere in an Infinite Liquid in Figure H4a:
Added Mass [From Lamb (1932)]:

$$a'_w = \frac{2}{3}\rho \pi R^3 \tag{H15}$$

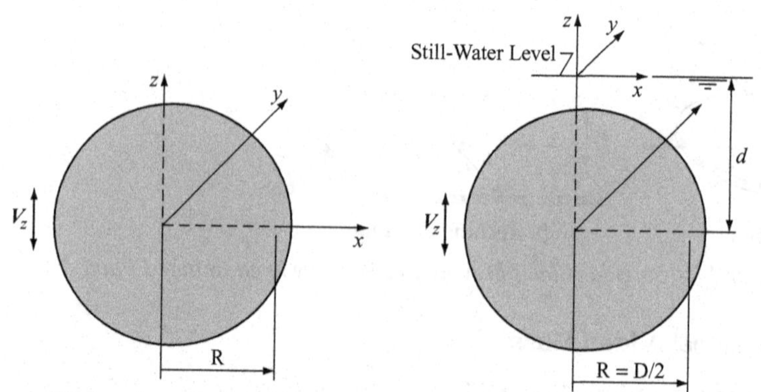

Figure H4. *Vertical Motions of a Sphere in an Infinite Liquid and Beneath a Free Surface.*

Sphere beneath a Free Surface in Figure H4b:
Added Mass per Unit Length [From Kennard (1967)]:

$$a'_w = C_a \frac{2}{3} \rho \pi R^3 \tag{H16}$$

The values of C_α are presented here for depth-to-diameter ratio, d/D:

$d/D = 0.00$	0.50	1.00	1.50	2.00	2.50	3.00	3.50	4.00	4.50
$C_A = 0.50$	0.88	1.08	1.16	1.18	1.18	1.16	1.12	1.04	1.00

These values are also reported by Brennan (1982). Comparing the C_α-value for $d/D = 4.50$, we see that the value is the same as it would be in eq. H15, that is, $C_\alpha = 1.00$. Hence, we conclude that the free-surface effect on the added mass is negligible for $d/D \geq 4.50$.

H3. Frequency-Dependent Added Mass

The previous sections are devoted to the added mass of basic shapes oscillating at infinite frequency. For the derivations of finite-frequency added-mass expressions, the book by Falnes (2002) is recommended.

References

CHAPTER 1

Bascom, Willard (1964), *Waves and Beaches*, Doubleday, Garden City, New York.

Bretschneider, Charles L. (1969), "Wave Forecasting," in Chapter 11 of *Handbook of Ocean and Underwater Engineering*, McGraw-Hill, New York. See Myers et al. (1969).

Craik, A. D. D. (2003), "G. G. Stokes and His Precursors on Water Wave Theory," in *Wind Over Waves II: Forecasting and Fundamental of Applications*, by S. G. Sajjadi and J. C. R. Hunt, Horwood Publishing, Chichester, U.K., pp. 3–22.

Earle, M. D. and J. M. Bishop (1984), "A Practical Guide to Ocean Wave Measurement and Analysis," Endeco, Inc., Marion, Massachusetts.

Myers, J. J., C. H. Holm, and R. F. McAllister (1969), *Handbook of Ocean and Underwater Engineering*, McGraw-Hill, New York.

U.S. Army (1984), *Shore Protection Manual*, U.S. Government Printing Office, Washington, DC, Vol. 1, 4th edition.

Wiegel, Robert L. (1964), *Oceanographical Engineering*, Prentice Hall, Englewood Cliffs, New Jersey.

CHAPTER 2

Achenbach, E. (1968), "Distribution of Local Pressure and Skin Friction Around a Circular Cylinder in Cross-Flow up to $R_e = 5 \times 10^6$," *Journal of Fluid Mechanics*, Vol. 34, Part 4, pp. 625–639.

Achenbach, E. (1971), "Influence of Surface Roughness on the Cross-Flow around a Circular Cylinder," *Journal of Fluid Mechanics*, Vol. 46, Part 3, pp. 321–335.

Avery, William H. and Chih Wu (1994), *Ocean Thermal Energy Conversion*, Oxford University Press, Oxford and New York.

Chadwick, Andrew and John Morfett (1986), *Hydraulics in Civil Engineering*, Allen and Unwin, London and Boston.

Courant, R. (1968), *Differential and Integral Calculus*, Volume II, Wiley-Interscience, New York, 29th edition.

Dawson, Thomas H. (1983), *Offshore Structural Engineering*, Prentice Hall, Englewood Cliffs, New Jersey.

Dettman, John W. (1969), *Mathematical Methods in Physics and Engineering*, McGraw-Hill, New York, 2nd edition.

Every, M., R. King, and D. Weaver (1982), "Vortex-Excited Vibrations of Cylinders and Cables, and Their Suppression," *Ocean Engineering* (Elsevier), Vol. 9, No. 2, pp. 138–157.

Garvey, R. E. (1990), "Composite Hull for Full-Ocean Depth," *Marine Technology Society Journal*, Vol. 24, No. 2, pp. 49–58.

Granger, Robert A. (1985), *Fluid Mechanics*, Holt, Rinehart and Winston, New York.

Jones, G., J. Cincotta, and R. Walker (1969), "Aerodynamic Forces on a Stationary and Oscillating Circular Cylinder at High Reynolds Numbers," National Aeronautics and Space Administration, Report TR R-300.

Karamcheti, K. (1966), *Principles of Ideal-Fluid Aerodynamics*, John Wiley & Sons, New York.

King, D. A. (1969), "Basic Hydrodynamics," in *Handbook of Ocean and Underwater Engineering*, McGraw-Hill, New York, Section 2, pp. 2–3 to 2–7.

McCormick, M. E. (1973), *Ocean Engineering Wave Mechanics*, Wiley-Interscience, New York.

McCormick, M. E. (1981), "On the Modeling of Ocean Moorings," U.S. Naval Academy Report EW-13–81, October.

McCormick, M. E. (2007), *Ocean Wave Energy Conversion*, Dover Publication, Long Island, New York. Originally published by Wiley-Interscience, New York, 1981.

Miller, B. L. (1976), "The Hydrodynamic Drag on Roughened Circular Cylinders," *Transactions*, Royal Institution of Naval Architects, Vol. 119, pp. 55–70.

Morkovin, M. V. (1964), "Flow around Circular Cylinder – A Kaleidoscope of Challenging Fluid Phenomena," *Proceedings*, Symposium on Fully Separated Flows (ASME), pp. 102–118.

Reynolds, A. J. (1974), *Turbulent Flows in Engineering*, John Wiley & Sons, New York.

Roshko, A. (1961), "Experiments on the Flow past a Circular Cylinder at Very High Reynolds Numbers," *Journal of Fluid Mechanics*, Vol. 10, Part 3, pp. 345–356.

Ross, C. T. F. (1990), *Finite Element Programs for Axisymmetric Problems in Engineering*, Halsted Press, Chichester, U.K.

Ross, David (1979), *Energy from Ocean Waves*, Pergamon Press, Oxford.

Sarpkaya, T. (1979), "Vortex-Induced Oscillations," *Journal of Applied Mechanics* (ASME), Vol. 46, June, pp. 241–258.

Schlichting, H. (1979), *Boundary Layer Theory*, McGraw-Hill, New York.

Shames, Irving H. (1962), *Mechanics of Fluids*, McGraw-Hill, New York.

Shaw, Ronald (1982), *Wave Energy – A Design Challenge*, Halsted Press, Chichester, U.K.

Thwaites, Bryan (ed.) (1987), *Incompressible Aerodynamics*, Dover, New York.

Vallentine, H. R. (1967), *Applied Hydrodynamics*, Butterworths, London, 2nd edition.

Van Dyke, Milton (1982), *An Album of Fluid Motion*, Parabolic Press, Stanford, California.

CHAPTER 3

Airy, George B. (1845), "On Tides and Waves," *Encyclopaedia Metropolitania*, Vol. 5, Article 192, pp. 241–396, London.

Brink-Kjær, O. and I. G. Jonsson (1973), "Verification of Cnoidal Shoaling: Putnam and Chinn's Experiments," Progress Report 28, 19–23, Institute of Hydrodynamics and Hydraulic Engineering, Technical Univ. of Denmark, Lyngby.

Bruun, Per (1985), *Design and Construction of Mounds for Breakwaters and Coastal Protection*, Elsevier, New York.

Carnahan, Brice (1969), *Applied Numerical Methods*, John Wiley & Sons, New York.

Charlier, Roger H. (1982), *Tidal Energy*, Van Nostrand Reinhold Publishing Company, New York.

Kinsman, Blair (1984), *Wind Waves*, Dover Publications, New York, 2nd edition.

Lamb, Horace (1945), *Hydrodynamics*, Dover Publications, New York, 7th edition.

Le Méhauté, Bernard (1969), "An Introduction to Hydrodynamics and Water Waves II: Wave Theories," U.S. Dept. of Commerce (Environmental Science Services) Report ERL 111-Pol 3–2, July.

Lighthill, James (1979), *Waves in Fluids*, Cambridge University Press, Cambridge, U.K.

McCormick, Michael E. (1973), *Ocean Engineering Wave Mechanics*, Wiley-Interscience, New York.

McCormick, M. E. (2007), *Ocean Wave Energy Conversion*, Dover Publications, Long Island, New York. Originally published by Wiley-Interscience, New York in 1981.

Medina, Josep R., C. Fassardi and R. T. Hudspeth (1990), "Effects of Wave Groups on the Stability of Rubble Mound Breakwaters," *Proceedings*, Coastal Engineering Conference, Delft, Chapter 116.

Phillips, Owen M. (1966), *The Dynamics of the Upper Ocean*, Cambridge University Press, Cambridge, U.K.

St. Denis, Manley (1969), "On Wind Generated Waves," Part 1 in *Topics in Ocean Engineering*, pp. 3–41.

Stoker, J. J. (1957), *Water Waves*, Interscience Publishers, New York.

Stokes, G. G. (1847), "On the Theory of Oscillatory Waves," *Transactions*, Cambridge Philosophical Society, Vol. 8 and Papers, Vol. I.

Stokes, G. G. (1880), Supplement to the 1847 paper under the same title, *Math. and Phys. Papers*, Cambridge Philosophical Society, Vol. I.

Svendsen, I. A. and I. G. Jonsson (1976), *Hydrodynamics of Coastal Regions*, Technical Univ. of Denmark, Lyngby.

Tucker, M. J and E. G. Pitt (2001), *Waves in Ocean Engineering*, Elsevier Science, Oxford.

Van Der Meer, J. W. (1988), "Deterministic and Probabilistic Design of Breakwater Armor Layer," *Journal of Waterway, Port, Coastal and Ocean Engineering* (ASCE), Vol. 114, No. 1, pp. 66–80.

Wiegel, Robert L. (1964), *Oceanographical Engineering*, Prentice Hall, Englewood Cliffs, New Jersey.

CHAPTER 4

Abramowitz, Milton and Irene A. Stegun (1965), *Handbook of Mathematical Functions*, Dover Publications, New York. Originally published by the U. S. Printing Office, Washington, DC, in 1964.

Bagnold, R. A. (1947), "Sand Movement by Waves," *Journal of the Institute of Civil Engineers*, Vol. 27, pp. 447–469.

Bhattacharyya, R. (1978), *Dynamics of Marine Vehicles*, Wiley-Interscience, New York.

Bhattacharyya, S. K. (1995), "On Two Solutions of the Fifth Order Stokes Waves," *Applied Ocean Research*, Vol. 17, pp. 63–68.

Boussinesq, J. (1872) – in French – "Theory of Waves and Swells Propagated in a Long Horizontal Rectangular Canal and Imparting to the Liquid Contained in this Canal Approximately Equal Velocities from the Surface to the Bottom," *Journal de Mathematiques Pures et Appliquees*, Vol. 17, Series 2, pp. 55–108.

Broeze, Jan (1992), "Computation of Breaking Waves with a Panel Method," *Proceedings*, 23rd Intl. Conf. Coastal Engineering, Venice, Vol. 1, pp. 89–102.

Crapper, G. D. (1984), *Introduction to Water Waves*, Ellis Harwood Ltd., Chichester, U.K.

Daily, J. W. and S. C. Stephan, Jr. (1952), "The Solitary Wave – Its Celerity, Profile, Internal Velocities and Amplitude Attenuation in a Horizontal Smooth Channel," *Proceedings*, 3rd Conf. on Coastal Engineering, Cambridge, MA, pp. 13–30.

Dawson, T. H. (1983), *Offshore Structural Engineering*, Prentice Hall, Englewood Cliffs, New Jersey.

De, S. C. (1955), "Contributions to the Theory of Stokes Waves," *Proceedings*, Cambridge Philosophical Society, Vol. 51, pp. 713–736.

Dean, R. G. (1965), "Stream Function Representation of Nonlinear Ocean Waves," *Journal of the Geophysical Research*, Vol. 70, No. 18, pp. 4561–4572.

Dean, R. G. (1967), "Relative Validities of Water Wave Theories," *Proceedings*, First Conference on Civil Engineering in the Oceans (ASCE), San Francisco.

Dean, R. G. (1974), "Evaluation and Development of Water Wave Theories for Engineering Application," U.S. Army Coastal Engineering Research Center Report SR-1, Vol. I, Vicksburg, Mississippi, November.

Dean, R. G. and R. A. Dalrymple (1984), *Water Wave Mechanics for Engineers and Scientists*, Prentice Hall, Englewood Cliffs, New Jersey.

Fenton, J. D. (1979), "A High-Order Cnoidal Wave Theory," *Journal of Fluid Mechanics*, Vol. 94, pp. 121–161.

Galvin, Cyril J. (1972), "Wave Breaking in Shallow Water," *Proceedings*, Seminar on *Waves on Beaches* at U. of Wisconsin (Madison), October, published by Academic Press, New York.

Gerstner, F. J. (1808), "Theorie der Wellen," *Ann. Physik*, Vol. 32, pp. 412–440.

Grilli, S. T., R. Subramenya, I. A. Svendsen, and J. Veeramony (1994), "Shoaling of Solitary Waves on Plane Beaches," *Journal of Waterway, Port, Coastal and Ocean Engineering* (ASCE), Vol. 120, No. 6, Nov./Dec., pp. 609–628.

Horikawa, K. (1978), *Coastal Engineering*, (Halsted) John Wiley & Sons, New York.

Khangaonkar, T. P. and Bernard Le Méhauté (1991), "Extended KdV Equation for Transient Axisymmetric Waves," *Ocean Engineering*, Vol. 18, No. 5, pp. 435–450.

Kinsman, Blair (1965), *Wind Waves*, Prentice Hall, Englewood Cliffs, NJ.

Korteweg, D. J. and G. de Vries (1895), "On the Change of Form of Long Waves Advancing in a Rectangular Canal, and on a New Type of Long Stationary Waves," *London, Edinburgh and Dublin Philosophical Magazine and Journal of Science*, Vol. 39, pp. 422–443.

Lamb, Horace (1945), *Hydrodynamics*, Dover Publishing, New York, 7th edition.

Le Méhauté, Bernard (1969), "An Introduction to Hydrodynamics and Water Waves," U.S. Department of Commerce, ESSA Report ERL 118-POL-3–2, Vol. II.

Lenau, C. W. (1966), "The Solitary Wave of Maximum Amplitude," *Journal of Fluid Mechanics*, Vol. 25, pp. 309–320.

Lighthill, James (1979), *Waves in Fluids*, Cambridge University Press, Cambridge, U.K.

Longuet-Higgins, M. S., E. D. Cokelet and M. J. H. Fox (1976), "The Calculation of Steep Gravity Waves," Proceedings, Conf. *Behavior of Offshore Structures*, Trondheim, Vol. 11, pp. 27–39.

Lord Rayleigh (1876), "On Waves," *London, Edinburgh and Dublin Philosophical Magazine and Journal of Science*, Ser. 5, Vol. 1, No. 4, pp. 257–279.

McCormick, Michael E. (1973), *Ocean Engineering Wave Mechanics*, Wiley-Interscience, New York.

McCowan, J. (1891), "On the Solitary Wave," *London, Edinburgh and Dublin Philosophical Magazine and Journal of Science*, Vol. 32, 5th Series, pp. 45–58.

McCowan, J. (1894), "On the Highest Wave of Permanent Type," *London, Edinburgh and Dublin Philosophical Magazine and Journal of Science*, Vol. 38, 5th Series, pp. 351–358.

Miche, M. (1944) – in French – "Undulatory Movements of the Sea in Constant or Decreasing Depth," Part 3, *Annales des Ponts et Chaussees*, Vol. 114, No. 1, p. 390.

Michell, J. H. (1893), "The Highest Waves in Water," *London, Edinburgh and Dublin Philosophical Magazine and Journal of Science*, Vol. 36, 5th Series, July–Dec., pp. 430–437.

Milne-Thomson, L. M. (1950), *Jacobian Elliptic Functions*, Dover Publications, New York.

Muga, Bruce J. and James F. Wilson (1970), *Dynamic Analysis of Ocean Structures*, Plenum Press, New York.

Munk, Walter H. (1949), "The Solitary Wave Theory and Its Application to Surf Problems," *Annals, New York Acad. Sci.*, Vol. 51, Art. 3, pp. 343–572.

Patel, M. H. (1989), *Dynamics of Offshore Structures*, Butterworths, London.

Penny, W. G. and A. T. Price (1952), "Finite Periodic Stationary Gravity Waves in a Perfect Liquid," *Philosophical Transactions, Royal Society, Series A*, Vol. 244, pp. 254–284.

Russell, J. Scott (1838), Report of the Committee on Waves to the British Association, Report 6, pp. 417–496.

Russell, J. Scott (1844), Report of rate Committee on Waves to the British Association, Report 13, pp. 311–390.

Russell, J. Scott (1881), "The Wave of Translation and the Work It Does as the Carrier Wave of Sound," *Proceedings, Royal Society of London*, Vol. 32, pp. 382–383, 4–43.

Sarpkaya, T. and M. Isaacson (1981), *Mechanics of Wave Forces on Offshore Structures*, Van Nostrand-Reinhold, New York.

Stokes, G. G. (1847), "On the Theory of Oscillatory Waves," *Transactions, Cambridge Philosophical Society*, Vol. 8, Part 4, pp. 441–455.
Stokes, G. G. (1880), *Mathematical and Physical Papers*, Cambridge University Press, Vol. 1, p. 227.
Svendsen, Ib. A. and J. Buhr Hansen (1978), "On the Deformation of Periodic Long Waves over a Gently Sloping Bottom," *Journal of Fluid Mechanics*, Vol. 87, Part 3, pp. 433–448.
Ursell, F. (1953), "The Long-Wave Paradox in the Theory of Gravity Waves," *Proceedings, Cambridge Philosophical Society*, Vol. 49, pp. 685–694.
U.S. Army (1984), *Shore Protection Manual*, U.S. Government Printing Office, Washington, D.C., Vol. 1, 4th edition.
Wiegel, R. L. (1960), "A Presentation of Cnoidal Wave Theory for Practical Application," *Journal of Fluid Mechanics*, Vol. 7, pp. 273–286.
Wiegel, R. L. (1964), *Oceanographical Engineering*, Prentice Hall, Englewood Cliffs, New Jersey.
Wilton, J. R. (1913), "On the Highest Waves in Deep Water," *Philosophical Magazine and Journal of Science* 7, Vol. 26, 6th Series, July–Dec., pp. 1053–1058.
Yamaguchi, Masataka (1992), "Interrelation of Cnoidal Wave Theories," *Proceedings* 23rd Intl. Conf. *Coastal Engineering*, Venice, pp. 737–750, 4–44.

CHAPTER 5

Abercromby, R. (1888), "Observations on the Height, Length and Velocity of Ocean Waves," *Philosophical Magazine*, Vol. XXV, 5th Series, April.
Abramowitz, M. and I. A. Stegun (1965), *Handbook of Mathematical Functions*, Dover Publications, New York.
Arthur, R. S. (1949), "Variability in Direction of Wave Travel," in *Ocean Surface Waves, Annulus, New York Academy of Science*, Vol. 51, pp. 511–522.
Barber, N. F. and F. Ursell (1948), "The Generation and Propagation of Ocean Waves and Swell: I. Wave Periods and Velocities," *Philiosophical Transactions of the Royal Society* (London), Vol. 240, No. S24, pp. 527–560.
Beal, R. C., editor (1987), "Measuring Ocean Waves from Space," *Proceedings*, Symposium, published in the *Technical Digest*, Johns Hopkins University Applied Physics Laboratory, Laurel, Maryland, Vol. 8, No. 1, January–March.
Bretschneider, C. L. (1959), "Wave Variability and Wave Spectra for Wind-Generated Gravity Waves," Beach Erosion Board (U.S. Army), Technical Memorandum 118.
Bretschneider, C. L. (1963), "A One-Dimensional Gravity Wave Spectrum," in *Ocean Wave Spectra* (U.S. Naval Oceanographic Office, National Academy of Sciences and National Research Council), Prentice Hall, Englewood Cliffs, NJ, pp. 41–65.
Bretschneider, C. L. (1977), "On the Determination of the Design Ocean Wave Spectrum," *Look Lab/Hawaii*, University of Hawaii, Vol. 7, No. 1.
Buckley, W. H., R. D. Pierce, J. B. Peters, and M. J. Davis (1984), "Use of the Half-Cycle Analysis Method to Compare Measured Wave Height and Simulated Gaussian Data Having the Same Variance Spectrum," *Ocean Engineering*, Vol. 11, No. 4, pp. 423–445.
Carter, D. J. T. (1982), "Prediction of Wave Height and Period for a Constant Wind Velocity Using JONSWAP Results," *Ocean Engineering*, Vol. 9, No. 1, pp. 17–33.
Cartwright, D. E. (1963), "The Use of Directional Spectra in Studying the Output of the Wave-Recorder on a Moving Ship Spectrum," in *Ocean Wave Spectra* (U.S. Naval Oceanographic Office, National Academy of Sciences and National Research Council), Prentice Hall, Englewood Cliffs, NJ, pp. 203–218.
Cornish, V. (1910), *Waves of the Sea and other Water Waves*, T. Fisher Unwin, London.
Cornish, V. (1934), *Ocean Waves and Kindred Geophysical Phenomena*, Cambridge University Press.
Darbyshire, J. (1952), "The Generation of Waves by Wind," *Proceedings, Royal Society (London)*, Series A, pp. 299–328.

Dean, R. G. (2003), *Beach Nourishment, Theory and Practice*, World Scientific Publishing Company, U.S.A.
Donelan, M. (1990), "Air-Sea Interaction," in *The Sea*, Vol. 9, Part A, Wiley-Interscience, New York, Chapter 7.
Ewing, J. A. (1971), "The Generation and Propagation of Directional Wave Data," in *Dynamic Waves in Civil Engineering*, edited by D. Hawells, I. Haigh and C. Taylor, Wiley-Interscience, New York, Chapter 4.
Ewing, J. A. (1986), "Presentation and Interpretation of Directional Wave Data," *Underwater Technology*, Society of Underwater Technology (U.K.), Vol. 12, No. 3, pp. 17–23.
Forristall, G. Z. (1978), "On the Statistical Distribution of Wave Heights in a Storm," *Journal of Geophysical Research*, Vol. 83, pp. 2353–2358.
Goda, Y. (1985), *Random Seas and Design of Marine Structures*, University of Tokyo Press.
Goda, Y. (1990), "Random Seas for Design of Marine Structures," in *Coastal Protection*, edited by K. W. Pilarcyzk, A. A. Balkema Press, Rotterdam, pp. 447–482.
Granger, R. A. (1995), *Fluid Mechanics*, Dover Publications, New York.
Gren, S. (1992), *A Course in Ocean Engineering*, Elsevier Science Publishers, Amsterdam.
Grosskopf, W. G. and N. C. Kraus (1994), "Guidelines for Surveying Beach Nourishment Projects," *Shore & Beach* (Amer. Shore and Beach Preserv. Assn.), Vol. 62, No. 2, pp. 9–16.
Hamada, T. (1964), "On the f^{-5} Law of Wind Generated Waves," Report of the Port and Harbour Technical Research Institute, Japan.
Hasselmann, K. T., P. Barnett, E. Bouws, H. Carlson, D. E. Cartwright, K. Enke, J. A. Ewing, H. Gienapp, D. E. Hasselmann, P. Kruseman, A. Meerburg, P. Muller D. J. Olbers, K. Richter, W. Sell and H. Walden (1973), "Measurements of Wind-Wave Growth and Swell Decay during the 'Joint North Sea Wave Project' (JONSWAP)," *Erganzungsheft zur Deutschen Hydrograph. Zeitschrift*, Reihe A(8), No. 12, pp. 1–95.
Hasselman, K. T., K. Ross, P. Müller and W. Sell (1976), "A Parametric Wave Prediction Model," *Journal of Physical Oceanography*, Vol. 6, pp. 200–228.
Hogben, N. (1990), "Long Term Wave Statistics," in *The Sea-Ocean Engineering Science*, Vol. 9, Part A, Chapter 8.
Isaacson, M. de St. Q. and N. G. MacKenzie (1981), "Long-Term Distribution of Ocean Waves: A Review," *Journal of Waterway, Port, Coastal and Ocean Engieering* (ASCE), Vol. 107, No. WW2, May, pp. 93–109.
ISSC (1967), Report of Committee 1, Environmental Conditions, International Ship Structures Committee, Oslo.
ITTC (1972), "Technical Decisions and Recommendations of the Seakeeping Committee," *Proceedings*, 12th and 13th International Towing Tank Conference, Rome (1969) and Berlin (1972).
Kinsman, B. (1965), *Wind Waves*, Prentice Hall, Englewood Cliffs, NJ.
Kitaigorodskii, S. A. (1962), "Application of the Theory of Similarity to the Analysis of Wind-Generated Wave Motion as a Stochastic Process," *Bulletin of the Academy of Science of the USSR*, Geophysical Series, 1, pp. 105–117.
Komen, G. J., L. Cavaleri, M. Donelan, K. Hasselmann, S. Hasselmann, and P. A. E. M. Janssen (1994), *Dynamics and Modelling of Ocean Waves*, Cambridge University Press.
Lamb, H. (1945), *Hydrodynamics*, Dover Publications, New York, 6th edition.
Le Méhauté, B. (1982), "Relationships between Narrow Band Directional Energy and Probability Densities of Deep and Shallow Water Waves," *Applied Ocean Research*, Vol. 4, No. 1, pp. 17–24.
Le Méhauté, B. and D. M. Hanes (1990), *Ocean Engineering Science*, Vol. 9, Part A of *The Sea*, Wiley-Interscience, New York.
Longuet-Higgins, M. S. (1952), "On the Statistical Distribution of the Heights of Sea Waves," *Journal of Marine Research*, Vol. 11, No. 3, pp. 245–266.
Longuet-Higgins, M. S., D. E. Cartwright, and N. B. Smith (1963), "Observations of the Directional Spectrum of Sea Waves using the Motions of a Floating Buoy," in *Ocean Wave*

Spectra (U.S. Naval Oceanographic Office, National Academy of Sciences and National Research Council), Prentice Hall, Englewood Cliffs, NJ, pp. 111–132.
McCormick, M. E. (1973), *Ocean Engineering Wave Mechanics*, Wiley-Interscience, New York.
McCormick, M. E. (1978), "Wind-Wave Power Available to a Wave Energy Conversion Array," *Ocean Engineering*, Vol. 5, No. 2, April, pp. 67–74.
McCormick, M. E. (1998a), "On the Use of Wind-Wave Spectral Formulas to Estimate Wave Energy Resources," *Journal of Energy Resource Technology* (ASME), Vol. 20, No. 2, December, pp. 314–317.
McCormick, M. E. (1998b), "A Generic Deep-Water Wave Spectral Formula," *Oceanic Engineering International* (ECOR), Vol. 2, No. 2, pp. 71–77.
McCormick, M. E. (1999), "Applcation of the Generic Spectral Formula to Fetch-Limited Seas," *Marine Technology Society Journal* (MTS), Vol. 33, No. 3, Fall, pp. 27–32.
McCormick, M. E. and Y. C. Kim (1997), "Ocean Wave-Powered Desalination," *Proceedings*, 27th Congress of the International Association for Hydraulics Research, San Francisco, June, Paper F19-D26–4.
McCormick, M. E. and S. E. Mouring (1995), "On the Hydrodynamic Design of Active Panels of Marine Vehicles," *Journal of Offshore Mechanics and Arctic Engineering* (ASME), Vol. 117, No. 4, pp. 290–294.
McCormick, M. E., R. Bhattacharyya, and S. E. Mouring (1997), "Hydroelastic Considerations in Ship Panel Design," *Naval Engineers Journal* (ASNE), Vol. 109, No. 5, September, pp. 61–66.
Mitsuyasu, H. (1970) "On the Growth of Spectrum of Wind-Generated Waves (2) – Spectral Shape of Wind Waves at Finite Fetch," *Proceedings*, 17th Conference on Coastal Engineering, pp. 1–7 (in Japanese).
Mitsuyasu, H., F. Tasai, T. Suhara, S. Mizuno, M. Ohkusu, T. Honda, and K. Rikiishi (1975), "Observations of the Directional Spectrum of Ocean Waves using a Cloverleaf Buoy," *Journal of Physical Oceanograpghy*, Vol. 5, pp. 750–760.
Mollison, D. (1982), "Ireland's Wave Power Resource," Report to the Natl. Board for Science and Technology and the Electricity Supply Board, ISBN 0-86282-023-5, June.
Moskowitz, L. (1964), "Estimates of the Power Spectrums for Fully Developed Seas for Wind Speeds of 20 to 40 Knots," *Journal of Geophysical Research*, Vol. 69, No. 24, pp. 5161–5179, December.
Neumann, G. (1952), "On Ocean Wave Spectra and a New Method of Forecasting Wind-Generated Seas," Technical Memorandum No. 43, Beach Erosion Board, U.S. Army Corps of Engineers.
Niedzwecki, J. M. and C. P. Whatley (1991), "A Comparative Study of Some Directional Sea Models," *Ocean Engineering*, Vol. 18, Nos. 1/2, pp. 111–128.
Phillips, O. M. (1957), "On the Generation of Waves by Turbulent Wind," *Journal of Fluid Mechanics*, Vol. 2, pp. 417–445.
Phillips, O. M. (1958), "The Equilibrium Range in the Spectrum of Wind Generated Waves," *Journal of Fluid Mechanics*, Vol. 4, No. 4, pp. 785–790.
Phillips, O. M. (1966), *The Dynamics of the Upper Ocean*, Cambridge University Press, London.
Phillips, O. M. (1985), "Spectral and Statistical Properties of the Equilibrium Range in Wind-Generated Gravity Waves," *Journal of Fluid Mechanics*, Vol. 156, pp. 505–531.
Pierson, W. J. (1952), "A Unified Mathematical Theory for the Analysis, Propagation and Refraction of Storm Generated Ocean Surface Waves," Parts I and II, Department of Meteorology and Oceanography, New York University.
Pierson, W. J. (1955), "Wind Generated Gravity Waves," *Advances in Geophysics*, Vol. 2, Academic Press, New York, pp. 93–178.
Pierson, W. J. (1964), "The Interpretation of Wave Spectrum in Terms of the Wind Profile Instead of Wind Measured at Constant Height," *Journal of Geophysical Research*, Vol. 69, No. 24, December, pp. 5191–5203.

Pierson, W. J. and L. Moskowitz (1964), "A Proposed Spectral Form for Fully Developed Wind Seas Based on the Similarity Theory of S. A. Kitaigorodskii," *Journal of Geophysical Research*, Vol. 69, No. 24, pp. 5181–5190, December.

Pilarczyk, K. W., editor (1990), *Coastal Protection*, A. A. Balkema, Rotterdam.

Putz, R. R. (1952). "Statistical Distribution for Ocean Waves," *Transactions*, American Geophysical Union, Vol. 33, No. 5, pp. 685–692.

Rice, S. O. (1944), "Mathematical Analysis of Random Noise," *Bell System Technical Journal*, Vol. 23, pp. 282–332.

Sarpkaya, T. and M. Isaacson (1981), *Mechanics of Wave Forces on Structures*, Van Nostrand Reinhold Company, New York, pp. 514–515.

Sorensen, R. M. (1973), "Ship-Generated Waves," in *Advances in Hydroscience*, Academic Press, New York, Vol. 9, pp. 49–83.

Sorensen, R. M. (1993), *Basic Wave Mechanics: For Coastal and Ocean Engineers*, Wiley-Interscience, New York.

St. Denis, M. (1969), "On Wind Generated Waves," in *Topics in Ocean Engineering*, Vol. 1, edited by C. L. Bretschneider, Gulf Publishing Company, Part 1, p. 4.

St. Denis, M. and W. J. Pierson (1953), "On the Motion of Ships in Confused Seas," *Transactions*, Society of Naval Architects and Marine Engineers, Vol. 61, pp. 280–357.

Strutt, J. W. (1880), "On the Resultant of a Large Number of Vibrations of the Same Pitch and of Arbitrary Phase," in *Scientific Papers*, Vol. I, Cambridge University Press, Chapter 68.

Sverdrup, H. U. and W. H. Munk (1947), "Empirical and Theoretical Relations between Wind, Sea and Swell," *Transactions*, American Geophysical Union, Vol. 27, No. VI, pp. 823–827.

Taylor, G. I. (1921), "Diffusion by Continuous Movement," *Proceedings*, London Mathematical Society, Ser. 2, Vol. XX, pp. 196–212.

Tucker, M. J. (1991), *Waves in Ocean Engineering*, Ellis Horwood Limited, Chichester, England.

Tucker, M. J. and E. G. Pitt (2001), *Waves in Ocean Engineering*, Elsevier, Amsterdam.

Tuckey, J. W. (1949), "The Sampling Theory of Power Spectrum Estimates," Symposium on Applications of Autocorrelation Analysis to Physical Problems. U.S. Office of Naval Research. NAVEXOS P-735, Woods Hole, Massachusetts, pp. 47–67.

U.S. Army (1984), "Shore Protection Manual," Vol. 1, Corps of Engineers, 4th Edition.

Weber, E. H. and W. Weber (1825), "Experimental Determination of Waves," Leipzig.

Weibull, W. (1951), "A Statistical Distribution Function of Wide Applicability," *Journal of Applied Mechanics* (ASME), September, pp. 293–297.

Wiegel, R. L., editor (1982), *Directional Wave Spectra Applications*, American Society of Civil Engineers, New York.

Wu, J. (1980), "Wind Stress Coefficients over Sea Surface near Neutral Conditions – A Revisit," *Journal of Physical Oceanography*, Vol. 10, No. 5, May, pp. 727–740.

Young, I. R. (1999), *Wind Generated Ocean Waves*, Elsevier Science, Oxford, U.K.

CHAPTER 6

Abramowitz, M. and I. A. Stegun (1965), *Handbook of Mathematical Functions*, Dover Publications, New York.

Abul-Azm, A. G. and A. N. Williams (1997), "Oblique Wave Diffraction by Segmented Offshore Breakwaters," *Ocean Engineering*, Vol. 24, No. 1, pp. 63–82.

Achenbach, J. D. and Z. L. Li (1986), "Reflection and Transmission of Scalar Waves by a Periodic Array of Screens," *Wave Motion* (North Holland, Amsterdam), Vol. 8, No. 2, March, pp. 225–234.

Adey, R. A. and C. A. Brebbia (1983), *Basic Computational Techniques for Engineers*, John Wiley & Sons, New York.

Bateman, H. (1964), *Partial Differential Equations of Mathematical Physics*, Cambridge University Press, London, Chapter XI.

Berkoff, J. C. W. (1972), "Computation of Combined Refraction-Diffraction," *Proceedings*, 13th Coastal Engineering Conference (ASCE), Vancouver, July, pp. 471–490.

Berkoff, J. C. W. (1976), "Mathematical Models for Simple Harmonic Linear Water Waves, Wave Diffraction and Refraction," Delft Hydraulics Laboratory, Publication No. 163, Delft, The Netherlands.

Blue, F. L. and J. W. Johnson (1949), "Diffraction of Water Waves Passing Through a Breakwater Gap," *Transactions, American Geophysical Union*, Vol. 30, No. 5, October, pp. 705–718.

Booij, N. (1983), "A Note on the Accuracy of the Mild-Slope Equation," *Coastal Engineering*, Elsevier Science, Amsterdam, Vol. 7, pp. 191–203.

Bourodimus, E. L. and A. T. Ippen (1966), "Wave Transformation in an Open Channel Transition," *Journal, Hydraulic Division* (ASCE), Vol. 94, No. HY5, pp. 1317–1329.

Brebbia, C. A. and S. Walker (1979), *Dynamic Analysis of Offshore Structures*, Butterworths, London.

Chandrasekera, C. N. and K. F. Cheung (1997), "Extended Linear Refraction-Diffraction Model," *Journal, Waterway, Port, Coastal and Ocean Engineering* (ASCE), Vol. 123, No. 5, pp. 280–286.

Copeland, G. J. M. (1985), "A Practical Alternative to the 'Mild-Slope' Wave Equation," *Coastal Engineering* (Elsevier), Vol. 9, pp. 125–149.

Dalrymple, R. A. and P. A. Martin (1990), "Wave Diffraction through Offshore Breakwaters," *Journal, Waterway, Port, Coastal and Ocean Engineering* (ASCE), Vol. 116, No. 6, pp. 727–741.

Dean, R. G. (1964), "Long Wave Modification by Linear Transformations," *Journal, Waterways, Harbors Division* (ASCE), Vol. 90, No. WW1, pp. 1–29.

Dickson, W. S., T. H. C. Herbers, and E. B. Thornton (1995), "Wave Reflection from Breakwater," *Journal, Waterway, Port, Coastal and Ocean Engineering* (ASCE), Vol. 121, No. 5, pp. 262–268.

Elmore, W. C. and M. A. Heald (1985), *Physics of Waves*, Dover Publications, New York.

Girolamo, P. De, J. K. Kostense and M. W. Dingemans (1988), "Inclusion of Wave Breaking in a Mild-Slope Model," *Computer Modeling in Ocean Engineering*, Balkema, Rotterdam.

Goda, Y. (1970), "A Synthesis of Breaker Indices," *Transactions, Japanese Society of Civil Engineers*, Vol. 2, Part 2, pp. 227–230.

Halliday, D. and R. Resnick (1978), *Physics*, Part II, John Wiley & Sons, New York, 3rd edition.

Healy, J. J. (1953), "Wave Damping Effect of Beaches," *Proceedings*, Minnesota International Hydraulics Convention (IAHR), pp. 213–220.

Johnson, J. W. (1952), "Generalized Wave Diffraction Diagrams," *Proceedings*, 2nd Conference on Coastal Engineering, Council on Wave Research, The Engineering Foundation, pp. 6–23.

Korn, G. A. and T. M. Korn (2000), *Mathematical Handbook for Scientists and Engineers*, Dover Publications, Mineola, New York. Originally published by McGraw-Hill in 1968.

Lamb, H. (1945), *Hydrodynamics*, Dover Publications, New York. Originally published by Cambridge University Press.

Lee, C. (1999), Discussion of "Extended Linear Refraction-Diffraction Model," *Journal, Waterway, Port, Coastal and Ocean Engineering* (ASCE), Vol. 125, No. 3, pp. 256–257.

Leenknecht, D. A., A. Szuwalski, and A. R. Sherlock (1992), "Automated Coastal Engineering System," Coastal Engineering Research Center, U.S. Army, Vicksburg, Mississippi, September.

Lindsay, R. B. (1960), *Mechanical Radiation*, McGraw-Hill, New York.

Liu, P. L.-F., S. B. Yoon, and R. A. Dalrymple (1986), "Wave Reflection from Energy Dissipation Region," *Journal, Waterway, Port, Coastal and Ocean Engineering* (ASCE), Vol. 112, No. 6, pp. 632–644.

Madsen, O. S. (1974), "Wave Transmission through Porous Structures," *Journal, Waterway, Port, Coastal and Ocean Engineering* (ASCE), Vol. 100, No. WW3, pp. 169–188.

Madsen, O. S. (1983), "Wave Reflection from a Vertical Permeable Wave Absorber," *Coastal Engineering* (Elsevier), Vol. 7, pp. 381–396.

Matthews, J. and R. L. Walker (1970), *Mathematical Methods of Physics*, Addison-Wesley, Reading, Massachusetts.

McCormick, M. E. and Cerquetti, J. (2003), "Empirical Formula for the Breaking Height Index," *Ocean Engineering International* (ECOR),Vol. 7, No. 2, pp. 90–95.

McCormick, M. E. and D. R. B. Kraemer (2002), "Polynomial Approximations for Fresnel Integrals in Diffraction Analysis," *Coastal Engineering*, Vol. 44, pp. 261–266.

Penny, W. G. and A. T. Price (1944), "Diffraction of Water Waves by Breakwaters," Misc. Weapons Development Technical History 26, Artificial Harbors, Sec. 3D.

Penny, W. G. and A. T. Price (1952), 'The Diffraction Theory of Sea Waves and the Shelter Afforded by Breakwaters," Part I of "Some Gravity Wave Problems in the Motion of Perfect Liquids" by J. C. Martin, W. J. Moyce, W. G. Penney, A. T. Price, and C. K. Thornhill, *Philosophical Transactions*, Royal Society of London, Series A, Vol. 244, pp. 236–253.

Putnam, J. A. and R. S. Arthur (1948), "Diffraction of Water Waves by Breakwaters," *Transactions, American Geophysical Union*, Vol. 29, No. 4, August, pp. 481–490.

Rey, V., M. Belzons, and E. Guazzelli (1992), "Propagation of Surface Gravity Waves over a Rectangular Submerged Bar," *Journal of Fluid Mechanics*, Vol. 235, pp. 453–479.

Saville, T. (1956), "Wave Runup on Shore Structures," *Journal, Waterways and Harbor Division* (ASCE) WW2, Vol. 82, No. WW2.

Shuto, N. (1972), "Standing Waves in front of a Sloping Dike," *Proceedings*, 13th Conference on Coastal Engineering (ASCE), pp. 1629–1647.

Sommerfeld, A. (1896), "Mathematische Theorie der Diffraction," *Mathematische Annalen*, Vol. 47, pp. 317–374.

Sommerfeld, A. (1949), *Partial Differential Equations in Physics* (translated by E. G. Strauss), Academic Press, New York.

Sommerfeld, A. (1954), *Optics* (translated by O. Laporte and P. A. Moldauer), Academic Press, New York.

Sollitt, C. K. and R. H. Cross (1972), "Wave Transmission through Permeables, Breakwaters," *Proceedings*, 13th Coastal Engineering Conference (ASCE), Vancouver, Vol. III, pp. 1872–1846.

Stoker, J. J. (1957), *Water Waves*, Interscience, New York.

Tsai, C.-P., J.-S. Wang, and C. Lin (1998), "Downrush Flow from Waves on Sloping Seawalls," *Ocean Engineering*, Vol. 10, Nos. 4–5, pp. 295–308.

Twu, S.-H. and C.-C. Liu (1999), "The Reflection Coefficient of Sloping Walls," *Ocean Engineering*, Vol. 26, No. 11, pp. 1085–1094.

U.S. Army (1984), "Shore Protection Manual," Coastal Engineering Research Center, U.S. Army, Vicksburg, Mississippi.

Wang, K.-H., and X. Ren (1992), "Wave Waves on Flexible and Porous Breakwaters," *Journal, Waterway, Port, Coastal and Ocean Engineering* (ASCE), Vol. 119, No. 5, pp. 1025–1047.

Wiegel, R. L. (1962), "Diffraction of Waves by Semi-Infinite Breakwaters," *Journal, Hydraulics Division, Proceedings* (ASCE), Vol. 88, HY1, January, pp. 27–44.

Wiegel, R. L. (1964), *Oceanographical Engineering*, Prentice Hall, Englewood Cliffs, NJ.

Williams, A. N. and W. W. Crull (1993), "Wave Diffraction by Array of Thin-Screen Breakwaters," *Journal, Waterway, Port, Coastal and Ocean Engineering* (ASCE), Vol. 119, No. 6, pp. 606–617.

Yokoki, H., M. Isobe, and A. Watanabe (1992), "A Method for Estimating Reflection Coefficient in Short-Crested Random Seas," *Proceedings*, 23rd International Conference on Coastal Engineering (ASCE), Venice, Chapter 57, pp. 765–776.

Zienkiewicz, O. C. and R. L. Taylor (2000), *Finite Element Method*, Oxford Univ. Press, 5th edition.

CHAPTER 7

Battjes, J. A. (1974), "Surf Similarity," *Proceedings*, Fourteenth Coastal Engineering Conference (ASCE), Copenhagen, June, pp. 466–480.
Bowen, A. J. (1969a), "Rip Currents, 1," *Journal of Geophysical Research*, Vol. 74, pp. 5467–5478.
Bowen, A. J. (1969b), "The Generation of Longshore Currents on a Plane Beach," *Journal of Marine Research*, Vol. 27, pp. 205–215.
Bowen, A. J., D. L. Inman, and V. P. Simmons (1968), "Wave Set-Down and Set-Up," *Journal of Geophysical Research*, American Geophysical Union, Vol. 73, No. 7, pp. 2569–2577, April.
Dean, R. G. (1974), "Evaluation and Development of Water Wave Theories for Engineering Application," Special Report No. 1, Coastal Engineering Research Center.
Dean, R. G. and R. A. Dalrymple (2002), *Coastal Processes*, Cambridge University Press, New York.
Fredsoe, J., editor (2002), "Surf and Swash Zone Mechanics," *Coastal Engineering, Special Issue*, Elsevier, Vol. 45, Nos. 3–4, May.
Galvin, C. J. (1968), "Breaker Type Classification on Three Laboratory Beaches," *Journal of Geophysical Research*, Vol. 73, No. 12, pp. 3651–3659.
Galvin, C. J. (1972), "Wave Breaking in Shallow Water," in *Waves on Beaches*, Academic Press, edited by R. E. Meyer, pp. 413–456.
Goda, Y. (1970a), "A Synthesis of Breaker Indices," *Proc. Japan Soc. Civil Engineers*, Vol. 180, August, pp. 39–49.
Goda, Y. (1970b), "A Synthesis of Breaker Indices," *Trans. Japan Soc. Civil Engineers*, Vol. 2, Part 2, pp. 227–230.
Goda, Y. (1985), *Random Seas and Design of Maritime Structures*, University of Tokyo Press.
Granger, R. A. (1995), *Fluid Mechanics*, Dover Publications, New York.
Herbich, J. B., editor (1999), *Handbook of Coastal Engineering*, McGraw-Hill Book Company, New York.
Horikawa, K. (1978), *Coastal Engineering*, John Wiley & Sons, New York.
Horikawa, K. and C. T. Kuo (1966), "A Study of Wave Transformation Inside Surf Zone," *Proceedings*, 10th Coastal Engineering Conference (ASCE), Vol. 1, pp. 217–233.
Hughes, S. A. (2004), "Estimation of Wave Run-up on Smooth, Impermeable Slopes Using the Wave Momentum Flux Parameter," *Coastal Engineering*, Vol. 51, Nos. 11–12, December, pp. 1085–1104.
Hunt, I. A. (1959), "Design of Seawalls and Breakwaters," *Journal of Waterways, Harbors and Coastal Engineering* (ASCE), Vol. 85, WW3, pp. 123–152. Originally published as a U.S. Army Corps of Engineers report, Lake Survey, Detroit.
Iribarren, C. R. and C. Nogales (1949), "*Protection des Ports*," Section II, Comm. 4, XVIIth Intl. Navigation Congress, Lisbon, pp. 31–80.
Jonsson, I. G. (1966), "Wave Boundary Layers and Friction Factors," *Proceedings*, 10th Coastal Engineering Conference (ASCE), pp. 127–148.
Kamphius, J. W. (1975), "Friction-Factor under Oscillatory Waves," *Journal of Waterways, Harbors and Coastal Engineering* (ASCE), Vol. 101, pp. 135–144.
Kamphius, J. W. (1991), "Alongshore Sediment Transport Rate," *Journal of Waterways, Harbors and Coastal Engineering* (ASCE), Vol. 117, pp. 624–640.
Kamphius, J. W. (2002), "Alongshore Sediment Transport Rate (II)," *Abstracts*, 28th International Conference on Coastal Engineering (Institute of Civil Engineers), Cardiff, Wales, July, Paper No. 4.
Kim, Y. C., editor (2009), *Handbook of Coastal and Ocean Engineering*, World Scientific Publications, Singapore.
Komar, P. D. and M. K. Gaughan (1972), "Airy Wave Theory and Breaker Height Prediction," *Proceedings*, 13th Coastal Engineering Conference (ASCE), Vol. 1, pp. 405–418.
Liu, P. L.-F. and R. A. Dalrymple (1978), "Bottom Frictional Stresses and Longshore Currents Due to Waves with Large Scales of Incidence," *Journal of Marine Research*, Vol. 36, pp. 357–475.

Longo, S., P. Petti, and I. J. Losada (2002), "Turbulence in the Swash and Surf Zones: A Review," *Coastal Engineering*, Vol. 45, Nos. 3–4, May, pp. 129–147.

Longuet-Higgins, M. S. (1970a), "Longshore Currents Generated by Obliquely Incident Sea Waves, 1," *Journal of Geophysical Research*, Vol. 75, No. 33, pp. 6778–6789.

Longuet-Higgins, M. S. (1970b), "Longshore Currents Generated by Obliquely Incident Sea Waves, 2," *Journal of Geophysical Research*, Vol. 75, No. 33, pp. 6790–6801.

Longuet-Higgins, M. S. (1972), "Recent Progress in the Study of Longhore Currents," in *Waves on Beaches*, R. E. Meyer, ed., Academic Press, New York, pp. 203–248.

Longuet-Higgins, M. S. and R. W. Stewart (1960), "Changes in the Form of Short Gravity Waves on Long Waves and Tidal Currents," *Journal of Fluid Mechanics*, Vol. 8, pp. 565–583.

Longuet-Higgins, M. S. and R. W. Stewart (1961), "The Changes in Amplitude of Short Gravity Waves on Steady Non-Uniform Currents," *Journal of Fluid Mechanics*, Vol. 10, pp. 529–549.

Longuet-Higgins, M. S. and R. W. Stewart (1962), "Radiation Stress and Mass Transport in Gravity Waves, with Application to 'Surf Beats'," *Journal of Fluid Mechanics*, Vol. 13, pp. 481–504.

Longuet-Higgins, M. S. and R. W. Stewart (1963), "A Note on Wave Set-Up," *Journal of Marine Research*, Vol. 21, No. 1, pp. 4–10.

Longuet-Higgins, M. S. and R. W. Stewart (1964), "Radiation Stresses in Water Waves: A Physical Discussion with Applications," *Deep-Sea Research*, Vol. 11, pp. 529–562.

Massel, S. R. (1989), *Hydrodynamics of Coastal Zones*, Elsevier, Amsterdam.

McCormick, M. E. (1973), *Ocean Engineering Wave Mechanics*, Wiley-Interscience, New York.

McCormick, M. E. and J. Cerquetti (2002), "Empirical Formula for the Breaking Wave Index," *Ocean Engineering International* (ECOR),Vol. 7, No. 2.

Miller, C. D. (1987), "Longshore Transport and Wave Decay in the Surf Zone," *Proceedings*, Coastal Hydrodynamics Conference (ASCE), University of Delaware, June–July, pp. 140–154.

Munk, Walter H. (1949), "The Solitary Wave Theory and Its Application to Surf Problems," *Annals*, New York Acad. Sci., Vol. 51, Art. 3, pp. 343–572.

Refaat, H. E. A. A., Y. Tsuchiya, and Y. Kawata (1990), "Similarity of Velocity Profiles in Non-Uniform Longshore Currents," *Proceedings*, Twenty-Second Coastal Engineering Conference (ASCE), Delft, July, Vol. 1, pp. 281–292.

Saville, T. (1961), "Experimental Determination of Wave Setup," *Proceedings*, 2nd Technical Conf. on Hurricanes, National Hurricane Research Project, Report No. 50, pp. 242–252.

Sorensen, R. M. (1997), *Basic Coastal Engineering*, Chapman and Hall, New York.

Wang, L. and H. Du (1993), "Approach to Calculating Methods of Wave Breaking Depth and Heights," *China Ocean Engineering* (Chinese Ocean Engineering Society), Vol. 7, No. 1, pp. 85–98.

Weggel, J. R. (1972), "Maximum Breaker Height," *Journal of Waterways, Harbors and Coastal Engineering Div.* (ASCE), Vol. 98, WW4, Nov., pp. 529–548.

U.S. Army (1984), *Shore Protection Manual*, U.S. Government Printing Office, Washington, DC, Vol. 1, 4th edition.

U.S. Army (1990), "Practical Considerations in Longshore Transport Rate Calculations," U.S. Army Corps of Engineers Coastal Engineering, Technical Note, II-24, December.

Zill, D. G. (1986), *A First Course in Differential Equations with Application*, Prindle, Webber & Schmidt Publishers, Boston, 3rd edition.

CHAPTER 8

CIRIA (1990), "Groynes in Coastal Engineerin: Data on Performance of Exisiting Groyne Systems," Construction Industry Research and Information Association (CIRIA), Technical Report No. 135, London.

Dean, R. G. and R. A. Dalrymple (2002), *Coastal Processes*, Cambridge University Press, New York.

Fleming, C. A. (1990), "Guide on the Uses of Groynes in Coastal Engineering," Construction Industry Research and Information Association (CIRIA), Technical Report No. 119, London.

Goda, Y. (1975), *Random Seas and Design of Maritime Structures*, University of Tokyo Press.

Herbich, J. B., editor (1999), *Handbook of Coastal Engineering*, McGraw-Hill, New York.

Horikawa, K. (1978), *Coastal Engineering*, John Wiley & Sons, New York.

Hudson, R. Y. (1953), "Wave Forces on Breakwaters," *Transactions*, American Society of Civil Engineers (ASCE), Vol. 118, p. 653.

Hudson, R. Y. (1959), "Laboratory Investigations of Rubble-Mound Breakwaters," *Proceedings*, American Society of Civil Engineers (ASCE), Waterways and Harbors Division, Vol. 85, WW3, Paper No. 2171.

Hudson, R. Y. (1961a), "Wave Forces on Rubble-Mound Breakwaters and Jetties," Miscellaneous Paper 2–453, U.S. Army Waterways Experiment Station, Vicksburg, MS.

Hudson R. Y. (1961b), "Laboratory Investigations of Rubble-Mound Breakwaters," *Transactions*, American Society of Civil Engineers (ASCE), Vol. 126, Part IV.

Iribarren Cavanilles, R. (1938), "A Formula for Calculation of Rock-Fill Dikes," *Revista de Obras Publicas* – Translation into English in *The Bulletin of the Beach Erosion Board*, Vol. 3, No. 1, January, 1949.

Iribarren Cavanilles, R. and C. Nogales *y* Olano (1950), "Generalization of the Formula for Calculation of Rock-Fill Dikes and Verification of Its Coefficients," *Revista de Obras Publicas* – Translation into English in *The Bulletin of the Beach Erosion Board*, Vol. 5, No. 1, January, 1951.

Kim, Y. C., editor (2009), *Handbook of Coastal and Ocean Engineering*, World Scientific Press, Hackensack, NJ.

Melby, J. A. (2002), "Damage Development on Stone Armored Breakwaters and Revetments," U.S. Army Corps of Engineers, Coastal Engineering Technical Note ERDC/CHL CHETN-111–64, June. Available at http://bigfoot.wes.army.mil/cetn.index.mil.

Ramakumar, R. (1993), *Engineering Reliability: Fundamentals and Applications*, Prentice Hall, Englewood Cliffs, NJ.

Sorensen, R. M. (1997), *Basic Coastal Engineering*, Chapman and Hall Publishers, New York.

Weibull, W. (1951), "A Statistical Distribution Function of Wide Applicability," *Journal of Applied Mechanics* (ASME), September, pp. 293–297.

U.S. Army (1984), *Shore Protection Manual*, U.S. Government Printing Office, Washington, DC, Vol. 1, 4th edition.

U.S. Army (1994), "Coastal Groins and Nearshore Breakwaters," American Society of Civil Engineers (ASCE), New York.

CHAPTER 9

Ablowitz, M. J. and A. S. Fokas (1997), *Complex Variables*, Cambridge University Press, Cambridge, U.K.

Abramowitz, M. and I. A. Stegun (1965), *Handbook of Mathematical Functions*, Dover Publications, New York.

Agarwal, A. K. and A. K. Jain (2003), "Dynamic Behavior of Offshore Spar Platforms under Regular Sea Waves," *Ocean Engineering* (Elsevier), Vol. 30, No. 4, pp. 487–516.

Akyildiz, H. (2002). "Experimental Investigation of Pressure Distribution on a Cylinder due to the Wave ns, Diffraction in Finite Water Depth," *Ocean Engineering* (Elsevier), Vol. 29, No. 9, August, pp. 1119–1132.

Borgman, L. E. (1965), "The Spectral Density for Ocean Wave Forces," *Proceedings*, Coastal Engineering Conference (ASCE), Santa Barbara, California, October, pp. 147–182.

Borgman, L. E. (1981), "Directional Spectrum Density of Forces on Platforms," *Proceedings*, Directional Wave Spectra Applications (ASCE), edited by R. L. Wiegel, Berkeley, California, September, 1981, pp. 315–332.

Bostrom, T. and T. Overvik (1986), "Hydrodynamic Force Coefficients in Random Wave Condition," *Proceedings*, 5th International Offshore Mechanics and Arctic Engineering Symposium (ASME), Tokyo, pp. 236–243.

Brennen, C. E. (1982), "A Review of Added-Mass and Fluid Inertial Forces," U.S. Navy, Naval Civil Engineering Laboratory Report CR82.010, January.

Bretschneider, C. L. (1963), "A One-Dimensional Gravity Wave Spectrum," in *Ocean Wave Spectra* (U.S. Naval Oceanographic Office, National Academy of Sciences and National Research Council), Prentice Hall, Englewood Cliffs, NJ, pp. 41–65.

Bretschneider, C. L. (1965), "On the Probability Distribution of Wave Force and an Introduction to the Correlation Drag Coefficient and the Correlation Inertial Coefficient," *Proceedings*, Coastal Engineering Conference (ASCE), Santa Barbara, California, October, pp. 183–217.

Bretschneider, C. L. (1967), "Probability Distribution of Wave Force," *Journal of the Waterways and Harbors Division* (ASCE), WW2, May, pp. 5–26.

Burrows, R., R. G. Tickell, D. Hames, and G. Najafian (1997), "Morison Wave Force Coefficients for Application to Random Seas," *Applied Ocean Research*, Elsevier Science, Vol. 19, pp. 183–199.

Cassidy, M. J., R. Eatock Taylor, and G. T. Houlsby (2001), "Analysis of Jack-Up Units Using a Constrained New Wave Methodology," *Applied Ocean Research*, Elsevier Science, Vol. 23, No. 4, August, pp. 221–224.

Chakrabarti, S. K. (1975), "Wave Forces for Fixed Offshore Structures," Preprint, Structural Engineering Conference (ASCE), New Orleans, April.

Crandall, S. H. and W. D. Mark (1963), *Random Vibrations in Mechanical Systesm*, Academic Press, New York.

Cook, G. R. (1987), "The Lighthill Correction to the Morison Equation," Ph.D. dissertation, Department of Civil Engineering, Johns Hopkins University, Baltimore, MD.

Dean, R G. and R. A. Dalrymple (1984), *Water Wave Mechanics for Engineers and Scientists*, Prentice Hall, Englewood Cliffs, NJ.

Faltinsen, O. M. (1990), *Sea Loads on Ships and Offshore Structures*, Cambridge University Press, Cambridge, U.K., 1st edition, Chapter 4.

Frank, W. (1967), "Oscillation of Cylinders in or below the Free Surface of Deep Fluids," Naval Ship Research and Development Center (U.S. Navy) Report 2375, October.

Garrett, C. J. R. (1971), "Wave Forces on a Circular Dock," *Journal of Fluid Mechanics*, Vol. 46, Part I, pp. 129–139.

Garrison, C. J. (1974), "Hydrodynamics of Large Objects in the Sea, Part I – Hydrodynamic Analysis," *Journal of Hydronautics* (AIAA), Vol. 8, pp. 5–12.

Garrison, C. J. (1975), "Hydrodynamics of Large Objects in the Sea, Part II – Motion of Free-Floating Bodies," *Journal of Hydronautics*, Vol. 9, pp. 58–63.

Garrison, C. J. (1978), "Hydrodyamic Loading of Large Offshore Structures: Three-Dimensional Source Distribution Methods," Chapter 3 of *Numerical Methods in Offshore Engineering*, edited by O. C. Zienkiewicz, R. W. Lewis, and K. G. Stagg, John Wiley & Sons, New York.

Gradshteyn, I. S. and I. M. Ryzhik (1980), *Tables of Integrals, Series and Products*, Academic Press, New York, 2nd edition.

Havelock, T. H. (1917), "Some Cases of Wave Motion due to a Submerged Obstacle," *Proceedings*, Royal Society, A, Vol. 93, pp. 520–532.

Havelock, T. H. (1926), "The Method of Images in Some Problems of Surface Waves," *Proceedings*, Royal Society, A, Vol. 115, pp. 268–280.

Havelock, T. H. (1940), "The Pressure of Water Waves upon a Fixed Obstacle," *Proceedings*, Royal Society of London, Series A, No. 963, Vol. 175, pp. 409–421.

Hudson, P. J. (2001), "Wave-Induced Migration of Grounded Ships," Doctoral dissertation, Dept. of Civil Engineering, Johns Hopkins University, Baltimore, MD.

Hudspeth, R. T. and J. H. Nath (1985), "High Reynolds Numbers Wave Force Investigation in a Wave Flume," Oregon State University, Civil Engineering Dept., Final Report Sep 82-Feb 85, March.

Ippen, A. T., editor (1966), *Estuary and Coastline Hydrodynamics*, McGraw-Hill, New York.

Isaacson, M. (1979), "Nonlinear Inertial Forces on Bodies," *Journal of Waterways, Harbors and Coastal Engineering Division* (ASCE), Vol. 105, No. WW3, pp. 213–227.

Isaacson, M. and D. J. Maull (1976), "Transverse Forces on Vertical Cylinders in Waves," *Journal of Waterways, Harbors and Coastal Engineering Division* (ASCE), Vol. 102, No. WW1, pp. 49–60.

Isaacson, M., J. Baldwin, and C. Niwinski (1991), "Estimation of Drag and Inertia Coefficients from Random Wave Data," *Journal of Offshore Mechanics and Arctic Engineering* (ASME), Vol. 113, May, pp. 128–136.

Keulegan, G. H. and L. H. Carpenter (1958), "Forces on Cylinders and Plates in an Oscillating Fluid," *Journal of Research of the National Bureau of Standards*, Washington, DC, Vol. 60, No. 5, May.

Korn, G. A. and T. M. Korn (2000), *Mathematical Handbook for Scientists and Engineers*, Dover Publications, Mineola, New York. Originally published in 1968 by McGraw-Hill, New York.

Lamb, H. (1932), *Theoretical Hydrodyamics*, Cambridge University Press, U.K., 6th edition. Subsequently published by Dover Publications, New York, in 1945.

Lewis, F. M. (1929), "The Inertia of the Water Surrounding a Vibrating Ship," *Transactions*, Society of Naval Architects and Marine Engineers, Vol. XXVII, pp. 1–20.

Lee, C.-H. (1995), "WAMIT Theory Manual," MIT Report 95–2, Department of Ocean Engineering, Massachusetts Institute of Technology, Cambridge, MA.

Lighthill, J. (1979), "Waves and Hydrodynamic Loading," *Proceedings*, Behavior of Offshore Structures Conference, London, Vol. 1.

Longuet-Higgins, M. S. (1952), "On the Statistical Distribution of the Heights of Sea Waves," *Journal of Marine Research*, Vol. 11, No. 3, pp. 245–266.

MacCamy, R. C. and R. A. Fuchs (1954), "Wave Forces on Piles: A Diffraction Theory," Beach Erosion Board, U.S. Army, Tech. Memo. No. 69, December.

McCormick, M. E. and J. Cerquetti (2004), "Alternative Wave-Induced Force and Moment Expressions for a Fixed, Vertical, Truncated, Circular Cylinder in Waters of Finite Depth," *Proceedings*, OMAE, Vancouver, June.

McCormick, M. E. and P. J. Hudson (2001), "An Analysis of the Motions of Grounded Ships," *International Journal of Offshore and Polar Engineering*, Vol. 11, No. 2, pp. 95–105.

McIver, P. and D. V. Evans (1984), "Approximation of Wave Forces on Cylinder Arrays," *Applied Ocean Research*, Vol. 6, No. 2, pp. 101–107.

Mei, C. C. (1995), *Mathematical Analysis in Engineering*, Cambridge University Press, Cambridge, U.K.

Miles, J. and F. Gilbert (1968), "Scattering of Gravity Waves by a Circular Dock," *Journal of Fluid Mechanics*, Vol. 34, Part 4, pp. 783–793.

Milgram, J. H. (1970), "Active Water-Wave Absorbers," *Journal of Fluid Mechanics*, Vol. 43, pp. 845–859.

Milne-Thomson, L. M. (1960), *Theoretical Hydrodynamics*, Macmillan, New York, 4th edition.

Mogridge, G. R. and W. W. Jamieson (1976), "Wave Forces on Large Circular Cylinders: A Design Method," Hydraulics Laboratory, National Research Council of Canada, Report MH-111.

Morison, J. R., J. W. Johnson, M. P. O'Brien, and S. A. Schaaf (1950), "The Force Exerted by Surface Waves on Piles," *Petroleum Transactions*, American Institute of Mining Engineers, Vol. 189, TP 2846, pp. 149–154.

Najafian, G., R. G. Tickell, R. Burrows, and J. R. Bishop (2000), "The UK Christchurch Bay Compliant Cylinder Project: Analysis and Interpretation of Morison Wave Force and Response Data," *Applied Ocean Research*, Vol. 22, No. 2, June, pp. 129–153.

Newman, J. N. (1962), "The Exciting Forces on Fixed Bodies in Waves," *Journal of Ship Research*, Vol. 6, No. 3, December, pp. 10–17.

O'Kane, J. J., A. W. Troesch, and K. P. Thiagaragan (2002), "Hull Component Interaction and Scaling for TLP Hydrodynamic Coefficients," *Ocean Engineering*, Vol. 29, No. 5, May, pp. 513–532.

Rahman, M. (1987a), "A Design Method of Predicting Second-Order Wave Diffraction Caused by Large Offshore Structures," *Ocean Engineering*, Vol. 14, No. 1, pp. 1–18.

Rahman, M. (1987b), "Approximate Wave Force Analysis on Rectangular Caissons," *Ocean Engineering*, Vol. 14, No. 1, pp. 71–78.

Robertson, J. M. (1965), *Hydrodynamics in Theory and Application*, Prentice Hall, Englewood Cliffs, NJ.

Sarpkaya, T. (1976), "In-Line and Transverse Forces on Smooth and Sand Roughened Cylinders in Oscillatory Flow at High Reynolds Numbers," U.S. Navy, Naval Postgraduate School Report NPS-69SL76062, Monterey, CA.

Sarpkaya, T. (1981), "Morison Equation and the Wave Forces on Offshore Structures," U.S. Navy Civil Engineering Laboratory Report CR82.008, Port Hueneme, CA.

Sarpkaya, T. (1986), "Forces on a Circular Cylinder in Viscous Oscillatory Flow at Low *KC* Numbers," *Journal of Fluid Mechanics*, Vol. 165, pp. 61–71.

Sarpkaya, T. and M. Isaacson (1981), *Mechanics of Wave Forces on Structures*, Van Nostrand Reinhold, New York.

Saunders, H. E. (1957), *Hydrodynamics in Ship Design*, Society of Naval Architects and Marine Engineers, New York.

Schinzinger, R. and P. A. A. Laura (1991), *Conformal Mapping*, Dover Publications, Mineola, New York.

Schrefler, B. A. and O. C. Zienkiewicz, editors (1988), *Computer Modeling in Ocean Engineering*, A. A. Balkema, Rotterdam.

Simon, M. J. (1982), "Multiple Scattering in Arrays of Axisymmetric Wave-Energy Devices. Part I. A Matrix Method using a Plane-Wave Approximation," *Journal of Fluid Mechanics*, Vol. 120, Part 1.

Söylemez, M. and O. Yilmaz (2004), "Hydrodynamic Design of a TLP Type Offloading Platform," *Ocean Engineering*, Vol. 30, No. 10, July, pp. 1269–1282.

Spring, B. H. and P. L. Monkmeyer (1974), "Interaction of Plane Waves with Vertical Cylinders," *Proceedings*, 14th Intl. Conf. on Coastal Engineering (ASCE), pp. 1828–1847.

Stokes, G. G. (1851), "On the Effect of the Internal Friction of Fluids on the Motions of Pendulums," *Transactions*, Cambridge Philosophical Society, Vol. 9, Part II. Also published in Stokes' *Mathematical and Physical Papers*, Cambridge University Press, Vol. III, pp. 1–141.

Sundaravadivelu, R., V. Sundar, and T. S. Rao (1999), "Wave Forces and Moments on an Intake Well," *Ocean Engineering*, Vol. 26, No. 4, pp. 363–380.

Taylor, R. Eatock and A. Rajagopalan (1983), "Load Spectra for Slender Offshore Structures in Waves and Currents," *Earthquake Engineering and Structural Dynamics* (Intl. Assn. For Earthquake Engineering), John Wiley & Sons, New York, Vol. II, pp. 831–842.

Vallentine, H. R. (1967), *Applied Hydrodynamics*, Butterworths, London, 2nd edition.

van Oortmerssen, G. (1971), "The Interaction between a Vertical Cylinder and Regular Waves," *Proceedings*, Symposium on Offshore Hydrodynamics, Netherlands Ship Model Basin, Publication No. 375, Wageningen, pp. XI.1–XI.24

von Kerczek, C. and E. O. Tuck (1969), "The Representation of Ship Hulls by Conformal Mapping," *Journal of Ship Research* (Society of Naval Architect and Marine Engineering), Vol. 13, pp. 284–298.

Weibull, W. (1951), "A Statistical Distribution Function of Wide Applicability," *Journal of Applied Mechanics* (ASME), September, pp. 293–297.

Wiegel, R. L. (1964), *Oceanographical Engineering*, Prentice Hall, Englewood Cliffs, NJ.

Wiegel, R. L., K. E. Beebe, and J. Moon (1957), "Ocean Wave Forces on Circular Cylindrical Piles," *Journal of the Hydraulics Division* (ASCE), Vol. 83, HY2, Paper 1199, April.

Wigley, C., editor (1960), "The Collected Papers of Sir Thomas Havelock on Hydrodynamics," U.S. Navy, Office of Naval Research publication ONR/ACR-103.

Woodward-Clyde Consultants (1980), "Assessment of the Morison Equation," U.S. Navy, Civil Engineering Laboratory Report CR 80.022, July.

Yeung, R. W. (1981), "Added-Mass and Damping of a Vertical Cylinder in Finite-Depth Waters," *Applied Ocean Research*, Vol. 3, No. 3, pp. 119–133.

Yilmaz, O. and A. Incecik (1998), "Analytical Solutions of Diffraction Problem of a Group of Truncated Vertical Cylinders," *Ocean Engineering*, Vol. 25, No. 6, pp. 385–394, June.

CHAPTER 10

Abramowitz, M. and I. A. Stegun (1965), *Handbook of Mathematical Functions*, Dover Publications, New York.

Baker, N. J., M. A. Mueller, and P. R. M. Brooking (2003), Electrical Power Conversion in Direct Drive Wave Energy Converters," *Proceedings*, Fifth European Wave Energy Conversion Conference, University College Cork, Ireland, pp. 197–204.

Bendat, J. S. and A. G. Piersol (1971), *Random Data: Analysis and Measurement Procedures*, Wiley-Interscience, New York.

Bhattacharyya, S. K., S. Sreekumar, and V. G. Idichandy (2002), "Coupled Dynamics of SeaStar Mini Tension Leg Platform," *Ocean Engineering*, Vol. 30, No. 6, pp. 709–737.

Bretschneider, C. L. (1963), "A One-Dimensional Gravity Wave Spectrum," in *Ocean Wave Spectra* (U.S. Naval Oceanographic Office, National Academy of Sciences and National Research Council), Prentice Hall, Englewood Cliffs, NJ, pp. 41–65.

Brooke, J., editor (2003), *Wave Energy Conversion*, Elsevier, Amsterdam.

Dean, R. G. and R. A. Dalrymple (1984), *Water Wave Mechanics for Engineers and Scientists*, Prentice Hall, Englewood Cliffs, NJ.

Falnes, J. (2002), *Ocean Waves and Oscillating Systems*, Cambridge University Press, Cambridge, U.K.

Hoerner, S. F. (1965), *Fluid-Dynamic Drag*, published by Dr.-Ing. S. F. Hoerner, Brick Town, NJ.

Ivanova, I. O. Ågren, H. Bernhoff, and M. Leijon (2003), "Simulation of a 100 kW Permanent Magnet Octagonal Linear Generator for Ocean Wave Conversion," *Proceedings*, Fifth European Wave Energy Conversion Conference, University College Cork, Ireland, pp. 197–204.

Joos, G. (1986), *Theoretical Physics*, Dover Publications, New York, 3rd edition.

Lighthill, J. (1996), *Waves in Fluids*, Cambridge University Press, Cambridge, U.K.

Lutes, L. D. and S. Sarkani (1997), *Stochastic Analysis of Structural and Mechanical Vibrations*, Prentice Hall, Upper Saddle River, NJ.

McCormick, M. E. (1983), "Analysis of Optimal Ocean Wave Energy Conversion," *Journal of Waterway, Port, Coastal and Ocean Engineering* (American Society of Civil Engineers), Vol. 109, No. 2, May, pp. 180–198.

McCormick, M. E. (2007), *Ocean Wave Energy Conversion*, Dover Publication, Long-Island, New York. Originally published by Wiley-Interscience, New York, 1981.

McCormick, M. E. and D. R. B. Kraemer (2006), "Long-Wave Approximations of the Added-Mass and Radiation Damping of a Heaving, Vertical, Circular, Cylindrical in Waters of Finite Depth," *Ocean Engineering International* (Natl. Research Council of Canada),Vol. 1, No. 2, pp. 59–66, 2006.

McCormick, M. E., D. G. Johnson, R. Hebron, and J. Hoyt (1981), "Wave Energy Conversion in Restricted Waters by a Heaving Cylinder/Linear-Inductance System," *Proceedings*, OCEANS 81 (IEEE Publication 81CH1685–7), Boston, September 16–18, pp. 898–901.

McCormick, M. E., J. P. Coffey, and J. B. Richardson (1982), "An Experimental Study of Wave Power Conversion by a Heaving, Vertical, Circular Cylinder in Restricted Waters," *Applied Ocean Research*, Vol. 4, No. 2, pp. 107–111.

McCormick, M. E., A. D. Lazarus, and C. Speight (2004), "Shore Protection by Point-Absorbers," *Proceedings*, Civil Engineering in the Oceans VI (ASCE), Baltimore, October 20–22, pp. 232–245.

Mueller, M. A., N. J. Baker, and E. Spooner (2000), "Electrical Aspects of Direct Drive Wave Energy Converters," *Proceedings*, Fourth European Wave Energy Conversion Conference, Aalborg, Denmark, December, Paper H4.

Newman, J. N. (1977), *Marine Hydrodynamics*, MIT Press, Cambridge, MA.

Omholt, T. (1978), "A Wave Activated Electric Generator," *Proceedings*, OCEANS 78 (Marine Technology Society), Washington, DC, pp. 585–589.

Patel, M. H. (1989), *Dynamics of Offshore Structures*, Butterworths, London.

Pierson, W. J. and L. Moskowitz (1964), "A Proposed Spectral Form for Fully Developed Wind Seas Based on the Similarity Theory of S. A. Kitaigorodskii," *Journal of Geophysical Research*, Vol. 69, No. 24, pp. 5181–5190, December.

Yeung, R. W. (1981), "Added Mass and Damping of a Vertical Cylinder in Finite-Depth Waters," *Applied Ocean Research*, Vol. 3, No. 3, pp. 119–133.

Zill, D. G. (1986), *A First Course in Differential Equations with Applications*, Prindle, Weber & Schmidt, Boston, 3rd edition.

CHAPTER 11

Ablowitz, M. J. and A. S. Fokas (1997), *Complex Variables*, Cambridge University Press, New York.

Abramowitz, M. and I. A. Stegun (1965), *Handbook of Mathematical Functions*, Dover Publications, New York.

Anderson, P. and H. Wuzhou (1984), "On the Calculation of Two-Dimensional Added-Mass and Damping Coefficients by Simple Green's Function Technique," Danish Center for Applied Math. and Mech., Technical University of Denmark, Report No. 287, June.

Arribas, F. P. and J. A. C. Fernádez (2006), "Strip Theories Applied to the Vertical Motions of High Speed Crafts," *Ocean Engineering*, Vol. 33, Nos. 8–9, pp. 1214–1229.

Bhattacharyya, R. (1978), *Dynamics of Marine Vehicles*, Wiley-Interscience, New York.

Bishop, R. E. D. and W. G. Price (1979), *Hydroelasticity of Ships*, Cambridge University Press, Cambridge, U.K.

Blagoveshchensky, S. N. (1962), *Theory of Ship Motions*, Dover Publications, New York. Translated by Louis Landweber.

Dean, R. G. and R. A. Dalrymple (1984), *Water Wave Mechanics for Engineers and Scientists*, Prentice Hall, Englewood Cliffs, NJ.

Faltinsen, O. M. (1974), "A Numerical Investigation of the Ogilvie-Tuck Formulas for Added-Mass and Damping Coefficients," *Journal of Ship Research* (Society of Naval Architect and Marine Engineering), Vol. 18, No. 2, June, pp. 73–84.

Faltinsen, O. M. (1990), *Sea Loads on Ships and Offshore Structures*, Cambridge University Press, Cambridge, U.K.

Faltinsen, O. M. (2005), *Hydrodynamics of High-Speed Marine Vehicles*, Cambridge University Press, Cambridge, U.K.

Fossen, T. I. and O. N. Smogeli (2004), "Nonlinear Time-Domain Strip Theory Formulation for Low-Speed," *Modeling, Identification and Control* (Norwegian Society of Automatic Control), Vol. 25, No. 4, pp. 201–221.

Frank, W. (1967), "Oscillation of Cylinders in or Below the Free Surface of Deep Fluids," Naval Ship Research and Development Center, Report 2375, Washington, DC, October.

Gradshteyn, I. S. and I. M. Ryzhik (1980), *Tables of Integrals, Series and Products*, Academic Press, New York, 2nd edition.

Grim, O. (1960), "A Method for More Precise Computation of Heaving and Pitching Motions Both in Smooth Water and in Waves," *Proceedings*, Third Symposium on Naval Hydrodynamics, Scheveningen, Netherlands.

Haskind, M. D. (1957), "Vozmushchayushchie Sili I Zalivaemostt," Sudov na Volnenii, Otdelenie Tekhnicheskikh Nauk, Izvestia Akademii Nauk S.S.S.R., No. 7, pp. 65–79.

Havelock, T. H. (1927), "The Method of Images in Some Problems of Surface Waves," *Proceedings, The Royal Society, Series A*, Vol. 115, pp. 268–280.

Havelock, T. H. (1942), "The Damping of the Heaving and Pitching Motion of a Ship," *Philosophical Magazine*, Ser. 7, Vol. 33, pp. 666–673, September.

Holstein, H (1937), "Uber dieVerwendung des Energiesatzes zur Losung von Operflachenwellenproblemen," *Ing, Arch.*, Vol. 8, pp. 103–111.

Jacobs, W. R. (1958), "The Analytical Calculations of Ship Bending Moment in Regular Waves," *Journal of Ship Research* (Society of Naval Architect and Marine Engineering), Vol. 1, No. 2, No. 1, June, pp. 20–30.

Jensen, J. J. (2001), *Load and Global Response of Ships*, Elsevier, Amsterdam.

Jensen, J. J. and P. T. Pedersen (1978), "Wave-Induced Bending Moments in Ships – a Quadratic Theory," Royal Institute of Naval Architects, pp. 151–165.

Katz, J. and A. Plotkin (2001), *Low-Speed Aerodynamics*, Cambridge University Press, Cambridge, U.K.

Korvin-Kroukovsky, B. V. (1955), "Investigation of Ship Motions in Regular Waves," *Transactions*, Society of Naval Architects and Marine Engineers, Vol. 63, pp. 386–435.

Korvin-Kroukovsky, B. V. (1961), *Theory of Seakeeping*, Society of Naval Architects and Marine Engineers, New York.

Korvin-Kroukovsky, B. V. and W. R. Jacobs (1957), "Pitching and Heaving Motions of a Ship in Regular Waves," *Transactions*, Society of Naval Architects and Marine Engineers, Vol. 65, pp. 590–632.

Krilov, A. (1896), "A New Theory for the Pitching and Heaving Motions of Ships in Waves," *Transactions*, Royal Institute of Naval Architects.

Lamb, H. (1932), Theoretical Hydrodyamics, Cambridge University Press, U.K., 6th edition. Subsequently published by Dover Publications, New York, in 1945.

Landweber, L. and M. Macagno (1957), "Added Mass of Two-Dimensional Forms Oscillating in a Free-Surface," *Journal of Ship Research* (*Society of Naval Architect and Marine Engineering*), Vol. 1, No. 3, November, pp. 20–30.

Landweber, L. and M. Macagno (1959), "Added Mass of a Three-Parameter Family of Two-Dimensional Forces Oscillating in a Free Surface," *Journal of Ship Research* (Society of Naval Architect and Marine Engineering), Vol. 2, No. 4, June, pp. 137–154.

Landweber, L. and M. Macagno (1967), "Added Mass of Two-Dimensional Forms by Conformal Mapping," *Journal of Ship Research* (Society of Naval Architect and Marine Engineering), Vol. 11, June, pp. 109–116.

Lewis, F. M. (1929), "The Inertia of the Water Surrounding a Vibrating Ship," *Transactions, Society of Naval Architects and Marine Engineers*, Vol. XXXVII, pp. 1–20.

Lloyd, A. R. J. M. (1989), *Seakeeping: Ship Behaviour in Rough Weather*, Ellis Horwood Ltd., Chichester, West Sussex, U.K.

Loukakis, T. A. and P. D. Sclavounos (1978), "Some Extensions of the Classical Approach to Strip Theory of Ship Motions, Including Calculation of the Mean Added-Forces and Moments," *Journal of Ship Research* (Society of Naval Architect and Marine Engineering), Vol. 22, March, pp. 1–19.

Macagno, M. (1968) "A Comparison of Three Methods of Computing the Added Mass of Ship Sections," *Journal of Ship Research* (Society of Naval Architect and Marine Engineering), Vol. 12, December, pp. 279–285.

McCormick, M. E. (1973), *Ocean Engineering Wave Mechanics*, Wiley-Interscience, New York.

McCormick, M. E. (2007), *Ocean Wave Energy Conversion*, Dover Publications, Long Island, New York. Originally published by Wiley-Interscience, New York in 1981.

McCormick, M. E. and P. J. Hudson (2009), "Theoretical High-Frequency Added-Mass for a Heaving Rectangular Strip," *Journal of Offshore Mechanics and Arctic Engineering* (ASME), in review.

McTaggart, K. A. (1996), "Improved Boundary Element Methods for Predicting Sectional Hydrodynamic Coefficients for Strip Theory Ship Motions Programs," Defence Research Establishment – Canada, Technical Memorandum 96/212, March.

Milne-Thomson, L. M. (1955), *Theoretical Hydrodynamics*, Macmillan Company, New York.

Motora, S. (1964), "Stripwise Calculation of Hydrodynamic Forces Due to Beam Seas," *Journal of Ship Research* (Society of Naval Architect and Marine Engineering), Vol. 8, No. 1, June, pp. 1–9.

Newman, J. N. (1962a), "The Exciting Forces and Wetting of Ships in Waves," David Taylor Model Basin (U.S. Navy) Translation, 307, November.

Newman, J. N. (1962b), "The Exciting Forces on Fixed Bodies in Waves," *Journal of Ship Research* (Society of Naval Architects and Marine Engineers), Vol. 6, No. 3, pp. 10–17.

Newman, J. N. (1977), *Marine Hydrodynamics*, MIT Press, Cambridge, Massachusetts.

Ogilvie, T. F. and E. O. Tuck (1969), "A Rational Strip Theory of Ship Motions: Part I," University of Michigan, Department of Naval Architecture and Marine Engineering Report No. 013.

Pedersen, T. (2000), "Wave Load Prediction – a Design Tool," Ph. D. Thesis, Department of Naval Architecture and Offshore Engineering, Technical University of Denmark, January.

Petersen, J. B. and L. Marnæs (1989), "Comparison of Non-Linear Strip Theory Predictions and Model Experiments," Technical University of Denmark, Report 387, April.

Phelps, B. P. (1997), "Determination of Ship Loads for Ship Structural Analysis," Defence Science and Technology Organization – Australia, Report No. DSTO-RR-0116.

Porter, W. R. (1960), "Pressure Distributions, Added-Mass and Damping Coefficients for Cylinders Oscillating in a Free Surface," University of California, Berkelely, Institute of Engineering Research, Report 82–16.

Price, W. G. and R. E. D. Bishop (1974), *Probabilistic Theory of Ship Dynamics*, Chapman and Hall (John Wiley & Sons), London.

Salvesen, N., E. O. Tuck, and O. M. Faltinsen (1970), "Ship Motions and Sea Loads," *Transactions*, SNAME, Vol. 78, pp. 250–287.

Schinzinger, R. and P. A. A. Laura (2003), *Conformal Mapping Methods and Applications*, Dover Publications, Mineola, New York.

SNAME (1973), "Seakeeping 1953–1974," Society of Naval Architects and Marine Engineering, Technical & Research Symposium S-3, June, 1974.

St. Denis, M. and W. J. Pierson (1953), "On the Motions of Ships in Confused Seas," *Transactions*, Society of Naval Architects and Marine Engineers, Vol. 61, pp. 260–357.

Stoker, J. J. (1957), *Water Waves*, Interscience Publishers, New York.

Tasai, F. (1959), "On the Damping Force an Added-Mass of Ships Heaving and Pitching," Reports of Research Institute for Applied Mechanics, Japan, Vol. VII, No. 26.

Thomson, W. (1848), "Notes on Hydrodynamics," *Cambridge and Dublin Mathematical Journal*, February.

Ursell, F. (1949), "On the Heaving Motion of a Circular Cylinder on the Surface of a Fluid," *Quarterly Journal of Mechanics and Applied Mathematics*, Vol. 2, pp. 218–231.

Ursell, F. (1954), "Water Waves Generated by Oscillating Bodies," *Quarterly Journal of Mechanics and Applied Mathematics*, Vol. 7, Part 4, pp. 427–437

Von Kerczek, C. and E. O. Tuck (1969), "The Representation of Ship Hulls by Conformal Mapping Functions," *Journal of Ship Research* (Society of Naval Architect and Marine Engineering), Vol. 13, No. 4, December, pp. 284–298.

Vugts, J. H. (1968), "The Hydrodynamic Coefficients for Swaying, Heaving and Rolling Cylinders in a Free Surface," *Proceedings*, Institute of Shipbuilding Progress, Vol. 5, No. 167, July, pp. 251–276.

Wang, H. T. and W. W. Miner (1989), "Numerical Evaluation of the Far Field Waves Pattern of the Radiation Green's Function," Naval Research Laboratory, NRL Memorandum Report 6478, Washington, DC, September 5.

Wendel, K. (1960), "Hydrodynamic Masses and Hydrodynamic Moments of Inertia," David Taylor Model Basin Translation 260, July. Translated by E. N. Labouvie and A. Borden.

Wigley, C. (1963), *The Collected Papers of Sir Thomas Havelock on Hydrodynamics*, Office of Naval Research, ONR/ACR-103, U.S. Government Printing Office, Washington, DC.

Yamamoto, Y., M. Fujino, and T. Fukasawa (1980), "Motion and Longitudinal Strength of a Ship in Head Sea and the Effects of Non-Linearities," *Naval Architecture and Ocean Engineering*, Society of Naval Architects of Japan, Vol. 18.

Yeung, R. W. (1981), "Added Mass and Damping of a Vertical Cylinder in Finite-Depth Waters," *Applied Ocean Research*, Vol. 3, No. 3, pp. 119–133.

Zill, D. G. (1986), *A First Course in Differential Equations with Applications*, Prindle, Weber & Schmidt, Boston, 3rd edition.

CHAPTER 12

Abramowitz, M. and I. A. Stegun (1965), *Handbook of Mathematical Functions*, Dover Publications, New York.

Agarwal, A. K. and A. K. Jain (2003), "Dynamic Behavior of Offshore Spar Platforms under Regular Sea Waves," *Ocean Engineering*, Vol. 30, No. 4, pp. 487–516.

Arockiasamy, M., D. V. Reddy, P. S. Cheema, and H. El-Tahan, "Stochastic Response of Compliant Platforms to Irregular Waves," *Ocean Engineering*, Vol. 10, No. 5, pp. 303–312.

Bayazitoglu, Y. O., G. Jones, and J. J. Hruska (1987), "Study Describes 2,600-ft CPT Platform Performance," *Ocean Industry*, Vol. 22, March, pp. 13–19.

Bhatta, D. D. and M. Rahman (2003), "On Scattering and Radiation Problem for a Cylinder in Water of Finite Depth," *International Journal of Engineering Science*, Vol. 41, pp. 931–967.

Bhattacharyya, S. K., S. Sreekumar, and V. G. Idichandy (2002), "Coupled Dynamics of SeaStar Mini Tension Leg Platform," *Ocean Engineering*, Vol. 30, No. 6, pp. 709–737.

Bishi, R. S. and A. K. Jain (1998), "Wind and Wave Induced Behavior of Offshore Guyed Tower Platforms," *Ocean Engineering*, Vol. 25, No. 7, pp. 501–519.

Brown, D. T. (2005), "Mooring Systems," Chapter 8, *Handbook of Offshore Engineering*, edited by S. K. Chakrabarti, Elsevier, Oxford, U.K.

Brown, D. T. and S. Mavrakos (1999), "Comparative Study on Mooring Line Dynamic Loading," *Marine Structures*, Vol. 12, No. 3, April, pp. 131–151.

Chakrabarti, S. K. (1975), "Wave Forces for Fixed Offshore Structures," Preprint, Structural Engineering Conference (ASCE), New Orleans, April.

Chakrabarti, S. K., editor (2005), *Handbook of Offshore Engineering*, Elsevier, Oxford, U.K.

Chakrabarti, S. K. and D. C. Cotter (1979), "Motion Analysis of Articulated Tower," *Journal of the Waterway, Port, Coastal and Ocean Division*, ASCE 105 (WW3), 281–292.

Chakrabarti, S. K. and D. C. Cotter (1980), "Transverse Motion of Articulated Tower," *Journal of the Waterway, Port, Coastal and Ocean Division*, ASCE 106 (WW1), 65–78.

Chakrabarti, S. K., C. Capanogla, and J. Halkyard (2005), "Historical Development of Offshore Structures," Chapter 1, *Handbook of Offshore Engineering*, edited by S. K. Chakrabarti, Elsevier, Oxford, U.K.

Chiou, R. B. and J. W. Leonard (1991), "Nonlinear Hydrodynamic Response of Curved, Singly-Connected Cables," *Proceedings*, 2nd Intl. Conf. on Computer Modeling in Ocean Engineering, Barcelona, Spain, Sept., pp. 412–424.

Choi, H. S. and J. Y. K. Lou (1991), "Nonlinear Behavior of an Articulated Offshore Loading Platform," *Applied Ocean Research*, Vol. 12, No. 2, pp. 63–67.

Crede, C. E. (1965), *Shock and Vibrations Concepts in Engineering Design*, Prentice Hall, Englewood Cliffs, NJ.

Dalrymple, R. A. and R. G. Dean (1972), "The Spiral Wavemaker for Littoral Drift Studies," *Proceedings*, 13th Conference on Coastal Engineering (ASCE), Vancouver.

Dawson, T. H. (1980), "Simplified Analysis of Offshore Piles under Cyclic Lateral Loads," *Ocean Engineering*, Vol. 7, No. 4, pp. 553–562.

Dean, R. G. and R. A. Dalrymple (1984), *Water Wave Mechanics for Engineers and Scientists*, Prentice Hall, Englewood Cliffs, NJ.

Edge, B. L. and P. G. Mayer (1969), "A Dynamic Structure-Soil-Wave Model for Deep Water," Preprint, Paper presented at the Ocean Engineering Conference II, Miami, Florida, December.

Ellis, E. A. and S. M. Springman (2001), "Modeling of Soil-Structure Interaction for a Piled Bridge Abutment in Plane Strain FEM Analysis," *Computers and Geotechnics*, Elsevier, Vol. 28, pp. 79–98.

Foster, E. T. (1967), "Predicting Wave Response of Deep Ocean Towers," *Proceedings*, Civil Engineering in the Ocean (ASCE), San Francisco, CA, September, pp. 75–98.

Garrett, C. J. R. (1971), "Wave Forces on a Circular Dock," *Journal of Fluid Mechanics*, Vol. 46, Part I, pp. 129–139.

Gere, J. M. (2001), *Mechanics of Materials*, Brooks/Cole, Pacific Grove, CA, 5th edition.

Gernon, B. J. and J. Y. K. Lou (1987), "Dynamic Response of a Tanker Moored to an Articulated Loading Platform," *Ocean Engineering*, Vol. 14, No. 6, pp. 489–512.

Gerwick, B. C. (1999), *Construction of Marine and Offshore Structures*, CRC Press, Boca Raton, FL, Chapter 12.

Grosenbaugh, M., S. Anderson, R. Trask, J. Gobat, W. Paul, B. Butman, and R. Weller (2002), "Design and Performance of a Horizontal Mooring for Upper-Ocean Research," *Journal of Atmospheric and Ocean Technology* (American Meteorological Society), Vol. 18, No. 9, pp. 1376–1389.

Han, S. and M. Grosenbaugh (2004), "Modeling of Seabed Interaction of Oceanographic Moorings in the Frequency Domain," *Proceedings*, Civil Engineering in the Oceans VI (ASCE), edited by M. J. Briggs and M. E. McCormick, pp. 396–407.

Hanna, S. Y. (1988), "Dynamic Response of a Compliant Tower with Multiple Articulations," *Proceedings*, 7th Offshore Mechanics and Arctic Engineering Conference (ASME), Vol. 1, pp. 257–269.

Hudson, P. J. (2001), "Wave-Induced Migration of Grounded Ships," Ph.D. dissertation (Dept. of Civil Engineering), Johns Hopkins University, Baltimore, MD.

Islamia, J. M. and S. Ahrnad (2003), "Nonlinear Seismic Response of Articulated Offshore Tower," *Defence Science Journal*, Indian Ministry of Defence, Vol. 53, No. 1, pp. 105–113, January.

Jordánr, M. A. and R. Beltrán-Aguedo (2004), "Nonlinear Identification of Mooring Lines in Dynamic Operation of Floating Structures," *Ocean Engineering*, Vol. 31, Nos. 3–4, pp. 455–482.

Kibbee, S. E. and D. C. Snell (2002), "New Directions in TLP Technology," *Proceedings*, Offshore Technology Conference, Houston, May, Paper 14175.

Kim, M. H. and W. Chen (1994), "Sender-Body Approximation for Slowly-Varying Wave Loads in Multi-Directional Waves," *Applied Ocean Research*, Vol. 16, No. 3, pp. 141–163.

Kokkinowrachos, K. and A. Mitzlaff (1981), "Dynamic Analysis of One- and Multi-Column Articulated Structures," *Proceedings*, International Symposium on Hydrodynamics in Ocean Engineering, Norwegian Institute of Technology, Trondheim, pp. 837–863.

Le Méhauté, Bernard (1969), "An Introduction to Hydrodynamics and Water Waves," U.S. Department of Commerce, ESSA Report ERL 118-POL-3–2, Vol. II.

Liaw, C. Y., N. J. Shankar, and K. S. Chua (1989), "Large Motion Analysis of Compliant Structures Using Euler Parameters," *Ocean Engineering*, Vol. 16, Nos. 5/6, pp. 646–667.

Lindsay, R. B. (1960), *Mechanical Radiation*, McGraw-Hill, New York.

Low, Y. M. and R. S. Langley (2008), "A Hybrid Time/Frequency Domain Approach for Efficient Coupled Analysis of Vessel/Mooring/Riser Dynamics," *Ocean Engineering*, Vol. 35, Nos. 5–6, pp. 433–446.

MacCamy, R. C. and R. A. Fuchs (1954), "Wave Forces on Piles: A Diffraction Theory," Beach Erosion Board, U.S. Army, Tech. Memo. No. 69, December.

Matlock, H. (1970), "Correlations for Design of Laterally Loaded Piles in Soft Clay," *Proceedings*, Offshore Technology Conference, Paper No. OTC 1204, Houston, TX, pp. 577–594.

Maus, L. D., L. D. Finn, and M. A. Danaczko (1986), "Exxon Study Shows Compliant Piled Tower Cost Benefits," Ocean Industry, March, pp. 20–25.

McClelland, B. (1969), "Foundations," in *Handbook of Ocean and Underwater Engineering*, edited by J. J. Myers, C. H. Holm, and R. F. McCallister, McGraw-Hill, New York, pp. 8–98 to 8–125.

McCormick, M. E. (1973), *Ocean Engineering Wave Mechanics*, Wiley-Interscience, New York.

McCormick, M. E. (1982), "An Analysis of Optimum Wave Energy Conversion," *Journal of Waterway, Port, Coastal and Ocean Engineering* (ASCE), Vol. 109, No. 2, May, pp. 180–198.

McCormick, M. E. and P. J. Hudson (2001), "An Analysis of the Motions of Grounded Ships," *International Journal of Offshore and Polar Engineering*, Vol. 11, No. 2, pp. 99–105.

McQuillan, J. (2002) "Modeling Dynamic Ground Reaction to Predict Motion of and Loads on Strand Ships in Waves," M.S. thesis (Ocean Engineering), Virginia Polytechnic Institute and State University, Blacksburg, Virginia.

Moan, T., K. Syvertsen, and S. Haver (1977), "Dynamic Analysis of Gravity Platforms Subjected to Random Wave Excitation," *Proceedings*, STAR Symposium (SNAME), Paper T4–1, San Francisco, May, pp. 119–146.

Modaressi, A. and P. Evesque (2001), "Is the Friction Angle the Maximum Slope of a Free Surface of a Non-Cohesive Material?" *Computerized Avalanching – 83 – Poudres & Grains*, Vol. 12, No. 5, pp. 83–102, June.

Molin, B. and J. L. Legras (1990), "Hydrodynamic Modeling of the Roseau Tower Stabilizer," *Proceedings*, 9th Intl. Conf. on Offshore Mechanics and Arctic Engineering (ASME), Houston, TX, February, Vol. I, Part B, pp. 329–338.

Myers, J. J., C. H. Holm, and R. F. Mcallister (1969), *Handbook of Ocean and Underwater Engineering*, McGraw-Hill, New York.

Nagamanit, I. and C. Ganapathy (1996), "Finite Element Analysis of Nonlinear Dynamic Response of Articulated Towers," *Computers & Structures*, Vol. 53, No. 2, pp. 214–223.

Ng, J. and B. J. Vickery (1989), "A Model Study of the Response of Compliant Tower to Wind and Wave Loads," *Proceedings*, Offshore Technology Conference, Houston, TX, Paper No. 6011.

Niedzwecki, J. W. and M. J. Casarella (1976), "On the Design of Mooring Lines for Deep Water Applications," *Journal of Engineering for Industry* (ASME), Vol. 98, pp. 514–522.

Patel, M. H. (1989), *Dynamics of Offshore Structures*, Butterworths, London.

Poulos, H. G and E. H. Davis (1980), *Pile Foundation Analysis and Design*, John Wiley & Sons, New York.

Pradnyana, G. and R. E. Taylor (1997), "The Second Order Response of Tension Leg Platforms," *Proceedings*, BOSS'97 (edited by J. H. Vugts), Elsevier Science, Oxford, U.K., pp. 303–317.

Reese, L. C. (1984), "Handbook of Design on Piled and Drilled Shafts under Lateral Load," Geotechnical Engineering Center, Bureau of Engineering Research, Univ. of Texas, Rept. No. FHWA-IP-84-11.

Reese, L. C. (2003), "Behavior of Piles Supporting Offshore Structures," in *Dynamics of Offshore Structures*, edited by J. F. Wilson, John Wiley & Sons, New York, Chapter 11.

Reese, L. C., W. R. Cox, and F. D. Koop (1974), "Analysis of Laterally Loaded Piles in Sand," *Proceedings*, Offshore Technology Conference, Paper No. OTC 2080, Houston, TX, pp. 473–485.

Regg, J., S. Atkins, B. Hauser, J. Hennessey, B. Kruse, J. Lowenhaupt, B. Smith, and A. White (2000), "Deepwater Development: A Reference Document for the Deepwater Environmental Assessment Gulf of Mexico OCS (1998 through 2007)," U.S. Dept. of Interior, Minerals Management Service, MMS 2000–015, May.

Sabuncu, T. and S. Calisal (1981), "Hydrodynamic Coefficients for Vertical Circular Cylinders of Finite Draft," *Ocean Engineering*, Vol. 8, No. 1, pp. 25–63.

Sannasiraj, S. A., V. Sundar, and R. Sundaravadivelu (1997), "Mooring Forces and Motion Responses of Pontoon-Type Floating Breakwaters," *Ocean Engineering*, Vol. 25, No. 1, pp. 27–48.

Sarpkaya, T. and M. Isaacson (1981), *Mechanics of Wave Forces on Structures*, Van Nostrand Reinhold Company, New York.

Sellers, L. L. and J. M. Niedzwecki (1992), "Response Characteristics of Multi-Articulated Offshore Towers," *Ocean Engineering*, Vol. 19, No. 1, pp. 1–20.

Smith, R. J. and C. J. MacFarlane (2001), "Static of a Three Component Mooring Line," *Ocean Engineering*, Vol. 28, No. 7, July, pp. 899–914.

Suneja, B. P. and T. K. Datta (1998), "Active Control of ALP with Improved Performance Function," *Ocean Engineering*, Vol. 25, No. 10, pp. 817–835.

Teng, B., D. Z. Ning, and X. T. Zhang (2004), "Wave Radiation by a Uniform Cylinder in Front of a Vertical Wall," *Ocean Engineering*, Vol. 31, No. 2, pp. 201–224.

Thomson, W. T. (1965), *Theory of Vibrations with Applications*, Prentice Hall, Englewood Cliffs, NJ, 1st edition.

Thomson, W. T. and M. D. Dahleh (1997), *Theory of Vibrations with Applications*, Prentice Hall, Englewood Cliffs, NJ, 5th edition.

Thomson, W. T. and M. D. Dahleh (1998), *Theory of Vibrations with Applications*, Prentice Hall, Englewood Cliffs, NJ, 5th edition.

Ursell, F. (1953), "The Long-Wave Paradox in the Theory of Gravity Waves," *Proceedings*, Cambridge Philosophical Society, Vol. 49, pp. 685–694.

Van Oortmerssen (1971), "The Interaction between a Vertical Cylinder and Regular Waves," *Proceedings*, Symposium on Offshore Hydrodynamics, Netherlands Ship Model Basin, Publication No. 375, Wageningen, pp. XI.1–XI.24.

Vandiver, J. K. (1981), "Prediction of the Damping-Controlled Response of Offshore Structures to Random Wave Excitation," *Marine Technology Society Journal*, Vol. 15, No. 1, pp. 31–41.

Vugts, J. H. and D. J. Hayes (1979), "Dynamic Analysis of Fixed Offshore Structures: A Review of Some Basic Aspects of the Problem," *Engineering Structures*, Vol. 1, pp. 114–120.

Will, S. A., D. G. Morrison, and D. E. Calkins (1988), "Composite Leg Platforms for Deep U.S. Gulf Waters," *Ocean Industry*, Vol. 23, March, pp. 23–28.

Wilson, J. F. (2003), *Dynamics of Offshore Structures*, John Wiley & Sons, New York, 2nd edition.

Wolf, J. P. (1988), *Soil-Structure Interaction Analysis in the Time Domain*, Prentice Hall, Englewood Cliffs, NJ.

Wolf, J. P. (1994), *Foundation Vibration Analysis Using Simple Physical Models*, Prentice Hall, Englewood Cliffs, NJ.

Wolf, J. P. and C. Song (1999), "The Guts of Dynamic Soil-Structure Interaction," Preprint, Keynote Lecture, International Symposium on Earthquake Engineering, Budva, Montenegro, Sept. 22–25.

Woodward-Clyde Consultants (1980), "Assessment of the Morison Equation," U.S. Navy, Civil Engineering Laboratory Report CR 80.022, July.

Yeung, R. W. (1981), "Added-Mass and Damping of a Vertical Cylinder in Finite-Depth Waters," *Applied Ocean Research*, Vol. 3, No. 3, pp. 119–133.

APPENDICES

Abramowitz, M. and I. A. Stegun (1965), *Handbook of Mathematical Functions*, Dover Publications, New York.

Adey, R. A. and C. A. Brebbia (1983), *Basic Computational Techniques for Engineers*, John Wiley & Sons, New York.

Brennen, C. E. (1982), "A Review of Added Mass and Fluid Inertial Forces," Naval Civil Engineering Laboratory Report CR82.010, January.

Dettman, J. W. (1962), *Mathematical Models in Physics and Engineering*, McGraw-Hill Book Co., New York. Now available from Dover Publications, Mineola, NY.

Falnes, J. (2002), *Ocean Waves and Oscillating Systems*, Cambridge University Press, Cambridge, U.K.

Ferrers, N. M., editor (1871), *The Mathematical Papers of the Late George Green*, Macmillan & Co., London.

Gradshteyn, I. S. and I. M. Ryzhik (1980), *Tables of Integrals, Series and Products*, Academic Press, New York, 2nd edition.

Havelock, T. H. (1929), "Forced Surface-Waves on Water," *Philosophical Magazine*, Vol. VIII, October, pp. 569–576.

Karamcheti, K. (1966), *Principles of Ideal-Fluid Aerodynamics*, John Wiley & Sons, New York.

Katz, J. and A. Plotkin (2001), *Low-Speed Aerodynamics*, Cambridge University Press, Cambridge, U.K.

Kennard, E. H. (1967), "Irrotational Flow of Frictionless Fluid, Mostly of Invariable Density," David Taylor Model Basin, U.S. Navy, Report 2299.

Lamb, H. (1932), *Theoretical Hydrodyamics*, Cambridge University Press, U.K., 6th edition. Subsequently published by Dover Publications, New York, in 1945.

Lewis, F. M. (1929), "The Inertia of the Water Surrounding a Vibrating Ship," *Transactions*, Society of Naval Architects and Marine Engineers, Vol. XXVII, pp. 1–20.
Lindsay, R. B. (1960), *Mechanical Radiation*, McGraw-Hill, New York.
Schinzinger, R. and P. A. A. Laura (2003), *Conformal Mapping*, Dover Publications, Mineola, NY.
Tolstov, G. P. (1962), *Fourier Series*, Prentice Hall, Englewood Cliffs, NJ. Now available from Dover Publications, Mineola, NY.
Wendel, K. (1956), "Hydrodynamic Masses and Hydrodynamic Moments of Inertia," David Taylor Model Basin Translation 260, July. Translated by E. N. Labouvie and A. Borden.
Wigley, C. (1963), *The Collected Papers of Sir Thomas Havelock on Hydrodynamics*, Office of Naval Research, ONR/ACR-103, U.S. Government Printing Office.

Index

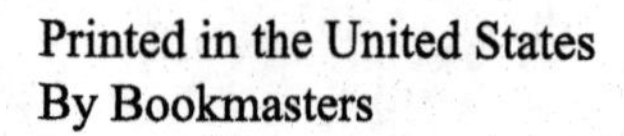
Printed in the United States
By Bookmasters